Brian A. Haughton

AF892170

Hepatotoxicology

Hepatotoxicology

Edited by

Robert G. Meeks
Steadman D. Harrison
Richard J. Bull

CRC Press
Boca Raton Ann Arbor Boston London

Library of Congress Cataloging-in-Publication Data

Cataloging information is on file with the Library of Congress

Developed by Telford Press

This book represents information obtained from authentic and highly regarded sources. Reprinted material is quoted with permission, and sources are indicated. A wide variety of references are listed. Every reasonable effort has been made to give reliable data and information, but the author and the publisher cannot assume responsibility for the validity of all materials or for the consequences of their use.

All rights reserved. This book, or any parts thereof, may not be reproduced in any form without written consent from the publisher.

Direct all inquiries to CRC Press, Inc., 2000 Corporate Blvd., N.W., Boca Raton, Florida 33431.

© 1991 by CRC Press, Inc.

International Standard Book Number 0-8493-8810-4

Printed in the United States

PREFACE

The idea for this book came from an initial contact made to the editors by Dr. Patrick Kelly of Telford Press, Inc., followed by a number of discussions with many individuals involved with the study of chemical-induced damage to the liver. It was the consensus of these individuals that a book that dealt with all aspects of liver damage would serve as a valuable addition to the literature and possibly could serve as a source book to which anyone interested in learning about or understanding this important topic could turn. In many respects, I believe we have accomplished what we set out to do. Researchers such as toxicologists, physiologists, physicians, biochemists, industrial hygienists, etc., interested in the effects of chemical agents on the structure and function of the liver, will find this book indispensable. This work will serve as a source to which investigators can turn for a historical perspective and basic information on the physiology, biochemistry, and pathology of the liver. An examination of the Table of Contents will show quickly that the range of topics covered in this book is broad. The first section deals with histological organization and histopathology of the liver, classification of chemical-induced injury, and the stages of cellular injury. Whereas, the second section deals with xenobiotic metabolism as well as various biochemical methods used to determine liver damage. The third section presents valuable information on the effects of various chemical agents on the liver and an up-to-date discussion of hepatocarcinogenesis including assessment of carcinogenesis and peroxisome proliferation.

The editors would like to deeply thank the many contributors to this book and the many individuals who assisted in the review and editing of the material presented. We also wish to acknowledge the many scientists in the field of hepatotoxicity that have contributed to the knowledge of chemical-induced liver damage. Just because they did not contribute directly to this volume does not minimize the tremendous importance that their contribution has made to the field of hepatotoxicity.

LIST OF CONTRIBUTORS

Robert S. Britton, M.S., Ph.D.
Assistant Research Professor
Department of Internal Medicine
St. Louis University Medical Center
St. Louis, Missouri

Richard J. Bull, B.S., Ph.D.
Professor and Chair
Pharmacology/Toxicology
College of Pharmacy
Washington State University
Pullman, Washington

Charles E. Cornelius, V.M.D., Ph.D.
Professor
Department of Physiological Sciences
University of California/Davis
Davis, California

John M. Cullen, V.M.D., Ph.D.
Associate Professor
Department of Microbiology, Pathology, and Parasitology
North Carolina State University
Raleigh, North Carolina

Francesco De Matteis, M.D., Ph.D.
Medical Director
Medical Research Council Toxicology Unit
Woodmansterne Road
Carshalton
Surrey, England

Mario U. Dianzani, Ph.D.
Department of Experimental Medicine and Oncology
Section of General Pathology
University of Torino
Torino, Italy

Robert B. Diasio, M.D.
Chairman
Department of Pharmacology
University of Alabama
UAB Station
Birmingham, Alabama

A. Jay Gandolfi, Ph.D.
Professor
Department of Anesthesiology and Pharmacology
University of Arizona
Tucson, Arizona

Eric A. Glende, Jr., Ph.D.
Associate Professor
Department of Physiology and Biophysics
School of Medicine
Case Western Reserve University
Cleveland, Ohio

Norio Hirota, M.D.
Department of Pathology
Jichi Medical School
Minamikawachi-machi
Kawachi-gun
Tochigi-ken, Japan

Andrea K. Hubbard, Ph.D.
Assistant Professor
Toxicology Program
University of Connecticut
Storrs, Connecticut

Prasada Rao S. Kodavanti, B.Sc., M.S., Ph.D.
Assistant Professor
Department of Pharmacology and Toxicology
University of Mississippi Medical Center
Jackson, Mississippi

Jeffrey L. Larson, Ph.D.
Pharmacology/Toxicology
College of Pharmacy
Washington State University
Pullman, Washington

Charles S. Leiber, M.D.
Director
Section of Liver Disease and Nutrition
Alcohol Research and Treatment Center
Veteran Affairs Medical Center
Bronx, New York

Charles Lindamood, III, Ph.D., D.A.B.T.
Head, Toxicology Section
Kettering-Meyer Laboratory
Southern Research Institute
Birmington, Alabama

Louis Marzella, M.D., Ph.D.
Associate Professor
Department of Pathology
University of Maryland
School of Medicine
Baltimore, Maryland

Harihara M. Mehendale, B.Sc., M.S., Ph.D.
Professor
Department of Pharmacology and Toxicology
University of Mississippi Medical Center
Jackson, Mississippi

Katsumi Miyai, M.D., Ph.D.
Professor
Department of Pathology
School of Medicine
University of California/San Diego
La Jolla, California

M. Sambasiva Rao, M.D.
Northwestern University
School of Medicine
Chicago, Illinois

Richard O. Recknagel, Ph.D.
Professor Emeritus
Department of Physiology and Biophysics
School of Medicine
Case Western Reserve University
Cleveland, Ohio

Janardan K. Reddy
Northwestern University
School of Medicine
Chicago, Illinois

Timothy P. Roth, B.S.
Research Assistant
Department of Anesthesiology
University of Arizona
Tucson, Arizona

Boris H. Ruebner, M.D.
Professor
Department of Medical Pathology
School of Medicine
University of California/Davis
Davis, California

David J. Sweeny, Ph.D.
Department of Pharmacology
University of Alabama
UAB Station
Birmingham, Alabama

Benjamin F. Trump, M.D.
Professor and Chairman
Department of Pathology
University of Maryland
School of Medicine
Baltimore, Maryland

Mary Vore, Ph.D.
Professor
University of Kentucky
Department of Pharmacology
College of Medicine
Lexington, Kentucky

Gary M. Williams, M.D.
Naylor Dana Institute of Disease Prevention
Valhalla, New York

TABLE OF CONTENTS

Chapter

1

Structural Organization of the Liver

Katsumi Miyai
University of California, San Diego, School of Medicine
Department of Pathology
La Jolla, California

INTRODUCTION

The structural organization of the liver reflects its function that is to serve as a guardian situated between the digestive tract and the rest of the body. Because of this interposition, it receives large amounts of nutrients and noxious substances entering the body through the digestive tract and portal vein. Hence, a major hepatic function involves efficient uptake of amino acids, carbohydrates, lipids and vitamins and their subsequent storage, metabolic conversion, and release into the blood and bile. Biotransformation is another major hepatic function. This process converts hydrophobic substances into water-soluble products that can be excreted into the bile or urine. Although the biotransformation system is also present in other organs (e.g., kidney, small intestine, and endocrine organs), that of the liver is most important quantitatively.

Formation of bile requires the liver to have an exocrine gland structure. In addition, the liver also plays an important role in the body's defense against macromolecules and particulate materials such as bacteria. Hepatic sinusoid-lining cells are quantitatively the most effective site of phagocytosis of particulate materials. Finally, the liver's large vascular capacity serves as a reservoir in the regulation of blood volume and flow through the body (Jones and Schmucker, 1977; Jones and Spring-Mills, 1984; Miyai, 1979).

Therefore, the structure of the liver is adapted to meet the functional requirements and has the following characteristic features. (1) The liver has a dual blood supply (portal vein and hepatic artery), and the splanchnic blood from the portal vein undergoes a second interaction with cells in a microvascular system. (2) The organization of the hepatic parenchyma and the microvascular system is highly efficient for exchange between the blood and hepatocytes, that is, (a) hepatocytes usually form single-cell-thick plates so that they have at least two surfaces exposed to the blood, (b) sinusoid-lining cells are highly porous and, unlike ordinary capillaries, are not lined by a continuous basement membrane, and (3) the hepatocyte excretes its products into blood as well as bite. Morphologically, this is expressed as the separate compartmentalization of the cell surface of hepatocytes into vascular and biliary spaces.

The large cell mass of the liver is supported by two tracts of connective tissue, the portal tract and that surrounding the branches of the hepatic vein. The former encloses afferent blood vessels (portal and hepatic arteries), bile ducts, lymphatic vessels, and nerves, whereas the latter surrounds tributaries of the hepatic vein. The two connective tissue tracts interdigitate but do not connect with each other, being separated by the intervening parenchyma. The terminal branches of portal veins and hepatic arteries are connected with a highly porous microcirculation system (sinusoids) traversing the parenchyma. Portal tracts and tributaries of hepatic veins serve as important landmarks to understand the microstructure and function of the liver (Elias, 1949; Healy, 1970; Jones and Schmucker, 1977; Jones and Spring-Mills, 1984; Miyai, 1979; Rappaport, 1980).

CELLULAR COMPOSITION

The liver constitutes approximately 2–5% of body weight in the adult man (1400–1600 g) and 5% at birth. Hepatocytes, estimated to number 250 billion in normal adult human liver, comprise about 80% of the total cellular population (Table 1–1) (Gates et al., 1961) and 80% of hepatic tissue volume (Rohr et al., 1976). In the rat, hepatocytes constitute about 60% of the total cellular population (Table 1–2) (Daoust, 1958) and 80% of hepatic tissue volume (Table 1–3) (Blouin et al., 1977; Weibel et al., 1969). Cells lining the sinusoid in the rat are estimated to be about 30% of the total cellular population (Table 1–2) but comprise only 6.3% of tissue volume (Table 1–3) (Blouin et al., 1977). Other cell types constitute small proportions of the cellular population and tissue volume. Biochemical studies have shown that, in normal human liver, 4% of the total protein is collagenous (Seyer et al.,

TABLE 1–1.
Distribution of Cell Types in Human Liver

Cell Types	Needle Biopsies (Germany) Apparent Distribution (% ± S.E.)	True Distribution (%)
Parenchymal	84.2 ± 2.7	86.0
Sinusoidal	14.7 ± 2.2	14.0
Bile duct	0.6	
Connective tissues	0.2	
Blood vessel	0.3	
Cell Types	**Surgical Biopsies (Michigan)* Distribution (%)**	
Parenchymal	79.8	
Sinusoid lining	16.6	
Others	3.6	
Binucleate parenchymal	7.8	

*Figures are mean of four specimens.
Source: Modified from Gates et al. (1961).

1977). Normal rat liver contains less collagen than does human liver, with only 0.55% of the total protein being collagenous (Seyer, 1980). Thus, despite its size, the liver normally contains relatively little connective tissue which provides an internal supporting framework for the parenchyma, and ensheathes blood and lymphatic vessels, bile ducts, and nerves.

BLOOD SUPPLY

The liver receives blood through the portal vein and hepatic artery. The former accounts for 75–80% and the latter 20–25% of the total hepatic blood flow (1500 ml/min), which is about one-quarter of cardiac output (Figure 1).

TABLE 1–2.
Distribution of Cell Types in Normal Rat Liver
(Percent of total cellular population)

Cell Types	Apparent Distribution (% ± S.E.)	True Distribution (% ± S.E.)
Parenchymal	68.8	60.6
Sinusoid	26.4	33.4
Bile Duct	1.7	2.0
Connective tissue	1.7	2.2
Blood vessel walls	1.4	1.8

Source: Modified from Daoust (1958).

TABLE 1–3.
Volumetric Composition of the Rat Liver Parenchyma*

Hepatocytes	77.8 ± 1.15
Nuclei	7.6 ± 0.50
Cytoplasm	70.2 ± 1.13
Sinusoidal cells	6.3 ± 0.49
Endothelial cells	2.8 ± 0.19
Nuclei	0.44 ± 0.08
Cytoplasm	2.36 ± 0.08
Kupffer cells	2.1 ± 0.31
Nuclei	0.39 ± 0.07
Cytoplasm	1.71 ± 0.07
Fat-storing cells	1.4 ± 0.19
Nuclei	1.12 ± 0.04
Cytoplasm	1.12 ± 0.04
Spaces	15.9 ± 0.75
Disse space	4.9 ± 0.35
Sinusoidal lumen	10.6 ± 0.45
Bile canaliculi	0.43 ± 0.05

*Data from Blouin (1977). Values are percentages % S.E.

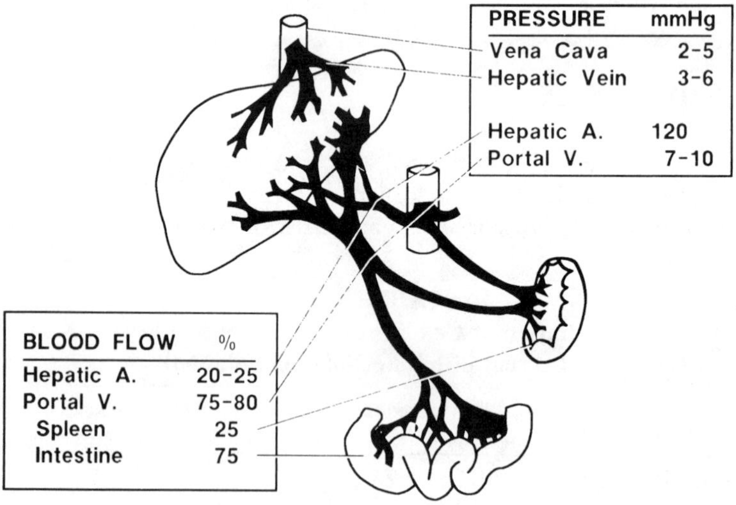

Figure 1. Blood supply of the liver.

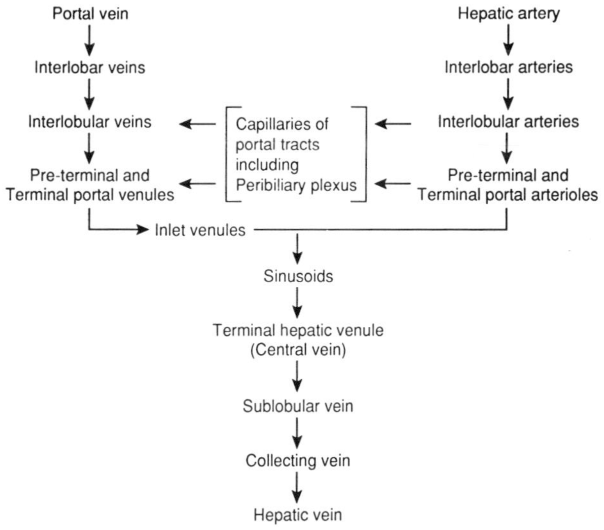

Figure 2. Organization of hepatic blood supply.

The portal vein conveys products of digestion to the liver after collecting the venous blood from the intestine. It also provides the liver with the venous return from the spleen. It is called a portal system because it conveys blood directly from one capillary bed (the intestine) to another (the liver) (Brauer, 1963; Healy, 1970).

The portal vein enters the liver at the porta hepatis and divides into interlobar and subsequently into interlobular branches that run in the connective tissue of the portal tracts. There is no valve anywhere in the portal vein. Interlobular portal veins branch repeatedly into smaller terminal divisions that eventually become continuous with sinusoids. Preterminal portal branches lie in small triangular portal tracts. Terminal portal venules are seen in the smallest round or oval portal tracts and form the vascular axis for simple acinus (see Hepatic Acinus). Terminal portal venules connect with sinusoids by way of short perpendicular inlet vennules, which pierce through the limiting plate of hepatocytes and lead directly into sinusoids (Figure 2). These terminal segments lack a muscular coat (Burkel, 1970; Healy, 1970). In certain species, inlet venules may contain contractile endothelial cells, showing sphincter activity that regulates portal blood flow into the sinusoids (McCuskey, 1966; Rappaport, 1972, 1980). The sinusoids converge towards a terminal hepatic venule (central vein). As the terminal hepatic venules run through the parenchyma, they receive sinusoids from all sides (Burkel and Low, 1966). They unite with other hepatic venules to form a sublobular vein. The sublobular veins join to form one of the hepatic veins that empty into the inferior vena cava (Healy, 1970).

The hepatic artery accompanies the portal vein. One to five arteries of variable sizes may be seen within a portal tract, anastomosing with one another. In their terminal portions, hepatic arteries branch to form arterioles (50–100 μm in diameter) with the media consisting of a single layer of smooth muscle cells and precapillaries arising from the terminal arterioles in the larger portal tracts, which constitute a transitional vessel encircled at their origin by cuffs of smooth muscle, and capillaries (less than 10 μm in diameter) forming a network within the portal tract (Burkel, 1970). These terminal branches of the hepatic arterial system form (1) general plexus within the portal tract, (2) peribiliary plexus surrounding bile ducts, and (3) arterial capillaries emptying directly into the sinusoids (Rappaport, 1980) (Figure 3). The general plexus supplies the structure within the portal tract other than the bile ducts, sending terminal branches into sinusoids. The peribiliary plexus plays an important role in secretion, absorption, and concentration of bile (Rappaport and Schneiderman, 1976). Terminal branches forming this plexus pass into the sinusoids or join the terminal portal venules (Rappaport, 1980).

Some authors maintain that terminal hepatic arterioles and capillaries traverse the hepatic parenchyma to join sinusoids near the terminal hepatic venules (central vein) (Elias, 1949; Wakim and Mann, 1942). This concept has not been supported by observations by others who have not found such an anastomosis (McKuskey, 1966; Nakata and Kanbe, 1966; Rappaport, 1980).

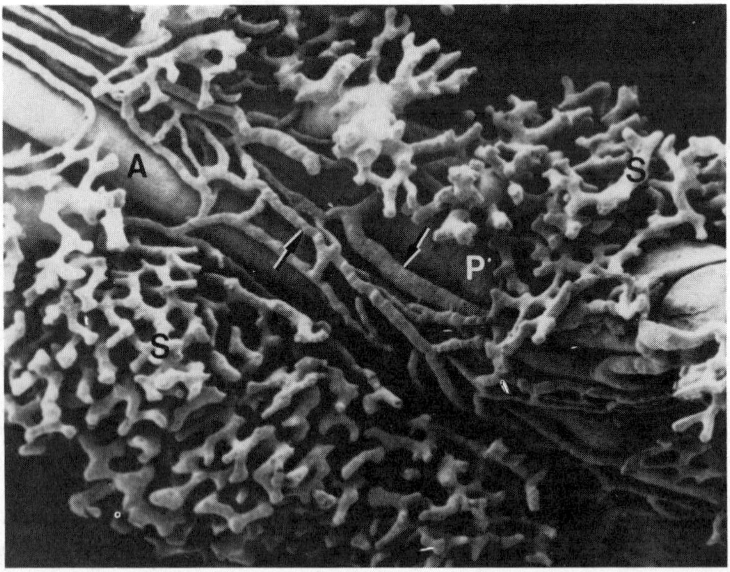

Figure 3. Scanning electron micrograph of a plastic cast of peribiliary plexus and its connecting blood vessels. It shows that capillaries comprising the peribiliary plexus are derived from hepatic arterioles (A) and that some of the capillaries drain into sinusoids (S). P = portal venule (×200). (Courtesy of Dr. T. Murakami.)

However, a recent study using scanning electron microscopy of corrosion casts has shown branches of intralobular arterioles terminating in the vicinity of the central veins, suggesting the presence of the aforementioned anastomosis (Kardon and Kessel, 1980). It seems that this controversy has not yet been settled.

As regards the control of blood flow and its distribution, the hepatic vascular system provides a number of possible sites for vasomotor mechanisms. Hepatic arterioles may constrict along their entire length. Sphincters have been identified at the level of precapillaries, at the entrance of the arteriolar branches into the sinusoids and into branches of the portal venules (McCuskey, 1966; Rappaport, 1972, 1980). In addition, McCuskey (1966) noted contractile structures consisting of endothelial cells located at the junction of sinusoids with portal venules (inlet sphincters) and with terminal hepatic venules (outlet sphincters). However, these latter observations seem to require further substantiation. Currently, blood flow and its distribution in the sinusoids are thought to be regulated primarily by arteriolar and precapillary sphincters. To a lesser degree, the large endothelial cells at the entry and exit of the sinusoids and arteriolar capillaries, by the contraction of cytoskeletal filaments, may contribute to "fine tuning" of the hemodynamic control (McCuskey, 1966).

ORGANIZATION OF HEPATIC PARENCHYMA

Microscopically, architecture of the hepatic parenchyma is rather simple. This lack of specific histological characteristic has led to difficulties in describing and defining a structural and functional unit of the liver. As a result, three concepts have been developed to describe the histological and functional units of the liver: hepatic (classic) lobule, portal lobule, and hepatic acinus. These concepts represent different ways to interpret particular aspects of liver structure and function, but they do not conflict with each other.

Hepatic Lobule

During the seventeenth century, Wepfer and Malpighi described small clumps of hepatic tissue adhering to minute blood vessels as the structural unit of the liver. Subsequently, Kiernan (1833) defined the hepatic lobule as a region of parenchyma, typically hexagonal in cross section, which surrounds a terminal hepatic venule (central vein) at its center (Figure 4). The boundary of the lobule is well defined in only a few species (e.g., pig) by interlobular septa of connective tissue (Figure 5). In most other species, including humans, connective tissue is sparse except in the portal tracts, which lie at the corners of the lobule and contain arteries, veins, bile ducts, and lymphatics. Within

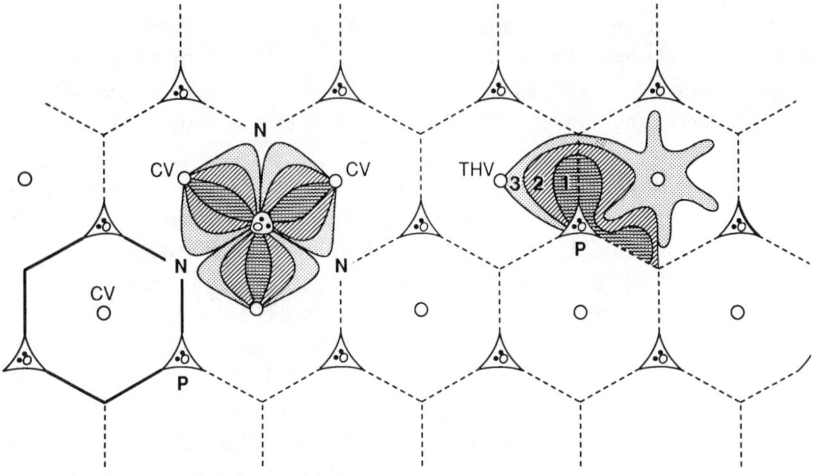

Figure 4. Schematic drawing of the three histologic units of the liver. Left: *Hepatic lobule* is typically hexagonal in shape and demarcated by portal tracts (P) in its periphery. Blood is drained by the central vein (CV). Middle: *Portal lobule,* the center of this unit, is formed by the portal tract. The periphery of the portal lobule is marked by central veins or by nodal points (N) which are water shed regions of blood supply. Right: *Hepatic acinus* the center of the acinus is formed by the terminal afferent vessels. Blood is drained by two or more terminal hepatic venules (ThV). The acinar parenchyma is arbitrarily divided into periportal zone (1), intermediate zone (2), and perivenular zone (3). The perivenular region is formed by the zone 3 of several adjacent acini and is shaped like a sea star.

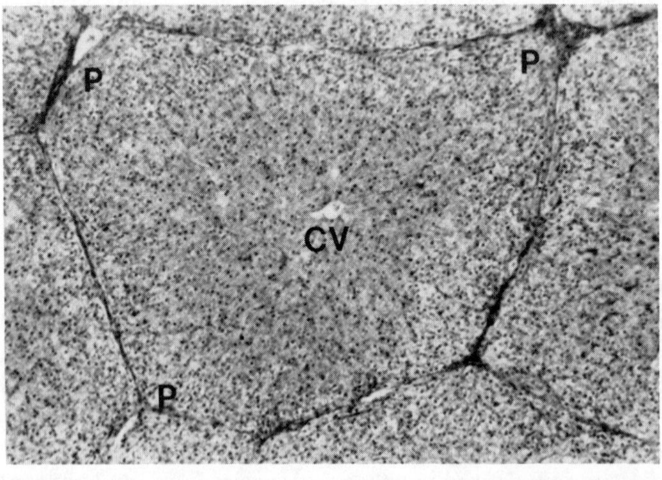

Figure 5. Light micrograph of pig liver. Lobules are demarcated by strands of connective tissue which connect portal tracts (P). Central vein (CV) is seen in the center of the lobule. (×214)

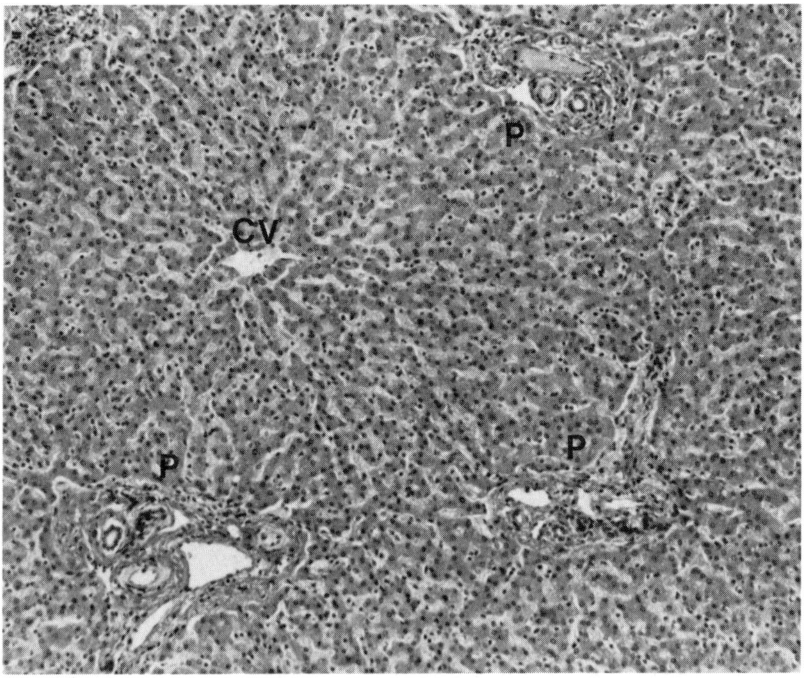

Figure 6. Light micrograph of human liver. In contrast with pig liver, lobule is not clearly demarcated. Portal tracts (P) located around a central vein (CV) serve as landmarks to delineate the hepatic lobule. Hepatocytes form single cell thick "cords" organized radially around the central vein (×360).

the lobule, the hepatocytes form cords arranged radically around the central vein, which is regarded as the structural center of the lobule. These cords are, in fact organized in the form of plates and sheets, one-cell thick, and branch and anastomose with one another to form a complicated system of walls between which the sinusoids interconnect in a continuous labyrinth (Elias, 1949) (Figures 6–8). According to the lobular concept, the blood flows from the periphery of the lobule, from the portal vein and hepatic artery, through sinusoids into the central veins. Since the classical lobule is easily defined by anatomical landmarks seen by light microscopy, it is convenient for histologic description and is, therefore, still most commonly used. Pathologic lesions are usually described as being pericentral (centrolobular), midzonal, or periportal (perilobular) in distribution (Jones and Spring-Mills, 1984).

Portal Lobule

Another concept of the structural unit of the liver was proposed by Mall (1906) who noted that in all other glands, the excretory duct is the center of the structural unit. In his view, everything radiates from the portal tract:

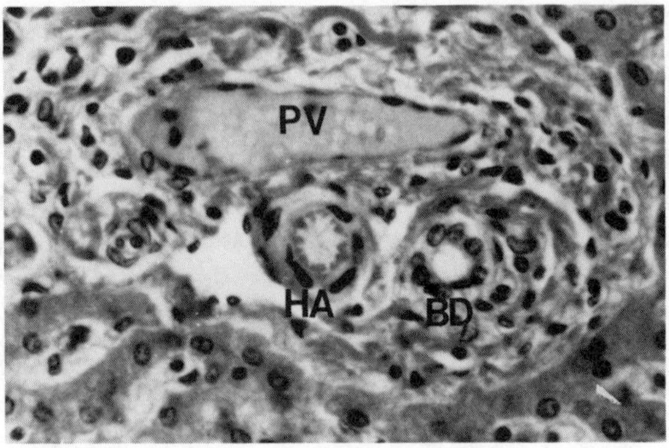

Figure 7. High-power micrograph of a portal tract showing a portal venule (PV), hepatic arteriole (HA), and bile duct (BD) (×1,300).

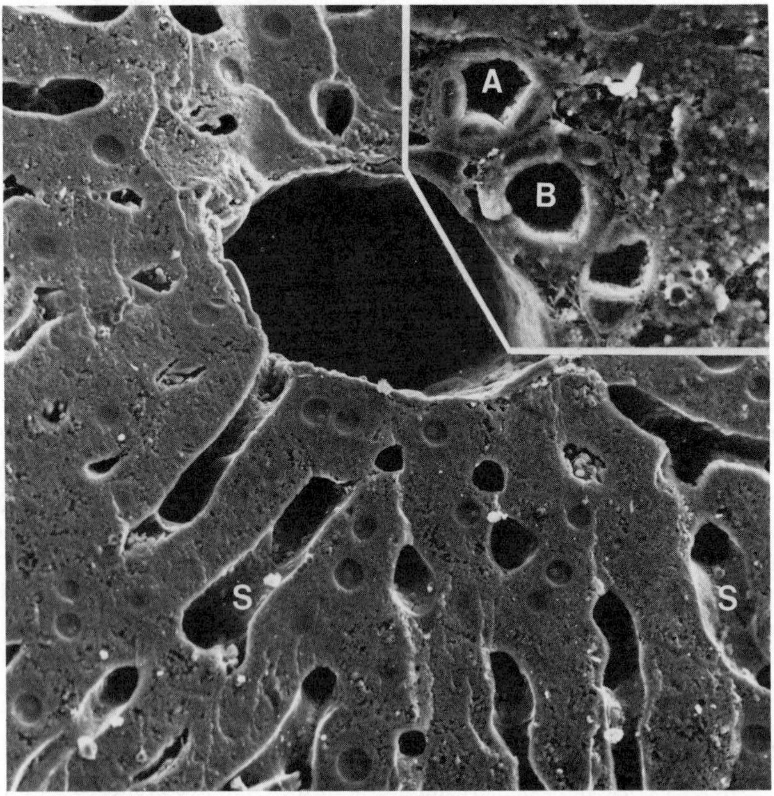

Figure 8. Scanning electron micrograph of a portal tract and adjacent lobular parenchyma. Single cell-thick liver cell plates and sinusoids (S) are seen. A = hepatic arteriole, B = bile duct (×570, inset 2,000).

arterial and portal blood vessels, bile ducts, lymphatic nerves, and connective tissue. Therefore, he defined the portal lobule as a triangular or wedge-shaped region of parenchyma that contains the interlobular bile duct in the portal tract in its center. Its periphery is marked by three different central veins (Figure 4). According to this concept, bile flows from the periphery to the centrally located bile duct, and blood flows from the center of the lobule to its periphery. Hence, the pathways for bile drainage and blood flow in the portal lobule are similar to most exocrine glands. Thus, the concept of the portal lobule enables better correlation between structure and function than does the classic lobule. However, those who criticize this concept argue that it unduly emphasizes the excretory function of the liver. Another argument against it is that the vascular supply to the portal lobule is preterminal so that it does not represent the smallest functional unit.

Hepatic Acinus

In vivo observations on the hepatic microcirculation after injection of dyes and the localization of ischemic damage in the hepatic parenchyma prompted Rappaport et al. (1954) to propose the hepatic acinus as the smallest functional unit related to the terminal branches of the afferent microcirculation. They noted that the blood supply to the parenchyma around each central vein was derived from several different sources, indicating that the lobular parenchyma was not one unit but consists of several different functional units or hepatic acini.

According to Rappaport et al. (1954) and Rappaport (1979) *simple hepatic acinus* (Figures 4 and 9) is defined as a parenchymal mass organized around a terminal portal venule (20–30 μm in diameter) and its accompanying hepatic arterioles, bile ducts, lymphatics, and nerve fibers. It lies between two or more terminal hepatic venules (central veins of the hepatic lobule). Between the portal tract and the terminal hepatic venule, there are about 20–22 hepatocytes. In a two-dimensional section, the acinus comprises only a portion of two adjacent hepatic lobules, and three-dimensionally it is shaped like a berry at the end of the terminal vascular and biliary stalk. Each simple acinus (supported by a microvascular framework consisting of the aforementioned blood vessels, sinusoids, and lymphatics together with nerves) forms the hepatic microcirculation unit. The simple acinus is also the unit for bile secretion; the bile produced is delivered into the terminal bile ductules, part of the corresponding axial portal tract. The watershed of bile flow is part of the dividing lines between two adjacent acini. In addition, the acinus is a metabolic unit as well (Rappaport, 1979).

The cells of the acinus surround the afferent vessels like layers of a bulb and are arbitrarily divided into three zones: 1, 2, and 3 (Figures 4 and 9). Zone 1 is roughly equivalent to the periportal zone, and zone 3 corresponds

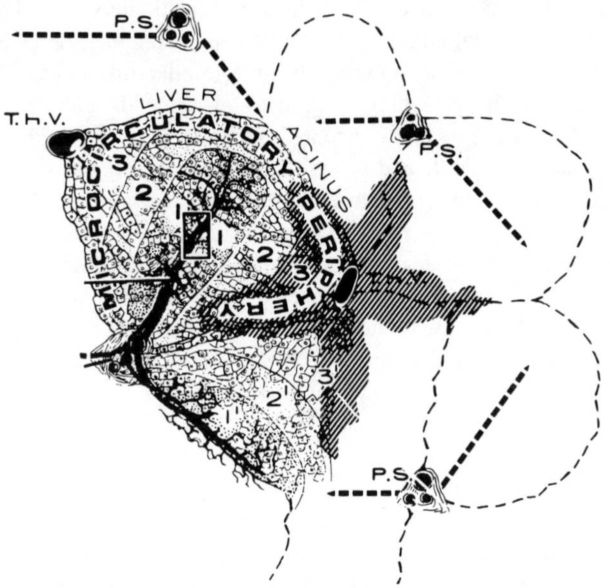

Figure 9. This diagram shows blood supply of the simple acinus, zonal arrangement and the microcirculatory periphery. The hepatic acinus occupies adjacent sectors of neighboring hexagonal lobules and it receives blood supply from the terminal branches of hepatic arterioles and portal venules and hepatic arterioles. Zones 1, 2, and 3, respectively, represent areas supplied with blood of first, second, and third quality with respect to oxygen and nutrient contents. The perivenular area consists of the microcirculatory periphery (zone 3) of several adjacent acini. In injuries involving this zone, the damaged area assumes the shape of a sea star (see Fig. 11). PS = portal space; ThV = terminal hepatic venule; 1,2,3 = acinar zones 1,2,3; heavy broken line = outline of lobular hexagons. (From Rappaport, 1979, with permission of the author and publisher.)

to the centrolobular zone of the hepatic lobule. The simple acini abbut against one another and, as shown in Figure 8, the sites where the terminal blood vessels from neighboring acini give out into their terminal capilleries constitute watershed areas and correspond to the "nodal points" of Mall (1906). It should be noted that not all periportal hepatocytes are equivalent to zone 1 hepatocytes in the acinus model. Only those cells that surround the terminal portal venule are identical with acinar zone 1 cells, whereas the periportal parenchyma around the preterminal portal venules contain cells from all three acinar zones. Since the cells in zone 1 are close to the afferent vessels, they receive blood rich in oxygen and nutrients, whereas the cells in zone 3 situated at the distal end of the microcirculation receive blood that has already exchanged gases and metabolites with cells in zones 1 and 2. It is thought that such an arrangement accounts for the greater sensitivity of the cells in zone 3 than those in other zones to damage from circulatory disturbances (ischemia, anoxia, or congestion) and nutritional deficiencies. Zone 2 is a transitional region between zones 1 and 3, and its dimension changes in response to the hemodynamic state within the acinus: with increased microcirculation, zone 1 expands over zone 2; when microcirculation decreases, zone 1 becomes part of zone 3 (Rappaport, 1979).

The *complex acinus* (Figure 10) is composed of at least three simple acini, and a sleeve of parenchyma around the preterminal vessels comprises small clumps, *acinuli,* which are supplied by portal venular and arteriolar branches from the preterminal vessels. The preterminal branches ramify in three directions, and their portal tracts, when cross sectioned, are triangular in shape. The terminal ramifications form the axis of simple acini that comprise the larger unit (complex acinus). The sinusoids drain into the terminal hepatic venule situated *between* the zone 3 of the acini forming the complex acinus. The *acinar agglomerate* (Figure 10) consists of at least three complex acini and the acini forming the sleeve of parenchyma around the large portal tract supplying the agglomerate. Thus, in summary, the hepatic parenchyma is divided into structural units of a first-order, second-order, and third-order nature, as shown in Figure 10.

Ordinarily one finds it difficult to identify the hepatic acinus in a histologic section because the acinus is not clearly demarcated by anatomical landmarks. When hepatocytes suffer an injury with zonal distribution, the injury usually

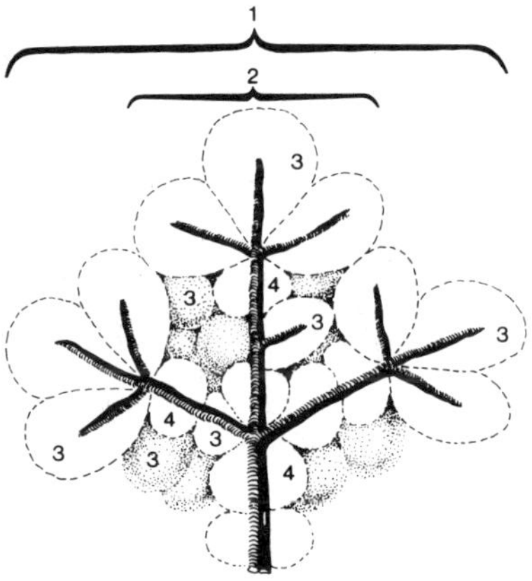

Figure 10. Schematic representation of hepatic acinar structure. This shows an acinar agglomerate, (1) three complex acini, (2) and a number of simple acini, (3). The acinar agglomerate is supplied by a branch of portal vein (and hepatic artery) which subdivides into three preterminal branches supplying each complex acinus. These branches in turn subdivide into terminal branches supplying the single acinus. The acinuli (4) are arranged as a sleeve around the preterminal and main supplying vessels. (From Rappaport et al., *Vitchow's Archives A* 1983; 402:107–137, with permission of the authors and publishers.)

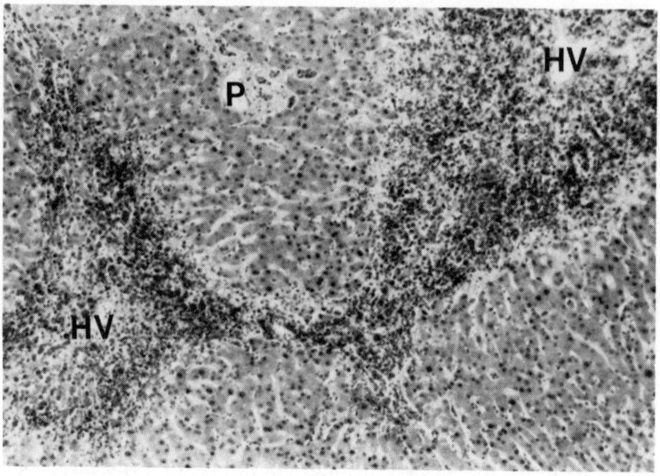

Figure 11. Injury due to severe passive congestion preferentially injures acinar zone 3 (perivenular zone), which is the microcirculatory periphery. Since this type of injury involves all the acini comprising the hepatic lobule, the damaged area encircles the hepatic venule (central vein), and when the injury is severe, it extends to water shed region, forming a sea-star-shaped area of injury. Compare the pattern of the injury with Figs. 4 and 9. P = portal tract, HV = hepatic venule (×300).

does not delineate an acinus. For example, ischemic or congestive injury to zone 3 hepatocytes affects those cells in each of the several acini surrounding a terminal hepatic venule (central vein) (Figure 11). Therefore, it is easier to recognize such an injury as "centrolobular" in the hepatic lobule. When the zone 3 injury is more advanced, it may extend to watershed zones of neighboring acini and bridge terminal hepatic veins (Figure 11). Such an injury corresponds to the seastar-shape pattern of circulatory periphery illustrated in Figures 4 and 9. However, the hepatic acinus may be delineated more clearly when only one or some of the acini comprising the lobule are selectively injured. For example, small thromboemboli occluding the afferent vessels supplying an acinus would cause ischemic injury primarily confined to this particular acinus and delineate its contour (Figure 12).

Functional Heterogeneity of Hepatic Parenchyma

Studies since the 1960s have accumulated a large body of evidence that support the notion that hepatocytes in different locations in the hepatic acinus differ with respect to their morphologic and functional characteristics (Gumucio and Chianale, 1988; Jungermann and Katz, 1982; Thurman and Kaufman, 1985). The ultrastructural differences between hepatocytes of the zone 1 (periportal zone) and zone 3 (pericentral zone) of the acinus are summarized in Table 1–4. The contradictory results with respect to the cell volume and

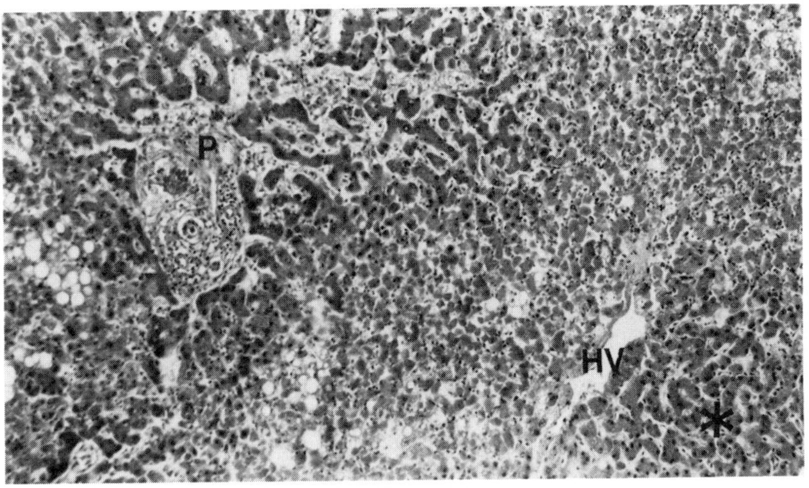

Figure 12. Ischemic injury caused by thromboemboli due to disseminated intravescular coagulation that impaired the blood supply only to some of the acini surrounding the hepatic venule. Since the ischemic injury is confined to the circulatory periphery of the affected acini, zone 3 of the unaffected acini (asterisk) remains intact. Compare the distribution of the injury with Figs. 4 and 9. P = portal tract; HV = hepatic venules (×330).

TABLE 1–4.
Hepatocyte Heterogeneity in Ultrastructure*

	Zone 1 > Zone 3	Zone 3 > Zone 1	No Zonal Predominance§
Cell volume†		4, 5	1, 2, 3
Mitochondria‡	3		1, 2, 5
SER‡		1, 2, 3, 4, 5	
RER‡	5		1, 2, 4, 5
Golgi-rich area‡	1, 2, 4, 5		
Lysosome‡		3, 4	1, 2, 5
Peroxisome‡	1, 2, 5	3, 4	
Lipid‡	1, 2, 3, 4, 5		
Bile canaliculi‡	1, 2, 5		

*Numbers refer to references.
†Expressed as μm^3/cell.
‡Expressed as volume density, $\mu m^3/\mu m^3$ mononuclear hepatocyte.
§References:

1. Jones et al., *Gastroenterology,* 1976; 71:1050–1060.
2. Jones et al., *Anat Rec* 1978; 192:277–288.
3. Loud J, *Cell Biol* 1968; 37:27–46.
4. Meihuizen and Blansjaar, *Mech Aging Developm* 1980; 13:111–118.
5. Schmucker et al., *J Cell Biol* 1978; 78:319–337.

TABLE 1–5.
Localization of Metabolic Processes in Hepatic Parenchyma*

Predominantly Acinar Zone 1 ("Periportal")	Predominantly Acinar Zone 3 ("Periportal")	Distributed Equally
Oxidative energy Metabolism		Metabolism of
Fatty acid oxidation		Ethanol
Citrate cycle		Acetaldehyde
Respiratory chain		
Glucose release	Glucose uptake	
Gluconeogenesis	Glycolysis	
Glycogen degradation to glucose	Glycogen synthesis from glucose	
Glycogen synthesis from lactate	Glycogen degradation to lactate	
Amino acid utilization	Ketogenesis	
Amino acid conversion to glucose	Lipogenesis including bile acid synthesis	
Amino acid degradation		
Urea formation	Biotransformation	
	Mixed function oxidation	
Secretion		
Bile Acids		
Bilirubin		

*Modified from Jungermann (1986); Thurman and Kaufman (1985); Traber et al. (1988).

some of the subcellular organelles observed by different investigators may have resulted from the differences in experimental conditions (e.g., feeding, diurnal rhythm, etc.). It is also conceivable that the investigators selected the area around the portal tracts as zone 1 without defining the size of the portal veins so that cells from all three acinar zones may have been taken as zone 1 cells.

The heterogeneity in enzyme content within the acinus has been demonstrated by a variety of techniques such as qualitative and quantitative enzyme histochemistry, microdissection of tissue samples from zones 1 and 3, followed by microchemical enzyme determinations or determination of enzyme activities after induction of selective acinar damage (Gumucio and Chianale, 1988; Jungermann and Katz, 1982; Thurman et al., 1986). These observations in combination with the structural heterogeneity mentioned above have given rise to the concept that hepatocytes in different regions in the acinus differ in their metabolism and function (Table 1–5) (Gumucio and Chianale, 1988; Jungermann and Katz, 1982; Thurman and Kauffman, 1985). Oxydative energy metabolism, amino acid utilization, and glucose-forming processes of gluconeogenesis and glycogen degradation to glucose predominate in zone 1. Amino acid utilization and bile secretion are also predominant in this region. On the other hand, glucose-utilizing processes of glycolysis and glycogen synthesis from glucose occur preferentially in zone 3 as do mixed function oxidation of biotransformation and lipogenesis. Some metabolic processes, such as ethanol and acetaldehyde metabolism, are equally distributed over different acinar zones (Gumucio and Chianale, 1988; Jungermann and Katz, 1982; Thurman and Kaufman, 1985).

Currently, the functional heterogeneity of hepatocytes in different acinar zones is believed to reflect mainly adaptation to differences in the microcirculatory environment (e.g., zonal concentration gradient of oxygen, substates, co-factors, hormones, etc.) which result from the unidirectional perfusion of the hepatic acinus. For example, when the direction of liver perfusion is changed (by retrograde perfusion), predominant contribution of hepatocytes in zones 1 and 3 to gluconeogenesis, glycolysis, ketogenesis, and oxygen consumption is reversed rapidly (Thurman and Kauffman, 1985). Differences in the capacity for bile acid excretion and biotransformation between zones 1 and 3 can be reduced within a few days by increasing serum bile acid concentration, which abolishes the sinusoidal concentration gradient of bile acids (Baumgartner et al., 1987). However, the predominant location of mixed function oxidation in zone 3 was not altered by a change in the direction of liver perfusion.

Therefore, there seems to be at least two types of regulation of hepatocyte heterogeneity. Some metabolic processes, as mentioned above, will occur predominantly in an acinar zone due to the availability of rate-limiting factors supplied through sinusoidal circulation. In this type of regulation, the direction of blood perfusion is crucial. A change in the direction of perfusion will result in a rapid change in the activity of a metabolic process in one zone. Carbohydrate metabolism is a typical example of this type of regulation. In contrast, the heterogeneity for other metabolic processes seems to be regulated differently. It is likely that the zonal heterogeneity of some enzymes in the mixed function oxidation system (e.g., cytochrome P450b and P450c) involves modulation of gene expression which is not readily altered by changes in the microcirculatory environment (Gumucio and Chinnale, 1988). As regards the physiologic and metabolic significance of the functional heterogeneity of hepatocytes within the acinus, it has been proposed that it is a necessary organization to achieve the regulation of an adequate concentration of solutes in hepatic venules and bile. This concept remains to be established.

HEPATOCYTE

The hepatocytes are the main functional unit of the liver. Human hepatocytes measure between 15 and 30 μm in greatest dimension with the average diameter of 25 μm. A single mononuclear hepatocyte has an average volume of 11 × 10^3 μm, which is twice as large as that of rodents, dogs, and cows

(4.77×10^3–5.54×10^3 μm^3) (Rohr et al., 1976; Weibel et al., 1969). According to Rohr et al. (1976), subcellular organelles subdivide the total volume of the human hepatocyte as follows: nucleus, 6.0%; rough endoplasmic reticulum (RER) 9.4%; smooth endoplasmic reticulum (SER) 5.9%; mitochondria 22%; "dense bodies" (lysosomes) 0.8%; peroxisomes (microbodies) 1.2%; cytosol (ground substance) 45% (Table 1–6). The volumetric composition of organelles in the rat hepatocyte (Table 1–6) (Weibel, 1969) is similar to that of the human hepatocyte. Morphometric data on the hepatic ultrastructure of the rats of different strains and age are available in the following references: Blouin (1983), Blouin et al. (1977), Jones and Schmucker (1977), Loud (1968), Meihuizen and Blahsjaar (1980), Schmucker et al. (1978), and Schmucker and Jones (1975).

Cell Surface

The individual hepatocyte is polyhedral in shape and has three different surfaces: canalicular, comprising the canalicular wall; lateral or intercellular, abutting directly upon adjacent hepatocytes; and sinusoidal, facing the sinusoid. The proportion of surface area belonging to each domain are estimated to be 13% (canalicular), 50% (lateral or intercellular), and 37% (sinusoidal), respectively (Weibel et al., 1969). Thus, the vascular and biliary surfaces have a ratio of 3:1 and constitute about half the cell surface. The *canalicular surface* comprises the wall of the bile canaliculus, which is a tubular space formed by the apposition of the edges of hemitubular grooves on the abutting surfaces of neighboring hepatocytes (Figure 13). The canalicular wall is studded with microvilli and the canalicular lumen is sealed by the tight junction (Zonula occludens). The bile canaliculus and tight junction will be discussed further in section 9. The *lateral surface* is closely approximated with that of an adjoining hepatocyte through desmosome, gap junctions, and studlike protrusions and their corresponding indentations, which loosely interlock with each other (Jones et al., 1980; Miyai, 1979). The *desmosomes* (Figure 14) are characterized by dense cytoplasmic underlining in thin sections. They are buttonlike points of contact that hold cells together. They also serve as anchoring sites for the intermediate filaments of cytokeratin type that extend from one side of the cell to the other across the cell interior. Desmosomes are apparently quite labile and their attachment plaques tend to become separted during the process of tissue dissociation (Staehelin, 1974). The *gap junctions* consist of arrays or plaques of intramembranous particles. As viewed in thin section by electron microscopy, they appear as bridges across the extracellular space with the segments of tightly apposed plasma membranes of adjacent hepatocytes (Revel and Karnovsky, 1967). In freeze-fracture replica, gap junctions are revealed to be round or oval-shaped plaques composed of tightly packed particles (8–9 nm in diameter) (Figure 15). When isolated

TABLE 1–6.
Volumetric Composition of Hepatic Ultrastructure

	Human liver* Percent		Rat Liver† Percent	
	Liver Tissue	Hepatocyte	Liver Tissue	Hepatocyte
Extrahepatocytic space	17.0		20.7	
Hepatocytes	83.0	100.0	79.3	100.0
Nuclei	5.0	6.0	5.6	7.3
Endoplasmic reticulum				12.1
Rough	8.0	9.4	9.6	7.2
Smooth	5.0	5.9	5.6	17.6
Mitochondria	18.0	22.0	14.0	1.8
Lysosomes	0.7	0.8	1.4	1.2
Peroxisomes	1.2	1.4	0.9	50.9
Cytosol	45.0	54.0	41.8	

*Data adapted from Rohr et al. (1976).
†Data adapted from Weibel et al. (1969).

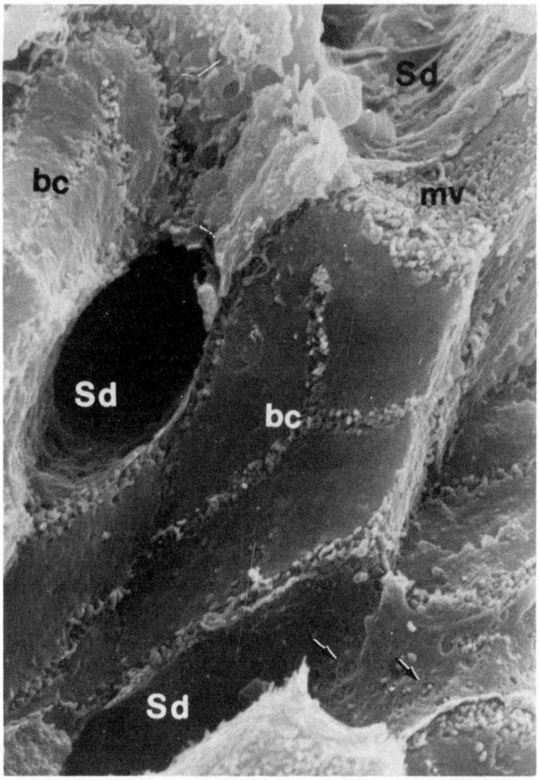

Figure 13. Scanning electron micrograph showing liver cell plates and sinusoides (Sd). Intercellular surface of hepatocytes, longitudinal profiles of bile canaliculi (bc) and sinusoids are revealed by fracturing the tissue prior to dehydration. Hepatocytes form single-cell-thick plates. Bile canaliculi are centrally located on the intercellular surface of hepatocytes. Sinusoidal surface of hepatocytes is studded with numerous microvilli (mv). Sinusoids are lined by fenestrated endothelial cells (×4,500).

and negatively stained, each of these particles has been shown to have an electron-opaque region (2 nm in diameter) believed to be the gap junction channel (Benedetti and Emmelot, 1968). In the normal hepatocyte, the gap junctions are estimated to occupy approximately 3% of the total cell surface area (Yee and Revel, 1978; Yancey et al., 1979). Functionally, the gap junctions form the channels for direct intercellular exchange of ions and small molecules. The assignment of the physiological functions of gap junctions in specific tissues is based on the relative importance of intercellular diffusion of these metabolites in the particular tissue. For cells such as hepatocytes that are involved in detoxification of small molecules, gap junctions are thought to increase the sink or buffer capacity of the cytoplasm (Spray et al., 1988).

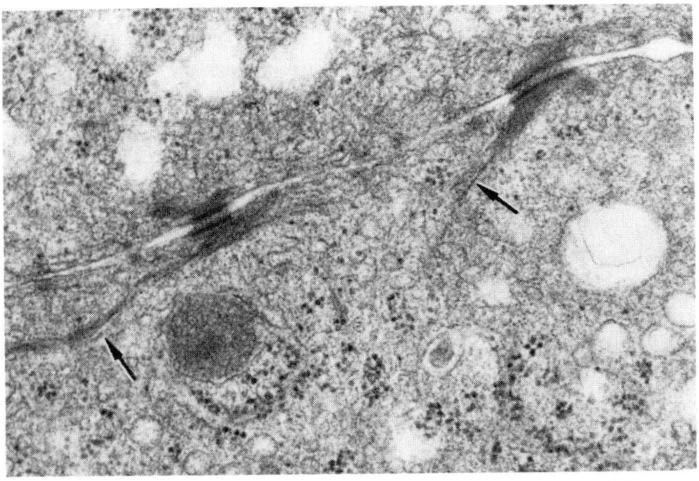

Figure 14. Transmission electron micrograph showing two desmosomes. Note intermediate filaments (arrows) are anchored to desmosomes (×36,000).

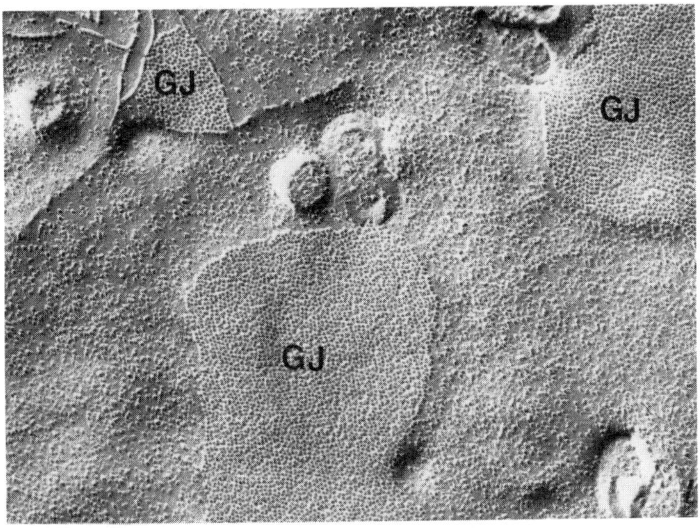

Figure 15. Transmission electron micrograph of a freeze-fracture replica showing gap junctions (GJ) which appear as plaques of tightly packed particles (×60,000).

Gap junctions are also believed to be involved in the control of tissue growth (Lowenstein, 1976). Reflecting their presumed roles in growth control, the cell surface area occupied by the gap junctions markedly decreases following partial hepatectomy and prior to cellular proliferation. It returns to normal state as regeneration proceeds (Meyer et al., 1981). Further details of the function and structure at molecular level of the hepatocyte gap junction are available in a recent review (Spray et al., 1988).

The *sinusoidal surface* is studded with microvilli and is the site across which metabolic exchanges between hepatocyte and the blood occur. The border between microvilli-studded and smooth domains of cell surface may shift depending on functional demands. For example, smooth-surfaced domains of the lateral surface decreases due to the expansion of microvilli-studded domain on the perisinusoidal border in cholestasis and regeneration (Desmet, 1972). In analogy to other epithelial cells, the sinusoidal and lateral surfaces are called the basolateral surface, and they differ from the canalicular surface in their biochemical composition, attached enzymes, and receptors (Evans, 1980; Maurice et al., 1985). For example, Na^+, K^+-ATPase, which serves as a sodium pump, is located in the basolateral, but not in the canalicular membrane (Blitzer and Boyer, 1978; Evans, 1980, whereas alkaline phophatase and Mg^{2+}-ATPase are localized to the canalicular but not basolateral membrane (Desmet, 1972). The functional roles of these two enzymes, however, are not understood. The distribution of these enzymes may change in response to altered functional states of the hepatocyte. For example, following bile duct obstruction, alkaline phosphatase activity spreads from the canalicular membrane to cover the entire cell surface within several hours (Hatoff et al., 1985; Ogawa et al., 1986). In addition, functionally active canalicular bile acid carrier protein becomes redistributed to and/or newly inserted into the basolateral domain. This reversal of the bile acid secretory polarity may represent a mechanism to prevent excessive accumulation of toxic bile acids within the hepatocyte (Fricker et al., 1987).

Endoplasmic Reticulum

What we now know as the endoplasmic reticulum was first recognized by Garnier between 1897 and 1899 as a basophilic component of the cytoplasm of glandular cells. He called this structure the ''ergastoplasm'' (''work plasm'' in Greek) because he presumed that it played a role in the elaboration of secretory products. During the 1930s it was found that the basophilia was due to RNA. By 1943, an isolated subcellular fraction called ''microsomes'' by Claude was found to contain most of the RNA of the cytoplasm. In 1953, Porter showed that his newly discovered membranous network, which he named the endoplasmic reticulum, could be identified with the highly basophilic and high RNA content of the ergastoplasm (for historical review, see Siekevitz and Zamecnik, 1981).

The endoplasmic reticulum (ER) consists of membranes arranged as parallel stacks of flattened cisternae and tubules. Two morphologically distinct regions of the ER can be distinguished (Figure 16): the rough ER (RER), which is studded with ribosomes on the cytoplasmic side of the membrane, and the smooth ER (SER), which is continuous with the membrane of the RER but lacks any attached ribosomes. These two regions also differ considerably in shape; whereas RER is organized in stacks of flattened cisternae, SER consists of a meshwork of tubules. The RER and SER occupy approximately 15% of the total cell volume, or 756 μm^3 per rat hepatocyte. The surface area of the ER is estimated to be approximately 63,000 μm^2 per hepatocyte or between 7.3 and 11 m^3/g liver, with 2.8 mg phospholipid and 4.7 mg protein per m^2. The ER of a hepatocyte has about 12.7×10^6 attached ribosomes. In addition to those attached to the membrane of the RER (membrane-bound ribosomes), many ribosomes are also present in the cytosol (free ribosomes) (Weibel et al., 1969; for additional data, see Blouin et al., 1977; Rohr et al., 1976).

The membrane of the ER undergoes constant renewal or turnover. The half-life of ER proteins ranges between a few hours and several days, with the average in rat liver being 60–70 hs. The turnover rates of phospholipids are more rapid than those of enzyme proteins. Despite the highly individual rates of turnover of practically all proteins and lipids in the ER, the basic composition, that is, the ratio between phospholipids and proteins, never changes significantly. This indicates that normal functions of the ER that are necessary for maintenance of the cell require a basic membrane structure that cannot be abolished or even changed significantly (De Pierre et al., 1988).

The hepatic ER is involved in (1) the synthesis of proteins for export, such as albumin, fibrinogen, prothrombin, and other plasma proteins, (2) the synthesis of enzymes associated with lysosomes and proxisomes, which remain within the cell, (3) the synthesis and metabolism of cholesterol, steroids, and bile acids, (4) the biotransformation of hydrophobic endogenous metabolites (e.g., bilirubin, steroids) and xenobiotics, (5) the metabolism of fatty acids, triglycerides, lipoproteins, and phospholipids, and (6) carbohydrate metabolism (glycogenolysis and release of glucose) (DePierre et al., 1988).

Reflecting the complexity of their function, both RER and SER are well developed in the hepatocytes. However, their relative quantities vary in cells in different acinar zones and may be markedly altered under different pathophysiologic conditions (Jones and Fawcett, 1966; Jones and Schmucker, 1977; Miyai, 1979). Proteins for export (e.g., albumin and other plasma proteins) are formed by the RER, enter its cisternae, and migrate through the SER and the Golgi complex and to the sinusoidal surface for release (Palade, 1975; Tavill, 1985). However, an immunohistochemical study of albumin synthesis (Lin and Chang, 1975) indicates that albumin is formed on the cytosol side of the RER and then released into the sinusoid apparently without passage

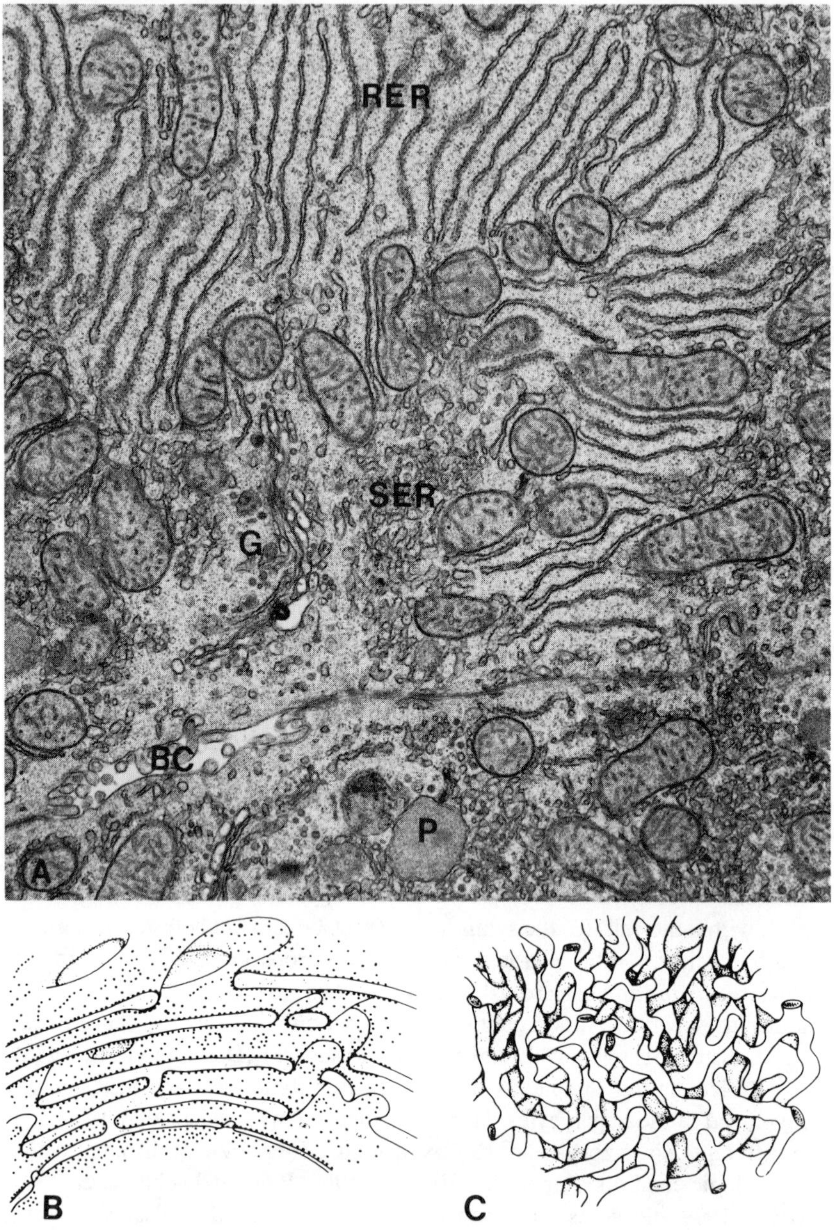

Figure 16. A: Transmission electron micrograph showing portions of two hepatocytes. It demonstrates appearance and relationship of organelles. RER = rough endoplasmic reticulum; SER = smooth endoplasmic reticulum; G = Golgi complex; BC = bile canaliculus; P = peroxisome ($\times$8,600). B: Diagrammatic representation of the RER which is organized in stacks of flattened cisternae with numerous ribosomes studded on the cytoplasmic side of the membrane. C: The SER, in contrast, consists of a meshwork of tubules without attached ribosomes. (B and C from Fig. 1–1 in *Visual Histology* by Moran and Rowley, published by Lea and Febiger, 1988, with permission of authors and publisher.)

through the ER cisternae or Golgi complex. The fact that albumin contains no carbohydrate in its molecules may render it unnecessary for albumin to be transported to the Golgi complex to acquire carbohydrates before its release into the cytosol. Presumably, albumin belongs to a class of proteins whose molecules are synthesized by bound polyribosomes and discharged directly into the cytosol. It should be remembered that not all protein synthesis in the hepatocyte involves the RER. It should be remembered that not all protein synthesis in the hepatocyte involves the RER. Many, including structural proteins of the cell itself, are formed by the free ribosomes. This process is particularly active during development and regeneration.

Many of the steps involved in cholesterol synthesis, including the rate-limiting enzyme 3-hydroxy-3-methylglutaryl coenzyme A reductase, are located in the microsomal fraction. The SER seems to be the main site of this synthetic activity since the hepatic capacity of cholesterol synthesis increases severalfold in association with phenobarbital-induced SER proliferation (Jones and Armstrong, 1965). Most steroids and lipid-soluble drugs and chemicals are metabolized into more water-soluble forms by the microsomal mixed-function oxidase system (a chain of enzymes such as cytochrome P450 and NADPH cytochrome C reductase) situated primarily in the SER. This enzyme system metabolizes lipid-soluble compounds of diverse chemical nature (e.g., phenobarbital, DDT, and some carcinogens), and when it is stimulated chronically by any of these compounds, its enzymes undergo induction with marked proliferation of the SER (Figure 17) (Jones and Fawcett, 1966; Remmer and

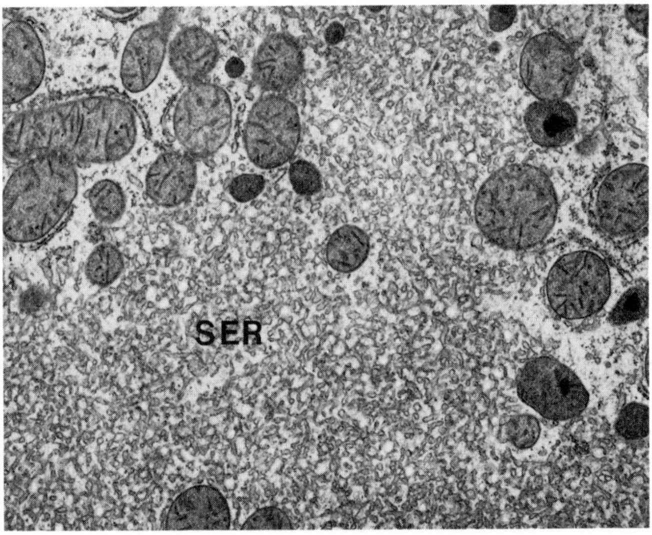

Figure 17. Proliferation of SER induced by phenobarbital (×22,000).

Merker, 1963). This adaptive response generally increases the liver's capacity to inactivate drugs, pesticides, and other noxious compounds as well as to metabolize steroid hormones. Induction of microsomal enzymes also enhances the activity of glucuronyl transferate, which may alleviate hyperbilirubinemia caused by the insufficiency of this enzyme. However, the induction of microsomal enzymes is not always beneficial to the host since the metabolites of some compounds are more toxic than their parent compounds. One such example is 3,4-benzpyrene, a relatively innocuous compound in cigarette smoke that becomes a potent carcinogen upon hydroxylation. Another example is carbon tetrachloride, which causes lethal cell injury upon the formation of free radicals during its biotransformation (Jones and Fawcett, 1966; Poli et al., 1985; Remmer and Merker, 1963).

Correlation between structure and function is also seen in the development of the SER in fetal and newborn livers. In the fetal liver, little SER is present and some of the enzymes for biotransformation are deficient. After birth, the SER increases in amount with a concomitant rise in the activity of the aforementioned enzymes that convert bilirubin and drugs into water-soluble metabolites capable of being excreted into the bile (Dallner et al., 1966a,b).

The ER is also involved in lipid metabolism. In the hepatocytes, triglyceride is formed from free fatty acids that are not used for energy production or membrane synthesis. Although a portion of triglyceride is stored as lipid droplets in the cytoplasm, the remainder is excreted into the blood as very low-density lipoprotein (VLDL). The VLDL droplets consist of a triglyceride core enclosed by a coat of apoprotein, phospholipid, and cholesterol. It is presumed that the apoprotein is produced by the RER and the lipid component by the SER. A portion of VLDL is then packaged in membrane vesicles budding off from the SER and releasing their contents into the blood. The remainder of VLDL is transported from the SER to the Golgi complex, where it acquires a carbohydrate component before it is released into the blood, also by exocytosis (Jones and Schmucker, 1977).

Golgi Complex

The discovery of the Golgi complex (apparatus) by Camillo Golgi, for whom it is named, took place in 1898. However, the controversy regarding whether the Golgi complex was a bona fide cell structure or an artifact continued until the mid-1950s. It stemmed from the question of whether one believed that the metallic impregnation methods (involving use of silver or O_sO_4) which Golgi and others used were staining a common structure with variable form and distribution in different cell types, or alternatively, that their methods produced artifactual deposition of heavy metals. With the introduction of electron microscopy in the early 1950s, those who believed in the existence of the Golgi complex began to outnumber those who did not,

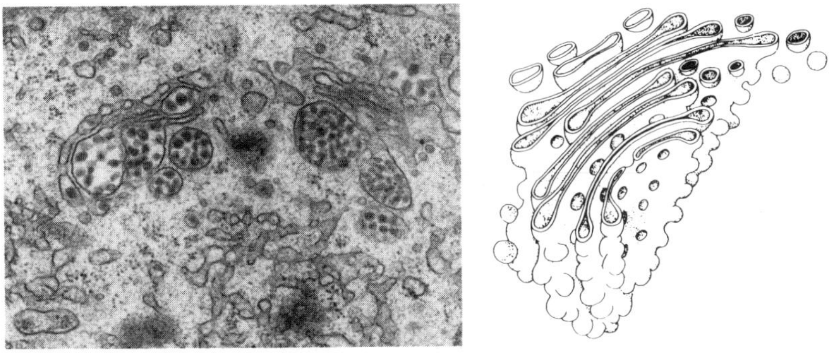

Figure 18. Transmission electron micrograph and diagrammatic representation of Golgi complex which consists of a stack of several disc-shaped cisternae and associated small vesicles. Electron micrograph shows electron-opaque secretory granules in cisternae and vesicles (×22,000). (The diagram is from Fig. 1–1 in *Visual Histology* by Moran and Rowley, published by Lea and Febiger, 1988, with permission of authors and publisher.)

and by the early 1960s even the most skeptical accepted the concept (Farquhar and Palade, 1981).

Morphologically, the Golgi complex consists of a stack of 4–6 dis-shaped cisternae, each with a diameter of 1 μm and a 30-nm lumen and associated small vesicles (20–30 nm in diameter) (Figure 18). The closely stacked cisternae (the Golgi stack) have two distinct faces: a convex, dis or forming face and a concave, trans or maturing face. The cis face is closely associated with the ER. The trans face is the face closest to the cell surface. The vesicles associated with the trans face include those containing protein or lipoprotein destined for secretion into the sinusoid after vesicular fusion with plasma membrane. Some of the vesicles in this region are primary lysosomes that recycle between the Golgi complex and secondary lysosomes (see section on Vasicular Transport System). In addition, an elaborate trans-tubular reticular membrane system has been demonstrated. This system is thought to be involved in terminal glycosylation, and in the sorting of endocytic and exocytic constituents of lysomosomes and plasma membrane (Griffiths and Simons, 1986; Farequhar, 1985). The Golgi complex are relatively numerous in the hepatocytes, and their number is estimated to be as many as 50 per cell. They tend to be more numerous in the proximity of the bile canaliculi. The membrane of the Golgi complex comprises approximate 2% of total cellular membrane surface area, or approximately 0.185 m^2 per gram of liver (Blouin, 1983).

Functionally, the Golgi complexes are involved in the transport of proteins and lipids for export and for uses within the cell. Sorting of proteins for selective export by modifying carbohydrate components added to proteins is one of the major functions of this organelle (Claude, 1970; Farquhar and Palade, 1981; Farquhar, 1985). The Golgi complex is also thought to participate in biliary secretion, but evidence for this concept is still circumstantial (Jones et al., 1980).

Vesicular Transport System

As mentioned above, proteins and lipoproteins are transported through the hepatocytes via membrane-bound vesicles and excreted into the blood. Vesicular transport is also involved in the endocytosis, intracellular transport, and exocytosis of a variety of other substances. Certain macromolecules are internalized specifically by *receptor-medicated endocytosis,* which differs from nonspecific fluid phase pinocytosis in that it involves binding of the ligand to be internalized to a specific receptor on the surface of the cell. Because endocytosis involves the formation of vesicles containing the ligand through pinching off from the plasma membrane, the ligand remains segregated from the cytoplasma by the surrounding vesicular membrane. Therefore, endocytosis differs from carrier-mediated transport systems, by catalizing the movement of small polar molecules and ions directly across the membrane barrier separating two aqueous compartment (Forgaç, 1988).

The macromolecules internalized by receptor-mediated endocytosis include serum proteins, such as low-density lipoprotein (LDL); asialoglycoproteins and transferin; hormones, such as insulin and epidermal growth factor (EGF); immunoglobulin A; and lysosomal enzymes. In addition to those substances normally taken up by cells, certain infectious agents such as many enveloped viruses and toxins seem to enter the cell through a receptor-mediated process (Forgac, 1988). Since proteins can be fixed and retained in the cell for morphological examination and their identity *in situ* can be confirmed immunologically and/or autoradiographically, or biochemically in isolated subcellular fractions, considerable progress has been made during the past decade in understanding the structure-function correlation of receptor-mediated endocytosis and related intracellular vesicular transport.

Figure 19 is a schematic diagram showing uptake, intracellular transport, and biliary secretion of some of the macromolecules internalized by receptor-mediated endocytosis. The first step in internalization of ligands by receptor-mediated endocytosis is binding of ligands to their corresponding cell surface receptors either prior to or following clustering of these receptors into specialized regions in the plasma membrane called *coated pits.* These pits are coated on their cytoplasmic surface with a protein called clathrin (Haigler et al., 1973; Wall et al., 1980). The involvement of receptors confers both specificity and the ability to concentrate ligands present at low concentrations in the serum (Forgac, 1988). Coated pits containing ligand-receptor complexes then pinch off from the plasma membrane to form clathrin-coated vesicles. Following this, internalized complexes appear in uncoated vesicular structures called *endosomes.* During an early phase (2 min) of endocytosis, these structures appear as small (0.1–0.2 μm) smooth-surfaced vesicles near the periphery of the cell, while at a later phase (5 min), ligand appears in larger (0.2–0.5 μm) vesicular structures with tubular extensions or those which contain small vesicles (i.e., multivesicular bodies) (Haigler et al., 1979; Geuze et al., 1983; Wall et al., 1980). The rapid appearance of ligands in uncoated

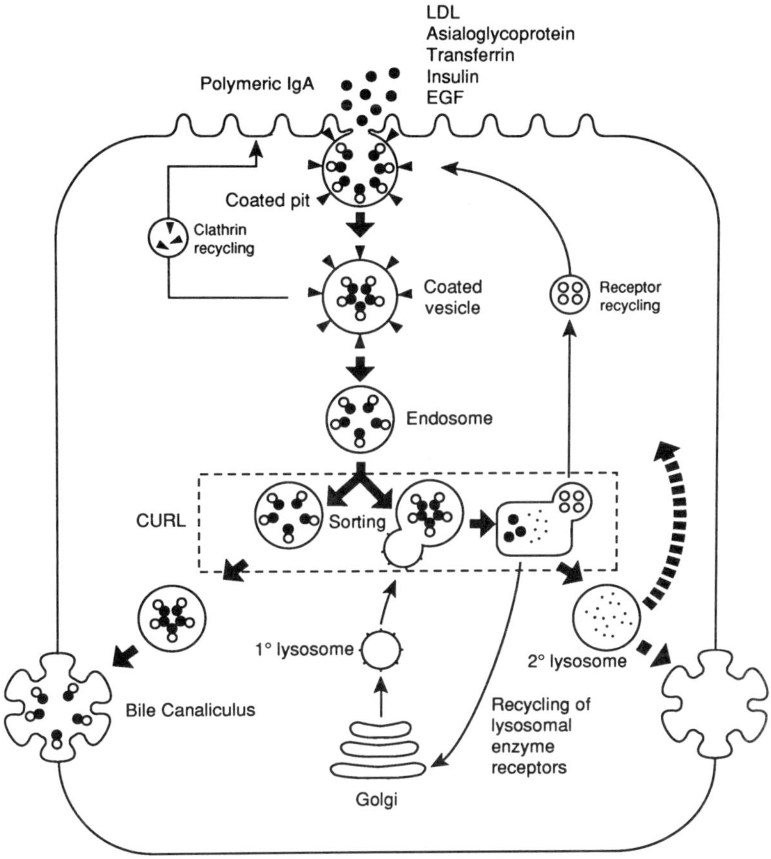

Figure 19. Schematic representation of the uptake, intracellular transport and biliary secretion of some of the macromolecules internalized by receptor-mediated endocytosis. Ligands (black dots) bind to their corresponding receptors (open circle) and endocytosed into clathrin (black triangles)-coated vesicles. They are subsequently sorted into appropriate vesicular compartments where receptors are uncoupled for recycling and ligands undergo digestion by lysosomal enzymes. The cytoplasmic region where sorting and digestion take place is called CURL (Compartment of Uncoupling of Receptor and Ligand). The products of lysosomal digestion of ligands are either secreted into bile or utilized within the hepatocytes (heavy dashed lines). Some macromolecules (e.g., polymeric IgA), however, are transported across the hepatocyte into bile without lysosomal degradation. In the case of IgA, a part of its receptor (secretory component) is secreted along with the bound IgA into bile.

endosomes shortly after their internalization through coated pits and coated vesicles suggests that some cellular mechanism is involved in removing the coat from such vesicles. According to recent studies, dissociation of clathrin is mediated by ATPase in the presence of ATP (Braell et al., 1984). Since internalized ligands rapidly move from small endosomal vesicles to larger ones, it has been suggested that endosomes undergo fusion either with each other or with preexisting vesicles within the cell. These larger vesicular-tubular structures have been termed CURL (compartment of uncoupling of receptor and ligand) (Geuze et al., 1983). Within CURL, exposure to low pH activates dissociation of the ligand-receptor complexes so that the unliganded receptor can return to the plasma membrane via recycling vesicles which bud off from CURL, and the ligand (located in the lumen of CURL) can be transported in a separate population of vesicles, which for most ligands destined for fusion with lysosomes (Forgac, 1988; Geuze et al., 1983, 1984). The process of receptor recycling is a property common to many receptor-mediated systems. Therefore, a given receptor may be utilized for multiple cycles of internalization and control may be exerted over the number of receptors present on the cell surface by regulating the efficiency of recycling (Brown et al., 1983). In some cases (e.g., EGF), however, the receptor continues with the ligand to be degraded in the lysosome (Stoscheck and Carpenter, 1984).

Not all ligands that are internalized by receptor-mediated endocytosis are directed to the lysosomes. An example is polymeric IgA, which is transported across the hepatocytes into bile without lysosomesal degradation. However, the IgA receptor is proteolytically cleaved and the ligand-binding domain (termed SC for secretory component) is secreted along with the bound IgA into bile. This proteolytic process does not seem to occur in lysosomes, indicating that sorting of IgA-receptor complex from other ligands destined for lysosomal degradation must occur (Geuze et al., 1984; Hoppe et al., 1985). A recent immunohistochemical study, using gold-conjugated antibodies against the receptor for asialoglycoprotein, lysosomal enzymes, and polymeric IgA, suggests that in addition to sorting ligands and receptor, CURL may also be involved in sorting of receptors from each other (Geuze et al., 1984). Whether receptor sorting is also triggered by acidification or by some other mechanism remains unknown.

It appears that receptor-mediated endocytosis shares many properties with other pathways of intracellular vesicular transport. One such example is that of newly synthesized lysosomal enzymes from the Golgi complex to the lysosome. As in the case of receptor-mediated endocytosis, this transport process is mediated by a specific receptor (the mannose-G-phosphatase receptor) which binds lysosomal enzymes in the Golgi complex and directs their packaging into vesicles destined to fuse with lysosomes. As with cell surface receptors, the mannose-G-phosphatase receptor is associated with coated pits in the trans Golgi cisternae which bud off to form coated vesicles

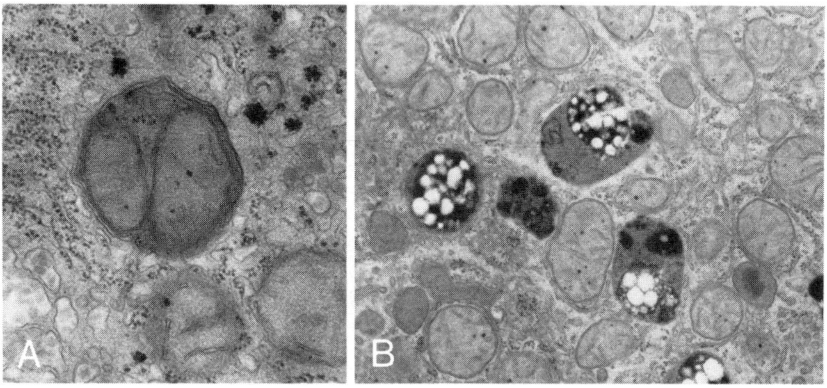

Figure 20. Lysosomes vary considerably in their morphology. A: This lysosome is involved in autophagocytosis of mitochondria and other organelles (×26,000). B: These lysosomes containing amorphous materials of variable density correspond to lipofuscin granules of light microscopy (×12,000).

(Geuze et al., 1985). In addition, lysosomal enzymes and their receptors dissociate at low pH, allowing the receptor to recycle to the Golgi complex. The structures in which this uncoupling occurs are morphologically similar to the CURL mentioned above suggesting that the same compartment may be used in both pathways (Geuze et al., 1985).

Lysosomes

The concept of lysosome as a sacklike structure surrounded by a membrane and containing acid phosphatase was first proposed on the basis of biochemical studies by deDuve and his associates in 1952. Between 1955 and 1956 combined biochemical and electron microscopic studies by Novikoff and others identified deDuve's lysosomes (lytic particles) with the "peribiliary dense bodies" in the hepatocyte which had first been described by Rouiller in 1954. Subsequent studies have established lysosome's role in intracellular digestion (for historical accounts, see Bainton, 1981; deDuve, 1975).

Lysosomes are defined as organelles specialized for intracellular digestion. They contain a wide variety of hydrolytic enzymes called *acid hydrolases* because they operate best at an acidic pH (about 5). Although it is simple to define lysosomes conceptually, it is difficult to identify them morphologically because they are highly pleomorphic. Lysosomal content may be homogeneous or heterogeneous and may include dense pigments, myelin figures, or partially digested organelles (Figure 20). Their positive identification in electron micrographs requires that the organelle be bound by a single membrane and show a positive histochemical reaction for acid phosphatase, a representative acid hydrolase (Bainton, 1981; deDuve, 1969, 1975; Novikoff et al., 1956; Novikoff, 1973). In addition, histochemistry has identified a region of SER and Golgi complex rich in acid phosphatase (Novikoff and Yam, 1978). It is thought that lysosomal enzymes synthesized in the RER and transported to the SER are subsequently transferred to the Golgi complex, where the enzymes are modified and packaged as lysosomes (Hasilik, 1980). The aforementioned group of acid-phosphatase-rich organelles (the Golgi complex, ER and lysosomes) has been called GERL (Novikoff and Yam, 1978).

Under physiologic conditions, lysosomes catabolize senescent organelles and inclusions. These "autophagic" processes maintain and rejuvenate the cells. When animals are starved, the hepatic mass decreases and lysosomes involved in autophagy increase in number, releasing energy-producing substances needed to sustain the entire body. Lysosomes also digest exogenous substances unnecessary for the cell or store some metabolites not needed immediately. For example, they participate in iron metabolism by storing ferritin-like compounds under normal conditions and by accumulating them in an iron storage disease (e.g., hemochromatosis) (deDuve, 1969, 1975; Novikoff, 1973).

Lysosomes seem to play important roles in a variety of pathologic processes, as suggested by their increase in size and number in viral hepatitis, anoxic injury, and cholestasis, to name a few. When a lysosomal enzyme is genetically deficient, it results in the accumulation of a compound normally metabolized by this enzyme (lysosomal storage diseases). An example is type II glycogenosis (Pompe's disease), in which glycogen accumulates in lysosomes because of the deficiency of α-glucosidase (deDuve, 1969; Novikoff, 1973). Detailed references on lysosomal disorders of the liver are available in an article by Ishak and Sharp (1987).

Peroxisomes

Biochemical studies on subcellular fractions that led to the discovery of lysosomes also identified another class of membrane-bound organelle containing enzymes that catalized reactions involving hydrogen peroxide, and hence have been termed *peroxisomes* (deDuve and Baudhuin, 1966). Three of these catalyzing enzymes produce hydrogen peroxide (Urate oxidase, d-amino acid oxidase and α-hydroxy acid oxidase) and one (catalase) destroys it. Electron microscopy of isolated peroxisomes proved that they were indeed the *microbodies* described earlier by Rouiller and Bernhard (1956).

Peroxisomes (*microbodies*) are ovoid-shaped particulate organelles bounded by a single membrane that contain a homogeneous matrix. Hepatic peroxisomes in many mammals other than humans have, within their matrix, a laminated structure called nucleoid that distinguishes them from lysosomes

(Figure 21) (Hruban and Rechcigl, 1969; Tsukada et al., 1966). Peroxisomes are about one-fourth as numerous as mitochondria in the hepatocyte (Loud, 1968; Weibel et al., 1969).

Functionally, peroxisomes are specialized for performing oxidative reactions using molecular oxygen. As stated above, they generate hydrogen peroxide, which they both use and destroy by means of the catalase they contain. They are involved in detoxifying various molecules; for example, oxidation of ethanol and acetaldehyde. It is estimated that peroxisomes normally account for about 10% of the total oxygen uptake of the liver. In addition to detoxification, peroxisomes catalize the β-oxidation of fatty acid to acetyl CoA which can be utilized by the mitochondria (to feed the citric acid cycle) or by biosynthetic reactions elsewhere. Peroxisomes are also thought to be involved in purine metabolism. (deDuve and Baudhuin, 1966; Lazarow, 1988).

Hepatic peroxisomes proliferate during embryologic development, during recovery from partial hepatectomy, and after the administration of a lipid-lowering agent, clofibrate, whereas they are absent or markedly reduced in number and size in certain congenital metabolic diseases such as Zellweger's cerebro-hepato-renal syndrome. However, the mechanism(s) regulating the biogenesis of peroxisomes is unknown. Several specific enzymatic defects of peroxisomes have been identified (Goldfischer, 1988).

Mitochondria

The earliest description of intracellular structures that probably represent mitochondria date back to the 1840s, only a few years after the discovery of the cell nucleus. The name *mitochondria* was first used in 1898 by Benda,

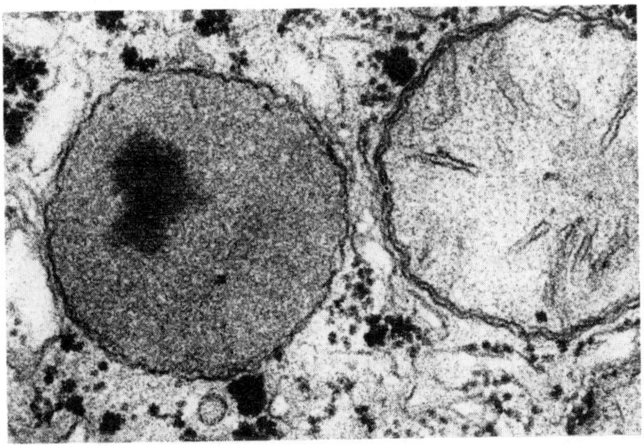

Figure 21. A peroxisome (left) and a mitochondrion (right) in a rat hepatocyte. A laminated, crystalloid core (nucleoid) seen in this peroxisome is normally absent in those of humans (× 50,000).

and derived from the Greek ''mitos'' (thread) and ''chondros'' (granule), referring to the appearance of these structures. In 1900, Michaelis found that the redox dye Janus Green B serves as a specific supravital stain of mitochondria, and this feature remained the ''official portrait'' of mitochondria until 1952, when the first high-resolution electron micrographs of mitochondria were published by Palade. Correlated morphologic and biochemical studies on isolated mitochondria initiated during the early 1940s have made a major contribution to our current knowledge of mitochondrial structure and function. For details of historical accounts, see Ernster and Schatz (1981).

Mitochondria in electron micrographs appear as elongated cylinders bounded by two layers of membrane, the outer and inner membranes. Together, they enclose and define two separate compartments, the matrix space and intermembrane space. The inner membrane is highly convoluted, forming many infoldings, known as *cristae,* projecting into the matrix space (Figure 16). According to recent studies by Candipan and Sjostrand (1984, a,b,c), however, some of the morphological features of the mitochondria, which are widely accepted as real, may be artifacts caused during preparation for electron microscopy. They suggest that (1) the inner and outer membranes are closely apposed and that there is no space within the cristae; (2) the cristae membrane is not simply an infolding of the inner membrane but differs structurally from it; (3) there is no evidence that the cristae undergo conformation changes in different metabolic states; and (4) two size classes of mitochondria, often referred to as ''condensed'' and ''orthodox,'' are preparatory artifacts.

As regards the functions and topographical localization of enzyme systems of the mitochondria, detailed description is beyond the scope of this chapter. It would suffice to state that the inner membrane and cistae contain respiratory enzyme complexes concerned with oxidative phosphorylation, whereas the matrix contains most of the components of the Krebs cycle and the enzymes involved in β-oxidation of fatty acids. The outer membrane is relatively unimportant as a locus for enxymes (for details, see Ontko and Dashti, 1976; Tzagoloff, 1982). Hepatic mitochondria are thought to be capable of replicating themselves with a half-life of about 10 days. This relatively rapid turnover seems to be carried out by autophagic sequestration of effete mitochondria by lysosomes; it is likely that this process is responsible for preserving the functional efficiency of mitochondria.

Mitochondria are numerous (about 800 to 1000 per cell) in the hepatocyte and vary in their number, size, shape, and enzymatic properties depending on the position of the cell within the hepatic acinus. They are smaller and more numerous in the hepatocytes in the acinar zone 3 than in those in zone 1, and occupy 13% of the total cell volume, in contrast to 20%, of cell volume in zone 1 cells (Loud, 1968). These differences may reflect an adaptation to the variation in the metabolic environment within the microcirculation.

The morphology of mitochondria varies greatly not only under pathologic but also under physiologic conditions, partly because of their mobility and plasticity in living cells. This makes it difficult to interpret mitochondrial morphology in relation to pathologic processes, except for a few conditions. During a deficiency of riboflavin, essential fatty acid, or dietary iron, giant mitochondria develop. Upon correction of the deficiency, the giant mitochondria return to normal size by means of division (for recent review, see Smith and Ord, 1983).

Cytoskeletal System

The cytoskeletal system consists of the microtubules and three different types of filaments: actin, myosin, and intermediate filaments. It is well established that virtually all nonmuscle cells have homologous cytoskeletal components so that much of what has been learned about the basic principles of cytoskeletal structure and function in nonmuscle cells is applicable to the hepatocyte (Phillips and Satir, 1988). *Microtubules* are long hollow, flexible tubules, about 24 nm in diameter, comprised of subunits of tubulin molecules that exist in a dynamic state of equilibrium with monomeric or oligomeric tubulin. Microtubules determine the distribution of intermediate filaments and are responsible for establishing and maintaining cell polarity. They are also implicated in the transport mechanism for secretory granules and their discharge from the cell. Microtubular functions may be impaired by colchicine, vinblastine, or vincristine which inhibit the formation of microtubules (Alberts et al., 1983; Soifer, 1986).

Little is known about the microtubules in the hepatocyte, in part because they are not abundant and their structure is not stable. Tubulin comprises about 1% of the soluble protein of the hepatocyte, and about 4% of the tubulin is assembled into microtubules (Reaven et al., 1977). There is good evidence that microtubules are involved in the secretion of proteins and lipoproteins by the hepatocytes (LeMarchand et al., 1974). Impaired vesicular transport and secretory function of the hepatocyte seen in alcoholic liver disease is attributed to depolymerization of microtubules induced by ethanol and its metabolite acetaldehyde (Baraona et al., 1984). Although the precise mechanism is not clearly elucidated, microtubules seem to play a role in bile secretion (Phillips and Stair, 1988).

Actin filaments (microfilaments) are the polymerized form of actin (F actin or fibrous actin) which measures 6 nm in diameter. They are formed from a pool of soluble subunits, G (globular) actin, and disassembled when no longer needed. This balance between assembly and disassembly is disturbed by specific chemicals that inhibit or enhance polymerization and lead to abnormal accumulation of either the polymer filaments or depolymerized subunits. Cytochalasins inhibit polymerization while phalloidin enhances it (Alberts et al., 1983).

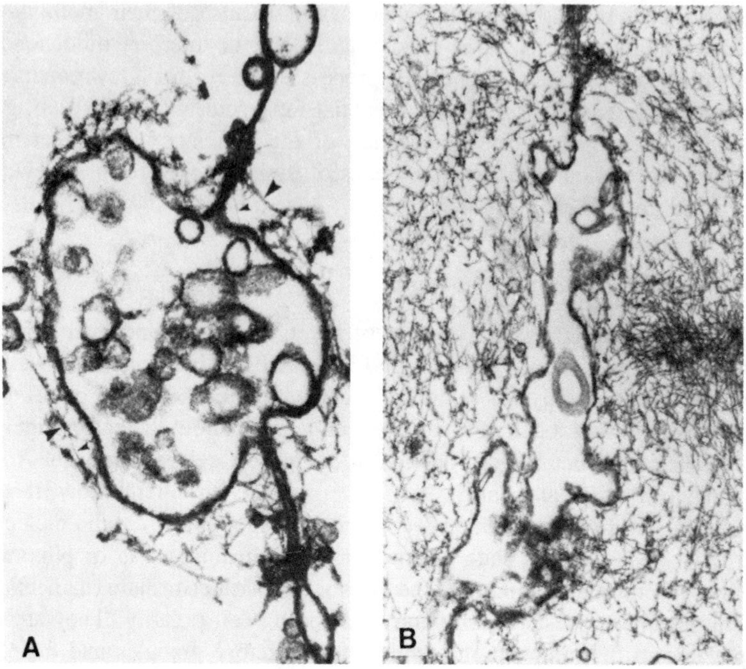

Figure 22. A: A bile canaliculus isolated in rat liver plasma membrane fraction. Note the network of filaments that remains around the canaliculus. Heavy meromycin decoration (not shown) reveals that many of the filaments are actin (×45,700). B: Similar preparation from an animal pretreated with phalloidin. Network around the canaliculus has increased (×45,700). (Courtesy of Dr. M.J. Phillips.)

In the hepatocyte, the estimated amount of actin varies from 1.5 (rat) (Gordon et al., 1977) to 5% (rabbit) (Brandon, 1976). Actin filaments form interconnecting network throughout the cytoplasm of the hepatocyte, but they are more conspicuous in the pericanalicular and perisinusoidal regions with attachment to the plasma membrane (French and Davies, 1975) and junctional complexes (Elias et al., 1980) (Figure 22). It has been shown that they are involved in contractile movement of bile canaliculi (Oshio and Phillips, 1981). This suggests that pericanalicular actin-myosin filament interaction may act as a pump to facilitate bile flow in the canalicular system (Oshio and Phillips, 1981). Alternatively, the observed canalicular motility may reflect a pressure valve phenomenon (Boyer, 1987). It is of interest that two types of actin inhibitors, cytocholasin and phalloidin, impair canalicular contractions and bile flow. Cytochalasin causes this by enhancing disassembly of actin filaments, whereas phalloidin does it by inhibiting their disassembly (Phillips and Poucell, 1987). In humans, ultrastructural changes similar to that caused by cytochalasm has been seen in the intrahepatic cholestasis caused by norethandrolone. Pericanalicular accumulation of actin filament resembling that caused by phalloidin has been noted in North American Indian Childhood cirrhosis and other types of cholestasis (Phillips and Poucell, 1987). It is, therefore, possible that actin filaments play a role in the pathogenesis of cholestasis in humans. Aside from canalicular motility, actin filaments may also be involved in the vesicular transport in the Golgi and pericanalicular regions (Oshio and Phillips, 1981) and that of receptor-mediated endocytosis (Salisbury et al., 1981). It is probable that vesicular transport in both the pericanalicular and perisinusoidal regions (and possibly elsewhere) of the hepatocytes is mediated by actin filaments.

Myosin is present in nonmuscle cells, but it differs from that in muscle cells in some respects. Its amount in nonmuscle cells is much smaller than that in skeletal muscle cells so that the myosin/actin ratio in the former is 1:100 in contrast to 1:6 in the latter. The dominant regulatory process for contraction in nonmuscle cells and smooth muscle is myosin-based, whereas that in skeletal muscle cells is actin-based (Phillips and Satir, 1988). Interaction between myosin and actin provides motility to a variety of subcellular events described above. The presence of myosin in the hepatocytes have been confirmed (Brandon, 1978), but little is known about it.

Intermediate filaments are a family of morphologically similar protein fibers with a diameter of 8–10 nm, which is intermediate between that of actin fibers (6 nm) and myosin fibers (15 nm). They are present throughout the cytoplasm, but are more numerous in the regions of the cell subjected to mechanical stress. Although their function is not known, they are thought to serve as scaffolding within the cytoplasm and as "mechanical integrater of cellular space" by providing support for actin filament/microtubule systems (Lazarides, 1980). Individual polypeptide subunits comprising intermediate filaments are arranged in an α-helical coil, which provides the filaments with high tensile strength and flexibility (Steinert, 1978). Although intermediate filaments are similar structurally, they are heterogeneous in their subunit composition and antigenicity. They are of five major types: cytokeratin, vimentin, desmin, glial fibrillary acidic protein, and neurofilament triplet protein (Wang et al., 1985). Cytokeratins are characteristic of epithelial cells including hepatocytes (Moll et al., 1982). They undergo alterations in their antigenicity as the differentiation of epithelial cells proceeds, so that epithelial cells can be classified according to the expression of their cytokeratin polypeptides (Sun et al., 1985).

In the hepatocyte, intermediate filaments form extensive networks in the cytoplasm which is revealed by electron microscopy with detergent extraction. They are especially concentrated around the nucleus, along cellular borders, and around bile canaliculi where they form a tubelike sheath (Ohta et al., 1988) (Figure 23). In addition, bundles of the filaments insert into desmosomes. Biochemically, hepatic intermediate filaments are of two types of cytokeratins which correspond to subtypes 8 and 18 in the scheme of Moll et al. (1982). Although intermediate filaments are structurally stable, the chemical composition changes rapidly when hepatocytes are placed in culture: not only does their polypeptide composition change but also vimentin appears within 2–3 days (Franke et al., 1979).

In some liver diseases, intermediate filaments form irregularly shaped aggregates which correspond to the Mallory body or alcoholic hyaline seen by light microscopy (Figure 24). Although the Mallory body is considered to be a hallmark of alcoholic liver disease, it is also seen in a variety of liver diseases, including chronic cholestasis, drug-induced liver injury, neoplasm, and even in extrahepatic locations (Phillips and Satir, 1988). Since Mallory bodies can be induced experimentally by microtubule inhibitors (Griseofulvin and colchicine), it is postulated that microtubular disassembly perturbs the normal structural relationship between microtubules and intermediate filaments with resultant collapse of intermediate filament network and aggregation of filaments (Denk, 1985).

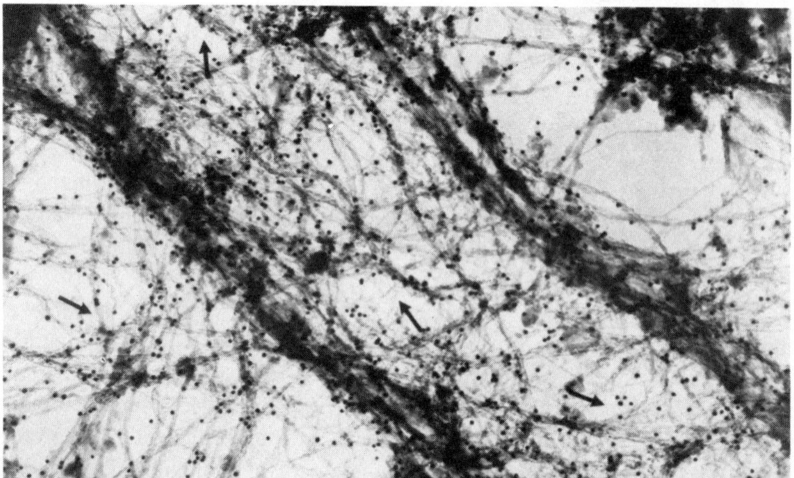

Figure 23. Pericanalicular sheath of intermediate filaments after extraction of cytosol and other organelles. Intermediate filaments are identified by the attachment of anti-CK55 antibody labeled with colloidal gold particles (dense dots). Note that most of the filaments stained except for actin (arrows) (×58,900). (From Ohta *et al.*, 1988 with permission of authors and publisher.)

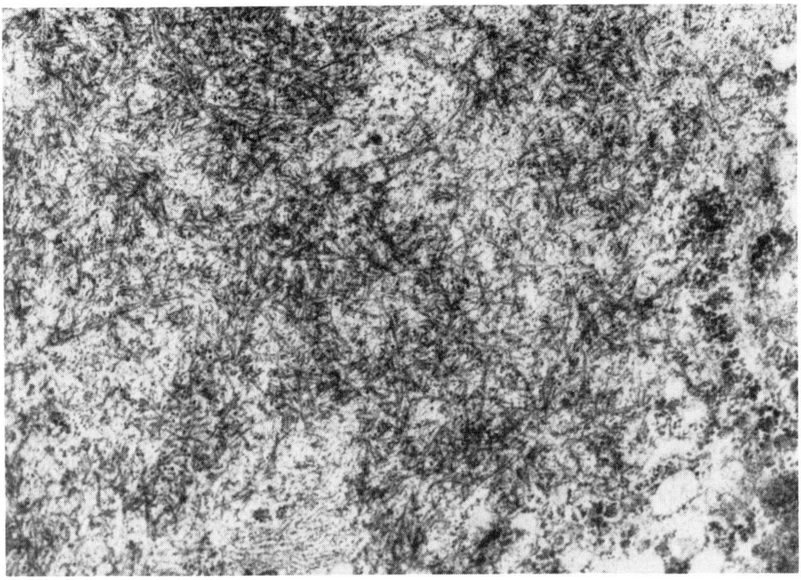

Figure 24. Transmission electron micrograph of Mallory body (alcoholic hyaline) in a hepatocyte from a patient with chronic alcoholic liver disease. The inclusion consists of aggregates of intermediate filaments lying free in the cytoplasm (×30,000).

Nucleus

The nucleus comprises 5–10% of the total volume of adult hepatocytes (Rohr et al., 1976; Weibel et al., 1969) and its size varies in correlation with the doubting of its DNA content. In the adult liver 30–80% of the nuclei are polyploid and 10–25% of hepatocytes are binucleate, whereas nearly all hepatocytes are mononucleate and diploid in the fetal or newborn liver. Polyploidy and binucleation show variations with respect to species, age, mitotic activity, endocrine regulation, and spatial distribution within the hepatic parenchyma (Alfert and Geschwind, 1958; Daoust, 1958; Geller, 1965; James et al., 1979). In areas around the terminal hepatic venules and in periportal regions, diploid nuclei and binucleate cells with two diploid nuclei are more numerous than elsewhere. Cells with tetraploid and octoploid nuclei are distributed in a parenchymal zone in a broad area between the two zones described above. These findings suggest that the functional heterogeneity of hepatocytes should be taken into account with regard not only to the metabolic zonation but also to the ploidy classes (Sasse, 1986). Adult hepatocytes are slowly self-replicating cells with a low level of DNA synthesis and mitosis. According to Carriere (1969), mitotic index in adult rat liver is about 0.03%. The nucleus is enclosed by the *nuclear envelope* (Figure 25), consisting of a pair of membranes. Since the outer nuclear membrane is continuous with the membrane of RER, the space formed between the outer and inner nuclear membrane is continuous with the cisternae of the ER. The inner and outer membranes frequently fuse to form the *nuclear pores*, which have an open channel through their center. Since RNA molecules and ribosomes are formed in the nucleus and proteins that function in the nucleus are synthesized in the cytoplasm, the nuclear pores are thought to be an important passage for the exchange of these macromolecules (Busch and Schildkraut, 1988).

The nucleus contains one or more prominent nucleoli. The *nucleolus* (Figure 25) consists of the granular and fibrillar components and its associated chromatin, and it plays a major role in the synthesis of all the cells' ribosome. Since almost all cellular growth or enlargement is closely correlated with the synthesis of new ribosomes, the functional state of the nucleolus is reflected in its morphology. With increased rate of ribosomal production (e.g., regenerating or neoplastic hepatocytes), the nucleolus becomes enlarged with a corresponding increase in the number of the granular component (Busch and Schildkraut, 1988; Bouteille et al., 1974). The *chromatin* (Figure 25) is condensed along the nuclear periphery and in association with the nucleolus, but it is generally more diffusely dispersed (hence, it stains less intensively with the hematoxylin of the hematoxylin and eosin stain) than that in the nuclei of other cellular types in the liver. Chromatin is a complex of nuclear

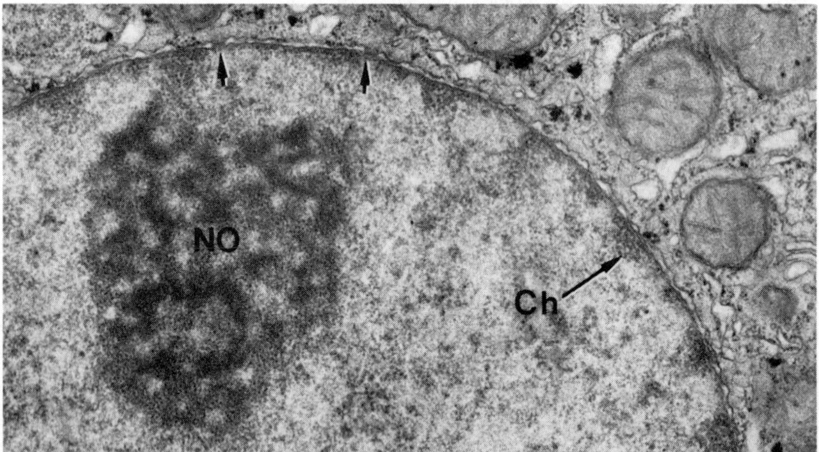

Figure 25. A portion of a nucleus of a hepatocyte. Nuclear chromatin (Ch) is condensed as ill-defined dense areas along the nuclear envelope. Nuclear pores (arrows) are seen between masses of peripheral chromatin. In a nucleolus (NO) granular component and ill-defined darker areas (which consists primarily of fibrillar component) are seen (×21,000).

DNA, its associated proteins (histones and nonhistone proteins), RNA, and small molecules (Busch and Schildkraut, 1988). It is a main constituent of the nucleus and the major repository of genetic information in the cell.

SINUSOID

The hepatic sinusoids form rich vascular networks with extensive anastomosis, and their lining cells are directly apposed to the parenchymal cells with essentially no intervening connective tissue. The lining of the sinusoid consists of a layer of highly porous endothelial cells and, unlike conventional capillaries, lacks a structurally defined basement membrane. Such a structure facilitates exchange between blood and hepatocytes. In addition, fixed macrophages (Kupffer cells) rest occasionally on the endothelial lining. (Fahimi, 1982; Widman et al., 1972; Wisse, 1972) (Figure 26).

Endothelial Cells

The endothelial cells are flattened and thin, and only their nuclei protrude into the sinusoidal lumen. Their cytoplasm is perforated with numerous fenestrae that are arranged in groups called sieve plates (Figure 27). The number and diameter of the fenestrae may vary according to the animal species (Fraser et al., 1986) and their localization in the hepatic lobule (Vidal-Vanaclocha and Barbera-Guillem, 1985; Wisse et al., 1985). Fenestration are larger in diameter in the periportal than in the centrolobular region, but the number of fenestrations per unit area is greater in the latter. The resulting porosity is greater in the centrolobular than in the periportal regions. The fenestrations are dynamic structures and undergo changes in their size and number in response to hormones and inhibitor of cytoskeletal fibers (Steffan et al., 1987). The neighboring endothelial cells are in contact with each other at their periphery, forming poorly defined cell contacts. As revealed in freeze-fracture preparations, these cell contacts consist of arrays of membrane particles, which remain separated from each other by particle-free intervals. This appearance probably represents an abortive form of tight junction formation, which provides mechanical adhesion without forming a permeability barrier between adjacent endothelial cells. Therefore, the only structurally open communication between the sinusoid and perisinusoidal space is through the fenestrations of endothelial cells (Yee and Revel, 1975).

One of the major functions of endothelial cells is thought to be a selective barrier between blood and hepatocytes. For example, chylomicrons newly generated in the digestive tract are too large to pass through endothelial fenestrations. The chylomicrons must first be reduced in size through metabolism in the bloodstream before they can pass through the sinusoidal lining

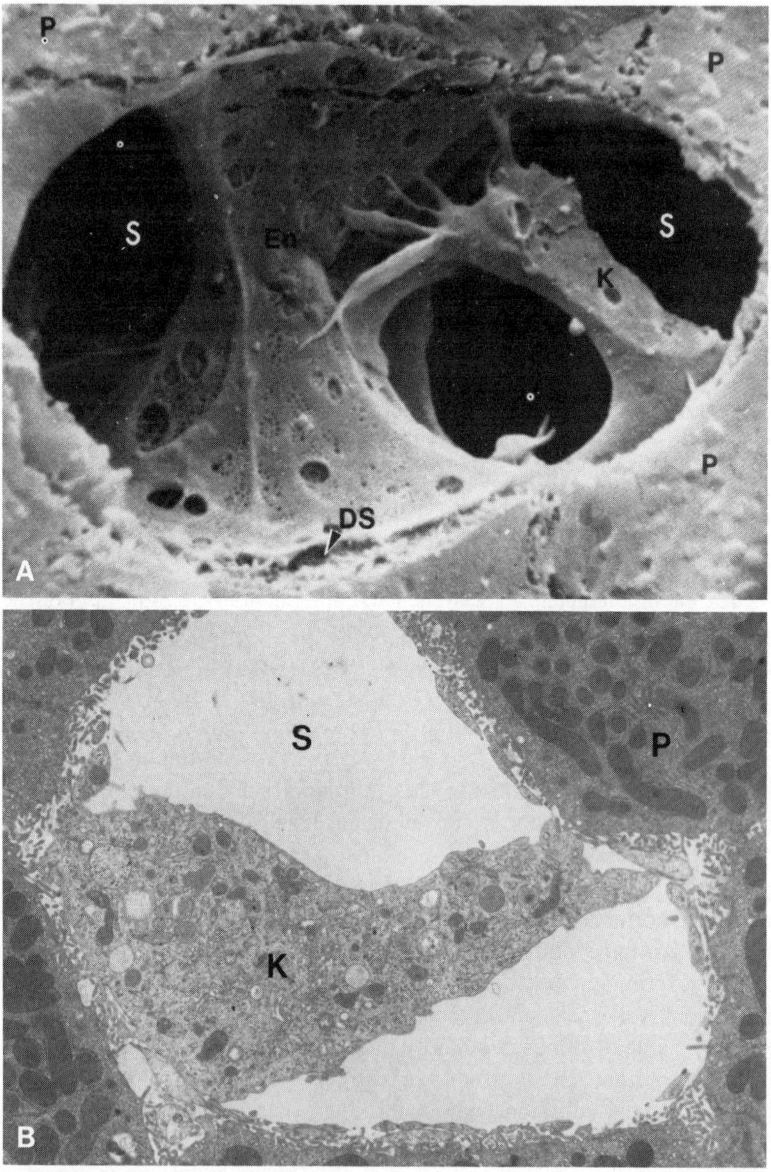

Figure 26. Scanning (A) and transmission (B) electron micrographs demonstrating the relationship of Kupffer cell (K) to the endothelium (En) and surrounding parenchyma (P). Note that the Kupffer cell is anchored to endothelial cells by cytoplasmic processes. DS = space of Disse. A (×15,000), B (×4,000).

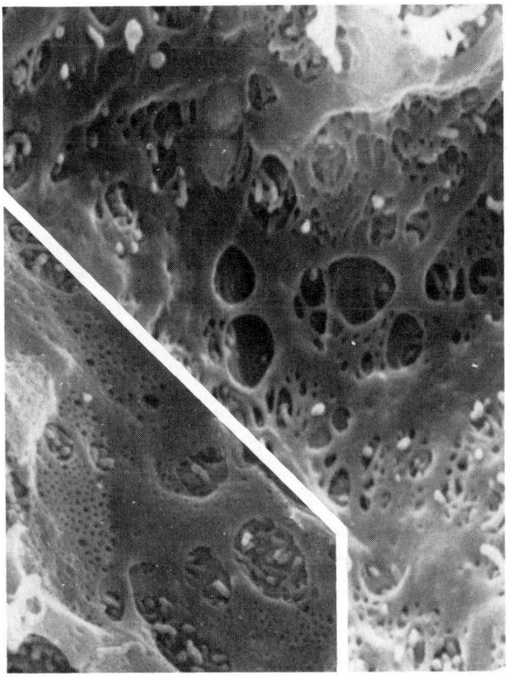

Figure 27. Scanning electron micrographs of the surface of sinusoidal endothelial cell. The cytoplasm of the endothelial cell is thin and perforated with numerous fenestrae of variable sizes (× 12,500 for both micrographs).

and be taken up by hepatocytes. Thus, endothelial cells protect the liver against exposure to excessive quantities of triglycerides (De Zanger and Wisse, 1982). Endothelial cells also play an important role in the regulation of lipoprotein metabolism. They have abundant scavenger receptors for modified low-density lipoproteins (LDL) and this enables them to take up such lipoproteins efficiently for lysosomal degradation (ban Berkel, 1982). They also possess receptors for liver lipase which is synthesized and secreted by hepatocytes and then bound to endothelial surface. Liver lipase breaks down circulating lipoproteins to cause the release of cholesterol and other components (van Berkel, 1982). Other functions of endothelial cells include turnover/catabolism of glycoproteins in the plasma through endocytosis and lysosomal digestion and synthesis of effector molecules (Brouwer et al., 1988).

Kupffer Cells (Fixed Macrophages)

These cells were first described by von Kupffer in 1899 as stellate cells with their processes traversing the sinusoids. Thus, they were thought to lie within the sinusoid but fixed to the endothelium. This was confirmed several decades later by electron microscopic studies (Widman et al., 1972; Wisse, 1972). Kupffer cells are thought to be end stage cells in the mononuclear phagocytic series. There is no evidence that they can be formed from endothelial cells (Bradfield, 1984). In normal adult rats, Kupffer cell population is maintained predominantly by self-renewal with a long life. Their proliferation under pathologic conditions results mainly from hyperplasia of local population and little from extrahepatic recruitment of monocytes (Jones and Summerfield, 1988). A bone marrow origin of Kupffer cells was suggested by an observation that a transplanted liver from a donor of opposite sex in a human became populated with Kuffer cells that had the sex karytype of the host (Gala et al., 1978). However, it seems likely that in adult animals Kupffer cell of bone marrow origin is quantitatively important only under conditions in which they are extensively lost. Kupffer cells comprise 80–90% of fixed macrophages of the reticuloendothelial system (Biozzi and Stiffel, 1965) and about 10% of all cells in the liver. Within the hepatic acinus, about 43% of Kupffer cells are located in the periportal zone, 29% in the perivenous, and 28% in the intermediate zone, respectively (Bouwens et al., 1986). The predominantly periportal distribution of Kupffer cells suggests that they monitor blood entering the sinusoid for materials that can be phagocytized and endocytosed.

Kupffer cells are anchored to endothelial cells by cytoplasmic processes (Figure 26). Their highly variable shape suggest that their form and their relations to the endothelium change continuously (Fahimi, 1982; Widman et al., 1972; Motta, 1984; and Wisse, 1972). They have morphologic features characteristic of macrophages. They do not form desmosomes or other intercellular junctions. They possess structures related to endocytosis: (1) bristle-coated vesicles (about 100 nm in diameter) and (2) peculiar sinous invaginations of the plasma membrane with a central dense line between parallel membranes. The latter are called "vermiform bodies" and have been seen in macrophages in extrahepatic sites, but are never seen in endothelial cells. The cytoplasm of Kupffer cells is not fenestrated and more abundant in organelles and more heterogenous in appearance than that of endothelial cells. Vacuoles of variable appearance presumed to be lysosomes are conspicuous. The cytoplasm also contains RER, Golgi complex, mitochondria, peroxisomes, and other organelles in modest amounts. Kupffer cells are distinguishable from other sinusoidal cells histochemically in that they stain for endogenous peroxidase and tartrate-resistant acid phosphatase (Fahimi, 1982; Widman et al., 1972; Wisse, 1972).

The main function of Kupffer cells is clearance of particulate and non-particulate materials from the blood. Particulate materials phagocytized by Kupffer cells include cellular debris (e.g., necrotic cells, organelles, and membranes) circulating tumor cells, microorganisms including viruses and exogenous particulate materials (e.g., colloidal gold, latex particles). Kupffer cells also endocytize a wide variety of macromolecules, e.g., immunoglobulins (IgA, IgG), immune complexes, enzymes, bacterial toxins, etc. Endotoxin appears to be taken up selectively by Kupffer cells, whereas many other nonparticulate macromolecules are taken up by endothelial cells as well (Jones and Summerfield, 1988; Plaaning-van Dalen et al., 1981). These functions primarily involve endocytosis and lysosomal digestion of endocytosed materials. Reflecting such a functional requirement, lysosomes comprise 14% of the cellular volume of Kupffer cells, in contrast to 7% in endothelial cells and hardly any space in fat storing cells. Kupffer cells account morphometrically for 17% of the lysosomal volume of the liver (Blouin, 1977).

In addition to the aforementioned functions, there is increasing evidence that Kupffer cells interact with hepatocyte and other cells. For example, Kupffer cells produce a factor that stimulate fibrinogen synthesis and secrete erythropoietin and colony stimulating factor. They also seem to influence fibroblast proliferation suggesting their role is fibrogenesis (Jones and Summerfield, 1988).

PERISINUSOIDAL SPACE OF DISSE

This is an extravascular space defined by the sinusoidal wall (i.e., endothelial cells) and the sinusoidal surface of the hepatocytes. As mentioned earlier, this space is freely permeable to blood plasma that comes in direct contact with the hepatocytes. A portion of plasma that has entered the perisinusoidal space is thought to flow in the direction opposite to that of the blood to the portal tract, where it enters lymphatic vessels. This subject will be discussed in the Lymphatic System section.

Lipocytes

Lipocytes (also called fat storing cells or Ito cells) are normally located in recesses between hepatocytes. They were initially recognized as a distinct type of cells by Ito using light microscopy (Ito and Nemoto, 1952) and further characterized by subsequent electron microscopic studies (Ito and Shibasaki, 1968; Fahimi, 1982; Wake, 1980). Lipocytes are characterized by conspicous lipid droplets in their cytoplasm (Figure 28). These lipid droplets contain vitamin A and their amount varies markedly, depending on the age of the animals and the dietary intake of vitamin A. The organelle composition of the remainder of the cell is reminiscent of that in fibroblasts with relatively well-developed RER and Golgi complex. Lipocytes have little, if any, manifestation of phagocytic activity. Functionally, they have been shown to be the major physiological site of vitamin A storage in the mammalian body (Hendriks et al., 1985). According to Brouwer et al. (1986), vitamin A metabolites are the major constituents of the lipid droplets. It is yet unclear if lipocytes play a role in the metabolism of other fat-soluble vitamins. As regards other functions, lipocytes have been implicated in perisinusoidal fibrogenesis because of their resemblance to fibroblasts and their close association with collagen fibers in the perisinusoidal space (Ballardini et al., 1983). In addition, they have been shown to produce various types of collagens (I, III, IC) in culture. However, hepatocytes in culture also secrete collagens (DeLeeuw et al., 1984). Therefore, the question as to whether lipocytes play a major role in fibrogenesis remains unresolved.

Pit Cells

Pit cells (large granular lymphocytes) were first recognized as a distinct cell type by Wisse et al. (1976). Initially, they were presumed to have an endocrine function because they contain cytoplasmic granules which resemble those of endocrine cell types (Wisse et al., 1976). However, subsequent studies suggest that they are the perisinusoidal equivalent of large granular lymphocytes and natural killer lymphocytes (Bouwens and Wisse, 1988). Although their morphology has been well characterized (Wisse et al., 1976; Kaneda and Wake, 1983), pit cells remain the least investigated sinusoidal cell type, presumably because they are seen only infrequently.

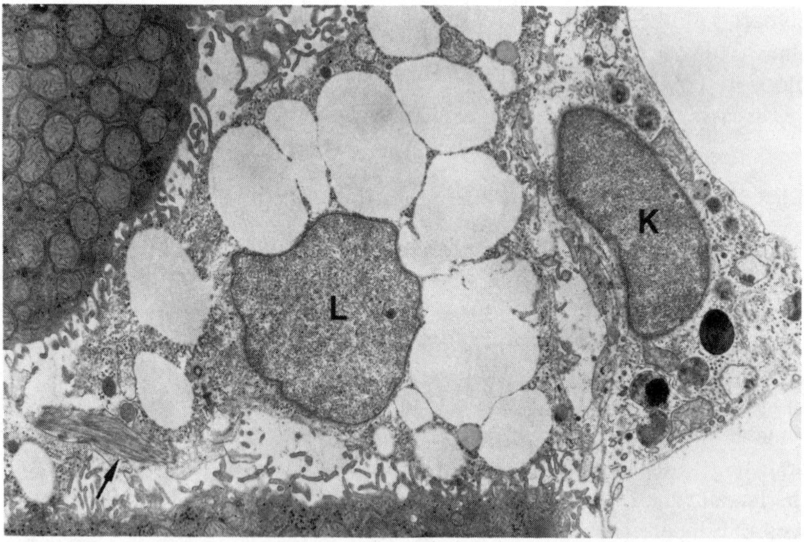

Figure 28. Lipocyte (Ito cell; L) with conspicuous lipid droplets in its cytoplasm is located in the perisinusoidal space. A bundle of collagen fibers (arrow) is present immediately adjacent to the lipocyte. A Kupffer cell (K) has several electron-dense lysosomes (×5,200).

Extracellular Matrix

Conventional electron microscopy revealed that the perisinusoidal space was empty except for the presence of the aforementioned cell types and occasional bundles of collagen-type fibers. Biochemical studies have shown that 53% of the total interstitial collagen is type I and 47% is type III in normal human liver (Seyer et al., 1977), whereas 60% is type I and 40% type III in normal rat liver (Seyer, 1980). Immunohistochemical studies have revealed that type I collagen is predominantly in the surface (Glisson's) capsule and portal connective tissue but only in modest amount in the perisinusoidal space. In contrast, type III collagen is found mainly in the perisinusoidal area but only small amounts in portal tracts and capsules (D'Ardenne et al., 1983; Martinez-Hernandez, 1984; Timple et al., 1977). Thus, type III is the predominant type of collagen present in the perisinusoidal space. However, type III collagen does not correlate exclusively with "reticulin fibers" seen by light microscopy because the latter seems to be a compound fibrous structure containing collagen type III fibronectin and additional noncollagenous glycoproteins (Unsworth et al., 1982). In addition, Martinez-Hernandez (1984) showed that perisinusoidal bundles of collagen-type fibers, especially those at branching and inflexion points of the sinusoids, contained type I collagen. Thus, precise correlation between the perisinusoidal bundles of fibers and their chemical composition remains to be established.

Using immunofluorescence, Hahn et al. (1980) reported the presence of a "basement membrane of different architecture" (presence of type IV collagen and absence of laminin). However, this "continuous" basement membrane seems to be an artifact due to the limited resolution of light microscopy since a subsequent electron immunohistochemical study (Martinez-Hernandez, 1984) revealed that type IV collagen is present as small discontinuous deposits. This latter study confirmed the absence of laminin. Immunohistochemistry has also demonstrated fibrorectin in the space of Disse, which forms irregular clusters, often attached to the nonvascular surface of the sinusoidal lining cells or to the hepatocyte plasma membrane (D'Ardenne et al., 1983; Martinez-Hernandez, 1984).

As regards the function of the extracellular matrix, bundles of collagen fibers are thought to form the intralobular structural frame anchoring and supporting cells in lobular parenchyma. The absence of continuous basement membrane in the space of Disse is conducive to facilitate the rapid metabolic exchange between hepatocytes and plasma. Although the functional significance of the fibronectin found in the perisinusoidal space is not clear, it may serve the same function as the fibronectin deposited in the early stage of development: that of a primary extracellular matrix, which is less restrictive than a basement membrane in terms of filtration (Martinez-Hernandez, 1984).

LYMPHATIC SYSTEM

In the liver, no lymphatic vessels are found in the parenchyma. Initial lymphatic vessels are found in the portal tracts and less prominently in association with hepatic veins (Comparini and Bastianini, 1965; Magari et al., 1981). Although the exact mechanism by which hepatic lymph is formed is not established, it is believed that a portion of the plasma resulting from interchange between the sinusoid and hepatocyte flows through the perisinusoidal space toward the portal tract where it percolates into the extracellular space. It then becomes the tissue fluid in the portal tract, and the hepatic lymph is drained by the lymphatic capillaries that form a complex plexsus around bile ducts and blood vessels. These portal lymphatic vessels converge at the porta hepatis and crain into hepatic lymph nodes and thence to the thoracic duct (Barrowman and Granger, 1981; Comparini and Bastianini, 1985; Magari et al., 1981). Approximately 80% of the hepatic lymph is estimated to flow through the aforementioned route. The remaining lymph appears to leave in vessels surrounding the hepatic vein and ascent to the neck along the internal thoracic arteries. Under normal circumstances, one-quarter to one-half of the total volume of the thoracic duct lymph comes from the liver and the remainder from the extrahepatic portal bed. The variation in the estimated hepatic lymph flow reflects the technical difficulty in measuring hepatic lymph output which is exquisitely sensitive to changes in hepatic venous pressure. Another observation of note is that the hepatic lymph differs from the lymph formed elsewhere in that its protein concentration is higher than that of the latter. This is attributed to the high porosity of the sinusoidal wall, but it is not understood why the concentration of some of the proteins in the hepatic lymph is higher than that in the plasma (Barrowman and Granger, 1981).

BILIARY SYSTEM

Bile Canaliculi

Bile canaliculi are smallest biliary spaces (1 ± 0.5 μm in diameter) located between two or more adjacent hepatocytes. Thus, in a single-cell-thick plate of hepatocytes, bile canaliculi form a network of polyhedral meshes with a single cell in each mesh. Because plates of hepatocytes branch and anastomose, the canaliculi form an extensive three-dimensional network with polyhedral meshes (Figure 29) (Daems, 1961; Elias, 1949; Jones et al., 1980). Although bile canaliculi were at one time thought to be separate structural entities surrounded by a cuticular wall, electron microscopy has shown that the limiting wall of the canaliculi consists of specialization of the plasma membrane of adjacent hepatocytes (Daems, 1961; Fawcett, 1955). The can-

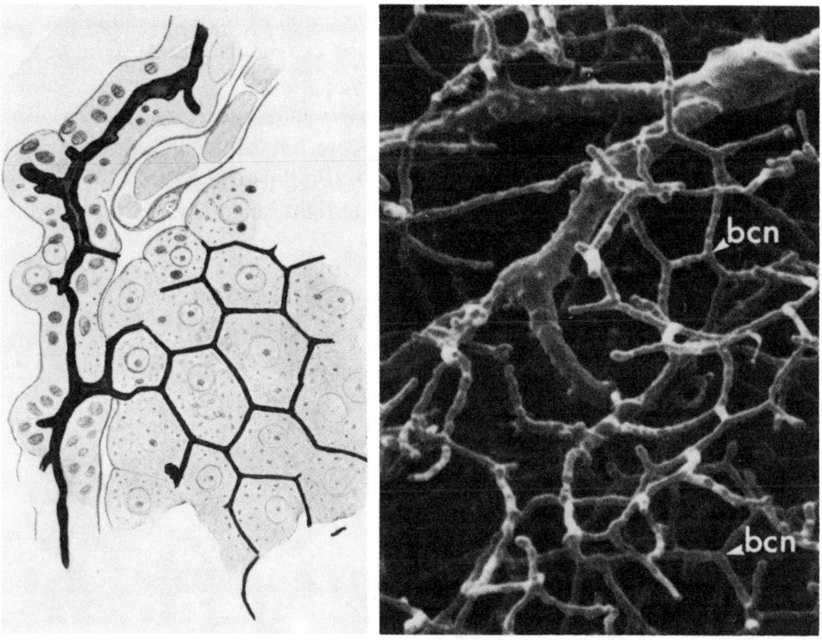

Figure 29. Left: Schematic diagram showing a network of bile canaliculi draining into an interlobular bile duct. Note that canalicular network forms polyhedral meshes with a single cell in each mesh. Right: Scanning electron micrography of a plastic cast of the biliary system. Note that bile canaliculi (bcn) form a network of polyhedral meshes resembling those illustrated in the diagram, and drain into bile ducts which are distinctly greater in diameter than canaliculi (×900). (Diagram from Elias, *Am J Anat* 1949; 85:379–456. Figure 47 with permission of the publisher. Scanning electron micrograph: Courtesy of Dr. T. Murakami.)

alicular wall is studded with numerous microvilli, and canalicular lumen is sealed by the tight junctions which appear as punctate regions of close apposition or fusion of membranes of adjacent cells in thin sections. When viewed in freeze-fracture replica preparations, the tight junctions appear as an interconnected series of strands of particles and grooves (Figure 30) (Farquhar and Palade, 1963).

According to Van Deurs and Luft (1979), the junctional elements are probably composed of particles and that strands are artifacts produced from cross linking during fixation. The particles are thought to represent intramembranous proteins that span adjacent plasma membrane bilayers. However, it has also been suggested that the strands may represent cylindrical inverted micelles which are interposed between two adjacent cell membranes at their point of fusion (Pinto da Silva and Kachar, 1982). Whichever view is correct, the tight junction provides a high resistance barrier to leakage of water and solutes into and from the bile canaliculus (Graf et al., 1984). As a result, the composition of the canalicular fluid can be quite different from that of the other pericellular fluid, allowing the accumulation of high concentrations of organic anions preferentially secreted into the canalicular lumen (Graf et al., 1984).

Actin microfilaments form a network surrounding the canaliculi and also insert into canalicular microvilli. They sustain the canalicular structure and presumably impart motility to the canaliculi (Phillips and Satir, 1988). They may also play a role in the regulation of the tight junctional permeability.

Bile Ductules

Bile ductules (terminal ductules, cholangioles, canals of Hering) form short channels that convey bile from bile canaliculi to the interlobular bile

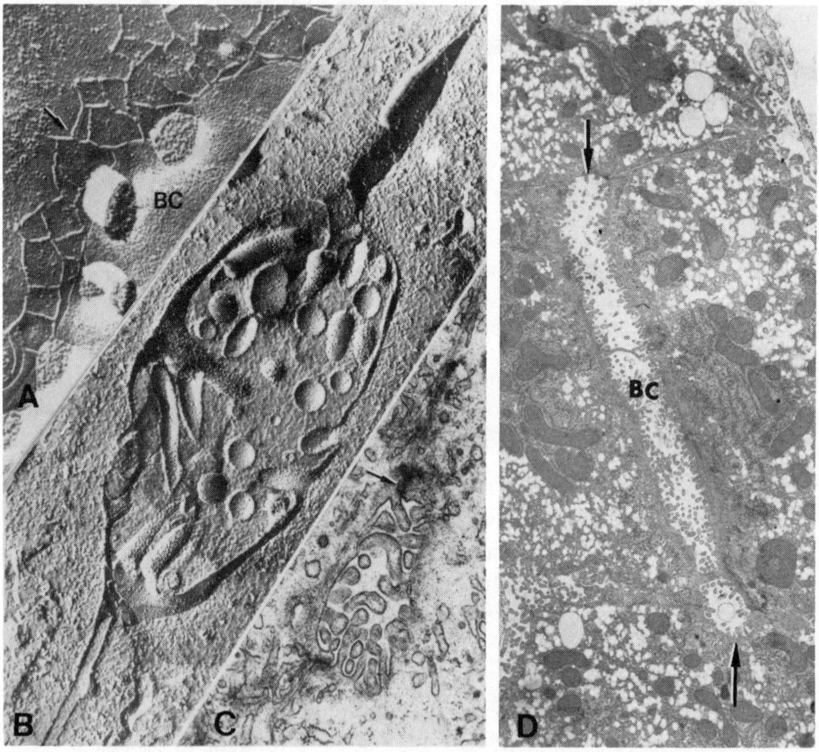

Figure 30. A: Freeze-fracture replica of a longitudinal profile of the tight junction (arrow) which seals the bile canaliculus (BC). B: Freeze-fracture replica of a bile canaliculus from a normal rat. Fractured surface of canalicular microvilli are studded diffusely with intramembranous granules. C: Thin section of a bile canaliculus. Tight junction is indicated by an arrow. D: A longitudinal profile of bile canaliculus (BC) which runs the entire length of a hepatocyte. This canaliculus seems to be continuous with another that runs perpendicularly at each end as indicated by the cross-sectional profile of the latter (arrows). A: ×36,000; B: ×35,000; C: ×16,000; D: ×4,100.

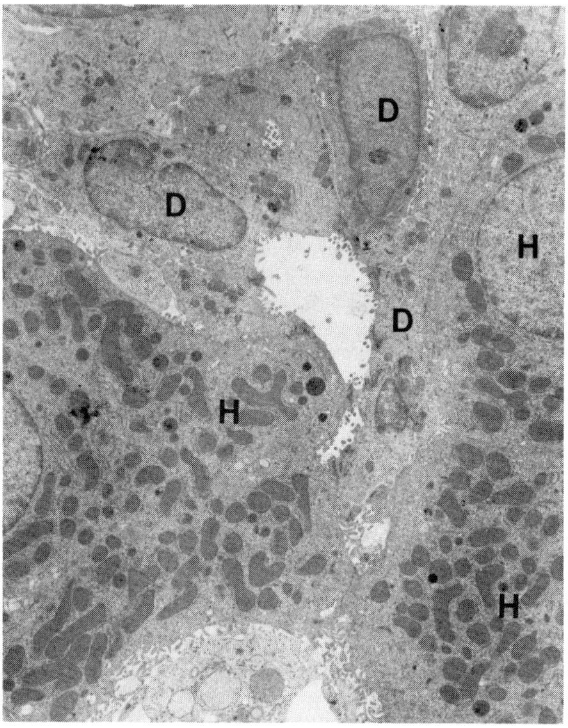

Figure 31. Junction between the bile canaliculus and bile ductule. The junction is formed by ductular cells (D) sharing a canalicular lumen with a hepatocyte (H) (×3,150).

ducts in the portal tract. They are lined by cuboidal cells and are often difficult to identify in the normal liver. At the junction with the bile canaliculus, bile ductular cells share a canalicular lumen with hepatocytes (Figure 31). The bile ductules are smaller than the interlobular bile ducts, usually measuring 15–20 μm in diameter. Their lumenal surface has short microvilli. The nuclei and mitochondria of the ductular cells are smaller than those of adjacent hepatocytes. The endoplasmic reticulum is poorly developed, but the Golgi complex and pinocytotic vesicles are well developed, suggesting that ductular cells are active, presumably in absorption and secretion (Jones and Spring-Mills, 1984; Steinlieb, 1972; Steiner and Carruthers, 1961).

Bile ductular cells proliferate under a variety of pathologic conditions, most notably in biliary obstruction and experimental chemical carcinogenesis. The origin of the bile ductular cells, however, is controversial. Some investigators believe that they develop from hepatocytes, whereas others maintain that they are derived from so-called oval cells which resemble biliary epithelial cells morphologically but possess biochemical features of fetal hepatocytes (Desmet, 1985). However, the embryologic development of the intrahepatic bile ducts suggests that biliary epithelial cells develop from primitive hepatocytes around branches of the portal vein (Bloom, 1926; Elias, 1955; Enzan et al., 1974; Shinjoji, 1984). First, a ring of primitive hepatocytes closest to the portal vein branches transforms into bile duct-type cells, followed by a second layer which undergoes the same process. Thus, a double-walled cylinder with a slitlike lumen called the ductal plate is formed (Jorgensen, 1977). Subsequently, this ductal plate is restructured resulting in a system of anastomosing bile ducts (Desmet, 1985; Jorgensen, 1977). In addition, immunohistochemical studies have shown that, during the remodeling of the ductal plates into bile ductules and ducts, the phenotypic expression of cytokeratin polypeptides in these transforming cells shift from the pattern characteristic of the hepatocyte to that of biliary epithelium (van Eyken et al., 1988a,b).

Interlobular Bile Ducts

Bile in the bile ductules flows into the interlobular bile ducts in the portal tract. These ducts measure 30–40 μm in diameter and form a continuous network for biliary passage. They are lined by a single layer of cuboidal or columnar epithelial cells with short microvilli on their lumenal surface and are encircled by a basal lamina.

The intracellular structure of the bile duct cells is similar to that of ductular cells in that the Golgi complex and pinocytotic vesicles are prominent, suggesting that bile duct cells are metabolically active (Jones and Spring-Mills, 1984; Steinlieb, 1972). Bile ducts increase in their size and join the *extrahepatic biliary system.* The diameter of the bile duct roughly corresponds to that of the adjacent artery.

Structural Aspect of Bile Secretion

According to current views, canalicular (hepatocytic) bile formation is mainly an osmotic process: water and inorganic electrocytes flow into the bile canaliculus along an osmotic gradient resulting from the excretion of solutes into the canalicular lumen (Erlinger, 1988; Reuben, 1984). The major component of the osmotic gradient is probably provided by bile acids and their counter ions. This bile acid-dependent fraction accounts for 30–60% of spontaneous basal bile flow, whereas another 30–60% is independent of bile acids. Although the mechanism of the latter is not established, it is probably due to the transport of inorganic ions (e.g., sodium and bicarbonate) by the hepatocyte. Although Na^+, K^+-ATPase, sinusoidal Na^+/H^+ antiporter, and a canalicular Cl^-/HCO_3^- antiporter are thought to be involved in the aforementioned processes, their precise roles remain to be defined. As bile flows through the biliary system, fluid and inorganic electrolytes are reabsorbed and/or secreted by the bile ductules and ducts. Secretion occurs mainly in response to secretin (Erlinger, 1988; Reuben, 1984).

As regards the respective roles of hepatocyte organelles in bile formation and secretion, very little is known. This lack of knowledge is understandable since bile contains no components (except for small amounts of proteins such as IgA and insulin) that can be visualized by electron microscopy. Although the Golgi complex and pericanalicular vesicles are presumed to be involved in the transport and secretion of bile acids, the evidence for this concept is circumstantial (Jones et al., 1980). With respect to Na^+, K^+-ATPase, biochemical and cytochemical studies have supported a basolateral (sinusoidal/lateral) localization (Blitzer and Boyer, 1978; Evans, 1980; Maurice et al., 1985). However, immunohistochemical studies using monoclonal (Schenk and Leffert, 1983) and polyclonal (Takemura et al., 1984) antibodies have suggested that the enzyme is also present on the canicular membrane. The subcellular localization and the mechanism by which this enzyme plays a role in bile secretion remain to be established (Erlinger, 1988).

Iron flux studies in isolated perfused liver suggest that nearly 95% of sodium in bile is secreted through the paracellular route (Graf and Peterlick, 1975). Since the tight junction regulates the paracellular transit of solutes and fluid, it must play an important role in bile formation. It has been proposed that the tight junctional permeability is inversely proportional to the number of parallel strands (and grooves) running parallel to the canaliculi per unit width of the tight junction (Bentzel et al., 1982; Claude and Goodenough, 1973). According to LaGarde et al. (1981), however, the structural organization of the hepatocellular tight junction is highly variable even over a 5–10-μm distance. In addition, recent studies in the intestinal epithelium suggest that focal discontinuities in junctional strands correlate with increased fluid transit through the paracellular route and that these discontinuities seem to be caused by contraction of the perijunctional actomyosin filaments (Madara and Pappenheimer, 1987). These observations suggest that the structure and function of the hepatocellular junction may not be easily correlated. The possible role of the cytoskeletal system in bile secretion has been discussed (see Cytoskeletal System Section).

NERVOUS SYSTEM

Extrinsic Innervation

Branches of the vagus, the splanchnic, and sometimes the phrenic nerves enter the liver primarily in association with blood and bile ducts. The anterior plexus, which surrounds the hepatic artery, is composed of fibers originating from the celiac ganglia and anterior vagus. Branches from the celiac ganglia and posterior vagus form the posterior plexus, which is located behind the hepatic portal vein and bile duct (Alexander, 1940; Sutherland, 1965). The innervation may vary among different species and individuals (Friedman, 1988).

Intrinsic Innervation

Most intrinsic nerves of the liver are associated with the vascular system and extend their branches to vascular smooth muscles, which is consistent with a vasomotor function (Alexander, 1940). Nerve fibers also extend to the endothelial cells in terminal branches of the hepatic artery, hepatocytes of the limiting plates, and to the perisinusoidal spaces where they may terminate on Kupffer cells and fat-storing cells of Ito (Alexander, 1940; Forssmann and Ito, 1977; Tsuneki and Ichihara, 1981). Although there was controversy, recent studies have shown neural supply to hepatocytes. The intralobular innervation is generally sparse and its extent of distribution varies among different species. The extent of intralobular innervation has also been found to correlate inversely with the frequency of occurrence of the gap junctions (Forssmann and Ito, 1977; Reilly et al., 1978), suggesting that tight electrical coupling between hepatocytes obviate the need for extensive innervations. The extent and distributions of adrenergic and cholinergic nerves within the liver differ considerably among different species. Little is known about the relative contribution of efferent and afferent fibers to the hepatic nerve supply (Freidman, 1988).

Functions of the Hepatic Nervous System

The efferent hepatic nerves, in particular those associated with the resistance vessels, regulate hepatic blood flow. α-Adrenergic stimuli reduce arterial blood flow and also increase vascular resistance in the portal veins without appreciable reduction in blood flow. β-Adrenergic stimuli may also play a part. However, cholinergic mechanisms do not seem to be involved in the hemodynamic control, since the vagal stimulation does not alter hepatic blood flow (Freidman, 1988; Lautt, 1980, 1983). Afferent nerves are also involved in the regulation of glucose and lipid metabolism (Friedman, 1988; Lautt, 1980, 1983). Although nerve plexus associated with the biliary system has been implicated in the regulation of biliary secretions, its precise role has yet to be established (Friedman, 1988; Lautt, 1980, 1983).

The afferent nerves are involved in detecting a range of physiological events. They are thought to serve as osmoreceptors, sodium receptors, baroreceptors, and metabolic receptors (Friedman, 1988; Lautt, 1980, 1983). It is likely that these receptors comprise an "early warning" system to signal changes in the osmolarity or ion concentration of ingested fluid or foodstuff so that appropriate gastrointestinal, endocrine, and renal responses can be made to accommodate fluctuations in body fluid balance (Friedman, 1988; Lautt, 1980, 1983).

REFERENCES

Alberts B, Bray D, Lewis J, Raff M, Roberts K, and Watson J: The cytoskeleton, in *Molecular Biology of the Cell.* New York, Garland Publishing, 1983, pp 550–609.

Alexander WF: The innervation of the biliary system. *J Comp Neurol* 1940; 72:357–370.

Alfert M and Geschwind JJ: Development of polysomaty in the rat liver. *Exp Cell Res* 1958; 15:230–232.

Bainton DF: The discovery of lysosomes. *J Cell Biol* 1981; 91:66s–76s.

Ballardini G, Degli Eposti S, Bianchi FB, Badiali de Giorgi L, Faccani A, Biolchini L, Busachi CA, and Pisi E: Correlation between Ito cells and fibrogenesis in an experimental model of hepatic fibrosis: a sequential stereological study. *Liver* 1983; 3:58–63.

Baraona E, Finkelman F, and Lieber CS: Re-evaluation of the effects of ethanol consumption on rat liver microtubules. *Clin Pathol Pharmacol* 1984; 44:262–278.

Barrowman JA and Granger DN: Hepatic lymph, in Lautt WW (eds): *Hepatic Circulation in Health and Disease.* New York, Raven Press, 1981, pp 137–152.

Baumgartner U, Miyai K, and Hardison WGM: Modulation of hepatic biotransformation and biliary secretion of bile acid by age and sinusoidal bile acid load. *J Physiol* 1987; 252 *Gastroent Liver Physiol:* G114–G119.

Benedetti EL and Emmelot P: Hexagonal array of subunits in tight junctions separated from isolated rat liver plasma membranes. *J Cell Biol* 1968; 38:15–24.

Bentzel CJ, Martinez M, Hainau B, Fromm MJ, and Hegel U: Morphological and physiological factors determining transjunctional fluxes, in Bradley SE and Purcell EF (eds): *The Paracellular Pathway.* New York, Josiah Macy Jr Foundation, 1982, pp 36–56.

Biozzi G and Stiffel C: The physiopathology of the reticuloendothelial cells of the liver and spleen. *Prog Liver Dis* 1965; 2:166–191.

Blitzer BL and Boyer JL: Cytochemical localization of Na^+, K^+-ALPase in the rat hepatocyte. *J Clin Invest* 1978; 62:1104–1108.

Bloom W: The embryogenesis of human bile capillaries and ducts. *Am J Anat* 1926; 36:451–462.

Blouin A: Morphomotry of liver sinusoidal cells, in Wisse E and Knook KL (eds): *Kuppfer Cells and Other Liver Sinusoidal Cells,* Amsterdam, Elsevier, 1977, pp 61–71.

Blouin A: Anatomy, ultrastructure and morphometry of the liver, in Glauman E et al. (eds): *Plasma Protein Secretion by the Liver,* London, Academic Press, 1983; pp 31–53.

Blouin A, Bolender RP, and Weibel, ER: Distribution of organelles and membranes between hepatocytes and non-hepatocytes in the rat liver parenchyma. *J Cell Biol* 1977; 72:441–445.

Bouteille M, Laval M, and DuPuy-Coin AM: Localization of nuclear functions as revealed by ultrastructural autography and cytochemistry, in Busch H (ed): *The Cell Nucleus,* New York, Academic Press, 1974, vol 1, pp 3–71.

Bouwens L, Baekeland M, De Zanger R, and Wisse E: Quantitation, tissue distribution and proliferation kinetics of Kupffer cells in normal rat liver. *Hepatology* 1968; 6:718–722.

Bouwens L and Wisse E: Tissue localization and kinetics of pit cells or large granular lymphocytes in the liver of rats treated with biological response modifiers. *Hepatology* 1988; 8:46–52.

Boyer JL: Contractile activity of bile canaliculi: contraction or collapse? *Hepatology* 1987; 7:190–192.

Bradfield JW: Liver sinusoidal cells. *J Pathology* 1984; 142:5–6.

Brael WA, Schlossman DM, Schmid SL, and Rothman JE: Dissociation of clathrin coats coupled to the hydrolysis of ATP: role of an uncoating ATPase. *J Cell Biol* 1984; 99:734–741.

Brandon D: Identification of myosin in rabbit hepatocytes. *Eur J Biochem* 1976; 65:139–146.

Brauer RW: Liver circulation and function. *Physiol Rev* 1963; 43:115–213.

Brouwer A, Hendriks HJF, Brekelmans P, Buijtenhek M, and Knook DL: Lipid analysis of different types of isolated liver cells and their lipid droplets, in Kirn A, Knook DL, and Wisse E (eds): *Cells of the Hepatic Sinusoid,* Rijswijk, Kupffer Cell Foundation, 1986, pp 175–181.

Brouwer A, Wisse E, and Knook DL: Sinusoidal endothelial cells and perisinusoidal fat-storing cells, in Arias M, Jacoby WB, Popper H, Schachter D, and Schafritz DA (eds): *The Liver: Biology and Pathobiology,* ed 2. New York, Raven Press, 1988.

Brown MS, Anderson RG, and Goldstein JL: Recycling receptors: the round-trip itinerary of migrant membrane proteins. *Cell* 1983; 32:663–667.

Burkel WE: The fine structure of the terminal branches of the hepatic arterial system of the rat. *Anat Rec* 1970; 167:329–349.

Burkel WE and Low FN: The fine structure of rat liver sinusoids, space of Disse and associated tissue space. *Amer J Anat* 1966; 118:769–775.

Busch H and Schildkraut C: Nucleus, DNA and chromatin, in Arias IM, Jacoby WB, Popper H, Schachter D, and Shafritz DA (eds): *The Liver: Biology and Pathobiology,* ed 2. New York, Raven Press, 1988, pp 47–70.

Candipan RC and Sjostrand FS: Water movement from intracrystal spaces in isolated liver mitochondria. *J Ultrast Res* 1984a; 89:249–260.

Candipan RC and Sjostrand FS: Freeze fracture analysis of isolated liver mitochondria in different metabolic states. *J Ultrast Res* 1984b; 89:274–280.

Candipan RC and Sjostrand FS: An analysis of contribution of preparatory techniques to the appearance of condensed and orthodox conformations of liver mitochondria. *J Ultrast Res* 1984c; 89:281–294.

Carriere R: The growth of liver parenchymal nuclei and its endocrine regulatin. *Int Rev Cytol* 1969; 25:201–277.

Claude A: Growth and differentiation of cytoplasmic membranes in the course of lipoprotein granule synthesis in the hepatic cell I. Elaboration of elements of the Golgi complex. *J Cell Biol* 1970; 47:745–766.

Claude P and Goodenough DA: Fracture faces of zonular occuludentes from ''tight'' and ''leaky'' epithelia. *J Cell Biol* 1973; 58:390–400.

Comparini L and Bastianini A: Graphic reconstructions in the morphological study of the hepatic lymphatic vessels. *Angiologica* 1965; 2:81–95.

Daems WTh: The microanatomy of the smallest biliary pathways in mouse liver tissue. *Acta Anat* 1961; 46:1–24.

Dallner G, Siekevitz P, and Palade GE: Biogenesis of endoplasmic reticulum membranes: Structural and chemical differentiation in developing rat hepatocyte. *J Cell Biol* 1966a; 30:73–96.

Dallner G, Siekevitz P, and Palade GE: Biogenesis of endoplasmic reticulum membranes II. Synthesis of constitutive microsomal enzymes in developing rat hepatocyte. *J Cell Biol* 1966b; 30:97–117.

D'Ardenne AJ, Burns J, Sykes BC, and Kirkpatrick P: Comparative distribution of fibronectin and type III collagen in human tissue. *J Pathol* 1983; 141:55–69.

Daoust R: The cell population of liver tissue and the cytological reference base. *Am Inst Biol Sci Publ* 1958; 4:3–10.

deDuve C: Lysosomes in retrospect, in Dingle JT and Fell HB (eds): *Lysosomes in Biology and Pathology,* Amsterdam; North-Holland, (1969); vol 2, pp 3–40.

deDuve C: Exploring cells with a centrifuge. *Science* 1975; 18:186–194.

deDuve C and Baudhuin P: Peroxisomes (microbodies and related particles). *Physiol Rev* 1966; 46:323–357.

DeLeeuw AM, McCarthy SP, Geerts A, and Knook DL: Purified rat liver fat-scoring cells in culture divide and contain collagen. *Hepatology* 1984; 4:392–493.

Denk H: Ethanol, Mallory bodies and microtubular system, in Seitz HK and Kommerell B (eds): *Alcohol Related Diseases in Gastroenterology,* Berlin, Springer-Verlag, 1985, pp 155–171.

DePierre JW, Anderson G, and Dallner G: Endoplasmic reticulum and Golgi complex, in Arias IM, Jacoby WB, Popper H, Schachter D, and Shafritz D (eds): *The Liver: Biology and Pathobiology,* ed 3. New York, Raven Press, 1988, pp 165–187.

Desmet VL: Morphologic and histochemical aspects of cholestasis, in Popper H and Schaffner F (eds): *Progress in Liver Diseases,* New York, Grune and Stratton, 1972, pp 97–125.

Desmet VL: Intrahepatic bile ducts under the lens. *J Hepatol* 1985; 1:549–559.

De Zanger R and Wisse E: The filtration effect of rat liver fenestrated sinusoidal endothelium on the passage of (remnant) chylomicrons to the space of Disse, Knook DL and Wisse E (eds): Amsterdam, Elsevier, North Holland, 1982, pp 69–72.

Elias H: A re-examination of the structure of the mamalian liver. I. Parenchymal architecture. II. The hepatic lobule and biliary systems. *Am J Anat* 1949; 84:311–333, 85:379–456.

Elias H: Origin and early development of the liver in various vertebrates. *Acta Hepatol* 1955; 3:1–56.

Elias E, Hruban E, Wade JB, and Boyer, JL: Phalloidin-induced cholestasis. A microfilament-mediated change in junctional complex permeability. *Proc Natl Acad Sci USA* 1980; 77:2229–2233.

Enzan H, Ohkita T, Fujita H, and Iijima S: Light and electron microscopic studies on the development of periportal bile ducts of the human embryo. *Acta Pathol Jpn* 1974; 24:427–447.

Erlinger S: Bile flow, in Arias IM, Jacoby WB, Popper H, Schacter D, and Shafritz DA (eds): *The Liver: Biology and Pathophysiology,* ed 2, New York, Raven Press, 1988, pp 643–661.

Ernster L and Schatz GT: Mitochondria: A historical review, in Gall JG, Porter KR, and Siekeviz P (eds): *Discovery in Cell Biology, (J Cell Biol* 1981; 91:no.3, part 2, 227s–255).

Evans WH: A biochemical dissection of the functional polarity of the plasma membrane of the hepatocyte. *Biochem Biophys Acta* 1980; 604:27–64.

Fahimi DH: Sinusoidal endothelial cells and perisinusoidal fat-storing cells: Structure and function, in Arias IM, Popper H, Schachter D, and Shafritz DA (eds): *The Liver: Biology and Pathobiology,* New York, Raven Press, 1982, pp 495–506.

Farquhar MG: Progress in unravelling pathways of Golgi traffic. *Annu Rev Cell Biol* 1985; 1:447–488.

Farquhar MG and Palade GE: Junctional complexes in various epithelia. *J Cell Biol* 1963; 77:375–412.

Farquhar MG and Palade GE: The Golgi apparatus (complex) (1954–1981) from artifact to center stage. *J Cell Biol* 1981; 91:77s–103s.

Fawcett DW: Observations on the cytology and electron microscopy of hepatic cells. *J Natl Cancer Inst* 1955; 15:1475–1502.

Forgac M: Receptor-mediated endocytosis, in Arias IM, Jacoby WB, Popper H, Schachter D, and Shafritz DA (eds): *The Liver: Biology and Pathophysiology,* ed 2. New York, Raven Press, 1988, pp 207–225.

Forssmann WG and Ito S: Hepatic innervation in primates. *J Cell Biol* 1977; 73:299–313.

Franke WW, Mayer D, Schmid E, Denk H, and Borenfreund E: Differences in expression of cytoskeletal proteins in cultured rat hepatocytes and hepatoma cells. *Exp Cell Res* 1979; 134:345–359.

Fraser R, Day WA and Fernando NS: Review: The liver sinusoidal cells. Their role in disorders of the liver, lipoprotein metabolism and atherogenesis. *Pathology* 1986; 18:5–11.

French SW and Davies PL: Ultrastructural localization of actin-like filaments in rat hepatocytes. *Gastroenterology* 1975; 68:765–774.

Fricker G, Landmann, and Meier PJ: Bile duct ligation reverses the bile salt secretory polarity of rat hepatocytes. *Hepatology* 1987; 7:1106.

Friedman MI: Hepatic nerve function, in Arias IM, Popper H, Schachter D, and Shafritz DA (eds): *The Liver: Biology and Pathobiology,* ed 2. New York, Raven Press, 1982, pp 949–959.

Gala RP, Sparkes RS, and Golde DW: Bone marrow origin of hepatic macrophages (Kupffer cells) in humans. *Science* 1978; 201:937–938.

Gates GA, Henley KS, Pollard HM, Schmidt E and Schmidt FW: The cell population of human liver. *J Lab Clin Med* 1961; 57:182–184.

Geller S: Intralobular distribution of polyploid cells in rat liver. *Anat Rec* 1965; 151:352–353.

Geuze HJ, Slot JW, Strous GJ, Hasilik A, and Von Figura K: Possible pathways for lysosomal enzyme delivery. *J Cell Biol* 1985; 101:2253–2262.

Geuze HJ, Slot JW, Strous GJ, Lodish HF, and Schwarz AL: Intracellular site of asialoglycoprotein receptor-ligand uncoupling: double-label immunoelectron microscopy during receptor-mediated endocytosis. *Cell* 1983; 32:277–287.

Geuze HJ, Slot JW, Strous GJ, Peppard J, von Figura K, Hasilik A, and Schwarz AL: Intracellular receptor sorting during endocytosis: comparative immunoelectron microscopy of multiple receptors in rat liver. *Cell* 1984; 37:195–204.

Goldfischer SL: Peroxisomal disease, in Arias IM, Jacoby WB, Popper H, Schachter D, and Shafritz DA (eds): *The Liver: Biology and Pathology,* ed 2. New York, Raven Press, 1988, pp 255–267.

Gordon DJ, Boyer JL, and Korn ED: Comparative biochemistry of nonmuscle actins. *J Biol Chem* 1977; 252:8300–8309.

Graf J, Gautam A, and Boyer JL: Isolated rat hepatocyte couplets: A primary secretory unit for electrophysiologic studies of bile secretory function. *Proc Natl Acad Sci USA* 1984; 81:6516–6520.

Graf J and Peterlick L: Mechanism of transport of inorganic ions into bile, in Taylor W (ed): *The Hepatobiliary System: Fundamental Pathological Mechanisms,* New York, Plenum, 1975.

Griffiths G and Simons K: The trans Golgi network: sorting at the exit site of the Golgi complex. *Science* 1986; 234:438–443.

Gumucio JJ and Chianale J: Liver cell heterogeneity and Liver function, in Arias IM, Jacoby WB, Popper H, Schachter D, and Schafritz DA (eds): *The Liver: Biology and Pathobiology,* ed 2. New York, Raven Press, 1988, pp. 931–947.

Hahn E, Wick G, Pencev D, and Timple R: Distribution of basement membrane proteins in normal and fibrotic human liver: collagen type IV, laminin and fibronectin. *Gut* 1980; 21:63–71.

Haigler HT, McKanna JA, and Cohen S: Direct visualization of the bining and internalization of a ferritin conjugate of epidermal growth factor in human carcinoma cells A-431. *J Cell Biol* 1979; 81:382–395.

Hasilik A: Biosynthesis of lysosomal enzymes trends. *Biochem Sci* 1980; 5:237–240.

Hatoff DE, Toyota N, Wong C, Miller AL, Takeya M, and Miyai K: Rat liver alkaline phosphatase: Evidence hepatocyte and portal triad enzymes differ. *Digest Dis Sci* 1985; 30:564–572.

Healy JE: Vascular anatomy of the Liver. *Ann NY Acad Sci* 1970; 170:8–17.

Hendriks HFJ, Verhoofstad WAMM, Boruwer A, DeLeeuw AM, and Knook DL: Perisinusoidal fat-storing cells are the main vitamin A storage sites in rat liver. *Exp Cell Res* 1985; 160:138–149.

Hoppe CA, Connolly TP, and Hubbard AL: Transcellular transport of polymeric IgA in the rat hepatocyte: Biochemical and morphological characterization of the transport pathway. *J Cell Biol* 1985; 101:2113–2123.

Hruban E and Rechcigl M: *Microbodies and Related Particles.* New York, Academic Press, 1969.

Ishak KG and Sharp HL: Metabolic errors and liver disease, in MacSween RN, Anthony PP, and Scheuer PJ (eds): *Pathology of the Liver,* ed 2. Edinburgh, Churchill Livingstone, 1987, pp 99–180.

Ito T and Nemoto M: Uber die Kupfferschen Sternzellen und die ''Fettspeicherungs zellen'' (''fat-storing cells'') in der Blut Kapillarenwand der menschlichen leber. *Fol Anat Jpn* 1952; 24:243–258.

Ito T and Shibasaki S: Electron microscopic study on the hepatic sinusoidal wall and fat-storing cells in the normal human liver. *Arch Hist Jpn* 1968; 29:137–192.

James J, Tas J, Bosch KS, DeMeese AJP, and Schuyt HC: Growth patterns of rat hepatocytes during postnatal development. *Eur J Cell Biol* 1976; 19:222–226.

Jones AL and Armstrong DT: Increased cholesterol biosynthesis following phenobarbital induced hypertrophy of agranular endoplasmic reticulum in liver. *Proc Soc Exp Biol Med* 1965; 119:1136–1139.

Jones AL and Fawcett DW: Hypertrophy of the agranular endoplasmic reticulum in human liver induced by phenobarbital (with a review of the functions of this organelle in liver. *J Histochem Cytochem* 1966; 14:215–232.

Jones AL and Schmucker DL: Current concepts of liver structure as related to function. *Gastroenterology* 1977; 73:833–851.

Jones AL, Schmucker DL, Renston RH, and Murakami T: The architecture of bile secretion. A morphological perspective of physiology. *Digest Dis Sci* 1980; 25:609–629.

Jones AL and Spring-Mills E: The liver and gallbladder, in Weiss L (ed): *Modern Concepts of Gastrointestinal Histology.* New York, Elsevier, 1984, pp 706–748.

Jones EA and Summerfield JA: Kupfer cells, in Arias IM, Jacoby WB, Popper H, Schachter D, and Shafritz, DA (eds): New York, Raven Press, 1988, pp 683–704.

Jorgensen MJ: The ductal plate malformation. A study of the intrahepatic bile duct lesion in infantile polycystic disease and congenital hepatic fibrosis. *Acta Pathol Microbiol Scand Section A* 1977; 257(Suppl):7–88.

Jungermann K and Katz N: Functional hepatocellular heterogeneity. *Hepatology* 1982; 2:385–395.

Kaneda K and Wake K: Distribution and morphologic characteristics of the pit cells in the liver of the rat. *Cell Tissue Res* 1983; 233:485–505.

Kardon RH and Kessel RG: Three dimensional organization of the hepatic microcirculation in the rodent as observed by scanning electron microscopy of corrosive casts. *Gastroenterology* 1980; 79:72081.

Kiernan F: The anatomy and physiology of the liver. *Phil Trans Roy Soc London B Biol Sci* 1833; 123:711–770.

LaGarde S, Elias E, Wade JB, and Boyer JL: Structural heterogeneity of hepatocyte ''tight'' junctions—A quantitative analysis. *Hepatology* 1981; 1:193–203.

Lautt WW: Hepatic nerves: A review of their functions and effects. *Can J Physiol Pharmacol* 1980; 58:105–123.

Lautt WW: Afferent and efferent neural roles in liver function. *Prog Neurobiol* 1983; 21:323–348.

Lazarides E: Intermediate filaments as mechanical integrators of cellular space. *Nature* 1980; 283:249–256.

Lazarow PB: Peroxisomes, in Arias IM, Jacoby WB, Popper H, Schachter D, and Shafritz DA (eds): *The Liver: Biology and Pathology,* ed 2. New York, Raven Press, 1988, pp 241–254.

LeMarchand Y, Patzelt G, Assimacopoulos-Jeannet F, Loten EG, and Jeanrenaud B: Evidence for a role of the microtubule system in the secretion of newly synthesized albumin and other proteins by the liver. *J Clin Invest* 1974; 53:1512–1517.

Lin C and Chang JP: Electron microscopy of albumin synthesis. *Science* 1975; 190:465–467.

Loud AV: A quantitative stereological description of the ultrastructure of normal rat liver parenchymal cells. *J Cell Biol* 1968; 37:27–46.

Lowenstein WR: Junctional communication and the control of growth. *Biochem Biophys Acta* 1979; 560:1–65.

Madara JL and Pappenheimer JR: Structural basis for physiological regulation of paracellular pathways in intestinal epithelia. *J Memb Biol* 1987; 100:149–164.

Magari S, Fujikawa K, and Nishi A: Form, distribution, fine structure and function of hepatic lymphatics with special reference to blood vessels and bile ducts. *Asian Med J* 1981; 24:254–270.

Mall FP: A study of the structural unit of the liver. *Am J Anat* 1906; 5:227–308.

Martinez-Hernandez A: The hepatic extracellular matrix. I. electron immunohistochemical studies in normal rat liver. *Lab Invest* 1984; 51:57–74.

Maurice M, Durand-Schneider AM, Garbaz M, and Feldmann G: Characterization of rat hepatocyte plasma membrane domains by monoclonal antibodies. *Eur J Cell Biol* 1985; 39:122–129.

McCuskey RS: A dynamic and static study of hepatic arterioles and hepatic sphincters. *Am J Anat* 1966; 119:455–478.

Meihuizen SP and Blansjaar N: Stereological analysis of liver parenchymal cells from young and old rats. *Mech Aging Develop* 1980; 13:111–118.

Meyer DJ, Yancey B, Revel JP, with an appendix by Peskott A: Intercellular communication in normal and regenerating rat liver. A quantitative analysis. *J Cell Biol* 1981; 91:505–523.

Miyai K: Ultrastructural basis for toxic liver injury, in Farber E and Fisher MM (eds): *Toxic Injury of the Liver,* pt A. New York, Marcel Dekker, 1979, pp 59–154.

Moll R, Franke WW, Schiller DL, Geiger B, and Krepler R: The catalog of human cytokeratins: Patterns of expression in normal epithelia, tumors and cultured cells. *Cell* 1982; 31:11–24.

Motta P: The three dimensional microanatomy of the liver. *Arch Histol Jpn* 1984; 47:1–30.

Nakata K and Kanbe A: The terminal distribution of the hepatic artery and its relationship to the development of focal liver necrosis following interruption of portal blood supply. *Acta Pathol Jpn* 1966; 16:313–321.

Novikoff AB: Lysosomes, a personal account, in Hers G and Van Hoof F (eds): *Lysosomes and Storage Diseases.* New York, Academic Press, 1973, pp 1–41.

Novikoff AB, Beaufay H, and DeDuve C: Electron microscopy of lysosome—rich fractions from rat liver. *J Biophys Biochem Cytol* 1956; 2:179–184.

Novikoff AB and Yam A: The cytochemical demonstration of GERL in rat hepatocytes during lipoprotein mobilization. *J Histochem Cytochem* 1978; 26:1–24.

Ogawa H, Miyai K, Hardison WG and Tamura J: Liver surface alkaline phosphatase activity correlates with bile acid load. *Gastroenterology* 1986; 90:1753.

Ohta M, Marceau N, and French SW: Pathologic changes in the cytokeratin pericanalicular sheath in experimental cholestasis and alcoholic fatty liver. *Lab Invest* 1988; 59:60–74.

Ontko JA and Dashti H: Enzymes of mitochondria, in Altman PI and Katz DD (eds): *Cell Biology.* Feb Am Soc Exp Biol, 1976.

Oshio C and Phillips MJ: Contractibility of bile canaliculi. Implication for liver function. *Science* 1981, 212:1041–1042.

Palade GE: Intracellular aspects of the process of protein synthesis. *Science* 1975; 189:347–358.

Phillips MJ and Poucell S: Cholestasis: surgical pathology, mechanisms and new concepts, in Farber E, Phillips MJ, and Kaufman N (eds): *Pathogenesis of Liver Diseases* (IAP Monograph No. 28), Baltimore, Williams and Wilkins, 1987, pp 65–94.

Phillips MJ and Satir P: The cytoskeleton of the hepatocyte: Organization, relationships and pathology, in Arias IM, Jacoby WB, Popper H, Schachter D, and Shafritz DA (eds): *The Liver: Biology and Pathology,* ed 2. New York, Raven Press, 1988, pp 11–27.

Pinto daSilva P and Kachar B: On tight junction structure. *Cell* 1982; 28:441–450.

Plaaning-van Dalen DP, Brouwer A, and Knook DL: Clearance capacity of rat liver Kupffer, endothelial and parenchymal cells. *Gastroenterology,* 1981; 81:1036–1044.

Poli G, Cheeseman KH, and Dianzani MU: *Free Radicals in Liver Injury.* Oxford, IRL Press, 1985.

Rappaport AM: Normal microcirculation of the mammalian liver: 16 mm color film, sound, 20 minutes duration. Toronto, Division of Instructional Media Services, Faculty of medicine, University of Toronto, 1972.

Rappaport AM: Physioanatomical basis of toxic liver injury, in Farber E and Fisher MM (eds): *Toxic Injury of the Liver,* New York, Marcel Dekker, 1979, pp 1–57.

Rappaport AM: Hepatic blood flow: Morphological aspects and physiologic regulation, in Javitt NB (ed): *Liver and Biliary Tract Physiology I.* (Internatl. Review of Physiology). University Park Press, 1980, pp 1–63.

Rappaport AM, Borowy ZJ, Longheed WM, and Lotto WN: Subdivision of hexagonal liver lobules into a structural and functional unit. Role in hepatic physiology and pathology. *Anat Recl* 1954; 119:11–33.

Rappaport AM and Schniederman JH: The functions of the hepatic artery. *Rev Physiol Biochem Pharmacol* 1976; 76:129–175.

Reaven EP, Cheng Y, and Miller MD: Quantitative analysis of tubulin and microtubule compartments in isolated rat hepatocytes. *J Cell Biol* 1977; 75:731–742.

Reilly FD, McCuskey PA, and McCuskey RS: Intrahepatic distribution of nerves in the rat. *Anat Rec* 1978; 191:55–68.

Remmer H and Merker HJ: Drug-induced changes in the liver endoplasmic reticulum: association with drug metabolizing enzymes. *Science* 1963; 142:1657–1658.

Reuben A: Bile formation: sites and mechanisms. *Hepatology* 1984; 4:15S–24S.

Revel JP and Karnovsky MJ: Hexagonal array of subunits in intercellular junctions of the mouse heart and liver. *J Cell Biol* 1967; 33:C7–C12.

Rohr HP, Luthy J, Gudat F, Oberholtzer M, Gysin C, Stadler G, and Bianchi L: Stereology: A new development to the study of human liver biopsy specimens, in Popper H and Schaffner F (eds): *Progress of Liver Diseases,* vol 5. New York, Grune and Stratton; 1976, pp 24–34.

Rouiller C and Bernhard W: ''Microbodies'' and the problem of mitochondrial regeneration in liver cells. *J Biophy Biochem Cytol* 1956; 2 (Suppl):355–359.

Salisbury JL, Condeelis JS, Maihle NJ, and Statir P: Receptor-mediated endocytosis by clathrin-coated vesicles: Evidence for a dynamic pathway. *Cold Springs Harbor Symp. Quart. Biol.* 1981; 46:733–741.

Sasse D: Liver structure and innervation, in Thurman RG, Kauffman FC, and Jungermann K (eds): *Regulation of Hepatic Metabolism Intra- and Intercellular Compartmentation.* New York, Plenum, 1986, pp 3–25.

Schenk DB and Leffert HL: Monoclonal antibodies to rat Na^+, K^+-ATPase block enzyme activity. *Proc Natl Acad Sci USA* 1983; 80:5281–5285.

Schmucker DL and Jones AL: Hepatic fine structure in young and aging rats treated with oxandrolone: A morphometric study. *J Lipid Res* 1975; 16:143–150.

Schmucker DL, Mooney JS, and Jones AL: Stereological analysis of hepatic fine structure in Fisher 344 rat. Influence of sublobular location and animal age. *J Cell Biol* 1978; 78:319–337.

Seyer JM: Interstitial collagen polymorphism in rat liver with C Cl_4-induced cirrhosis. Biochem. *Biophys Acta* 1980; 629:490–498.

Seyer JM, Hutcheson ET, and Kang AH: Collagen polymorphism in normal and cirrhotic human liver. *J Clin Invest* 1977; 59:241–248.

Shinjoji N: The origin of intrahepatic bile duct cells in the mouse. *J Embryol Exp Morphol* 1984; 79:25–39.

Shiratori Y, Geerts A, Ichida T, and Wisse E: Collagen production and Kupffer cell-modulated proliferation of fat-storing cells in culture, in Kirn A, Knook DL, Wisse E (eds): *Cells of the Hepatic Sinusoid.* Rijswijk, Kupffer Cell Foundation, 1986, pp 239–245.

Siekevitz P and Zamecnik PC: Ribosomes and protein synthesis, in Gall JG, Porter KR, Siekevits P (eds): *Discovery in Cell Biology (J Cell Biol* 1981; 91:no 3, part 2, 535–655).

Smith RA and Ord MJ: Mitochondrial form and function relationships in vivo: their potential in toxicology and pathology. *Int Rev Cytol* 1983; 83:63–134.

Soifer D: Factors regulating the presence of microtubules in cells. *Ann NY Acad Sci* 1986; 466:1–7.

Spray DC, Saez JC, and Hertzberg EL: Gap junctions between hepatocytes: Structural and regulatory features, in Arias IM, Jacoby WB, Popper H, Schachter D, and Shafritz DA (eds): *The Liver: Biology and Pathology,* ed 2. New York, Raven Press, 1988, pp 851–866.

Staehelin A: Structure and functions of intercellular junctions. *Int Rev Cytol* 1974; 39:191–283.

Steffan AM, Gendrault JL, and Kirn A: Increase in the number of fenestrae in mouse endothelial liver cells by altering the cytoskeleton with cytochalasin B. *Hepatology* 1987; 7:1230–1238.

Steiner JW and Carruthers JS: Studies on the fine structures of the terminal branches of the biliary tree. I. The morphology of normal bile canaliculi, bile preductules (ducts of Hering) and bile ductules. *Am J Path* 1961; 38:639–661.

Steinert PM: Structure of three-chain unit of the bovine epidermal keratin filament. *J Mol Biol* 1978; 123:49–70.

Steinlieb I: Functional implications of human portal and bile ductular structure. *Gastroenterology* 1972; 63:321–327.

Stoscheck CM and Carpenter G: Down regulation of FGF receptors: direct demonstration of receptor degradation in human fibroblasts. *J Cell Biol* 1984; 98:1048–1053.

Sun TT, Tseng SCG, Haung AJW, Cooper D, Schermer A, Lynch MH, Weiss R, and Eichner R: Monoclonal antibody studies of mannalian keratins: a review. *Ann NY Acad Sci* 1985; 455:307–329.

Sutherland SD: The intrinsic innervation of the liver. *Rev Int Hepatol* 1965; 15:569–578.

Takemura S, Omori K, Tanaka K, Omori K, Matsunra S, and Tashiro Y: Quantitative immunoferritin localization of (Na^+, K^+) ATPase on canine hepatocyte cell surface. *J Cell Biol* 1984; 99:1502–1510.

Tavill AS: Intracellular pathways of protein synthesis and secretion in the hepatocyte. *Seminars in Liver Disease* 1985; 5:95–109.

Thurman RG and Kauffman FC: Sublobular compartmentation of pharmacolohic events (SCOPE): Metabolic fluxes in periportal and pericentral regions of the liver lobule. *Hepatology* 1985; 5:144–151.

Thurman RG, Kauffman FC, and Baron J: Biotransformation and zonal toxicity, in Thurman RG, Kauffman FC, and Jungermann K (eds): *Regulations of Hepatic Metabolism. Intra- and Intercellular compartmentation.* New York, Plenum, 1986, pp 321–382.

Timple R, Wick G, and Gray S: Antibodies to distinct types of collagens and procollagens and their application in immunohistology. *J Immunol Methods* 1977; 18:165–182.

Tsukada H, Mochizuku Y, and Fujiwara S: The nucleoids of rat liver cell microbodies. Fine Structure and enzymes. *J Cell Biol* 1966; 28:449–460.

Tsuneki K and Ichihara K: Electron microscope study of vertebrate liver innervation. *Arch Histol Jpn* 1981; 44:1–13.

Tzagoloff A: *Mitochondria.* New York, Plenum Press, 1982.

Unsworth DJ, Scott DL, Almond TJ, Beard HK, Hollborow EJ and Walton KW: Studies on reticulin L. Stereological and immunohistological investigation of the occurrence of collagen type III, fibronectin and the non-collagenous glycoprotein of Pras and Glynn in reticulin. *Br J Exp Pathol* 1982; 63:154–166.

van Berkel TJC: Functions of hepatic non-parenchymal cells, in Sies H et al. (eds): *Metabolic Compartmentation.* London, Academic Press, 1982, pp 437–481.

Van Deurs B and Luft JH: Effects of glutaraldehyde fixation on the structure of tight junctions. 1979; *J Ultrast Res* 1979; 68:160–172.

Van Eyken P, Sciot R, and Desmet VJ: Intrahepatic bile duct development in the rat: a cytokeratin-immunohistochemical study. *Lab Invest* 1988a; 59:52–59.

Van Eyken P, Sciot R, Callea F, Van Der Steen K, Moerman P, and Desmet VJ: The development of the intrahepatic bile ducts in man: A keratin-immunohistochemical study. *Hepatology* 1988b; 8:1586–1595.

Vidal-Vanaclocha F and Barbera-Guillem F: Fenestration patterns in endothelial cells in rat liver sinusoids. *J Ultrast Res* 1985; 90:115–123.

Wake K: Perisinusoidal stellate cells (fat-storing cells, interstitial cells, lipocytes), their related structure in and around the liver sinusoids and vitamin A-storing cells in extrahepatic organs. *Int Rev Cytol* 1980; 66:303–353.

Wakim KG and Mann FC: The intrahepatic circulation of blood. *Anat Rec* 1942; 82:233–253.

Wall DA, Wilson G, and Hubbard AL: The galactose-specific recognition system of the mammalian liver: The route of ligand internalization in rat hepatocytes. *Cell* 1980; 21:79–93.

Wang E, Fischman D, Liem RKH, and Sun TT: Intermediate filaments. *Ann NY Acad Sci* 1985; 455:832.

Weibel ER, Staubli W, Gnagi HR, and Hess FA: Correlated morphometric and biochemical studies on the liver cell I. Morphometric model, stereologic methods, and normal morphometric data for rat liver. *J Cell Biol* 1969; 42:63–91.

Widman JJ, Cotran RS, and Fahimi HD: Mononuclear phagocytes (Kupffer cells) and endothelial cells: identification of two functional cell types in rat liver sinusoids by endogenous peroxidase activity. *J Cell Biol* 1972; 52:159–170.

Wisse E: An ultrastructural characterization of the endothelial cell in the rat liver sinusoid under normal and various experimental conditions, as a contribution to the distinction between endothelial and Kupffer cells. *J Ultrast Res* 1972; 38:528–562.

Wisse E, De Zanger RB, Charels K, Vander Smissen P, and McCuskey RS: The liver sieve: Considerations concerning the structure and function of endothelial fenestrae, the sinusoidal wall and the space of Disse. *Hepatology* 1985; 5:683–692.

Wisse E, Van't Noordende JM, and van der Meulen J: The pit cell: description of a new type of cell occurring in rat liver sinusoids. *Cell Tissue Res* 1976; 173:432–435.

Yancey SB, Easter E, and Revel JP: Cytological changes in gap junctions during liver regeneration. *J Ultrast Res* 1979; 67:229–242.

Yee AG and Revel JP: Endothelial cell junctions. *J Cell Biology* 1975; 66:200–204.

Yee AG and Revel JP: Loss and reappearance of gap junctions in regenerating liver. *J Cell Biol* 1978; 78:554–564.

Chapter

2

A Histopathologic Classification of Chemical-Induced Injury of the Liver

John M. Cullen
Department of Microbiology, Pathology and Parasitology
North Carolina State University
Raleigh, North Carolina

Boris H. Ruebner
Department of Medical Pathology
School of Medicine
University of California at Davis
Davis, California

GENERAL PRINCIPLES

The most frequent histologic pattern of acute hepatic injury produced by chemicals is a characteristic combination of hepatocytic swelling and steatosis. The model compound which is associated with this type of liver injury is carbon tetrachloride. The relative proportions of cytoplasmic swelling to steatosis vary with different agents. If the injury is relatively mild, it may regress; if severe, it may progress to necrosis and become irreversible. Apart from this pattern of injury, a great variety of nuclear and cytoplasmic changes may be produced by different chemicals. Some chemicals are associated with cholestasis or more complex injuries such as granulomas and vascular-based injuries. Fibrosis usually suggests a more chronic injury resulting from repeated injury or, rarely, a single major insult. The pattern of injury often gives some useful clues to etiology and may narrow the number of compounds considered possibly responsible for a particular type of injury.

Perhaps the best recognized classifications of hepatotoxic chemicals are those causing injury by predictable (intrinsic) reactions and those causing injury by idiosyncratic (hypersensitivity) reactions (Zimmerman, 1978). Intrinsic toxins produce injury to the majority of species exposed in a dose-related fashion. A great deal of work has been done on the mechanisms responsible for chemical-induced injuries and several classification systems exist (Zimmerman and Maddrey, 1987). Those relatively few intrinsic hepatotoxins, such as elemental phosphorus, which are directly injurious to particular cellular constituents, are classified as direct hepatotoxins. Target organelles may differ depending on the particular chemical. Classified as indirect hepatotoxins are those large numbers of chemicals which interfere with specific metabolic pathways leading to hepatocellular injury. Indirect hepatotoxins may be further subdivided into cytotoxic or cholestatic types. Chemical injury of the idiosyncratic or hypersensitivity type is relatively rare and is seen virtually only in humans as a reaction to certain therapeutic drugs. A compound or group of compounds may produce not only a characteristic cytologic change affecting hepatocytes, but the injury may also have a characteristic localization in the hepatic lobule (Figure 1). Lobular localization may give a clue to the mechanism responsible for toxic injury. Carbon tetrachloride, for example, injures predominantly the central zone of the lobule. Typically, chemicals of the type which have to be activated by the mixed-function oxidases (MFO) of the endoplasmic reticulum, a group of enzymes which are more abundant in the central rather than peripheral part of the lobule, produce centrilobular injury. However, the relative anoxia of the pericentral zones may also be a localizing factor for hepatocellular injury by chemicals, which are activated by other means. Centrilobular injury is produced by the great majority of hepatotoxic compounds and, therefore, is of limited value in identifying the particular agent responsible in a case of toxic liver injury. Stimulation or inhibition of the MFO system by administration of certain chemicals may alter the lobular localization of chemical-induced injury. For example, the hepatotoxic phytotoxin ngaione may produce midzonal or centrilobular injury in mice depending on their prior exposure to phenobarbital which induces MFO enzymes (Seawright and Hrdlicka, 1972).

Chemical compounds may also produce more than one pattern of injury depending on the dose and its manner of administration, in particular whether the chemical has been administered acutely or chronically. The route of administration may also make some difference since the cell types or organs which the chemical encounters first may be injured most severely. However, although toxic compounds reach the periportal hepatocytes first, pericentral localization of hepatic injury is much more common than periportal injury. This would strongly suggest that metabolic zonation may be more responsible

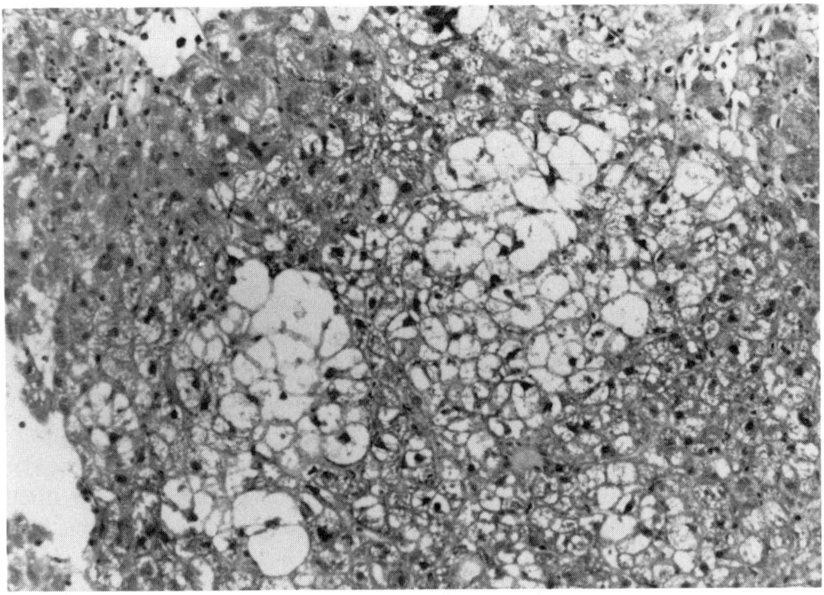

Figure 1. Diagram depicting both the lobular and acinar concept of hepatic organization. The classical hexagonal hepatic lobule is bordered by six portal radicles (P) and centered on the central vein (C). The lobule is divided somewhat arbitrarily into central zones surrounding central veins, midzones, and peripheral or periportal zones. An alternate concept is that of Rappaport's acinus. The acinus (shaded) is diamond shaped and centered on the terminal distributing branches of the portal radicle vasculature. Centroacinar areas (zone 1) have the best blood supply. Periacinar areas (or zone 3) are the apex of the shaded areas.

for lobular localization than circulatory factors. Protein deficiency is another factor which may diminish certain liver enzymes and alter the severity of hepatocellular injury (Hartroft, 1963). Differences between the sexes in the susceptibility of the liver to injury are quite common and probably are based partially on hormonally dependent increased MFO activity in males making them generally more susceptible to agents activated by these enzymes (Rouiller, 1964). Other endocrine factors such as thyroid function may also alter chemical toxicity (McLean et al., 1965). It should also be noted that some agents cause different patterns of injury in different species; for example, aflatoxin B_1 causes periportal necrosis in rats and pericentral necrosis in mice (Goldblatt, 1964). Pyrrolizidines produce hepatocellular necrosis in rats and veno-occlusive disease in humans (Bras, 1973). No doubt these different patterns of injury are related to metabolic differences between the different species. The relationship between a group of compounds and the lesion produced may thus be complex.

ACUTE HEPATOCELLULAR INJURY AND NECROSIS

Cytoplasmic Alterations

Hydropic change is characterized by swelling of the cytoplasm of hepatocytes, frequently with some vacuolization. This type of reaction may be seen particularly in centrilobular zones, as for instance in carbon tetrachloride injury (Jennings, 1955). Peripheral hydropic change is seen less commonly in such compounds as elemental phosphorus and allyl formate (Rouiller, 1964). In order to confirm a histologic diagnosis of hydropic change, one has to exclude other causes of swelling of hepatocytes such as increased glycogen content, which is a rare consequence of chemical injury and large-droplet fatty change which, in fact, is commonly associated with hydropic change. This makes the distinction between these two patterns of injury sometimes difficult and occasionally impossible. In fact, their frequent association represents the most common histologic type of liver injury. The association of hepatocellular swelling with large-droplet fatty change often tends to be a precursor to necrosis which is the most severe stage of this type of injury and is described in more detail below.

Accumulation of excessive glycogen in hepatocytes is seen diffusely in the glycogen storage diseases and focally in certain preneoplastic hepatocellular lesions (Bannasch, 1968). Excessive glycogen deposition develops rarely as a result of chemical injury. One example of this type of change appears to be the lesions induced in dogs by administration of corticosteroids where the hepatocytes are swollen as the result of excessive cytoplasmic glycogen accumulation (Figure 2) (Fittschen and Bellamy, 1984).

Fatty change unassociated with hydropic swelling of hepatocytes or necrosis may be divided into small-droplet fatty change and large-droplet fatty change (Table 2–1) (Hoyumpa et al., 1975). Small-droplet or microvesicular fatty change which is seen particularly in Reye's syndrome in humans has also been described after ingestion of aflatoxin B_1 in monkeys and after injection of excessive amounts of intravenous tetracycline in humans (Figure 3) (Lloyd-Still et al., 1974). In this type of hepatocellular injury the nucleus remains in the center of the hepatocyte and the small droplets are seen in the cytoplasm and do not displace the nucleus. Relatively small fat droplets tend to be seen in those conditions which are not associated with necrosis. Phosphorus and ethionine-induced toxicity cause microvesicular steatosis in addition to those conditions mentioned above (Hoyumpa et al., 1975). Large-droplet fatty change may be seen sometimes without necrosis but tends to be accompanied by necrosis more frequently than microvesicular steatosis (Figure 4). This type of fatty change is seen much more often than microvesicular steatosis and occurs in carbon tetrachloride toxicity, in rats fed diets choline deficient and low in methionine, as well as in hepatic injury by *Amanita phalloides,* and, of course, in alcoholic liver injury (Desmet, 1985; Jennings,

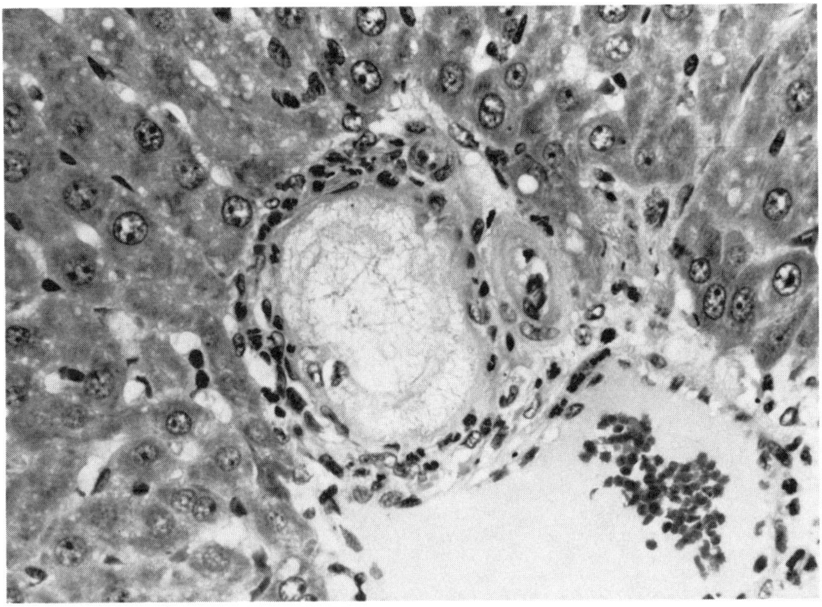

Figure 2. Distended hepatocytes in a dog due to excessive corticosteroid administration. Accumulation of glycogen in cytoplasm is responsible for the ballooned appearance of the cells.

TABLE 2–1.
Agents That Produce Macrovesicular Steatosis*

Acetamide	Ethyl bromide
Antimony	Ethyl chloride
L-Asparaginase	Hydrazine
Azacytidine	Hypoglycin
Azaserine	Methotrenate
Azauridine	Methyl bromide
BAL	Methyl dichloride
Barium salts	Orotic acid
Bleomycin	Phosphorus†
Borates	Rare earths‡
Carbon disulfide	Safrole
Chromates	Thallium compounds
Dichloroethylene	Uranium compounds
Dimethylhydrazine	Warfarin
Ethanol	

*After Zimmerman (1978).
Partial list of agents that produce fatty liver in rats:
† Phosphorus causes mainly fatty liver; may be some peripheral necrosis as well.
‡ Only rare earths of low atomic number (<61) produce fatty liver.

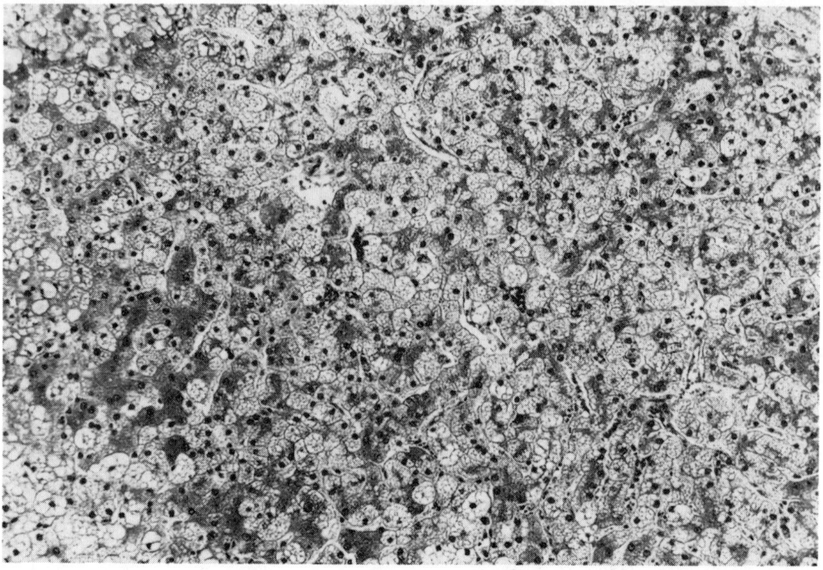

Figure 3. Microvacuolar steatosis is characterized by multiple small round vacuoles in the hepatocellular cytoplasm which usually do not displace the nucleus.

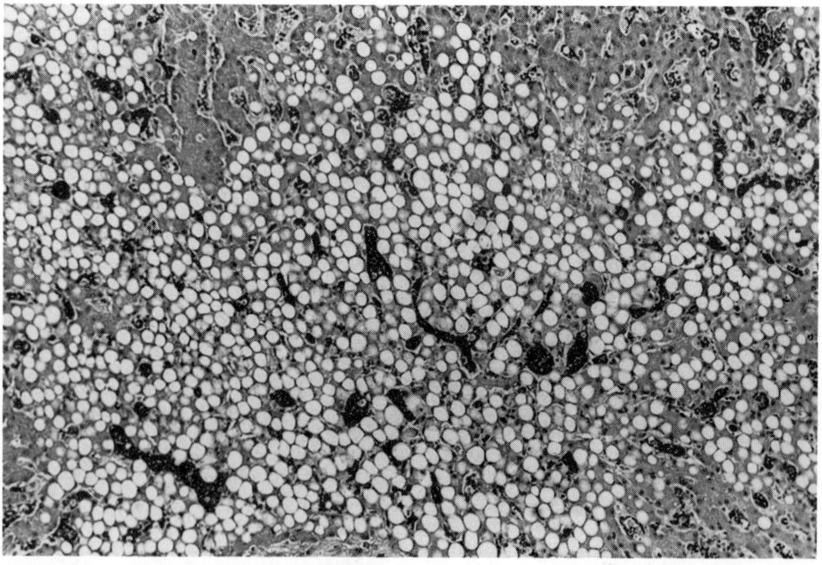

Figure 4. Macrovesicular steatosis is characterized by large rounded cytoplasmic vacuoles which often expand the hepatocyte and displace the nucleus.

1955, Wepler and Opitz, 1972). However, microvesicular steatosis may occur in alcoholic liver injury (Uchida et al., 1983). Methotrexate also produces this kind of liver injury. Chronic methotrexate administration may lead to fibrosis (Tolman et al., 1985). A partial listing of agents that cause steatosis is found in Table 2–2.

Proliferation of the smooth endoplasmic reticulum is a response not only to phenobarbital but also to the PCBs and 3-methyl-cholanthrene as well as DDT (Zimmerman, 1978). Classically, the cytoplasm of the hepatocytes showing this type of response has a ground glass appearance by light microscopy and the cell appears somewhat enlarged. However, this reaction may not always be easily recognized by light microscopy and may require electron microscopy and even quantitative electron microscopy for its demonstration. This change may be considered an adaptive response since evidence of injury of the hepatocyte is usually not seen. Generalized enlargement of the liver can also occur following lead injection (Columbano, 1983, 1984).

Phospholipidosis is another kind of cytoplasmic alteration. The hepatocytes are enlarged and have a foamy appearance by light microscopy and highly characteristic laminated inclusions are seen by electron microscopy. Classically, this picture has been produced by drugs such as Coralgil and Amiodarone (Rigas et al., 1986).

Mallory bodies are eosinophilic cytoplasmic inclusions by light microscopy and represent a proliferation of intermediate-type filaments. They were first described as a response of the hepatocyte to chronic ethanol ingestion (Lieber, 1975). However, we know now that Mallory bodies are a reaction which may be seen in a variety of different types of injury. Among these are cholestatic conditions in humans such as Wilson's disease, primarily biliary cirrhosis, and Indian childhood cirrhosis. Griseofulvin in rats (Denk et al., 1975) and Dieldrin in mice (Meirhenri et al., 1981) have also been found to produce these structures.

Hepatocellular Necrosis

Hydropic swelling and necrosis frequently have a characteristic zone localization. Fatty change is a frequent accompaniment of toxic hepatic injury, but its lobular localization is more variable than that of necrosis. Central necrosis is particularly seen after acetaminophen ingestion, and in carbon-tetrachloride injury (Figure 5) (Ferguson, 1977; Jennings, 1955). Halothane anesthesia also is associated with a predominantly central type of injury (Kenna et al., 1987). Midzonal necrosis is rarely seen. It has been described for instance after ngaione, a phytotoxin, and has been attributed to hypoxia associated with free-radical formation (Seawright, 1972). Peripheral necrosis, however, is seen after ferrous sulfate poisoning as well as phosphorus poisoning and after aflatoxin B_1 exposure in rats (Fletcher and Galambos, 1963;

TABLE 2–2.
Partial List of Agents that Produce Hepatic Necrosis and Steatosis*

Agent	Site of Necrosis[†] CZ	MZ	PZ	M	St	Agent	CZ	MZ	PZ	M	St
Acetaminophen	+	—	—	—	—	Dinitrophenol	—	—	—	(+)	+
Aflatoxins	+	—	+	+	+	Dinotrotoluene	+	—	—	(+)	+
Albitocin	—	—	+	—	—	Divinyl ether	+	—	—	—	+
Allyl compounds	—	—	+	—	—	Ethylene dibromide	+	—	—	—	+
Alloxan	—	—	+	—	—	Ethylene dichloride	+	—	—	—	—
α-Amanitin	+	—	—	—	+	Ferrous sulfate	—	—	+	—	—
Aniline	—	—	—	+	—	Fluoroacetate	+	—	—	D	+
ANIT	(+)	(+)	(+)	—	—	Galactosamine	—	—	—	+	(+)
Anthrapyrimidine	+	+	—	—	—	Iodobenzene	+	—	—	—	—
Antimony	—	—	—	—	+	Iodoform	+	—	—	—	+
Arsenic (inorg.)	+?	—	+	—	—	Islandicin	+	—	—	—	+
Arsenicals (org.)	+	—	—	+	—	Leteoskyrin	+	—	—	—	—
Beryllium	—	+	—	—	—	Manganese compounds	—	—	+	—	+
Botulinus toxin	+	—	—	—	—	Methylchloroform	+	—	—	(+)	(+)
Bromobenzene	+	—	—	—	(+)	Methylene chloride	(+)	—	—	—	(+)
Bromotrichloromethane	+	—	—	—	+	Naphthalene	+	—	—	—	—
CCl_4	+	—	—	—	+	Ngaione	—	+	—	—	—
Chlorobenzenes	+	—	—	(+)?	—	Paraquat	+	+	—	—	—
Chloroform	+	—	—	—	+	Phalloidin	+	—	—	—	—
Chloroprene	+	—	—	—	+	*P. vulgaris* endotoxin	—	—	+	—	—
Chlorinated biphenyls	(+)	—	—	+	+	Pyridine	+	—	—	—	(+)
Chloronaphthalenes	(+)	—	—	+	+	Pyrrolidizine alkaloids	+	—	—	—	—
Chloropropane	+	—	—	—	+	Rubratoxin	+	—	—	+	—
Dichloropropane	+	—	—	—	+	Selenium	—	—	—	—	—

Dioxane	+	—	—	—	—	Sporodesmin	—	—	+	—	+
Dioxins	+	—	—	(+)	—	Tannic acid	+	—	—	—	—
DDT	+	—	—	—	+	Thioacetamide	+	—	—	+	(+)
Diphtheria toxin	+	—	—	—	—	TNT	(+)	—	—	—	—
Dimethylnitrosamine	+	—	—	—	+	Urethane	+	—	—	—	+
Dinitrobenzene	+	—	—	(+)	+	Xylidine	+	—	—		

*After Zimmerman (1978).

† CZ = centrizonal necrosis; MZ = midzonal; PZ = peripheral zone; M = massive; St = steatosis; D = diffuse. Parenthesis indicates the lesion to be a nonprominent component of the injury.

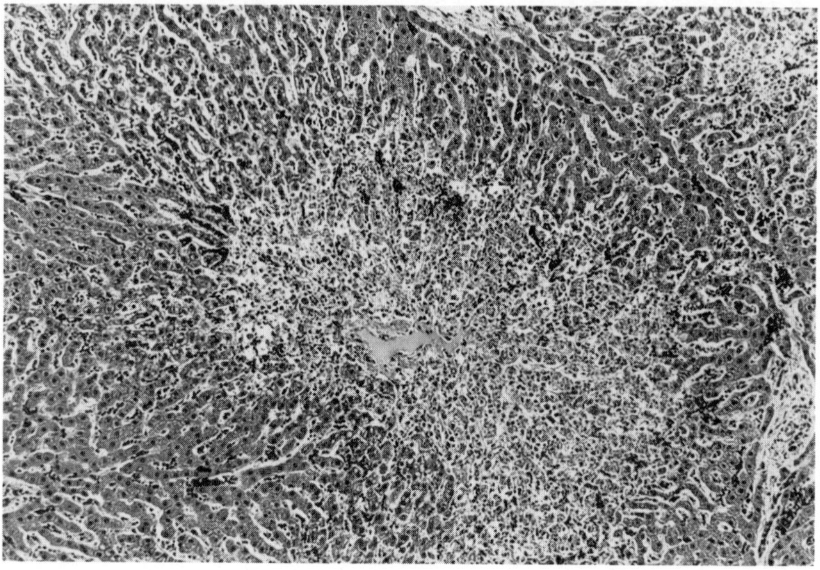

Figure 5. The most common pattern of chemical-induced hepatocellular necrosis is centrolobular necrosis seen here. Periportal hepatocytes are intact.

Goldblatt, 1969; Rodriquez-Iturbe, 1971; Robotham et al., 1974). Massive necrosis occurs in animals exposed to large doses of chemicals and has been reported in humans and animals exposed to aniline, trinitrotoluene, chlorinated biphenyls, and chloronapthalenes among others (Zimmerman, 1978). Diffuse focal necrosis is most often associated with viral infections but may occur following ingestion of galactosamine (Decker et al., 1972).

Lytic necrosis, characterized by nuclear pyknosis, and karyorrhexis, associated with cellular disintegration, is a more advanced stage of hydropic degeneration which has already been described and is frequently associated with fatty change in adjacent hepatocytes, as, for example, in carbon tetrachloride poisoning (Jennings, 1955). It is in fact indicative of irreversible injury produced by the same agents which cause hydropic change when they are given in doses that are associated with reversible injury.

Formation of acidophilic bodies or apoptosis is a different type of cell death which is differentiated clearly morphologically from that type of cell death associated with lytic necrosis (Figure 6) (Duvall and Wyllie, 1986). In this type of cell death the cytoplasm becomes dehydrated and more dense. This type of cell death is associated with nuclear pyknosis. Apoptosis is seen in a great variety of types of cell injury, particularly those of viral origin and also in atrophy as seen, for instance, in the myometrium after pregnancy. As far as chemical liver injury is concerned, apoptosis is seen in alcoholic liver injury and may also be found to some extent in injury by a great variety of toxic agents. Formation of apoptotic bodies is particularly striking in livers which are returning to normal size following lead-injection-induced hepatomegaly (Columbano et al., 1984).

Nuclear Changes

Hepatocyte necrosis is, of course, associated not only with cytoplasmic but also with nuclear changes and these have been described above. Nuclear changes not associated with necrosis are relatively uncommon. They include the inclusions seen in lead poisoning (Beattie et al., 1979). However, lead inclusions are seen much more frequently in the kidney than in the liver and do not appear to have any great pathologic significance.

Pyrrolizidine alkaloids produce hepatocellular injury and vascular injury in different animals under different circumstances (Ruebner et al., 1970). They have also been reported to produce megalocytosis, or karyomegaly (Figure 7). This is seen only in certain species and has to be distinguished from the nuclear enlargement sometimes seen in old age associated with increase in nuclear ploidy which is a regular accompaniment of aging.

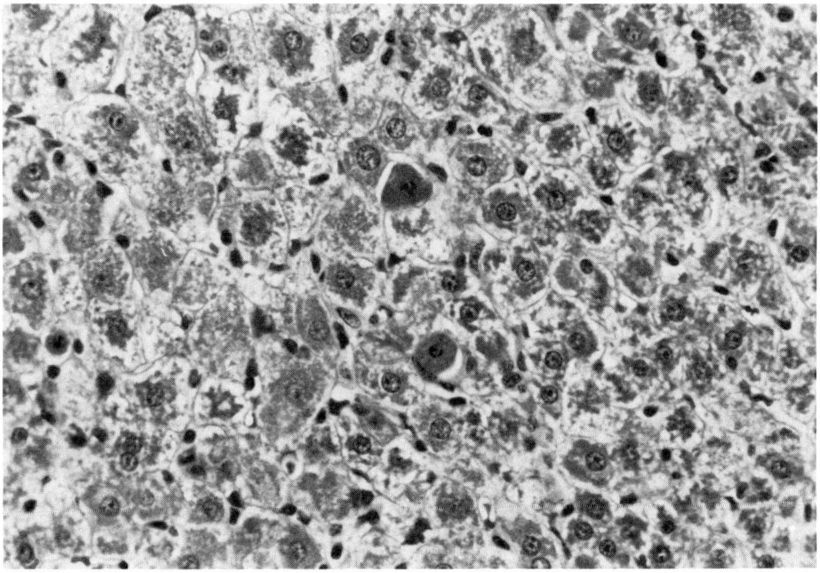

Figure 6. Individual cell necrosis or apoptosis occurs infrequently as a response to chemical-induced injury. It is characterized by condensation of cytoplasm, pyknosis of nuclei, and rounding of the cell outline.

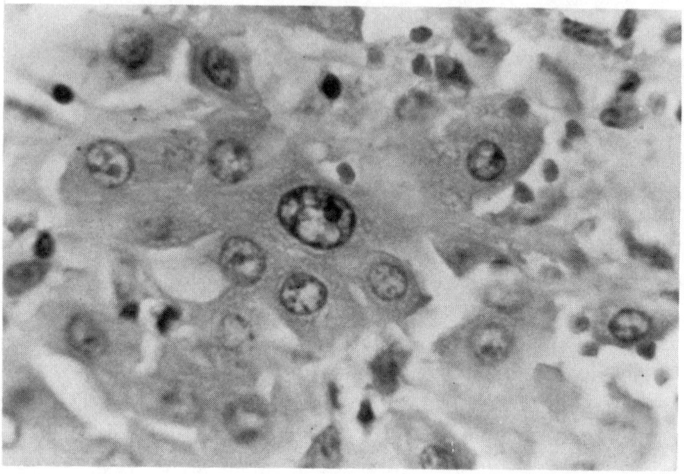

Figure 7. Karyomegaly associated with pyrrolizidine alkaloid ingestion is characterized by enlarged nuclei.

Chronic Hepatocellular Injury

Continuous or repetitive liver injury causes chronic liver injury associated with fibrosis and, if continued long enough, may lead to cirrhosis (Figure 8) (Cameron and Karunaratne, 1936). A variety of chemicals administered in this fashion may have this effect, particularly chlorinated hydrocarbons such as carbon tetrachloride, PCBs, dinitrobenzene, and trinitrotoluene (Bridge et al., 1942; Himsworth, 1947; Klatskin, 1975; Miller 1920) (Table 2–3). Pyrrolizidine alkaloids can also cause this type of response as do nitrosamines, dibutyltin salts, and excessive vitamin A (Klatskin, 1975; Rouiller, 1964; Zimmerman, 1969; Barnes and Magee, 1958; Barnes and Stoner, 1961). A particularly interesting group is that including inorganic arsenicals (Franklin et al., 1950; Weir, 1930), vinyl chloride (Plaa, 1980), and thorotast. While all these compounds produce fibrosis if given chronically, the resulting lesion in this group of compounds appears to be a more diffuse fibrosis, which has been termed hepatoportal sclerosis. Although cirrhotic nodules usually do not develop, these patients often have portal hypertension. Hemangiosarcoma is a late complication.

A granulomatous response is, of course, most characteristic of such chronic infections as tuberculosis and fungal infections, as well as of sarcoidosis. A sarcoidosis-like picture of a granulomatous response to a great variety of drugs is seen in association with drugs, such as quinine (McMaster and Hennigar, 1981). However, it may occasionally be seen as a response to certain metals such as beryllium (Stoeckle et al., 1969) and copper which may be contained in copper spray used by vineyard workers (Pimentel and Menezes, 1975, 1977).

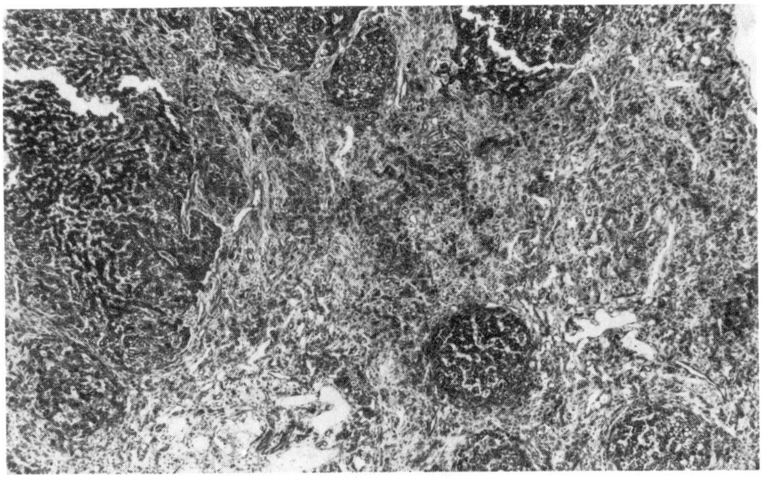

Figure 8. Chronic administration of hepatotoxic substances may result in architectural distortion of liver anatomy due to fibrosis, and regenerative nodule formation.

TABLE 2–3.
Drugs Which May Produce Cirrhosis*

Aflatoxins	Indospicine
Amanita phalloides	Lithocholate
Arsenic (inorg.)	Luteoskyrin
Azo dyes	Manganese
Bromobenzene	α-Methyldopa
Cadmium ion	α-Naphthylisothiocyanate
Carbon tetrachloride	Phosphorus
Chlorinated biphenyls	Pyrrolidizines
Chloroform	Rubratoxin
Cinchophen	Safrole
Cyclochloritine	Sporodesmin
Dimethylnitrosamine	Stilbamidine
1,2,5,6-Dibenzanthracene	Tannic acid
DDT	Tetrachlorethane
Dimethylnitrosamine	Thioacetamide
Dioxin	Thiourea
Ethanol	Trinitrotoluene
Ethylene dichloride	Trypan blue
Galactosamine	Urethane
Griseofulvin	Vinyl chloride

*After Zimmerman (1978).

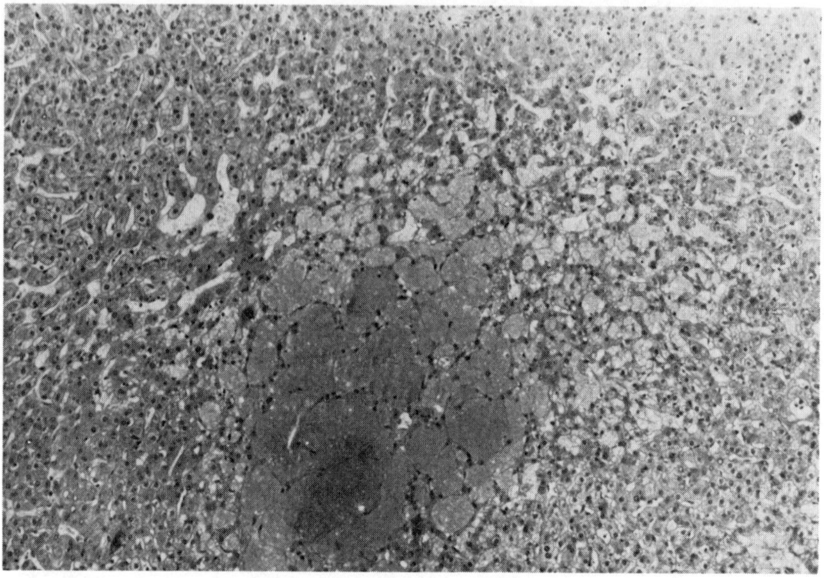

Figure 9. Peliosis formation of large vascular spaces within the liver may result from ingestion of contraceptive or anabolic steroids and occurs in cattle which ingest *Pimella* sp.

Vascular Injury

Peliosis hepatis is a striking hepatic lesion which may or may not be associated with symptoms. It is seen most characteristically in humans as the result of ingestion of contraceptive or androgenic steroid hormones (Taxy, 1978). It consists of the development of large irregular vascular spaces in the liver which may or may not be lined by endothelium (Figure 9). These spaces have no preferential lobular localization. Whether these vascular spaces are associated with necrosis remains debated. Most similar to the human disease is probably St. George's disease of cattle caused by a poisonous plant, *Pimella* spp (Kelly and Seawright, 1978). Somewhat similar lesions were seen after administration of lasiocarpine in mice by Ruebner (Ruebner et al., 1970) and after phalloidin exposure in the rat (Tuchweber et al., 1973).

Veno-occlusive disease was originally reported in Jamaican children as a result of ingestion of pyrrolidizine alkaloids in bush tea (Bras, 1973). Nonhuman primates exposed to the pyrrolidizine alkaloid monocrotaline have developed central vein occlusion (Allen and Carstens, 1971). It consists of the development of fibrovascular connective tissue in the region of the central veins and leads to portal hypertension and consequently liver failure. More recently it has been observed particularly in individuals exposed to urethane or patients treated with immunosuppressive compounds such as azathioprine and 6-thioguanine (Brodsky et al., 1961; Weitz et al., 1982). Portal hypertension without cirrhosis characterized by portal fibrosis has been reported in workers exposed to arsenic-containing insecticidal sprays (Huet, 1975). Hypervitaminosis A may also result in noncirrhotic portal hypertension due to perisinusoidal fibrosis (Russel et al., 1974).

The Budd-Chiari syndrome is seen characteristically after obstruction of the blood flow through the hepatic veins or the inferior vena cava, usually by thrombosis. It develops when there is an increased tendency for thrombosis, as for example, in polycythemia. It has also been reported in women taking contraceptive steroids (Lalonde et al., 1982).

Pigment Deposits

Accumulations of pigments such as lipofuscin, hemosiderin, copper, porphyrins, or bile may be found in hepatocytes and Kupffer cells following exposure to a variety of chemicals. Lipofuscin may increase as a normal aging change or as a result of hepatocellular necrosis. Bile accumulation is considered in section 3. The iron-containing pigment hemosiderin is found in excess most often in hemochromatosis and after hemolysis or multiple transfusions. However, excessive hemosiderin can be deposited in the hepatocytes after exposure to hexachlorobenzene, excessive dietary ingestion of iron, and in patients who ingest excessive iron supplements (Figure 10) (Hennigar et al., 1979). This was quite common in the African Bantu population due to iron contamination of their diet (Bothwell, 1962). However, this lesion appears to be becoming less frequent (MacPhail et al., 1979).

Porphyrin crystals, derived from disturbed heme synthesis, are deposited

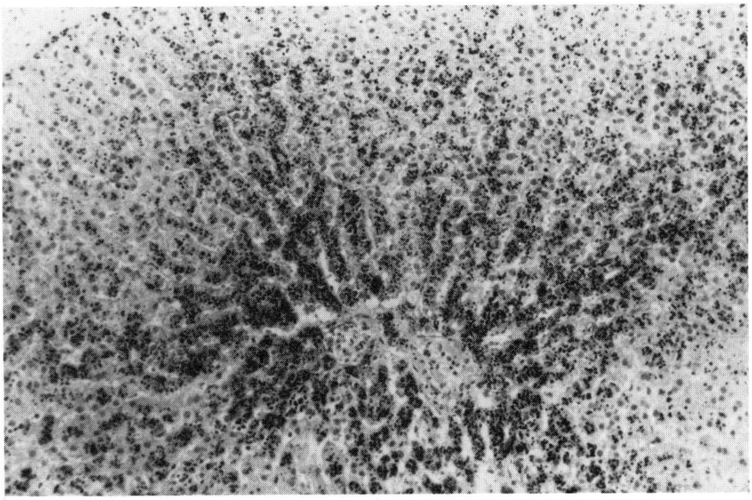

Figure 10. Accumulation of hemosiderin in hepatocytes and Kupffer cells may be appreciated histologically as granular cytoplasmic deposits when the liver is stained with appropriate histochemical stain which discloses iron (Perls stain).

in the liver in inherited porphyrias. In porphyria cutanea tarda, an inherited disease, which is exacerbated by ethanol ingestion, needle-shaped phorphyrin crystals are found in the liver (Cortes et al., 1980). Acquired porphyria may be associated with the ingestion of several compounds including hexachlorobenzene and contraceptives (Behm, 1974; Schmid, 1960; Francis and Smith, 1987). Griseofulvin in mice is associated with deposition of brown doubly refractile protoporphyrin crystals, mostly in bile canaliculi (Matilla and Molland, 1974). Diethoxycarbonyl, dihydrocollidine and trimethyl phenyl thioethylmethylsidnone also have this effect (Bloomer, 1988). These compounds inhibit ferrochetolase which results in accumulation of protoporphyrin in the liver.

Accumulated copper may be detected as refractile granules in hepatocellular cytoplasm and demonstrated more readily histochemically. Chemicals that produce prolonged cholestasis and excessive ingestion of copper salts may lead to deposition of copper in hepatocytes or Kupffer cells.

CHOLESTATIC INJURY

Cholestasis may be defined as an interruption in the flow of bile due to excretory failure of hepatocytes with accumulation of bile constituents in the blood (Desmet, 1972). Cholestasis may occur through impairment of uptake of bile constituents from the blood into hepatocytes, interruption of the movement of these substances through the hepatocyte into the bile canaliculus, or injury of bile ducts. Many chemicals may produce an interruption of bile flow as their only or primary action (Plaa and Preistly, 1977). Obviously, necrosis of hepatocytes will also impair bile secretion, but this cause of cholestasis will not be considered in this section.

Cholestasis is recognized histologically by the presence of green-brown plugs or casts in dilated canaliculi. Accumulated bile may also be found within the lumen of bile ducts and as brownish granules in the cytoplasm of hepatocytes and Kupffer cells. Most often bile accumulation occurs in central zones, although bile may be present throughout the entire lobule in severe cases (Dubin and Peterson, 1958). The histologic pattern of chemical-induced cholestasis can be divided into three categories: (1) *canalicular* or ''bland'', which is characterized by the absence of histologic evidence of injury, (2) *hepatocanalicular* or ''pericholangitic,'' associated with periportal chronic inflammation and mild hepatocellular injury, and (3) a third type proposed by Zimmerman termed cholangiodestructive or ductal, which is characterized by damage to larger interlobular bile ducts (Zimmerman and Maddrey, 1987). However, it should be kept in mind that different species often have dissimilar histologic responses to the same chemical so that classification of chemicals varies with the species examined. A partial listing of chemicals which cause

TABLE 2–4.
Chemicals That Produce Cholestasis

Type of Cholestasis	Agent
Canalicular (bland)	Ictergenin
	Manganese excess
	Cytochalasin B
	Phalloidin
	Endotoxins
	C-17 alkylated steroids
	Dinitrophenol
	4,4-Diaminodiphenylmethane
	4,4-Diaminodiphenylamine
	Flavaspidic acid
	Lithocholic acid
Hepatocanalicular (pericholangitic)	Chlorpromazine
	Erythromycin estolate
	Organic arsenicals
Cholangiodestructive	α-Naphthylisothiocyanate
	Sporodesmin
	Paraquat

cholestasis and the histologic pattern associated with each chemical is presented in Table 2–4.

"Bland" cholestasis is most often seen in humans after administration of certain drugs such as contraceptive or anabolic steroids (Figure 11). However, a similar pattern is also produced by several naturally occurring compounds such as icterogenin, a triterpene pentacyclic alkaloid found in the plant *Lippia rehmani* (Arias, 1963). Cattle, sheep, rabbits, and rats are susceptible to this compound (Bullock et al., 1974). Triterpenes, lantadene A and B, which occur in the plant *Lantana camara* can also produce a similar pattern of cholestasis in ruminants (Seawright, 1972; Gopinath and Ford, 1969). Cytochalasin B, a fungal alkaloid, can, like anabolic steroids, produce canalicular cholestasis through its disruptive action on cytoplasmic microtubules which appear necessary for movement of bile through canaliculi (Phillips et al. 1978, 1983). In low doses, phalloidin, another fungal toxin produced by *Amanita phalloides,* can also produce cholestasis (Watanabe and Phillips, 1986). Other agents capable of causing cholestasis without associated inflammatory lesion include endotoxins, and manganese in excessive quantities (Zimmerman et al., 1979; Lustig et al., 1982; Witzleben, 1972; Phillips et al., 1978).

Hepatocanalicular cholestasis, the second histologic type of cholestasis, is characterized by the presence of mild inflammatory infiltrates associated with interlobular bile ducts or scattered through the parenchyma (Figure 12). This type of injury is generally associated with reactions to drugs. Chlor-

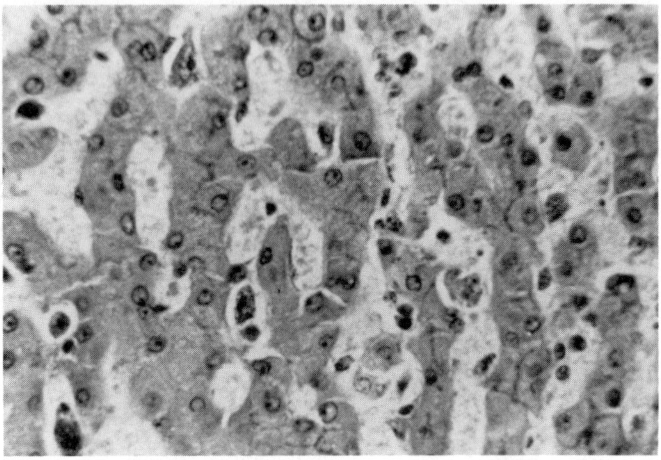

Figure 11. Accumulation of bile within canaliculi without associated hepatic inflammation is characteristic of canalicular or bland cholestasis.

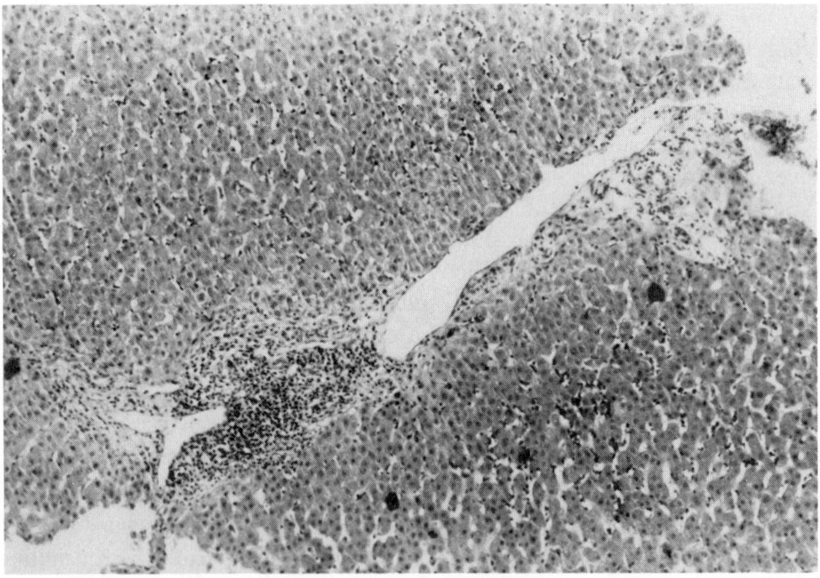

Figure 12. Cholestasis accompanied by portal tract inflammation (hepatocanalicular cholestasis) is more common in humans exposed to certain drugs, such as chlorpromazine as in this case.

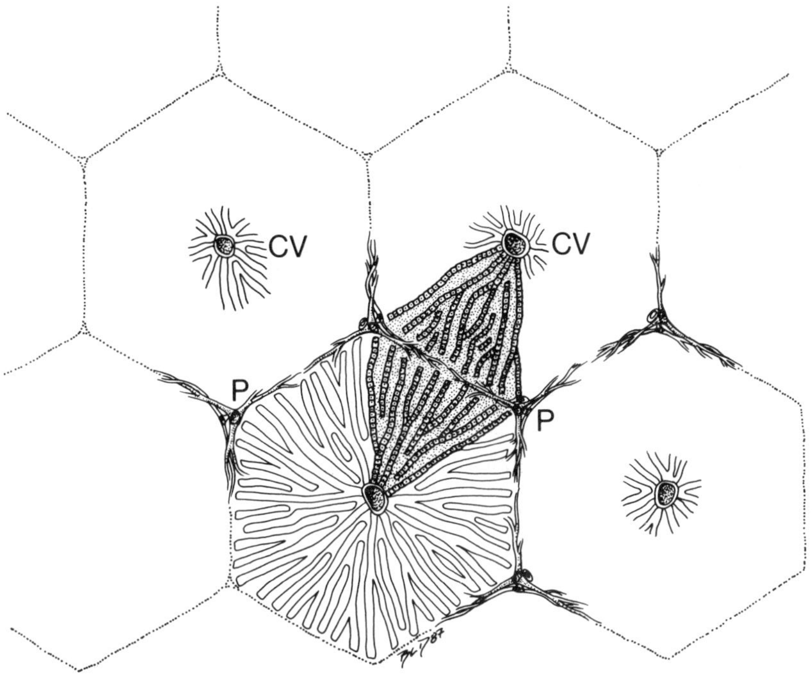

Figure 13. Cholangiodestructive cholestasis is characterized by destruction of bile ducts due to injury of biliary epithelium, as in this case of α-napthyl isothiocyanate toxicity in the mouse.

promazine, a sedative, and the antibiotic erythromycin estolate are typically associated with this type of lesion in humans (Levine et al., 1966). Synthetic chemicals such as organic arsenicals have been associated with this type of lesion. However, it is quite possible that the lesions produced by organic arsenicals were really the result of viral hepatitis.

Cholangiodestructive cholestasis may be best exemplified by α-napthylisothiocyanate (ANIT), a synthetic chemical which produces cholestasis through destruction of biliary epithelium in bile ducts (Goldfarb et al., 1962). Depending on the dose, duration of exposure, and species affected, biliary proliferation, cirrhosis, and variable degrees of parenchymal injury may occur. Rats, guinea pigs, hamsters, dogs, sheep, and cattle are affected (Lopez and Mazzanti, 1955; McClean, 1958; Indacochea-Redmond, 1971; Gopinath, 1970). Rats and mice are more susceptible than guinea pigs. Rabbits and hamsters, however, do not develop cholestasis following ANIT exposure, but do incur mild hepatocellular injury. Ruminants and dogs also develop hepatocellular rather than biliary injury (Gopinath, 1970; Indacochea-Redmond and Plaa, 1971).

Initial lesions following a single dose in susceptible species include swollen biliary epithelium, necrotic cell debris within ductal lumens, and neutrophilic periportal infiltrates (Figure 13). Subsequent biliary proliferation is maximal four days following exposure and has resolved to a virtually normal appearance within seven days. Chronic administration of cholangiodestructive toxins may lead to significant biliary proliferation and fibrosis resembling biliary cirrhosis (Figure 14).

Cholangiodestructive cholestasis is also produced by sporodesmin a metabolite of the fungus *Pithomyces chartarum,* which infests various plant materials (Mortimer and Taylor, 1963). Cholestasis develops in sheep, rats, and rabbits (Mortimore and Taylor, 1963; Slater et al., 1964). While this compound selectively injures bile ducts in rats, its effects vary depending on the species affected, as sheep also develop periportal hepatocellular injury (Mortimore, 1962, Slater, 1962). Bile duct injury in sheep consists of progressive destruction of bile ducts produced and subsequent periductal fibrosis. Toxic doses of the herbicide, paraquat, aniline-oil-contaminated rapeseed oil, 4-4′diamino-diphenyl-methane, and several drugs have produced similar lesions in humans as have certain organic-tin-containing compounds in rodents (Baptista et al., 1983; Kopelman et al., 1966; Matsumoto et al., 1980; Mullick et al., 1981; Yermakoff et al., 1979).

Crystalline debris accumulated in portal bile ducts and larger intra- and extrahepatic bile ducts in association with cholestasis and photosensitization is reported in sheep from South Africa grazing the plant *Tribulus terrestrus* (Kellerman and Coetzer, 1985). Additional exposure to the cholangiodestructive mycotoxin sporodesmin may also play a role in the evolution of this lesion.

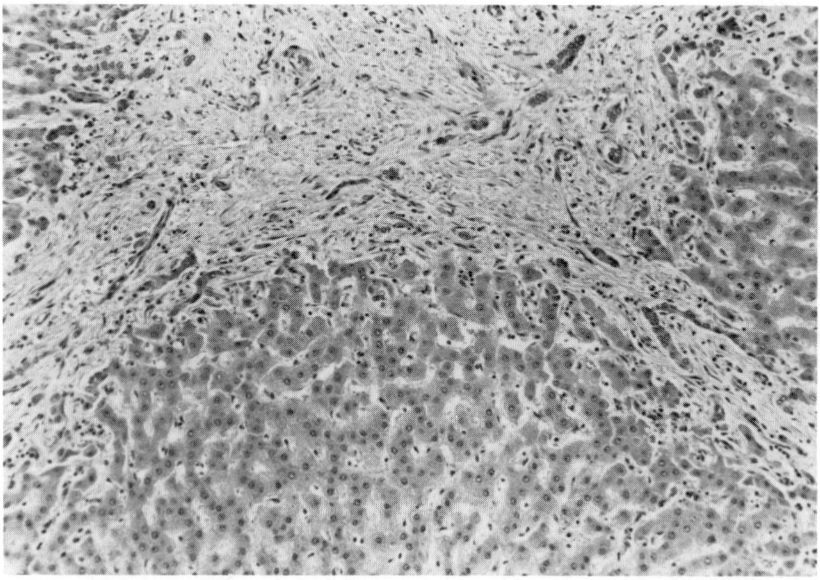

Figure 14. Chronic exposure to cholangiodestructive toxins can result in a pattern of portal fibrosis and biliary proliferative which resembles primary biliary cirrhosis, as in this case of chronic sporodesimin ingestion in a sheep.

REFERENCES

Allen JR and Carstens LA: Monocrotaline-induced Budd-Chiari syndrome in monkeys. *Am J Dig Dis* 1971; 16:111–121.

Arias IM: Effects of a plant acid (icterogenin) and certain anabolic steroids on the hepatic metabolism of bilirubin and sulfobromophtalein (BSP). Ann *NY Acad Sci* 1963; 104:1014.

Baptista A, Bianchi L, deGroote J, Desmet VJ, Gedigk P, Ishak KG, et al. Histopathology of the intrahepatic biliary tree. *Liver* 1983; 3:161–175.

Bannasch P: The cytoplasm of hepatocytes during carcinogenesis electron and light microscopical investigations of the nitrosomorpholine-intoxicated rat liver, in *Recent Results in Cancer Research,* New York, Springer, 1968.

Barnes JM and Magee PN: The biliary and hepatic lesion produced experimentally by dibutyltin salts. *J Pathol Bacteriol* 1958; 75:267–279.

Barnes JM and Stoner HB: The toxicology of tin compounds. *Pharmacol Rev* 1961; 18:211–231.

Beattie AD, Mullin PJ, Baxter RH, and Moore MR: Acute lead poisoning—an unusual cause of hepatitis. *Scot Med J* 1979; 24:318–321.

Behm AR: Oral contraceptives and porphyria cutanea tarda. *Can Med Assoc J* 1974; 110:1052–1054.

Bloomer JR: The liver in protoporphyria. *Hepatology* 1988; 8:402–407.

Bothwell TH: Siderosis in the Bantu: A comparison of the incidence in males and females. *Br Med J* 1962; 1:522–524.

Bourgeois CH, Olson L, Comer D, et al. Encephalopathy and fatty degeneration of the viscera: A clinicopathologic analysis of 40 cases. *Am J Clin Pathol* 1971; 56:558–571.

Bourgeois CH, Shank RC, Grossman RA, Johsen DO, Wooding WL, and Chandavimol P: Acute aflatoxin B_1 toxicity in the macaque and its similarities to Reye's syndrome. *Lab Invest* 1971; 24:206–216.

Bras G: Aspects of hepatic vascular diseases in Gall EA, Mostofi FK (eds): *The Liver*, Baltimore, Williams and Wilkins, 1973, pp 406–430.

Bridge JC, et al. Discussion on trinitrotoluene poisoning. *Proc R Soc Med* 1942; 35:553.

Brodsky I, Johnson H, Killman S-A, and Cronkite EP: Fibrosis of central and hepatic veins, and perisinusoidal spaces of the liver following prolonged administration of urethane. *Am J Med* 1961; 30:976–980.

Bullock G, Eakins MN, Sawyer BC and Slater TF: Studies on bile secretion with the aid of the isolated perfused rat liver. I. Inhibitory action of sporodesmin and icterogenin. *Proc R Soc Lond Biol* 1974; 186:333–356.

Cameron GR and Karunaratne WAE: Carbon tetrachloride cirrhosis in relation to liver regeneration. *J Pathol Bacteriol* 1936; 42:1–13.

Clark R, Borirakchanyavat V, Davidson AR, et al. Hepatic damage and death from overdose of paracetamol. *Lancet* 1973; 1:66–70.

Columbano A, Ledda GM, Sirigu P, Perra T, and Pani P: Liver cell proliferation induced by a single dose of lead nitrate. *Am J Pathol* 1983; 110:83–85.

Columbano A, Ledda-Columbano GM, Coni PP, Vargiu M, Gavino F, and Pani P: Liver hyperplasia and regression after lead nitrate administration. *Tox Path* 1984; 12:89–95.

Cortes JM, Oliva H, Paradinas FJ, and Hernandez-Guio C: The pathology of the liver in porphyria cutanea tarda. *Histopathology* 1980; 4:471–485.

Decker K and Keppler D: Galactosamine induced liver injury, in Popper H and Schaffner F (eds): *Progress in Liver Disease*, vol IV. New York, Grune and Stratton, 1972, pp 183–199.

Denk H, Gschnait F, and Wolff K: Hepatocellular hyalin (Mallory bodies) in long term griseofulvin-treated mice: A new experimental model for the study of hyalin formation. *Lab Invest* 1975; 32:773–776.

Desmet VJ: Morphologic and histochemical aspects of cholestasis, in Popper H and Schaffner F (eds): *Progress in Liver Disease*, vol. IV. New York, Grune and Stratton, 1972; pp 97.

Desmet VJ: Alcoholic liver disease, histological features and evolution. *Acta Med Scand* 1985; 703(Suppl):111–126.

Dubin IN and Peterson L: An explanation for the centrolobular localization of intrahepatic bile stasis in acute liver diseases. *Am J Med Sci* 1958; 336:45–52.

Duvall E and Wyllie AH: Death and the cell. *Immunol Today* 1986; 7:115–119.

Ferguson DR, Snyder SK, Cameron AJ: Hepatotoxicity in acetaminophen poisoning. *Mayo Clinic Proceedings* 1977; 52:246–248.

Fletcher GF and Galambos JT: Phosphorus poisoning in humans. *Arch Intern Med* 1963; 112:846–852.

Fittschen C and Bellamy JEC: Prednisone-induced morphologic and chemical changes in the liver of dogs. *Vet Pathol* 1984; 21:399–406.

Franklin M, et al. Fowler's solution as an etiologic agent in cirrhosis. *Am J Med Sci* 1950; 219:589.

Francis JE and Smith AG: Polycyclic aromatic hydrocarbons cause hepatic porphyria in iron-loaded C57BL/10 mice: comparison of uroporphyrinogen decarboxylase inhibition with induction of alkoxyphenoxazone dealkylations. *Biochem Biophys Res Commun* 1987; 146:13–20.

Goldblatt LA: *Aflatoxin: Scientific Background, Control and Implications*. New York, Academic Press, 1969.

Goldfarb S, Singer EJ, and Popper H: Experimental cholangitis due to alpha-naphthylisothiocyanate (ANIT). *Am J Pathol* 1962; 40:685–698.

Gopinath C and Ford EJH: The effect of *Lantana camara* on the liver of sheep. *J Path* 1969; 99:75–85.

Gopinath C and Ford EJH: The effect of alpha-napthylisothiocyanate on the liver of sheep and calves. *J Pathol* 1970; 100:269–280.

Hartroft WS: Experimental hepatic injury, in Schiff L (ed): *Diseases of the Liver*, ed 2. Philadelphia, Lippincott, 1963, pp 101–141.

Hennigar GR, Greene WB, Walker EM, and de Saussure C: Hemochromatosis caused by excessive vitamin iron intake. *Am J Pathol* 1979; 96:611–623.

Himsworth HP: *Lectures on the Liver and Its Disease*. Cambridge, Harvard University Press, 1947.

Hoyumpa AM Jr, Greene HL, Dunn GD, et al. Fatty liver: biochemical and clinical considerations. *Am J Dig Dis* 1975; 20:1142–1170.

Huet PM, Guilleaume E, Cote J, et al: Noncirrhotic presinusoidal portal hypertension associated with chronic arsenical intoxication. *Gastroenterology* 1975; 68:1270–1277.

Indacochea-Redmond N and Plaa GL: Functional effects of α-napthylisothiocyanate in various species. *Toxicol Appl Pharmacol* 1971; 19:71–80.

Jennings RB: Fatal fulminant carbon tetrachloride poisoning. *Arch Pathol* 1955; 55:269–284.

Kellerman TS and Coetzer JAW: Hepatogenous photosensitivity diseases in South Africa. *Onderstepoort J Vet Res* 1985; 52:157–173.

Kelly WR and Seawright AA: Pimelea poisoning of cattle, in Keller RF, VanKampen KR, and James LF (eds): *Effects of Poisonous Plants on Livestock.* New York, Academic Press, 1978, pp 293–300.

Kenna JG, Neuberger J, Mieli-Vergani G, Mowat AP, and Williams R: Halothane hepatitis in children. *Br Med J (Clin Res)* 1987; 294:1209–1211.

Klatskin G: Toxic and drug-induced hepatitis, in Schiff L (ed): *The Liver,* ed 4. Philadelphia, JB Lippincott, 1975, pp 604–710.

Kopelman H, Scheuer PJ, and Williams R: The liver lesion of the Epping Jaundice. *Q J Med* 1966; 35:553–567.

Lalonde G, Theoret G, Daloze P, Bettez P, and Katz SS: Inferior vena cava stenosis and Budd-Chiari syndrome in a woman taking oral contraceptives. *Gastroenterology* 1982; 82:1452–1456.

Levine RA, Briggs GW, and Lowell DM: Chronic chlorpromazine cholangiolitic hepatitis: Report of a case with immunofluorescent studies. *Gastroenterology* 1966; 50:665–670.

Lieber CS: Alcohol and the liver. Transition from adaptation to tissue injury, in Khanna JM, Israel Y, and Kalant H (eds): *Alcoholic Liver Pathology.* Ontario, Addiction Research Foundation, 1975, pp 171–189.

Lloyd-Still JD, Grand RJ, and Vawter GF: Tetracycline hepatotoxicity in the differential diagnosis of postoperative jaundice. *J Pediatrics* 1974; 84:366–370.

Lopez M and Mazzanti L: Experimental investigation on alpha-napthyl-isothiocyanate as a hyperplastic agent on the biliary ducts in the rat. *J Pathol Bact* 1955; 69:243–250.

Lustig S, Pitlik SD, and Rosenfeld JB: Liver damage in self-induced hyper-manganemia. *Arch Intern Med* 1982; 142:405–406.

MacPhail AP, Simon MO, Torrance JD, Charlton RW, Bothwell TH, and Issacson C: Changing patterns of dietary iron overload in black South Africans. *Am J Clin Nutri* 1979; 32:1271–1278.

Matilla A and Molland EA: A light and electron microscopic study of the liver in a case of erythrohepatic protoporphyria and in griseofulvin-induced porphyria in mice. *J Clin Pathol* 1974; 27:698–709.

Matsumoto T, Matsumori H, Kuwabara N, Fukuda Y, and Ariwa R: A histopathological study of the liver in paraquat poisoning. An analysis of fourteen autopsy cases with emphasis on bile duct injury. *Acta Pathol Jpn* 1980; 30:859–870.

McLean AE, McLean EK, and Judah JD: Cellular necrosis in the liver induced and modified by drugs. *Internat Rev Exp Path* 1965; 4:127–157.

McMaster KR and Hennigar GR: Drug-induced granulomatous hepatitis. *Lab Invest* 1981; 44:61–73.

Meierhenri E, Ruebner B, Gershwin ME, Hsieh LS, and Franch SW: Mallory body formation in hepatic nodules of mice ingesting dieldrin. *Lab Invest* 1981; 44:392–396.

Miller J: Atrophy of the liver. *Br Med J* 1920; 2:581–590.

Mortimore PH and Taylor A: The experimental intoxication of sheep with sporodesmin, a metabolic product of *Pithomyces chartarum*. IV. Histological and histochemical examination of orally dosed sheep. *Res Vet Sci* 1963; 4:166–185.

Mullick FG, Ishak KG, Mahabir R, and Stromeyer FW: Hepatic injury associated with paraquat toxicity in humans. *Liver* 1981; 1:209–221.

Phillips MJ, Oda M, and Funatsu K: Evidence for microfilament involvement in norethandrolone-induced intrahepatic cholestasis. *Am J Pathol* 1978; 93:729–744.

Phillips MJ, Oshio C, Miyairi M, and Smith CR: Intrahepatic cholestasis as a canalicular motility disorder: evidence using cytochalasin. *Lab Invest* 1983; 48:205–211.

Phillips MJ, Poucell S, and Oda M: Biology of disease: mechanisms of cholestasis. *Lab Invest* 1986; 54:593–608.

Pimentel JC and Menezes AP: Liver disease in vineyard sprayers. *Gastroenterology* 1977; 72:275–283.

Pimentel JC and Menezes AP: Liver granulomas containing copper in vineyard sprayer's lung: A new etiology of hepatic granulomatosis. *Am Rev Respir Dis* 1975; 111:189–195.

Plaa GL: Toxic responses of the liver, in Doull J, Klassen CD, and Amdur MO (eds): *Casarett and Doull's Toxicology: The Basic Science of Poisons*. New York, Macmillan, 1980, pp 206–231.

Plaa GL and Preistly BG: Intrahepatic cholestasis induced by drugs and chemicals. *Pharmacol Rev* 1977; 28:207–273.

Reid WD, Christie B, and Krishna G: Bromobenzene metabolism and hepatic necrosis. *Pharmacology* 1971; 6:41–55.

Rigas B, Rosenfeld LE, Barwick KW, Enriquez R, Helzberg J, Batsford WP, Josephson ME, and Riely CA: Amiodarone hepatotoxicity: a clinicopathologic study of five patients. *Ann Intern Med* 1986; 104:348–351.

Rimington C, Quinn JI, and Roets GCS: Studies on the photosensitization of Animals in South Africa. The icterogenic factor in gelldikkop: Isolation of the active principle from *Lippia rehemani. Onderstepoort J Vet Sci* 1937; 9:225–255.

Robotham JL, Troxler RF, and Leitman PS: Iron poisoning: Another energy crisis. *Lancet* 1974; 2:664–665.

Rodriquez-Iturbe B: Acute yellow phosphorous poisoning. *N Engl J Med* 1971; 284:157.

Rouiller C: Experimental toxic injury of the liver, in Rouiller C (ed): *The Liver,* vol II, New York, Academic Press, 1964, pp 335–476.

Ruebner BH, Watanabe K, and Wand JS: Lytic necrosis resembling peliosis hepatis produced by lasiocarpine in the mouse liver. A light and electron microscopic study. *Am J Pathol* 1970; 60:247–271.

Russel RM, Boyer JL, Bagheri SA, and Hruban Z: Hepatic injury from chronic hypervitaminosis A resulting in portal hypertension and ascites. *N Engl J Med* 1974; 291:435–440.

Searle JW, Kerr JF, Halliday JW, and Powell LW: Iron storage disease, in MacSween RNM, Anthony PP, Scheuer PJ (eds): *Pathology of the Liver,* ed 2. New York, Churchill-Livingstone, 1987, pp 181–201.

Seawright AA and Allen JG: Pathology of the liver and kidney in *Lantana* poisoning of cattle. *Aust Vet J* 1972; 48:323–331.

Seawright AA and Hrdlicka J: The effect of prior dosing with phenobarbitone and diethylaminoethyl diphenylpropyl acetate (SKF 525-A) on the toxicity and liver lesion caused by ngaione in the mouse. *Br J Exp Pathol* 1972; 52:242–252.

Scheuer PJ: Drugs and toxins, in Scheuer PJ (ed): *Liver Biopsy Interpretation,* Baltimore, Williams & Wilkins, 1980, pp 88–101.

Schmid R: Cutaneous prophyria in Turkey. *N Engl J Med* 1960; 263:397–398.

Slater TF, Strauli UD, and Sawyer BC: Sporidesmin poisoning in the rat. I. Chemical changes. *Res Vet Sci* 1964; 5:450–472.

Solis-Herruzo JA, Castellano G, Colina F, Morillos JD, Munoz-Yaqüe MT, Cola MC, and Jelavic D: Hepatic injury in the toxic epidemic syndrome caused by ingestion of adulterated cooking oil (Spain, 1980). *Hepatology* 1984; 4:131–139.

Stoeckle JE, Hardy HL, and Weber AL: Chronic beryllium disease. Long-term follow-up of sixty cases and selective review of the literature. *Am J Med* 1969; 46:545–561.

Takeuchi T, Kivirikko KI, and Prockop DJ: Increased protocollagen hydroxylase activity in the livers of rats with hepatic fibrosis. *Biochem Biophys Res Commun* 1967; 28:940–944.

Taxy JB: Peliosis: a morphologic curiosity becomes an iatrogenic problem. *Hum Pathol* 1978; 9:331–340.

Tolman KG, Clegg DO, Lee RG, and Ward JR: Methotrexate and the liver. *J Rheumatol Suppl* 1985; 12:29–34.

Tuchweber B, Kovacs K, Khandekar JD, and Garg BD: Peliosis-like changes induced by phalloidin in the rat liver: a light and electron microscopic study. *J Med* 1973; 4:327–345.

Uchida T, Kao H, Quispe-Sjogren M, and Peters RL: Alcoholic foamy degeneration a pattern of acute alcoholic injury of the liver. *Gastroenterology* 1963; 84:683–692.

Watanabe S and Phillips MJ: Acute phalloidin toxicity in living hepatocytes. Evidence for a possible disturbance in membrane flow and for multiple functions for actin in the liver cell. *Am J Pathol* 1986; 122:101–111.

Weir JF: Cirrhosis associated with chronic inorganic arsenical poisoning. *Proc Staff Meet Mayo Clin* 1930; 5:173–175.

Weitz H, Gokel JM, Loeshke K, Possinger K, and Eder M: Veno-occlusive disease of the liver in patients receiving immunosuppressive therapy. *Virchows Arch (Pathol Anat)* 1982; 395:245–256.

Wepler W and Opitz K: Histologic changes in the liver biopsy in amanita phalloides intoxication. *Hum Pathol* 1972; 3:249–254.

Witzleben CL: Physiologic and morphologic natural history of a model of intrahepatic cholestasis (manganese-bilirubin overload). *Am J Pathol* 1972; 66:577–588.

Yermakoff JK, Fuller GC, and Rodil JV: An experimental model of hepatic fibrosis induced by the administration of dibutyltin dichloride. *Toxicol Appl Pharmacol* 1979; 49:31–40.

Zimmerman HJ: Papaverine revisited as a hepatotoxin. *N Engl J Med* 1969; 281:1364–1365.

Zimmerman HJ: *Hepatotoxicity: The Adverse Effects of Drugs and Other Chemicals on the Liver.* New York, Appleton-Century-Crofts, 1978.

Zimmerman HJ, Fang M, Utili R, Seeff LB, and Hoofnagle J: Jaundice due to bacterial infection. *Gastroenterology* 1979; 77:362–374.
Zimmerman HJ and Maddrey WC: Toxic drug-induced hepatitis, in Schiff L and Schiff ER (eds): *Diseases of the Liver,* ed 6. Philadelphia, Lippincott, 1987, pp 591–667.

Chapter 3

Pathology of the Liver: Functional and Structural Alterations of Hepatocyte Organelles Induced by Cell Injury

Louis Marzella and Benjamin F. Trump
Maryland Institute for Emergency Medical Services Systems and Department of Pathology
School of Medicine
University of Maryland
Baltimore, Maryland

INTRODUCTION

In this chapter, we will review the structural and functional alterations induced by acute cell injury in hepatocytes. We will begin by reviewing common etiologies of cell injury, review the role of mediators released by parenchymal and nonparenchymal cells and consider the occurrence of hepatocyte adaptations to injury. We will then review the manifestations of hepatocyte injury in the subcellular organelles. We will discuss and illustrate the main structural alterations and correlate them with parallel functional alterations. Finally we will discuss the temporal evolution (stages) of cell injury and review the role of ion shifts in the evolution of cell injury towards cell death.

ACUTE AND CHRONIC HEPATOCYTE INJURY

Definition of Cell Injury

The concept of cell injury is fundamental to the study of the pathogenesis of disease. Disease must be understood in terms of the biologic events which change the function and the structure of the constituent cells of an organ. Cellular injury is defined as a state of pathologically altered cellular function. Normal cells in physiologic conditions are at homeostasis. When an injurious stimulus is applied, cells expend metabolic energy to maintain homeostasis. If the stimulus is severe enough or lasts sufficiently long, the ability of the cells to adapt may be exceeded and cellular functions become altered; the cell is said to be injured. If the injury is sublethal, the cell survives even though it may function abnormally. Chronic sublethal injury or altered environmental factors may induce cells to undergo stable, structural and functional alterations that restore homeostasis. Sublethally injured cells may also continue to deteriorate structurally and functionally to reach a point from which recovery is no longer possible. The so-called point of no return is not yet precisely defined either ultrastructurally or functionally. However, it has been possible to define the point of no return empirically. Tissues are, for example, subjected to injury of reproducible severity for progressively longer periods of time. In this manner, the duration of injury (e.g., anoxia) needed to induce cell death in a particular tissue is established. It therefore becomes possible to study the cells to determine what distinctive structural or functional alterations have taken place at or near the point of no return. These alterations are likely to be important determinants of the death of cells.

Causes of Hepatocellular Injury: Ischemia, Anoxia, Free Radicals, Toxins

Damage to liver induced by *in vivo* ischemia (Marzella and Glaumann, 1981) or by perfusion with nitrogen-saturated buffer (Anundi et al., 1987) begins after 30 min of oxygen deprivation. Increases in intracellular calcium by mechanisms to be discussed may contribute to the injury and calmodulin antagonists, and calcium channel blockers appear to confer some degree of protection against anoxia (Farber and Young, 1981). Provision of substrates, such as fructose, that are readily metabolized to lactate protects against the hypoxic injury if intravascular flow is maintained. The presence of an abundant cellular supply of glycogen also offers some protection against injury (Anundi et al., 1987). Free radicals and other toxic oxygen species generated during reperfusion have also been proposed to contribute to the cell damage induced by ischemia. During ischemia, a reducing environment is created. When oxygen is reintroduced at reflow, the nonenzymatic reduction of oxygen to superoxide anion may occur. During ischemia, degradation of adenine nucleotides occurs and products are generated which can be oxidized enzymatically at reflow with the generation of superoxide.

In conditions where low flow states occur, the centrilobular hepatocytes are anoxic. Hepatocytes in the midzonal regions are hypoxic and more susceptible to free-radical-mediated damage (Marotto et al., 1988). Superoxide and other toxic oxygen species cause cell injury by causing strand breaks in DNA and by oxidation of thiol groups and scission of proteins and lipids. No specific target molecule of superoxide whose damage is responsible for cell death has been identified. Cellular enzymatic and nonenzymatic defenses are active in detoxification of superoxide and other toxic oxygen species generated during normal cellular metabolism and during cell injury. Ablation of free radicals with enzymes or radical scavengers has not, however, consistently afforded protection to the ischemic liver.

Free radicals generated by hepatic metabolism or by drug biotransformation reactions can also cause cell damage by oxidation of unsaturated fatty acids of phospholipids. Lipid peroxidation is a feature of hepatocytes injured by toxins and by the accumulation of iron (Comporti, 1985; Hulcrantz et al., 1984). These reactions cause the autocatalytic breakdown of unsaturated fatty acids and the generation of fatty aldehydes and other carboxyl compounds which can in turn react with SH groups and covalently bind to proteins (Comporti, 1985). Altered membrane permeability or inactivation of critical membrane pumps can result from lipid peroxidation (Lowrey et al., 1981). The quantitative importance of lipid peroxidation in the evolution of cell death is undetermined. The prevailing opinion is that peroxidation is not an early event in cell injury such as that induced by ischemia (Silver and Szabo, 1983).

The ultrastructural manifestations of hepatocyte injury due to these various causes are similar. In *in vitro* systems, the absence of mechanical factors and inflammatory cells delays the evolution of cell injury. In hypothermic conditions (4°C), ultrastructural signs of cell injury also develop more slowly (Myagkaya et al., 1987).

Injury Caused by Mediators Released by Parenchymal and Nonparenchymal Cells

The resident liver macrophages (Kupffer cells) (Figure 1) are important in the clearing of necrotic hepatocytes by phagocytosis and rapid degradation of cellular fragments present in the sinusoids or in the space of Disse (Glaumann and Marzella, 1981). Kupffer cells are also active in the uptake and killing of bacteria and in the phagocytosis and in the lysis of tumor cells (Xu et al., 1984). The distribution and activity of Kupffer cells over the liver lobule is heterogeneous. Cells near the portal veins are more abundant, larger, and have higher phagocytic capacity (Sleyster and Knook, 1982). If the phagocytic capacity of the Kupffer cells is enhanced, the ability of the liver

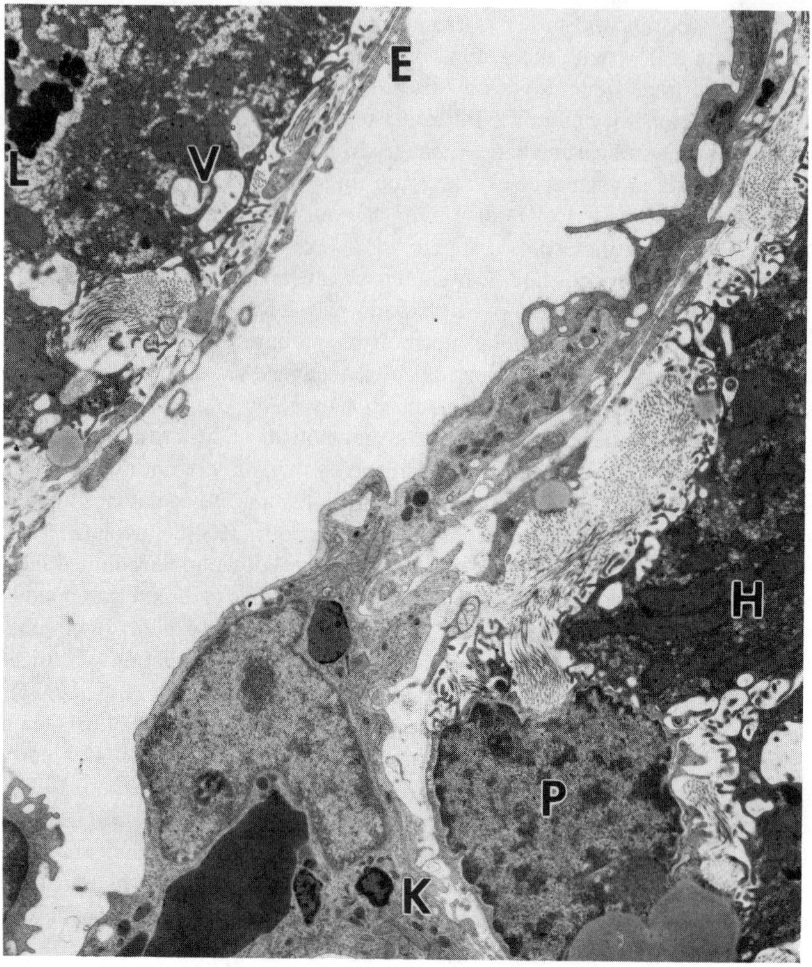

Figure 1. Ultrastructure of nonparenchymal and parenchymal liver cells. Perisinusoidal cells (P) are located within the space of Disse and contain variable amounts of vitamin A. They are thought to play a role in the production of hepatic collagen, particularly following liver injury. Basement membrane-like material may be found between perisinusoidal cells and endothelial cells. Bundles of collagen are found in the space of Disse in contact with the hepatocyte plasma membrane. This collagen reacts primarily with pro-III collagen antibodies and to a lesser extent with type I collagen antibodies in both humans and rats (Clement et al., 1986). An erythrocyte is enveloped by the Kupffer cells (K) plasma membrane. Both complement and Fc receptors on the Kupffer cells participate in the phagocytosis of erythrocytes. Fenestrated endothelial cells (E) line the sinusoids. Apical vacuoles (V) and lysosomes (L) are seen in the hepatocytes (H). Magnification: ×2,800.

to resist many types of insults increases (Loegering and Blumenstock, 1985). However, activation of the Kupffer cells may also increase the susceptibility of the liver to some types of injury such as that induced by endotoxin. It has been proposed that the release of toxic oxygen species and lysosomal enzymes from the Kupffer cells may participate in the induction of hepatocyte injury. It has been shown that phagocytizing macrophages release lysosomal enzymes to the extracellular milieu (Dean et al., 1979; Nichols, 1982). In sum the Kupffer cells perform a very important function, but the potential for hepatocyte damage exists because of release of enzyme and of toxic oxygen species from the Kupffer cells.

The hepatocytes are also a potential source of damage to sinusoidal cells. It has been shown in cultured hepatocytes that stable intermediates of drug biotransformation are released to the medium. *In vivo* these toxic intermediates can diffuse into other types of liver cells or reach other distant organs. Here they can induce damage directly or after further biotransformation (Jones et al., 1985; Anders et al., 1988). A case in point is vinyl chloride, which is biotransformed inside the hepatocyte to a toxic (epoxide) intermediate, which can diffuse into nonparenchymal liver cells where it induces DNA alkylation. The ability of nonparenchymal cells to repair adducts is lower than that of hepatocytes. These phenomena may explain the localization of vinyl chloride-induced neoplastic lesions in the nonparenchymal liver cells despite their very low ability to activate this drug to its toxic reactive intermediate (Guengerich and Liebler, 1985). Nonresident cells and circulating toxins derived from the gut or other systemic sites can also induce liver injury. Increased numbers of neutrophils and macrophages accumulate in liver sinusoids in lethal and some forms of sublethal injury (Ogawa et al., 1985). This influx is, for example, seen in septic states and is associated with hepatocyte injury (Sato et al., 1982). It is possible that free radicals generated by neutrophils deposited in the sinusoids contribute to hepatocyte injury.

Hepatocyte Adaptations

In the early stages of acute cell injury, some of the functional alterations may have beneficial effects and help the cell to survive. For example, in ischemia in the heart, loss of Ca^{2+} regulation by the sarcoplasmic reticulum is a very early event and may be caused by inactivation of a key regulatory protein. Yet this inactivation causes immediate loss of contractility of the injured myocyte, with consequent preservation of metabolic energy. Similarly, inactivation of transcriptional and/or translocational activity may be beneficial initially as it decreases ATP consumption. During injury states, active repair and detoxification processes are also activated to restore homeostasis. Excision and repair of damaged macromolecules, proteolytic degradation of denatured proteins, reacylation of lysophosphatides in membranes, endocytosis, and lysosomal degradation of cell surface membranes bearing complement lesions are some examples of cellular repair processes.

More stable cellular adaptations may create morphological and biochemical alterations of cells which restore homeostasis by meeting increased functional demands. Growth of liver parenchyma is such an adaptive response. Both an increase in the number of hepatocytes (hyperplasia) and in the size and numbers of subcellular organelles (hypertrophy) occur. Two well-known examples are the administration of phenobarbital and other xenobiotics which induce the proliferation of smooth endoplasmic reticulum (SER), and the chronic intake of ethanol or of lipid-lowering drugs which induce the proliferation of peroxisomes (Reddy et al., 1984). In both these examples, the growth is in response to increased levels of substrates and results in enhanced capacity of the hepatocytes to oxidize these substrates (Reddy et al., 1984). When the stimulus is removed, the hypertrophied hepatocytes regress in size by a decrease in the synthesis of specific proteins and by turning on lysosomal degradation pathways. In the case of hyperplasia, it has been shown that a regression of the liver size occurs when the stimulus for growth is removed; corresponding decreases in liver DNA content occur and increased numbers of isolated necrotic cells are found in the liver (Bursch et al., 1984; Columbano et al., 1985). It has therefore been proposed that programmed cell death (apoptosis) may be responsible for the regression of hyperplastic cells in the liver. Loss of liver mass by focal coagulative necrosis of hepatocytes without compensating regeneration is another adaptive response to altered hormonal or nutritional environment that leads to hepatic atrophy. Atrophy, for example, has been observed in liver lobes deprived of portal blood flow (Rozga et al., 1986).

Cell Turnover

An ongoing process of degradation and resynthesis constantly renews the subcellular organelles and molecular components of the hepatocytes. Focal or even extensive liver injury is usually followed by the typical inflammatory responses followed by the proliferation of hepatic parenchymal and nonparenchymal cells and restitution of lobular architecture (Ogawa et al., 1985). Isolated hepatocyte necrosis is yet another means for turnover of hepatic proteins and lipids. Morphologically, this type of physiologic cell death is associated with condensation of the nuclear chromatin and with breakup of the cell into membrane-bound cytoplasmic fragments that are readily phagocytosed by Kupffer cells. Frequently, these fragments have an overall increase in electron density and become surrounded by adjacent hepatocytes. However, true phagocytic uptake of these fragments by the hepatocytes is an uncommon event. Single-cell necrosis has also been termed apoptosis (Wyllie, 1986). The morphologic features of cells dying because of termination of life-span are not sufficiently characteristic to readily permit this type of "physiologic" cell death to be distinguished from cell death caused by cell injury. Isolated single-cell necrosis is augmented in some pathologic states. Little or nothing is known about the mechanism of increased single-cell necrosis. Endonuclease activation may play a role (McConkey et al., 1988). It has also been hypothesized that enhanced protein degradation concomitantly with inadequate biosynthesis may be a mechanism leading to isolated cell necrosis. This process is, for example, postulated to occur in cultured cells that have achieved high population density (Amenta et al., 1986).

Susceptibility of Hepatocytes to Injury

The localization of the hepatocyte within the liver lobule is an important determinant of susceptibility to injury. Differences in enzyme composition, organelle distribution, and oxygen tensions are the main determinants of susceptibility to injury. For example, the enhanced susceptibility of centrilobular hepatocytes to hypoxic injury such as occurs in shock is well known. Oxygen tensions are important in the regulation of metabolic reactions in hepatocytes (Marzella et al., 1986). By reversing the flow of perfusate to the perfused liver, the oxygen gradient across the liver lobule can be reversed; rates of some metabolic reactions can be altered and the localization of hepatocyte necrosis within the hepatic lobule can be shifted (Marotto et al., 1988). The susceptibility of the liver to cell injury caused by toxins metabolized by reductive or oxidative metabolism can be, for example, altered by increasing the tissue oxygen tensions with hyperbaric hyperoxia (Bernacchi et al., 1984; Marzella et al., 1986).

FUNCTIONAL AND STRUCTURAL ALTERATIONS INDUCED BY CELL INJURY IN HEPATOCYTES

In this section, we will describe the structural and functional alterations induced by cell injury on the hepatocyte organelles. The maintenance of structural and functional integrity of membranes of the cell surface and of intracellular organelles is essential for maintenance of cellular function. The capacity of cellular membranes to segregate proteins and substrates and to maintain ionic gradients is essential for such fundamental cell functions as generation of ATP, signaling, regulation of biosynthetic and catabolic reactions, and motility of organelles and cells.

At the ultrastructural level, diverse types of insults induce similar ultrastructural alterations in hepatocytes. At the functional level, ischemic, toxic, immunologic insults induce diverse biochemical alterations. In time, the diverse biochemical alterations affect common integrated basic organelle functions. We will focus on the most common functional alterations of injured hepatocytes in this section and will discuss them along with the corresponding structural alterations. In the next section we will define how these alterations evolve in time and how they can lead to loss of homeostasis and of regulated cell functions. Finally, we will present current theories of how these structural and functional alterations ultimately may lead to cell death.

We will begin our discussion of injury with the organelles responsible for the generation of energy and for the steady-state maintenance of ionic levels; namely, mitochondria, plasma membranes, and endoplasmic reticulum. Disruptions of these functions occur early in cell injury and figure prominently in current hypotheses on susceptibility of cells to injury and on the evolution of cell injury towards cell death.

Mitochondria

We begin our review of manifestations of injury in subcellular organelles by discussing alterations seen in the mitochondria. The appearance of the mitochondria is profoundly affected by the fixative(s) used and the agents used for dehydration and for embedding. Ultrastructural alterations affecting matrix density, presence and number of mitochondrial granules, width of spacing between cristae and inner and outer membranes, and overall volume of the organelles have been described after different fixation protocols and must be distinguished from alterations induced by cell injury (Figures 2 and 3) (Candipan and Sjostrand, 1984c). The presence of spaces between cristae in isolated rat liver mitochondria is markedly influenced by the composition of the medium (Candipan and Sjöstrand 1984a; Schwerzmann et al., 1986). Uncoupling of mitochondria or inhibition of respiration causes the widening of the space between the cristae and an effective shrinkage of the mitochondria (Candipan and Sjöstrand, 1984a). Addition of ADP or substrate to isolated mitochondria causes a decrease in the volume of the organelle without, however, widening the space between the cristae. The change in appearance of this space does not seem to be due to variation in preparative technique (Candipan and Sjöstrand, 1984b). There is general agreement that mitochondrial ultrastructural appearance reflects the ability of the organelle to synthesize ATP (Laiho and Trump, 1975; Smith and Ord, 1983). Orthodox shape is associated with active oxidative phosphorylation; condensation (Figure 4) reflects a low-energy state in the cell due to a decrease in the ATP levels. Various degrees of swelling are functionally associated with altered permeability of mitochondrial membrane, ionic shifts, osmotic swelling, and loss in ability to generate ATP.

The most abundant cations present in mitochondria are K^+ and Mg^{2+}; Na^+ and Ca^{2+} are present in lower concentrations. The major anions are phosphate, ATP, ADP, and other metabolites. Isolated mitochondria can

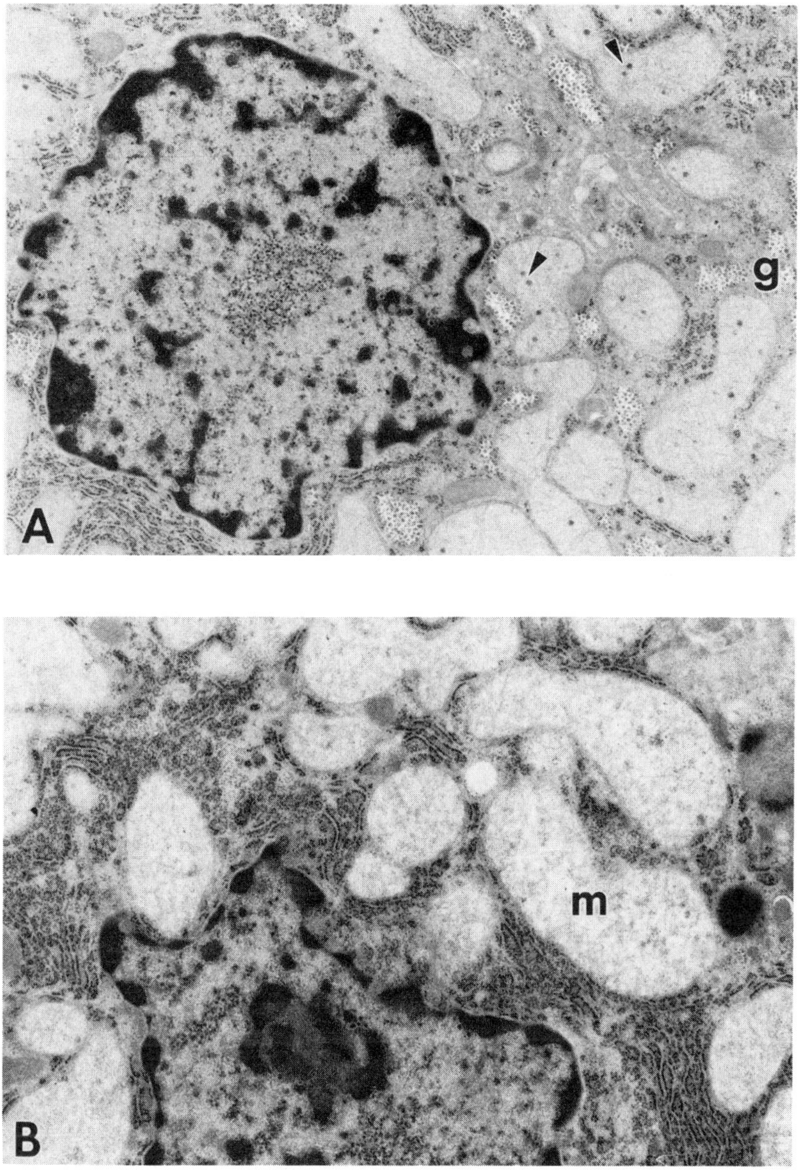

Figure 2A, B. Ultrastructure of mitochondria. (A) Mitochondria are in orthodox configuration, the matrix granules (arrowheads) and cristae are distinct. (B) Mitochondria of the hepatocyte are swollen. The matrix (m) is poorly preserved. In both micrographs, prominent nuclear alterations are present. Namely, indentation of nuclear envelope and slight condensation of chromatin. Abundant chromatin and perichromatin granules are present. Note the absence of glycogen (g) in Figure 2B. From dog liver 1 week after a 1-h episode of hypovolemic shock. Magnification: A: ×5,800; B: ×4,600.

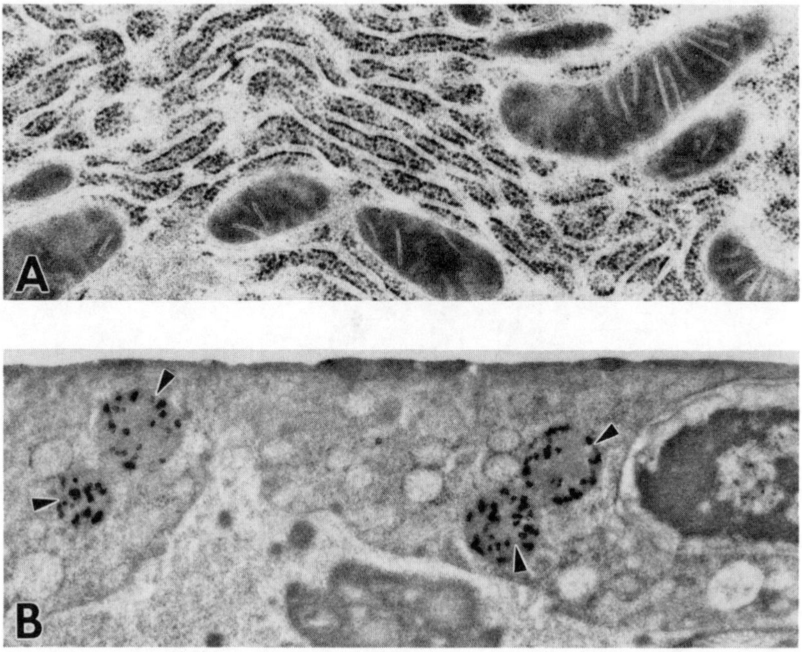

Figure 3A, B. Ultrastructure of mitochondria. Physical fixation by rapid freezing and primary chemical fixation with osmium tetroxide enhance the preservation of electron dense deposits in the matrix of mitochondria. These deposits may represent increased accumulations of calcium. Ultrastructurally, these mitochondrial densities are similar to those that form in injured tissues under conditions of calcium accumulation (Jennings et al., 1975). (A) The liver specimen was quick frozen in liquid nitrogen, freeze-substituted, and fixed in tetrahydrofuran containing osmium tetroxide. Stained with uranyl acetate and lead citrate. Normal ER and mitochondria are seen. (B) Fixed as above, staining was omitted to enhance the appearance of the mitochondrial matrix densities (arrowheads). Magnification A: ×11,200; B: ×8,600.

rapidly become depleted of K^+ and phosphates. The presence of Ca^{2+} in the isolation medium results in increased mitochondrial Ca^{2+} content and impairment of mitochondrial respiration (Arnold et al., 1985). In the presence of inorganic phosphate, isolated mitochondria can accumulate large amounts of calcium (Carafoli, 1986). The Ca^2/Pi molar ratio in these injured mitochondria suggests that calcium hydroxyapatite deposits are formed. *In vivo,* mitochondria and other cellular compartments of necrotic cells also are noted to have hydroxyapatite crystals. In contrast to some tissues such as myocardium, these deposits are seldom seen at the histological and ultrastructural levels in the liver with conventional aldehyde fixation.

Mitochondrial injury characterized by swelling, decreased number of cristae, and decreased density of the matrix can be induced in isolated hepatocytes by altering permeability of cellular membranes through solvent effects (Berger et al., 1987). In this type of injury, relative sparing of the endoplasmic reticulum (ER) occurs in the early stages. The ultrastructural appearance of this cellular change is illustrated in Figures 2A, B. Enlarged, pale mitochondria were also observed by Walker et al. (1983) in livers of mice treated with acetaminophen.

The preservation of potential and proton gradients by the mitochondria is an important determinant of cell survival. In the case of hypoxic injury, exposure of hepatocytes to 30 min of hypoxia causes a 20% decrease in

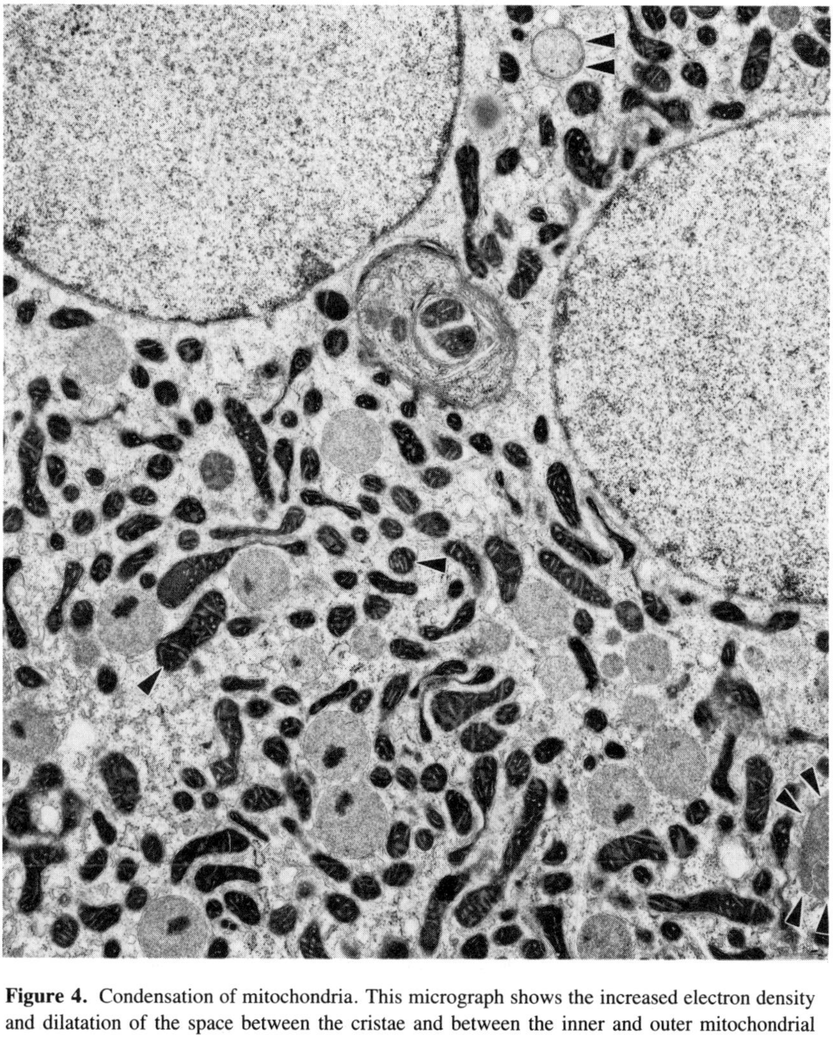

Figure 4. Condensation of mitochondria. This micrograph shows the increased electron density and dilatation of the space between the cristae and between the inner and outer mitochondrial envelope (arrowheads) which are associated with decreased levels of ATP in the cell. Compare to the "orthodox" appearance of two of the mitochondria (double arrowheads) in this field. From a cultured hepatocyte 24 h after isolation. Magnification: ×5,000.

mitochondrial membrane potential. This decrease induces a 40–50% drop in mitochondrial calcium levels. The endoplasmic reticulum calcium pool decreases more slowly and to a lesser degree during this time (Aw et al., 1987b). The pH of mitochondria is preserved after 30 min of hypoxia. The preservation of membrane potential and pH gradient may be achieved by inhibition of ion transport across the inner mitochondrial membrane. For example, specific proteins can inhibit ATP synthase and H^+ conductance; H^+-K^+ transporters because of their voltage dependence can be inhibited by slight decreases in membrane potential; translocation of ADP into the mitochondria can also be inhibited (Aw et al., 1987a). The activation of these processes during hypoxia preserves mitochondrial integrity by maintaining ion equilibrium across the mitochondria membrane. In response to toxins which oxidize SH groups, increased permeability of the inner mitochondrial membrane to protons has been proposed to occur. Increased permeability leads to a decrease in membrane potential and to efflux of calcium from the mitochondria down its electrochemical gradient (Rizzuto et al., 1987). Alterations in mitochondrial volume occur by the net transport of solutes into the mitochondria followed by rapid water equilibration. Mitochondrial swelling, driven by osmotic forces caused by net solute transport can occur in the absence of altered membrane permeability (Garlid and Beavis, 1985); Mg^{2+} may function in the regulation of solute uptake by mitochondria *in vivo*.

Increase in the size of the mitochondria unrelated to swelling can be induced relatively rapidly *in vivo* (in as little as two days) by the intraperitoneal injections of some toxins. It is thought that the increase in mitochondrial size may occur at least in part through a process of fusion of two or more organelles. This alteration in size appears to be reversible and is associated with little functional impairment of the mitochondria (Wakabayashi et al., 1983).

Massive swelling of mitochondria (Figure 5) begins with uptake of calcium and activation of phospholipase A_2 (Broekemeir et al., 1985). Simultaneous inhibition of lysophospholipid-acyltransferase activity occurs from a fall in the mitochondrial GSH:GSSG ratio. The accumulation of lipid degradation products within the inner-membrane bilayer causes a generalized increase in permeability to small molecules (less than 1500 molecular weight). Consequently, there is release of calcium and magnesium, loss of membrane potential, and swelling of the mitochondria. The activation of phospholipase A_2 and the accumulation of lysophospholipids with a corresponding increase in the levels of free fatty acids have been demonstrated to occur in ischemic mitochondria (Okayasu et al., 1985). These changes are associated with a loss of respiratory control and correlate with increase in membrane permeability and high-amplitude swelling.

It has been shown that by ultrastructural criteria, the increase in permeability and swelling occurs in a heterogeneous fashion in isolated *in vitro* mitochondria (Beatrice et al., 1982). This heterogeneity may be due to the

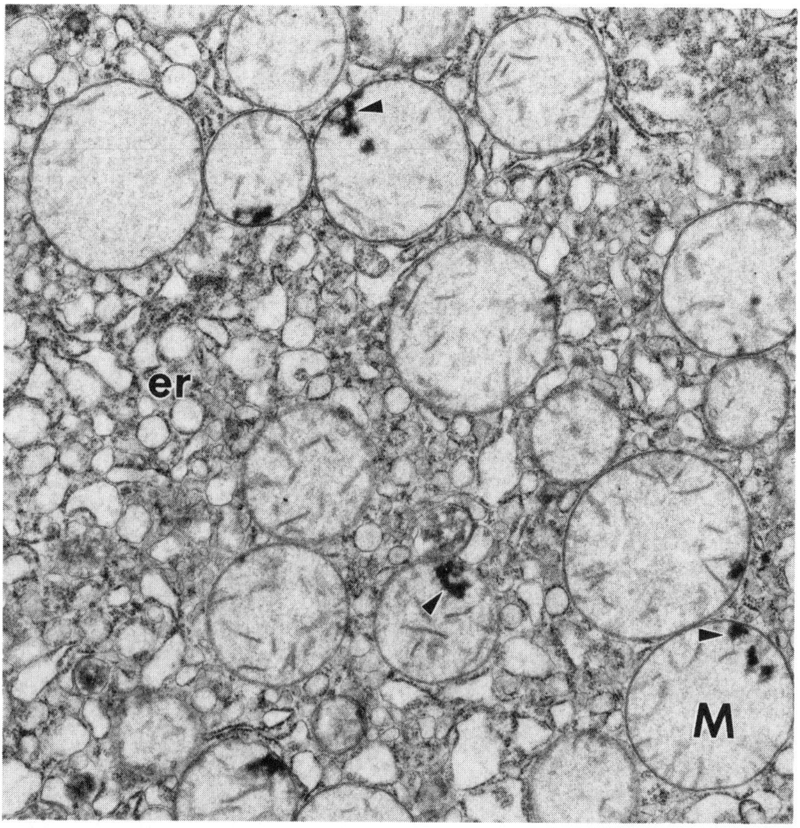

Figure 5. Uniform high-amplitude swelling of mitochondria. In cells which are at or past the "point of no return" mitochondria (M) show high-amplitude swelling and amorphous (flocculent) electron dense deposits (arrowheads). Note also the presence of dilatation and vesiculation of the endoplasmic reticulum (ER). From human liver 3.5 h after death. Magnification: ×6,900.

presence of subpopulations of mitochondria derived from different regions of the hepatic lobules or it may reflect functional heterogeneity of mitochondria at the subcellular level. In injured cells, mitochondria showing various degrees of swelling can be seen (Figure 6). It is possible that this nonuniformity in degree of swelling is caused by the presence of focal increase in calcium levels in injured cells, which may cause mitochondrial injury in discrete subcellular sites. Spread of injury to the remainder of the cell may evolve as a "wave front" phenomenon involving release of Ca^{2+} by the injured mitochondria followed by uptake from contiguous uninjured mitochondria. Nonuniform development of mitochondrial injury by means of increased permeability through "cooperative" interaction between less stable and more stable mitochondria has been postulated to occur in mitochondrial fractions injured *in vitro* (Beatrice et al., 1982).

In addition to pathologic release of cations and swelling of mitochondria, physiologic cycles of release and accumulation of cations have been described in the absence of loss of mitochondria membrane permeability and membrane potential. In the liver mitochondria, these cycles are dependent on carriers which appear to release calcium in exchange for protons and perhaps also sodium (Nicholls and Ackerman, 1982). It has been proposed that this cycle is dependent on the opposing effects of phospholipase A_2 and lysophospholipid acyltransferase (Broekemeier et al., 1985). Functionally, these cycles would result in only slight changes in proton electrochemical gradient, and ultrastructurally the mitochondria would appear to have a decreased matrix space

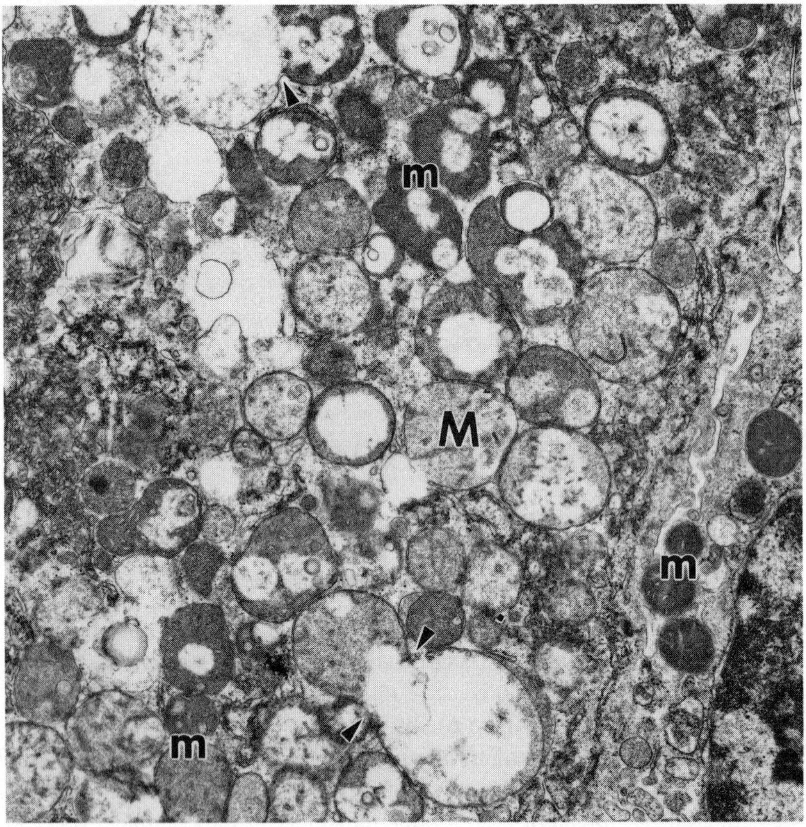

Figure 6. Heterogeneity of mitochondrial appearance in injured hepatocytes. The swollen mitochondria (M) are enlarged and more electron lucent. Rupture of the outer mitochondrial membrane appears to cause unfolding of the inner membrane (arrowheads). Mitochondria in orthodox or slightly condensed configurations are present in the severely injured and in the sublethally injured hepatocyte. Discrete zones of decreased electron density and extraction of matrix contents are seen in many mitochondrias. From rat liver 24 h after CCl_4 administration. Magnification: ×6,900.

(condensed). Agents which decrease glutathione levels cause a decline in the activity of the acyltransferase and increased permeability of mitochondrial membranes to ions and small molecules. Ultrastructurally, these mitochondria would have either an ''orthodox'' or a slightly swollen appearance.

In isolated hepatocytes, the volume of mitochondrial matrix increases by 20–40% in the presence of cyclic AMP-dependent or Ca^{2+}-mobilizing hormones such as glucagon or vasopressin. This volume change is dependent on the presence of extracellular calcium and is associated with marked stimulation of oxidation of several mitochondrial substrates and elevation of ATP concentrations (Halestrap et al., 1986a,b). In these conditions, mitochondrial swelling is due to entry of K^+ into the matrix compartment. It has been proposed that a K^+/H^+ antiport which pumps K^+ out of mitochondria and an electrogenic uniporter which carries K^+ both regulate the content of K^+ and the volume of the mitochondrial matrix (Jung and Brierly, 1984; Brierly et al., 1984; Halestrap et al., 1986a,b).

Plasma Membranes

The integrity of the plasma membrane is essential for maintaining the normal distribution of ions in the cell (see section 5.1). The volume of cells is regulated by the plasma membrane by actively extruding sodium to balance the colloid osmotic pressure of large intracellular molecules. When the ability of the Na^+ pump to maintain volume regulation is exceeded, colloid osmotic swelling of the cell begins. Alteration of permeability leads to ionic shifts and to loss of plasma membrane potential. Progressive increases in plasma membrane permeability lead to leakage of cellular metabolites. Eventually larger-molecular-weight constituents such as proteins leak out of the cell. Several types of injurious agents induce plasma membrane leaks by a variety of mechanisms. Some toxic agents, such as mercurials, oxidize thiol groups of membrane proteins. Oxidation and cross-linking of proteins and disruption of the membrane bilayer can also be caused by free-radical attack or by the action of cationic proteins such as polylysine and mellitin (Dufton et al., 1984). Activation of phospholipases by altered calcium levels also leads to altered plasma membrane permeability due to the detergentlike action of lipid degradation products. Some proteins such as bacterial toxins, and proteins released from T lymphocytes and eosinophils also induce leakage of plasma membrane. These proteins create channels through the lipid bilayer through which ionic leaks can occur. The insertion of viral envelope proteins into the plasma membrane also enhances permeability (Wyke et al., 1980). Immunological injury to the hepatocytes occurs if antibodies to components of the plasma membrane are generated (Sato et al., 1985). These antibodies can bind complement and induce the formation of complement channels in the plasma membrane followed by loss of ion equilibrium and passive influx of water along osmotic gradients. The ability of the cell to initiate compensatory ionic shifts and to remove surface membrane lesions by blebbing or endocytosis of membrane patches are important determinants in the survival of cells (Bashford et al., 1986; Carney et al., 1986).

The cell responds to increase in volume by an increased efflux primarily of K^+ and Cl^-. Ion transport pathways that protect against cellular swelling have been described in many cell types and involve either separate K^+ and Cl^- conduction, K-Cl co-transport, or parallel K^+-H^+ and Cl^--HCO_3^- exchanges. Osmotically driven loss of water from the cell accompanies these ionic shifts (Eveloff and Warnok, 1987). Compensatory ionic shifts in response to cell shrinkage due to hyperosmotic stress also occur. In this case, either co-transport of Na-Cl or Na-K-2Cl is involved; Na^+-H^+ and Cl^--HCO_3^- exchange systems are also operative in some cell types (Eveloff and Warnok, 1987). Calcium plays an important role in the compensatory ionic shifts that follow increased cell volume. In some cell types, both K^+ as well as Cl^- conductance has been found to be dependent upon calcium.

Cellular swelling is accompanied by simplification of the plasma membrane contour with loss of cell surface specializations such as microvilli. Unattached cells also assume a more rounded shape. Some degree (5–10%) of elastic stretching of membranes also accounts for a small increase in cell volume (Li et al., 1986). Swelling is also accompanied by the formation of blisters (blebs) on the apical portion of the hepatocytes that borders upon the space of Disse. The formation of blebs is more prominent on the surface of hepatocytes and other cells in culture (Lemasters et al., 1987). In these conditions, blebbing is facilitated by the diminution of cell to cell attachments and can be readily induced by a variety of insults (Figure 7).

Hypoxia in perfused livers causes the formation of hepatocyte blebs in the plasma membrane bordering the space of Disse (Figure 8). By scanning microscopy, these blebs can be seen occasionally to extend into the sinusoidal space through endothelial fenestrations (Lemasters et al., 1983). Agents that interfere with actin filaments by destabilizing their structure, increasing their polymerization, or altering their distribution also induce hepatocyte blebbing (Prentiki et al., 1979). Toxic cell injury induced by compounds such as acetaminophen (Walker et al., 1983) or $HgCl_2$ (Smith et al., 1987) also results in blebs of hepatocytes. Severe liver hypoxia induces the formation of large vacuoles which are thought to be formed by invagination of the plasma membrane of the hepatocytes (Figure 9). These vacuoles can be seen to be connected with the cell surface in suitably oriented sections and to contain plasma components (French et al., 1984).

More subtle alterations of the plasma membrane are induced by diet and by drugs such as ethanol and chlorpromazine. These compounds alter the fluidity of hepatocytes plasma membrane and cause alterations in the activity of integral membrane proteins. Some functional consequences are, for ex-

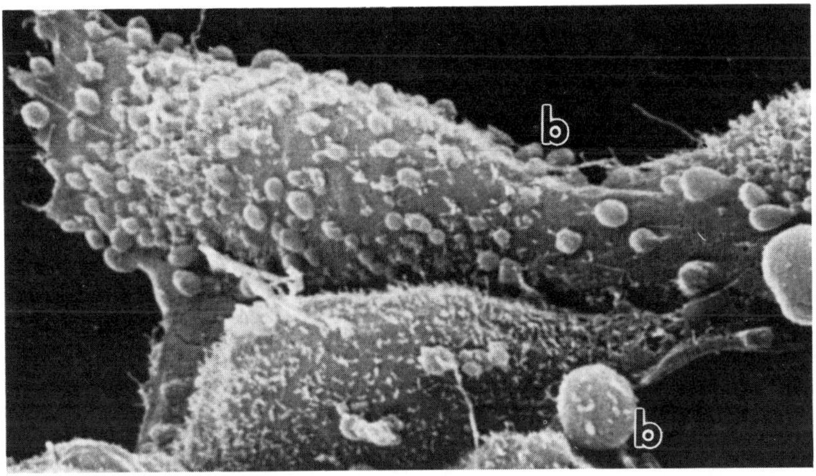

Figure 7. Blebbing of plasma membrane in cultured rabbit renal proximal tubular cells exposed to $HgCl_2$ (50 μM) for 7 min at 37°C. $HgCl_2$ oxidizes SH groups of proteins and alters the permeability of cellular membranes. The blebs (B) are of various sizes. By phase contrast microscopy, the blebs may be observed to detach from the cell surface or to be reabsorbed. Magnification: ×860. (Scanning micrograph courtesy of Patricia P. Phelps, Department of Pathology, University of Maryland School of Medicine.)

ample, inhibition of bile acid transport (Schachter, 1984; Mills et al., 1987). Physical alterations of lipid bilayers may have regulatory as well as pathologic significance.

Endoplasmic Reticulum—Golgi Apparatus

The endoplasmic reticulum (ER) is made up of an intracellular (endoplasmic) reticulum (network) of membrane sheets and tubules. The ER functions in the (1) biosynthesis of protein, phospholipid, cholesterol, and oligosaccharide moieties, (2) the integration of these biosynthetic products into membrane bilayers, and (3) translocation of membranes and luminal contents to other intracellular compartments or to the extracellular space. In addition, the ER functions in the regulation of cytosolic Ca^{2+} and in the detoxification of drugs. An early and universal alteration induced by cell injury is the dilatation of the ER (Figure 10). The precise mechanism is unknown. It is thought that disorganization of cytoskeletal and integral proteins on ER membrane and accumulation of isoosmotic fluid in the cisternae play a role. In severe states of acute cell injury, vesiculation of the ER and loss of polysomes in ER and cytosol occur.

The synthesis of all secretory proteins and of some membrane proteins occurs in the RER. Hepatocyte proteins are constantly being synthesized and degraded. In the case of the ER, estimates of protein turnover are in the range of one to two days (Shiraki and Guengerich, 1984) without marked differences in turnover rates of specific proteins. These data suggest that the ER membranes are degraded in bulk by uptake into the lysosomes by autophagy. High rates of lipid turnover have also been found for membrane phospholipids in the liver. The enzymes that assemble phospholipid molecules are located on the cytoplasmic face of the ER cisternae. Enzymes which synthesize cholesterol in the hepatocytes are also located in the ER. Experimental evidence indicates that two mechanisms participate in the translocation of lipids from ER to other subcellular compartments. The first mechanism is transport by

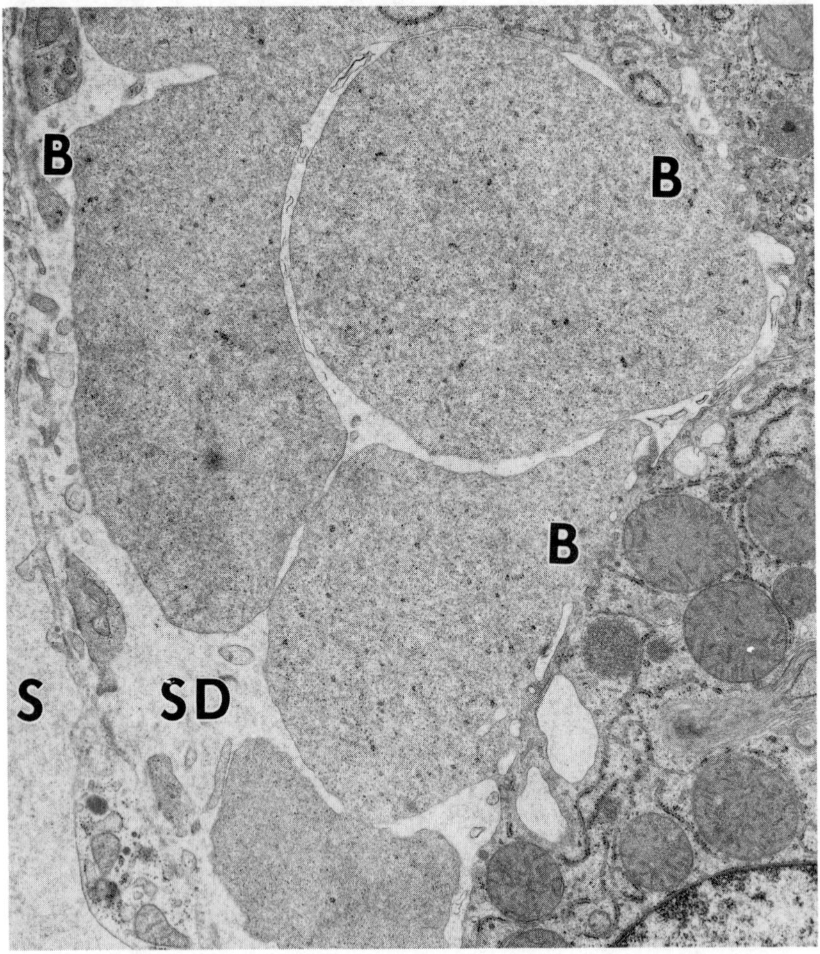

Figure 8. Plasma membrane blebs in space of Disse (SD). The blebs (B) are forming from the surface membrane of two adjacent hepatocytes. The sinusoidal (S) space is also seen. From dog liver during acute hypovolemic shock. Magnification ×5,300.

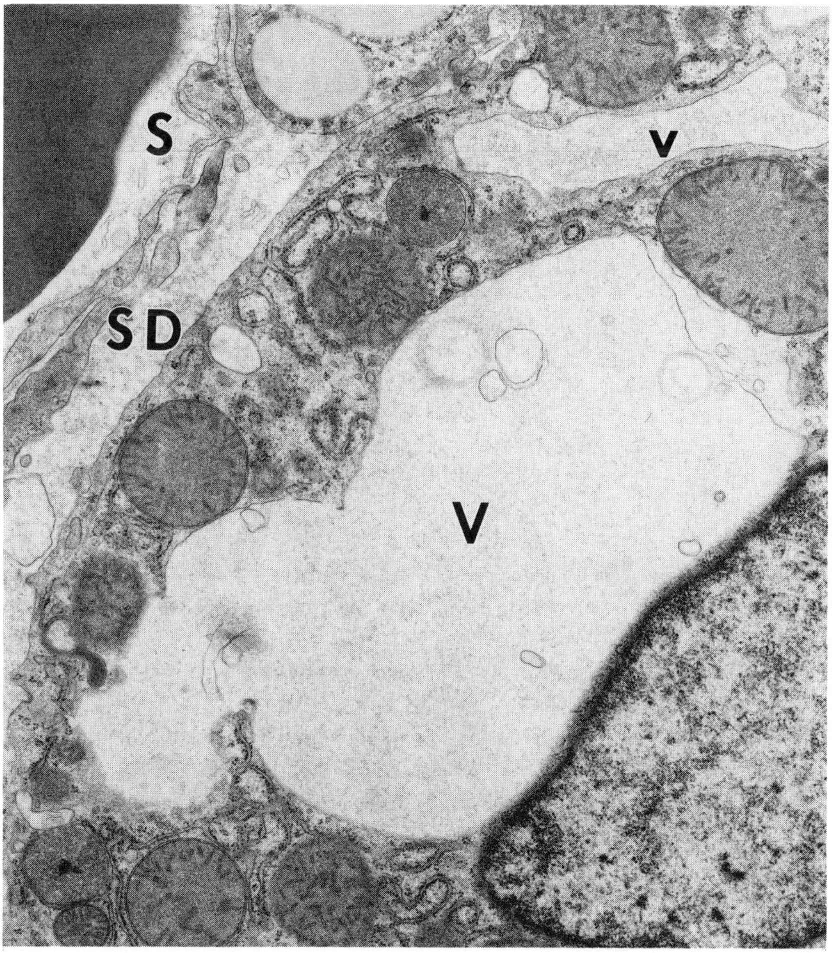

Figure 9. Plasma membrane-derived hypoxic vacuoles. Vacuoles of various sizes (V) are present below the space of Disse (SD). The mitochondria are swollen. There is no dilatation of ER or nuclear envelope. Same preparation as Figure 8. Magnification: ×9,000.

membrane vesicles which also ferry proteins to the Golgi apparatus, lysosomes, and plasma membranes. The existence of this vesicular pathway is suggested by the need of metabolic energy for cholesterol transport (Kaplan and Simoni, 1985). Protein-catalyzed transfer has also been proposed to function in the translocation of membrane lipids because protein and lipid transport differ in kinetics and in susceptibility to inhibitors (Sleight and Pagano, 1983).

The liver is a major site for biosynthesis of lipoproteins. Both the protein and lipid components are synthesized by the ER. Morphologically, the lipoprotein particles are seen in smooth and ribosome-poor cisternae and in transitional vesicles which transport them to the Golgi apparatus. The size of

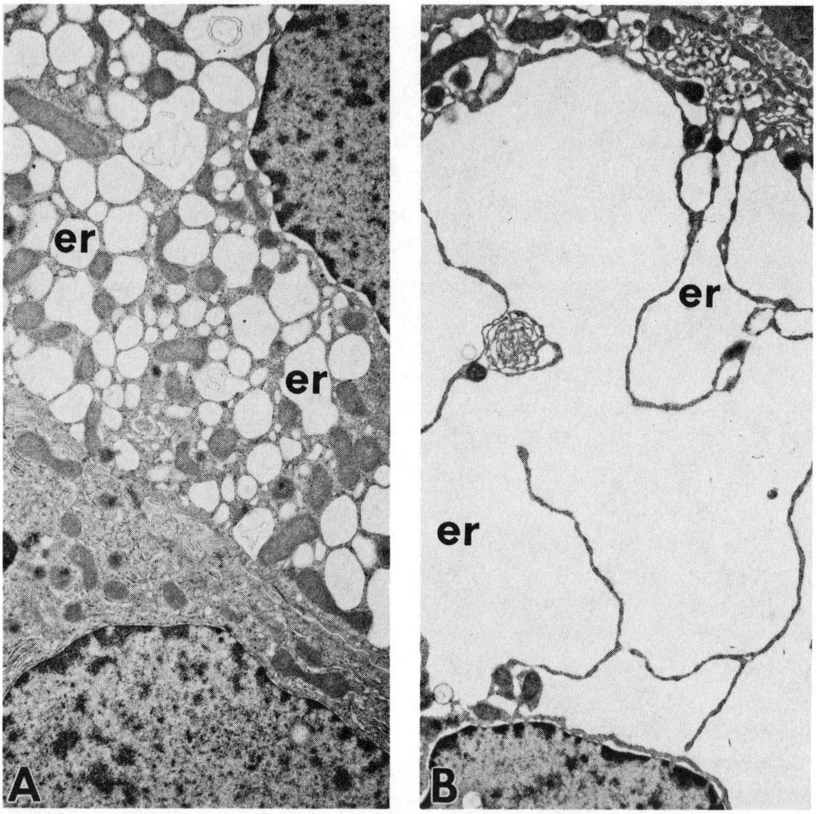

Figure 10A, B. Dilatation of endoplasmic reticulum (ER). Dilatation is moderately severe in (A) and extreme in (B). Note that the nuclear envelope is also dilated. Contrast the ER dilatation with the normal appearance of ER in contiguous hepatocytes. The mitochondria are in orthodox configuration (A) or are slightly condensed (B). When extreme, dilatation of ER corresponds to ballooning seen by light microscopy. From rat liver 4 h (A) or 24 h (B) after CCl_4 administration. Magnification: A: ×3,400; B: ×4,200.

the particles increases by 50–100% in the Golgi apparatus and the ratio of triglyceride, cholesterol, and phospholipid components is modified (Higgins and Hutson, 1984). Several types of injurious agents block the transport of newly synthesized lipid from the hepatocyte and cause its accumulation in ER (Figure 11A) and other compartments of the secretory pathway (Poli et al., 1985).

The ER is an important regulator of cytosolic Ca^{2+} homeostasis. The liver ER contains a high-affinity ATP-dependent system for transporting cytosolic calcium into the lumen. Inositol 1,4,5-triphosphate generated by the signal transduction pathway causes rapid release of calcium from the lumen of the ER (Dawson and Irvine, 1984). The ER ATP-dependent transport

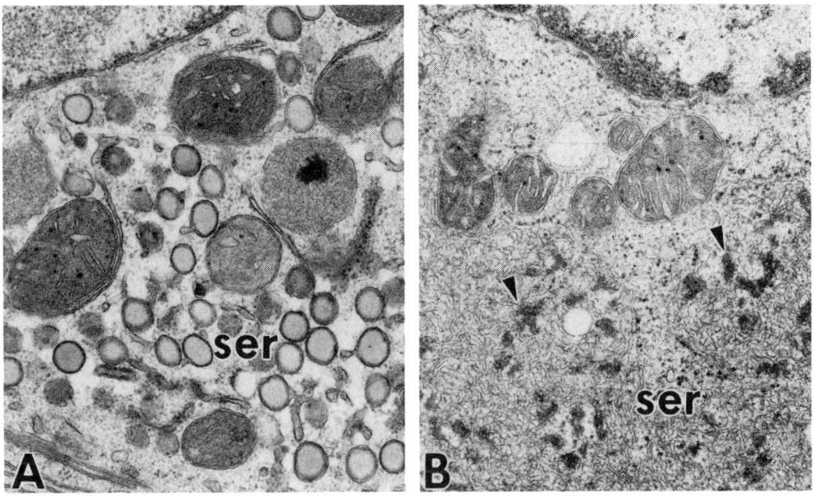

Figure 11A, B. (A) Lipid accumulation in smooth endoplasmic reticulum (SER) in a cultured hepatocyte from rat liver. (B) Proliferation of SER in rat liver by phenobarbital. The administration of CCl_4 before sacrifice also induced aggregation of SER and formation of electron dense deposits (arrowheads) on the SER clusters. Similar densities on smooth endoplasmic reticulum membranes have been seen in hepatocytes exposed to various toxins (Reynolds et al., 1975). Magnification: A: ×7,700; B: ×8,500.

system has high affinity for Ca^{2+} and can quickly take up Ca^{2+} and restore normal cytosolic Ca^{2+} levels. ATP-dependent pumps in the plasma membrane are also activated by elevation of cytosolic Ca^{2+} and extrude calcium from the cell. Decreased levels of cellular ATP can permit sustained elevation of cytosolic Ca^{2+}. In the presence of ATP, massive influx of extracellular calcium into the cell can saturate the ER transport. At this point, the lower affinity but higher-capacity mitochondrial Ca^{2+} transport system is activated. Concentrations of Ca^{2+} at steady state are estimated to be 8 nmole/kg in the RER (Somlyo et al., 1985). This figure corresponds to approximately 20% of total hepatocyte calcium. In physiologic conditions, liver mitochondria only contain approximately 5% of total hepatocyte calcium and do not participate in the regulation of cytosolic calcium levels (Somlyo et al., 1985). Uptake of Ca^{2+} into hepatocyte ER and extrusion by the plasma membrane into the extracellular space are the normal mechanisms for cytosolic Ca^{2+} regulation because of the high affinity of the ATPase located in these membranes for calcium. It has been proposed that mitochondria can accumulate calcium during pathologic conditions in which altered permeability of the plasma membrane causes saturation of ER calcium uptake capacity and elevation of cytosolic ionized calcium to micromolar concentrations. In these conditions, mitochondria accumulate calcium (Laiho and Trump, 1975) and can buffer it by precipitating in the matrix with phosphate (Carafoli, 1986).

The formation of insoluble Ca^{2+}-phosphate salts can be seen by electron microscopy. Since normal mitochondrial energy transduction is essential to maintain the potential that drives calcium uptake, such calcium deposits may theoretically still form in viable cells. Another possibility is that calcium deposits may form in nonviable cells as a result of equilibration of cytoplasmic and plasma milieu in necrotic cells. Calcium deposits would then appear at cellular locations which serve as nucleation sites for formation of calcium salts.

The smooth endoplasmic reticulum (SER) is active in the biotransformation of nonpolar drugs via the cytochrome P-450 mixed-function oxidase system (Figure 11B). In the first stage of drug biotransformation, reactive drug intermediates may be created by oxidation through cytochrome P-450. These reactive drug intermediates induce cell injury by covalent binding to proteins, nucleic acids, and lipids.

Lysosomes

The ultrastructural appearance of the lysosomes in hepatocytes is variable and depends to a large extent on variation in the rates of uptake and degradation of cellular organelles and other degradable substrates by the hepatocyte. The substrates can be taken up by autophagy (into autophagosomes) during normal turnover of cellular proteins and lipids (Figure 12) as well as by heterophagy (into endosomes or phagosomes) of extracellular lipids and proteins via the plasma membrane (Marzella and Glaumann, 1987). Phagocytosis of cellular debris present in the sinusoids in the space of Disse is performed by the Kupffer cells (Glauman and Marzella, 1981). Convincing evidence of (hetero)phagocytosis by hepatocytes is seldom seen. The process of degradation of substrates segregated into autophagosomes and endosomes begins when these vacuoles merge with acid hydrolase-containing vesicles (*primary lysosomes*) or vacuoles (*secondary lysosomes*) and acquire digestive enzymes. In the case of autophagosomes, frequently two or more membranes completely surround the segregated cytoplasm. After the autophagosome fuses with a lysosome, degradation begins. At this stage, the vacuole is called an *autophagolysosome*. During the process of degradation, only the outer membrane of the autophagolysosome is preserved. All other macromolecular components are degraded to the level of individual molecules (amino acids, fatty acids, glycerol, cholesterol, monosaccharides) which permeate or are transported back into the cytosol for reutilization.

The size of the autophagolysosome decreases by fission into smaller vesicles that bud outwardly into the cytosol for recycling or inwardly into the vacuolar matrix for degradation. In the case of endosomes, a prelysosomal compartment has been defined structurally and functionally. This prelysosomal compartment has an acid pH. It is in this compartment that dissociation of ligands from receptors and recycling of membranes back to the surface of the cell occurs. The ligands are then transported to the lysosomes for degradation.

In addition to (macro)autophagic and heterophagic uptake of materials, pathways for the lysosomal uptake of secretory materials (crinophagy) and cytosol (microautophagy) have been outlined. Extralysosomal pathways for degradation of proteins also exist. These pathways are thought to participate primarily in the degradation of abnormal proteins and proteins that turn over at high rates (Marzella and Glaumann, 1987).

The biogenesis of lysosomal enzymes follows the RER-SER-Golgi route that is taken by lipids and proteins destined for export to the extracellular

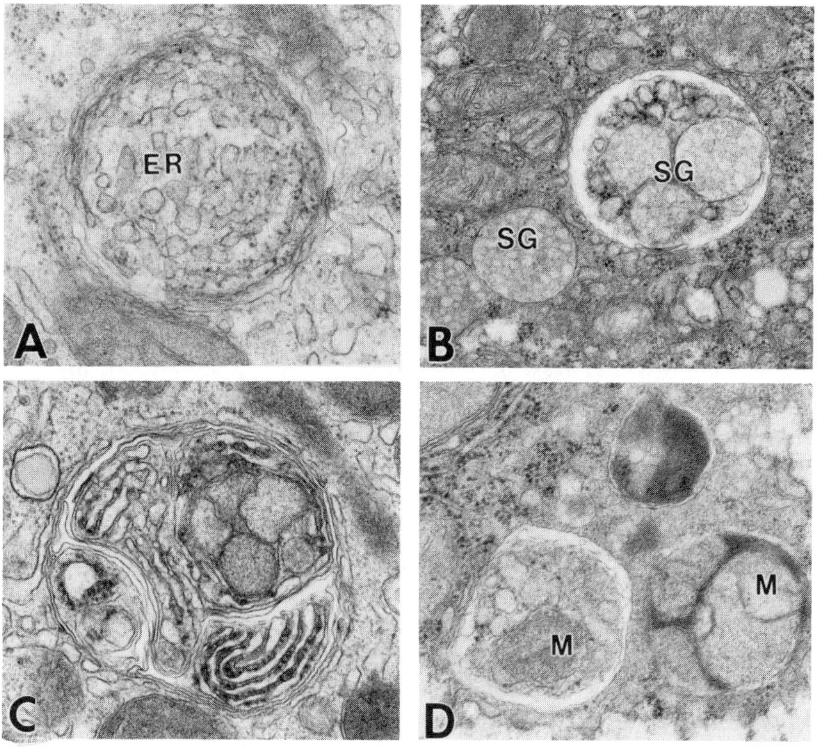

Figure 12A-D. Stages of autophagic vacuole formation. (A) Multiple bounding membranes surround rough and smooth ER cisternae which appear ultrastructurally normal. This represents the earliest (nascent) stage of AV formation. (B) In the next stage, the inner and outer bounding membranes of the AV are more distinct and are separated by a wide electron lucent space. Some secretory granules (SG) and ER cisternae are sequestered in the AV. (C) The cytoplasmic contents of this AV have begun to show ultrastructural signs of degradation. (D) This micrograph illustrates the progressive degradation of mitochondria (M) segregated inside AVs. A third vacuole (lysosome) contains degraded contents that are morphologically unrecognizable. Magnification: A: ×6,800; B-D: ×8,500.

space (Marzella and Glaumann, 1983). A "signal" amino acid sequence on the nascent lysosomal enzyme determines the insertion of the oligopeptide chain into the lumen of the ER. Proteolytic removal of the signal sequence and addition of carbohydrate chains to the nascent oligopeptide also occur in the ER (Nishimura et al., 1988). In the Golgi apparatus, processing of the carbohydrate chains of the lysosomal enzymes occurs. The lysosomal enzymes bind to mannose-6-phosphate receptors in the Golgi region and are transported within vesicles to the lysosomes either directly or, as has recently been proposed, via an endosomal compartment. Transport of lysosomal enzymes that is independent of mannose-6-phosphate receptors also occurs. The various enzymes are transported through the ER-Golgi route at different rates possibly because of specific interactions with other membrane proteins that may regulate rates of transport (Cardelli et al., 1986). Apparently physiologic secretion of lysosomal enzymes from hepatocytes has also been described (Nishimura et al., 1988).

The lysosomal cathepsins play a major role in degradation of cellular proteins. These proteinases have acid optima (pH 3–5) and they become inactive and unstable at neutral or alkaline pH. In addition, cytosolic inhibitors bind and inactivate cathepsins that may leak out of the lysosomes of injured cells. The lysosome is the major intracellular site for the degradation of lipids as can be inferred by the accumulation of lipids that occurs in lysosomal enzyme-deficiency states (Glew et al., 1985). Phospholipid turnover in the liver is rapid. The estimates of half-lives for phospholipids range from approximately 10 h to 2 days depending on the labelled precursor used to measure the turnover (Van Den Bosch, 1980). Phospholipases are also found in the cytosol and in all the various cellular membranes where they act in concert with lysophospholipid acyltransferases to remodel phospholipids (Van Den Bosch, 1980).

Phospholipid inclusions in the form of myelin figures are found in the lysosomes of patients with drug-induced lipidosis. These agents retard the degradation of lipid which normally enters the lysosomes by autophagy and heterophagy. Lysosomal degradation is inhibited by alkalinization of the lysosomal milieu by drugs (such as chloroquine) that are weak bases (Poole and Ohkuma, 1981). Some drugs such as aminoglycosides may inhibit lysosomal phospholipases (Hostetler and Hall, 1982; Regec et al., 1986). A variety of other drugs with amphiphilic structure are thought to bind to phospholipids and prevent their degradation by lysosomal enzymes (de la Iglesia et al., 1974; Dake et al., 1985). Phospholipids may become chemically altered before entry into the lysosomes. Formation of lipid peroxides may occur in large amounts during oxidative stress or slowly as cells age. These molecules are resistant to digestion and accumulate in the lysosomes as lipofuscin.

The size of the lysosomal compartment is variable and it reflects at steady state the rates of autophagic segregation and degradation of hepatocyte cy-

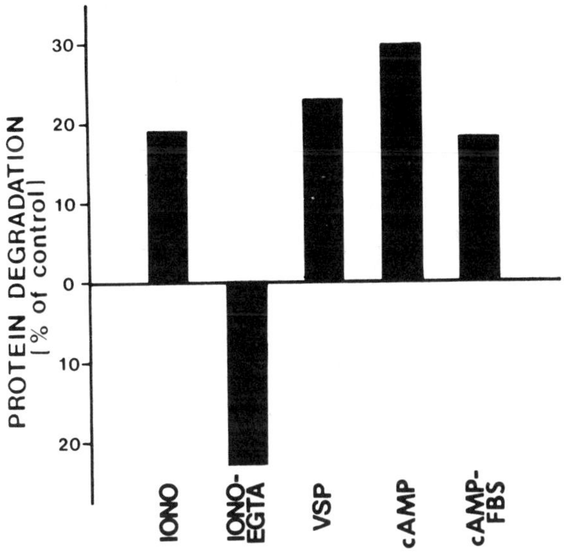

Figure 13. Regulation of proteolysis in cultured rat hepatocytes. Deprivation of nutrients and other serum factors and the addition of agents that increase intracellular levels of cAMP and Ca^{2+}, and stimulate proteolysis in the hepatocytes. Proteolysis in control cultured hepatocytes is normalized to a value of 0. Ionomycin (IONO), vasopressin (VSP), and dibutyryl cyclic AMP (cAMP) all stimulate proteolysis. Ionomycin, in the presence of a divalent ion chelator, depletes intracellular calcium and decreases proteolysis (IONO-EGTA). In the presence of serum, dibutyryl cyclic AMP is less effective in stimulating proteolysis (cAMP-FBS). (Reproduced with permission from Williams and Wilkins: *Lab Invest* 1988; 58:643–652. Yu and Marzella, p. 648, Figure 5.)

toplasm. When protein degradation is stimulated, the size of the lysosomal compartment expands. Amino acid deprivation, glucagon, and cyclic-AMP are potent stimulators of hepatocyte autophagic protein degradation; insulin is inhibitory (Figures 13 and 14) (Pfeifer, 1978; Poso et al., 1982; Seglen et al., 1980; Yu and Marzella, 1988). In physiologic conditions such as in the postabsorptive state, autophagic protein degradation in the hepatocytes can be stimulated to provide amino acids to maintain amino acid homeostasis in the blood and to provide substrates for gluconeogenesis and maintenance of blood glucose levels. During prolonged starvation, hepatocyte lysosomal protein degradation decreases and skeletal muscle proteolysis is turned on (Mortimore et al., 1983). During hepatic regeneration, lysosomal proteolysis is suppressed to enhance cell growth. The propensity for cell growth is, in general, associated with lowered rates of cellular protein degradation. Immortalized cells manifest lower rates of protein degradation than their normal cells of origin which manifest a finite life-span. Neoplastic transformation of immortalized cells lowers even further steady-state protein degradation and also diminishes the ability of physiologic stimuli to stimulate protein degradation in the transformed cells. It has been proposed that the lowered degradation capacity may confer a growth advantage to the transformed cells (Seglen et al., 1986; Ahlberg et al., 1987; Tayek et al., 1988; Lee et al., 1989).

In injured cells, an increase in the numbers and size of lysosomes occurs. Inhibition of lysosomal hydrolases by enzyme inhibitors or by alkalinization of the lysosomal milieu retards degradation and causes an expansion of the lysosomal compartment due to the accumulation of undegraded materials (Henell and Glaumann, 1984; Marzella and Glaumann, 1987). As far as can be judged from cultured hepatocytes, toxic cell injury causes a decrease in protein degradation and a reduction in the volumes of lysosomes (Yu and Marzella, 1988). The expansion of the lysosomal compartment in livers during

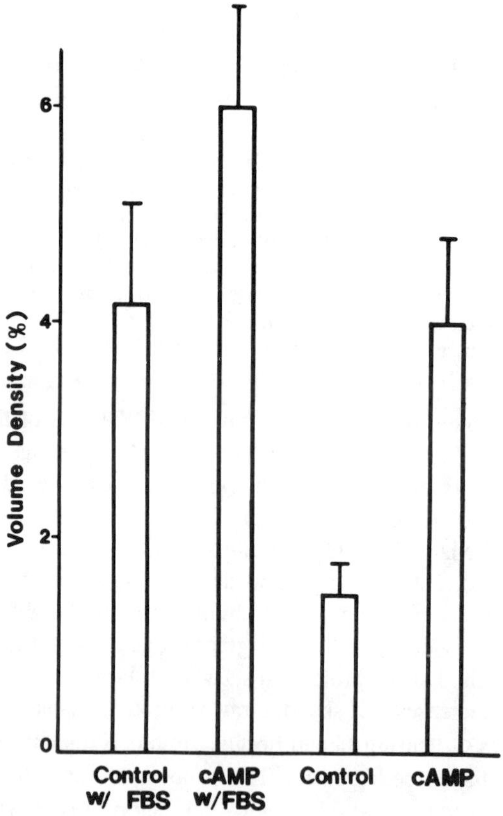

Figure 14. Levels of cAMP and of nutrients regulate proteolysis in hepatocytes through the lysosomal degradation pathway. The ordinate shows the volume density of the lysosomal compartment in hepatocytes. The addition of dibutyryl cyclic-AMP (cAMP) and the absence of serum (W/FBS) cause an increase in the volume density of the lysosomes. (Reproduced with permission from Williams and Wilkins: *Lab Invest* 1988; 58:643–652, Yu and Marzella, p 648, Figure 5.)

acute injury *in vivo* is, therefore, not caused directly by toxins. Changes in the levels of hormones and nutrients (particularly amino acids) occurring from systemic effects of toxic injury are responsible for the increase in the lysosomal compartment (Yu and Marzella, 1988).

Undergradable substances introduced into the lysosomes by autophagy and heterophagy accumulate over time and can be seen ultrastructurally as electron dense deposits. In the case of metals such as copper, excessive accumulation may cause labilization of the lysosomal membrane. It has also been suggested that hepatocyte lysosomes can sequester metals from the cytosol and can extrude them from the cell (Helman et al., 1985). In the case of iron overload, increased deposition of ferritin occurs in the liver lysosomes and is associated with increased labilization of the lysosomes (Peters and Seymour, 1976). The presence of Fe^{2+} induces leakage of enzymes from lysosomes by free-radical reactions that can be blocked by lipid-soluble free-radical scavengers (Mak and Weglicki, 1985). During autophagy, uptake of cytosolic ferritin and of hemoproteins occurs. Degradation of these iron-containing proteins by the lysosomes occurs slowly (Glaumann and Marzella, 1981). The iron is released to the cytosol where it participates in peroxidative reactions that can aggravate cell injury (Starke et al., 1985).

The number of lysosomes varies in hepatocytes located in different regions of liver lobules. Lysosomal shape is also influenced by the presence of undigestible residue. Typical examples are the rounded shapes of lipofuscin-containing lysosomes and tubule-shaped lysosomes caused by the presence of crystalline deposits or other residues (Naito et al., 1988). Age causes an increase in the number of lysosomes and an alteration in the distribution of lysosomal enzyme activity within the liver lobule (DePriester et al., 1984). Age-related changes in membrane composition and functional activity of other hepatocyte organelles have been described (Schmuckler et al., 1984).

In the hepatocyte, the lysosomes are dynamic structures in motion through the cellular cytoplasm along microtubular tracks (Yu and Marzella, 1990). Clustering near Golgi apparatus and smooth ER and in pericanalicular regions is frequently seen. The Golgi-ER is the site where biogenesis of lysosomal vesicles (primary lysosomes) occurs through budding of vesicles from *trans* regions of the Golgi apparatus. The clustering of lysosomes near bile canaliculi has been interpreted as indication of release of lysosomal contents into bile by fusion with canalicular plasma membrane. The heterogeneous morphological appearance of lysosomes and the functional relationships between the various forms are summarized diagrammatically in Figure 15.

Nuclei

Changes in the structure of the nuclei are classic indicators of cell necrosis at the light microscopic level. Alterations of nuclear envelope, chromatin,

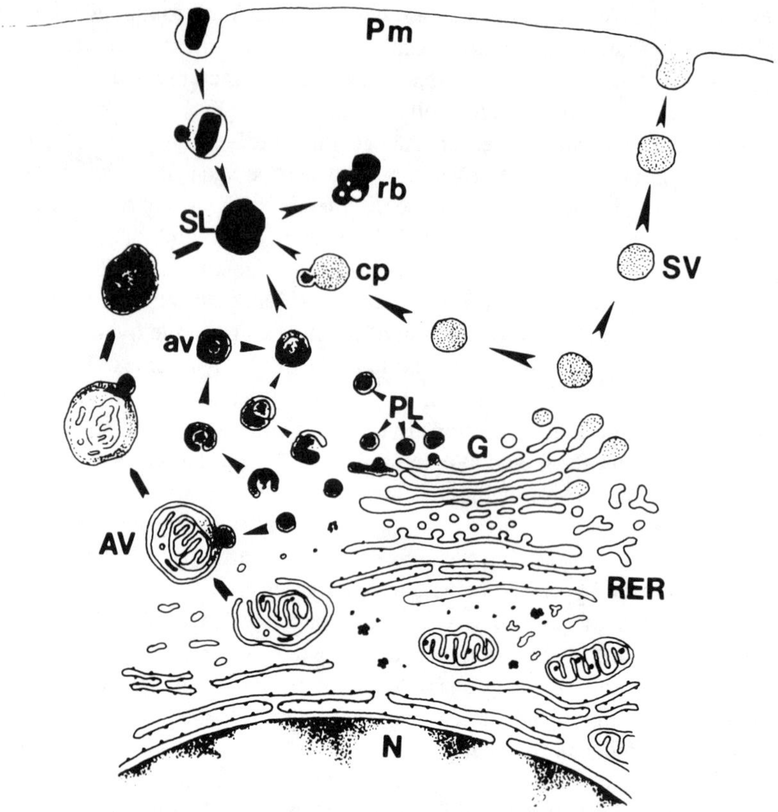

Figure 15. Lysosomal degradation pathways in hepatocytes. The autophagic and heterophagic pathways for the segregation of intracellular and extracellular material are shown converging to the main cellular locus of degradation, namely the secondary lysosome (SL). (Macro)autophagosomes (AV) sequester entire cellular organelles and cytosol. (Micro)autophagosomes (av) sequester cytosolic components and are formed by invagination of the lysosomal membrane. Crinophagic vacuoles (CV) degrade secretory proteins and lipids. Primary lysosomes (PL) are seen to form in the trans region of the Golgi (G) apparatus. PM: plasma membrane; RB: residual body; SV: secretory vacuole; RER: rough endoplasmic reticulum; N: nucleus.

and nucleoli are also useful markers of sublethal cell injury at the ultrastructural level (Figure 16). Moreover, physiologic events also affect the appearance of the chromatin and nucleoli. The shape of the nuclear envelope is maintained by a protein network, the lamina, which is also the site for the attachment of regions of chromatin to the nuclear envelope. These attachment sites are spaced regularly along the chromosomes and maintain a stable orientation of the chromosomes within the nuclei (Newport and Forbes, 1987). The size of the nucleus and the density of the chromatin changes when transcriptional activity is altered. Changes in DNA content during various stages of the cell cycle also affect nuclear size. Increases in nuclear size and alterations in nuclear shape are characteristic of aging cells. In addition to increasing ploidy, an accumulation of macromolecules, such as RNA and altered proteins, occurs in aged cells and is indicated by increases in measurable nuclear dry mass (Mukherjee and Weinstein, 1986).

Polyamines bind to nucleic acids and influence the appearance of the chromatin. Changes in the levels of polyamines in the nuclei are associated with condensation and decondensation of chromatin (Hougaard et al., 1987). In addition, binding of polyamines to DNA protects against thermal or irradiation induced damage, methylation, and enzymatic cleavage. The concentration of Ca^{2+}, Mg^{2+}, and H^{+} also have profound influences on the morphological appearance of the chromatin. In isolated hepatocyte nuclei, the presence of high calcium (3 mM) and magnesium (11 mM) causes a marked condensation of the chromatin. Chelation of the cations restores the granular-fibrillar appearance of the chromatin. Decrease in pH also causes condensation of the chromatin and a decrease in the amount of extractable water (Miseta et al., 1987). These considerations suggest that the chromatin condensation seen in the early stages of cell injury is caused by shifts in ions and in pH.

Hepatocyte nucleoli are of the reticular (nucleolonema) type composed of networks of dense fibrillar and granular components. Fibrillar centers are interspersed in this network. Fibrillar centers are structures containing proteins involved in ribosomal DNA transcription (Scheer and Rose, 1984). When

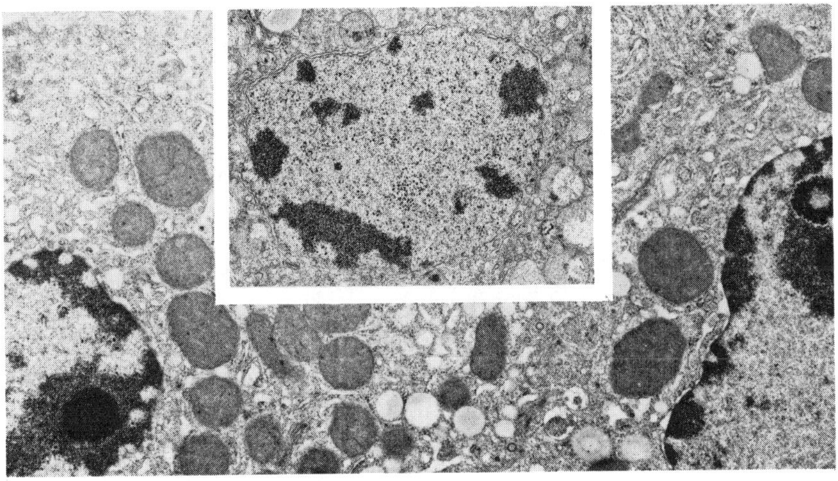

Figure 16. Nuclear alterations in cell injury. Compaction and segregation of nucleoli caused by aflatoxin B administration can be seen (arrowheads). The inset shows the occurrence of coarse chromatin clumps and incipent loss of electron density of nuclear contents caused by cell necrosis. Magnification: $\times 5{,}000$. Inset: $\times 2{,}000$.

protein synthesis is inhibited, the fibrillar centers decrease in size (Wachtler et al., 1987) and become surrounded by a coating of dense fibrillar components surrounded by an accumulation of granular material. The centers tend to become smaller in size. Inhibitors of RNA synthesis cause decreases in size of nucleoli and nucleolar segregation (separation of fibrillar and granular components) (Cole et al., 1986; Almeida et al., 1987; Wachtler et al., 1987).

Cytoskeleton

The prominent alterations of the hepatocyte plasma membrane illustrated in Figures 8 and 9 are thought to be related to changes in cytoskeletal components, particularly actin. It is not known whether actin induces localized deformation of the plasma membrane by contraction or whether loss of anchorage points between the plasma membrane and the cytoskeleton induces the pits and blebs shown in Figures 8 and 9. These alterations are considered sublethal because blebs can disappear after they form. Restoration of normal cell-surface contour as well as pinching off and detachment of the blebs from the cell can be observed in cultured cells. The addition of cytochalasin to cultured anoxic hepatocytes retards the time course of loss of cell viability but does not affect the ultimate number of dead cells. Depolymerization of microtubules with colchicine also had no effect on cell death (Okayasu et al., 1984). These results support the proposal that cytoskeletal disorganization is not a lethal event. Intermediate filaments are important for the structural support and organization of hepatocytes and other hepatic structures. Changes in numbers, localization, and organization of these filaments occur commonly in sublethal and chronic types of injury in all tissues including the liver (Jahn, 1982; Ohta et al., 1988).

OVERVIEW OF STRUCTURAL ALTERATIONS OF INJURED CELLS

The morphological alterations of the various organelles we have reviewed and illustrated are summarized in Table 3–1. These ultrastructural features are reliable indicators of cell injury in hepatocytes. Analogous alterations occur in cell injury in most tissues.

Stages of Cell Injury

The ultrastructural manifestations of cell injury that we have summarized in Table 3–1 have also been categorized on the basis of severity for each organelle. The type and extent of alterations of the various organelles to injury of graded severities has also been defined. For the purpose of standardization,

TABLE 3–1.

Cell Injury	Ultrastructural Characteristics
Reversible:	Dilated endoplasmic reticulum Condensed mitochondria Condensed nuclear chromatin Increased cell cytoplasmic volume (swelling) Blebs on plasma membrane Formation of apical vacuoles Decreased numbers of polysomes Altered cytoskeletal organization
Transition Point:	All of the above features plus swelling (low amplitude) and amorphous (*flocculent*) densities of mitochondria
Point of No Return:	Empirically defined in the various tissues. However, *no* structural or biochemical definition is yet available
Necrosis:	High-amplitude swelling of mitochondria Large flocculent densities of mitochondria Massive swelling of cell, electron lucent cytosol Marked clumping of nuclear chromatin Discontinuities and vesiculation of membranes Formation of cell fragments Phagocytosis of cells by resident macrophages and/or inflammatory cells

it is helpful to classify these ultrastructural alterations of injured cells into stages. It should be recognized that this classification is arbitrary because it creates transition points in a sequence of alterations that evolves gradually. The classification also does not imply that all cells or even all the organelles within a cell move synchronously through the various stages of injury. Histological and subcellular microenvironments exist through which functional and structural alterations may evolve at various rates. At some point during this continuum of changes, the alterations became irreversible and the cell dies. This state has been designated the "point of no return." The specific biochemical and structural correlate of this transition point has not been identified. Dead cells undergo various changes which at the morphologic level are defined as necrosis.

The ultrastructural manifestations of cell injury are classified by stages in Figure 17. Some of the earliest manifestations of cell injury are seen in mitochondria where loss of matrix granules and matrix condensation occur (Figure 4). As injury progresses, swelling of mitochondria begins. In dead cells mitochondria are very enlarged, the matrix is pale, the cristae are diminished in numbers and are shortened. Amorphous flocculent densities are

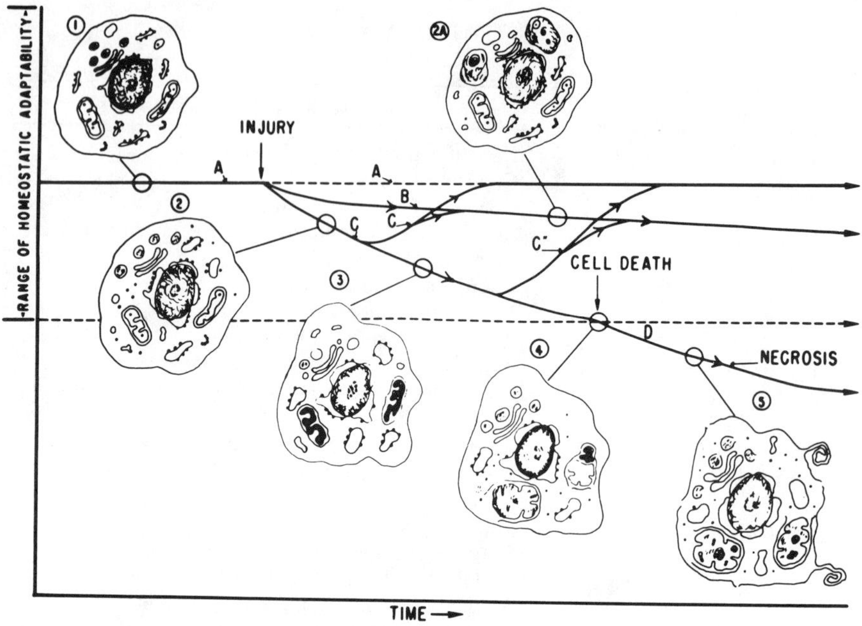

Figure 17. Stages of cell injury. Plotted along the abscissa is time, and along the ordinate is range of homeostatic ability. Stage 1 shows a cell in a normal steady state (curve A). At the arrow, an injury is applied which may be acutely lethal or sublethal. In the case of the lethal injury, the cell loses homeostatic ability along curve C. Before the point of cell death, recovery can occur if the injurious stimulus is removed. Such recovery might proceed along curve C′ or curve C″. In stage 2, the principal change consists of dilatation of the ER and slight clumping of the nuclear chromatin; some ribosomes may be detached from the endoplasmic reticulum, and the entire cell may be slightly swollen. In stage 3, the mitochondria are condensed, the cell is more swollen, and blebs begin to appear on the cell membrane. After the point of cell death, recovery cannot occur, even if the injurious stimulus is removed the cell is said to enter the phase of necrosis. During this transitional period, stage 4 is seen, and in it changes similar to those in stage 3 occur, except that some mitochondria are markedly swollen, while in others, portions of their inner compartments are condensed and other portions are swollen. The lysosomes remain intact; there is increased ribosomal scattering and increased cell swelling. During the phase of necrosis, the cell undergoes degradation by autolysis and denaturation, and a typical morphologic picture occurs as in stage 5. (Reproduced with permission. Trump BF, McDowell EM, and Arstila AU: Cellular reaction to injury, in LaVia MF and Hill RB Jr., eds: *Principles of Pathobiology,* ed 3. Oxford University Press, New York, 1980, pp 20–111.)

commonly seen in the mitochondria at this stage (Figure 5). Rupture of the outer membrane (Figure 6) and unfolding of the inner membrane follow.

Features of cell surface membrane such as microvilli and endocytotic pits tend to decrease on the surface of injured cells. Blebs of various sizes (Figures 7 and 8) protrude into the extracellular space and may seal and detach from the cell. In severely injured cells, the surface membrane becomes smoother and apical vacuoles occasionally of large size may form (Figure 9). In irreversibly injured cells, membrane breaks may be seen.

In the early stages of cell injury dilatation of the endoplasmic reticulum occurs (Figure 10A) and is followed by vesiculation and detachment of ribosomes. Accumulation of lipidlike densities may occur intraluminally (Figure 11A). The Golgi apparatus becomes dilated and ultrastructurally unrecognizable in early stages of injury. Clustering and loss of polarity of smooth and rough cisternae is often seen in subacute forms of injury.

Nuclei of hepatocytes normally show evenly dispersed chromatin, the nucleoli show interspersed fibrillar and granular components. Condensation of the chromatin, particularly near the nuclear envelope, occurs in early stages of cell injury. Separation of nucleolar components into distinct structural regions also occurs and is very distinct in some types of injury. More severe injury causes dilatation of nuclear envelope. Clumping of the chromatin and the appearance of electron lucent areas occurs after cell death and is due to the activation of degradative enzymes. These ultrastructural manifestations of injury are depicted in Figure 16.

These stages of cellular injury defined by transmission electron microscopy have been validated by various other morphologic techniques, such as light microscopy, scanning, x-ray microanalysis, and immunocytochemistry. The evolution of functional alterations following the onset of cell injury has also been followed analytically. As discussed previously, characteristic biochemical and biophysical alterations have been identified, which temporally parallel the ultrastructural stages defined by electron microscopy. Figure 17 summarizes in diagrammatic form the concept of evolution of cell injury through morphologically definable stages towards cell death. This diagram also illustrates the concept of reversibility of sublethal types of cell injury.

ROLE OF ION SHIFTS IN CELL INJURY

Ion Regulation

Altered membrane permeability and dysfunction of energy metabolism lead to loss of homeostasis of ions such as Na^+, Ca^+, K^+, and H^+ (Figure 18). Loss of ability to regulate ionic homeostasis is thought to be an important event in cell injury.

The ionic composition of cells is markedly different from that of the extracellular space; in addition, cells contain large amounts of negatively charged proteins. In the case of calcium, for example, the extracellular concentration is in the millimolar range, whereas the cytosolic concentration of free calcium is four orders of magnitude lower. The plasma membrane of cells maintains these ionic gradients by the action of membrane pumps and exchangers. The plasma membrane constitutes a permeability barrier that restricts the free movement of ions and metabolites between the cellular interior and the extracellular space. Metabolic energy is expended to maintain

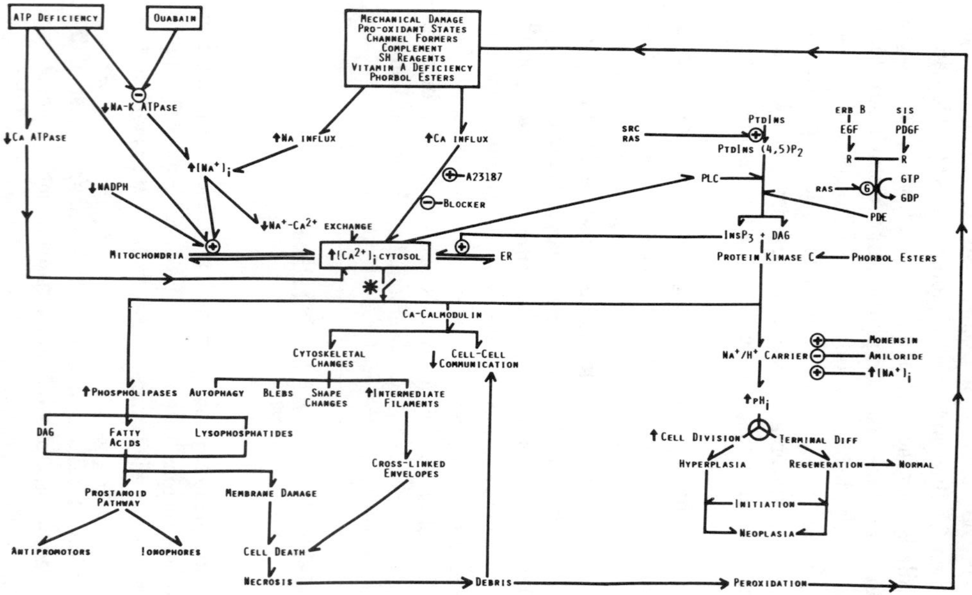

Figure 18. Ion homeostasis and cell injury. The diagram emphasizes the role of plasma membrane, mitochondria and ER in ion equilibrium. Several types of insult may lead to ion deregulation. Loss of ion regulation, in particular Ca^{2+}, may constitute a common final pathway towards cell death. This diagram also illustrates the role of ions such as Ca^{2+} and H^+ as intracellular signals that modulate cellular responses essential for adaptation, repair and expression of neoplastic transformation. (Reproduced with permission. Trump BF and Berezesky IK: Cell injury, ion regulation, and tumor promotion, in *Banbury Report 25: Nongenotoxic Mechanisms in Carcinogenesis*. Cold Spring Harbor Laboratory, 1987, pp 69–82.)

the concentration of ions and metabolites in the two compartments at different levels. Na^+ and Ca^{2+} are extruded from the cells by energy-dependent pumps and by exchanger systems. Efflux of potassium from the cell occurs because of concentration gradients. Physiologically altered fluxes of Na^+, K^+, and H^+, due to changes in the activity of the Na^+/H^+ exchanger and of the Na^+/K^+ ATPase, can directly or through changes in intracellular Ca^{2+}, induce differentiation of cells, as well as promote cell proliferation.

In cell injury, loss of ATP (needed to drive Na^+/K^+ ATPase) may cause a rise in cytosolic Na^+ levels. The presence of Na^+/K^+ ATPase inhibitors such as digitalis or ouabain will have an analogous effect on cytosolic Na^+ levels. Increased Na^+ levels cause influx of Ca^{2+}, dependent on the Na^+/Ca^{2+} exchanger; the exchanger normally functions to regulate intracellular Ca^{2+} by utilizing the downhill movement of Na^+ to provide energy for the uphill extrusion of Ca^{2+} (Schanne and Moore, 1986). Membrane depolymerization may occur due to the influx of sodium. The change in membrane potential can cause voltage-dependent calcium channels to open, increasing the influx of Ca^{2+}. The ability to extrude protons from the cells in exchange for the uptake of Na^+ is also impaired in injured cells. Since protons are generated continuously from cellular metabolism, the intracellular pH declines as the intracelullar buffering capacity is exceeded. Loss of ATP also prevents Ca^{2+} extrusion by the Ca^{2+} ATPase. This enzyme is one of the major plasma membrane systems for ejecting CA^{2+} in nonexcitable cells. Loss of ion homeostasis in cell injury can be aggravated if the permeability of the surface membrane is perturbed by the detergentlike activity of Ca^{2+}-dependent phospholipases in the plasma membrane. Toxins such as heavy metals which oxidize thiol groups alter the permeability of plasma membrane.

Loss of ion homeostasis and in particular sustained elevations of cytosolic calcium are deleterious to cell function (Figure 18). Increased cytosolic calcium levels function as a second messenger system that modulates a great variety of cellular activities. By virtue of the great importance of calcium in the integration of cell functions, it has been hypothesized that calcium deregulation is a common mechanism in the evolution of cell injury towards cell death. Increased intracellular calcium can cause profound cellular alterations by modulation of enzyme activity. Nucleases, lipases, proteases, and protein kinases can be activated. Increased expression of oncogenes may occur, with far reaching effects on regulation of cellular metabolism.

Proton Regulation

The pH of a variety of hepatocyte subcellular organelles is acidic. Maintenance of pH involves the activity of an ATP-dependent proton pump (H^+-ATPase). The lysosomes were the first organelles to be shown to be acidified following the realization that the pH optimum of all the lysosomal hydrolases is in the acid range (Ohkuma, 1987). This feature is thought to be advantageous to the cell as it would minimize degradative activity of the lysosomal enzymes if leakage out of the lysosomes into the cytosol would occur during cell injury. Subsequently, acidification of vesicles, functionally coupled with the lysosomes (namely endosomes, multivesicular bodies, clathrin-coated vesicles) was demonstrated (Maxfield, 1982; Van Dyke et al., 1985). The acidification of these vesicles permits the dissociation of endocytosed ligands from receptors. The ligands are then ferried to the lysosomes by vesicular transport for degradation. The receptors recycle to the surface of the cell for multiple rounds of receptor binding, and internalization through clathrin-coated vesicles. Alkalinization of the acidic endocytic compartments interferes both with receptor recycling as well as with degradation of ligands in the lysosomes by the mechanisms explained above. Disruption of proton gradients also interferes with cellular secretory pathways (Strous et al., 1983). The Golgi apparatus of hepatocytes and a variety of secretory granules have been demonstrated to be acidic (Glickman et al., 1983; Schwartz et al., 1985). Maintenance of proton gradients across the mitochondrial membrane is essential for maintenance of oxidative phosphorylation. Loss of this gradient is an important feature of cell injury.

The pH of hepatocyte cytoplasm is maintained by a Na^+-H^+ exchange which is primarily localized in the basolateral domain of the hepatocyte plasma membrane. A Cl^--HCO_3^- exchange may also participate in pH regulation as well as in bile formation in the hepatocyte (Scharschmidt and Van Dyke, 1987). The activity of the Na^+-H^+ exchange is primarily regulated by levels of cytoplasmic pH. However, levels of cytosolic calcium and protein kinase C activity may also participate in regulation (Mahnensmith and Aronson, 1985). This exchanger, therefore, protects the cells from H^+ loads generated from intermediary metabolism. Activation of the exchanger can maintain anerobic glycolysis since phosphofructokinase, the rate-limiting enzyme in glycolysis, is inhibited at low pH. In some cell types, the exchanger also functions in cell volume regulation, particularly in response to exposure to hypertonic solutions. The activity of the exchanger can also be affected by growth factors. The consequent changes in cellular pH are thought to play a role in the initiation of proliferative responses (Busa and Nuccitelli, 1984). Decrease in extracellular pH appears to delay cell death in isolated hepatocytes, probably by enhancing calcium efflux from the cells (Pentilla and Trump, 1974). However, the combination of low pH and hypoxia has synergistic effects in the killing of cultured normal and neoplastic cells (Rotin et al., 1986).

Cell Swelling

In addition to altered cell physiology, loss of ion regulation also causes an influx of water and other solutes into the cell. The cell undergoes osmotic swelling. In conditions where blood flow is interrupted, the intracellular environment may become hypertonic due to the accumulation of small-molecular-weight metabolites. Upon blood reflow, osmotic swelling may occur. Epithelial cells can respond to osmotic stress by increasing the number of ionic channels or transport proteins in the plasma membrane. This increased capacity for ion transport is associated with fusion of cytoplasmic vesicles and tubular structures with the plasma membrane. Decreases in ionic transport correlated with decreased surface areas of plasma membrane are also known to occur. In so far as it is possible, cells may also alter their shape, and the organization of the plasma membrane to adapt to severe increases in cell volume. The movement of water across biological membranes is sufficiently rapid that it does not play any rate-limiting role in the development of changes in cellular volume. Even in polarized cells with water-impermeable domains of plasma membrane, it is primarily the solute load that determines cellular volume. Cells swell rapidly in response to an increased osmotic load until the influx of water abolishes the differences in colloid osmotic pressure across the plasma membrane. Loss of intracellular water activates the uptake of solute from the extracellular space. Rapid adjustments occur to restore cellular volume. In addition, the intracellular generation of low-molecular-weight substances tends to restore osmolality to normal and normalize cell volume.

Shifts in ions such as Na^+, K^+, Ca^{2+}, and small molecules such as amino acids and other organic components are primarily responsible for cell swelling. The volume of cells is regulated by the plasma membrane by actively extruding sodium to balance the colloid osmotic pressure of large molecules. At the ultrastructural level, swelling is manifested by increased electron lucency of the cytosol. As swelling continues and the cell volume increases, the plasma membrane begins to lose its surface microvilli.

MECHANISMS OF CELL DEATH: A FINAL COMMON PATHWAY?

Despite differences in function and in susceptibility to injury, it is remarkable that at the ultrastructural level, quite diverse types of insults, such as ischemic, toxic, immunologic, and infectious, all induce similar ultrastructural alterations in cells. At the biochemical level, these insults induce more distinct alterations by virtue of modification of specific biochemical pathways or modification of a specific organelle. However, with time, the initial biochemical alteration inevitably effects integrated diverse cellular functions. As a consequence, despite differences in the initial insult, injured cells manifest common structural and functional alterations. These common functional alterations reflect a loss of homeostasis. It has been hypothesized that loss of ion equilibrium, particularly calcium, may represent a common final pathway towards cell death (Figure 18).

Cells that pass the point of no return are irreversibly injured, i.e., they are dead. Dead cells lose structural integrity due to osmotic and other physical factors and to the degradative action of intracellular and extracellular enzymes (the process of necrosis). Factors such as blood flow (which allows the influx of inflammatory cells), adjacent tissue cells, and extracellular matrices influence the rapidity of necrosis. At the light microscopic level, dead cells are morphologically indistinguishable from living cells until the process of necrosis begins to induce changes. However, at the ultrastructural level readily identifiable alterations permit the identification of irreversibly injured cells several hours before the necrotic process begins and even before histologic alterations become apparent in the tissue by light microscopy.

In this review, we have stressed that maintenance of structural and functional integrity of the plasma membrane and of specialized intracellular membrane compartments is critical to maintain ionic equilibrium. The integrity of intracellular membranes is also essential to separate the subcellular compartments with their specialized functions. These membrane-bound microenvironments facilitate enzymatic reactions for biosynthesis and oxidative phosphorylation, regulate availability of ions and substrates, and segregate lytic reactions. The ability to generate ATP needed to drive chemical reactions and maintain chemical equilibrium is another key function whose loss may lead to irreversible injury. The earliest and most important consequence of dysfunction of energy metabolism is thought to be the loss of homeostasis of ions such as Na^+, Ca^{2+}, K^+, and H^+. Loss of ion regulation, energy loss, and altered membrane permeability enhance each other in a deleterious way. Thus, after a specific initiating event, cellular injury may evolve into a common final pathway. This concept is illustrated in Figure 18.

Other cellular functions are also affected by acute cell injury but are less likely to directly cause cell death because, in many instances, they are reversible. These functions are (1) cytoskeletal organization required to maintain cell shape, integrity of plasma membrane, and locomotion of cell and subcellular organelles, (2) transcription of DNA, translation of mRNA transcripts, and translocation and export of macromolecules, (3) intermediary metabolism of fuels, (4) intercellular communication and adherence to other cells and to the extracellular matrix, and (5) uptake, processing and degradation of macromolecules derived from outside or inside cells.

ACKNOWLEDGMENTS

We wish to thank Dr. Thomas Jones for the critical review of this manuscript. We gratefully acknowledge the expert assistance of Messrs. Seung Chang, Robert Pendergrass, and Perry Comegys and Mrs. Janet Folmer of the Electron Microscopy Laboratories. We also thank Mr. Wayne J. Ivusich, Ms. Millie Michalisko, and Ms. Diane B. Dix of the Publications Division for their assistance.

REFERENCES

Ahlberg J, Yucel T, Ericksson L, and Glaumann H: Characterization of the proteolytic compartment in rat hepatocyte nodules, *Virchows Arch B* 1987; 53:79–88.

Almeida H, Magalhães M, and Magalhães MC: Nucleolar alterations induced by 4-aminopyrazolo (3,4-d)-pyrimidine in adrenal cortex and liver cells of rat. *Cell Tissue Res* 1987; 248:231–234.

Amenta JS, Brocher SC, Mehta J, Manjunath D, and Baccino FM: Evidence for a special relationship between proteolysis and single cell necrosis. *Toxicol Pathol* 1986; 14:335–341.

Anders MW, Lash L, Dekant W, Elfarra AA, and Dohn DR: Biosynthesis and biotransformation of glutathione *S*-conjugates to toxic metabolites. *CRC Crit Rev Toxicol* 1988; 18:311–341.

Anundi I, King J, Owen DA, Schneider H, Lemasters JJ and Thurman RG: Fructose prevents hypoxic cell death in the liver. *Am J Physiol* 1987; 253:G390–G396.

Arnold PE, Lumlertgul D, Burke TJ, and Schrier RW: *In vitro* versus *in vivo* mitochondrial calcium loading in ischemic acute renal failure. *Am J Physiol* 1985; 248:F845–F850.

Aw TY, Andersson BS, and Jones DP: Mitochondrial transmembrane ion distribution during anoxia. *Am J Physiol* 1987a; 252:C356–C361.

Aw TY, Andersson BS, and Jones DP: Suppression of mitochondrial respiratory function after short-term anoxia. *Am J Physiol* 1987b; 252:C362–C368.

Bashford CL, Alder GM, Menestrina G, Micklem KJ, Murphy JJ, and Pasternak CA: Membrane damage by hemolytic viruses, toxins, complement and other cytotoxic agents. A common mechanism blocked by divalent cations. *J Biol Chem* 1986; 261:9300–9308.

Beatrice MC, Stiers DL, and Pfeiffer DR: Increased permeability of mitochondria during Ca^{2+} release induced by *t*-butyl hydroperoxide or oxalacetate. The effect of ruthenium red. *J Biol Chem* 1982; 257:7161–7171.

Berger ML, Reynolds RC, and Combes B: Carbon tetrachloride-induced morphologic alterations in isolated rat hepatocytes. *Exp Mol Pathol* 1987; 46:245–257.

Bernacchi A, Myers RAM, Trump BF, and Marzella L: Protection of hepatocytes with hyperbaric oxygen against carbon tetrachloride-induced injury. *Toxicol Pathol* 1984; 12:315–323.

Brierly GP, Jurkowitz MS, Farooqui T, and Jung DW: K^+/H^+ antiport in heart mitochondria. *J Biol Chem* 1984; 259:14672–14678.

Broekemeier KM, Schmid PC, Schmid HH, and Pfeiffer DR: Effects of phospholipase A_2 inhibitors on ruthenium red-induced Ca^{2+} release from mitochondria. *J Biol Chem* 1985; 260:105–113.

Bursch W, Lauer B, Timmermann-Trosiener I, Barthel G, Schuppler J, and Schulte-Hermann R: Controlled death (apoptosis) of normal and putative preneoplastic cells in rat liver following withdrawal of tumor promoters. *Carcinogenesis* 1984; 5:453–458.

Busa WB and Nuccitelli R: Metabolic regulation via intracellular pH. *Am J Physiol* 1984; 246:R409–R438.

Candipan RC and Sjöstrand FS: Water movement from intracristal spaces in isolated liver mitochondria. *J Ultrastruc Res* 1984a; 89:249–260.

Candipan RC and Sjöstrand FS: Freeze-fracture analysis of isolated liver mitochondria in different metabolic states. *J Ultrastruct Res* 1984b; 89:274–280.

Candipan RC and Sjöstrand FS: An analysis of the contribution of the preparatory technique to the appearance of condensed and orthodox conformations of liver mitochondria. *J Ultrastruct Res* 1984c; 89:281–294.

Carafoli E: Membrane transport in cellular homeostasis of calcium. *J Cardiovasc Pharmacol* 1986; 8:S3–S6.

Cardelli JA, Golumbeski GS, and Dimond RL: Lysosomal enzymes in *Dictyostelium discoideum* are transported to lysosomes at distinctly different rates. *J Cell Biol* 1986; 102:1264–1270.

Carney D, Hammer C, and Shin ML: Elimination of terminal complement complexes in the plasma membrane of nuclear cells: influence of extracellular Ca^{2+} and association with cellular Ca^{2+}. *J Immunol* 1986; 137:263–270.

Clement B, Grimaud JA, Campion J-P, Deugnier Y, and Guillouzo A: Cell types involved in collagen and fibronectin production in normal and fibrotic human liver. *Hepatology* 1986; 6:225–234.

Cole KE, Hsu IC, and Trump BF: Comparative ultrastructural effects of aflatoxin B_1 on mouse, rat and human hepatocytes in primary culture. *Cancer Res* 1986; 46:1290–1296.

Columbano A, Ledda-Columbano GM, Coni PP, Faa G, Liguori C, Santa Cruz G, and Pani P: Occurrence of cell death (apoptosis) during the involution of liver hyperplasia. *Lab Invest* 1985; 52:670–675.

Comporti M: Lipid peroxidation and cellular damage in toxic liver injury. *Lab Invest* 1985; 53:599–623.

Dake MD, Madison JM, Montgomery CK, Shellito JE, Hinchcliffe WA, Winkler ML, and Bainton DF: Electron microscopic demonstration of lysosomal inclusion bodies in lung, liver, lymph nodes, and blood leukocytes of patients with amiodarone pulmonary toxicity. *Am J Med* 1985; 78:506–512.

Dawson AP and Irvine RF: Inositol 1,4,5-triphosphate-promoted Ca^{2+} release form microsomal fractions of rat liver. *Biochem Biophys Res Comm* 1984; 120:858–864.

Dean RT, Hylton W, and Allison AC: Lysosomal enzyme secretion by macrophages during intracellular storage of particles. *Biochim Biophys Acta* 1979; 584:57–65.

DePriester W, Van Manen R, and Knook DL: Lysosomal activity in the aging rat liver. II. Morphometry of acid phosphatase positive dense bodies. *Med Ageing Dev* 1984; 26:205–216.

Dufton MJ, Hider RC and Cherry RJ: The influence of melittin on the rotation of band 3 protein in human erythrocyte membrane. *Eur Biophys J* 1984; 11:17–24.

Eveloff JL and Warnock DG: Activation of ion transport systems during cell volume regulation. *Am J Physiol* 1987; 252:F1–F10.

Farber JJ and Young EE: Accelerated phospholipid degradation in anoxic rat hepatocytes. *Arch Biochem Biophys* 1981; 211:312–320.

French SW, Benson NC, and Sun PS: Centrilobular liver necrosis induced by hypoxia in chronic ethanol-fed rats. *Hepatology* 1984; 4:912–917.

Garlid KD and Beavis AD: Swelling and contraction of the mitochondrial matrix. II. Quantitative application of the light scattering technique to solute transport across the inner membrane. *J Biol Chem* 1985; 260:13434–13441.

Glaumann H and Marzella L: Degradation of membrane components by Kupffer cell lysosomes. *Lab Invest* 1981; 45:479–490.
Glew RH, Basu A, Prence EM, and Remaley AT: Lysosomal storage diseases. *Lab Invest* 1985; 53:250–269.
Glickman J, Croen K, Kelly S, and Al-Awqati Q: Golgi membranes contain an electrogenic H^+ pump in parallel to a chloride conductance. *J Cell Biol* 1983; 97:1303–1308.
Guengerich FP and Liebler DC: Enzymatic activation of chemicals to toxic metabolites. *CRC Crit Rev Toxicol* 1985; 14:259–307.
Halestrap AP and Dunlop JL: Intramitochondrial regulation of fatty acid β-oxidation occurs between flavoprotein and ubiquinone. A role for changes in the matrix volume. *Biochem J* 1986a; 239:559–565.
Halestrap AP, Quinlan PT, Whipps DE, and Armston AE: Regulation of the mitochondrial matrix volume *in vivo* and *in vitro*. The role of calcium. *Biochem J* 1986b; 236:779–787.
Helman RG, Adams LG, Pierce KR, Bridges CH, and Bailey EM: The role of lysosomes in the pathogenesis of copper-induced hepatotoxicity: morphological studies. *J Comp Pathol* 1985; 95:25–35.
Henell F and Glaumann H: Effect of leupeptin on the autophagic vacuole system of rat hepatocytes. Correlation between ultrastructure and degradation. *Lab Invest* 1984; 51:46–56.
Higgins JA and Hutson JL: The role of Golgi and endoplasmic reticulum in the synthesis and assembly of lipoprotein lipids in hepatocytes. *J Lipid Res* 1984; 25:1295–1305.
Hostetler KY and Hall LB: Inhibition of kidney lysosomal phospholipases A and C by aminoglycoside antibiotics: possible mechanism of aminoglycoside toxicity. *Proc Natl Acad Sci USA* 1982; 79:1663–1667.
Hougaard DM, Bolund L, Fujiwara K, and Larsson L-I: Endogenous polyamines are intimately associated with highly condensed chromatin *in vivo*. A fluorescence, cytochemical and immunocytochemical study of spermine and spermidine during the cell cycle and in reactivated nuclei. *Eur J Cell Biol* 1987; 44:151–155.
Hulcrantz R, Ericsson JLE, and Hirth T: Levels of malondialdehyde production in rat liver following loading and unloading with iron. *Virchows Archiv [Cell Pathol]* 1984; 45:139–146.
de la Iglesia FA, Feuer G, Takada A, and Matsuda Y: Morphologic studies on secondary phospholipidosis in human liver. *Lab Invest* 1974; 30:539–549.
Jahn W: Electron microscopic characterization of intermediate filaments in rat liver parenchymal cells. *Eur J Cell Biol* 1982; 26:259–264.
Jennings RB, Ganote CE, and Reimer KA: Ischemic tissue injury. *Am J Pathol* 1975; 81:179–198.
Jones TW, Gerdes RG, Ormstad K, and Orrenius S: The formation of both a mono- and bis-substituted glutathione conjugate of hexachlorobutadiene by isolated hepatocytes and following *in vivo* administration to the rat. *Chem Biol Interact* 1985; 56:251–257.
Jung DW and Brierly GP: The permeability of uncoupled heart mitochondria to potassium ion. *J Biol Chem* 1984; 259:6904–6911.

Kaplan MR and Simoni RD: Transport of cholesterol from the endoplasmic reticulum to the plasma membrane. *J Cell Biol* 1985; 101:446–453.

Laiho KU and Trump BF: Studies on the pathogenesis of cell injury. Effects of inhibitors of metabolism and membrane function on the mitochondria of Ehrlich ascites tumor cells. *Lab Invest* 1975; 32:163–182.

Lee HK, Meyers RAM, and Marzella L: Stimulation of autophagic protein degradation by nutrient deprivation in a differentiated murine teratocarcinoma (F9 12–1a) cell line. *Exptl Mol Pathol* 1989; 50:139–146.

Lemasters JJ, Stemkowski CJ, Sungchul JI and Thurman RG: Cell surface changes and enzyme release during hypoxia and reoxygenation in the isolated, perfused rat liver. *J Cell Biol* 1983; 97:778–786.

Lemasters JJ, DiGiuseppi J, Nieminen AL, and Herman B: Blebbing, free Ca^{2+} and mitochondrial membrane potential preceding cell death in hepatocytes. *Nature* 1987; 325:78–81.

Li W, Aurora TS, Haines TH, and Cummins ZH: Elasticity of synthetic phospholipid vesicles and submitochondrial particles during osmotic swelling. *Biochemistry* 1986; 25:8220–8229.

Loegering DJ and Blumenstock FA: Depressing hepatic macrophage complement receptor function causes increased susceptibility to endotoxemia and infection. *Infect Immun* 1985; 47:659–664.

Lowrey K, Glende EA, and Recknagel RO: Destruction of liver microsomal calcium pump activity by carbon tetrachloride and bromotrichloro-methane. *Biochem Pharmacol* 1981; 30:135–140.

McConkey DJ, Hartzell P, Duddy SK, Håkansson H, and Orrenius S: 2,3,7,8-Tetrachlorodibenzo-*p*-dioxin kills immature thymocytes by Ca^{2+}-mediated endonuclease activation. *Science* 1988; 242:256–259.

Mahnensmith RL and Aronson PS: The plasma membrane sodium-hydrogen exchanger and its role in physiological and pathophysiological processes. *Circ Res* 1985; 56:773–788.

Mak IT and Weglicki WB: Characterization of iron-mediated peroxidative injury in isolated hepatic lysosomes. *J Clin Invest* 1985; 75:58–63.

Marotto ME, Thurman RG, and Lemasters JJ: Early midzonal cell death during low-flow hypoxia in the isolated, perfused rat liver: protection by allopurinol. *Hepatology* 1988; 8:585–590.

Marzella L and Glaumann H: Effects of *in vivo* liver ischemia on microsomes and lysosomes. *Virch Arch (Cell Pathol)* 1981; 36:1–25.

Marzella L and Glaumann H: Biogenesis translocation and function of lysosomal enzymes. *Int Rev Exp Pathol* 1983; 25:239–278.

Marzella L and Glaumann H: Autophagy, microautophagy and crinophagy as mechanisms for protein degradation, in Glaumann H and Ballard FJ (eds): *Lysosomes: Their Role in Protein Breakdown.* London, Academic Press, 1987, pp 319–370.

Marzella L, Muhvich K, and Myers RAM: Effects of hyperoxia on liver necrosis induced by hepatotoxins. *Virchows Archiv (Cell Pathol)* 1986; 51:497–507.

Maxfield FR: Weak bases and ionophores rapidly and reversibly raise the pH of endocytic vessels in cultured mouse fibroblasts. *J Cell Biol* 1982; 95:676–681.

Mills PR, Meier PJ, Smith DJ, Ballatori N, Boyer JL, and Gordon ER: The effect of changes in the fluid state of rat liver plasma membrane on the transport of taurocholate. *Hepatology* 1987; 7:61–66.

Miseta A, Ludani A, Kellermayer M, Trombitas K, and Hazlewood CF: Hydrogen ion dependent reversible fibrillarization and hydration of chromatin in divalent cation-depleted isolated rat liver nuclei. *Physiol Chem Phys Med NMR* 1987; 19:3–8.

Mortimore GE, Hutson NJ, and Surmacz CA: Quantitative correlation between proteolysis and macro- and microautophagy in mouse hepatocytes during starvation and refeeding. *Proc Natl Acad Sci USA* 1983; 80:2179–2183.

Mukherjee AB and Weinstein ME: Culture media variation as related to *in vitro* aging of human fibroblasts. I. Effects on population doubling, nuclear volume and nuclear morphology. *Mech Aging Dev* 1986; 37:55–67.

Myagkaya GL, van Veen HA, and James J: Ultrastructural changes in rat liver during Euro-Collins storage, compared with hypothermic *in vitro* ischemia. *Virchows Arch B* 1987; 531:176–182.

Naito M, Takahashi K, and Hiroshi H: An ultrastructural and experimental study on the development of tubular structures in the lysosomes of Gaucher cells. *Lab Invest* 1988; 58:590–598.

Newport JW and Forbes DJ: The nucleus: structure, function and dynamics. *Annu Rev Biochem* 1987; 56:535–565.

Nichols BA: Uptake and digestion of horseradish peroxidase in rabbit alveolar macrophages. Formation of a pathway connecting lysosomes to the cell surface. *Lab Invest* 1982; 47:235–246.

Nicholls D and Akerman K: Mitochondrial calcium transport. *Biochim Biophys Acta* 1982; 638:57–88.

Nishimura Y, Furuno K, and Kato K: Biosynthesis and processing of lysosomal cathepsin L in primary cultures of rat hepatocytes. *Arch Biochem Biophys* 1988; 263:107–116.

Ogawa K, Suzuki J-I, Narasaki M, and Mori M: Healing of focal injury in rat liver. *Am J Pathol* 1985; 119:158–167.

Ohkuma S: The lysosomal proton pump and its effect on protein breakdown, in Glaumann H and Ballard FJ (eds): *Lysosomes: Their Role in Protein Breakdown.* London, Academic Press, 1987, pp 115–148.

Ohta M, Marceau N, and French SW: Pathologic changes in cytokeratin pericanalicular sheath in experimental cholestasis and alcoholic fatty liver. *Lab Invest* 1988; 59:60–74.

Okayasu T, Curtis MT, and Farber JL: Cytochalasin delays but does not prevent cell death from anoxia. *Am J Pathol* 1984; 117:163–166.

Okayasu T, Curtis MT, and Farber JL: Structural alterations of the inner mitochondrial membrane in ischemic liver cell injury. *Arch Biochem Biophys* 1985; 236:638–645.

Pentilla A and Trump BF: Extracellular acidosis protects Ehrlich ascites tumor cells and rat renal cortex against anoxic injury. *Science* 1974; 185:277–278.

Peters TJ and Seymore CA: Acid hydrolase activities and lysosomal integrity in liver biopsies from patients with iron overload. *Clin Sci Mol Med* 1976; 50:75–78.

Pfeifer U: Inhibition by insulin of the formation of autophagic vacuoles in rat liver. *J Cell Biol* 1978; 78:152–167.

Poli G, Chiarpotto E, Albano E, Cottalasso D, Nanni G, Marinari VM, Bassi AM, and Dianzani MV: Carbon tetrachloride-induced inhibition of hepatocyte lipoprotein secretion: functional impairment of Golgi apparatus in early phases of such injury. *Life Sci* 1985; 36:533–539.

Poole B and Ohkuma S: Effect of weak bases on the intralysosomal pH in mouse peritoneal macrophages. *J Cell Biol* 1981; 90:665–669.

Poso AR, Wert JJ, and Mortimore GE: Multifunctional control by amino acids of deprivation-induced proteolysis in liver: role of leucine. *J Biol Chem* 1982; 257:12114–12120.

Prentki M, Chaponnie C, Jeanrenaud B, and Gabbiani G: Actin microfilaments, cell shape and secretory processes in isolated rat hepatocytes. Effect of phalloidin and cytochalasin D. *J Cell Biol* 1979; 81:592–607.

Reddy JK, Lalwani ND, Qureshi SA, Reddy MK, and Moehle CM: Induction of hepatic peroxisome proliferation in non-rodent species, including primates. *Am J Pathol* 1984; 114:117–183.

Regec AL, Trifillis AL, and Trump BF: The effect of gentamicin on human renal proximal tubular cells. *Toxicol Pathol* 1986; 14:238–241.

Reynolds ES, Moslen MT, Szabo S, Jaeger RJ, and Murphy SD: Hepatotoxicity of vinyl chloride and 1,1-dichloroethylene. Role of mixed function oxidase system. *Am J Pathol* 1975; 81:219–236.

Rizzuto R, Pitton G, and Azzone FG: Effect of Ca^{2+}, peroxides, SH reagents, phosphate and aging on the permeability of mitochondrial membranes. *Eur J Biochem* 1987; 162:239–249.

Rozga J, Jeppsson B, and Bengmark S: Portal branch ligation in the rat. Re-evaluation of a model. *Am J Pathol* 1966; 125:300–308.

Rotin D, Robinson B, and Tannock IF: Influence of hypoxia and an acidic environment on the metabolism and viability of cultured cells: potential implications for cell death in tumors. *Cancer Res* 1986; 46:2821–2826.

Sato K, Ikeda T, Katami K, and Ogawa H: Preparation of monoclonal antibody to hepatocellular membranes and its application to induction of liver cell membrane damage. *Acta Pathol Jpn* 1985; 35:1375–1383.

Sato T, Tanaka J, Kono Y, Jones RT, Cowley RA, and Trump BF: Hepatic cellular injury following lethal *Escherichia coli* bacteremia in rats. *Lab Invest* 1982; 47:304–310.

Schachter D: Fluidity and function of hepatocyte plasma membranes. *Hepatology* 1984; 4:140–151.

Schanne FA and Moore L: Liver plasma membrane calcium transport. Evidence for a Na^{+}-dependent Ca^{2+} flux. *J Biol Chem* 1986; 261:9886–9889.

Scharschmidt BF and Van Dyke RW: Proton transport by hepatocyte organelles and isolated membrane vesicles. *Annu Rev Physiol* 1987; 49:69–85.

Scheer U and Rose KM: Localizations of RNA polymerase I in interphase cells and mitotic chromosomes by light- and electron microscopic immunocytochemistry. *Proc Natl Acad Sci USA* 1984; 81:1431–1435.

Schmuckler DL, Vessey DA, Wang RK, James JL, and Maloney A: Age-dependent alterations in the physicochemical properties of rat liver microsomes. *Mech Aging Dev* 1984; 27:207–217.

Schwartz AL, Strous GJAM, Slot JW, and Geuze HG: Immunoelectron microscopic localization of acidic intracellular compartments in hepatoma cells. *EMBO J* 1985; 4:899–904.

Schwerzmann K, Cruz-Orive LM, Eggman R, Sänger A, and Weibel ER: Molecular architecture of the inner membrane of mitochondria from rat liver: a combined biochemical and stereological study. *J Cell Biol* 1986; 102:97–103.

Seglen PO, Gordon PB, and Poli A: Amino acid inhibition of the autophagic/lysosomal pathway of protein degradation in isolated rat hepatocytes. *Biochim Biophys Acta* 1980; 630:103–118.

Seglen PO, Schwarze PE, and Saeter G: Changes in cellular ploidy and autophagic responsiveness during rat liver carcinogenesis. *Toxicol Pathol* 1986; 14:342–348.

Shiraki H and Guengerich FP: Turnover of membrane proteins: kinetics of induction and degradation of seven forms of rat liver microsomal cytochrome P-450, NADPH-cytochrome P-450 reductase and epoxide hydrolase. *Arch Biochem Biophys* 1984; 235:86–96.

Silver EH and Szabo S: Role of lipid peroxidation in tissue injury after hepatic ischemia. *Exp Mol Pathol* 1983; 38:69–76.

Sleight RG and Pagano RE: Rapid appearance of newly synthesized phosphatidyl ethanolamine at the plasma membrane. *J Biol Chem* 1983; 258:9050–9058.

Sleyster ECh and Knook DL: Relation between localization and function of rat liver Kupffer cells. *Lab Invest* 1982; 47:484–490.

Smith MW, Ambudkar IS, Phelps PC, Regec AL, and Trump BF: $HgCl_2$-induced changes in cytosolic Ca^{2+} of cultured rabbit renal proximal tubular cells. *Biochim Biophys Acta* 1987; 931:130–142.

Smith RA and Ord MJ: Mitochondrial form and function relationships *in vivo:* their potential in toxicology and pathology. *Intern Rev Cytol* 1983; 83:63–134.

Somlyo AP, Bond M, and Somlyo AV: Calcium content of mitochondria and endoplasmic reticulum in liver frozen rapidly *in vivo. Nature* 1985; 314:622–625.

Starke PE, Gilbertson JD, and Farber JL: Lysosomal origin of the ferric iron required for cell killing by hydrogen peroxide. *Biochem Biophys Res Commun* 1985; 133:371–379.

Strous GJAM, Willemsen R, van Kerkhof P, Slot JW, Geuze HJ, and Lodish HF: Vesicular stomatitis virus glycoprotein, albumin and transferrin are transported to the cell surface via the same Golgi vesicles. *J Cell Biol* 1983; 97:1815–1822.

Tayek JA, Blackburn GL, and Bistrian RR: Alterations in whole body, muscle, liver and tumor tissue protein synthesis and degradation in Novikoff hepatoma and Yoshida sarcoma tumor growth studied *in vivo. Cancer Res* 1988; 48:1554–1558.

Trump BF, Berezesky IK: Cell injury, ion regulation, and tumor promotion, in Butterworth BE and Slaga TJ (eds): *Banbury Report 25: Nongenotoxic Mechanisms in Carcinogenesis.* New York, Cold Spring Harbor Laboratory, 1987, pp 69–82.

Trump BF, McDowell EM, and Arstila AV: Cellular reaction to injury in LaVia MF and Hall RB Jr (eds): *Principles of Pathology.* New Yorki, Oxford University Press, 1980, pp 20–111.

Van Den Bosch H: Intracellular phospholipases A. *Biochem Biophys Acta* 1980; 604:191–246.

Van Dyke RW, Scharschmidt BF, and Steer CJ: ATP-dependent proton transport by isolated brain clathrin-coated vesicles. Role of clathrin and other determinants of acidification. *Biochim Biophys Acta* 1985; 812:423–436.

Wachtler F, Popp W, and Schwarzacher HG: Structural changes in nucleoli during inhibition of protein- and RNA-biosynthesis. *Cell Tissue Res* 1987; 247:583–589.

Wakabayashi T, Horiuchi M, Sakaguchi M, Omata H, and Misawa K: Induction of megamitochondria in the mouse and rat livers by hydrazine. *Exp Mol Pathol* 1983; 39:139–153.

Walker RM, McEllijott TF, Massey TE, and Racz WJ: Ultrastructural effects of acetaminophen in isolated mouse hepatocytes. *Exp Mol Pathol* 1983; 39:165–175.

Wyke AW, Impraim CC, Knutton S, and Pasternak CA: Components involved in virally mediated membrane fusion and permeability changes. *Biochem J* 1980; 190:625–638.

Wyllie AH: What is apoptosis? *Histopathology* 1986; 10:995–998.

Xu ZL, Bucana CD, and Fidler IJ: In vitro activation of murine Kupffer cells by lymphokines or endotoxins to lyse syngeneic tumor cells. *Am J Pathol* 1984; 117:372–379.

Yu QC and Marzella L: Pathobiology of lysosomes, in Trump BF and Mergner WJ (eds): *Mechanisms of Cell Damage and Cell Death.* Boca Raton, CRC Press, 1990, pp 143–178.

Yu QC and Marzella L: Response of autophagic protein degradation to physiologic and pathologic stimuli in rat hepatocyte monolayers. *Lab Invest* 1988; 58:643–652.

Chapter

4

Xenobiotic Biotransformation

Charles Lindamood, III, Ph.D., D.A.B.T.
Kettering-Meyer Laboratory
Southern Research Institute
Birmingham, Alabama

INTRODUCTION

Liver is the primary site of biotransformation, followed by lung, kidney, intestine, intestinal flora, skin, and endocrine organs, such as the gonads, placenta, and adrenal. Liver, which receives 25% of the cardiac output, has a primary function of receipt and processing of chemicals absorbed in the gastrointestinal tract. The liver, therefore, has a major role in biotransformation. Such capabilities are believed to have evolved in lung, intestine, and skin since these are portal routes of entry. Capabilities in kidney and endocrine organs apparently evolved to biotransform xenobiotics into cellular sources of fatty acids and steroids, respectively. Most phase I enzymes are located in the endoplasmic reticulum, a site where lipophilic xenobiotics partition.

Biotransformation plays a key role in toxicity and carcinogenicity. It is not a "detoxification" process, but a mechanism for converting poorly excretable lipophilic xenobiotics to readily excretable water-soluble compounds. Many electrophilic, reactive species are generated during the biotransformation process. These electrophiles are chemically reactive with the basic cellular constituents: protein, RNA, and DNA. These chemical interactions can lead to disruption of normal cellular function and result in toxicity and carcinogenicity. The occurrence, degree, and extent of toxicity in a particular cell or organ depends on several factors. Of prime importance is whether the biotransformation process produces a compound more or less toxic than the parent compound and the relative activities of enzymes generating the electrophiles, biotransforming the electrophilic intermediates to less reactive compounds, and repairing the damage to cellular constituents. In turn, the relative activities of the enzymes of these competing pathways are influenced by factors such as genetics, i.e., species and strain (Kato, 1979; Walker, 1980), age (Kitani, 1984), endocrine status, including sex and pregnancy (Gustafsson et al., 1983; Stock, 1984), disease states and diet (Guengerich, 1984; Bidlack et al., 1986). Dose, route of administration, pharmacokinetic considerations (Wilkinson and Shand, 1975), and diurnal and seasonal variations (Walker, 1980) also influence the relative activities of these pathways. Concurrent exposure to inhibitors (Testa and Jenner, 1981; Netter, 1982), stimulators (Guengerich, 1984), and inducers (Conney, 1986) of xenobiotic biotransformation are major influencing factors on the relative activities of the competing pathways for xenobiotic biotransformation. Since biotransformation is a major function of the liver, this organ is, in many instances, one of the major targets of toxic or carcinogenic electrophiles generated by biotransformation of xenobiotics.

GENERAL PRINCIPLES

Research monographs devoted to the general principles of xenobiotic biotransformation include those of Jakoby (1980), Jakoby et al. (1982), Caldwell and Paulson (1984), and Gorrod et al. (1988). The individual chapters within these monographs address the complete aspects of the enzymology of xenobiotic biotransformation as well as the multiple pathways for biotransformation of specific chemical classes and functional groups. A detailed discussion of the role of prostaglandin endoperoxide synthetase and peroxidases in xenobiotic biotransformation and bioactivation is considered outside the scope of this chapter since these enzymes act predominately at extrahepatic sites [see review by Eling et al. (1983) and Larson et al. in Gorrod et al. (1988)].

Phase I Biotransformation

Quantitatively, 90% of the phase I oxidation of xenobiotics is catalyzed by the cytochrome P450 monooxygenase enzyme system. The remaining 10% of the phase I oxidations of xenobiotics can be accounted for by the flavin-containing monooxygenase and the molybdenum hydroxylases.

Cytochrome P450 Monooxygenase

The enzyme system playing the predominant role in phase I biotransformation is the class of enzymes known as cytochrome P450 monooxygenases [P450; see reviews by Whitlock (1986), Wolf (1986), Guengerich (1987),

Nebert et al. (1987), Nebert and Gonzalez (1987), Guengerich (1988), Gonzalez (1989), Guengerich (1989), and Nebert et al. (1989)]. This system is believed to have evolved at an early time since its activity is detected in all types of organisms, including bacteria. [See Waterman and Estabrook (1983) for a review of P450 history.] Its initial function may have been to allow utilization of molecules in the environment by incorporation of oxygen as a first step in synthesis of fatty acids and steroids from alkane molecules and utilization of environmental chemicals as energy sources. The associated peroxidase activity may have evolved to deactivate reactive oxygens in the environment. Later activities developed most likely in response to toxic components in food sources and hydrophobic substances in the general environment, which, once absorbed, could not be readily eliminated. This enzyme system, discovered independently in the laboratories of Klingenberg and Garfinkel in 1958, was described as an hepatic microsomal binding pigment. Omura and Sato, in the period 1962 to 1964, named this pigment cytochrome P450 due to its biochemical properties resembling a cytochrome and its unusual spectral characteristics. The term P450 is derived from the absorption of light at 450 nm of the carbon monoxide adduct of the reduced hemeprotein. Based upon this property, *in vitro* assays for P450 utilize CO binding to the reduced form to measure its activity. The physiological role of P450 in steroid synthesis was described in 1964 and 1965 by Omura and Sato and Cooper and co-workers. Although it was not known at that time, its role in the biotransformation of drugs and xenobiotics was demonstrated in 1949 by Mueller and Miller and in 1955 by Brodie and co-workers.

All tissues, except striated muscle and erythrocytes, possess P450 activity. It is also found in several subcellular organelles, of which microsomal and mitochrondrial activities are most prominent. Microsomal P450 is a two-enzyme complex of NADPH-cytochrome P450 reductase and P450. The reductase functions to transfer reducing equivalents from electron donors to P450-substrate complexes and has preference for NADPH as the source of electrons. P450 contains heme and is a *b*-type cytochrome with the oxidized enzyme containing Fe^{3+} and the reduced enzyme Fe^{2+}. Cytochrome b_5 and cytochrome b_5 reductase are capable of participation in electron transfer in P450 reactions. For the mitochrondrial system, a flavoprotein reductase and an iron-sulfur moiety (adrenodoxin) that resembles the prokaryote P450 electron-transport chain is involved in electron transfer.

P450 plays important roles in both the synthesis and catabolism of molecules critical for cellular function [see Guengerich (1987) for individual references]. Synthetic reactions include conversion of cholesterol to bile acids (cholesterol 7-α-hydroxylation), steroids (various hydroxylation reactions), and vitamin D (vitamin D_3 25-hydroxylation), alkanes to fatty acids; and fatty acids to eicosanoids (arachidonic acid epoxidation, prostacyclin and thromboxane syntheses). Catabolic reactions include those for fatty acids (ω- and ω-1 hydroxylations, β-oxidations), steroids (various hydroxylations), and eicosanoids (ω-, ω-1, ω-2 hydroxylations).

Several general classes of xenobiotics are biotransformed by P450. Generic reactions include aliphatic hydroxylation at ω-1 carbons of straight-chain alkanes or alkyl functional groups, aromatic hydrocarbon hydroxylation at carbon-hydrogen bonds, alkene epoxidation at carbon-carbon double bonds, dealkylation via hydroxylation of carbon atoms alpha to nitrogen, oxygen, or sulfur atoms, deamination via hydroxylation of carbons alpha to primary amino groups, oxidation of sulfur- and nitrogen-containing groups by insertion of oxygen to a free-electron pair, and dehalogenation of alkyl halides by insertion of an oxygen atom at the carbon-halogen bonds. Reduction reactions are catalyzed by the P450 system, but are rare relative to oxidations. These readily occur only under conditions of low oxygen tension in which substrates compete with oxygen for electron transfer from NADPH-cytochrome P450 reductase. Reduction reactions are important biotransformation/bioactivation pathways for xenobiotics containing nitrogen functional groups such as nitro, azo, arene oxides, *N*-oxides, and alkyl halides, however. Other enzymes for catalysis of nitrogen group reduction include xanthine oxidase, aldehyde oxidase, and DT-diaphorase of liver cytosol and bacteria reductases of the intestinal microflora.

All of the oxidative reactions involve insertion of oxygen into a carbon - hydrogen bond, as represented by the following equation:

$$RH + O_2 + NADPH + H^+ \rightleftharpoons ROH + NADP^+ + H_2O_2$$

The basic reaction mechanism of microsomal P450 involves binding of a substrate to oxidized P450 (Fe^{3+}). NADPH provides reducing equivalents to NADPH P450 reductase, which catalyzes the reduction (Fe^{3+} to Fe^{2+}) of the substrate-oxidized P450 complex. The reduced complex then combines with molecular oxygen. (It is at this step that carbon monoxide competes with molecular oxygen for binding.) A second electron and two hydrogens are acquired for NADPH-cytochrome P450 reductase, and the reduced P450-substrate-oxygen ternary complex rearranges by an unknown mechanism with transfer of electrons to molecular oxygen. Subsequent products are the oxidized (hydroxylated) substrate, water and regenerated, oxidized P450.

To date, 34 distinct P450 proteins have been characterized in the rat and 16 distinct P450 proteins in humans. A proposed nomenclature for the P450 system is given in Table 4–1 (Nebert et al., 1989). Currently there are nine designated mammalian P450 gene families with 71 genes in total. Four are believed to have evolved to biotransform xenobiotics and five to metabolize endogenous compounds. The CYP1A gene or polycyclic aromatic hydrocarbon (PAH)-inducible family consists of one subfamily of two genes. The CYP1A1 gene is distinguished by induction by β-napthoflavone and the

TABLE 4–1.
P450 Genes and Their Products*

Locus Symbol	Protein Name	Trivial Name	Species and Source
CYP1A1	IA1	c	Rat
		P_1	Mouse
		P_1	Human
		form 6	Rabbit
		IA1	Trout
CYP1A2	IA2	d	Rat
		P_3	Mouse
		P_2	Mouse
		P_3	Human
		form 4	Human
		LM4	Rabbit
CYP2A1	IIA1	a1	Rat
CYP2A2	IIA2	a2	Rat
CYP2A3	IIA3	a3	Rat
		15α	Mouse
		P450(1)	Human
		IIA3	Human
CYP2B1	IIB1	b	Rat
CYP2B2	IIB2	e	Rat
CYP2B3	IIB3	IIB3	Rat
CYP2B4	IIB4	LM2	Rabbit
		p54	
		BO	
		p-450_1, b15	
		B1	
		b46	
CYP2B4P		(pseudogene)	Rabbit
CYP2B5	IIB5	b52	Rabbit
		B2	
		HP1	
CYP2B6	IIB6	LM2	Human
CYP2B7	IIB7		Human
CYP2B8	IIB8		Human
CYP2B9	IIB9	pf26	Mouse
CYP2B10	IIB10	pf3/46	Mouse
CYP2C1	IICV1	PBc1	Rabbit
CYP2C2	IIC2	PBc2, K	Rabbit
		pHP2	
CYP2C3	IIC3	PBc3	Rabbit
		3b	
CYP2C4	IIC4	1-88	Rabbit
		PBc4	
CYP2C5	IIC5	form 1	Rabbit

Continued

TABLE 4–1 (cont.).
P450 Genes and Their Products*

Locus Symbol	Protein Name	Trivial Name	Species and Source
CYP2C6	IIC6	PB1	Rat
		pTF2	
CYP2C6P	(pseudogene)		Rat
CYP2C7	IIC7	f	Rat
		pTF1	
CYP2C8	IIC8	form 1	Human
		IIC2	
		mp-12, mp-20	Human
CYP2C9	IIC9	IIC1	Human
		mp-4	
CYP2C10	IIC10	mp	Human
		mp-8	Human
CYP2C11	IIC11	h, M-1, 16α	Rat
CYP2C12	IIC12	i, 15β	Rat
CYP2C13	IIC13	g	Rat
CYP2C14	IIC14	pHP3	Rabbit
CYP2C15	IC15	b32-3	Rabbit
CYP2D1	IID1	db1	Rat
		CMFa	Rat
		db2	Rat
CYP2D2	IID2	CMF2	Rat
CYP2D3	IID3	db3	Rat
		CMF3	Rat
CYP2D4	IID4	db4	Rat
CYP2D5	IID5	db5	Rat
		CMF1b	Rat
CYP2D6	IID6	db1	Human
CYP2D7	IID7		Human
CYP2D8	IID8		Human
CYP2D9	IID9	16α	Mouse
CYP2D10	IID10	cb	Mouse
CYP2E1	IIE1	j	Human
			Rat
		3a	Rabbit
CYP2E2	IIE2	IIE2	Rabbit
CYP2F1	IIF1		Human
CYP2G1	IIG1	olf1	Rat
CYP2H12	IIH1	PB15	Chicken
CYP3A1	IIIA1	pcn1	Rat
CYP3A2	IIIA2	pcn2	Rat
CYP3A3	IIIA3	HLp	Human
CYP3A4	IIIA4	nf-25	Human
		pcn1	
		nf-10	

Continued

TABLE 4–1 (cont.).
P450 Genes and Their Products*

Locus Symbol	Protein Name	Trivial Name	Species and Source
CYP3A5	IIIA5	pcn3	Human
CYP3A6	IIIA6	3c	Rabbit
CYP4A1	IVA1	LAω 1	Rat
CYP4A2	IVA2	LAω 2	Rat
CYP4A3	IVA3	LAω 3	Rat
CYP4A4	IVA4	p-2	Rabbit
CYP4A5	IVA5	LAω 1	Rabbit
CYP4A6	IVA6	LAω 2	Rabbit
CYP4A7	IVA7	LAω 3	Rabbit
CYP4B1	IVB1	Lung P450	Human
		form 5	Rat
		form 5	Rabbit
CYP6A1	VIA1		House fly
CYP11A1	XIA1	scc	Human
			Cow
CYP11B1	XIB1	11β	Cow
CYP17	XVIIA1	17α	Cow
			Human
			Pig
			Rat
			Chicken
CYP19	XIXA1	arom	Human
			Chicken
CYP21A1	XXIA1	c21A	Mouse
		c21	Cow
		c21	Pig
CYP21A1P		(pseudogene c21A)	Human
CYP21A2	XXIA2	c21B	Human
CYP21A2P		(pseudogene c21B)	Mouse
CYP26	XXVIA1	26-ohp	Rabbit
CYP51	P450LI	14DM	*S. cerevisiae*
			C. tropicalis
CYP52	P450LII	alk	*C. tropicalis*
CYP101	P450CI	cam	*Ps. putida*
CYP102	P450CII	BM-3	*B. megaterium*

*Modified from Nebert et al. (1989). Used with the permission of Dr. Daniel C. Nebert.

CYP1A2 gene by induction with safrole or isosafrole. [*Note:* For individual references on specific inducers and enzyme activities of the P450 gene families discussed below, see Ryan et al. (1982); Guengerich et al. (1982a,b), Dannan et al. (1983), and reviews by Nebert and Gonzalez (1987), Guengerich (1987)]. Also note that all P450s have overlapping substrate activities, and activities attributed to a specific P450 are those for which it is most active relative to other P450s. The descriptions are those for the rat since the rat P450 has been most fully characterized. In some cases, activities associated with a P450 orthologue in one species are absent in another species. [Guengerich (1989) provides a description of the human P450.] Activity of this P450 gene is induced by polycyclic aromatic hydrocarbons (PAH; e.g., 3-methylcholanthrene, 3-MC), dioxins [2,3,7,8-tetrachlorodibenzo-*p*-dioxin (TCDD) is the prototype] and polyhalogenated biphenyls (Arochlor 1254, a mixture of polyhalogenated biphenyls, is the prototype), with these compounds inducing CYP1A1 to a greater extent than CYP1A2. CYP1A1 has high activity for hydroxylation and epoxidation of benzo(*a*)pyrene and other PAH substrates, whereas CYP1A2 has high activity for [*N*-hydroxylation of aromatic amines. Substrate activities associated with CYP1A1 include 7-ethoxyresorufin *O*-deethylation, and warfarin 6- and 8-hydroxylation. Enzyme activities associated with CYP1A2 include 17-β-estradiol 2- and 4-hydroxylation, acetanilide (acetaminophen) 4-hydroxylation, and *N*-hydroxylation of aromatic amines, including 2-aminofluorene, 1- and 2-naphthylamine, 4-aminobiphenyl, and amino acid pyrolysates. Polymorphic variation in gene expression of CYP1A2 is believed to be a determinant of individual risk to carcinogenicity of certain aromatic amines [see Gonzalez et al. (1986) for a review of P450 polymorphism and susceptibility to toxicity and carcinogenicity].

The CYP2 gene or major family consists of eight subfamilies (A - H). Subfamily CYP2A (Major) consists of three genes. Activity of the CYP2A1 in the rat is induced by phenobarbital (PB) or 3-MC. CYP2A1 and CYP2A2 demonstrate positional specificity toward hydroxylation of prototype substrate testosterone with CYP2A1 active for 7-α-hydroxylation and CYP2A2 active for 15-α-hydroxylation of testosterone. CYP2A3 in the rat is expressed in the lung and is inducible by 3MC. The B subfamily consists of 10 genes. In the rat, subfamily B has two genes that are inducible by PB (CYP2B1 and CYP2B2). This P450 gene subfamily is the primary P450 responding to induction by PB. CYP2B1 and CYP2B2 have similar enzyme activities, with catalytic activities for common substrates higher for CYP2B1 than for CYP2B2. Enzyme activities include 7-pentoxyresorufin *O*-dealkylation, warfarin 4′-hydroxylation, d-benzphetamine *N*-demethylation, aminopyrine *N*-demethylation, testosterone 16-α, 16-β, and 17-β-hydroxylation, and androstendione 16-β-hydroxylation.

Subfamily C consists of 15 genes. Most of these genes are constitutively expressed, whereas others are under developmental and sex-specific regulation. Some of these are inducible by PB, but others have no known inducers or have activities reduced by inducers. Some activities are actually induced by inhibitors of other P450s. Enzyme activities within this P450 gene family include testosterone 16-α, and 2-α-hydroxylation, 17-β-estradiol 2- and 4-hydroxylation, nefedipine oxidation, ethylmorphine *N*-demethylation, warfarin 4′-, 6-, and 7-hydroxylation, procarbarbazine methyl-*N*-oxidation, *S*-mephenytoin 4-hydroxylation, hexobarbital 3′-hydroxylation, and tolbutamide methyl hydroxylation.

Subfamily D consists of ten genes. In rats and humans, this subfamily is associated with debrisoquine-4-hydroxylase activity. Additional catalytic activities include sparteine δ-5-oxidation, bufuralol 1′-hydroxylation, propranol 4-hydroxylation and de-isopropylation, and encainide *O*-demethylation. In humans, polymorphism in the gene expression of this P450 subfamily has been related to individual variation of therapeutic and toxic responses to a variety of drugs. High levels of genetic expression of this subfamily have been associated with liver, gastrointestinal, and cigarette-induced lung cancer.

Subfamily E is inducible by ethanol, imidazole, isoniazid, acetone, trichloroethylene, and pyrazole. The most potent inducers, however, appear to be aniline, alcohols, and nitrosamines. Aniline *p*-hydroxylase activity is associated with this P450 gene family (Liebler et al., 1987). Since bioactivation of nitrosamines is the highest activity of this P450 gene family, high genetic expression of this gene family is proposed to increase individual risk to cancer. Inhibition and induction of this P450 activity by ethanol is believed to explain reduced half-lives of drugs biotransformed by P450 and potentiation of hepatotoxicity of xenobiotics [see review by Liebler et al. (1987) for references]. Ethanol potentiation of carbon tetrachloride, xylene, isoniazid, phenylbutazone, and acetaminophen hepatotoxicity is believed to be due to ethanol-induced biotransformation of the compounds.

The CYP3 gene family is inducible by steroids, the prototype of which is pregnelone 16-α-carbonitrile (PCN). Two genes of this subfamily in the rat are designated CYP3A1 and CYP3A2, which are distinguished by PCN and dexamethasone induction of CYP3A1, but not CYP3A2. PB induces both genes. Additional inducers of CYP3A1 include dexamethasone, hydantoins, polyhalogenated biphenyls, and macrolide antibiotics, such as triacetyloleandromycin and griseofulvin. Enzyme activities for CYP3A1 include ethylmorphine *N*-demethylase, warfarin 10-hydroxylation, testosterone 6-β-hydroxylation, estradiol 2- and 4-hydroxylation, and mephenytoin 4-hydroxylation. Activities of CYP3A2 are not defined.

The CYP4 gene family is clofibrate-inducible. Clofibrate is the prototype of a class of hypolipidemic drugs that are peroxisome proliferators and hepatocellular tumor promoters in rodents (Reddy et al., 1980). The endogenous substrates are unknown, but may be eicosanoids, such as prostaglandins and leukotrienes (Bains et al., 1985; Hardwick et al., 1987). This P450 gene family has high specificity for lauric acid 11- and 12-ω-hydroxylation (Orton and Parker, 1982) and C-20 ω-hydroxylation of arachidonic acid (Bains et al., 1985).

Four cytochrome P450 monooxygenases are involved in stereoidgenesis. CYP17, CYP21A2, CYP11A1, and CYP19 code for 17-α-hydroxylation, C21 hydroxylation, 11-β-hydroxylation and cholesterol side-chain cleavage, and aromatase, respectively. The side-chain cleavage and 11-β-hydroxylase enzymes are mitochondrial, and the 21-hydroxylase, 17-α-hydroxylase and aromatase are microsomal. A genetic defect in expression of the P450 for C21 hydroxylation is associated with hyperandrogenism and cortisol deficiency.

P450 activities are sex-, tissue-, and age-dependent and differentially regulated at the constitutive level and in response to xenobiotic exposure and endogenous mediators such as hormones. The mechanisms by which these factors regulate P450 are not fully understood, but in general involve control of gene expression at the transcriptional level [see Adesnick and Atchinson (1986) and Whitlock (1986) for detailed reviews and individual references]. Other mechanisms of regulation involve increased stability or increased rate of translation of cytochrome P450 mRNA or decreased rate of turnover of the P450 protein (Waterman and Estabrook, 1983; Guengerich, 1987; Nebert and Gonzalez, 1987).

Regulation of gene transcription of CYP1A1 (PAH-inducible) is believed to be by two *trans*-acting factors which function to control constitutive and inducible activities (reviewed by Nebert and Jones, 1989). Induction involves positive control and results from binding of PAH-type inducers to a cytosolic receptor encoded by the regulatory gene of the *Ah* locus. (*Ah* signifies aromatic hydrocarbon in reference to inducibility by these agents.) Structural genes that are part of the *Ah* locus include CYP1A1, CYP1A2, UDP-glucuronsyl-transferase, NAD(P)H: menadione oxidoreductase, and glutathione transferase. The inducer-receptor complex interacts with the *cis*-acting genomic regulatory element to increase transcription of these structural genes. Two models have been proposed for regulation of constitutive activity. Regulation may involve an endogenous ligand for the *Ah* receptor that is a substrate for the CYP1A1 protein product and rendered inactive by metabolism. Inactivation of the protein leads to accumulation of endogenous ligand and stimulation of transcription. Alternatively, metabolism of an endogenous substrate by the CYP1A1 protein product leads to activation of an apo-repressor protein to an active repressor. The repressor binds a *cis*-acting genomic element to inhibit transcription. Allelic variation in the genes coding for the inducer receptor as well as for the putative repressor protein are believed to result in polymorphic variation in both constitutive activities and response to induction and determine individual susceptibility to toxicity and carcinogenicity to drugs and xenobiotics biotransformed by these proteins.

Transcription of genes encoding P450 for biosynthesis of steroids (androgens, estrogens, glucocorticoids, and mineral corticoids) is responsive to ACTH or FSH, depending on cell type. cAMP is the intracellular messenger mediating the increased transcription. To increase transcription of the P450, cAMP itself acts indirectly by increasing the transcription of a hypothetical steroid hydroxylase-inducing protein. This protein presumably increases transcription by binding to a specific *cis*-acting genomic control element for the appropriate P450 gene.

The mechanisms by which the CYP2 gene families are transcriptionally activated is unknown. For the CYP2B1 and CYP2B2 (PB-inducible) gene families, it is not clear if a receptor is involved with PB induction since no receptor has been identified, induction requires high doses or exposures, and there are no structural-activity relationships for PB-type induction of P450. Several mechanisms may be involved in the increased transcription of the PB-inducible P450. A hierarchy of *cis*-acting genomic control elements may influence transcription upon interaction with trans-acting factors, such as PB or other PB-type inducer, heme and/or some other tissue-specific factor. For families CYP3 (steroid or PCN-inducible type) and CYP4 (clofibrate-inducible), similarly to the PB-inducible family, increased gene transcription is believed to be involved in induction (Hardwick et al., 1987). The species-, developmental-, sex-, and tissue-specific expression of P450 activities involve additional regulatory mechanisms. The most likely mechanism(s) for this regulation of P450 activities is transcriptional activation via receptors, *trans*- and *cis*-acting genomic control elements, and positive- and negative-acting enhancer elements.

The mechanisms regulating P450 activities are important determinants of the selectivity and sensitivity to toxicity resulting from P450-mediated bioactivation. As mentioned previously, polymorphic variation in the constitutive activities of the P450 gene families is believed to determine individual sensitivity to toxicity and carcinogenicity to drugs and xenobiotics. Similarly, species, developmental, sex- and tissue-specific sensitivity to toxicity can, in many cases, be attributed to the differential regulation of P450 bioactivation pathways at these levels. Induction is a major modifying factor for manifestation of toxicity and carcinogenicity as a result of xenobiotic exposure, with both phase I and II biotransformation pathways being independently and coordinately regulated. In the rat, P450s of high constitutive activities include CYP2C6, CYP2C11 in males, CYP2C12 in females, and CYP3A and CYP4A (Guengerich et al., 1982b; Bains et al., 1985). These activities are repressed or respond marginally to inducers. For example, PB induces not only the activities of the CYP2B subfamily, but also the activities of the CYP2A, CYP3A (PCN or steroid-inducible), epoxide hydrolase and UDP-glucuronyltransferases. 3-MC and TCDD induce the activities of P450 subfamilies CYP1A (PAH-inducible), CYP2A (Major) but not CYP2B (PB-inducible). Arochlor 1254 induces the activities of P450 subfamilies CYP1A, CYP2A, and CYP2B. Dexamethasone induces the activity of P450 subfamily CYP3A (PCN-inducible) while repressing the transcription of epoxide hydrolase. PB, PAH, PCN, and Arochlor 1254 all repress the activity of CYP2C11, the enzyme of highest constitutive activity in male rat liver (Guengerich et al., 1982b). PAH and TCDD, in addition to inducing CYP1A, increase transcription of UDP-glucuronyltransferase, NAD(P)H:menadione oxidoreductase and glutathione transferase genes, all of which are part of the *Ah* locus or gene battery (see above). These relatively specific changes in activities of both phase I and phase II enzymes of biotransformation have led to the concept that induction has evolved as an adaptive mechanism to assure survival in response to toxicity and mutagenicity from xenobiotics in the environment. Although, for acute toxicity, induction can result in an increased or decreased effect depending on whether the induced activities are bioactivating or detoxifying [see Conney (1982) for examples], for carcinogenesis, induction generally results in protection against electrophiles generated during biotransformation by an overall increase in activities of detoxifying pathways relative to bioactivating pathways (Conney, 1982; Guengerich, 1988).

Flavin-Containing Monooxygenase

Next to P450-linked enzymes, flavin-containing monooxygenase (FMO) is considered to be the most active enzyme in phase I biotransformation [reviewed by Ziegler (1988)]. Historically, it has been referred to as amine oxidase or as *N,N*-dimethylaniline *N*-oxidizing enzymes (EC 1.14.13.8.). It is a flavoprotein with FAD as co-factor and is localized in the endoplasmic reticulum. Liver, lung, and kidney have the highest activity, with all tissues containing activity except muscle. The liver and lung enzymes are immunologically distinct, and differ in thermal stability, pH optimum, and sensitivity to anionic detergents. FMO catalyzes the oxidation of nucleophilic nitrogen, sulfur, and phosphate atoms. In contrast to P450, it lacks carbon hydroxylation activity. Tertiary amine substrates are oxidized to amine oxides, secondary amines to hydroxyl amines and nitrones, and primary amines to hydroxyl amines and oximes. Sulfides, thiols, thioamides, and thiocarbamates are sulfur-containing substrates that give rise to various sulfoxides, sulfones, thiols, sulfinic, and sulfenic acids as products. Organophosphates, trivalent phosphorus-containing and phosphines are phosphate-containing substrates. The enzyme has higher activity for *S* and *P*-oxygenation relative to *N*-oxygenation; compounds bearing nucleophilic sulfur, phosphate, and nitrogen atoms are preferentially *S*- and *P*-oxygenated. Methimazole (*N*-methyl-2-mercaptoimidazole) and dimethylaniline are substates commonly used to measure activity. This enzyme is believed to have the endogenous function of oxidizing the sulfur-containing compound cysteamine to cystamine. This serves to control the endogenous cystamine-cysteamine balance and hence to regulate disulfide generation during peptide synthesis.

The enzyme is characterized by large species differences in activity; activities are high in hogs, dogs, guinea pigs, and humans. The reaction mechanism involves a hydroperoxyflavin as oxygenating intermediate with oxygen transfer from the hydroperoxyflavin to the substrate. FMO only catalyzes the oxidation of nitrogen compounds bearing functional groups that are readily oxidized by organic peroxides. The hydroperoxyflavin will not attack nitrogen atoms in amides, carbamides, imines, nitrones, oximes, isocynates, nitriles, or heterocyclic aromatic amines. Amines, hydroxylamines, and hydrazines are excellent substrates. The thermal lability of the liver enzyme has been utilized to distinguish it from P450-linked enzyme. There are no known inhibitors of the FMO; however, methimazole (a substrate) has been utilized as a competitive inhibitor to differentiate between P450 and FMO activities. The purified liver enzyme appears to have one binding site for activators, which appear to be positively charged lipophilic compounds such as primary alkylamines (*n*-octylamine is the prototype) or alkyl guanidines. The enzyme is not inducible by xenobiotics; activity is actually reduced by PB, β-NF, or PAH treatment. Hormones appear to exert an important regulatory control on FMO activity.

P450-linked enzymes and FMO compete for oxidation of organic nitrogen, phosphate, and sulfur compounds. For organic nitrogen compounds, primary amines are better substrates for P450 oxidation, whereas secondary and tertiary amines are better substrates for FMO oxidation. The p*Ka* of the substrate nitrogen appears to determine the relative activities of each enzyme system for oxidation of the amine with basic amines oxidized preferentially by FMO and acidic amines by P450-linked enzymes. For organic sulfur compounds, structure can be the primary determinant of selectivity of biotransformation by the two processes. In most cases, final oxidation products for organic sulfur compounds result from the simultaneous and/or sequential action of both P450-linked enzymes and FMO as well as other enzymes. Organophosphates are substrates for both P450 and FMO, while trivalent phosphorus-containing compounds and phosphines are preferentially oxygenated by FMO. Along with P450-linked enzymes, FMO participates in the bioactivation of several xenobiotics, including aminofluorenes, aromatic amines, azo dyes, hydrazines, thiocarbonyls, and thiophospho compounds (Guengerich and Liebler, 1985).

Molybdenum Hydroxylases

Molybdenum-containing enzymes important for carbon oxidation include aldehyde oxidase and xanthine oxidase dehydrogenase. Aldehydes and *N*-heterocyclics are typical substrates. Because specificity is much broader than implied by the common names of the enzymes, molybdenum hydroxylases is now considered the preferred nomenclature. The enzymes catalyze nucleophilic oxidation of carbon via hydroxyl ion attack with the incorporated oxygen atom ultimately derived from water. Heterocyclic compounds are preferred substrates relative to aldehydes. Aliphatic aldehydes are better substrates for aldehyde oxidase than xanthine oxidase with aromatic and heterocyclic substituents resulting in increased substrate affinities for both enzymes. For heteroaromatic rings, carbon atoms of the carbocyclic ring would be the preferred sites for P450 oxidation, while carbon atoms adjacent to the ring nitrogen atom would be the preferred sites for molybdenum hydroxylase oxidation since these carbon atoms are the most electropositive.

The enzymes are predominantly cytosolic with some aldehyde oxidase activity in microsomes. The enzymes are ubiquitious in nature. In mammals, the highest levels of activities of aldehyde oxidase are in liver, and for xanthine oxidase, the highest levels of activities are in small intestine, and milk and mammary gland. Each enzyme has a similar dimeric structure of two identical subunits. Each subunit contains one atom molybdenum, one atom flavin, and two atoms of Fe_2S_2. Substrates interact at the molybdenum atom with the molybdenum atom reduced from Mo^{6+} to Mo^{4+}. The substrate-molybdenum complex rearranges to a thiol complex with oxidized product ultimately released by hydrolysis. Mo^{4+} is reoxidized to Mo^{6+} by intramolecular transfer of electrons from Mo^{6+} to FAD and Fe_2S_2.

The endogenous functions of these enzymes is not clearly understood. Xanthine oxidase participates in purine catabolism, and aldehyde oxidase catalyzes the oxidation of vitamins such as pyridoxal and *N*-methyl nicotinamide; however, neither enzyme is required for these physiological processes to proceed. Xanthine oxidase may be important for storage and mobilization of iron. There is, however, an increasing appreciation for their role in the biotransformation of xenobiotics such as aldehydes, uncharged bases, quaternary *N*-heterocyclics, and carbocyclics. They are believed important in the bioactivation and detoxification of aromatic nitro compounds, azo dyes, nitrosamines, *N*-oxides, and sulphoxides via reduction reactions and to possibly play a role in the generation of reactive oxygen species during oxidation of both endogenous and exogenous compounds.

Epoxide Hydrolase

Epoxide hydrolase (EC 3.3.2.3.) catalyzes the hydration of arene oxides and aliphatic epoxides to their corresponding *trans*-1,2-dihydrodiols. This enzyme system, along with glutathione conjugation, plays an important role in minimizing the toxicity of epoxides generated by the oxidation of aromatic and aliphatic hydrocarbons by P450-linked enzymes [see reviews by Glatt et al. (1984) and Glatt and Oesch (1987)].

The enzyme is present in several tissues. One form is in close proximity to P450 in the endoplasmic reticulum, and this may reflect its role in detoxification of electrophilic epoxides generated by P450. Epoxide hydrolase activities are also present in the plasma membrane, golgi apparatus, nuclear membrane, mitochondria, and cytosol. Because of differences in immunogenicity, pH optimum, molecular weight, and 1,1,1-trichloropropene oxide inhibition, the cytosolic and microsomal enzymes are believed to be two distinct forms. Epoxides that are good substrates for the microsomal epoxide hydrolase appear to be poor substrates for cytosolic epoxide hydrolase and vice versa. For example, styrene oxide, a common substrate for determination of enzyme activity, is a poor substrate for the cytosolic enzyme. Therefore, low substrate activity of one form appears to be compensated by the other. There are no known specific inducers of epoxide hydrolases, but PB, 3-MC, PCN, Aroclor 12154, *trans*-stilbene oxide, lindane, antioxidants, alcohols, ketones, and imidazoles all induce the microsomal form of the enzyme. 1,1,1-Trichloropropene oxide and cyclohexene oxide are the prototype enzyme substrates. The cytosolic form has PB as the only known inducer. On the other hand, the nuclear and microsomal forms are immunologically similar but have dissimilar substrate specificities and inducibility. Species and sex differences are not particularly noteworthy. The mouse has low activity but responds to induction to a greater degree. The low activity of the enzyme in mice could not be related to the sensitivity of this species to hepatocarcinogenicity. The principal example of epoxide hydrolase participation in a bioactivation reaction is generation of reactive vicinal dihydrodiol-epoxides, the major ultimate carcinogens of PAH.

The cytosolic glutathione transferases and microsomal epoxide hydrolase enzyme systems display both regio- and stereoselectivity for alkene and arene epoxide biotransformation. Their activities for alkene and arene epoxide biotransformation are complementary. For example, in the case of polycyclic arene oxides, biotransformation by glutathione transferases is favored by increasing size; and K-region epoxides are better substrates than non-K-region epoxides. The relative tissue concentrations of epoxide hydrolase and glutathione transferase and their differential activities toward non-K and K-region epoxides of polycyclic arene oxides are believed to be important determinants of tissue susceptibility to PAH toxicity/carcinogenicity.

Esterases and Amidases

Esterases and amidases are enzymes that tend to be nonspecific for hydrolysis of ester and amide linkages [see review by Satoh (1987)]. Substrates include aliphatic and aromatic esters and thioesters. Esters are hydrolyzed to carboxylic acids and alcohols, and amides are converted to amines and carboxylic acids. The resulting carboxylic acid, alcohol, and amine compounds are substrates for phase II biotransformation. Esterase activities are highest in the liver, kidney, and small intestine. Appreciable activities are present also in adipose tissue, central nervous system, lung, nasal muscosa, and testes. Activity is predominately microsomal, but there is activity also in cytosol and mitochondria. Because of the broad and overlapping substrate specificity of these enzymes, classification has been difficult. Based on inhibition by the organophosphate, paraoxon, there are four main groups. As preferred substrates, A-esterases (arylesterases) have aromatic esters, B-esterases (carboxylesterases) have aliphatic esters, C-esterases (acetyl-esterases) have acetyl esters, and cholinesterases have choline esters. Based on differences in immunological and chromatographic behavior, substrate specificity, differences in hormonal regulation, and responses to inhibitors and inducers, however, several isozymes are believed to exist within each group. Inducers of the liver enzymes are PB, DDT, polychlorinated biphenols, *trans*-stilbene oxide and 3-MC. The exact physiological role of these esterases is unknown, but since these enzymes have significant lipase activity, lipid metabolism may be one of their functions. The enzymes play an important role in drug and xenobiotic biotransformation. Examples of substrates include isocarboxazid, aspirin, organophosphates, procaine, clofibrate, phenacetin, acetaminophen, and benzodiazepines. Their primary role in bioactivation is hydrolysis of acyl groups of aromatic amines and acetanilides, leading to methemoglobinemia or allergic reactions. Generally, inhibition of these enzymes, for example with organophosphates, results in increased toxicity of xenobiotics because hydrolysis of esters and amides generally results in increased hydrophophilicity and accelerated excretion. Examples of xenobiotics bioactivated by these enzymes to cause hepatotoxicity include aromatic amines, phenacetin, propanil, procaine, and esters of allyl alcohol.

Alcohol, Aldehyde, Ketone Oxidation-Reduction

Several different enzymes are involved in the biotransformation of alcohols, aldehydes, and ketones. Since aldehydes and ketones contain reactive functional groups, these enzymes play an important role in biotransformation of these groups to less reactive, and thus less toxic, intermediates.

Alcohol dehydrogenase (EC 1.1.1.1.) is ubiquitous in nature. The enzyme is predominately cytosolic. Its highest activity is in liver, but there is appreciable activity in stomach, lung, intestine, and kidney. Alternate biotransformation pathways for alcohols include glucuronide conjugation and catalase-dependent and P450-linked enzyme (ethanol-inducible subfamily) oxidation. Alcohol structure determines the extent of biotransformation by a particular pathway. Primary and secondary aliphatic and aromatic alcohols are good substrates, but hindered, and tertiary alcohols are poor substrates. Selected aldehydes and ketones can also be reduced by alcohol dehydrogenase.

In humans, several isozymes that differ in substrate specificity are known to exist. Three gene loci, the expression of which is developmentally regulated, code for distinct subunits that can form up to 15 dimeric isozymes from random combination of subunits. A sixteenth form that is characterized by insensitivity to prototype alcohol dehydrogenase inhibitors, i.e., pyrazole and

4-methoxypyrazole, may be coded by a separate gene locus. Co-factor supply (physiologic NAD^+/NADH ratio) is an important regulatory factor of enzyme activity, and polymorphic variation in subunit gene expression is an important determinant of substrate specificity and affinity.

The classic examples of alcohol dehydrogenase-mediated bioactivation are in the toxicity of methanol and ethylene glycol. Formate, a metabolite of methanol, causes toxicity to the retina leading to blindness; and oxalate, a metabolite of ethylene glycol, forms a precipitant in kidney tubules leading to stone formation. Leading to hepatotoxicity, alcohol dehydrogenase converts allyl formate and allyl alcohol to the reactive aldehyde, acrolein (Rees and Tarlow, 1967; Reid, 1972). Since alcohol dehydrogenase has a periportal distribution in the liver, allyl formate and allyl alcohol induce periportal necrosis.

Aldehyde dehydrogenases (EC 1.2.1.3.) catalyze the oxidation of aldehydes to acids by using NAD as co-factor. The physiological substrates for the enzymes are unknown; the substrate specificity is broad. Classical substrates include acetaldehyde, formaldehyde, and glycolaldehyde, which are the biotransformation products of ethanol, methanol, and ethylene glycol, respectively. The enzymes are present in all organs and in cytosol, mitochondria and microsomes. Liver has the highest activity, but kidney also possesses high activity. The enzymes exist as several isozymes within the cytosolic, mitochrondrial, and microsomal compartments. One cytosolic isozyme is specific for the oxidation of formaldehyde complexed with glutathione and is referred to as formaldehyde dehydrogenase. Additional isozymes within the cytosol are differentiated by selective induction with PB or TCDD and 3-MC. The isozyme induced by TCDD and 3-MC has received special study due to its increased activity in tumor cells [see Marselos and Lindahl (1988)]. Since acetaldehyde is preferentially oxidized in mitochondria, an isozyme for its oxidation may be localized in these structures. Additional mitochrondrial isozymes can be differentiated by selective inhibition with the prototype aldehyde dehydrogenase inhibitor, disulfiram.

Aldehyde reductase (EC 1.1.1.2.) catalyzes the reduction of carbonyl compounds. The enzyme is predominately cytosolic, with highest activity in kidney and significant activities in liver and brain. Carbonyl-containing substrates include aromatic aldehydes and ketones, aliphatic ketones, methyl, and unsaturated ketones. The metabolic functions of these enzymes are unknown, as are the physiological roles of aldehydes generated during intermediary metabolism. Since many aldehydes, particularly aldehyde derivatives of neurotransmitters, appear to be physiologically active, these enzymes may function to regulate endogenous aldehyde activity. At least three izoenzymes have been characterized. These include aldehyde reductase (the same isozyme as glucuronate reductase, mevaldate reductase, lactaldehyde reductase, and daunorubicin reductase), aldose reductase, and glycerol dehydrogenase.

Ketone reductases catalyze the reduction of ketones to alcohols with NADPH as a co-factor. The enzymes have an ubiquitous tissue distribution and are principally cytosolic. Liver and kidney have the highest activities. Substrates include aromatic, aliphatic, alicyclic, and unsatured ketones as well as aldehydes. The model ketone for assay of activity is *p*-nitroacetophenone. Although both aldehyde and ketone reductases act on aldehyde substrates, the two enzyme families can be differentiated based on substrate and inhibitor specificity and stereospecificity toward co-factor.

For the alcohol, aldehyde, ketone oxidation-reduction enzymes, aldehyde dehydrogenase is most important in bioactivation. Inhibition of aldehyde dehydrogenase during biotransformation generally results in toxicity due to accumulation of electrophilic aldehydes. Also, alcohol, aldehyde, ketone oxidation-reduction enzymes are a few of several enzymes involved in bioactivation of nitro polycyclic aromatic hydrocarbons by one election reduction (Beland et al., 1985).

Phase II Biotransformation

The apparent physiological function of phase II biotransformation reactions is regulation, via termination, of the biological activities of endogenous compounds. This is the case for steroid hormones. Another possible role for such reactions is to form conjugates that serve as transport forms of steroid hormones, prolonging the half-life of the hormone and allowing selective uptake by target cells through specific transport systems that recognize the conjugating moiety. The most important phase II biotransformation reactions in mammals are glucuronidation, sulfation, glutathione conjugation, acetylation, methylation, and amino acid conjugation. For xenobiotic biotransformation, glucuronidation, sulfation, glutathione conjugation, and acetylation are of primary importance.

Glucuronyltransferases

Reactions catalyzed by glucuronyltransferases [EC 2.8.2.1.; see reviews by Bock et al. (1986) and Bock and Schirmer (1987)] are both qualitatively and quantatively the most important phase II biotransformation reactions due to the broad range and chemical diversity of substrates. UDP-glucuronic acid is the co-factor for these enzymes. The transferases catalyze the translocation of glucuronic acid from UDP-glucuronic acid to an appropriate receptor to form the β-D-glucuronide. The enzyme location is predominately microsomal. Activity is highest in the liver; measurable activities are present also in kidney, intestine, skin, brain, and spleen. Substrates are aliphatic alcohols and carboxylic acids (converted to *O*-glucuronides), primary and secondary aromatic and aliphatic amines (conjugated to *N*-glucuronides), free sulfydryl-containing compounds (conjugated to *S*-glucuronides), and compounds containing nucleophilic carbon atoms (conjugated to *C*-glucuronides). Glucuronidation contributes a carboxyl group, which exists primarily in ionized form at physiologic pH. This ionized carboxyl group promotes excretion by increasing the xenobiotic's water solubility and substrate affinities for transport by biliary and renal organic anion systems.

Several isozymes of glucuronyltransferases are believed to exist, but the exact number is unknown. Some separation has been achieved based on preferential selectivity toward phenols, bilirubin, testosterone, chloramiphenol, and morphine as substrates. In rat liver, there are at least five isozymes; four are specific for endogenous bilirubin and steroids. These four isozymes are bilirubin-GT (substrates include androsterone, etiocholanolone and lithocholic acid), 17-β-hydroxysteroid-GT (substrates include testosterone and estradiol at C17), and aromatic steroid-GT (substrates include estrone and estradiol). Xenobiotic biotransformation appears to involve a fifth isozyme termed phenol-GT_{MC} (MC signifies inducibility by 3-MC) or GT_1. However, 17-OH-steroid-GT also conjugates exogenous planar phenols, such as 1-naphthol and 4-nitrophenol. For GT_1, no endogenous substrate is known. Phenols and polyphenols of PAH are substrates; 3-hydroxybenzo(*a*)pyrene, 4-nitrophenol, 1-naphthol, and 2-aminophenol are prototypes. PB is selective for induction of UDP-glucuronyltransferases that have bilirubin and steroids as substrates and also induces GT_1 at higher doses. UDP-glucuronyltransferase isozymes are also differentially regulated at the developmental level, with the PB-inducible isozymes appearing early in fetal life and the 3-MC-inducible isozymes in late fetal life. GT_1 is believed to be the UDP-glucuronyltransferase isozyme that undergoes coordinate induction with CYP1A1 (PAH-inducible). Co-factor supply is another important regulatory factor; UDP-glucuronic acid availability is dependent on local concentrations of glucose, UDP, and NAD^+ (Bidlack et al., 1986).

Although glucuronide conjugation is considered to be primarily a detoxification reaction, there are examples where glucuronide conjugates are intermediates leading to toxic metabolites. The most notable is the induction of bladder cancer in workers in the dye industry exposed to arylamines such as 1- and 2-naphthylamine, 4-aminobiphenyl, and benzidine. After oxidation and then conjugation with glucuronide in the liver, the product is subsequently hydrolyzed by the acidic environment in the kidney, resulting in release of an electrophilic nitrinium compound. A similar example is the colon carcinogenicity of 2′,3-dimethyl-4-aminobiphenyl. After oxidation and glucuronide conjugation in liver, the conjugate is secreted in the bile to the intestine, where bacterial β-glucuronidase liberates the glucuronide group to produce an electrophile. For the liver, glucuronidation participates in the bioactivation of *N*-hydroxy-arylamines.

Glutathione Transferases

Glutathione transferases [GSH transferase; EC 2.5.1.18.; reviewed by Ketterer et al. (1983), Mannervik (1985); Ketterer (1988), and Pickett and Lu (1989)] are a family of enzymes that catalyze the initial step in the formation of *N*-acetylcysteine (mercapturic acid) derivatives of xenobiotics. Glutathione (GSH), a tripeptide of the three amino acids, glycine, glutamine, and cysteine, is the co-factor for the enzymes. The reaction mechanism involves nucleophilic attack of the sulfydryl atom of glutathione on electrophilic atoms. This results in formation of a thioether bond between the xenobiotic and glutathione. The conjugates are ultimately cleaved to cysteine derivatives in the kidney, the cysteine moiety is acetylated, and the compound is then excreted in the urine as the mercapturic acid derivative. Substrate specificity is extremely broad; rates of reaction are governed by the hydrophobic and electrophilic nature of the substrate. Substrates can be any xenobiotic with an electrophilic carbon, nitrogen, sulfur, or oxygen atom. Common xenobiotic substrates include aryl and alkyl halides, olefins, organic peroxides, quinones, and sulfate esters.

This enzyme activity is detectable in almost every organ; high activities are in liver and testes. In certain organs, GSH transferases may be localized or absent in certain cell types. Likewise, certain isozymes may be predominant in a particular organ. These enzyme activities are predominately cytosolic; but activities are also present in the nucleus, microsomes, and mitochondria. Several cytosolic isozymes in rat liver have been characterized. They are homo- and heterodimers of at least seven subunits (Y_a, Y_{b1}, Y_{b2}, Y_c, Y_k, Y_n, Y_p) that are distinguished by mobility on sodium dodecyl sulphate polyacrylamide gel electrophoresis. 1-Chloro-2,4-dinitrobenzene is the reference substrate to assay GSH transferase activities. Other substrates relatively selective for assay of the various GSH transferase subunits include 1,2-dichloro-4-dinitrobenzene (subunits Y_b and Y_n), *trans*-4-phenyl-3-buten-2-one (Y_{b2}), and 4-hydroxynon-2-enal (Y_k). Cumene and linoleate hydroperoxide are standard substrates to measure the peroxidase activity of GSH transferase; peroxidase activities are high for subunits Y_a, Y_c, Y_k, Y_p, particularly Y_k. GSH transferase/GSH peroxidase activities are high in testes, relative to other tissues, and in the nucleus, relative to the cytoplasm. Inducers of rat GSH transferase include antioxidants, such as ethoxyquin, butylated hydroxytoluene and butylated hydroxyanisole, and PB, 3-MC, TCDD, and *trans*-stilbene oxide. Activities of subunits Y_a, Y_{b1}, and Y_{b2} appear to be selectively induced by these agents, with Y_c only marginally affected.

The precise biological functions of GSH transferase are unknown. Endogenous substrates are steroids, quinones, leukotrienes, and prostaglandins. One function may be to serve as a storage depot for compounds, such as bilirubin, in order to minimize toxicity. The most important biological functions appear to be protection of membranes and DNA from peroxidation, and provision or a cellular means for directing reaction of electrophiles away from other cellular constituents. Endogenous electrophiles detoxified by GSH transferases include cytotoxic and/or genotoxic products of lipid and nucleic acid peroxidation, such as cholesterol-5,6-oxide, hydroxyalkenals (e.g., 4-hydroxynon-2-enal), linoleic acid hydroperoxide, and 5-hydroperoxymethyluracil. The efficiency of detoxification of these electrophiles by GSH transferases appears to be a significant factor for relative risk for toxicity and carcinogenicity. For example, although cholesterol-5,6-oxide is a weak bacterial mutagen and hydroxyalkenals are strong bacterial mutagens, cholesterol-5,6-oxide is considered a potentially greater hazard because it is a poor substrate and hydroxyalkenals are excellent substrates for GSH transferases. Similar considerations apply to electrophiles generated during xenobiotic biotransformation. High-polar compounds are poor substrates for GSH transferases. For example, *N*-sulfonyloxy-*N*-acetyl-2-aminofluorene and *N*-hydroxy-2-aminofluorene are highly polarized and are not efficiently detoxified; they react with GSH only nonezymatically. For xenobiotic electrophiles reacting enzymatically with GSH transferases, the relative efficiency of detoxification can vary with substrate affinity of the GSH transferase subunits. Since aflatoxin-8,9-oxide is only a low-affinity substrate for subunits Y_a and Y_c, detoxification by GSH transferases is inefficient. Epoxides of benzo(*a*)pyrene (BP) are substrates for all GSH transferase subunits; they are high-affinity substrates for subunits Y_{b2} and Y_p. Further, subunit Y_{b2} displays stereoselectivity toward the (+) enantiomers of BP oxides. All these factors contribute to the high efficiency of detoxification of benzo(*a*)pyrene oxide by GSH transferase.

Although GSH conjugation is in general a detoxification pathway for xenobiotic biotransformation, an increasing number of xenobiotics are known to be bioactivated by this pathway [reviewed by Anders et al. (1988) and Lock (1988)]. γ-Glutamyltransferase (EC 2.3.2.2.) and dipeptidases (EC 3.4.13.6., 3.4.11.2.), respectively, catalyze the removal of the glutamyl and glycine moieties of the glutathione *S*-conjugates to yield the corresponding cysteine *S*-conjugates. Cysteine *S*-conjugates are the penultimate toxic metabolites of nephrotoxic halogenated alkenes. These conjugates may form electrophiles nonenzymatically, be detoxified by acetylation to a mercapturic acid by cysteine *S*-conjugate *N*-acetyltransferase (EC 2.3.1.80.), or be biotransformed by cysteine conjugate β-lyase (EC 2.6.1.64.) to toxic and/or nontoxic thiols. Some of the thiols formed by cysteine *S*-conjugate β-lyase can be further biotransformed to toxic and/or nontoxic methylthio derivatives by *S*-methyltransferases (see below). The selectivity for kidney toxicity is due to the active transport of the *S*-cysteine conjugates by the renal anion transport system, urinary excretion of the conjugates, and kidney localization of cysteine *S*-conjugate β-lyase. For hepatotoxicity, GSH conjugation plays a role in the bioactivation of vicinal dihaloethanes, such as 1,2-dibromoethane. The gluthione-*S*-conjugate mustards of these compounds nonenzymatically form electrophilic episulfonium ions.

Sulfotransferases

Sulfotransferases (EC 2.8.2.1.) are cytosolic enzymes present in liver, kidney, intestinal tract, and lungs. These enzymes play an important role for conjugation, with inorganic sulfate, of compounds with hydroxyl group moieties. The intermediates formed are ionized organic sulfates, which are readily excretable. Substrates are typically phenols and aliphatic alcohols, which are biotransformed to sulfate esters or ethereal sulfates. Aromatic amines and hydroxylamines can also be substrates for sulfotransferases; sulfamates and *N,O*-sulfates are the respective products. Numerous endogenous, low-molecular-weight compounds, including catecholamines, hydroxysteroids, and bile acids, are substrates for sulfotransferases.

The co-factor for these enzyme reactions is 3′-phosphoadenosine-5′-phosphosulfate (PAPS), the source of sulfate. In the cell, the major source of sulfate for PAPS synthesis is derived from cysteine oxidation. Since the free-cysteine concentration in the cell is limited, PAPS concentration is rate-limiting for sulfate conjugation. Other than the regulatory effects of free-cysteine concentration on PAPS availability, dietary sulfate or local concentrations of sulfate are important determinants of sulfotransferase activities. Sulfotransferases can be classified into four classes: phenols, catecholamines, and organic hydroxylamines are substrates for aryl sulfotransferases; hydroxylated steroids, primary alcohols, and secondary alcohols are substrates for hydroxysteroid sulfotransferase; phenols of aromatic rings of steroids are substrates for estrone sulfotransferase; and conjugated and unconjugated bile acids are substrates for bile salt sulfotransferase.

The primary type of sulfotransferase involved in xenobiotic biotransformation is that reacting with aryl compounds. At least three forms of aryl sulfotransferase and three forms of hydroxysteroid sulfotransferase have been purified to homogeneity from rat liver. The enzymes are developmentally and hormonally regulated, but apparently are not inducible by PB or 3-MC exposure.

During xenobiotic biotransformation, sulfate conjugation generally results in the formation of less toxic intermediates. Some sulfate conjugates, however, are unstable and spontaneously rearrange to electrophilic species. 2-Acetylaminofluorene, *N*-methyl-4-aminoazobenzene, safrole, and estragole are hepatocarcinogens in which the ultimate carcinogens are believed to be sulfate esters (Miller and Miller, 1981a).

N-Acetyltransferases

Xenobiotics or endogenous compounds containing amino, hydroxyl, or sulfhydryl groups are substrates for acetyltransferases [reviewed by King and

Glowinski (1983) and Weber and Hein (1985)]. Amines, including aromatic and aliphatic primary amines, hydrazines, hydrazides, and sulfonamides, have been the most widely studied substrates. Endogenous substrates are 5-hydroxy-tryptamine and histamine. The enzymes are cytosolic and are present in several tissues; liver has the highest activity. Acetyl coenzyme A is the co-factor. The enzymes have been designated acetyl CoA: amine acetyl transferases. Sulfamethazine *N*-acetyltransferase is another term for the enzyme activity; sulfamethazine is the prototype substrate. The activity is characterized by polymorphism across many species, and individuals (or species and strains) are characterized as "slow" or "fast" acetylators. This polymorphism is determined by a single gene locus containing two co-dominant alleles: R = rapid, and r = slow. This polymorphism is an important determinant of susceptibility to toxicity and carcinogenicity of drugs (e.g., isoniazid) or xenobiotics, particularly aromatic amines, which are bioactivated or detoxified by acetylation.

For aromatic amines, such as 2-aminofluorene, benzidine, 4-aminobiphenyl, and 2-napthylamine, acetyl CoA-dependent acetylation is catalyzed by *N*-acetyl-transferase (NAT; EC 2.3.1.5.). This enzyme is the same as that catalyzing the *N*-acetylation of sulfamethazine (Flammang et al., 1987). Acetyl CoA-dependent acetylation also occurs with the aromatic amines after P450 biotransformation to *N*-hydroxy derivatives. For the resulting arylhydroxamic acid derivatives of aromatic amines, intramolecular transfer of the acetyl group of the hydroxamic acid from the amine to the hydroxyl group of the nitrogen is catalyzed by *N,O*-arylhydroxamic acid acyltransferase (AHAT) in an acetyl CoA-independent reaction. AHAT is believed to be the same protein as NAT, with the *N,O*-intramolecular transfer of the acetyl group occurring because the arylhydroxyamic acids are recognized as and substitute for acetyl CoA as acetyl donor for the enzyme reaction. The resulting product of the *N,O*-intramolecular acetyl group transfer is an unstable *N*-acyloxyarylamine, which spontaneously degrades to an electrophilic arylnitrenium ion.

Methylation

Methylation reactions include *N*-, *O*-, and *S*-methylation. In general, these reaction pathways are primarily of quantitative importance for endogenous compounds, but xenobiotics such as aliphatic and aromatic amines are methylated; *N*-heterocyclics, mono- and polyhydric phenols, and sulfhydryl-containing xenobiotics can be substrates. These enzymes have a strict requirement for *S*-adenosyl-L-methionine as a co-factor and methyl group donor. All the enzymes are subject to end-product inhibition by demethylated *S*-adenosyl-L-methionine (*S*-adenosylhomocysteine). Sulfhydryl groups apparently are required for the catalytic activity of these enzymes since sulfhydryl group reagents inhibit enzyme activity.

N- and *O*-methylation reactions are catalyzed by catechol *O*-methyltransferase (EC 2.11.6.), histamine *N*-methyltransferase (EC 2.11.8.), and indolethylamine *N*-methyltransferase. Catechol *O*-methyltransferase is ubiquitous in nature; the highest activity in mammals is in liver. This enzyme is primarily cytosolic. In rat liver, two forms of the enzyme have been purified and are designated COMT I (molecular weight 23,000) and COMT II (molecular weight 47,000). Mg^{2+} is required for activity; enzyme activity is inhibited by Ca^{2+}. Phenolic and polyphenol compounds are inhibitors of the enzyme. The enzyme has strict specificity for the catechol nucleus as a methyl group acceptor, but broad specificity in regard to additional substituents on the aromatic nucleus. Its primary function appears to be extraneuronal inactivation of catecholamines by methylation. Catecholamine-type drugs are the most important nonphysiological substrates that are methylated.

Histamine *N*-methyltransferase is ubiquitous in nature; its activity is principally cytosolic. The enzyme has strict specificity for histamine as a methyl acceptor. It is sensitive to inhibition by high substrate concentrations of histamine.

Indolethylamine *N*-methyltransferases are enzymes, present in various tissues, that are both specific and nonspecific for *N*-methylation of amines. There is a broad substrate specificity for the methyl donor. Enzyme activity is generally assayed with tryptamine, *N*-methylserotonin, *N*-methyltryptamine, or β-phenylethylamine as a substrate. The enzymes are subject to end-product inhibition by *N*-methylated indoleamines. Substrates include endogenous biogenic amines, amine drugs such as amphetamine, and xenobiotic amines such as aniline.

Thiol *S*-methyltransferase is predominately a microsomal enzyme. Activities are high in liver, kidney, and lung; lower activities are in testis, spleen, and intestine. Thioacetanilide is the standard thiol substrate used to assay activity *in vitro*. The enzyme has the important function of detoxifying mercaptan xenobiotics, particularly hydrogen sulfide produced by anaerobic bacteria in the intestinal tract. Additional thiol or sulfhydryl group substrates include drugs such as dialkylcarbamates (e.g., Antabuse), thiosubstituted pyridines and purines (antithyroid drugs), substrates from disulfide interchange reactions involving glutathione, and, perhaps most importantly for xenobiotic biotransformation, thioethers of cysteine produced during glutathione conjugation of xenobiotics (see above).

Amino Acid Conjugation

One pathway for elimination of xenobiotics with carboxylic acid functional groups is formation of acyl conjugates with amino acids. Quantitatively, conjugation by glucuronidation is more important, and conjugation with sugars to form glycosides can also occur. Xenobiotic substrates generally are cyclic and polycyclic carboxylic acids such as aryl acids (e.g., benzoic acid) and arylacetic acids (e.g., phenylacetate). Physiological substrates include fatty acids and bile acids. For bile acids, conjugation functions to minimize potential toxicity and to prolong biological activity by increasing detergent properties, which facilitate intestinal reabsorption. In mammals, glycine is the most common amino acid acyl acceptor; glutamine and taurine are also common acceptors. Taurine is generally the acceptor for bile acid conjugation. Structure of the compound containing the carboxylic acid moiety, the species of animal, and the specific amino acid utilized as the acyl acceptor determine the extent of amino acid conjugation.

The numbers of enzymes and isozymes, and the mechanisms, distribution, and regulation of enzymes involved in amino acid conjugation are not fully characterized. Conjugation involves two coupled enzyme systems located in cytosol, mitochondria and microsomes. The first class of enzymes are ATP-dependent acid: CoA ligases, which require ATP and coenzyme A as cofactors to catalyze the conversion of the carboxylic acid - containing substrate to a thioester derivative of coenzyme A. This reaction is reversible. The second class is acyl-CoA: amino acid *N*-acyltransferases, which catalyze the transfer of the acyl moiety of the coenzyme A thioester derivative to the amino group of the acceptor amino acid. This reaction is irreversible.

Benzoates and arylacetates are the prototype substrates for characterization of the enzymes catalyzing amino acid conjugation of the carboxylic acid moieties on xenobiotics. Kidney and liver mitochrondria are known sites of localization of the amino acid conjugation enzymes for these substrates. In mammals, there are at least two forms of CoA ligases for benzoic acid as substrate; the two forms are distinguished by differential specificity for salicylate conjugation. There are also at least two *N*-acyltransferases, one with benzyl CoA as the preferred substrate and one with arylacetyl CoA as the preferred substrate. Additional forms of CoA ligase and *N*-acyltransferase are specific for bile acid conjugation. Liver microsomes have high activities of these enzymes.

XENOBIOTIC BIOTRANSFORMATION, HEPATOTOXICITY, AND HEPATOCARCINOGENICITY

As stated previously, biotransformation functions as a mechanism for modifying xenobiotics such that they acquire increased polarity and hence become excretable. This process can occur by one or more of several competing pathways, all of which lead to products that are excretable but differing in the relative degrees of electrophilicity of the intermediates in the pathways. Whether bioactivation or detoxification occurs is influenced by the relative activities of the competing pathways for biotransformation. Table 4–2 (Guen-

TABLE 4–2.
Xenobiotics Biotransformed to Hepatotoxins/Hepatocarcinogens*

Chemical Class	Examples	Enzymes Involved	Type of Transformation	Ultimate Toxic Metabolites
Alkyl carbon systems				
Substituted alkanes	Ethyl carbamate, aflatoxin B2	?	Desaturation to olefins (see below)	Expoxides?
Haloalkanes	CHC13,CC14, halothane	P450	Heteroatom release, reduction	Acyl halides, radicals
Vicinal dihaloalkanes	1,2-Dibromoethane, 1,2-dichloroethane, 1,2-dibromo-3-chloropropane	P450, GSH transferase	Heteroatom release, conjugation	Alpha-Halocarbonyls, half-mustards/epi-sulfonium ions
Heteroatom-substituted cyclopropyls	Benzylcycloproplylamine protein	P450, peroxidases, flavovation	One-electron oxidation, suicide inacti-	Methylene radicals
Olefinic carbon systems Olefins	Styrene, acrylonitrile	P450	Epoxidation, oxidative group transfer; suicide inactivation	Epoxides
Haloalkenes	Vinylidene chloride, trichloroethylene	P450	Epoxidation, oxidative group transfer, suicide inactivation	Epoxides, acyl halides, alpha-halocarbonyls
Dihydrofurans	Aflatoxins	P450	Epoxidation	Epoxides?
Aromatic carbon systems				
Polycyclic aromatic hydrocarbons	Benzo(a)pyrene, 7,12-dimethylbenzanthracene, napthalene	P450	Epoxidation	Epoxides, diol epoxides
Halobenzenes	Bromobenzene, pentachlorophenol	P450	Epoxidation, oxidative group transfer	Epoxides radicals (?), quinones (?)
Alkenylbenzenes	Safrole, estragole	?	C-oxidation, sulfation	Epoxides, sulfate esters

Acetanilides (paracetamol, Tylenol)	Acetaminophen dases oxidation	P450, peroxione- electron	Formal dehydration, qui- none radicals (?)	Quinoneimines, semi-
Furans ketone 3-methylfuran	Pyrrole, 3,4-dimethyl-	P450	?	?
Pyrroles pyrrole				
Thiophenes	Thiophene	?	?	?
Nitrogen-based-systems *N*- Nitrosamines	Dimethylnitrosamine	P450 eroatom release	C-Hydroxylation, hetsamines	α-Hydroxy N-nitro-
Pyrrolines alkaloids (lasiocarpine, monocrotalline)	Pyrrolizidine	P450	Heteroatom release	Pyrroles
Aminofluorenes acetylaminofluorene	2-Aminofluorene, 2-peroxidases	P450, FMO, elec- tron oxidation	*N*-Oxidation, one esters, radicals, nitroso compounds	Hydroxylamines, and
Aromatic amines benzidines, Trp-2	2-Naphthylamine, peroxidases	P450, FMO,	N-Oxidation esters	Hydroxylamines and
Azo dyes aminoazo benzene	*N,N*-Dimethyl-4-	P450, FMO sulfation	*N*-Oxidation, esters	Hydroxylamines and
Hydrazines dimethylhydrazine	Procarbazine,	P450, FMO *N*-oxidation, *C*- oxidation, car- bene formation	Dehydrogenation, thane?	CH3, CH4* diazome-
Nitro compounds	Ronidazole P450 reductase, others	P450, NADPH- oxygen radicals	Reduction	Nitro anion radicals,
Sulfur based systems Thiocarbonyls	Thioacetamide	FMO	S-Oxygenation	Sulfenes, sulfines
Thiophosphono compounds	Parathion, other insecticides	FMO, P450	S-Oxygenation	P=O compounds (paraoxon)
Metal-based systems Alkyl tins	Tetramethyl tin	P450	Dealkylation	Triethyl tin ?

*Modified with permission of Guengerich and Liebler (1985). See that publication for references. Reprinted with permission from *CRC Critical Reviews in Toxicology*. Copyright CRC Press, Inc., Boca Raton, Florida.

gerich and Liebler, 1985) lists xenobiotics for which the liver is a major target for toxicity and/or carcinogenicity and for which toxicity and carcinogenicity require biotransformation. Previous reviews by Gillette et al. (1974), Miller and Miller (1981b), Neal and Halpert (1982), Nelson (1982), Macdonald (1983), Ortiz de Montellano and Correia (1983), Guengerich and Liebler (1985), Dipple et al. (1985), and Kadlubar and Hammons (1987) provide detailed descriptions of biotransformation pathways and intermediates leading to toxicity and carcinogenicity for these compounds.

P450 plays a major role in the bioactivation of organic halides [reviewed by Macdonald (1983) and Kadlubar and Hammonds (1987); see also Chapter 13], which includes haloalkanes (chloroform, carbon tetrachloride, halothane), dihaloalkanes (1,2-dibromoethane, 1,2-dichloroethane, 1,2-dibromo-3-chloropropane), haloalkenes (vinyl chloride, vinyl bromide, vinylidene chloride, trichloroethylene), and halobenzenes (bromobenzene, chlorobenzene, pentachlorophenol). Haloalkanes can be biotransformed by oxidation to halohydrins that nonezymatically form reactive acyl halides, haloaldehydes, or haloacroleins; they can also be transformed by reduction to reactive free radicals and electrophilic halogen. Haloalkenes are converted to epoxides, vicinal haloaldehydes, and acyl halides and halobenzenes to epoxides, which are alkylating agents.

For the haloalkanes such as chloroform, oxidation of the carbon-halogen bond by P450 yields a halohydrin intermediate that is unstable and decomposes nonenzymatically to a carbonyl-containing compound with release of hydrogen halide. The carbonyl species are electrophilic and react nonenzymatically with cellular nucleophiles. In the case of chloroform, phosgene ($COC1_2$) is believed to be the electrophile mediating hepatotoxicity and hepatocarcinogenicity. CYP2B α-hydroxylates chloroform to the halohydrin, trichloromethanol, which dehydrochlorinates to phosgene. Phosgene can react with critical cellular nucleophiles, hydrolyze to carbon dioxide and HCl or form cysteine and glutathione conjugates.

Reductive biotransformation plays a role in the bioactivation of haloalkanes such as carbon tetrachloride and halothane. With P450, either one- or two-electron transfer can occur. NADPH P450 reductase can catalyze one-electron transfer. Because the haloalkane must compete with molecular oxygen for electron transfer, bioactivation of haloalkanes proceeds primarily by P450-mediated carbon-halogen oxidation, with reductive bioactivation taking precedence under hypoxic or anaerobic conditions or when the haloalkane structure precludes carbon-halogen oxidation (i.e., carbon tetrachloride). Enzymes of the CYP2B subfamily are able to catalyze one-electron reduction of carbon tetrachloride to the trichloromethyl radical. This radical induces lipid peroxidation and suicide inactivation of CYP2B. The trichloromethyl radical undergoes a hydrogen ion abstraction to form chloroform, a one-electron reduction to dichlorocarbene and electrophilic chlorine, or oxidation to phosgene. It has been proposed that two forms of P450 biotransform carbon tetrachloride to electrophiles (Kadlubar and Hammons, 1987). CYP2B may mediate reduction to the trichloromethyl radical, chloroform, and dichloromethylcarbene, and then CYP2E1 may catalyze the oxidation of trichloromethyl radical to phosgene and electrophilic chlorine. P450 reductive biotransformation of halothane (1,1,1-trifluoro-2-bromo-2-chloroethane) yields a 1,1,1-trifluoro-2-chloroethyl radical, which induces lipid peroxidation or covalently binds critical cellular nucleophiles. Oxidative biotransformation by P450 generates an unstable dihalocarbinol, which nonenzymatically decomposes, via dehydrobromination, to the electrophile, trifluoroacetyl chloride.

Haloalkenes are bioactivated by P450-mediated oxidation of the vinyl moiety. Oxidation of vinyl chloride or vinyl bromide yields the ultimate carcinogenic electrophiles, vinyl halogen oxides. These intermediates can nonenzymatically form an additional electrophile haloacetaldehyde, epoxides or covalently bind heme, and inactivate P450. Detoxification reactions include hydration of the vinyl halogen oxide by epoxide hydrolase and reduction of the haloacetaldehyde by aldehyde and alcohol dehydrogenases. CYP2B is the P450 subfamily most active for catalyzing epoxide formation from vinyl chloride and vinyl bromide. P450-generated acyl halides and haloaldehydes are also bioactivation intermediates in the biotransformation pathways for structurally related vinylidene chloride and trichloroethylene. The haloacetaldehyde is the major electrophilic metabolite. An oxygenated intermediate from P450 oxidation leads to nonenzymatic formation of a haloaldehyde or an epoxide or to P450 inactivation by heme alkylation. Epoxides are relatively minor bioactivation products. For vinylidene chloride, both CYP2A and CYP2B are most active for oxidation and most susceptible to inactivation.

Both P450 and glutathione transferase play a role in the bioactivation of dihaloalkanes. The major pathway for 1,2-dichloroethane involves glutathione transferase conjugation and subsequent formation of electrophilic episulfonium derivatives. In addition, P450-mediated carbon-halogen oxidation and halogen oxidation play a role in bioactivation. Carbon-halogen oxidation results in an unstable gem-chlorohydrin intermediate that nonenzymatically dehydrohalogenates to form the electrophile, 2-chloroacetaldehyde. Halogen oxidation by P450 yields 1-chloroso-2-chloroethane, which is electrophilic. This product nonenzymatically forms additional electrophiles. Alternatively, it may be conjugated by glutathione transferases, with ultimate formation of the episulfonium electrophile. Additional, but minor, microsomal bioactivation pathways include initial carbon-halogen reduction and dehydrohalogenation to electrophilic species.

Halobenzenes, such as bromobenzene, are similarly bioactivated by P450 to epoxides (arene oxides) [reviewed by Jollow and Smith (1977)]. Chlorobenzene, bromobenzene, iodobenzene, and *o*-dichlorobenzene are biotransformed to arene oxides. All induce hepatotoxicity, whereas fluorobenzene and *p*-chlorobenzene are not biotransformed to arene oxides and do not cause hepatic necrosis. Arene oxides are detoxified by spontaneous rearrangement to a phenol or by biotransformation by epoxide hydratase and/or glutathione transferases. Bromobenzene is biotransformed by two P450 pathways: a 3,4-bromobenzene oxide synthetase (CYP2B) and a 2,3-bromobenzene oxide synthetase (CYP1A). Although both arene oxides are electrophilic, the former is believed to be primarily responsible for hepatotoxicity. This compound illustrates how both dose and induction modify toxicity resulting from xenobiotic biotransformation. As the dose of administered bromobenzene increases, arene oxide formation increases as does the severity of necrosis due to P450 induction exceeding epoxide hydrolase induction and limited availability of glutathione. Although the CYP2B subfamily is predominant for bioactivation of halobenzenes, activation by the P450 family can reduce toxicity due to coordinate induction of epoxide hydrolase and glutathione transferase. In the case of bromobenzene, PB pretreatment at low doses results in reduced hepatic necrosis due to the induction of epoxide hydrolase and glutathione transferase; but at higher doses, PB pretreatment results in increased hepatic necrosis because the induction of P450 activity exceeds the increased epoxide hydrolase and glutathione transferase activities. If P450s are induced with 3-MC, bromobenzene hepatic necrosis is reduced because the pattern of epoxide formation is altered (increased proportion of 2,3-bromobenzene epoxide relative to 3,4-bromobenzene epoxide) and epoxide hydrolase is induced.

As is the case with halocarbons, suicide inactivation from covalent modifications of the enzymes for which the xenobiotics are the preferred substrates is a common mechanism of hepatotoxicity. Olefins, acetylenes, and heteroatom-substituted cyclopropyls are examples [reviewed by Ortiz de Montellano and Correia (1983)]. These latter compounds likewise lead to inactivation of P450 by alkylation of the prosthetic heme group after P450 biotransformation to a radical. The P450 gene family preferentially inactivated by compounds in these chemical classes also appears to be that which is PB-inducible. For fluorinated ether anesthetics such fluoroxene, hepatic necrosis is believed to result from alkylation of critical macromolecules; lethality is due to trifluoroethanol production after P450-mediated biotransformation [reviewed by Guengerich and Liebler (1985) and Kaminsky and Fraser (1988)]. This chemical class is biotransformed by at least three subfamilies of P450: CYP2B, CYP3A, and CYP1A1. Trifluoroethanol production is via epoxide formation by P450 or from heteroatom release after alkylation of the prosthetic heme group. CYP2B and CYP3A biotransformation is believed to favor trifluoroethanol production, whereas CYP1A1 is most susceptible to suicide inactivation by alkylation of the prosthetic heme group.

P450 biotransforms several other chemical classes to electrophiles. P450 biotransformation of pyrrolizidine alkaloids yields *N*-oxides, pyrroles, and demethylated or deesterified derivatives [reviewed by Guengerich and Liebler,

(1985)]. *N*-Oxide and pyrole formation are independent pathways. *N*-Oxidation is detoxifying and may be catalyzed by FMO; P450 biotransformation yields electrophilic pyrroles by dehydrogenation. This latter pathway involves hydroxylation at a carbon adjacent to the heterocyclic nitrogen to given an unstable carbinolamine intermediate, as in P450-mediated *N*-dealkylation. Aflatoxins are hepatotoxins that are bioactivated to a 2,3-epoxide [reviewed by Kadlubar and Hammons (1987)]. Aflatoxin B1 is the prototype and CYP1A2, CYP2A, and CYP2B are all active for oxidation to the epoxide. In humans, CYP3A4 bioactivates aflatoxin B1 (Shimada and Guengerich, 1989). In the rat, the CYP2B appears to be most active for epoxidation; however, exposure to inducers of P450 results in detoxification from increased proportions, relative to the epoxides, of ring-hydroxylated and *O*-demethylated derivatives.

Acetanilides, such as acetaminophen and phenacetin, are nonprescription analgesics that cause hepatic necrosis in overdosage. For acetaminophen, sulfation and glucuronidation are the major pathways for biotransformation [reviewed by Jollow and Smith (1977) and Nelson (1982)]. Minor bioactivation pathways for acetaminophen include direct oxidation by P450, peroxidases, or prostaglandin synthetase to the electrophile *N*-acetyl-*p*-benzoquinoneimine (NAPQI). Massive doses are required for toxicity, because glutathione transferase efficiently detoxifies this electrophile until glutathione is depleted to 15% of normal levels. NAPQI can also undergo hydrolysis to acetamide or be reduced by P450 back to acetaminophen. The P450 most active for bioactivation of acetaminophen is CYP2A. CYP2B and CYP2A1 produce the electrophilic metabolite to a lesser extent. Clinical toxicity is believed to require activation of the minor pathway by prior or concurrent exposure to P450 inducers; for example, the use of PB for anticonvulsant therapy. Induction of PB-inducible P450, however, can modify toxicity by coordinate induction of acetaminophen glucuronidation and reduction of NAPQI back to acetaminophen.

Polyaromatic hydrocarbons (PAH) are bioactivated by the sequential action of P450 and epoxide hydrolase [reviewed by Gelboin (1980) and Conney (1982)]. Initial biotransformation by P450 results in epoxide, phenol, and quinone intermediates. The oxygenated products undergo further biotransformations involving primarily P450, epoxide hydrolase and glutathione, glucuronide or sulfate conjugation. BP has been the prototype compound for the study of the pathways for PAH bioactivation. Formation of epoxides is believed to be the initial step. Three epoxides, BP 4,5-oxide, BP 7,8-oxide, and BP 9,10-oxide, are formed. In general, these epoxides are efficiently detoxified by rearrangement to a phenol, hydration through the action of epoxide hydrolase or glutathione conjugation. The BP 7,8-oxide is the epoxide that is biotransformed to the ultimate carcinogen. The biotransformation pathway is characterized by stereoselectivity. P450 biotransforms BP to the proximate carcinogen (+)-BP 7,8-oxide; very little (−)-BP 7,8-oxide is formed.

Epoxide hydrolase is stereoselective for hydration of (+)-BP 7,8-oxide to the optically pure (−)-*trans*-BP-7,8-diol. This BP-diol can then be biotransformed to four possible stereoisomeric 9,10-epoxides by CYP1A1. These are the enantiomeric pairs (1 and 2) of the cis- and trans-isomers. CYP1A1 is stereoselective for the formation of the ultimate carcinogen, (+)-*trans*-BP 7,8-diol-9,10-epoxide 2, the selection of which is based upon tumorigenicity data in mice. These latter epoxides are extremely electrophilic because, while they can be detoxified by glutathione conjugation, they are refractory to hydration by epoxide hydrolase.

The potential for toxicity and carcinogenicity, therefore, is believed to be related to the extent of formation of (+)-BP 7,8-diol-9,10-epoxide 2. The various P450 families exhibit positional specificity in the initial oxygenation of BP and the amount of (−)-*trans*-7,8-diol formed and stereoselectivity in biotransformation of BP(−)-*trans*-7,8-diol to the diol epoxides 1 and 2. The degree of enantiomeric purity of the BP metabolites that are formed depends on the relative activities of the P450 subfamilies. For BP in the rat, the CYP1A1 and CYP1A2 subfamilies appear to be predominant for bioactivation, whereas the CYP2A and CYP2B subfamilies have less involvement. In humans, CYP3A4 is predominant for bioactivation of the diol (Shimada et al., 1989). Although PAH-inducible P450 is involved in bioactivation, induction tends to result in detoxification from increased P450 activities that oxygenate PAH at positions resulting in less formation of epoxides and diols, from a minimal increase in epoxide hydrolase activity, and from increased UDP-glucuronyl transferase activity. Activation of PB-inducible P450 results overall in detoxification; but relative to PAH-inducible P450 activation, induction results in a greater increase in epoxide hydrolase activity, which both bioactivates and detoxifies PAH epoxides and diols.

Studies on the bioactivation and carcinogenicity of PAH has led to the bay-region theory of PAH carcinogenesis [Jerina and Daly (1977); reviewed by Conney (1982) and Dipple et al. (1984)]. This theory states that the amount of bay-region diol-epoxides formed from the parent compound is the primary determinant of potential for carcinogenicity. The ultimate carcinogens are believed to be diol epoxides on saturated benzo rings, with the epoxide oxygen *trans* to a benzylic hydroxyl group. Therefore, similar to BP, other PAH are bioactivated to diol epoxides by a sequence of three biotransformations by P450 and epoxide hydrolase. In general the (+)-*trans* epoxide diol is the ultimate carcinogen [reviewed by Dipple et al. (1984)]. Benzo(*e*)phenanthrene is an exception in that both cis- and trans-isomers are equally involved in carcinogenicity. The PAH structure determine the bay-region epoxide formed and the P450 family that is predominant for bioactivation. For example, in the rat, both CYP2B1 and CYP1A1 are believed to be most active in bioactivation of BP; CYP2B1, CYP1A1, and CYP1A2 for 2-aminoanthracene and 1,2,3,4-dibenzanthracene; and CYP1A1 for 7,12-dimethylbenz (*a*)anthracene and 3-MC (Shimada and Nakamura, 1987; Guengerich, 1988).

Several nitro polyaromatic hydrocarbon compounds are hepatocarcinogens. These include mono- and dinitropyrenes, 6-nitrochrysene, 7-nitrobenz(*a*)-anthracene, and nitrobenz(*a*)anthracenes. Bioactivation involves nitroreduction to nitrosoarenes, *N*-hydroxyarylamines, and arylamines; nitroaromatic ring oxidation and *N*-hydroxy arylamine *O*-acetylation [reviewed by Beland et al. (1985) and Kadlubar and Hammons (1987); see references therein]. Nitroreduction is believed to involve sequential one-electron reactions that can be catalyzed by several enzymes, including bacterial nitroreductases and mammalian enzymes such as P450 reductase, NADH-cytochrome b_5 reductase, NADH-dependent hydroxylamine reductase, cytosolic DT diaphorase, xanthine oxidase, aldehyde oxidase, alcohol dehydrogenase, and mitochondrial succinate dehydrogenase and dihydrolipoyl dehydrogenase. Nitroreduction can also occur nonenzymatically by endogeneous reducing agents, such as pyridine and flavin nucleotides. The nitroreduction products are electrophilic per se, but can be further bioactivated by *O*-esterification by acetyl CoA-dependent *O*-acetyltransferases. For dinitropyrenes, reduction to amines by P450 and NADPH-P450-reductase is inactivating. Both ring oxidation and nitroreduction are believed to be important for the bioactivation for nitro polyaromatic hydrocarbons such as 6-nitrochrysene, 3- and 6-nitrobenz-(*a*)pyrene, and 1-nitropyrene. Ring oxidation precedes nitroreduction; the nitro group determines the regioselectivity of ring oxidation. The nitro substituent inhibits periregion oxidation. The stereoselectivity of further biotransformation of the ring-oxidized compounds to dihydrodiol and diol epoxides is not effected by the nitro group, but the quasidiequatorial dihydrodiol conformation is favored over the quasidiaxial, which inhibits biotransformation to diol epoxides. After ring oxidation, nitroreduction results in ring-oxidized *N*-hydroxyamino-PAH, which are electrophilic and which may undergo further bioactivation by *O*-acetylation as mentioned above.

Nitrosamines are potent carcinogens in several organs and species. Liver is a primary target for hepatotoxicity and hepatocarcinogenicity. This chemical class can be divided into four subclasses: nitrosodialkylamines, cyclic nitrosamines, β-oxidized nitrosamines, and nitroarylalkylamines. P450-mediated biotransformation is believed to be an obligatory step in bioactivation, with the nitrosamine structure determining which P450 form is predominant for bioactivation [reviewed by Dipple et al. (1985) and Kadlubar and Hammons (1987)]. Most nitrosamines are thought to be bioactivated by P450-mediated *N*-dealkylation and α-carbon hydroxylation. *N N*-Dimethylnitrosamine (DMN) is structurally the simplest nitrosamine, and its biotransformation has been the most intensively studied. P450-mediated α-carbon hydroxylation is believed to ultimately result in formation of the putative carcinogenic electrophile, the alkyldiazonium ion. For DMN, at least two forms of rat P450 mediate the *N*-dealkylation reaction. One is a high-affinity and the other is a low-affinity system. In the rat, the high-affinity system appears to be CYP2E1 and the low-affinity system CYP2B. Enzymes other than P450, that is, FMO and a liver cytosolic enzyme, are believed to mediate α-carbon hydroxylation of DMN. This latter enzyme has a low K_M value for DMN α-carbon hydroxylation. *N*-Nitroso derivatives with longer alkyl side chains, cyclic structures, or aryl carbon substituents can be oxidized by various forms of P450 at different positions and at varying rates. Oxidation can occur at β-, ω-, ω-1, or aryl carbons and be a major determinant of the ultimate electrophile formed as well as the tissue-specific carcinogenicity. Nitrosamine structure can also result in bioactivation by pathways other than α-hydroxylation.

Hydrazine and its alkylated derivatives are carcinogens; liver is one of the primary target organs. Some hydrazines occur naturally in mushrooms. Procarbazine is a hydrazine derivative used as an antitumor agent. A series of oxidation steps results in the formation of carcinogenic electrophiles (Kadlubar and Hammons, 1987). Dimethylhydrazine and procarbazine have been the prototype agents for the study of the biotransformation of this chemical class. Azodialkanes are the first oxidation products formed. This pathway appears to be mediated by the CYP2B gene family and by monoamine oxidase. Initial oxidation by FMO results in a monoalkylated derivative and an aldehyde. Additional oxidations to azoxydialkanes and diazenes are separate bioactivation pathways. The electrophilic diazenes result from further oxidations of the primary oxidation products by P450, peroxidases and FMO. Formation of azoxydialkanes from the azodialkanes requires the sequential action of P450. *N*-Oxidation is catalyzed by CYP2C6 and CYP1A1. Next, α-carbon oxidation similar to that for nitrosamine bioactivation appears to be necessary for formation of the ultimate carcinogenic metabolite of azoxydialkanes, methylazoxymethanol (MAM) from DMH. CYP2E1 may catalyze this reaction. From MAM, the methyldiazonium ion, which is the same putative carcinogenic electrophilic as for nitrosamines, can form spontaneously or from catalysis by choline or alcohol dehydrogenase.

Aromatic amines, such as aminofluorene, benzidine, 2-napthylamine, and 2-acetylaminofluorene are carcinogens for rodent liver [reviewed by Miller and Miller (1981b), Dipple et al. (1985), Weber and Hein (1985), and Kadlubar and Hammons (1987)]. Three separate, but parallel, metabolic pathways result in electrophilic intermediates during biotransformation. The amino group undergoes hydroxylation by P450 and acetylation by acetyl CoA-dependent *N*-acetyltransferase. The resulting hydroxamic acid can then be bioactivated by PAPS-dependent sulfotransferase to the electrophilic *N*-hydroxy sulfate ester. Sulfotransferase is believed to be the primary enzyme of bioactivation for hepatocarcinogenicity since species, tissue, sex, and age sensitivity to arylamine hepatocarcinogenicity parallel the activity of this enzyme and since increases in dietary sulfate enhance the hepatocarcinogenicity of these compounds. A second bioactivation pathway involves rearrangement of the hydroxamic acid by *N*-acetyltransferase to yield *N*-acetoxyarylamine, an electrophile that spontaneously forms an additional electrophile, the arylnitrenium ion. This pathway is believed to be important for mammary gland tumorigenesis since sulfotransferase activity is absent in mammary gland. A third bioactivation pathway is responsible for urinary bladder carcinogenesis. After *N*-hydroxylation by P450, the *N*-hydroxyarylamine can be glucuronidated to form *N*-hydroxylamine-*N*-glucuronide. In the acidic environment of the urine, this glucuronide derivative nonenzymatically forms the electrophilic arylnitrenium ion. Another bioactivation pathway involves production of nitroxyl radicals by the action of peroxidases.

Arylamine bioactivation is the primary example of the modifying effect of genetics on biotransformation and, ultimately, the type and quantity of electrophilic metabolites formed. Rates of *N*-acetylation, *N*-hydroxylation, sulfation, and deacylation are dependent on genetic factors. P450 can ring hydroxylate (C-1, C-3, C-5, C-7, and C-9 positions) as well as *N*-hydroxylate arylamines. Ring hydroxylation is considered to be detoxifying since these intermediates do not result in phase II bioactivation products. Activities for ring versus *N*-hydroxylation vary among the P450 subfamilies; *N*-hydroxylation is catalyzed predominately by the CYP1A2 family. Induction of the CYP1A2 activity, however, results in inhibition of hepatocarcinogenesis because total oxidation is increased with the relative proportions of *N*-hydroxylation to ring-hydroxylated intermediates remaining unchanged. Species susceptibility to arylamine hepatocarcinogenesis has been related to activities of *N*- versus ring-hydroxylation. The guinea pig is resistant to arylamine hepatocarcinogenesis, and of several species studied (rabbit, mouse, rat, hamster, guinea pig), it has the highest activity for ring hydroxylation and the lowest activity for *N*-hydroxylation. Similarly, the acetylator phenotype influences bioactivation of arylamines. For example, dogs lack acetylation enzyme activity and do not bioactivate arylamines by this pathway.

Arylamine biotransformation provides an additional example of how dose and duration of exposure can influence the degree of toxicity. Sulfation, glucuronidation, and acetylation all result in electrophilic intermediates (Kadlubar and Beland, 1985). The acetylated and glucuronidated conjugates however, are more stable than the sulfated conjugate; therefore, acetylation and glucuronidation are considered detoxifying (Mulder, 1979). Acute exposures tend to lead to conjugation with the sulfate since it is a low-affinity, high-capacity system. As the dose increases, acetylation and glucuronidation tend to predominate since acetylation is an intermediate affinity and capacity system and glucuronidation is a low-affinity, high-capacity system. With chronic exposures, glucuronidation is favored since it is inducible and sulfation and acetylation are not.

Several additional compounds are bioactivated by phase II biotransformation after obligatory hydroxylation by P450. Alkenylbenzenes, such as safrole (1-allyl-3,4-methylenedioxybenzene) and estragole (1-allyl-4-methoxybenzene), are rodent hepatocarcinogens. P450 hydroxylates allylarenes at the 1′-carbon, resulting in a 1′-hydroxy metabolite, which is a proximate carcinogen [reviewed by Miller and Miller (1981a) and Miller and Miller (1983)]. This metabolite is weakly electrophilic per se and can be further biotransformed by P450 to electrophilic epoxides, but the major ultimate carcinogenic electrophile is believed to be the 1′-sulfuric acid ester resulting from sulfation at the 1′-hydroxyl group by PAPS-dependent sulfotransferase. The P450 family specific for alkenylbenzene biotransformation has not been characterized, but activation of the CYP2B subfamily enhances hepatocarcinogenicity by safrole. Aminoazo dyes, such as *N*-methyl-4-aminoazobenzene, undergo a similar bioactivation pathway. P450 biotransforms the parent compound to a *N*-hydroxy derivative, which is then bioactivated to the ultimate electrophilic metabolite, a sulfuric acid ester, by PAPS-dependent sulfotransferase. For *N*-methyl-4-aminoazobenzene, the P450 family most active for *N*-hydroxylation is CYP1A. Induction of this P450 enzyme activity, however, inhibits aminoazo dye carcinogenesis through increasing biotransformation by P450-mediated *N*-demethylation and azo reduction. Therefore, overall, there is less *N*-hydroxylated product for sulfation. Aminoazo dyes were one of the first examples of the dietary influence on biotransformation and the resulting manifestation of toxicity. Since flavin adenine dinucleotide is a cofactor for the enzymecatalyzed azo reduction, increases in dietary riboflavin have a protective effect on hepatocarcinogenesis by aminoazo dye.

CONCLUSIONS

A understanding of xenobiotic biotransformation is one of science's most notable examples of the practical applications and societal benefits derived from basic research. Studies on xenobiotic biotransformation have defined the roles of genetics, age, diet, and various environmental factors, and have demonstrated the importance of induction and inhibition in xenobiotic/drug exposures in individual sensitivity to chemicals [see Conney (1982) for numerous examples]. Perhaps the most important concept derived from research on xenobiotic biotransformation is that electrophiles generated during the process initiate toxicity and carcinogenicity by binding to critical cellular nucleophiles (Miller and Miller, 1981b). This concept has been the driving force behind a significant portion of research in carcinogenesis, toxicology, and pharmacology and has been utilized to assess the toxic and carcinogenic potential of drugs and xenobiotics [see Hammons et al. (1985) for an example] and to design chemotherapeutic regimens with enhanced tissue selectivity and drug analogs with reduced toxicity. Studies on the mechanisms of genetic control of the *Ah* locus (Nebert and Gonzalez, 1987) have contributed to current concepts on genetic regulation of enzyme and cellular function and to understanding the genetic basis for adaptive responses that provide cells' selective survival advantage in response to environmental insult.

Several central questions remain for future investigations on xenobiotic biotransformation. Many of the endogenous functions and substrates for enzymes of biotransformation are not yet defined. For example, despite intensive investigation the endogenous substrate for the *Ah* locus and its precise role in the genetic control of cellular differentiation and proliferation is not known. The consequences of xenobiotic exposure on anabolism and catabolism of endogenous substrates likewise is not known. The prime example is the possible adverse consequences of competitive and noncompetitive inhibition and induction of P450-catalyzed anabolism and catabolism of fatty acids and steroids (Conney, 1982). The role of electrophiles generated during xenobiotic biotransformations in promotion of neoplasia has received little attention, although there are examples where electrophiles play significant roles in promotion (Guengerich, 1988). PB is both a classic promoter of neoplasia and inducer of P450, but definition of a causal relationship between tumor promotion and P450 induction remains elusive. In fact, despite years of investigation, the molecular mechanisms by which PB mediates increased gene transcription of CYP2B1 and CYP2B2 remains a mystery. Similarly, a causal relationship between CYP4 induction by hypolipidemic agents, lipid lowering, and peroxisomal proliferation remains a central unanswered question. Although great progress has been made toward understanding the molecular mechanisms of the genetic control of enzymes of biotransformation, further progress is essential not only for the complete understanding of genetic regulation of these enzymes, but for understanding the basis for the genetic control of cell proliferation, differentiation, and neoplastic transformation. Not only are selected enzymes of biotransformation part of, or coordinately regulated with, the *Ah* gene battery, but enzyme activities involved in cell proliferation and differentiation, such as aldehyde dehydrogenase, phosphatidylcholine biosynthesis, arachidonic acid metabolism, lipid deacylation, phospholipase A_2, γ-glutamyltranspeptidase, and protein kinase C (Nebert and Gonzalez, 1987), may also be coordinately regulated with the *Ah* locus. Likewise, neoplastically transformed cells are believed to be resistant to cytotoxic therapy due to altered genetic regulation of enzymes of biotransformation such that detoxification is enhanced relative to bioactivation [for example, reduced P450 and increased epoxide hydrolase and phase II enzyme activities; Bock and Schirmer (1987)].

The vast majority of knowledge on xenobiotic biotransformation has come from rodent studies. The applicability of these findings to humans needs further definition (Walker, 1980; Guengerich, 1988), particularly in light of reliance of rodent bioassays for human safety and risk assessment in toxicity and carcinogenicity. Long-term goals include continued reductions in animal use for safety testing and risk assessment through prediction of toxicity from known biotransformation pathways and intermediates; routine phenotyping for biotransformation polymorphisms in individuals prior to job placement or drug therapy (Conney, 1982), and dietary (Wattenberg, 1978), lifestyle, and genetic interventions that favor detoxification relative to bioactivation during xenobiotic exposure. Research on xenobiotic biotransformation will continue to impact not only on the direction of basic science but on society as well.

ACKNOWLEDGMENTS

The author would like to thank Dr. Donald L. Hill of Southern Research Institute for editing the manuscript, and Dr. Daniel W. Nebert of NIH, Dr. F. Peter Guengerich of Vanderbilt University, and Dr. Daniel C. Liebler of the University of Arizona for reviewing the manuscript.

REFERENCES

Due to the extensive literature base of this discipline, references have been limited to review articles. These articles can be consulted for individual references of the subject area.

Adesnick M and Atchinson M: Genes for cytochrome P-450 and their regulation. *CRC Crit Rev Biochem*, 1986; 19:247–305.

Anders MW, Lash L, Dekant W, Elfarra AA, and Dohn DR: Biosynthesis and biotransformation of glutathione *S*-conjugates to toxic metabolites. *CRC Crit Rev Toxicol* 1988; 18:311–341.

Bains SK, Gardiner SM, Mannweiler K, Gillett D, and Gibson GG: Immunochemical study on the contribution of hypolipidaemic-induced cytochrome P-452 to the metabolism of lauric acid and arachidonic acid. *Biochem Pharmacol* 1985; 34:3221–3229.

Beland FA, Heflich RH, Howard PC, and Fu PP: The *in vitro* metabolic activation of nitro polycyclic aromatic hydrocarbons, in *Polycyclic Hydrocarbons and Carcinogenesis*. Harvey RG (ed): Washington, D.C., American Chemical Society, 1985, pp 371–396.

Bidlack WR, Brown RC, and Mohan C: Nutritional parameters that alter hepatic drug metabolism, conjugation, and toxicity. *Fed Proc* 1986; 45:142–148.

Bock KW, Bock-Hennig BS, Fischer G, Lilienblum W, and Schirmer G: UDP-Glucuronyltransferases and their toxicological significance. *Adv Exp Med Biol* 1986; 197:171–184.

Bock KW and Schirmer G: Species differences in glucuronidation and sulfation in relation to hepatocarcinogenesis. *Arch Toxicol Suppl* 1987; 10:125–135.

Caldwell J and Paulson GD (eds): *Foreign Compound Metabolism*. London, Taylor and Francis, 1984.

Conney AH: Induction of microsomal enzymes by foreign chemicals and carcinogenesis by polycyclic aromatic hydrocarbons: G.H.A. Clowes Memorial Lecture. *Cancer Res* 1982; 42:4875–4917.

Conney AH: Induction of microsomal cytochrome P-450 enzymes: the first Bernard B. Brodie Lecture at Pennsylvania State University. *Life Sci* 1986; 39:2493–2518.

Dannan GA, Guengerich FP, Kaminsky LS, and Aust SD: Regulation of Cytochrome P450. *J Biol Chem* 1983; 258:1282–1288.

Dipple A, Michejda CJ, and Weisburger EK: Metabolism of chemical carcinogens. *Pharmac Ther* 1985; 27:265–296.

Dipple A, Moschel RC, and Bigger CAH: Polynuclear aromatic carcinogens, in Searle CE (ed): *Chemical Carcinogens,* Am. Chem. Soc. Monogr. 182, Washington, DC, American Chemical Society, 1984, pp 41–163.

Eling T, Boyd J, Reed G, Mason R, and Swarajah K: Xenobiotic metabolism by prostaglandin endoperoxide synthetase. *Drug Metabolism Rev* 1983; 14:1023–1053.

Flammang TJ, Yamazoe Y, Guengerich FP, and Kadlubar FF: The *S*-acetyl coenzyme A-dependent metabolic activation of the carcinogen *N*-hydroxy-2-amino-fluorene by human liver cytosol and its relationship to the aromatic amine *N*-acetyltransferase phenotype. *Carcinogenesis* 1987; 8:1967–1970.

Gelboin HV: Benzo(*a*)pyrene metabolism, activation and carcinogenesis: role and regulation of mixed-function oxidases and related enzymes. *Physiol Rev* 1980; 60:1107–1166.

Gillette JR, Mitchell JR, and Brodie BB: Biochemical mechanisms of drug toxicity. *Annu Rev Pharmacol* 1974; 14:271–288.

Glatt HR, Mertes I, Wolfel T, and Oesch F: Epoxide hydrolases in laboratory animals and in man, in Greim H, Jung R, Kramer M, Marquardt H, and Oesch F (eds): *Biochemical Basis of Chemical Carcinogenesis.* New York, Raven Press, 1984.

Glatt HR and Oesch F: Species differences in enzymes controlling reactive epoxides. *Arch Toxicol Suppl* 1987; 10:111–124.

Gonzalez FJ: The molecular biology of cytochrome P450s. *Pharmacol Rev* 1989; 40:244–288.

Gonzalez FJ, Jaiswal, AK, and Nebert, DW: P-450 genes: evolution, regulation, and relationship to human cancer and pharmacogenetics. *Cold Spring Harbor Symp Quant Biol* 1986; 51:879–890.

Gorrod JW, Oelschläger H, and Caldwell J (eds): *Metabolism of Xenobiotics*, London, Taylor and Francis Ltd., 1988.

Guengerich FP: Effects of nutritive factors on metabolic processes involving bioactivation and detoxification of chemicals. *Annu Rev Nutri* 1984; 4:207–231.

Guengerich FP: Cytochrome P-450 enzymes and drug metabolism, in Bridges JW, Chasseaud LF, and Gibson GG (eds): *Progress in Drug Metabolism,* vol 10. London, Taylor and Francis, Ltd., 1987, pp 1–54.

Guengerich FP: Roles of cytochrome P-450 in chemical carcinogenesis and cancer chemotherapy. *Cancer Res* 1988; 48:2946–2954.

Guengerich FP: Characterization of human microsomal cytochrome P-450 enzymes. *Annu Rev Pharmacol Toxicol* 1989; 29:241–264.

Guengerich FP, Danna, GA, Wright STG, Martin MV, and Kaminsky LS: Purification and characterization of rat liver microsomal cytochrome P-450: electrophoretic, spectral, catalytic, and immunochemical properties and inducibility of eight isozymes treated with phenobarbital or beta-naphthoflavone. *Biochemistry* 1982; 21:6019–6030.

Guengerich FP, Danna, GA, Wright ST, Martin MV, and Kaminsky LS: Purification and characterization of microsomal cytochrome P-450s. *Xenobiotica* 1982b; 12:701–716.

Guengerich FP and Liebler DC: Enzymatic activation of chemicals to toxic metabolites. *CRC Crit Rev Toxicol* 1985; 14:259–307.

Gustafsson JA, Mode A, Norstedt G, and Skett P: Sex steroid induced changes in hepatic enzymes. *Annu Rev Physiol* 1983; 4:51–60.

Hammons GJ, Guengerich FP, Weis CC, Beland FA, and Kadlubar FF: Metabolic oxidation of carcinogenic arylamines by rat, dog, and human hepatic microsomes by purified flavin-containing and cytochrome P-450 monooxygenases. *Cancer Res* 1985; 45:3578–3585.

Hardwick JP, Song B-J, Huberman E, and Gonzalez FJ: Isolation, complementary DNA sequence, and regulation of rat hepatic lauric acid-w-hydroxylase (cytochrome P-450LAW): identification of a new cytochrome P-450 gene family. *J Biol Chem* 1987; 262:801–810.

Jakoby WB (ed): *Enzymatic Basis of Detoxification,* vols 1 and 2. New York, Academic Press, 1980.

Jakoby WB, Bend JR, and Caldwell J (eds): *Metabolic Basis of Detoxification of Functional Groups,* New York, Academic Press, 1982.

Jerina DM and Daly JW: Oxidation at carbon, in Parke DV and Smith RL (eds): *Drug Metabolism. From Microbe to Man.* London, Taylor and Francis, Ltd., 1977.

Jollow DJ and Smith C: Biochemical aspects of toxic metabolites: formation, detoxification, and covalent binding, in Jollow DJ, Kocsis JJ, and Snyder R (eds): *Biology Reactive Intermediates, Proceedings International Conference*, New York, Plenum, 1977.

Kadlubar FF and Beland FA: Chemical properties of ultimate carcinogenic metabolites of arylamines and arylamides, in *Polycyclic Hydrocarbons and Carcinogenesis*, vol 283, Washington, DC, American Chemical Society, 1985, pp 341–370.

Kadlubar FF and Hammons GJ: The role of cytochrome P-450 in the metabolism of chemical carcinogens, in Guengerich FP (ed): *Mammalian Cytochromes P-450,* vol 2. Boca Raton, CRC Press, 1987, pp 81–130.

Kaminsky LS and Fraser JM: Multiple aspects of the toxicity of fluroxene and its metabolite 2,2,2-trifluroethanol. *CRC Crit Rev Toxicol* 1988; 19:87–112.

Kato R: Characteristics and differences in the hepatic mixed function oxidases of different species. *Pharmac Ther* 1979; 6:41–98.

Ketterer B: Protective role of glutathione and glutathione transferages in mutagenesis and carcinogenesis. *Mut Res* 1988; 202:343–361.

Ketterer B, Coles B, and Meyer DJ: The role of glutathione in detoxification. *Environ Health Perspect* 1983; 49:59–69.

King CM and Glowinski IB: Acetylation, deacylation and acyltransfer. *Environ Health Perspect* 1983; 49:43–50.

Kitani K: The effect of aging on hepatic metabolism of xenobiotics, in Caldwell J and Paulson GD (eds): *First Proceedings International Symposium on Foreign Compound,* London, Taylor and Francis Ltd., 1984, pp 275–287.

Liebler CS, Lasker JM, Alderman J, and Leo MA: The microsomal ethanol oxidizing system and its interaction with other drugs, carcinogens, and vitamins. *Ann NY Acad Sci* 1987; 492:11–24.

Lock EA: Studies on the mechanisms of nephrotoxicity and nephrocarcinogenecity of halogenated alkenes. *CRC Crit Rev Toxicol* 1988; 19:23–42.

Macdonald TL: Chemical mechanisms of halocarbon metabolism. *CRC Crit Rev Toxicol* 1983; 11:85–120.

Mannervik B: The isoenzymes of glutathione transferase, in Meister A (ed). *Advances of Enzymology,* vol 57. New York, John Wiley and Sons, 1985.

Marselos M and Lindahl R: Substrate preference of a cytosolic aldehyde dehydrogenase inducible in rat liver by treatment with 3-methylcholanthrene. *Toxicol Appl Toxicol* 1988; 95:339–345.

Miller JA and Miller EC: Mechanisms of chemical carcinogenesis. *Cancer* 1981b; 47:1055–1064.

Miller JA and Miller EC: The metabolic activation and nucleic acid adducts of naturally-occurring carcinogens: Recent results with ethyl carbamate and the spice flavors safrole and estragole. *Br J Cancer* 1983; 48:1–15.

Miller EC and Miller JA: Searches for ultimate chemical carcinogens and their reactions with cellular macromolecules. *Cancer* 1989; 47:2327–2345.

Ortiz de Montellano PR and Correia MA: Suicidal destruction of cytochrome P-450 during oxidative drug metabolism. *Annu Rev Pharmacol Toxicol* 1983; 23:481–503.

Mulder GJ: Detoxification or toxification? Modification of the toxicity of foreign compounds by conjugation in the liver. *Trends Biochem Sciences* 1979; 4:86–90.

Neal RA and Halpert J: Toxicology of thiono-sulfur compounds. *Annu Rev Pharmacol Toxicol* 1982; 22:321–339.

Nebert DW, Adesnik M, Coon MJ, Estabrook RW, Gonzalez FJ, Guengerich FP, Gunsalus IC, Johnson EF, Kemper B, Levin W, Phillips IR, Sato R, and Waterman MR: The P-450 gene superfamily: recommended nomenclature. *DNA* 1987; 6:1–11.

Nebert DW and Gonzalez FJ: P-450 genes. Structure, evolution and regulation. *Annu Rev Biochem* 1987; 56:945–993.

Nebert DW and Jones JE: Minireview: regulation of the mammalian cytochrome P_1-450 (CYP1A1) gene. *Int J Biochem* 1989; 21:243–252.

Nebert DW, Nelson DR, Adesnik M, Coon MJ, Estabrook RW, Gonzalez FJ, Guengerich FP, Gunsalus IC, Johnson EF, Kemper B, Levin W, Phillips IR, Sato R, and Waterman MR: The P-450 superfamily: updated listing of all genes and recommended nomenclature for the chromosomal loci. *DNA* 1989; 8:1–13.

Nelson SD: Metabolic activation and drug toxicity. *J Med Chem* 1982; 25:753–765.

Netter KJ: Inhibition of oxidative drug metabolism in microsomes, in Schenkman JB and Kupfer D (eds): *International Encyclopedia of Pharmacology and Therapeutics*, vol 108: *Hepatic Cytochrome P-450 Monooxygenase System.* Oxford, Pergamon, 1982, pp 741–761.

Orton TC and Parker GL: The effect of hypolipidemic agents on the hepatic microsomal drug-metabolizing enzyme system of the rat. *Drug Metab Disp* 1982; 10:110–115.

Pickett CB and Lu AYH: Glutathione *S*-transferases: gene structure, regulation, and biological function. *Annu Rev Biochem* 1989; 58:743–764.

Reddy JK, Azarnoff DL, and Hignite CE: Hypolipidemic hepatic peroxisome proliferators form a model class of chemical carcinogens. *Nature* 1980; 283:397–398.

Rees KR and Tarlow MJ: The hepatotoxic action of allyl formate. *Biochem J* 1967; 104:757–761.

Reid WD: Mechanism of allyl alcohol-induced hepatic necrosis. *Experientia* 1972; 28:1058–1061.

Ryan DE, Thomas PE, Reik LM, and Levin W: Purification, characterization, and regulation of five rat hepatic microsomal cytochrome P-450 isozymes. *Xenobiotica* 1982; 12:727–744.

Satoh T: Role of carboxylesterases in xenobiotic metabolism. *Rev Biochem Toxicol* 1987; 8:155–181.

Shimada T and Guengerich FP: Evidence for cytochrome $P450_{NF}$, the nifedipine oxidase, being the principal enzyme involved in the bioactivation of aflatoxins in human liver. *Proc Natl Acad Sci USA* 1989; 86:462–465.

Shimada T, Iwasaki M, Martin MV, and Guengerich FP: Human liver microsomal cytochrome P-450 enzymes involved in the bioactivation of procarcinogens detected by umu gene response in *salmonella typhimurium* TA 1535/pSK1002. *Cancer Res* 1989; 49:3218–3228.

Shimada T and Nakamura S-I: Cytochrome P-450-mediated activation of procarcinogens and promutogens to DNA-damaging products by measuring expression of UMU gene in Salmonella typhimurium TA1535/p5K 1002. *Biochem Pharmacol* 1987; 36:1979–1987.

Sipes IG and Gandolfi A Jay: Biotransformation of toxicants, in Klassen CD, Amdur, MO, and Doull J (eds): *Casarett and Doull's Toxicology, The Basic Science of Toxicology,* ed 3. New York, MacMillan, 1986, pp 64–98.

Stock BH: Hepatic drug metabolism in pregnancy. *Rev Drug Metab Drug Interactions* 1984; 5:53–81.

Testa B and Jenner P: Inhibitors of cytochrome P-450s and their mechanism of action. *Drug Metab Rev* 1981; 12:1–117.

Walker CH: Species variations in some hepatic microsomal enzymes that metabolize xenobiotics, in Bridges JW and Chasseaud LF (eds): *Progress in Drug Metabolism*, vol. 5, New York, Wiley, 1980, pp 113–164.

Waterman MR and Estabrook RW: The induction of microsomal electron transport enzymes. *Mol Cell Biochem* 1983; 53/54:267–278.

Wattenberg LW: Inhibitors of chemical carcinogenesis. *Adv Cancer Res* 1978; 26:197–226.

Weber WW and Hein DW: *N*-Acetylation pharmacogenetics. *Pharmacol Rev* 1985; 37:25–79.

Wilkinson GR and Shand DG: A physiological approach to hepatic drug clearance. *Clin Pharmac Ther* 1975; 18:377–390.

Whitlock JP Jr: The regulation of cytochrome P-450 gene expression. *Annu Rev Pharmacol Toxicol* 1986; 26:333–369.

Wolf CR: Cytochrome P-450s: polymorphic multigene families involved in carcinogen activation. *Trends in Genetics* 1986; 2:209–214.

Ziegler DM: Flavin-containing monooxygenases. *Drug Metabol Rev* 1988; 19:1–32.

Chapter

5

Liver Function Tests in the Differential Diagnosis of Hepatotoxicity

Charles E. Cornelius
Department of Physiological Sciences
and
California Primate Research Center
University of California at Davis
Davis, California

INTRODUCTION

A multitude of tests are currently available to detect and diagnose hepatobiliary disorders in laboratory animals. Most biochemical tests generally fall into two categories: (1) tests which measure *specific liver functions* and/or hepatic blood flow, or (2) *serum markers* to detect necrosis, cholestasis, progressive fibrogenesis, or hepatoma. Most investigators select a battery of tests (a liver profile) since if one or more tests are abnormal, there is little probability of missing an occult hepatic disorder. It is imperative to understand what each test indicates with regard to morphologic, circulatory, or metabolic alterations. Examples are elevated serum enzyme activity suggesting necrosis and/or cholestasis; delayed hepatic clearance of foreign dyes or galactose indicating decreased hepatic blood flow or portasystemic shunts; elevated levels of serum bilirubin conjugates suggesting intra- or extrahepatic cholestasis; decreased plasma caffeine clearance indicating defective demethylation and function of the microsomal enzyme system; elevated serum levels of procollagen-peptide III levels suggesting a progressive fibrosis; and elevated serum α-1-fetoprotein levels suggesting hepatocellular carcinoma (Cornelius, 1987).

Most recognized laboratory tests are quite useful since they are noninvasive and can be easily performed with only one or a few blood samples required. Normal ranges of test values in each species under study must be established since values may vary due to strain, age, sex, circadian rhythm, nutrition, and, at times, from unique stresses present during the handling and housing of animals. A variety of other quality controls must be considered such as sample preservation and site of sample collection, animal health, anesthetic used, method of and speed of euthanasia, and the analytical method selected (Bartosek and Guaitani, 1982).

It is well known that land mammals can range in size from that of shrews at 3 g to elephants near 3000 kg and have many similar gross anatomic and histologic structures, as well as physiologic mechanisms (Mordenti, 1986). When values for liver weight and hepatic blood flow in a variety of terrestrial and aquatic mammals are plotted against body weight and submitted to linear (log-log) regression analyses, the slope of regression lines are 0.89 and 0.91, respectively (Prothero, 1982). Over this range ($1–10^8$ g) of body weights, the percentage of the body which is liver drops from 7.6% in the shrew to 0.9% in the whale. It is not clearly understood why larger mammals ''invest'' less in liver size, but this observation does pose questions regarding interspecies scaling of physiologic properties. Analyses of the pharmacokinetics of drugs used in liver function tests in numerous species has indicated that drug elimination among the species generally obeys the power equation $Y = aW^b$ (Mordenti, 1986). In this equation, Y is the compound measured, W the body weight, log a the y intercept, and b the slope (plot of log Y versus log W). Interspecies pharmacokinetic scaling may be approached from either an allometric (pharmacokinetic parameters) or physiologic (physiologic flow models) approach. In these ways, the more rapid elimination of compounds in small mammals can be scaled and more easily compared, if needed, with data observed in the larger species.

In this review, information on the evaluation of hepatic integrity will be primarily confined to known tests on mice, rats, rabbits, dogs, and nonhuman primates. A few interesting differences will also be discussed about certain other domestic animals, which are also used occasionally in biomedical research. A variety of new advanced techniques are now available to assess hepatobiliary disorders in humans. Examples are ultrasonography, CT scan angiography, and nuclear magnetic resonance imaging. Only a few studies have been performed as yet on laboratory and domestic animals.

INDICATIONS FOR AND CLASSIFICATION OF TESTS

Indications for Tests

Liver function tests and serum markers allow for the detection of a variety of specific hepatic disorders such as necrosis, cholestasis, decreased organ perfusion, and defective metabolic functions. Animals to be utilized in highly complex metabolic or toxicological studies should be first checked to ascertain if any primary or secondary hepatic disorders are present; in this way, normal and trusted control values can be established for a laboratory. If liver abnormalities are present, tests can aid in the specific diagnosis of the type of lesions present, in prognosis and for evaluating approaches and responses to therapy.

Tests are particularly useful in the detection of hepatotoxicities occurring in animals following the administration of drugs. Since drugs administered *per os* must first pass through the liver after intestinal absorption, most toxicological studies include a liver profile. Any detection of hepatic injury must be assessed as to the types and reversibility of lesions, margins of safety with regard to dosage, and ultimately their benefits versus risks. Drugs producing hepatotoxicity in more than one animal species are usually rejected for clinical trials. For an animal to be useful as a model for humans, it should possess an analogous metabolism of the drug administered. Although small rodents are used due to their ease of handling and cost effectiveness, nonhuman primates have been shown, in many cases, to more closely resemble humans in their metabolism of a variety of drugs and may be more ideal animals to use in measuring drug hepatotoxicities. In studies with dogs and monkeys concerning the prediction of the clinical toxicity of cancer chemotherapeutic agents, the combined use of dogs and monkeys may result in a favorable overlap in positive data and reduce the number of false-negative predictions (Schein and Anderson, 1973).

Collection of Blood Samples

In mice, blood is most conveniently collected by heart puncture, but is not without risk (0.7 ml of blood from 25-g mouse). Anesthesia should be used. Blood can also be obtained from the retro-ocular sinus or by tail laceration. In rats, blood (up to 10 ml) can be obtained from the orbital sinus, cardiac puncture, or from jugular or femoral veins. In rabbits, blood can be obtained from ear veins, by cardiac or femoral vein puncture; while in the dog, commonly used sites are the external jugular, saphenous, or antebrachial cephalic veins. In monkeys, either the cubital vein at the elbow or great saphenous vein are easily sampled following anesthesia. The use of heparinized plasma for most tests in small rodents eliminates hemolysis and ensures ample sample for determinations.

Classification of Tests

Laboratory tests to assess hepatic disorders in animals and humans can be arbitrarily classified as either (1) tests that truly measure a liver function or (2) those measuring abnormal levels of serum constituents (markers) resulting from hepatopathy (Cornelius, 1987). Some overlap does exist between classifications.

Function Tests

1. Hepatic clearance of organic anions: bilirubin, sulfobromophthalein (BSP); indocyanine green (ICG); and bile acids.
2. Hepatic blood flow: galactose and IGG clearance.
3. Hepatic microsomal function; aminopyrine breath test and caffeine clearance test.

Serum Markers

1. "Liver-specific" hepatic enzymes to measure hepatocellular necrosis: alanine aminotransferase in small animal species (ALT); glutamic dehydrogenase (GD), sorbitol (iditol) dehydrogenase, ornithine carbamyltransferase (OCT), arginase and guanase in all mammals.
2. "Liver-specific" hepatic enzymes to measure hepatic cholestasis (intrahepatic or extrahepatic biliary obstruction); alkaline phosphatase hepatic isoenzyme (AP); γ-glutamyl transferase (GGT); 5′ nucleotidase (5′NT); and leucine aminopeptidase (LAP).
3. Other enzymes present in high concentrations in both liver and other tissues and not as "liver-specific": aspartate aminotransferase (AST); isocitric dehydrogenase (ICD) and lactic dehydrogenase (LDH).
4. Procollagen type III peptide levels (PIIIP) to assess if hepatic fibrogenesis is active.
5. Ammonia levels (blood) in hepatoencephalopathies.
6. Ligandin levels in necrosis and hepatoma.
7. Hyaluronate levels due to hepatic endothelial cell damage.
8. α-1-Fetoprotein (AFP) levels to detect hepatoma.
9. Carcinoembryonic antigen (CEA) levels to detect cancer metastasis to the liver.
10. Elevations of antibodies against a variety of cellular components: mitochondrial, nuclear and specific liver membrane proteins.
11. Albumin, globulin, amino acids, cholesterol, and other lipids.

Biochemical analysis of a variety of minerals, metabolites, and enzymes obtained from liver biopsies can also be quite useful in studying specific biochemical defects in inherited, acquired, and experimentally induced hepatobiliary disorders. The multitude of tests listed in the above list have not all been well-tested in laboratory animals. Only those which have been studied in depth or are promising for future investigations will be covered in this review.

FUNCTION TESTS

Serum Bilirubin Levels or Clearance

The hepatic clearance of large organic anions such as bilirubin involve (1) hepatic uptake at the sinusoid involving a facilitated diffusion mechanism utilizing specific membrane proteins; (2) intracellular binding to ligandin, followed by enzymatic conjugation of bilirubin in the endoplasmic reticululum with either glucuronic acid, glucose, or xylose; and (3) subsequent excretion into the bile against a concentration gradient (Cornelius, 1987). Unconjugated bilirubin must be conjugated to a polar sugar conjugate in order to be excreted into the bile. Such bilirubin conjugates in mice, rats, and monkeys are primarily diglucuronides, while canine bile also contains appreciable amounts of glucose and xylose conjugates. The rabbit excretes equal amounts of biliverdin and bilirubin conjugates into bile due to the existence of only trace amounts of biliverdin reductase in the liver and spleen (George et al., 1988).

Bilirubin tolerance tests are no longer used; however, the level of unconjugated and conjugated bilirubin in the serum are reliable markers and generally indicate the type of jaundice present. Bilirubin is produced in the body primarily from heme turnover in erythrocytes (75–80%) with lesser amounts originating from hepatic hemoproteins (P450, etc.) and ineffective erythropoiesis. Nearly all bilirubin except for minor amounts originating from hepatic heme turnover enters the plasma compartment prior to its excretion by the liver (Blanckaert and Schmid, 1982). The serum unconjugated bilirubin level, under steady-state conditions, is inversely related to its hepatic clearance rate and directly related to its production rate. Conjugated bilirubin is found in the serum due to its regurgitation into the serum if cholestasis is present.

Most laboratories exclusively use some modification of the *Van den Bergh test.* This test will many times indicate the existence of trace amounts of "direct-reacting" conjugated bilirubin in animals; this may be due to false-positive reactions and not detected by chromatographic procedures. Increased "indirect-reacting" bilirubin in the serum suggests the presence of hemolysis or defective hepatic uptake or conjugation of bilirubin. "Direct-reacting" pigment points towards hepatobiliary cholestasis. It is not uncommon to have both unconjugated and conjugated bilirubin in both hemolytic and cholestatic syndromes. Such measurements are useful clinically but only estimate and do not accurately reveal the exact amounts of bilirubin as available from newer and more elaborate TLC or HPLC procedures (Blanckaert,

1980; Blanckaert et al., 1980). These new techniques provide extremely sensitive methods which are capable of even measuring small increases in the level of conjugated bilirubin in normobilirubinemic sera in moderate liver disorders (Scharschmidt et al., 1982). The recent finding of an additional direct-reacting bilirubin fraction in the serum during cholestasis has finally explained the occasional presence of major amounts of direct-reacting serum pigment when bilirubinuria is not present. This "delta" bilirubin or "biliprotein" is not cleared into the urine as are bilirubin conjugates, since delta bilirubin is covalently bound to albumin (Weiss et al., 1983).

A substantial neonatal hyperbilirubinemia is only present during the first few days of life in nonhuman primates, like in humans. Serum unconjugated bilirubin levels may be elevated in neonates up to 6 mg/dl and present for 3–4 days. Such persistent hyperbilirubinemia in simian neonates can be prevented by administering heme oxygenase inhibitors such as Sn or Zn protoporphyrin (Cornelius and Rodgers, 1984).

Serum Bilirubin Levels

Total bilirubin values in normal sera is over 95% unconjugated pigment. All experimental animals have less than 0.5 mg/dl (0.85 μmol/dl) of bilirubin except for Bolivian and Peruvian squirrel monkeys (and horses) which exhibit higher total serum unconjugated bilirubin values up to 4 mg/dl (5.5 μmol/dl) upon a 24-h fast. If these squirrel monkeys are used in experimental studies, it must be understood that they exhibit a "Gilbert-like" syndrome with fasting hyperbilirubinemia (Cornelius et al., 1985). Representative and selected values which favorably agree with other data in the literature are presented in Table 5–1.

Urinary Bile Pigments

Since unconjugated bilirubin does not pass the renal filter, the urine contains only conjugated (noncovalently bound) bilirubin. The presence of bilirubin conjugates in the urine suggests that a cholestatic syndrome is present, either of an intrahepatic or extrahepatic obstructive process. Most dogs, unlike other species, normally excrete small amounts of bilirubin conjugates into the urine. In severe obstructive processes, large amounts may be present in the urine due to its regurgitation to the serum and is a sensitive test for cholestasis. The presence of urobilinogen in the urine indicates that total bile duct obstruction is not present, while increased levels can occur from either excessive hemolysis or hepatocellular diseases. Erythrocyte hemolysis results in increased quantities of heme available for bilirubin and urobilinogen production; in hepatic disorders, less is recycled in the enterohepatic circulation and lost to the serum and urine (Cornelius, 1987). Following phototherapy and irradiation of unconjugated bilirubin, its photooxidation products are efficiently removed by the kidney (Elias et al., 1987).

TABLE 5–1.
Total Serum Bilirubin

Species	X ± SD	X ± SD	Reference
	mg/dl	μmol/dl	
Mice	0.25 ± 0.16	(0.43 ± 0.27)	Harrison et al. (1978)
Rats	0.2 ± 0.01	(0.34 ± 0.02)	Lazar et al. (1974)
Rabbits	0.21 ± 0.26	(0.36 ± 0.44)	Melby and Altman (1974)
Dogs	0.23 ± 0.11	(0.39 ± 0.19)	Mitruka and Rawnsley (1977)
Chimpanzees	0.2 ± 0.1	(0.34 ± 0.02)	Wisecup et al. (1969)
Baboons	0.2 ± 0.2	(0.34 ± 0.03)	Kruckenberg et al. (1972)
Rhesus monkeys	0.2 ± 0.3	(0.34 ± 0.05)	Evans et al. (1953)
Brazilian squirrel monkeys			
fed	0.28 ± 0.03	(0.48 ± 0.05)	Cornelius et al. (1985)
fasted	0.44 ± 0.03	(0.75 ± 0.08)	
Bolivian squirrel monkeys			
fed	0.40 ± 0.03	(0.68 ± 0.05)	Cornelius et al. (1985)
fasted	2.0 ± 0.20	(3.4 ± 0.3)	
Howler monkeys	0.7 ± 0.1	(1.2 ± 0.2)	Katz et al. (1968)
Marmosets	0.20	(0.34)	Deinhardt et al. (1967)

Sulfobromophthalein (BSP) Clearance

The hepatic transport of BSP is similar to that of bilirubin; however, BSP is partly conjugated to glutathione and excreted either as free BSP or as a conjugate. BSP retention can be observed due to low hepatic glutathione levels, e.g., as observed in hyperthyroidism (Sir et al., 1987). BSP utilizes the same sinusoidal-membrane facilitating protein, also binds to ligandin, and competes with conjugated bilirubin for excretion (Cornelius, 1987). This foreign dye has been utilized for some time to detect hepatobiliary disorders in animals. More recently, other tests such as the measurement of serum markers (enzymes and bile acids) have proved equally sensitive and require only collecting a heparinized blood sample. BSP clearance is usually measured in large animal species (horses, cattle) as the BSP $T_{1/2}$ since its removal from the plasma is by a single exponential process during the first 15 min. In smaller animals, its removal rate is usually measured from one blood sample as a percent retention of the original dose since the dye is rapidly cleared and not by one exponential "first-order" process (Cornelius, 1988). In smaller animals in which BSP is cleared from the plasma by two exponential processes (K_i and K_2), the transport maximum (T_m) can be estimated from the slower exponential component K_2. The clearance of dye by the liver can be altered by a variety of factors: the serum albumin concentration, volume of distribution, hepatic uptake from plasma, excretion into the canaliculus, and obstructive processes within the ductular/ductal system.

In the mouse, serum BSP concentration was found to be 1.15 ± 2.3 mg/dl (1.37 ± 2.74 μmol/dl) at 15 min after the injection of 100 mg/kg (Kutob and Plaa, 1962). Following a 75-mg/kg dose via the tail vein in another study, serum contained 1.0 ± 0.7 mg/dl (1.37 ± 0.83 μmol/dl) at 30 min after injection (Becker and Plaa, 1967). Estimates of $T_{1/2}$ for BSP clearance in mice were slightly over 15 min. Following the administration of halogenated hepatotoxins, a good correlation was obtained between BSP retention and histomorphologic evidence of hepatic injury.

Rats have been studied extensively as animal models. BSP plasma clearance is slightly higher in females than in males (Orzes et al., 1985); it is apparently due to a higher affinity of and not the number of BSP affinity sites on hepatocytes. Phenol-3,6-dibromophthalein disulfonate has been used extensively as an analog of BSP and is not biotransformed prior to biliary excretion (Klaassen and Strom, 1978). BSP plasma disappearance increases following phenobarbital administration in rats (Klaassen, 1970). Impaired BSP biliary excretion, hepatic storage, and plasma retention have been observed as early as 3 h after CCl_4 administration (Priestly and Plaa, 1970). Lineweaver-Burk plots of the BSP biliary excretion rates versus the BSP serum concentration curve in rats indicate a higher apparent K_m is present in aflatoxin-treated rats (Carrillo et al., 1985).

Following the injection of 60 mg/kg of BSP, plasma levels in rats were less than 5 mg/dl (5.97 μmol/dl) at 30 min (Priestly and Plaa, 1970). Others (Mancini and Kocsic, 1974) observed plasma BSP levels of 1.05 ± 0.26 mg/dl (1.25 μmol/dl) at 30 min after the injection of 60 mg/kg. In this study, DMSO doubled the hepatotoxic potential of CCl_4.

In the dog, the rate of BSP disappearance (and percent retention) is independent of intravenous dosage between 5 and 20 mg BSP per kilogram (Cornelius, 1988). The use of 5 mg/kg is generally accepted among veterinary practitioners and usually detects hepatobility disorders involving necrosis and cholestasis. Less than 5% retention at 30 min has been accepted as within normal limits. BSP retention also detects congenital portasystemic shunts in puppies where little BSP is presented in the liver. Increased BSP retention has been observed in a variety of canine hepatopathies. Any disorders effecting hepatic blood flow will also result in delayed BSP clearance due to diminished hepatic perfusion.

In nonhuman primates, BSP clearance has been utilized extensively in the past, although newer tests are now used more commonly (Warren et al., 1984). BSP percent retention and fractional clearance (In $2/T_{1/2}$) in rhesus monkeys is reported to be <5% in 30min and 0.2 ± 0.04% per minute, respectively (Vogin et al., 1966). BSP retention in chimpanzees is reported as 5 ± 3% in 20 min (Wisecup et al., 1969) within the range of 2–8% (Deinhardt et al., 1962, 1967). In Howler monkeys, fractional clearance (K) is reported as 0.3 (0.2–0.5%)/min (Katz et al., 1968). Brazilian and Bolivian squirrel monkeys have BSP K values of 0.13 ± 0.14 and 0.19 ± 0.21%/min, respectively (Cornelius et al., 1985). No delay in BSP clearance was observed in Bolivian and Peruvian squirrel monkeys with a Gilbert-like syndrome as in a minority of the cases of Gilbert's syndrome in humans (Cornelius et al., 1985). Delayed BSP clearance has been observed in a multitude of hepatopathies in nonhuman primates (Kruckenberg et al., 1972).

Indocyanine Green (ICG) Clearance

ICG clearance has been used for some time to estimate circulation time, its plasma clearance, and hepatic blood flow (Keiding and Skak, 1985). In this section, its use in measuring hepatic clearance will be covered. The most simple approach is to measure its $T_{1/2}$ for hepatic clearance from the peripheral blood after the injection of 0.5 mg/kg (Warren et al., 1984). ICG is cleared much like BSP, however, it is not conjugated prior to excretion into bile and not reabsorbed in the gut into the enterohepatic circulation. A more complex method uses the constant infusion of ICG and measurement of steady-state ICG levels in a peripheral vessel (A) and the hepatic vein (V). Blood flow = infusion/A – V. From this same data, hepatic V_{max} and K_m of ICG can be calculated. Since the plasma clearance of ICG follows Michaelis-Menten kinetics, the use of Lineweaver-Burk plots allows for the calculation of V_{max} and K_m for its removal (Paumgartner et al., 1970).

In mice, a special external measuring device revealed ICG $T_{1/2}$'s of 2.6 and 3.7 min after a 4 mg/kg dose (Danielson et al., 1971). $T_{1/2}$'s increased to between 3.9–4.4 min after a 24 h fast and deprivation of water. ICG $T_{1/2}$ in rats has been recorded as 2.2 min (Vonk et al., 1974). Initial fractional clearances (K) decreases progressively in rats when the ICG dose is increased from 0.25–2.5 mg/100 g body wt ($T_{1/2} = 0.692/K$); $T_{1/2}$ values correspondingly increase to 1.5 to 6.9 min (Paumgartner et al., 1970). It is recommended that levels of 5.0 mg/kg furnishes the best level of dye to inject to measure hepatic function. Low concentration of injected dye are affected primarily by the liver blood flow. Since it is not possible to use a large enough dose of dye to measure the hepatocyte's capacity to remove ICG independent of blood flow, one must assess ICG V_{max} from the kinetic analysis of several submaximal values. By this method, V_{max} and K_m have been determined in normal rats to be 6.5 (5.7–7.5) mg/100 g liver/min and 0.44 (0.33–0.56) mg/kg, respectively (Paumgartner et al., 1970).

Clearance studies in dogs has revealed ICG $T_{1/2}$'s of 8.4 ± 2.3 min during the first 12 min after the intravenous injection of 0.5 mg/kg (Bonasch and Cornelius, 1964). Plasma volumes can also be determined by extrapolating the plasma ICG log concentration plot to zero time (mg injected/0 time concentration). Similar $T_{1/2}$ values were reported by Banaszak et al. (1960). ICG plasma $T_{1/2}$'s in cats receiving 1.0 mg/kg are reported to be approximately 5.0 min (Burczynski et al., 1987).

$T_{1/2}$ values for ICG clearance have been reported in a variety of nonhuman primates (minutes): rhesus monkeys at 0.5 mg/kg, 1.9 ± 0.11 (Vogin et al., 1966), 1.7 ± 0.5 (Warren et al., 1984); squirrel monkeys at 5.0 mg/kg, 15.4 (Cornelius et al., 1985); and baboons at 0.25 mg/kg, 6–8 (Grainger et al., 1983). Average V_{max} and K_m values for ICG clearances using a Lineweaver-Burk plot were 0.42 mg/kg/min and 0.541 mg/kg (Warren et al., 1984). Delayed ICG clearance occurs in a variety of hepatobiliary disorders which effect the hepatic transport of the dye into the bile (Cornelius, 1988).

Serum Bile Acids (SBA)

There are a number of methods currently used in measuring bile acids. Radioimmunoassay (RIA); 3-α-hydroxy steroid dehydrogenase; HPLC; and luminescent procedures (Cornelius, 1987). A recent review comprehensively details the use of SBA's in evaluating hepatic disorders in domestic animals (Parraga and Kaneko, 1985). The new simplified RIA test for cholylglycine is easily performed; however, it does not measure all the bile acids and their conjugates, which vary among the species. One can, however, obtain normal values for most species using this test. The enzymatic method has recently been adapted for the "random-access" analyzer RA1000 (Technicon Instruments Corp.).

Most bile acid studies have been performed in the fasting state; normal fasting SBA levels are lower than postprandial levels. There remains a controversy over whether fasted or postprandial levels are more "diagnostic"; however, most investigators are using fasting SBA levels (Cornelius, 1987). Although they may not offer more information than other tests about hepatic necrosis or obstruction, they are used by many investigators as part of the liver profile. SBA levels are equally as sensitive as serum enzyme tests to detect hepatic necrosis and fibrosis. They quantitatively have a 100-fold greater flux in the enterohepatic circulation than bile pigments; elevations occur more rapidly than bilirubin in the serum in hepatobiliary disorders. Normal values have been established for most species (μmol/l): rats, 6.5 ± 1.1; dogs, 2.3 ± 0.4 (Center et al., 1985) and 2.6 ± 0.4 (Meyer, 1986); cats, 1.7 (Center et al., 1985); minipigs, 16.4 (Krocker and Romer, 1984); and Bolivian and Brazilian squirrel monkeys, 10 ± 3 and 16 ± 3, respectively (Cornelius et al., 1985).

Hepatic Blood Flow

The liver has two afferent blood supplies: (1) the portal veins which drain the intestine and spleen (70–75% of blood supply) and (2) the hepatic artery (25–30% of blood supply). Under resting conditions, total hepatic flow is approximately 25% of cardiac output, ranging 100–130 ml/min/100 g liver in dogs, cats, and humans. Approximately 30% of hepatic volume is blood, representing 12% of the total blood volume (Lautt and Greenway, 1987). Food ingestion increases flow while in exercise, and flow is decreased as blood is shunted to the muscles and brain. A prime regulator and mover of blood in the sinusoids of normal animals comes from the hepatic anterioles, whose tone is regulated in part by the local concentration of adenosine secreted into the space of Mall (Lautt and Greenway, 1987). Portal pressure and flow depend greatly upon the degree of constriction or dilatation of both the mesenteric arterioles and the hepatic veins. Hepatic fibrosis generally results in increased portal pressure, decreased hepatocellular perfusion, and ultimately the shunting of blood around the liver via anastomoses with other venous systems.

Although hepatic blood flow can be accurately measured by using implanted electromagnetic flow meters, thermostromuhrs, transonic flow probes, and direct blood collections, these techniques are complex and suffer from the effects of invasive procedures. Indirect measurements using indicator dilution methodology can be used, but the hepatic clearance of substances from the plasma lend themselves more easily to studies using a number of experimental animals (Campra and Reynolds, 1982). Indicator-dilution tests involve the injection of a tracer bolus such as ^{131}I-albumin upstream in the portal vein and the measurement of its concentration continuously downstream in the catheterized hepatic vein. The volume of flow is measured by the dilution of a known quantity of tracer over a specific time period.

The easiest techniques to use, however, are based upon the Fick principle involving the selective removal of dyes or metabolites by the liver. Substances usually utilized are galactose and indocyanine green (ICG) which are exponentially removed from the plasma primarily by the liver as a first-order process. Although a *continuous infusion method* can be used, the easiest approach on a number of animals is by the *single injection method.* Sequential tests are also possible by using this method. ICG can be used; however, its extraction efficiency must be determined by measuring simultaneous hepatic venous and peripheral blood concentrations. Galactose has an advantage since the extraction ratio is near one at plasma concentrations below 1.25 mM/l. If plasma galactose concentrations are above 2 mmol/L (which is above V_{max}), its plasma disappearance or galactose elimination capacity (GEC) is accounted for by the quantity of hepatic galactokinase present (Zaramello et al., 1985). At lower plasma galactose concentrations, the determination of hepatic plasma flow can be determined by multiplying its fractional clearance, k (min^{-1}) by the initial distribution volume, where $k = \ln 2/T_{1/2}$. Plasma flow is next converted to blood flow by adjusting with hematocrit values. The galactose clearance test should become an attractive addition to more sensitive liver profiles since both GEC and the hepatic blood flow can both be estimated with minor adjustments (Tvedegaard et al., 1985). It is critical that investigators establish their own normal values by using either ICG or galactose. References to selected articles using experimental animals are mice (Danielson et al., 1971); rats (Paumgartner et al., 1970); Rabinovici and Shapira, 1966; Denis et al., 1975); dogs (Banaszak et al., 1960; Bonasch and Cornelius, 1964); cats (Burczynski et al., 1987); baboons (Grainger et al., 1983); and rhesus monkeys (Vogin et al., 1966).

Microsomal Function

The liver's ability to rapidly complete specific biotransformations can be assessed by using drugs as probes (Kaplovitz et al., 1982). Drugs with low hepatic intrinsic clearance can be used to measure the functional liver mass independent of the effects of hepatic blood flow. By using drugs such as *aminopyrine* and *caffeine,* hepatic microsomal demethylation activity can be assessed. ^{14}C-aminopyrine has been used as a breath test in rats (Lauterburg and Bircher, 1976), horses (Kupfer et al., 1977), and humans (Hofmann, 1982). The radioactive methyl groups of ^{14}C-aminopyrine are removed by microsomal demethylation, liberating ^{14}C-formate which is ultimately converted to and measured as $^{14}CO_2$ in the expired breath (Goldberg and Brown, 1987). The rate of $^{14}CO_2$ correlates particularly well with more advanced hepatocellular disorders with diminished microsomal function. Additional studies are needed to assess the value of this complex test in laboratory animals.

A more simple test to measure microsomal function (*N*-demethylation) is the *caffeine clearance test* (Burg, 1975). Human and canine neonates clear caffeine slowly from the plasma; this reflects both their liver's lack of metabolizing capacity and ability to excrete it into the urine (Warszawski et al., 1977). Mean $T_{1/2}$ values for plasma elimination in day-old puppies and adult dogs after a 50-mg/kg dose is 47 ± 5 and 6.6 ± 0.8 min, respectively. $T_{1/2}$ values reach normal adult levels by 2 weeks of age. Considerable variation exists in the kinetics of plasma caffeine clearance in various animals (Bonati et al., 1984). In comparative toxicological tests which assume linear kinetics, it was necessary to administer 9 mg/kg to rabbits and mice, 5 mg/kg to rats and 2 mg/kg to monkeys to mimic a 1-mg/kg dose of human blood exposure. The plasma fasting caffeine level is quite sensitive and will even detect the severity of impairment of microsomal function in chronic caffeine drinkers (Wahllander et al., 1985). Its use in children has been proposed since its clearance from plasma can easily be measured in saliva (Jost et al., 1985). The test correlates well with other tests such as ICG clearance, the aminopyrine breath test, and galactose elimination capacity. Such a noninvasive test may be useful in animals and replace more cumbersome and invasive techniques (Cornelius, 1987). The proper physical care of laboratory animals must be carefully monitored if microsomal function is to be measured (Vesell et al., 1973). Reductions in aniline hydroxylase, ethylmorphine *N*-demethylase activity, and cytochrome P450 content in hepatic microsomes has been reported in rats housed in a dirty environment.

SERUM MARKERS TO MEASURE HEPATOBILIARY DISORDERS

General

Toxicologists are having a difficult time with the "enzyme explosion" since their number has doubled every 10 years since 1930 (Bartosek and Guaitani, 1982). Elevations in serum enzyme activity have proved to be among the most sensitive diagnostic indicators of hepatocellular necrosis and/or intrahepatic and extrahepatic obstructions in the biliary system. Increases in serum activity usually relate to either a greater enzyme release from cells or decreased clearance from the serum (Boyd, 1983). The duration of elevations in serum enzyme activities varies and is dependent upon a multitude of factors such as the concentration gradient from cell to serum, their intracellular location, molecular size, rate of plasma disappearance or inactivation, and rate of hepatic overproduction (e.g., alkaline phosphatase). Plasma half-life of enzymes may vary tenfold (Bergmayer, 1974). During alterations in (1) hepatocellular permeability and necrosis or (2) cholestasis (intrahepatic or extrahepatic), different liver-specific enzymes are released into the serum. Liver-specific enzymes are defined as those located primarily in the liver and not usually found to be elevated in the serum in diseases involving other organs. Detailed investigations on the enzyme activities in various mammalian tissues (humans, marmoset, baboon, Patas monkey, dog, cat, pig, rabbit, guinea pig, ferret, mouse, and rat) have revealed which enzymes are primarily concentrated in hepatocytes (Boyd, 1983; Clampitt, 1978). Some enzymes may be elevated in the serum due to liver disorders (e.g., lactate dehydrogenase (LDH); aspartate aminotransferase (AST); isocitrate dehydrogenase (ICD, etc.), but are not useful diagnostically since they are in high concentration in other tissues and not liver-specific. Elevations in the activity of liver-specific enzymes in the serum provide little aid in predicting the reversibility of lesions, but with experience one can estimate the severity of a necrotic or cholestatic process.

Serum liver-specific enzymes which are elevated during *hepatocellular necrosis* are (1) alanine aminotransferase (ALT) in only rodents, dogs, cats, and nonhuman primates; and (2) sorbitol (iditol) dehydrogenase (SDH); (3) glutamic dehydrogenase (GDH); (4) ornithine carbamyltransferase (OCT); (5) arginase; and (6) guanase in all ureotelic mammalian species. Livers of large animal species (e.g., horses, cattle) contain only small amounts of ALT activity (Cornelius, 1963). Liver-specific enzymes which confirm intrahepatic ductular cholestasis (can be secondary to necrosis) or extrahepatic bile duct obstructions are (1) gamma glutamyl transferase (GGT); (2) alkaline phosphatase (AP—the hepatic isoenzyme); (3) 5′ nucleotidase (5′ND); and (4) leucine aminopeptidase (LAP).

Combinations of serum enzyme analysis can also be rewarding. If one measures both serum arginase and ALT activities in the dog with hepatic necrosis, it can be determined prognostically if necrosis is subsiding with regeneration or the necrosis is progressing (Cornelius et al., 1963; Mia and Koger, 1979). When both ALT and arginase remain elevated, prognosis is poor; while if arginase subsides and ALT remains elevated, regeneration and recovery is likely. One can also estimate the degree of necrosis and/or cholestasis that is present if both ALT (or SDH) and GGT) activities are measured. If both GGT and ALT activities are elevated, lesions can be assumed to be both cholestatic (e.g., elevated GGT) and necrotic (e.g., elevated ALT or SDH). If one observes only GGT or ALT are primarily elevated, it can be assumed that either cholestasis or necrosis, respectively, is the principle hepatobiliary lesion (Cornelius, 1988).

Serum Enzymes to Measure Hepatocellular Necrosis

Mice

Alanine aminotransferase (ALT) activities in normal mice have been reported to be X $\pm$ SD (IU/l) 13.1 $\pm$ 4.8 (Mitruka and Rawnsley, 1977) and 40 ± 21 (Harrison et al., 1978). ALT values in albino mice range 2.1–23.8 IU/l (Bonilla, 1972; Nomura et al., 1975). Serum aspartate aminotransferase (AST) values normally vary in the range 23–48 IU/l but the enzyme is present in many tissues and not liver specific.

Rats

ALT activities in normal rats have been reported to be (IU/l): 27 ± 5 (Benirschke et al., 1978), 32 ± 11 (Melby and Altman, 1974), 24 ± 2 (Mitruka and Rawnsley, 1977), and 26 ± 6 (Teschke et al., 1983). Normal serum ALT values range 17–30 IU/l in albino rats (Burns et al., 1971; Vondruska and Greco, 1973; Nomura et al., 1975). Serum AST activities vary in the range 46–81 IU/l. Mean activities of arginase are 21 IU/l (Boyd, 1983), while serum GDH activities are 3.9 ± 1.4 IU/l (Teschke et al., 1983).

Rabbits

ALT activities in normal rabbits are reported as (IU/l): 33 ± 13 (Yu et al., 1979), 51 ± 11 (Benirschke et al., 1978), 49 ± 20 (Melby and Altman, 1974) and 64 ± 6.2 (Mitruka and Rawnsley, 1977). Mean serum AST activities are in the range 33–64 IU/l, they can be measured but are not "liver specific."

Dogs

Serum ALT activities (IU/l) in normal dogs have been reported as 3–61 (Mia and Koger, 1979), 14 ± 6.6 (Lumsden et al., 1979), 64 ± 27 (Benirschke et al., 1978), and 20–45 (Bunch et al., 1985). Serum AST activities were 19 ± 4 (Lumsden et al., 1979) and 46 ± 14 IU/l (Benirschke et al., 1978). Serum GDH activities fall in the range 1–9 IU/l (Adbelkader and Hauge, 1986). Serum arginase activities are in the range 0–14 IU/l (Mia and Koger, 1979). Mean serum enzyme activities for GDH and SDH have been reported at 26 and 28 IU/l, respectively (Boyd, 1983).

Cats

Serum ALT levels are similar to that in dogs; numerous authors report activities within a range of 2–60 IU/l (Cornelius, 1988). Serum arginase levels range 0–12 IU/l (Mia and Koger, 1979). Mean serum activity of SDH in cats has been reported as 3.9 IU/l (Boyd, 1983).

Nonhuman Primates

Ranges of serum ALT, SDH, and GDH activities in normal marmosets have been reported as 0–14, 1–9, and 3–42 IU/l, respectively (Davy et al., 1984). Serum ALT and ASP activities in owl monkeys are reported at 52 ±

36 and 166 ± 108 IU/l, respectively (Mrema et al., 1986). Serum ALT values in baboons and rhesus monkeys are reported to be 20 ± 13 (Burns et al., 1967), 23 ± 10 (Anderson, 1966), or 0–60 (Warren et al., 1984) IU/l.

Serum Enzymes to Measure Cholestasis

Cholestasis may be present due to either (1) intrahepatic ductal/ductular obstructions associated with or without hepatic necrosis or fibrosis and (2) extrahepatic common duct obstruction due to tumor, biliary "stones," etc. The term *cholestasis* is also occasionally used more broadly to describe insufficient bile production or biochemical defects resulting in diminished excretory transport of organic ions into the bile. Among the serum enzymes to measure cholestasis, alkaline phosphatase (AP) and gamma glutamyltransferase (GGT) are used most commonly and parallel each other in obstructive disorders. The use of leucine aminopeptidase (LAP) and 5′ nucleotidase (5′ND) offer no apparent advantages over AP or GGT; however, one study concluded that 5′ND levels in dogs discriminated biliary tract disturbances better than AP or GGT (Abdelkader and Hauge, 1986). Such enzymes are primarily bound to canalicular membranes in the liver; in cholestasis, they regurgitate into the serum either via the paracellular shunt pathway or directly through the plasma membrane (DeBroe et al., 1985).

Unlike other "cholestatic" enzymes, a variety of AP isoenzymes are present in other tissues and can be elevated in other organ disorders, e.g., rickets. Intestinal and hepatic "steroid-induced" AP isoenzymes have been identified in the dog (Wellman et al., 1982). Isoenzymes can be identified by a variety of techniques such as starch and polyacrylamide electrophoresis or their rates of denaturation by heat and urea (Kaplan, 1986). The increase in serum AP activity in obstructive disorders is due to a unique overproduction of the hepatic AP isoenzyme (Kaplan and Righetti, 1970) which results from enhanced RNA translation and not from increased transcription (Seetharam et al., 1986). The prevalence of increased bile acid concentrations in cholestasis is apparently necessary for the release and transport of hepatic AP to the serum pool (Schlaeger et al., 1982). Unfortunately, a variety of methods are used to measure either the total serum AP activities or specific isoenzymes which use different substrates, buffers, temperatures, pH's, and co-factors. The use of AP as a serum marker will no doubt be continually used; however, GGT has certain advantages as being more liver-specific and does not increase as markedly as AP in hepatic necrosis or in bony disorders (Meyer, 1983). Urinary GGT levels have also been utilized to detect kidney damage (Ford, 1974).

Mice

Serum AP values decline from one month of age at 140 ± 30 IU/l to as low as 30 ± 10 IU/l at 9 months (Bailey et al., 1976). Others reported levels of 175 ± 50 at 1–3 months, declining to 73 ± 13 at 6–12 months (Frith et al., 1980). Elevated levels in young growing mice no doubt reflects the presence of the osteogenic isoenzyme.

Rats

Serum AP levels in albino rats have been reported to range 128–313 IU/l (Burns et al., 1971; Vondruska and Greco, 1973; Nomura et al., 1975). Serum levels in other normal rat colonies were reported at 98 ± 35 (Melby and Altman, 1974) and 88 ± 16 (Mitruka and Rawnsley, 1977) IU/l. GGT levels in rat sera have been reported to average 0.6 IU/l (Boyd, 1983) and 2.2 ± 0.3 IU/l (Teschke et al., 1983).

Rabbits

Serum AP levels in normal rabbits have been reported at 40 ± 26 (Yu et al., 1979), 11 ± 0.6 (Benirschke et al., 1978), and 11 ± 5 (Melby and Altman, 1974) IU/l.

Dogs

Serum AP levels in normal dogs have been reported at 32 ± 12 (Lumsden et al., 1979), 22 ± 0.6 (Benirschke et al., 1978), and 31 ± 10 (Meyer and Noonan, 1981) IU/l. Serum GGT levels have been reported to be 0–11 (Abdelkader and Hauge, 1986), 3 ± 1 (Meyer and Noonan, 1981), 2–4 (Bunch et al., 1982), and 0–10 (Shull and Hornbuckle, 1979). The use of GGT levels of dogs is particularly useful as GGT activity, unlike serum AP, does not increase in the serum in early hepatic necrosis.

Nonhuman Primates

Serum AP levels in owl monkeys are reported to be quite high at 614 ± 589 IU/l (Mrema et al., 1986), while activities in rhesus and squirrel monkeys average and range at 48 (28–68) and 40 (0–90) IU/l, respectively (Cornelius et al., 1983). Serum AP and GGT activities in marmosets range 160–790 and 0–19 IU/l, respectively (Davy et al., 1984).

Other Serum and Blood Markers

Serum Procollagen-Peptide Type III Levels

A new noninvasive RIA test has recently been developed for use in humans to detect the fibrotic activity present in cirrhotic liver. The presence and amount of liver fibrosis is easily established through biopsy of the liver, but proves to detect the level of fibrogenesis have been previously lacking in our diagnostic armamentarium.

Procollagen peptide-III (PIIIP) is deposited along with collagen in the fibrotic matrix of the liver. PIIIP is released to the serum from the helical portion of the procollagen molecule when it matures into thin collagen fibrils (Fleischmajer et al., 1983). This test first requires the identification of the liver as the affected organ since serum PIIIP levels are also elevated in active pulmonary fibrosis and chronic renal failure (Hahn and Schuppan, 1983). Its elevation in serum of human patients with viral hepatitis remains high only if a chronic active hepatitis persists. When elevated in human serum, it has proved to be a good index of fibrogenesis, but not the quantity of hepatic collagen present (Surrenti et al., 1985). PIIIP levels have also proven, unlike serum bilirubin levels, to be excellent prognostic indicators of human survival in primary biliary cirrhosis (Erickson and Zetterval, 1985). Serum levels are closely correlated with the histologic grade of active lobular fibrosis (Galambos et al., 1985). Currently, antibodies are produced in rabbits against collagens isolated from fetal goat and bovine skin. It is expected that this RIA test will be soon developed for use in laboratory animals.

Ligandin

This hepatic cytosolic protein is a glutathione transferase isoenzyme and constitutes about 2% of soluble protein; it binds large organic anions such as BSP, ICG, bilirubin, steroid hormones, etc. (Bass et al., 1977). Ligandin has been purified from hepatic cells of all air-breathing vertebrates (Fleischner et al., 1975). The RIA test is quite specific for each species and requires its isolation from each species (except nonhuman primates) for subsequent antibody production. It is also present in intestinal epithelium and proximal renal tubules (Hobig et al., 1974). The RIA of ligandin in serum (Bass et al., 1977; Bass et al., 1979) and urine (Bass et al., 1978) in rats and humans suggests its usefulness in monitoring hepatic necrosis and for detecting the presence of hepatoma (Ohmi and Arias, 1981). Serum ligandin levels correlate well with the severity of hepatic necrosis and histologic lesions (Sherman et al., 1983), although they may not always correlate closely with serum ALT activities. Serum levels may be elevated 40-fold in human patients with hepatocellular carcinoma. Following the implantation of hepatoma cells into livers of nu/nu athymic mice, elevated serum ligandin levels originated from tumor cells and not from normal murine hepatic cells (Ohmi and Arias, 1981). The measurement of serum or urinary ligandin levels will offer a new approach to detecting hepatocellular carcinoma in animals.

Alpha-1-Fetoprotein (AFP) and Carcinoembryonic Antigen (CEA)

Such proteins are produced by proliferating malignant cells which apparently recapitulate certain characteristics of developing fetal tissues. This may occur due to a derepression of fetal genomes which should be repressed as tissue differentiation proceeds. AFP is normally synthesized by embryonic yolk sac and liver. Eighty to ninety percent of hepatocellular carcinomas (HCC) synthesize AFP; it can be detected by RIA in 90% of human patients with HCC (Price and Alberti, 1985). CEA can be detected in the serum by a variety of techniques. About 80% of patients with hepatic metastases have elevated CEA levels; levels correlated better with metastases than with primary HCC. Both AFP and CEA can also be elevated in the serum in hepatic necrosis and regeneration and are not specific only for hepatic cancers (Price and Alberti, 1985). Although these two tests are not currently used in laboratory animal testing, they could provide, in the future, useful information for cancer research. Studies are needed to establish their usefulness.

Blood Ammonia

Animals with diffuse hepatobiliary disease have reduced capacities to metabolize protein nitrogen and synthesize urea. Serum levels of urea precursors such as ammonia and amino acids are many times elevated and associated with hepatoencephalopathy (Cornelius, 1988). Blood urea values may also be lower in terminal cases since less is excreted into the urine (Rudman et al., 1973). Since measurement of the rate of urea production is complex, indirect tests are used such as measuring the fasting blood ammonia level or performing an ammonium tolerance test. It is necessary for each laboratory to establish its own normal blood ammonia levels since methods vary and require careful scrutiny. Although arterial ammonia levels are slightly higher, venous samples can also be used. Mean concentrations of venous and arterial levels in dogs are 60 and 75 μg/dl, respectively (Strombeck et al., 1975). The dog has been studied extensively; blood ammonia values above 120 μg/dl are suspicious for hepatic disease and can increase tenfold in serious cases. Elevated levels can also occur in hemorrhagic shock or renal failure. In these cases, more urea enters the intestine where it is easily degraded to ammonia. Elevated ammonia levels are common in puppies and kittens with congenital portasystemic shunts (Meyer et al., 1978; Berger et al., 1986).

An oral ammonia tolerance test has been used in the dog (100 mg NH_4 Cl/kg) and is quite effective in detecting hepatobiliary disorders and portasystemic shunts (Meyer et al., 1978). In one study, dogs with shunting exhibited ammonia venous levels averaging 1050 μg/dl at 30 min as compared to fasting levels of 236 μg/dl. Normal dogs at 30 min after NH_4 Cl administration had levels averaging 156 μg/dl as compared to fasting levels of 80 μg/dl. An ammonia tolerance test using the rectal route of administration has been reported to be preferable to the oral test (Rothuizen and Van Der Ingh, 1982).

Blood Glucose

Although blood glucose levels can be depressed in advanced hepatic disorders, it is important to appreciate that fasting blood glucose values may be normal in the vast majority of instances where chronic hepatobiliary disorders are present. Hypoglycemia is only observed in fulminating and terminal hepatic dysfunction or in association with hepatocellular carcinoma in humans (Feingold and Siperstein, 1982). Both the liver (with its large reserve capacity) and the kidney (through gluconeogenesis) both contribute to the maintenance of normal blood glucose levels. Although the oral or intravenous glucose tolerance test and blood levels of pyruvic and lactic acid may be abnormal in hepatopathies, such tests have little specific diagnostic value. It is at times even difficult to assess whether glucose intolerance is due to diabetes mellitus or its associated complications of microvascular disease. The complex metabolic alterations that lead to minor glucose intolerance in chronic liver disorders are not yet established, but it is postulated that they may result from exaggerated peripheral insulin concentrations, decreased hepatic degradation of insulin, insulin resistance, and/or hepatic hyposecretion of insulin (Feingold and Siperstein, 1982). Since the method used and other physiologic factors such as age, sex, and activity all effect the reported normal values in animals, it is recommended to establish normal well-controlled values for your own laboratory. It must be stressed that blood glucose levels are low only in the most serious hepatopathies and other tests are far more sensitive to detect necrosis or fibrosis.

Serum Proteins and Amino Acids

Serum albumin concentrations are controlled by a variety of factors: the extent, nature, and duration of the hepatopathy; nutritional status; circulatory disorders; any kidney losses to the urine; digestion and intestinal amino acid absorption; and the presence of any inflammatory bowel disorders. Serum albumin levels are obviously not specific for hepatic disorders, but the combination of an absolute low albumin concentration and/or an elevated gamma globulin level are quite typical. The separation of serum albumin and other proteins are best performed by using cellulose acetate or other electrophoretic methods (Rothkopf-Ischebeck, 1980). Changes in albumin levels in early hepatic necrosis are minimal or not present; levels only fall significantly after extensive and chronic hepatic synthetic malfunction. Serum albumin levels may be static and not reflect the turnover or synthesis of albumin in the animal.

Changes in the levels of *serum immunoglobulins* are not specific for hepatocellular function but reflect the ability of the reticuloendothelial system (e.g., Kupffer cells) in the liver to remove antigens flowing from the intestinal tract in the portal blood. Any inefficient hepatic uptake or shunting of antigens past the liver results in an overproduction of antibodies by the extrahepatic RE system (Triger and Wright, 1973). Elevated Ig M, Ig G, and Ig A levels are common in chronic hepatopathies. The primary value of observing a hyperglobulinemia is only its use as a clue to the presence of a chronic hepatic disorder (Kaplowitz et al., 1982).

Since normal serum proteins levels range in concentration greatly due to water balance, it is recommended to measure the total protein level and then estimate the concentrations of each serum protein fraction (e.g., γ, β, α, and albumin) by paper or gel electrophoresis. Each laboratory is advised to establish normal control values for the laboratory animal to be used in an investigation. The following selected references are provided for various normal serum protein values in animals:

Mice: Harrison et al. (1978), Melby and Altman (1974), Mitruka and Rawnsley (1977), Caisey and King (1980), and Frith et al. (1980).

Rats: Benirschke et al. (1978), Melby and Altman (1974), and Mitruka and Rawnsley (1977).

Rabbits: Yu et al. (1979), Benirschke et al. (1978), and Mitruka and Rawnsley (1977).

Dogs: Moreland (1974), Caisey and King (1980), Lumsden et al. (1979), and Benjamin (1978).

Nonhuman Primates: Kruckenberg et al. (1972), Mitruka and Rawnsley (1977), Robinson and Ziegler (1968), Altman and Dittmer (1974), Moreland (1974), and Beland et al. (1979).

Abnormal patterns in the concentration of *serum amino acids* during the occurrence of hepatic disorders are well documented (Fischer, 1982). These pattern alterations usually involve either those responsive to insulin and glucagon derangements (branched-chain amino acids) or those which are cleared by the liver (aromatic amino acids). Most information comes from human studies (Records et al., 1976). In cirrhosis, there is an increase in serum aromatic amino acids and decreased branched-chain amino acids; while in acute hepatitis, all amino acids are elevated in association with normal levels of branched-chain amino acids. Serum phenylalanine and tyrosine levels increase and branched-chain amino acids decrease in concentration in dogs with portasystemic shunts or chronic active hepatitis (Strombeck and Rogers, 1978). Attempts to correlate the ratios of various serum amino acids to the existence of clinical symptoms in heptoencephalopathy have been inconclusive. A number of amino acid tolerance tests have been used (Gornall and Bardawill, 1952), but more simple liver-specific procedures are now available.

Serum Lipids

It is not surprising that a variety of changes in the levels of various lipids and lipoproteins occur in hepatobiliary disorders since the liver is so intimately involved in lipid metabolism. Serum triglycerides, phospholipids, fatty acid acids, and total cholesterol and its ester are occasionally used to differentiate cholestatic from necrotic processes (Bass et al., 1976). Although changes in triglycerides and total cholesterol levels are not liver-specific, the percent of cholesterol ester (CE) or the occurrence of a unique ''lipoprotein-X'' (LP-X) can suggest hepatocellular deficiencies or cholestasis, respectively, may be present (Kaplowitz et al., 1982). In hepatocellular disorders, less CE is formed due to low lecithin cholesterol acyltransferase (LCAT) activity in the liver. Elevated levels of both total cholesterol and LP-X are observed in animals with biliary obstruction (Sabesin, 1982). Normal lipoproteins are spherical bodies, while the abnormal LP-X is spherical in structure (Monzato et al., 1976). After the elimination of extrahepatic obstruction in animals, LP-X quickly disappears from the plasma and hepatic LCAT activity returns (Ritland and Bergan, 1975). Although LP-X is not more sensitive than other tests to detect cholestasis, it is another useful test that could be used. The density classes of lipoproteins (VLDL; LDL; HDL; HDL_2) are also occasionally measured and are altered significantly in hepatobiliary disorders (Cooper, 1982). There is a need for investigations on such lipids in laboratory animals.

Normal total *serum cholesterol* values vary greatly in literature and are no doubt markedly effected by diet composition (Beher et al., 1963). Total cholesterol (mg/dl) values in mice have been reported as 107 ± 5.6 (Beher et al., 1963), 64 ± 12 (Mitruka and Rawnsley, 1977), 208 ± 42 (Frith et al., 1980), 150 ± 40 (1% cholesterol diet, Beher et al., 1963), and 350 ± 63 (1% cholesterol + 0.5% cholic acid, Beher et al., 1963). Serum total cholesterol (mg/dl) levels in rats are reported as 87 ± 11 (Beher et al., 1963), 50 ± 10 (Benirschke et al., 1978), 49 ± 18 (Melby and Altman, 1974), 46 ± 10 (Mitruka and Rawnsley, 1977). Following the ingestion by rats of 1% cholesterol or 1% cholesterol plus 0.5% cholic acid, levels of 110 ± 18 and 640 ± 200 mg/dl, respectively, were observed (Beher et al., 1963). In rabbits, serum total cholesterol values (mg/dl) are reported as 39 ± 21 (Yu et al., 1979); 20 ± 4 (Benirschke et al., 1978), 25 ± 10 (Melby and Altman, 1974), and 26 ± 12 (Mitruka and Rawnsley, 1977). Total serum cholesterol levels (mg/dl) in dogs are also considerably higher on high-lipid diets (260 ± 40) than on kennel rations (190 ± 10) (Hoe and Harvey, 1961). Others have reported levels (mg/dl) for dogs at 195 ± 72 (Melby and Altman, 1974), 181 ± 12 (Mitruka and Rawnsley, 1977), 128 ± 110 (Caisey and King, 1980), and 143 ± 29 (Benirschke et al., 1978). The percent of the total cholesterol that was esterified was similar at 63 ± 12 and 62 ± 3, respectively, in normal dogs receiving low- and high-fat content in their diets (Hoe and Harvey, 1961). Total serum cholesterol values (mg/dl) in nonhuman primates are reported as the following: rhesus monkeys: 130 ± 35 (Benirschke et al., 1978), 138 ± 25 (Melby and Altman, 1974), and 140 ± 6 (Mitruka and Rawnsley, 1977); squirrel monkeys: 155 ± 31 (Benirschke et al., 1978) and 140 ± 6 (Mitruka and Rawnsley, 1977). The percent ester ranges 60–80% in most species but should be established for the experimental animal in question on its specific dietary regime.

Miscellaneous Serum Markers

An overwhelming number of serum markers are available for use in specific investigations. Some selected tests will be briefly mentioned in this section. Most of these tests have been used very little in laboratory animals but may become useful diagnostic tools once they are investigated more fully in a variety of animal species.

Elevation in *serum guanase activity* has recently been acclaimed as more liver-specific in dogs and humans with liver necrosis than are elevations in serum ALT activities (Ito et al., 1982). Blood *uric acid* levels above 1 mg/dl have long been known to suggest the presence of diffuse hepatic disease in dogs since their end product of purine metabolism is allantoin and not uric acid as in other mammals. In hepatic disorders, hepatic uricase activity is diminished in the liver. One canine breed, the Dalmatian coach hound, is unlike other dogs and excretes uric acid since its hepatic cells do not extract uric acid from the blood (Gresecke and Tiermeyer, 1984). Elevated *hyaluronate* levels in serum has also recently been recommended as a test which discriminates the presence of cirrhosis more sensitively than serum transaminase activity, prothrombin time, galactose elimination capacity, or serum bilirubin levels (Engstrom-Laurent et al., 1985). This high-molecular-weight polysaccharide is produced by mesenchymal cells and normally removed from the circulation by endothelial cells in the liver sinusoids. A variety of *antibodies* can also be used as serum markers for hepatobiliary disorders. Elevations in various antimitochondrial, antinuclear, and antiplasma membrane antibodies are currently used at university hospitals to differentiate types of cirrhosis and probe specific alterations at the cellular level (Cornelius, 1987). Studies to date have primarily been confined to assessing human disorders; future studies will no doubt be extended to animals models.

Hemostatic Function

The liver synthesizes nearly all of the coagulation factors: prothrombin, fibrinogen, and factors V, VII, IX, and X (Suttie and Jackson, 1977). Clotting problems occur only when substantial impairment of the liver's synthetic ability is compromised (Kaplowitz et al., 1982). Most factors are normally present in excess of the required quantities in the serum. Factor levels can be measured by direct assay or by tests assessing the interplay of many factors. The Quick, one-stage prothrombin time, is commonly used (Quick, 1966). This conversion of prothrombin to thrombin in the presence of calcium ions and thromboplastin (tissue extract) measures the extrinsic coagulation pathway. Since prothrombin time is increased when a variety of factors are lacking (factors I, II, V, VII, and X), it is particularly sensitive in detecting impaired hepatic synthetic abilities. The observance of prothrombin levels less than 50% of normal in humans suggest that a risk for bleeding may exist (Spector and Corn, 1967). Increases in prothrombin time in vitamin K deficiencies due to obstructive jaundice occur due to less binding of calcium ions to phospholipids. This is due to the presence of abnormal amounts of γ-carboxyglutamic acid residues which are necessary for calcium binding (Stenflo et al., 1974).

Prothrombin time serves as a good prognostic indicator since if it is prolonged in hepatic necrosis, there is a likelihood of progressive and fulminant disease (Kaplowitz, 1982). Bleeding associated with liver disorders in animals can also be caused by anatomic lesions associated with portacaval shunts or esophageal varices. The coagulation process is complex; the reader is referred to detailed information on general hemostasis by Martinez and Palascok (1982) and a review on animals by Dodds (1980). In hepatic disease, problems can be arbitrarily divided into the following groups: abnormal or depressed synthesis and/or decreased synthesis of clotting factors, defective blood vessels, inadequate removal of procoagulants; thrombocytopenia and/or platelet dysfunction, increased fibrinolysis, and deviations in a variety of fibrinolytic processes (Rock, 1984).

Hemostatic disorders are usually studied in animals by using only a limited profile of tests. For an adequate and complete study, it is recommended that practical screening tests should always include (1) a one-stage prothrombin time, (2) a platelet count, (3) whole blood clot retraction and lysis, (4) a fibrin-fibrinogen split product assay, (5) thrombin clotting time, and (6) an activated partial thromboplastin time (Dodds, 1980). This profile will allow for a definitive decision concerning the type of hematostatic defect that is present in a variety of inherited and metabolic diseases. It must be understood that clarifying the specific coagulation defect associated with a liver disease is usually quite complex. Some abnormalities in hemostasis may be effected simultaneously in varying degrees; however, special patterns in the "hemostatic" profile do occur in hepatopathies. Inadequate therapeutic responses of vitamin K to low levels of factors V, VIII, and IX would suggest an uncertain prognosis (Rock, 1984).

Measuring the prothrombin time is probably the easiest and most simplistic approach to measure hepatic synthetic capacity of hemostatic factors in animals. Normal values should be established on control animals in laboratories by using specific tests which are needed for a study. Examples of hemostatic values reported in rats and nonhuman primates are shown in the following:

Rats

Here we see prothrombin times of 14 ± 0.4 s (Colvin and Wang, 1974); bleeding times (Creskoff et al., 1949); partial thromboplastin times of 21 ± 4 s (Tschopp and Zucker, 1972); and mean plasma fibrinogen levels of 184 mg/dl (158–214) (Colvin and Wang, 1974).

Monkeys

Mean prothrombin times in most macaques range 11–14 s (Kruckenberg et al., 1972). Most mammals fall into these ranges; the reader is referred to a recent review by Dodds (1980), which compares the majority of tests in animals with established human values.

LIVER BIOPSY

Although biopsy of the liver is not usually covered in reviews on liver function tests, its discussion should not be omitted. Many times, determination of the type of hepatic lesions present or the biochemical mechanisms responsible for jaundice may still remain in doubt (Brobst and Schall, 1972; Dalton and Hill, 1972). Biopsy techniques, although simple, are not without risk. Obvious advantages of histologic examination of the liver are to detect metabolic disorders such as extensive lipidosis, amyloidosis, or glycogen storage diseases; or to point to obscure hepatomegaly and suspected hepatic lesions associated with nearly normal liver function tests. One can also perform biochemical determinations to measure heavy metal accumulations, abnormal hepatocellular enzyme activities, vitamin levels, and the cytochemical determination of certain disorders, e.g., inherited copper storage in Bedlington terriers (Johnson et al., 1984; Cornelius, 1987).

Hepatic biopsy allows for a multitude of analyses of specific enzyme activities that may relate to hepatobiliary disorders. A few selected examples are (1) relating the levels of both uridine diphosphate glucuronyl transferase and its substrate, UDPGA, to altered hepatic bilirubin conjugation and clinical jaundice (Trotman et al., 1983; Portman et al., 1984); (2) correlating heme oxygenase (HO) activities to bilirubin production rates and jaundice following therapy with HO inhibitors (Simionatto et al., 1985); and (3) relating hyperammonemia with hepatic levels of arginosuccinate synthetase deficiency rather than to portasystemic shunts or cirrhosis (Strombeck et al., 1975).

CONCLUSIONS

It is hoped that this brief introduction into the assessment of hepatic function in hepatobiliary disorders in laboratory animals will introduce the investigator to the literature on this subject. Because of the rapid regeneration of the liver, one must always ask one question first: What function or pathological process in the liver am I measuring at this moment in time? It is hoped that extensive investigations on new sophisticated hepatic function tests currently used in human medicine will be accomplished in the years ahead on laboratory animals. The significance of quality control studies used in the safety evaluation of drugs greatly depends upon such future investigations.

REFERENCES

Abdelkader SV and Hauge JG: Serum enzyme determination in the study of liver disease in dogs. *Acta Vet Scand* 1986; 27:59–70.

Altman NH and Dittmer R: *Blood and Other Body Fluids,* Federation of the American Society for Experimental Biology, 1974, pp 1701–1711.

Anderson DR: Normal values for clinical blood chemistry tests for *Macaca mulatta* monkeys. *Am J Vet Res* 1966; 27:1484–1489.

Bailey JM, Hill WD, Fiscus AG, Reilly CA, and Finkel MP: Plasma alkaline phosphatase in mice with experimentally-induced osteosarcomas. *Lab Anim Sci* 1976; 26:66–69.

Banaszak EF, Steikiel WJ, Grace RA, and Smith JJ: Estimation of hepatic blood flow using a single dye clearance method. *Am J Physiol* 1960; 198:877–880.

Bartosek I and Guaitani A (1982). Biochemical and enzymatic control of animal quality, in Bartosek I (ed): *Animals in Toxicological Research,* New York, Raven Press, pp 103–114.

Bass BD, Hoffmann WE, and Dorner JL: Normal canine lipid profiles and effects of experimentally induced pancreatitis and hepatic necrosis on lipids. *Am J Vet Res* 1976; 37:1355–1357.

Bass NM, Kirsch RE, Tuff SA, Campbell JA, and Saunders SJ: Radioimmunoassay measurement of urinary ligandin in nephrotoxin treated rats. *Clin Sci* 1979; 56:419–426.

Bass NM, Kirsch RE, Tuff SA, and Saunders SJ: Radioimmunoassay of ligandin. *Biochem Biophys Acta* 1977; 494:131–143.

Bass NM, Kirsch RE, Tuff SA, and Saunders SJ: Radioimmunoassay of plasma ligandin. *Gastroenterol* 1978; 75:589–594.

Becker BA and Plaa GL: Assessment of liver function in mice. *Lab An Care* 1967; 17:267–272.

Beher WT, Baker GD, and Penny DE: A comparative study of the effects of bile acids and cholesterol on cholesterol metabolism in the mouse, rat, hamster and guinea pig. *J Nutr* 1963; 79:523–530.

Beland MF, Sehgal PK, and Peacock WC: Baseline blood chemistry determinations in the squirrel monkey. *Lab Anim Sci* 1979; April, pp 195–199.

Benirschke K, Garner FM, and Jones TC (eds): *Pathology of Laboratory Animals,* vol I. New York, Springer-Verlag, 1978, pp 1–1050.

Benjamin M: *Outline of Veterinary Clinical Pathology.* Ames, Iowa, Iowa State Univ. Press, 1978, pp 1–250.

Berger B, Whiting PG, Breznock EM, Bruhl-Day R, and Moore PF: Congenital feline portasystemic shunts. *J Am Vet Med Assoc* 1986; 188:517–521.

Bergmeyer HU: *Methods of Enzymatic Analysis,* vol I. New York, Academic Press, 1974, pp 6–14.

Blanckaert N: Analysis of bilirubin and bilirubin mono- and diconjugates—determination of their relative amounts in biological samples. *Biochem J* 1980; 185:115–128.

Blanckaert N, Kabra P, and Farina F: Measurement of bilirubin and its monoconjugates and diconjugates in human serum by alkaline methanolysis and high-performance liquid chromatography. *J Lab Clin Med* 1980; 96:198–212.

Blanckaert N and Schmid R: Physiology and pathophysiology of bilirubin metabolism, in Zakim D and Boyer TD (eds): *Hepatology—A Textbook of Liver Disease*. Philadelphia, WB Saunders Co., 1982, pp 246–296.

Bonasch H and Cornelius C: Indocyanine green clearance—a liver function test for the dog. *Am J Vet Res* 1964; 25:254–259.

Bonati M, Latini R, Tognoni G, Young JF, and Garrattini S: Interspecies comparison of *in vivo* caffeine pharmacokinetics in man, monkey, rabbit, rat and mouse. *Drug Metab Rev* 1984; 15:1355–1383.

Bonilla CA: Use of automated instrumentation to establish normal serum parameters in experimental animal models. *Adv Automated Analyt Technicon Int Congr* 1972; 7:85–89.

Boyd JW: The mechanisms relating to increases in plasma enzymes and isoenzymes in diseases of animals. *Vet Clin Path* 1983; 12(2):9–24.

Brobst DF and Schall WD: Needle biopsy of the canine liver and correlation of laboratory data with histopathologic observations. *J Am Vet Med Assoc* 1972; 161:382–388.

Bunch SE, Castleman WL, Hornbuckle WE, and Tennant BC: Effects of long-term primidone and phenytoin administration on canine hepatic function and morphology. *J Am Vet Med Assoc* 1982; 46(1):105–115.

Burczynski FJ, Pushka KL, Sitar DS, and Greenway CV: *Am J Physiol* 1987; 252:H953–H962.

Burg AW: Physiological disposition of caffeine. *Drug Metab Rev* 1975; 4:199–228.

Burns KR, Ferguson FG, and Hampton SH: Compendium of normal blood and valves for baboons, chimpanzees and marmosets. *Am J Clin* 1967; 48:484–494.

Burns EL, Usher GS, and Taylor TW: Individual biorange patterns. *J Occup Med* 1971; 13:138–146.

Caisey JD and King DJ: Clinical chemical values for some common laboratory animals. *Clin Chem* 1980; 26:1877–1879.

Campra J and Reynolds T: The hepatic circulation, in Arias IM, Popper H, Schacter D, and Shafitz DA (eds): *The Liver—Biology and Pathobiology*. New York, Raven Press, 1982, pp 627–646.

Carrillo MC, Rodriguez JV, Monti JA, and Garay EAR: Hepatic transport and biliary excretion of sulfobromophthalein in aflatoxin B_1-treated rats. *Tox Appl Pharmacol* 1985; 77:353–357.

Center SA, Baldwin BH, deLahunta A, Dietz AE, and Tennant B: Evaluation of serum bile acid concentrations for the diagnosis of portasystemic venous anomalies. *J Am Vet Med Assoc* 1985; 186:1090–1094.

Clampitt RB: An investigation into the value of some clinical biochemical tests in the detection of minimal changes in liver morphology and function in the rat. *Arch Toxicol Suppl* 1978; 1:1–13.

Colvin H and Wang W: Toxic effects of warfarin in rats fed different diets. *Toxicol Appl Pharmacol* 1974; 28:337–348.

Cooper AD: Hepatic lipoprotein and cholesterol metabolism, in Zakim D and Boyer TD (eds): *Hepatology—A Textbook of Liver Disease*. Philadelphia, WB Saunders, 1982, pp 109–137.

Cornelius CE: Relation of body-weight to hepatic glutamic pyruvic transaminase activity. *Nature* 1963; 200:580–581.

Cornelius CE: A review of new approaches to assessing hepatic function in animals. *Vet Res Commun* 1987; 11(5):423–441.

Cornelius CE: Liver function, in Kaneko J (ed): *Clinical Biochemistry of Domestic Animals*. New York, Academic Press, 1988.

Cornelius CE, Douglas GH, Gronwall RR, and Freedland RA: Comparative studies on plasma arginase and transaminases in hepatic necrosis. *Cornell Vet* 1963; 53:181–191.

Cornelius CE and Rodgers PA: Prevention of neonatal hyperbilirubinemia in rhesus monkeys by tin-protoporphyrin. *Ped Res* 1984; 18:728–730.

Cornelius CE, Rodgers PA, Bruss ML, and Ahlfors CE: Characterization of a Gilbert-like syndrome in squirrel monkeys. *J Med Primatol* 1985; 14:59–74.

Cornelius CE, Rodgers PA, and Warren DL: The assessment of hepatic function in rhesus and squirrel monkeys. *Vet Med/Small An Clin* 1983; December 1885–1888.

Creskoff AJ, Fitz-Hugh T, and Farris EJ: Hematology of the rat—methods and standards, in Farris EJ and Griffiths JQ (eds): *The Rat in Laboratory Investigation*. Philadelphia, Lippencott, 1949, pp 406–420.

Dalton JRF and Hill FWG: A procedure for the examination of the liver and pancreas in the dog. *J Small Anim Pract* 1972; 13:527–530.

Danielson B, Eriksson K, Karlmark B, and Schildt B: Estimation of nutrient liver blood flow in mice by atraumatic densitometry. *Acta Chir Scand* 1971; 137:621–629.

Davy CW, Jackson MR, and Walker J: Reference intervals for some clinical parameters in the marmoset: effect of age and sex. *Lab Anim* 1984; 18:135–142.

DeBroe ME, Roels F, Nouven EJ, Claeys L, and Wieme RJ: Liver plasma membrane: the source of high molecular weight phosphatase in human serum. *Hepatol* 1985; 5:118–128.

Deinhardt F, Courtois G, Dherte P, Osterrieth P, Ninane G, Henle G, and Henle W: Studies of liver function tests in chimpanzees after innoculation with human hepatitis virus. *Am J Hyg* 1962; 75:a311–321.

Deinhardt F, Holmes AW, Capps RB, and Popper H: I. Transmission of disease by serial passages, and description of liver lesions. *J Exp Med* 1967; 125:673–688.

Denis P, Ossenberg FW, and Benhamou J: Hepatic blood flow and enzyme induction in the rat. *Biochem Pharmacol* 1975; 24:249–251.

Dodds J: Hemostasis and coagulation, in Kaneko JJ (ed): *Clinical Biochemistry of Domestic Animals*. New York, Academic Press, 1980, pp 671–718.

Elias MM, Comin EJ, Ochoa JE, and Rodriguez Garay EA: Renal handling of bilirubin photoderivatives. *Experimentia* 1987; 43:875–878.

Engstrom-Laurent A, Loof L, Nyberg A, and Schroder T: Increased serum levels of hyaluronate in liver disease. *Hepatol* 1985; 5:638–642.

Erickson S and Zetterval O: The *N*-terminal propeptide of collagen type III in serum as a prognostic indicator in primary biliary cirrhosis. *J Hepatol Suppl* 1985; 2:S225.

Evans AS, Evans BK, and Sturtz V: Standards for hepatic and hematologic tests in monkeys. *Proc Soc Exp Biol Med* 1953; 82:437–440.

Feingold KR and Siperstein MD: Abnormalities of glucose metabolism in liver disease, in Zakim D and Boyer TD (eds): *Hepatology—A Textbook of Liver Disease*. Philadelphia, WB Saunders, 1982, pp 499–507.

Fischer JE: Amino acids in hepatic coma. *Dig Dis Sci* 1982; 27:97–102.

Fleischner G, Kamisaka K, Hobig WH, Jakoby WB, and Arias IM: Human ligandin: characterization and quantitation. *Gastroenterology* 1975; 69:A21/821.

Fleishchmajer R, Olsen BR, Timpe R, Perlish JS, and Lovelace O: Collagen fibril formation during embryogenesis. *Proc Natl Acad Sci (USA)* 1983; 80:3354–3358.

Ford EJH: Activity of gamma-glutamyl transpeptidase and other enzymes in the serum of sheep with liver or kidney damage. *J Comp Path* 1974; 84:231–243.

Frith CH, Suber RL, and Umholtz R: Hematologic and clinical chemistry findings in control BALB/c and C57BL/6 mice. *Lab Anim Sci* 1980; 30:835–840.

Galambos MR, Collins DC, and Galambos JT: A radioimmunoassay procedure for type III procollagen: its use in the detection of hepatic fibrosis. *Hepatology* 1985; 5:38–42.

George JW, Nulk K, Weiss A, Bruss ML, and Cornelius CE: Biliverdin reductase activity in cattle, sheep, rabbits and rats. *FASEB J* 1988; 2(4):A738.

Goldberg DM and Brown J: Advances in the application of biochemical tests to diseases of the liver and biliary tract: their role in diagnosis, prognosis and the elucidation of pathogenetic mechanisms. *Clin Biochem* 1987; 20:127–148.

Gornall AG and Bardawill CJ: The study of liver function in dogs. *Can J Med Sci* 1952; 30:256–271.

Grainger SL, Keeling PW, Brown IM, Marigold JH, and Thompson RP: Clearance and non-invasive determination of the hepatic extraction of indocyanine green in baboons and man. *Clin Sci* 1983; 64:207–212.

Gresecke D and Tiermeyer W: Defect of uric acid uptake in Dalmation dog liver. *Experientia* 1984; 40:1415–1416.

Hahn EG and Schuppan D: Collagen metabolism in liver disease, in Bianchi L (ed): *Liver in Metabolic Diseases*. Lancaster, MTP Press, 1983, pp 309–323.

Harrison SD, Burdeshaw JA, Crosby RE, Cusic AM, and Denine EP: Hematology and clinical chemistry reference values for C57B46 × DBA/2F$_1$ mice. *Cancer Res* 1978; 38:2636–2639.

Hobig WH, Pabst MJ, Fleischner G, Gatmaitan G, Arias IM, and Jacoby WB: The identity of glutathione-*S*-transferase B with ligandin, a major binding protein in liver. *Proc Natl Acad Sci (USA)* 1974; 71:3879–3882.

Hoe CM and Harvey DE: An investigation into liver function tests in dogs, pt 1: serum transaminases. *J Small Anim Pract* 1961; 2:22–31.

Hofmann AF: The aminopyrine demethylation breath test and the serum bile acid level: nominated but not yet elected to join the common liver tests. *Hepatology* 1982; 2:512–517.

Ito S, Takaoka T, Nakaya Y, Hiasa Y, Mori H, Tanaka K, and Ichihara A: Clinical value of the determination of serum guanase activity. *Gastroenterology* 1982; 83:1102–1108.

Johnson GF, Gilbertson SR, Goldfischer S, Grushoff PS, and Sternlieb I: Cytochemical detection of inherited copper toxicosis of Bedlington terriers. *Vet Path* 1984; 21:57–60.

Jost G, Gambon R, Rossi E, and Preisig R: Noninvasive quantification of liver function in children using salivary caffeine clearance. *J Hepatol Suppl* 1985; 2:S259.

Kaplan MM: Serum alkaline phosphatase—another piece is added to the puzzle. *Hepatology* 1986; 6:526–528.

Kaplan MM and Righetti A: Induction of rat liver alkaline phosphatase: the mechanism of the serum elevation in bile duct obstruction. *J Clin Invest* 1970; 49:508–516.

Kaplowitz N, Eberle D, and Yamada T: Biochemical tests for liver disease, in Zakim D and Boyer TD (eds): *Hepatology: A Textbook of Liver Disease*. Philadelphia, WB Saunders, 1982, pp 583–612.

Katz S, Gilardoni A, Genovesse N, Cornelius C, and Malinow MM: Liver function studies in free-ranging Howler monkeys with hepatic pigmentation. *Lab Anim Care* 1968; 18:626–630.

Keiding S and Skak C: Intrinsic hepatic clearance of ICG—a new quantitative measurement of liver function in man. *J Hepatol Suppl* 1985; 2:S262.

Klaasen CD: Effects of phenobarbital on the plasma disappearances and biliary excretion of drugs in rats. *J Pharm Exp Therap* 1970; 175:289–300.

Klaassen CD and Strom SC: Comparison of biliary excretory function and bile composition in male, female and lactating female rats. *Drug Met Disposit* 1978; 6:120–124.

Krocker R and Romer C: The significance of serum bile acid concentrations as indicators of hepatic function in the mini-pig. *Zbl Vet Med* 1984; 31:287–295.

Kruckenberg SM, Cornelius CE, and Cook JE: Liver function and disease in primates, in Fiennes RNT (ed): *Pathology of Simian Primates,* pt I. Basel, Switzerland, S Karger, 1972, pp 711–755.

Kupfer A, Gronwall R, and Presig R: Hyperbilirubimia and slow acetylation in the horse: a parallel to Gilbert's syndrome. *Digestion* 1977; 16:353.

Kutob SD and Plaa GL: Assessment of liver function in mice with bromsulfalein. *J Appl Physiol* 1962; 17:123–125.

Lauterburg BH and Bircher J: Expiratory measurement of maximal aminopyrine demethylation in vivo. *J Pharmacol Exp Therap* 1976; 196:501–509.

Lautt WW and Greenway CV: Conceptual review of the hepatic vascular bed. *Hepatology* 1987; 7:952–963.

Lazar G, Serra D, and Tuchweber B: Effect on cadmium toxicity of substances influencing reticuloendothelial activity. *Toxicol Appl Pharmacol* 1974; 29:367–376.

Lumsden JH, Mullen K, and McSherry BJ: Canine hematology and biochemistry reference values. *Canad J Comp Med* 1979; 43:125–130.

Mancini RE and Kocsis JJ: Dimethylsulfoxide increases the lethality of CCl_4 in rats but decreases its hepatotoxicity. *Tox Appl Pharmacol* 1974; 27:206–209.

Martinez J and Palascok J: Hemostatic alterations in liver disease, in Zakim D and Boyer TD (eds): *Hepatology—A Textbook of Liver Diseases*. Philadelphia, WB Saunders, 1982, pp 546–580.

Melby EC and Altman NH: *Handbook of Laboratory Animal Science,* Vols 1–3. Boca Raton, CRC Press, 1974.

Meyer DJ: Serum gamma-glutamyltransferase as a liver test in cats with toxic and obstructive hepatic disease. *J Am An Hosp Assoc* 1983; 19:1023–1030.

Meyer DJ: Liver function tests in dogs with portasystemic shunts: measurement of serum bile acid concentration. *J Am Vet Med Assoc* 1986; 188:168–169.

Meyer DJ and Noonan NE: Liver tests in dogs receiving anticonvulsant drugs (Diphenylhydantoin and Primidone). *J Am An Hosp Assoc* 1981; 17:261–264.

Meyer DJ, Strombeck DR, Stone EA, Zenoble RD, and Buss DD: Ammonia tolerance test in clinically normal dogs and in dogs with portasystemic shunts. *J Am Vet Med Assoc* 1978; 173:377–386.

Mia AS and Koger BS: Comparative studies on serum arginase and transaminases in hepatic necrosis in various species of domestic animals. *Vet Clin Path* 1979; 8:9–15.

Mitruka BM and Rawnsley HM: *Clinical Biochemical and Hematological Reference Values in Normal Experimental Animals.* New York, Masson Publ Co, 1977, pp 1–272.

Monzato E, Fellin R, and Baggeo G: Formation of lipoprotein X—its relation to bile components. *J Clin Invest* 1976; 57:1248–1260.

Mitruka BM and Rawnsley HM: *Clinical Biochemical and Hematological Reference Values in Normal Experimental Animals.* New York, Masson Publ Co, 1977, pp 1–272.

Monzato E, Fellin R, and Baggeo G: Formation of lipoprotein X— its relation to bile components. *J Clin Invest* 1976; 57:1248–1260.

Mordenti J: Man versus beast: pharmaco-kinetic scaling in mammals. *J Pharmaceut Sci* 1986; 75:1020–1040.

Moreland AF: Biological values for various laboratory animals. *Lab Anim Digest* 1974; 9:41–42.

Mrema JE, Heidari MB, Schmidt DA, and Green TJ: Selected serum biochemical values for reference F-1 colony—born owl monkeys. *Vet Clin Path* 1986; 15:6.

Nomura G, Tamura M, Hirohashi K, and Nagasi S: Comparison of constituents of blood of several experimental animals with human blood. *Exp Animals (Tokyo)* 1975; 22(suppl):321–355.

Ohmi N and Arias IM: Ligandinemia in primary liver cell cancer in rat and man. *Hepatology* 1981; 1:316–318.

Orzes N, Bellentani S, Alden R, Simon P, Ferretti I, Lunazzi GC, Sottocasa G, and Tiribelli C: Sex differences in the hepatic uptake of sulfobromophthalein in the rat. *Clin Sci* 1985; 69:587–593.

Parraga ME and Kaneko JJ: The serum bile acids and the bile acid profile as tests of liver function. *Vet Res Commun* 1985; 9:79–88.

Paumgartner G, Probst P, Kraines R, and Leevy CD: Kinetics of Indocyanine green removal from the blood. *NY Acad Sci* 1970; 170:134–147.

Portman OW, Roy-Chowdhury J, Roy-Chowdhury N, Alexander M, Cornelius CE, and Arias IM: A non-human primate model for Gilbert's syndrome. *Hepatology* 1984; 4:175–179.

Price CP and Alberti KG: Biochemical assessment of liver function, in Wright R, Millward-Sadler G, Alberti K, and Karran S (eds): *Liver and Biliary Disease.* London, WB Saunders, 1985, pp 455–493.

Priestly BE and Plaa GL: Temporal aspects of carbon tetrachloride-induced alteration of sulfobromophthalein excretion and metabolism. *Tox Appl Pharmacol* 1970; 17:786–794.

Prothero JW: Organ scaling in mammals: the liver. *Comp Biochem Physiol* 1982; 71A:567–577.

Quick AJ: *Hemorrhagic Diseases and Thrombosis,* ed 3. New York, Lea and Febiger, 1966, pp 391–407.

Rabinovici N and Shapira N: The intrahepatic distribution of portal venous and arterial blood. *J Surg Res* 1966; 6:74–79.

Records CO, Buxton B, and Chase RA: Plasma and brain amino acids in fulminant hepatic failure and their relationship to hepatic encephalopathy. *Eur J Clin Invest* 1976; 6:387–394.

Ritland S and Bergan A: Plasma concentration of lipoprotein–X (LP-X) in experimental bile duct obstruction. *Scand J Gastroenterol* 1975; 10:17–24.

Robinson FR and Ziegler RF: Clinical laboratory values of Beagle dogs. *Lab Anim Care* 1968; 18:39–49.

Rock WA: Laboratory assessment of assessment of coagulation disease. *Clin Lab Med* 1984; 4:419–441.

Rothkopf-Ischebeck M: Differentiation of various experimentally-induced pathological processes in Wistar rats by serum protein electrophoresis using cellulose-acetate membranes. *Lab Animals* 1980; 14:153–165.

Rothuizen J and Van Der Ingh I: Rectal ammonia tolerance test in the evaluation of portal circulation in dogs with liver disease. *Res Vet Sci* 1982; 33:22–25.

Rudman P, DiFolco TJ, and Galambos JT: Maximal rates of excretion and synthesis of urea in normal and cirrhotic subjects. *J Clin Invest* 1973; 52:2241–2249.

Sabesin SM: Cholestatic lipoproteins—their pathogenesis and significance. *Gastroenterology* 1982; 83:704–709.

Scharschmidt BF, Blanckaert N, and Farina FA: Measurement of serum bilirubin and its mono- and diconjugates—applications to patients with hepatobiliary disease. *Gut* 1982; 23:643–649.

Schein P and Anderson T: The efficacy of animal studies in predicting clinical toxicity of cancer chemotherapeutic drugs. *Int J Clin Pharmacol* 1973; 8:228–238.

Schlaeger R, Haux P, and Kattermann P: Studies on the mechanism of the increase in serum alkaline phosphatase activity in cholestasis. *Enzyme* 1982; 28:3–13.

Seetharam S, Sussman NL, Komoda T, and Alpers DH: The mechanism of elevated alkaline phosphatase activity after bile duct ligation in the rat. *Hepatology* 1986; 6:374–380.

Sherman M, Bass NM, Campbell JA, and Kirsch RE: Radioimmunoassay of human ligandin. *Hepatology* 1983; 1:75–85.

Shull RM and Hornbuckle W: Diagnostic use of serum gamma-glutamyltransferase in canine liver disease. *Am J Vet Res* 1979; 40:1321–1324.

Simionatto CS, Anderson KE, Drummond ES, and Kappas A: Studies on the mechanism of SN-protoporphyrin suppression of hyperbilirubinemia. *J Clin Invest* 1985; 75:513–521.

Sir T, Wolff C, Soto J, Perez G, and Merino R: Relation between hepatic levels of glutathione and sulphobromophthalein retention in hyperthyroidism. *Clin Sci* 1987; 73:235–237.

Spector I and Corn M: Laboratory tests of hemostasis. *Arch Int Med* 1967; 119:577–582.

Stenflo J, Fernland P, and Egan W: Vitamin K dependent modifications of glutamic acid residues in prothrombin. *Proc Nat Acad Sci (USA)* 1974; 71:2730–2733.

Strombeck DR, Meyer DJ, and Freedland RA: Hyperammonemia due to a urea cycle enzyme deficiency in two dogs. *J Am Vet Med Assoc* 1975; 166:1009–1111.

Strombeck DR and Rogers Q: Plasma amino acid concentrations in dogs with hepatic disease. *J Am Vet Med Assoc* 1978; 173:93–97.

Surrenti C, Casini A, Milani S, Pasquinelli I, and Banchetti E: Is the determination of serum *N*-terminal procollagen type III peptide a reliable index of hepatic fibrosis? *J Hepatol Suppl* 1985; 2, S335.

Sutti JW and Jackson CM: Prothrombin, structure, activation and biosynthesis. *Physiol Rev* 1977; 57:1–70.

Teschke R, Neuefeind M, Nishimura M, and Strohmeyer G: Hepatic gamma-glutamyltransferase activity in alcoholic fatty liver: comparison with other liver enzymes in man and rats. *Gut* 1983; 24:625–630.

Triger DR and Wright R: Hypergammaglobulinemia in liver disease. *Lancet* 1973; 1:1494.

Trotman BW, Shaw L, Roy-Chowdhury J, Malet PF, and Rosato EF: Effect of phenobarbital on serum and biliary parameters in a patient with Crigler-Najjar syndrome, type II and acquired cholestasis. *Dig Dis Sci* 1983; 28:753–762.

Tschopp TB and Zucker MB: Hereditary defect in platelet function in rats. *Blood* 1972; 40:217–226.

Tvedegaard E, Poulson HE, Vilstrup H, and Thomsen HK: Functional status of the liver during chronic renal failure: an experimental study in the rabbit. *Liver* 1985; 5:183–188.

Vesell ES, Lang CM, White WJ, Passananti GT, and Tripp SL: Hepatic drug metabolism in rats: impairment in a dirty environment. *Science* 1973; 179:896–897.

Vogin EE, Moreno OM, Brodie DA, and Mattis PA: Indocyanine green clearance in the rhesus monkey. *J Appl Physiol* 1966; 21:1880–1882.

Vondruska JF and Greco RA: Certain hematologic and blood chemical values in Charles River CD albino rats. *Bull Am Soc Vet Clin Path* 1973; 2:3–10.

Vonk RJ, Veen H, Prop G, and Meyer DKF: The influence of taurocholate and dehydrocholate choleresis on plasma disappearance and biliary excretion of indocyanine green in the rat. *Naunyn-Schmiedeberg's Arch Pharmacol* 1974; 282:401–410.

Wahllander A, Renner E, and Preisig R: Fasting caffeine concentration. *Scand J Gastroenterol* 1985; 20:1133–1141.

Warren D, Cornelius C, and Ford B: Liver function studies on rhesus monkeys following the administration of hydrazine sulfate. *Vet Hum Tox* 1984; 26:295–299.

Warszawski D, Gorodischer R, Moses S, and Bark H: Caffeine pharmacokinetics in young and adult dogs. *Biol Neonate* 1977; 32:138–142.

Weiss JS, Gautam A, Lauff JJ, Sundberg MW, Jatlow P, Boyer JL, and Seligson D: The clinical importance of a protein-bound function of serum bilirubin in patients with hyperbilirubinemia. *New Engl J Med* 1983; 309:147–150.

Wellman ML, Hoffman WE, Dorner JL, and Mock RE: Comparison of the steroid-induced, intestinal, and hepatic isoenzymes of alkaline phosphatase. *Am J Vet Res* 1982; 43:1204–1207.

Wisecup WG, Hodson HH, Hanly WC, and Felts PE: Baseline blood levels of the chimpanzee. Liver function tests. *Amer J Vet Res* 1969; 30:955–962.

Yu L, Pragay DA, Chang D, and Wicher K: Biochemical parameters of normal rabbit serum. *Clin Biochem* 1979; 12:83–87.

Zaramello ME, Vallini R, and Tiribelli C: Estimation of hepatic blood flow by galactose plasma clearance in patients with liver disease. *Liver* 1985; 5:129–133.

Chapter

6

The Isolated Hepatocyte and Isolated Perfused Liver as Models for Studying Drug- and Chemical-Induced Hepatotoxicity

David J. Sweeny and Robert B. Diasio
Division of Clinical Pharmacology
and Department of Pharmacology
University of Alabama at Birmingham
Birmingham, Alabama

INTRODUCTION

The capacity of a drug or chemical to produce hepatotoxicity *in vivo* is often determined by the interaction of a series of complex cellular processes that are involved in the uptake, metabolism, and elimination of these potentially toxic compounds. Because of the importance of these processes in determining drug- or chemical-induced hepatotoxicity, a better means of evaluating the potential of a compound to produce toxicity *in vivo* can be obtained by using an *in vitro* model in which these processes are preserved. The present chapter discusses the use of two such *in vitro* models, the isolated hepatocyte and the isolated perfused liver, for studying drug- and chemical-induced hepatotoxicity.

THE USE OF ISOLATED HEPATOCYTES AS A MODEL FOR STUDYING DRUG- OR CHEMICAL-INDUCED HEPATOTOXICITY

The parenchymal cell (hepatocyte) is the main functional unit of the liver, constituting 60% of the liver cell population and 80% of the total liver volume. Because of the vast number and size of hepatocytes relative to other cell types (i.e., Kupffer and endothelial) in the liver, the isolation of hepatocytes from whole liver has become a fairly simple and routine procedure. Hepatocytes can be isolated from practically any species (e.g., chicken, fish, human), but those from rodent, especially the rat, are the most widely used in drug metabolism and toxicity studies. The present chapter therefore describes the use of isolated rat hepatocytes as a model for studying drug- and chemical-induced hepatotoxicity. While isolated rat hepatocytes in culture have been suggested to be useful for certain toxicological studies (McQueen and Williams, 1987), the present chapter deals exclusively with the use of freshly isolated suspensions of hepatocytes.

Preparation of Isolated Hepatocytes: A Historical Perspective

In the earliest procedures (Kaltenback, 1954; Branster and Morton, 1957), mechanical means, such as homogenization or the forcing of livers through a series of successively finer steel meshes, were used for the isolation of hepatocytes. Invariably these procedures resulted in a very low yield (below 10%) and a high proportion of damaged hepatocytes. In an effort to improve the yield and reduce damage caused to hepatocytes during their isolation, Howard et al. (1967) developed a procedure in which digestive enzymes were used to aid in the dispersion of the liver, thereby reducing the amount of mechanical treatment needed for the isolation of hepatocytes. In this procedure, the liver was first quickly perfused with ice cold Ca^{2+}-free Hank's buffer containing a mixture of collagenase and hyaluronidase. The liver was subsequently removed from the animal, cut into small slices, and incubated in a shaking water bath at 37°C under an atmosphere of 95% O_2–5% CO_2. Incubation of liver slices with hyaluronidase alone, using enzyme concentrations of up to 0.5%, failed to disperse the slices into hepatocytes even after 2 h of incubation. Collagenase by itself resulted in some cell dispersion. However, it was found that collagenase was more effective when used in combination with hyaluronidase. A concentration of 0.15% collagenase and 0.15% hyaluronidase, with an incubation time of 1 h was found to produce the best liver digestion. Incubation of liver slices with 0.15% hyaluronidase and 0.25% collagenase also resulted in adequate liver digestion; however, this concentration of collagenase resulted in a greater number of damaged hepatocytes.

The technique employing digestive enzymes for the isolation of hepatocytes was further improved by Berry and Friend (1969), who found that increased cell yield and viability could be attained by perfusing the liver *in situ* (via the portal vein) with a Ca^{2+}-free Hank's solution containing collagenase (0.05%) and hyaluronidase (0.10%). During the first 5 min of this procedure the perfusate was allowed to flow to waste, after which the perfusate was recirculated. Following 5–10 min of perfusion with the enzyme solution, oozing of fluid from the surface of the liver into the thoracic and abdominal cavities indicated that enzymatic dispersion of the liver was occurring. After 10–30 min of perfusion with the enzyme solution, the liver became soft and disintegrated easily upon touching. Subsequent incubation of the liver at 37°C for 15 min in buffer containing the digestive enzymes (with gentle shaking) resulted in a high yield of hepatocytes. Even higher cell yields could be obtained by perfusing the liver for 10 min with a Ca^{2+}- and Mg^{2+}-free Hank's solution containing 2 *mM* EDTA prior to incubation of the liver with the enzyme solution. With this procedure, yields sixfold higher than those reported by Howard et al. (1967) could be obtained. Moreover, the viability of hepatocytes, as judged by trypan blue exclusion, prepared by this method was consistently greater than 95%.

Further improvement of the enzymatic digestion technique was made by Seglen (1972), who developed a two-step perfusion method for the isolation of hepatocytes. During the initial step, the liver was perfused for 10 min in a nonrecirculating manner with either a Ca^{2+}-free or a Ca^{2+}-free buffer containing the Ca^{2+} chelator EGTA (5 *mM*). This initial perfusion facilitated the dispersion of the liver, presumably by removing the calcium salts that serve as part of the extracellular matrix (Anderson, 1953). After a 5–10 min perfusion with the Ca^{2+}-free buffer, the liver was perfused in a recirculating fashion with a buffer containing a mixture of collagenase (0.73%) and hyaluronidase (0.17%). In contrast to the method of Berry and Friend (1969) where no calcium was added to the buffer containing the digestive enzymes, calcium (4.0 *mM* resulted in the best digestion of liver) was included in the perfusate mixture. The addition of calcium facilitated the digestion of the liver, presumably by enhancing the activity of the collagenase since calcium is required for optimal activity of this enzyme (Seifter et al., 1970). With this two-step perfusion procedure, cell yields of close to 100%, with a viability of greater than 95% (judged by trypan blue), could be obtained after a 10-min perfusion. The use of this two-step procedure, with the modification that collagenase (0.12%) is used without any hyaluronidase, continues to be the most widely used method for liver digestion in the preparation of isolated hepatocytes (Moldeus et al., 1978).

Following enzymatic digestion, the liver is usually placed in an incubation buffer (e.g., Krebs-Henseleit bicarbonate) containing 1–2% bovine serum albumin. The bovine serum albumin not only helps protect the hepatocytes from mechanical damage, but also reduces the clumping of hepatocytes. After a period (usually 2–15 min) of gentle shaking at 37°C, the cell mixture is filtered through nylon mesh. With optimal digestion of the liver, only the connective and vascular tissue should remain in the nylon mesh. The cell suspension is subsequently centrifuged (50 g) at 4°C for 2 min. This centrifugation pellets the viable hepatocytes, while the nonparenchymal (e.g., Kupffer cells) and nonviable hepatocytes remain in the supernatant fluid. A hepatocyte suspension with greater than 95% viability is usually obtained with 2–3 centrifugations. With this procedure 2–3 $\times$ 10^8 viable cells are usually isolated from a rat liver (10–12 g).

Incubations are usually performed in round bottom flasks at 37°C under an atmosphere of 95% O_2–5% CO_2 using a cell concentration of 1–2 $\times$ 10^6 hepatocytes/ml. Hepatocytes can be incubated under these conditions for as long as 5 h without any significant decrease in cell viability (Thomas and Reed, 1988a).

Techniques Used in the Assessment of Isolated Hepatocyte Viability

Following the isolation procedure and during the course of the experiment, some means must be available for determining the viability of the hepatocyte preparation. Ideally the technique used for assessment of hepatocyte viability should be easy, rapid, and sensitive. Unfortunately, no one method currently available fulfills all of these criteria.

The most commonly used methods for assessment of hepatocyte viability are based on determination of the structural integrity of the cell membrane. These methods frequently examine either the uptake of a normally nonpermeable dye (e.g., trypan blue) into the hepatocyte or the release of intracellular enzymes (e.g., lactate dehydrogenase, LDH) into the incubation medium.

The basis for the trypan blue exclusion method is that this dye is usually incapable of entering the hepatocyte because the high molecular weight (M.W. = 961) and negative charge of the trypan blue molecule prevents it from crossing the cell membrane. Hepatocytes with a damaged cell membrane, however, will take up trypan blue. Therefore, the viability of a hepatocyte preparation can be rapidly determined by incubating an aliquot of the hepatocyte suspension with a solution of trypan blue (0.14%) and determining, by the use of a microscope, the percentage of cells (damaged) whose nucleus are stained blue.

Assaying for leakage of intracellular enzymes is not as rapid a procedure as trypan blue exclusion, but is generally regarded as being a quantitatively more reliable measure of membrane integrity than dye exclusion methods. The activity of the cytosolic enzyme LDH is most frequently determined. LDH activity is typically assayed spectrophotometrically by monitoring the oxidation of NADH in the presence of pyruvate (Bergmeyer and Bernt, 1974). The assessment of cell viability by measuring LDH activity can be determined in one of two ways. First, the activity of LDH can be measured by using the incubation medium from which the hepatocytes have been removed by gentle centrifugation; or by a second approach, frequently referred to as LDH latency or NADH penetration assay, in which LDH activity is measured by using a suspension of isolated hepatocytes (Moldeus et al., 1978). With both approaches, the rate of oxidation of NADH is compared to that obtained using a sample of hepatocytes that have been completely lysed by treatment with detergents such as Triton X-100.

Although trypan blue exclusion and LDH leakage are the most widely used methods for assessing hepatocyte viability, the use of these techniques is often criticized in that they only identify cells that are extensively damaged. Additional methods (e.g., succinate stimulation of respiration or the measurement of intracellular levels of K^+, Na^+, and Ca^{2+}) are available that are more sensitive measures of cell membrane integrity than the trypan blue and LDH methods. However, these methods are more time consuming and involve the use of more sophisticated equipment than the trypan blue and LDH methods (Baur et al., 1975). Moreover, while succinate stimulation of respiration has been suggested to be a useful method for ascertaining the integrity of the cell membrane, Pfaff et al. (1980) have indicated that the use of succinate may not accurately reflect the integrity of the cell membrane, since the stimulation of respiration can be blocked by endogenous substrates (e.g., malate) that compete with succinate for uptake into the mitochondria. Similarly, while the measurement of intracellular K^+, Na^+, and Ca^{2+} levels can be used as sensitive determinant of cell membrane integrity, the intracellular concentrations of these ions have been shown to be temperature dependent, and therefore during the period (10–20 min) of temperature equilibration that occurs at the beginning of the incubation (Pfaff et al., 1980), measurement of intracellular Na^+, K^+, and Ca^{2+} may not accurately reflect the viability of the hepatocyte preparation.

The appearance of the isolated hepatocyte can also be used to assess the viability of the hepatocyte preparation. Under the electron microscope, viable hepatocytes appear spherical and have an intact cell membrane and well-preserved intracellular organelles (Howard et al., 1973). In contrast, damaged hepatocytes demonstrate numerous outpouches of the cell membrane, commonly referred to as blebs or bullae. The blebbing of the cell membrane appears to result from the disruption of the cytoskeletal structure of the hepatocyte (Mirabelli et al., 1988), and the development of blebs on the hepatocyte surface is regarded as a common event during the course leading to irreversible cell injury (Lemasters et al., 1987).

Advantages of Isolated Hepatocytes as a Model for Studying Drug- or Chemical-Induced Hepatotoxicity

Isolated hepatocytes offer a number of advantages over broken cell (e.g., microsomes), intact liver, and *in vivo* models for studying drug- or chemical-induced hepatotoxicity. Compared to broken cell preparations, where homogenization and fractionation of the liver often result in loss or removal of certain pathways of drug metabolism, isolated hepatocytes retain the capacity to metabolize compounds by both monooxygenase and conjugation reactions. This is an important advantage of isolated hepatocytes compared to broken cell preparations, since compounds are often metabolized to reactive metabolites by monooxygenation and detoxified by conjugation pathways. More importantly, since the pathways involved in activation and detoxification are usually located in different cellular compartments (Orrenius and Moldeus, 1982), a more accurate picture of whether or not a compound will produce hepatotoxicity can be seen by using isolated hepatocytes than with broken cell preparations.

Isolated hepatocytes also have the advantage of maintaining the capacity to synthesize the required co-factors for both monooxygenase and conjugation reactions. This is important, since under physiological conditions the activity of drug metabolism pathways is often controlled by the availability of co-factors (Thurman and Kauffman, 1980). Therefore, isolated hepatocytes should more closely resemble the metabolism that occurs *in vivo,* since with broken cell preparations co-factors must be added back (often in supersaturating concentrations) to the incubation mixture. Moreover, the availability of certain cellular co-factors (e.g., glutathione) are often critical factors in the detoxification of reactive drug metabolites. Therefore, hepatocytes should provide a better insight on the capacity of a drug or chemical to produce hepatotoxicity *in vivo* than broken cell preparations.

Although preparation of isolated hepatocytes disrupts the zonal architecture of the liver and the polarity of the hepatocyte, isolated hepatocytes still have many distinct advantages over intact liver preparations. One important advantage is that a great number of hepatocytes are typically isolated from each liver, and because of this the use of isolated hepatocytes allows for a large number of different experiments to be performed from each animal. This also allows each liver to serve as its own control. Isolated hepatocytes also allow for sampling throughout the experiment so that intracellular events (e.g., depletion of GSH) occurring during the development of hepatotoxicity can be determined. Moreover, since incubation conditions can be easily controlled, the use of isolated hepatocytes is more applicable to determining the mechanism by which a drug or chemical produces hepatotoxicity than *in vivo* studies, where factors such as blood flow, extrahepatic metabolism, and protein binding may influence the ability to determine the mechanism by which a drug or chemical produces hepatotoxicity.

Use of Isolated Hepatocytes in Examining Drug- and Chemical-Induced Hepatotoxicity

Isolated hepatocytes have been extensively used for studying the metabolism and toxicity produced by a wide variety of drugs and chemicals. While it is beyond the scope of this chapter to provide a comprehensive review of the use of isolated hepatocytes in studying hepatotoxicity, the following section utilizes specific compounds as examples to demonstrate the usefulness of isolated hepatocytes in the study of drug- and chemical-induced hepatotoxicity. In addition, this section helps illustrate how isolated hepatocytes have been useful in clarifying the mechanism(s) by which drugs or chemicals can produce hepatotoxicity.

Examination of Acetaminophen-Induced Hepatotoxicity Using Isolated Hepatocytes

One of the more widely studied compounds that illustrates the usefulness of isolated hepatocytes in studying drug-induced hepatotoxicity is acetaminophen. As is seen *in vivo,* acetaminophen produces a dose-related toxicity in isolated hepatocytes (Moldeus, 1978). For this reason, isolated hepatocytes have been used extensively to examine the metabolism of acetaminophen and the mechanism(s) by which acetaminophen produces hepatotoxicity.

A diagram of the metabolism of acetaminophen is shown in Figure 1. At low concentrations, acetaminophen is primarily metabolized by sulfation and glucuronidation to unreactive metabolites (Moldeus, 1978b; Sweeny and Weiner, 1985). The capacity to metabolize acetaminophen by sulfation and glucuronidation is limited, and as the concentration of acetaminophen is increased a greater proportion of the dose is metabolized by cytochrome P450 mono-oxygenation (Jollow et al., 1973). The product of the cytochrome (P450-mediated oxidation is a reactive metabolite that has been putatively identified as *N*-acetyl-*p*-benzoquinoneimine (NAPQI) (Dahlin et al., 1984). Normally, the reactive metabolite is detoxified by conjugation with glutathione. Because cellular levels of glutathione are limited and the capacity for resynthesis minimal, the formation of the acetaminophen-glutathione conjugates leads to a stoichiometric reduction in cellular glutathione levels (Moldeus, 1978). Following depletion of glutathione, the reactive metabolite has been shown to bind to cellular proteins, and hepatotoxicity subsequently develops.

The interrelationship between the activation of acetaminophen in the endoplasmic reticulum by cytochrome P450 monooxygenation and the detoxification of the NAPQI by conjugation with glutathione in the cytosol is a critical factor in determining whether or not acetaminophen-induced hepatotoxicity will develop. Since hepatocytes retain the capacity to metabolize acetaminophen by both of these pathways, isolated hepatocytes have proven

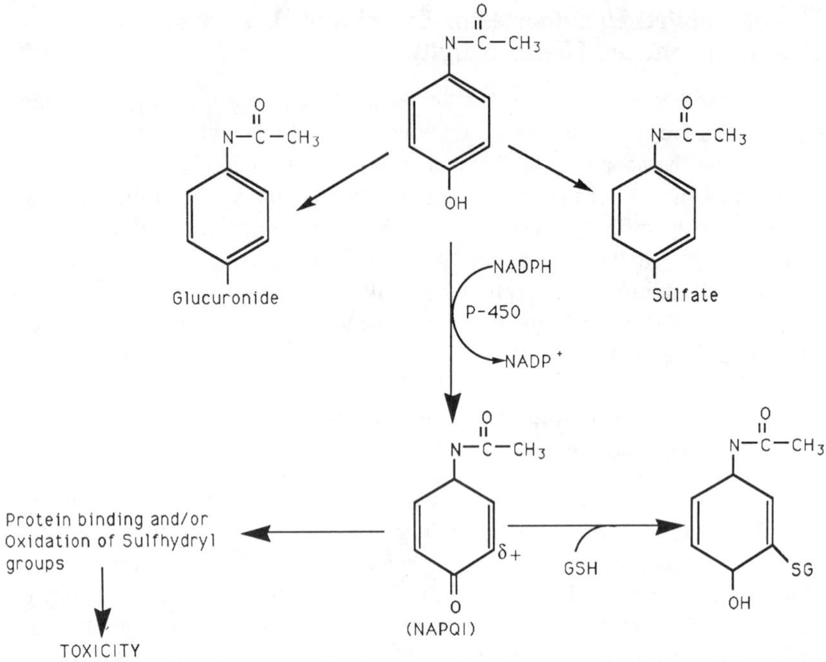

Figure 1. Acetaminophen metabolism and toxicity. At low doses, acetaminophen is metabolized to unreactive metabolites by sulfation and glucuronidation. The capacity for these pathways is limited, and as the concentration of acetaminophen increased a greater proportion of the dose is metabolized by cytochrome P450 to a reactive metabolite (NAPQI). This reactive metabolite is normally detoxified by conjugation with reduced gluthathione (GSH). However, cellular GSH is limited, and following GSH depletion NAPQI can produce cell toxicity by removal (either by alkylation or oxidation) of protein sulfhydryl groups.

to be an excellent system for examining factors that will influence the balance between these activating and detoxifying reactions. For example, hepatocytes isolated from rats are normally resistant to the hepatotoxicity produced by acetaminophen because of the low inherent capacity of the cytochrome P450 system to metabolize acetaminophen (Moldeus, 1978b). However, hepatocytes isolated from rats in which the activity of the cytochrome P450 system has been induced by compounds such as phenobarbital (Pessayre et al., 1980) form an increased amount of the reactive metabolite and are sensitive to the toxicity produced by acetaminophen. Similarly, hepatocytes from mice and hamsters, which readily metabolize acetaminophen by the P450 system, are quite sensitive to the hepatotoxicity produced by acetaminophen (Moldeus, 1978a; Boobis et al., 1986).

The capacity for detoxification of the reactive metabolite by conjugation with glutathione is critical for preventing the acetaminophen-induced hepatotoxicity. The importance of this detoxification pathway can be clearly seen in isolated hepatocytes. For example, hepatocytes isolated from animals pretreated with the glutathione-depleting agent diethyl maleate are more sensitive to the toxicity produced by acetaminophen (Moldeus, 1978a). While, on the other hand, incubation of hepatocytes with compounds, such as *N*-acetylcysteine, that increase intracellular glutathione levels, has been shown to reduce the toxicity produced by acetaminophen (Massey and Racz, 1981). The reduced toxicity produced by incubation of hepatocytes with precursors of glutathione was suggested to result from an increased capacity to detoxify the reactive metabolite by conjugation with glutathione, since incubation with these compounds resulted in the increased formation of the acetaminophen-glutathione conjugate.

One of the most important uses for isolated hepatocytes in examining acetaminophen-induced hepatotoxicity is in helping to determine the events that ultimately lead to cell death. The depletion of intracellular glutathione levels and the binding of the acetaminophen-reactive metabolite with cellular proteins have been suggested to be important in the development of the acetaminophen-induced hepatotoxicity. However, the capacity of *N*-acetylcysteine to prevent acetaminophen-induced liver damage *in vivo,* even up to 15 h after administration of acetaminophen (Boobis et al., 1986), suggests that other critical events may also be important in the development of acetaminophen-induced hepatotoxicity. In this regard, studies with isolated hepatocytes have recently demonstrated a dissociation between the binding of the reactive metabolite with cellular proteins and cell death (Boobis et al., 1986). In these studies, isolated hamster hepatocytes were incubated with acetaminophen (2.5 *mM*) for 90 min, after which the cells were washed to remove all but bound acetaminophen. The hepatocytes were placed in fresh medium (in the absence of acetaminophen), and over the course of incubation a dramatic decrease in viability was observed. However, hepatocytes incubated with dithiothrietol or *N*-acetylcysteine during this second incubation remained viable over the course of the experiment. The prevention of cell toxicity produced by *N*-acetylcysteine, and dithiothreitol did not result from a reduction of the binding of acetaminophen to cellular proteins, but was suggested to result from the capacity of these reducing agents to repair the oxidative damage caused by the reactive metabolite of acetaminophen. In support of the role of an oxidative process in the development of acetaminophen-induced hepatotoxicity, studies in isolated rat hepatocytes using chemically synthesized NAPQI have demonstrated the oxidative effects of this reactive metabolite on cellular proteins (Albano et al., 1985). Moreover, the potentiation of the toxicity of NAPQI in these studies by inhibition of glutathione reductase with 1,3-bis-(2-chloroethyl)-1-nitrosourea (BCNU) appears to further support the role for oxidative events in the toxicity of acetaminophen.

Examination of Oxidative Mechanisms of Hepatotoxicity Using Isolated Hepatocytes

In recent years, it has become increasingly evident that the metabolism of a number of drugs and chemicals can lead to the production of reactive oxygen species that are potentially toxic to the liver. Normally, these reactive oxygen species are detoxified by a variety of cellular mechanisms (Sies, 1986). However, under conditions of increased oxidative stress the capacity of the liver to detoxify these reactive oxygen species may become limited; consequently, toxicity to the liver may result. The pathways responsible for the formation and detoxification of the reactive oxygen species are located in different cellular compartments. Moreover, the activities of these pathways rely on the availability of cellular reducing equivalents (e.g., NADPH). Therefore, isolated hepatocytes have proven to be an excellent model for examining the interaction between these activating and detoxifying pathways and also for studying the mechanism by which oxidative stress produces hepatotoxicity.

Reactive oxygen species can be formed as a consequence of the one-electron reduction of drugs and chemicals by enzymes located in the mitochondria and endoplasmic reticulum (Sies, 1986). In this reaction (Figure 2), the compound (X) undergoing one-electron reduction can be reoxidized by molecular oxygen (O_2), leading to the formation of the superoxide radical ($O_2^{\cdot -}$) and reformation of the parent compound. Following this cycle, the parent compound can repeat this process (redox cycling), the net result being the formation of the superoxide radical at the expense of cellular reducing equivalents (e.g., NADPH).

The superoxide radical is capable of producing a number of cytotoxic effects. Normally, however, the superoxide radical rapidly dismutates (either spontaneously or enzymatically by superoxide dismutase) to hydrogen peroxide, which is also quite toxic to the cell. The formation of hydrogen peroxide can lead to the production of more reactive oxygen species (e.g., hydroxide radical, $^{\cdot}OH$). However, hydrogen peroxide is normally detoxified enzymatically by either glutathione peroxidase or catalase (Sies, 1986).

The napthoquinone menadione is a widely studied compound that can illustrate the usefulness of isolated hepatocytes in examining the metabolism and toxicity of compounds that can undergo redox cycling (Thor et al., 1982; Orrenius et al., 1985). In the intact cell, menadione can either undergo a two-electron reduction by the cytoplasmic enzyme DT-diaphorase to the hydroquinone or undergo a one-electron reduction by several flavoprotein reductases to the semiquinone radical (Figure 3). Metabolism by the cytoplasmic DT-diaphorase is considered a detoxification process, since the hydroquinone is subsequently metabolized by sulfation and glucuronidation and eliminated from the cell. The importance of DT-diaphorase in detoxifying menadione is demonstrated by the potentiation of menadione toxicity in isolated hepatocytes produced by dicumarol, which inhibits the conversion of menadione to the hydroquinone (Thor et al., 1982).

Formation of the semiquinone radical by the one-electron reduction of menadione can potentially produce hepatotoxicity by one of two mechanisms. First, the semiquinone radical is capable of reacting directly with cellular thiols, and secondly, the semiquinone radical can reduce molecular oxygen to the superoxide radical. While both mechanisms have been shown to contribute to the depletion of cellular thiols, the major loss of cellular thiols appears due to the formation of reactive oxygen species. This is suggested by the increased levels of GSSG and increased actinomycin A–insensitive oxygen consumption that is observed in hepatocytes incubated with menadione (Thor et al., 1982). Moreover, the capacity of the reducing agent dithiotheritol to restore the majority of sulphydryl groups that were lost in the presence of menadione is further support that the loss of sulphydryl groups is due to an oxidative rather than an alkylating process (DiMonte et al., 1984).

The formation of the reactive oxygen species by menadione-stimulated

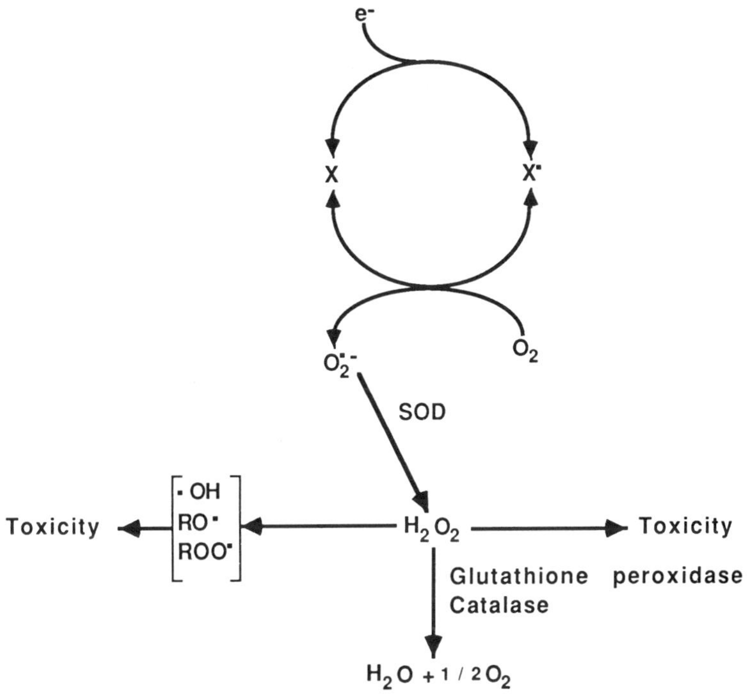

Figure 2. Formation of reactive oxygen species during redox cycling. The one-electron reduction of a compound (X) in the mitochondria or endoplasmic reticulum leads to the formation of a radical (X$^{\cdot}$). The radical intermediate can be reoxidized by molecular oxygen, resulting in the formation of the superoxide radical ($O_2^{\cdot -}$). The superoxide radical dismutates to hydrogen oxide, which can be subsequently detoxified enzymatically or lead to the formation of more powerful oxygen radicals (e.g., hydroxide radical, ·OH).

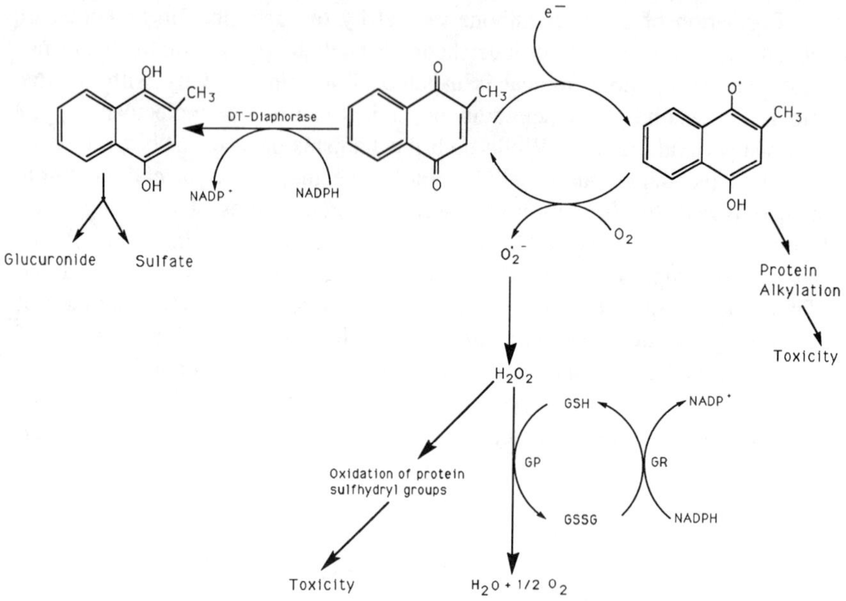

Figure 3. Metabolism and toxicity of menadione. The hydroquinone menadione can be metabolized by the cytoplasmic enzyme DT-diaphorase in a two-electron reduction or undergo a one-electronc reduction by a variety of flavoprotein reductases in the mitochondria or endoplasmic reticulum. Metabolism by DT-diaphorase is considered a detoxification process since the hydroquinone can be eliminated from the cell following sulfation or glucuronidation. The product of the one-electron reduction (semiquinone radical) can be reduced back to menadione by molecular oxygen (O_2), during which the superoxide radical ($O_2^{\cdot -}$) is formed. $O_2^{\cdot -}$ can dismutate to H_2O_2, which can be detoxified by glutathione peroxidase at the expense of reduced glutathione (GSH). Cellular GSH levels are usually maintained by glutathione reductase (GR), which catalyzes the reduction of oxidized glutathione (GSSG) using reducing equivalents supplied by NADPH. In case of extreme oxidative stress, GSH levels can be depleted and toxicity to the cell can result following oxidation of critical cellular proteins. The alkylation of cellular proteins by the semiquinone radical may also contribute to the loss of cell viability.

redox cycling can potentially lead to the development of hepatotoxicity. However, under normal conditions the hepatocyte is capable of detoxifying these reactive oxygen species. The superoxide radical can be rapidly dismutated to H_2O_2, which can be subsequently metabolized by mitochondrial or cytoplasmic glutathione peroxidase. The reduction of H_2O_2 by glutathione peroxidase occurs at the expense of GSH, leading to the formation of GSSG. The loss of GSH is minimized by the reduction of GSSG by glutathione reductase, utilizing reducing equivalents supplied by NADPH. The capacity of the hepatocyte to reduce GSSG by glutathione reductase appears to be an important means of maintaining a supply of GSH under conditions of oxidative stress, since the toxicity of menadione to isolated hepatocytes is increased in the presence of BCNU, an inhibitor of glutathione reductase (Ross et al.,

1986). Moreover, the capacity to supply NADPH for the glutathione reductase system appears to be an important factor in the prevention of oxidative injury. This is suggested by the observation that hepatocytes isolated from rats in which the activity of the pentose pathway (i.e., supply of NADPH) was decreased (e.g., by fasting) were more susceptible to menadione-induced toxicity (Smith et al., 1987).

Under conditions of low oxidative stress the hepatocyte is capable of detoxifying the reactive oxygen species. However, with increased oxidative stress, cellular GSH levels become rapidly depleted, thereby severely affecting the capacity of the hepatocyte to detoxify the reactive oxygen species. The continued production of reactive oxygen species in the presence of a compromised detoxification process subsequently results in the oxidation of critical cellular proteins and ultimately cell death.

Examination of the Course of Events Leading to the Development of Hepatotoxicity Using Isolated Hepatocytes

For a number of compounds (e.g., acetaminophen, menadione), the depletion of soluble thiols and the subsequent removal of protein thiols, either by alkylation or oxidation, are clearly important initial events in the development of hepatotoxicity. However, while the importance of these initial changes have been well documented, the course of events that subsequently follows have not been fully established.

Orrenius and coworkers (1984) have suggested that an increase in cytosolic free Ca^{2+} may be a critical event in the development of the hepatotoxicity produced by a variety of drugs and chemicals. Normally, cytosolic free Ca^{2+} levels are maintained in the range of $1\text{–}2 \times 10^{-7}$ *M* by the sequestering of Ca^{2+} in the mitochondria and endoplasmic reticulum, and by the action of the plasma membrane Ca^{2+}-ATPase (Campbell, 1987). However, during the course of development of hepatotoxicity, cytosolic free Ca^{2+} levels may be increased as a result of depletion of Ca^{2+} from mitochondrial and extramitochondrial compartments, and by inhibition of plasma membrane Ca^{2+}-ATPase. Inhibition of plasma membrane Ca^{2+}-ATPase appears especially critical in causing an increase in cytosolic free Ca^{2+} levels, since this enzyme functions in maintaining the large transmembrane Ca^{2+} gradient by efficiently transporting Ca^{2+} out of the cell. Inhibition of plasma membrane Ca^{2+}-ATPase appears to be important in the toxicity produced by cystamine in isolated hepatocytes (Nicotera et al., 1985), since restoration of plasma membrane Ca^{2+}-ATPase activity produced by incubating isolated hepatocytes with dithiothreitol correlated well with the ability of this agent to prevent the cystamine-induced hepatotoxicity. Moreover, the role of extracellular Ca^{2+} in the development of cystamine toxicity was suggested by the capacity of the Ca^{2+} channel blocker verapamil to prevent this toxicity.

The increase in cytosolic free Ca^{2+} produced by these hepatotoxins (e.g., cystamine) results in the stimulation of a number of cellular phospholipases and proteases. The increase in phospholipase activity could be prevented by the preincubation of isolated hepatocytes with chlorlpromazine or dibucaine (Nicotera et al., 1986). However, while these compounds inhibited the increase in phospholipase activity they were unable to prevent the loss of cell viability. On the other hand, the preincubation of isolated hepatocytes with the inhibitors of Ca^{2+} activated proteases (antipain and leupeptin) prevented the increase in protease activity as well as the loss of cell viability. These findings may suggest that the activation of proteases by the sustained increase in cytosolic free Ca^{2+} may lead to the loss of viability as a result of the increased catabolism of proteins that are critical in maintaining the cytoskeletal structure of the hepatocyte (Nicotera et al., 1986).

While an increase in cytosolic free Ca^{2+} levels has been suggested by a number of researchers to be important in the mechanisms leading to hepatotoxicity (Reynolds 1963; Farber and El-Mofty, 1975), others have questioned the role Ca^{2+} may have in the development of hepatotoxicity. For example, Stacey and Klaassen (1982) demonstrated that omission of Ca^{2+} from the incubation buffer had no effect on the toxicity produced by cadmium and copper in isolated hepatocytes, whereas Smith et al. (1981) demonstrated that the toxicity produced by carbon tetrachloride, bromobenzene, and ethylmethanesulfonate was actually increased when isolated hepatocytes were incubated in Ca^{2+}-free buffer. Furthermore, no increase in cytosolic free Ca^{2+} levels was observed during the "chemical hypoxia" produced in isolated hepatocytes by cyanide and iodoacetate, despite the fact that these conditions lead to bleb formation and loss of cell viability (Lemasters et al., 1987).

Recently, Thomas and Reed (1988a) demonstrated that incubation of isolated hepatocytes in the presence of reduced ($< 2.0\ mM$) levels of Ca^{2+} resulted in an increased loss of cell viability. This toxicity was suggested to be a consequence of an oxidative process, because of the correlation of malondialdehyde formation (indicator of lipid peroxidation) with the loss of cell viability. Moreover, the role of an oxidative process in this toxicity was further suggested by the finding that the antioxidants vitamin E and chlorpromazine, as well as the iron chelators desferrioxamine and EDTA prevented the production of malondialdehyde and the loss of cell viability. Further studies (Thomas and Reed, 1988b) suggested that the toxicity produced in isolated hepatocytes incubated in the presence of reduced extracellular Ca^{2+} may have resulted from mitochondrial Ca^{2+} cycling, since ruthenium red and lanthanides, which inhibit Ca^{2+} uptake into the mitochondria, prevented this toxicity. Therefore, while an increase in cytosolic free Ca^{2+} appears important in the development of the toxicity produced by a number of compounds, an increase in cytosolic free Ca^{2+} may not be essential to the development of drug- and chemical-induced hepatotoxicity.

THE ISOLATED PERFUSED LIVER AS A MODEL FOR STUDYING DRUG- AND CHEMICAL-INDUCED HEPATOTOXICITY

The isolated perfused liver is an additional *in vitro* intact cell preparation that is commonly used for examining drug- and chemical-induced heptotoxicity. This technique was first described by Claude Bernard around 1855 (Miller, 1972). However, it was not until the 1950s that this model became extensively used for studying the physiology and biochemistry of the liver. Isolated perfused liver experiments have been described utilizing a variety of species. However, the rat liver model is the most widely used, and therefore this part of the chapter will deal with the use of the isolated perfused rat liver as a model for studying drug- and chemical-induced hepatotoxicity.

The Isolated Perfused Rat Liver Technique

The surgical procedure used in the isolation and removal of the liver from the animal is similar to that utilized in the preparation of isolated hepatocytes. Briefly, this procedure typically involves the cannulation of the hepatic portal vein with a polyethylene catheter through which perfusate is pumped into the liver (anterograde perfusion). In some instances the "in-flow" cannula may be placed into the inferior vena cava (above the diaphram) to achieve flow opposite to that occurring under physiological conditions (retrograde perfusion). In addition to cannulation of the portal vein, the common bile duct may be cannulated (i.e., with PE-10 tubing), thereby facilitating the collection of bile during the experiment. The surgical procedure should be performed as quickly as possible (5–7 min) to minimize the duration of liver hypoxia. Once removed from the carcass, the liver is placed in a heated and humidified chamber and the liver perfused in either a recirculating or nonrecirculating fashion. Generally, when red cells are used (20–40% hematocrit), the perfusate is recirculated. The inclusion of red cells in the perfusate medium facilitates the transport of oxygen to the liver, and typical flow rates with buffer containing red cells is 1–2 ml/g liver/min. When no red cells are included, the perfusate (Krebs-Henseleit bicarbonate buffer) must be delivered to the liver at higher flow rates (2–4 ml/g liver/min) to compensate for the lower-oxygen-carrying capacity of the buffer (Meijer et al., 1981). With Krebs-Henseleit bicarbonate buffer as perfusate medium, the liver is usually perfused in a nonrecirculating fashion.

Despite the higher flow rates that are required when using the Krebs-Henseleit bicarbonate buffer, utilization of this buffer offers certain advantages over the use of perfusates containing red cells. For example, with red cells problems (e.g., nonuniform liver perfusion) may be encountered due to hemolysis or clotting of red cells. Moreover, use of red cells may affect the hepatic extraction of a drug, due to binding of drug to the red cell. Another disadvantage of perfusates containing red cells is that during the course of the experiment recirculation of perfusate leads to accumulation of metabolic end products (e.g., lactate) which may have an effect on the liver. Krebs-Henseleit bicarbonate buffer is usually not recirculated, thereby circumventing the buildup of metabolic end products in the perfusate.

Because of the low-oxygen-carrying capacity of the Krebs-Henseleit bicarbonate buffer and the problems that may be encountered with using red cells, some researchers have utilized buffers containing fluorocarbon emulsions to increase the oxygen-carrying capacity of the perfusate. While the use of fluorocarbon emulsions has been suggested to have many benefits (Krowe et al., 1974), the use of these emulsions has not become widespread, possibly due to the high cost of the fluorocarbons and the uncertainty as to the effect of these agents on liver cell function.

Assessment of Viability of the Isolated Perfused Liver Preparation

As with isolated hepatocytes, it is important to have some means of determining viability of the isolated perfused liver preparation. The most direct and probably easiest way of assessing the viability of the perfused liver is by monitoring the oxygen consumption of the preparation. This can be done by allowing the effluent perfusate to flow past a Clark-type oxygen electrode and calculating oxygen consumption by recording the difference between the oxygen concentration entering and leaving the liver. Typically the isolated perfused rat liver consumes approximately between 1.5 and 1.8 μmoles O_2/g liver/min (Badr et al., 1989; Matsumura et al., 1986). The metabolic state of the liver also can be determined by measuring rates of release of glucose, lactate, and pyruvate. From these two later measurements the cytosolic NAD^+/NADH ratio can be calculated (Veech et al., 1969). Similarly, the cellular redox state can be examined using microprobes, placed on the surface of the liver, that monitor the relative fluorescence of reduced cellular pyridine nucleotides (NADH and NADPH) (Ji et al., 1979). Typically the cellular pool of pyridine nucleotides should be highly oxidized. However, under anoxic conditions pyridine nucleotides are predominantly in the reduced state (Ji et al., 1980).

In assessing the development of hepatotoxicity, the release of intracellular enzymes into the effluent perfusate is probably the easiest and most straightforward approach. As with isolated hepatocytes, lactate dehydrogenase is the enzyme activity typically measured. Trypan blue exclusion can also be used (Belinsky et al., 1984). However, this method is less applicable to the perfused liver model, since the liver must first be fixed (with 1% paraformaldehyde), sectioned, and then examined by light microscopy. This procedure thereby prevents the trypan blue exclusion method from being used routinely for assessing the viability of the isolated perfused liver preparation. However, the trypan blue exclusion procedure is very applicable for pinpointing the region of the liver lobule that is the site of toxicity (e.g., allyl alcohol–induced periportal toxicity) (Belinsky et al., 1984).

In addition to permitting easy access to perfusate samples, the isolated perfused liver allows for the collection of bile samples. Bile flow can be monitored throughout the perfusion period. A gradual decrease in bile flow may not always indicate hepatotoxicity. However, in instances where bile flow decreases rapidly, this may be a sign of hepatotoxicity. Moreover, when combined with analysis of biliary constituents (e.g., oxidized glutathione) a decrease in bile flow can provide useful information in assessing the development of hepatotoxicity.

Advantages of the Isolated Perfused Liver as a Model for Examining Drug- and Chemical-Induced Hepatotoxicity

The isolated perfused liver offers many advantages over other *in vitro* models (e.g., microsomes, isolated hepatocytes) for studying drug- and chemical-induced hepatotoxicity. Like isolated hepatocytes, the isolated perfused liver is an intact cell model and therefore offers all the same advantages over broken cell preparations. However, the isolated perfused liver offers several distinct advantages over isolated hepatocytes. First, while the lobular organization has been lost in the isolated hepatocyte model as a result of dispersion of the liver with collagenase, the lobular architecture is maintained in the isolated perfused liver; therefore, the toxicity produced by a variety of compounds in this model should more closely resemble that observed *in vivo*. For example, it has been shown that the toxicity produced by a number of compounds *in vivo* is specific for hepatocytes located in either periportal (e.g.,allyl alcohol) or pericentral (e.g., acetaminophen) regions of the liver lobule. A similar toxicity can be produced by these agents in the isolated perfused liver; therefore, this model has been used to determine the basis for these zone-specific toxicities. Second, while the isolated hepatocyte model makes use of only one of cell type (i.e., parenchymal cell) found in the intact liver, the isolated perfused liver maintains all cell types (i.e., parenchymal, Kupffer, and endothelial cells) that are normally present in the liver. Since nonparenchymal cells also have the capacity to metabolize drugs and exogenous compounds (Steinberg et al., 1987), the isolated perfused liver should closely resemble the metabolism and toxicity of compounds that are seen *in vivo*.

An additional advantage of the isolated perfused liver over isolated hepatocytes is that the polarity (i.e., sinusoidal and biliary surfaces) of the hepatocyte is preserved in this model. This allows for a determination of the hepatic extraction of a compound, and for an examination of the elimination of a compound and its metabolites into perfusate and bile over time (Sweeny and Reinke, 1987; Sweeny et al., 1988). Moreover, since bile flow is maintained, the isolated perfused liver permits the determination of the potential of a drug or chemical to produce cholestasis (i.e., decrease bile flow).

The isolated perfused liver also offers many advantages over *in vivo* models (e.g., intact animal) for studying drug- and chemical-induced hepatotoxicity. With the isolated perfused liver, the perfusate medium is well defined. There are no hormones or other endogenous substances (e.g., plasma proteins) normally present *in vivo* which may compromise our understanding of the mechanism by which a drug or chemical produces hepatotoxicity. With the isolated perfused liver, perfusate and bile samples can be easily obtained so that the time course of changes leading to hepatotoxicity can be followed.

One deficiency of the isolated perfused liver model is that compared to an *in vivo* model or cultured isolated hepatocytes there is a limited period (usually 3–4 h) during which the isolated perfused liver remains viable. Therefore, the isolated perfused liver would not be an appropriate model for studying the development of hepatotoxicity following chronic or subchronic exposure to a drug or chemical.

Use of the Isolated Perfused Liver Preparation as a Model for Studying Drug- and Chemical-Induced Hepatotoxicity

The isolated perfused liver has been used extensively as a model to examine the hepatic metabolism and toxicity of a variety of agents. While it is beyond the scope of this section to provide a comprehensive review of the use of the isolated perfused rat liver in studying hepatotoxicity, this section utilizes specific drugs or chemicals as examples to demonstrate the usefulness of the isolated perfused rat liver as a model for studying drug- and chemical-induced hepatotoxicity.

Examination of the Mechanism of Zonal Toxicity Using the Isolated Perfused Rat Liver

A major application of the isolated perfused liver is that the model can be used in studying the mechanism by which a variety of compounds produce toxicity to specific zones of the liver lobule. As an example, allyl alcohol has been shown to produce periportal toxicity following *in vivo* administration to rats (Piazza, 1915). This toxicity is dependent on the metabolism of allyl alcohol by alcohol dehydrogenase, since the allyl alcohol–induced hepatotoxicity can be prevented by the prior administration of pyrazole, an inhibitor of alcohol dehydrogenase (Reid, 1972). It has been suggested that allyl alcohol–induced periportal toxicity results from preferential metabolism in this region of the liver lobule, since histochemical evidence indicated that periportal hepatocytes contained higher levels of alcohol dehydrogenase than pericentral hepatocytes (Reid, 1972). However, analysis of the zonal distribution of alcohol dehydrogenase utilizing a microchemical assay had earlier demonstrated that alcohol dehydrogenase activity was greater in hepatocytes located in centrilobular rather than periportal regions of the liver lobule (Morrison and Brock, 1967). Moreover, the contention that periportal toxicity results from preferential metabolism allyl alcohol in this region of the liver lobule may be further questioned, since in the intact cell alcohol dehydrogenase activity may be regulated by the availability of NAD^+ (Kashiwagi et al., 1982), which is uniformly distributed across the liver lobule (Matschinsky et al., 1978).

Using the isolated perfused rat liver Thurman and colleagues have extensively examined the metabolism of allyl alcohol in an effort to determine the reason for the regional specific toxicity. Employing microlight guides, which are small enough to be placed on the surface of periportal or pericentral regions of the liver lobule and which detect changes in NADH fluorescence associated with the metabolism of allyl alcohol, Thurman and colleagues demonstrated that the metabolism of allyl alcohol occurs at slightly higher rates in hepatocytes located in the pericentral region of the liver lobule (Belinsky et al., 1984). However, despite the lower rates of allyl alcohol metabolism in periportal hepatocytes, only hepatocytes located in this region demonstrated signs of hepatotoxicity (i.e., by trypan blue uptake) following a 60-min infusion of allyl alcohol.

While it is well recognized that cellular enzymes are unevenly distributed across the liver lobule (Jungermann and Katz, 1982), there also exists an oxygen gradient across the liver lobule of between 40 and 60 Torr (Lautt, 1976) which may contribute to the development of regional-specific hepatotoxicities. Using microoxygen probes placed on either periportal or pericentral regions of the liver lobule, Thurman and co-workers have demonstrated that oxygen uptake by hepatocytes is dependent on the local oxygen tension (Matsumura et al., 1986). That is, when the oxygen concentration is greater in periportal regions, as is observed during anteriograde perfusion, hepatocytes in this region consume two to three times the amount of oxygen as do hepatocytes "downstream" in the pericentral region of the liver lobule. However, when the direction of perfusate flow was reversed, the pattern of oxygen consumption shifted so that pericentral hepatocytes consumed two to three times as much oxygen as periportal hepatocytes. By reversing the direction of flow through the liver (i.e., retrograde infusion) the toxicity produced by allyl alcohol could be shifted from hepatocytes primarily in the periportal region to hepatocytes primarily in the pericentral region of the liver lobule (Badr et al., 1986). These findings indicate that the oxygen gradient across the liver lobule is an important factor in determining the site of toxicity of allyl alcohol.

The toxicity produced by the redox cycling agent menadione could also be shifted from the periportal region to the pericentral region of the liver lobule by reversing the direction of perfusate flow through the liver (Badr et al., 1989). The toxicities produced by both allyl alcohol and menadione could be prevented by the free-radical scavenger (+)-cyanidol and the iron-trapping agent desferrioxamine, indicating that the menadione- and allyl alcohol–induced periportal toxicities may result as a consequence of an oxidative process that is "fueled" by the high oxygen tension normally present in this region of the liver lobule.

Examination of Drug- or Chemical-Induced Cholestasis Using the Isolated Perfused Rat Liver

Since the isolated perfused liver maintains the capacity to produce bile, this model can be used to examine the mechanism by which a drug or chemical produces cholestasis. The phenothiazine chlorpromazine can serve to illustrate this point. Chlorpromazine has been shown to cause overt cholestatic hepatitis in approximately 1–2% of patients receiving this agent (Ishak and Irey, 1972). Initially it had been suggested that the chlorpromazine-induced hepatotoxicity was due to a drug-hypersensitivity reaction. However, the finding that chlorpromazine administration produced elevations of serum alkaline phosphatase activity in as many as 50% of the patients receiving this medication suggested a high intrinsic toxicity of this agent (Dickes et al., 1957). This contention was supported by the observation that chlorpromazine produced toxicity in liver slices (Dujovne et al., 1968), and inhibited bilirubin secretion and bile formation *in vivo* (Hargreaves, 1965; Stefko et al., 1963).

In the isolated perfused rat liver, chlorpromazine has been shown to produce a rapid, dose-related decrease in bile flow (Kendler, 1971). Hepatic perfusate flow was also decreased chlorpromazine. However, the decrease in bile flow did not appear to result from a decreased perfusate flow rate since a similar mechanical reduction in perfusate flow rate had no significant effect on bile flow (Hrulsan et al., 1978; Tavaloni et al., 1979). The decrease in bile flow produced by chlorpromazine appeared to result from an effect of this agent on the bile acid–independent component of bile flow, since bile acid secretion was not affected during the chlorpromazine-induced cholestasis. Electron microscopic examination of livers perfused with cholestatic doses of chlorpromazine demonstrated a marked increase in the appearance of intracellular vacuoles, dilation of bile canaliculi, and evagination of canalicular diverticuli (Hruben et al., 1978). The activity of the plasma membrane–associated enzymes Mg^{2+}-ATPase and 5′-nucleotidase were also decreased in livers perfused with cholestatic concentrations of chlorpromazine, whereas plasma membrane Na^{+}-K^{+}-ATPase activity was not affected (Tavaloni et al., 1979). The electron microscopic and biochemical changes observed during the chlorpromazine-induced cholestasis suggest that the cholestatic properties of chlorpromazine may result from an effect of this agent on the canalicular membrane function.

Summary

The isolated hepatocyte and isolated perfused liver have been widely shown to be useful models for studying drug- and chemical-induced hepatotoxicity. These models retain the capacity to metabolize compounds by monooxygenase and conjugation pathways. In addition, the capacity for cofactor synthesis is preserved. Thus, the pattern of metabolism and toxicity oberved with these models will more closely resemble *in vivo* than other *in vitro* preparations. The isolated hepatocyte and perfused liver models also permit easy sampling throughout the course of the experiment, thereby allowing for the determination of cellular changes that occur during the development of hepatotoxicity. Moreover, because incubation conditions can be easily controlled, the isolated hepatocyte and isolated perfused liver permit a better insight into the mechanism of toxicity than can be obtained *in vivo*.

REFERENCES

Albano E, Rundgren M, Harvison PJ, Nelson SD, and Moldeus P: Mechanism of *N*-acetyl-*p*-benzoquinone imine cytotoxicity. *Mol Pharamcol* 1985; 28:306–311.

Anderson NG: The mass isolation of whole cells from rat liver. *Science* 1953; 117:627–628.

Badr MZ, Belinsky SA, Kauffman FC, and Thurman RG: Mechanism of hepatotoxicity to periportal regions of the liver due to allyl alcohol: role of oxygen and lipid peroxidation. *J Pharmacol Exp Ther* 1986; 238:1138–1142.

Badr MZ, Ganey PE, Yoshihara H, Kaffman FC, and Thurman RG: Hepatotoxicity of menadione predominate in oxygen-rich zones of the liver lobule. *J Pharmacol Exp Ther* 1989; 248:1317–1322.

Baur H, Kasperek S, and Pffaf E: Criteria of viability of isolated liver cells. *Hoppe-Seyler's Z Physiol Chem* 1975; 356:827–838.

Belinsky SA, Popp JA, Kauffman FC, and Thurman RG: Trypan blue uptake as a method to investigate hepatotoxicity in periportal and pericentral regions of the liver lobule: studies with allyl alcohol in the perfused liver. *J Pharmacol Exp Ther* 1984; 230:755–760.

Bergmeyer HU and Bernt E: Lactate dehydrogenase UV-assay with pyruvate and NADH, in Bergmeyer HU (ed): *Methods of Enzymatic Analysis*. New York, Academic Press, 1974, pp 574–579.

Berry MN and Friend DS: High-yield preparation of isolated rat liver parenchymal cells. A biochemical and fine structural study. *J Cell Biol* 1969; 43:506–520.

Boobis AR, Tee LBG, Hampden CE, and Davies DS: Freshly isolated hepatocytes as a model for studying the toxicity of paracetamol. *Fd Chem Toxic* 1986; 24:731–736.

Branster MV and Morton RK: Isolation of intact liver cells. *Nature* 1957; 180:1283–1284.

Campbell AK: Intracellular calcium: Friend or Foe? *Clin Sci* 1987; 72:1–10.

Dahlin DC, Miwa GT, Lu AYH, and Nelson SD: *N*-Acetyl-*p*-benzoquinone imine: a cytochrome *P*-450-mediated oxidation product of acetaminophen. *Proc Natl Acad Sci USA* 1984; 81:1327–1331.

Dickes R, Schenker V, and Deutsch L: Serial liver-function and blood studies in patients receiving chlorpromazine. *N Engl J Med* 1957; 256:1–7.

DiMonte D, Ross D, Bellomo G, Eklow L, and Orrenius S: Alterations in intracellular thiol homeostasis during the metabolism of menadione by isolated rat hepatocytes. *Arch Biochem Biophys* 1984; 235:334–342.

Dujovne CA, Levy R, and Zimmerman HJ: Hepatotoxicity of phenothiazines *in vitro* as measured by loss of aminotransferases to surrounding media. *Proc Soc Exp Biol Med* 1968; 128:561–563.

Farber JL and El-Mofty SK: The biochemical pathology of liver necrosis. *Am J Pathol* 1975; 81:237–250.

Hargreaves T: Cholestatic drugs and bilirubin metabolism. *Nature* 1965; 206:154–156.

Howard RB, Christen AK, Gibbs FA,and Pesch LA: The enzymatic preparation of isolated intact parenchymal cells from rat liver. *J Cell Biol* 1967: 35:675–684.

Howard RB, Lee JC, and Pesch LA: The fine structure, potassium content and respiratory activity of isolated liver parenchymal cells prepared by improved enzymatic techniques. *J Cell Biol* 1973; 57:642–658.

Hruban Z, Tavoloni N, Reed JS, and Boyer JL: Ultrastructural changes during cholestasis induced by chlorpromazine in the isolated perfused rat liver. *Virchows Arch B Cell Path* 1978; 26:289–305.

Ishak KG and Irey NS: Hepatic injury associated with the phenothiazines. *Arch Pathol* 1972; 93:283–304.

Ji S, Chance B, Nishiki K, Smith T, and Rich T: Micro-light guides: a new method for measuring tissue fluorescence and reflectance. *Am J Physiol* 1979; 236:C144–156.

Ji S, Lemasters JJ, and Thurman RG: A non invasive method to study the metabolic events within sublobular sregions of hemoglobin-free perfused liver. *FEBS Lett* 1980; 113:37–41.

Jollow DJ, Mitchell JR, Potter WZ, Gillette JR, and Brodie BB: Acetaminophen-induced hepatic necrosis. II. Role of covalent binding *in vivo*. *J Pharmacol Exp Ther* 1973; 187:195–202.

Jungermann K and Katz N: Functional hepatocellular heterogeneity. *Hepatology* 1982; 2:385–395.

Kaltenbach JP: The preparation and utilization of whole cell suspensions obtained from solid mammalian tissues. *Expt Cell Res* 1954; 7:568–571.

Kashiwagi T, Ji S, Lemasters JJ, and Thurman RG: Rates of alcohol dehydrogenase-dependent ethanol metabolism in periportal and pericentral regions of the perfused rat liver. *Mol Pharmacol* 1982; 21:438–443.

Kendler J, Bowry S, Seef LB, and Zimmerman HJ: Effect of chlorpromazine on the function of the perfused isolated liver. *Biochem Pharmacol* 1971; 20:2439–2445.

Krone W, Huttner WB, Kampf SC, Rittich B, Seitz HJ, and Tarnowski W: Long-term perfusion of the isolated rat liver. Maintenance of its functional state by use of a fluorocarbon emulsion. *Biochim Biophys Acta* 1974; 372:55–71.

Lautt WW: Method for measuring hepatic uptake of oxygen or other blood substances *in situ*. *J Appl Physiol* 1976; 40:269–274.

Lemasters JJ, DiGuiseppi J, Nieminen A-L, and Herman B: Blebbing, free calcium and mitochondrial membrane potential preceding cell death in hepatocytes. *Nature* 1987; 325:78–81.

Lemasters JJ, Ji S, and Thurman RG: Centrilobular injury following hypoxia in isolated, perfused rat liver. *Science* 1981; 213:661–663.

Lemasters JJ, Stemkowski CJ, Ji S, and Thurman RG: Cell surface changes and enzyme release during hypoxia and reoxygenation in the isolated, perfused rat liver. *J Cell Biol* 1983; 97:778–786.

Massey TE and Racz WJ: Effects of *N*-acetylcysteine on metabolism, covalent binding and toxicity of acetaminophen in isolated mouse hepatocytes. *Tox Appl Pharmacol* 1981; 60:220–228.

Matschinsky FM, Hintz CS, Reichlmeier K, Quistorff B, and Chance B: *The Intralobular Distribution of Oxidized and Reduced Pyridine Nucleotides in the Liver of Normal and Diabetic Rats.* New York, Academic Press, 1978.

Matsumura T, Kauffman FC, Meren H, and Thurman RG: Oxygen uptake in periportal and pericentral regions of liver lobule in perfused liver. *Am J Physiol* 1986; 250:G800–G805.

McQueen CA and Williams GM: Toxicology studies in cultured hepatocytes from various species, in Rauckman EJ and Padilla GM (eds): *The Isolated Hepatocyte: Use in Toxicology and Xenobiotic Biotransformation.* New York, Academic Press, 1987, pp 51–67.

Meijer DKF, Keulemans K, and Mulder GJ: Isolated perfused rat liver technique, in Jakoby WB (ed): *Methods in Enzymology.* New York, Academic Press, 1981, pp 81–94.

Miller LL: History of isolated liver perfusion and some still unsolved problems, in Bartosek I, Guaitani A, and Miller LL (eds): *Isolated liver perfusion and its applicaqtion.* New York, Raven Press, 1972, pp 1–9.

Mirabelli F, Salis A, Marinoni V, Finardi G, Bellomo G, Thor H, and Orrenius S: Menadione-induced bleb formation in hepatocytes is associated with the oxidation of thiol groups in actin. *Arch Biochem Biophys* 1988; 264:261–269.

Moldeus P: Paracetamol metabolism and toxicity in isolated hepatocytes from mouse, in Aitio A (ed): *Conjugation Reactions in Drug Biotransformation.* Amsterdam, Elsevier/North Holland, 1978a, pp 293–302.

Moldeus P: Paracetamol metabolism and toxicity in isolated hepatocytes from rat and mouse. *Biochem Pharmacol* 1978b; 27:2859–2863.

Moldeus P, Hogberg J, and Orrenius S: Isolation and use of liver cells, in Fleisher S and Packer L (eds): *Methods in Enzymology.* New York, Academic Press, 1978, pp 60–71.

Morrison GR and Brock FE: Quantitative measurement of alcohol dehydrogenase in the lboule of normal livers. *J Lab Clin Med* 1967; 7:116–120.

Nicotera P, Hartzell P, Baldi C, Svensson SA, Bellomo G, and Orrenius S: Cystamine induces toxicity in hepatocytes through the elevation of cytosolic calcium and the stimulation of a non-lysozomal proteolytic system. *J Biol Chem* 1986; 261:14628–14635.

Nicotera P, Moore M, Mirabelli F, Bellomo G, and Orrenius S: Inhibition of hepatocyte plasma membrane calcium-ATPase activity by menadione metabolism and its restoration by thiols. *FEBS Lett* 1985; 181:149–153.

Orrenius S and Moldeus P: Compartmentation of detoxification reactions, in Sies H (ed): *Metabolic Compartmentation.* New York, Academic Press, 1982, pp 485–520.

Orrenius S, Rossi L, Eklow-Lastbom L, and Thor H: Oxidative stress in intact cells. A comparison of the effects of menadione and diquat in isolated hepatocytes, in Poli G, Cheeseman KH, Dianzani MU, and Slater TF (eds): *Free Radicals in Liver Injury.* Washington, DC, IRL Press, 1985, pp 99–106.

Orrenius S, Thor H, and Bellomo G: Alterations in thiol and calcium-ion homeostasis during hydroperoxide and drug metabolism in hepatocytes. *Trans Biochem Soc* 1984; 12:23–28.

Pessayre D, Wandscheer J-C, Cobert B, Level R, Degott C, Batt AM, Martin N, and Benhamou J-P: Additive effects of inducers and fasting on acetaminophen hepatotoxicity. *Biochem Pharmacol* 1980; 29:2219–2223.

Pfaff E, Schuler B, Krell H, and Hoke H: Viability control and special properties of isolated rat hepatocytes. *Arch Toxicol* 1980; 44:3–21.

Piazza JG: Toxicity of allyl formate. *Z Exp Pathol Ther* 1915; 17:318–325.

Reid WD: Mechanism of allyl alcohol-induced hepatic necrosis. *Experentia* 1972; 28:1058–1061.

Reynolds JL: Liver parenchymal cell injury. I. Initial alterations of the cell following poisoning with carbon tetrachloride. *J Cell Biol* 1963; 19:139–157.

Ross D, Thor H, Threadgill MD, Sandy MS, Smith MT, Moldeus P, and Orrenius S: The role of oxidative processes in the cytotoxicity of substituted 1,4-napthoquinones in isolated hepatocytes. *Arch Biochem Biophys* 1986; 248:460–466.

Seifter S, Takahashi S, and Harper E: Further demonstration that cysteine reacts with metal component of collagenase. *Bioc Biop A* 1970; 214:559–570.

Seglen PO: Preparation of rat liver cells. I. Effect of calcium on enzymatic dispersion of isolated, perfused liver. *Exp Cell Res* 1972; 74:450–454.

Sies H: Biochemistry of oxidative stress. *Angew Chem Int Ed Engl* 1986; 25:1058–1071.

Smith PF, Alberts DW, and Rush GF: Menadione-induced oxidative stress in hepatocytes isolated from fed and fasted rats: the role of NADPH-regenerating pathways. *Tox Appl Pharmacol* 1987; 89:190–201.

Smith MT, Thor H, and Orrenius S: Toxic injury to isolated hepatocytes is not dependent on extracellular calcium. *Science* 1981; 213:1257–1259.

Stacy N and Klaassen CD: Lack of protection against chemically induced injury to isolated hepatocytes by omission of calcium from the incubation medium. *J Toxicol Environ Health* 1982; 9:267–276.

Stefko PL and Zbinden G: Effect of chlorpromazine, chlordiazepoxide, diaepam and chlorprothixene on bile flow and intrabiliary pressure in cholecystectomized dogs. *Am J Gastroenterol* 1963; 39:410–417.

Steinberg P, Lafranconi WM, Wolf CR, Waxman DJ, Desch F, and Friedberg T: Xenobiotic metabolizing enzymes are not restricted to parenchymal cells in rat liver. *Mol Pharmacol* 1987; 32:463–470.

Sweeny DJ, Barnes S, and Diasio RB: Formation of conjugates of 2-fluoro-β-alanine and bile acids during the metabolism of 5-fluorouracil and 5-fluoro-2-deoxyuridine in the isolated perfused rat liver. *Cancer Res* 1988; 48:2010–2014.

Sweeny DJ and Reinke LA: Metabolism of benzo(*a*)pyrene in the isolated perfused rat liver. Factors affecting the release of phenolic metabolites into bile and perfusate. *Carcinogenesis* 1987; 8:779–783.

Sweeny DJ and Weiner M: Metabolism of acetaminophen in hepatocytes isolated from mice and rats of various ages. *Drug Metab Dispos* 1985; 13:377–379.

Tavoloni N, Reed JS, Hruban Z, and Boyer JL: Effect of chlorpromazine on hepatic perfusion and bile secretory function in the isolated perfused rat liver. *J Lab Clin Med* 1979; 94:726–741.

Thomas CE and Reed DJ: Effect of extracellular calcium omission on isolated hepatocytes. I. Induction of oxidative stress and cell injury. *J Pharmacol Exp Ther* 1988a; 245:493–500.

Thomas CE and Reed DJ: Effect of extracellular calcium omission on isolated hepatocytes. II. Loss of mitochondrial membrane potential and protection by inhibitors of uniport calcium transduction. *J Pharmacol Exp Ther* 1988b; 245:501–507.

Thor H, Smith MT, Hartzell P, Bellomo G, Jewell SA, and Orrenius S: The metabolism of menadione (2-methyl-1,4-napthoquinone) by isolated hepatocytes. A study of the implications of oxidative stress in intact cells. *J Biol Chem* 1982; 257:12419–12425.

Thurman RG and Kauffman FC: Factors regulating drug metabolism in intact hepatocytes. *Pharmacol Rev* 1980; 31:229–252.

Trush MA, Mimnaugh EG, and Gram TE: Activation of pharmacological agents to radical intermediates. Implications for the role of free radicals in drug action and toxicity. *Biochem Pharmacol* 1982; 31:3335–3346.

Veech RL, Eggleston LV, and Krebs HA: The redox state of free Nicotinamide-Adenine Dinucleotide Phosphate in the cytoplasm of rat liver. *Biochem J* 1969: 115:609–619.

Chapter

7

Biochemical Methods of Studying Hepatotoxicity

Prasada Rao S. Kodavanti and Harihara M. Mehendale
Department of Pharmacology and Toxicology
University of Mississippi Medical Center
Jackson, Mississippi

INTRODUCTION

The methods normally employed for the detection of hepatotoxicity vary with the circumstances of their use. Tests and methods employed in experimental hepatotoxicity must cover a wide range of experimental circumstances since they involve both *in vivo* and *in vitro* models (Zimmerman, 1976, 1978). *In vivo* studies are essential to demonstrate a toxic agent that has in fact a demonstrable adverse effect on the liver in a setting of physiological significance. Whole-animal models must be used to verify that the effects of various factors and manipulations on hepatotoxicity are indeed related to the mechanism of the toxicity. *In vitro* models, however, can be used to examine specific aspects of the mechanisms of hepatotoxicity. The *in vitro* aspects have investigational value, but only when they are supported by *in vivo* models. In this chapter, we discuss biochemical methods to identify specific aspects of hepatotoxicity using both *in vivo* and *in vitro* models. A variety of biochemical methods are available to assess hepatotoxicity, but they are scattered in the literature. In this chapter, we attempt to compile the biochemical methods, useful not only to assess hepatotoxicity, but also to understand the mechanistic aspects of hepatotoxicity. Before describing the specific methodology, a brief consideration of the biochemical homeostasis of the liver might be useful.

The liver is very complex in nature and has a pivotal role in several vital processes in the animal. For example:

1. Majority of proteins found in plasma are synthesized by the hepatocyte and secreted into systemic circulation. Several proteins are metabolized by transamination and deamination reactions, and excess ammonia is finally excreted as urea.
2. Liver acts as a major store for carbohydrate in the form of glycogen and plays a key role in the maintenance of blood glucose by glycogenolysis and gluconeogenesis.
3. Liver is a major site for cholesterol and triglyceride synthesis. The triglycerides are further modified to form other complex lipids and lipoproteins.
4. Liver synthesizes carbohydrates from fats and proteins, and also ATP from glucose via Embden-Meyerhof pathway.
5. Liver also synthesizes complex and critical organic molecules such as heme, cytochrome P450, and certain coenzymes.
6. Liver displays significant regenerating capacity unlike many other organs. When 70% of the liver is removed, entire liver can be regenerated within 7 days after resection.
7. Liver provides an effective site for phagocytosis of solid materials and is responsible for the removal of many endogenous metabolic end products and unwanted exogenous materials.
8. Liver is also the principal organ for elimination of biochemical waste in the form of bilirubin, excretion of bile acids, and secretion of bile.
9. Liver plays a major role in the detoxification and metabolic elimination of toxic chemicals.

It is clear from the above that biochemistry of the liver is very complex and affects the physiology and biochemistry of all other tissues in the body. Definitive and reliable methods are needed to assess perturbation of the delicate homeostatic balance between many simultaneously and sequentially operating processes in this metabolic storehouse. Liver is often the first and important organ to be exposed to variety of toxic chemicals. The target organ of toxicity may vary depending on the chemical nature of a toxicant, but the liver plays a major role in determining whether a chemical would be metabolically eliminated. A compound is classified as toxic when it causes an unfavorable change in the biochemical homeostasis of the animal system. This toxicity can lead to cellular injury, culminating in organ damage. Recovery from toxic injury may occur by virtue of intrinsic cytoprotective and other cellular homeostatic mechanisms. Bridges et al. (1983) have proposed an illustrative scheme for chemically induced cytotoxicity, which occurs in a progression of phases eventually resulting in either cell death or in recovery owing to cellular cytoprotective mechanisms.

Chemical ⟶	Initiating factor ⟶	Early molecular changes ⟶	Irreversible changes
	↓	↓	↓
	Detoxication	Recovery	Cell injury

Among these phases, early biochemical changes signify initiation of injury, and the nature and degree of these changes are often the determinants of whether recovery from toxicity can occur. This scheme also illustrates another important point. The extent of injury is a function of the net balance between irreversible changes and early molecular changes that lead to repair or replacement of cells destroyed by injury. Therefore, biochemical parameters indicative of these two separate events are important in the assessment of the outcome of any initial injury. In this chapter, we address both the recovery and the progressive phases of liver injury and have described the methodology employed for assessing these aspects.

HOW DO WE KNOW WHETHER A COMPOUND IS HEPATOTOXIC?

Hepatotoxicity of a compound can be detected by several means. The first step is to look for the chemical nature of a compound and make a reasonable guess based on the knowledge of how the liver's detoxification systems can handle this particular toxicant. The route of exposure is not very important in the sense that even if the compound is applied topically, it can exert hepatotoxic effect by reaching the liver via circulation. Many biochemical tests for hepatotoxicity have been used clinically for the last several years and they are still considered as very popular markers of hepatotoxicity. The importance of such tests lies in their simplicity and one does not have to sacrifice the experimental animals. Such tests are particularly very useful with larger test animal species. Histopathological examination of the liver is the next logical step in identifying and quantitating the nature and extent of liver injury. Indeed, a correlation of the biochemical markers of injury itself or the mechanisms underlying liver injury is often essential in experimental toxicology before conclusive interpretation of either findings. However, the discussion in this chapter will be limited to the biochemical methodology.

Biochemically, serum enzyme analyses have become the standard measure of hepatotoxicity during the past 25 years (Zimmerman and Seeff, 1970). Measurement of enzyme activities of serum permit detection of hepatotoxicity with far less labor than that required for other tests. The rationale for the use of serum transaminases and other enzymes is that these enzymes, normally contained in the liver cells, gain entry into the general circulation when liver cells are injured. This is useful for detection of early damage, demonstration of hepatic injury without sacrificing the animal, and for serial measurement in larger animals (Zimmerman, 1974). This experimental tool has been widely used to study the possible hepatotoxicity of unknown agents, to the comparison of the toxicity of known agents, to the demonstration of onset of injury, and to the study of the potentiation or inhibition of the toxic phenomena by physiologic manipulation or administration of various agents (Zimmerman, 1976). However, there are certain limitations for this use. Serum enzyme activities may increase as a result of leakage from cells with altered permeability of membranes due to systemic effects of an agent rather than specific hepatic injury caused by a chemical (Zimmerman, 1974). For example, anoxia or vascular collapse due to general systemic effects of a toxic agent may elevate serum enzyme activities which may be misinterpreted as evidence of specific chemical-induced hepatotoxicity (Zimmerman and Seeff, 1970). In reality, these enzymes might have leaked from other tissues or organs such as the myocardium and kidneys. Furthermore, some forms of hepatotoxicity, such as induced by ethionine and phosphorus, may be accompanied by little or no elevation of serum enzyme activities (Zimmerman, 1974). Accordingly, the interpretations about hepatotoxicity of a test chemical derived from the measurements of serum enzymes must be corroborated by other markers of injury such as histologic observations of the liver tissue, or hepatic function.

The serum enzymes as markers of liver injury can be divided into four groups based on specificity and sensitivity to liver damage (Zimmerman and Seeff, 1970).

Group 1: These enzymes indicate more selectively hepatic cholestasis, when elevated, e.g., alkaline phosphatase (AP), 5′-nucleotidase (5′-ND), γ-glutamyl transpeptidase (γ-GT).

Group II: These enzymes indicate parenchymal injury, when elevated, e.g., aspartate transaminase (AST), alanine transaminase (ALT), fructose-1,6-diphosphate aldolase (ALD), lactate dehydrogenase (LDH), isocitrate dehydrogenase (ICDH), ornithine-carbamoyl-transferase (OCT), and sorbitol dehydrogenase (SDH).

Group III: These enzymes represent injury of other tissues, when elevated, e.g., creatine phosphokinase (CPK).

Group IV: Enzymes depressed in hepatic injury, e.g., cholinesterase (ChE).

In general, enzymes which fall in groups I and II are most useful in the study of experimental hepatotoxicity. Hence, the details of measuring these enzymes have been described in order to aid in the quantitation of the degree of hepatotoxicity.

ENZYMES INDICATIVE OF CHOLESTASIS

ALKALINE PHOSPHATASE

This family of enzymes is also known as orthophosphoric monoester phosphohydrolase, ALP, and EC 3.1.3.1. Alkaline phosphatase is a prototype of those enzymes that reflect pathological reductions in bile flow. This enzyme has been extensively employed in experimentally induced hepatic dysfunction. However, species variation exists. While this enzyme has been extremely useful for detecting biliary dysfunction in the dog, it may not be useful for detecting cholestatic changes in rat, since the normal serum enzyme activities are exceptionally high and are usually susceptible to fluctuations due to dietary factors unrelated to cholestasis (Gutman, 1959). Alkaline phosphatase refers not to a single enzyme but to a family of enzymes with different physicochemical properties and broad overlapping substrate specificities. Initially, questions were raised regarding the physiologic significance of ALP, since its pH optimum is so high. However, at low substrate concentrations, such as might be found *in vivo,* the pH optimum for the enzyme falls within the physiological range (Ross et al., 1951).

Bessey et al. (1946) introduced the first diagnostically useful application of 4-nitrophenylphosphate as substrate for alkaline phosphatase assay. The study group of the American Association for Clinical Chemistry on alkaline phosphatase has published a standardized method for measuring this enzyme (Tietz et al., 1983). The procedure described here is the method of Tietz et al. (1983) with slight modifications.

Sample Preparation

Draw the blood into a tube either with an anticoagulant (plasma) or without an anticoagulant (serum). Care should be taken to avoid hemolysis. EDTA and citrate are inhibitors of alkaline phosphatase and, therefore, are unsuitable as anticoagulants. Centrifuge the blood at 700 × g for 10 min to separate serum or plasma from the clot or from packed cells.

Reagents

1. *2-Amino-2-methyl-1,3-propanediol buffer (0.35 M):* Weigh 18.4 g of this compound (FW 105.14) and dissolve in distilled water, adjust the pH to 10.4 at 30°C by adding 1 *N* HCl, and make up the volume to 500 ml.
2. *Metal ion buffer solution (5 mM magnesium acetate, 2.5 mM zinc sulfate, and 5 mM HEDTA):* Dissolve 0.071 g of magnesium acetate (FW 142.4), 0.041 g of zinc sulfate (FW 161.43), and 0.139 g of *N*-(2-hydroxyethyl)ethylenediaminetriacetic acid (HEDTA; FW 278.26) in a final volume of 100 ml 2-amino-2-methyl-1,3-propanediol buffer.

This buffer should be prepared carefully by dissolving HEDTA first, followed by zinc sulfate, and, finally, add magnesium acetate after complete dissolution of the first two salts.

3. *4-Nitrophenyl phosphate (160 mM):* Dissolve 5.9 g of 4-nitrophenyl phosphate (FW 371.2) in 2-amino-2-methyl-propanediol buffer and make up the volume to 100 ml.

Procedure

Pipet 0.02 ml of serum or plasma into a cuvette. To this, add 0.5 ml of 2-amino-2-methyl-propanediol buffer followed by 0.4 ml of metal ion buffer and 0.1 ml of 4-nitrophenyl phosphate. Monitor the reaction at 30°C over a period of 10 min in a spectrophotometer at A_{405}. During this reaction, colorless 4-nitrophenyl phosphate is hydrolyzed by alkaline phosphatase to form brilliantly yellow 4-nitrophenol. The rate at which the color is generated is proportional to the enzyme activity.

Prepare a standard curve by taking different concentrations of 4-nitrophenol (0.5 to 5 μmol/ml) and calculate the sample values from this. The unit of enzyme activity can be represented as μmol 4-nitrophenol formed/min/ml serum or plasma.

In the liver, ALP is associated with both sinusoidal and canalicular membranes (Hagerstrand, 1975) and its elevations in serum are indicative of cholestasis. Industrial chemicals are not usually associated with this type of response, although a large number of drugs (including amitriptyline, chlorpromazine, chlorthiazide, diazepam, fluphenazine, imipramine, methimazole, nitrofurantoin, penicillamine, promazine, thiouracil, and triflupromazine) do cause intrahepatic cholestasis (Popper and Schaffner, 1959; Schaffner and Raisfeld, 1969; Zimmerman, 1978).

5′-NUCLEOTIDASE

This enzyme is also known as 5′-ribonucleotide phosphohydrolase, 5′-ND, and EC 3.1.3.5. 5′-Nucleotidase is localized in the membranes of hepatocyte and bile ductular cells and used as an indicator to assess obstructive liver injury (Song et al., 1969). This enzyme specifically hydrolyzes 5′-nucleotides, such as adenosine 5′-monophosphate (AMP) to their nucleosides and inorganic phosphorus. Procedures for the assay of 5′-ND are based on measuring one of these end products produced by the hydrolytic action of the enzyme.

UV Method

The UV method for measuring 5′-ND activity has been described by

Arkesteijn (1976). The principle in this method is that 5′-ND catalyses the hydrolysis of AMP to yield adenosine and inorganic phosphate (Pi) as depicted in Eq. (1). The auxillary enzyme, adenosine deaminase (ADA), deaminates adenosine [Eq. (2)], producing inosine and ammonium ion (NH_4^+). In a coupled reaction catalyzed by L-glutamate dehydrogenase (GDH), the NH_4^+ reacts with 2-oxoglutarate in the presence of NADH to form glutamate and NAD [Eq. (3)]. The rate of decrease in absorbance at 340 nm because of NADH oxidation is directly proportional to the rate of adenosine formation and hence, 5′-ND activity.

$$\text{AMP} \xrightarrow{5'\text{-ND}} \text{Adenosine} + \text{P}_i \tag{1}$$

$$\text{Adenosine} \xrightarrow{\text{ADA}} \text{Inosine} + \text{NH}_4^+ \tag{2}$$

$$\text{NH}_4^+ + \text{2-Oxoglutarate} + \text{NADH} \longrightarrow \text{L-Glutamate} + \text{NAD} \tag{3}$$

Sample Preparation

Serum or plasma collected in heparin, citrate or oxalate can be used. Hemolysis should be avoided (Zygoweiz et al., 1977).

Reagents

1. *β-Glycerophosphate buffer (0.05 M):* Dissolve 1.08 g of glycerol-2-phosphate, disodium salt (FW 216.04) in distilled water, adjust pH to 9.0, and make up the volume to 100 ml.
2. *Adenosine monophosphate, sodium salt (16 mM):* Dissolve 55.55 mg of Na-AMP (MW 347.2) in 10 ml of β-glycerophosphate buffer.
3. *NADH (2 mM):* Dissolve 14.2 mg of NADH (MW 709.4) in 10 ml of β-glycerophosphate buffer.
4. *2-Oxoglutarate (37 mM):* Dissolve 62.2 mg of 2-oxoglutaric acid, sodium salt (FW 168.08) in 10 ml of β-glycerophosphate buffer.
5. *Glutamate dehydrogenase (GDH):* From a commercial source, 220 U/ml.
6. *Adenosine deaminase (ADA):* From a commercial source, 8 U/ml.

Procedure

Place 0.05 ml of the sample in a semimicrocuvette. To this, add 0.5 ml glycerophosphate buffer, 0.2 ml AMP, 0.1 ml NADH, 0.1 ml 2-oxoglutarate, 0.05 ml GDH, and 0.05 ml ADA. Monitor the reaction over a period of 10 min at a constant temperature (20–30°C) at A_{340} in a spectrophotometer. The rate of decrease in absorbance owing to the oxidation of NADH to NAD is recorded and this is proportional to the 5′-ND activity. The unit of enzyme activity can be represented as μmol NADH oxidized/min/ml serum or plasma.

Colorimetric Method

The colorimetric method has been described by Arnonson and Touster (1974). In this colorimetric method, the inorganic phosphate (P_i) liberated during hydrolysis of AMP by the enzyme is measured. The amount of P_i released is proportional to the enzyme activity [Eq. (4)].

$$\text{AMP} \xrightarrow{5'\text{-ND}} \text{Adenosine} + P_i \tag{4}$$

Reagents

1. *50 mM Na-AMP, pH 7.0 (MW 347.2):* Dissolve 173.6 mg of Na-AMP in 10 ml of distilled water.
2. *0.5 M Glycine-NaOH buffer, pH 9.1 (MW 75.07):* Dissolve 3.753 g of glycine in 90 ml of distilled water, adjust the pH to 9.1 with NaOH, and make up the volume to 100 ml.
3. *0.1 M $MgCl_2$* (FW 203.31): Dissolve 2.033 g of $MgCl_2$ in distilled water and make up to a final volume of 100 ml.
4. *50% Trichloroacetic acid (TCA):* Dissolve 50 g of TCA in distilled water and the volume is made up to 100 ml.

Procedure

Mix the reagents 1, 2, 3, and water in the volume ratio of 1:2:1:5. This mixture can be stored frozen and used for at least a year. Add 0.1 ml of serum or plasma to 0.9 ml of the above mixture and incubate at 37°C for 15–30 min (needs standardization) in a shaking incubator bath. After incubation, stop the reaction by adding 0.1 ml of 50% TCA. Centrifuge the contents and estimate P_i in the supernatant by using malachite green (Carter and Karl, 1982). The unit of enzyme activity can be represented as μmol P_i/mg protein/h.

Inorganic Phosphate (Pi) Assay With Malachite Green

Inorganic phosphate (P_i) assay with malachite green is one of the various methods available to measure P_i (Fiske and Subbarow, 1925; Lowry and Lopez, 1946; Carter and Karl, 1982). This method is adapted from Carter and Karl (1982), which has several advantages over the other methods. The principle involved in this method is that the P_i liberated during substrate hydrolysis combines with molybdate in acidic media forming phosphomolybdate complex. This reacts with malachite green forming a colored complex. Sulfuric acid reacts with the uncomplexed malachite green, reducing the background color and eliminating the contribution of P_i produced from any subsequent nonenzymatic hydrolysis of substrate. This method has several advantages over the other methods reported in the literature. For example, (1) the color developed is stable for several hours; (2) the sensitivity is approximately 1 nmol P_i, and (3) proteins up to 0.1 mg/ml do not affect the linearity of the standard curve.

Reagents

Solution A: Mix 4 volumes of 2 *N* HCl and 3 volumes of 0.1 *M* Na_2MoO_4 just before use.

2 *N* HCl: Take 100 ml of concentrated HCl and add to 500 ml of distilled water.

0.1 *M* sodium molybdate: Dissolve 2.42 g Na_2MoO_4 (FW 241.95) to a final volume of 100 ml in distilled water.

Solution B: Dissolve 42 mg of malachite green dye in 100 ml of 1% polyvinyl alcohol (Aldrich Co.).

Solution C: 7.8% sulfuric acid.

Procedure

To 1 ml of sample, add 0.7 ml of solution A and 0.3 ml of solution B. Mix the contents and allow it to stand for 2 min for color development between malachite green and phosphomolybdate complex, then add 2 ml of solution C and mix thoroughly. Allow the tubes to stand for 1 h at room temperature to complete the color development. Read the color at A_{625} in a spectrophotometer.

Prepare the standard curve by taking different concentrations of P_i (5–50 nmol/ml) and calculate the sample values from this.

Elevations of 5′-ND in serum are considered mainly of hepatobiliary origin (Hill and Sammons, 1967), despite the widespread localization of the enzyme in other body tissues (Goldberg, 1973). The reason for this phenomenon is not known yet. The postulation regarding this phenomenon is that hepatobiliary tissue is the only tissue suited to release 5′-ND into circulation, since this enzyme is attached to plasma membranes and is solubilized only by treatment with detergents. Liver and biliary tree have a high local concentration of bile salts, which are detergents. Since 5′-ND is so specific for liver disease, it has been used to confirm that elevations of ALP reflect liver disease. On the other hand, measurements of serum 5′-ND are particularly useful in diagnosing liver disease in childhood (Belfield and Goldberg, 1971) and in pregnancy (Seitanidis and Moss, 1969). An additional use for 5′-ND assays is in following the progression or regression of metastatic tumors in the liver (Van der Slick et al., 1970). Moreover, 5′-ND has a higher predictive value and lower rate of false positives in screening for metastatic liver cancer than ALP and γ-GT (Kim et al., 1977).

γ-GLUTAMYLTRANSPEPTIDASE

This enzyme is also known as γ-GT, γ-glutamyl-peptide, amino acid γ-glutamyltransferase, and EC 2.3.2.2. This enzyme is a very good marker for cholestasis. It catalyzes the transfer of γ-glutamyl group from a γ-glutamyl peptide to an amino acid or another peptide. Orlowski and Meister (1963) first introduced the substrate, L-γ-glutamyl-*p*-nitroanilide for the assay of γ-GT, which, after transfer of the γ-glutamyl group to glycylglycine, releases the chromogenic product *p*-nitroaniline. The procedure described here is essentially similar to the one described by Szasz (1969).

Sample Preparation

Serum or plasma, collected using heparin or EDTA, is suitable for γ-GT assay. This enzyme activity is unaffected by hemolysis.

Reagents

1. *Tris buffer, pH 8.25 (100 mM):* Dissolve 1.211 g of tris base (MW 121.1) in distilled water and make up the volume to 100 ml.
2. L-*γ-Glutamyl-p-nitroanilide (17.6 mM):* Dissolve 50.5 mg of glutamyl-*p*-nitroanilide (MW 287) in 10 ml of tris buffer.
3. *Glycylglycine (280 mM): Dissolve 0.37 g of glycylglycine (FW 132.12) in a final volume of 10 ml tris buffer.*

Procedure

Add 0.05 ml of the sample to a semimicrocuvette. To this, add 0.5 ml of tris buffer followed by 0.25 ml glutamyl-*p*-nitroanilide (GNA) and 0.25 ml glycylglycine (GG). Carry the reaction at 30°C or at some other constant temperature. The reaction is

$$\text{GNA} + \text{GG} \xrightarrow{\gamma\text{-GT}} \text{p-Nitroaniline} + \text{Glutamylglycylglycine} \quad (5)$$

The substrate is colorless, while *p*-nitroaniline absorbs strongly at A_{405}. Thus, measure the amount of *p*-nitroaniline liberated kinetically at A_{405} and this is proportional to the γ-GT activity.

Prepare the standard curve by taking different concentrations of *p*-nitroaniline (0.05 to 2 μmol/ml) and calculate sample values from this. The unit of enzyme activity can be represented as μmol *p*-nitroaniline/min/ml serum or plasma.

By histochemical techniques, γ-GT has been localized to the whole hepatobiliary tree, from hepatocytes to bile duct, in the liver (Naftalin et al., 1969). In liver disease, serum γ-GT elevation correlates well with serum ALP and is the most sensitive indicator of cholestasis (Whitfield et al., 1972). However, a finding of elevated γ-GT has limited usefulness, because the enzyme is ubiquitous and is elevated under a variety of pathological states in addition to hepatobiliary disease (Goldberg and Martin, 1975). Additionally, this enzyme is inducible by alcohol, in the absence of hepatic disorder (Zien and Discombe, 1970). One potential advantage for γ-GT use as a diagnostic tool is in its identification by histochemical means at early hepatocellular carcinoma. This is so because 40-fold increases of γ-GT have been found in loci of preneoplastic hepatocytes in an animal model for liver cancer (Cameron et al., 1978). Therefore, histochemically γ-GT is the most often used parameter for the detection and quantitation of the preneoplastic nodules.

ENZYMES THAT ARE MORE SPECIFIC MARKERS OF HEPATIC INJURY

These enzymes are further subdivided into different categories based on their specificity.

(a) Enzymes that are somewhat nonspecific and can reflect injury to extrahepatic tissue. For example, AST and LDH levels may rise in plasma or serum upon myocardial or renal injury.

(b) Enzymes found mainly in the liver. For example, ALT elevation in plasma or serum is indicative of liver injury.

(c) Enzymes that are almost exclusively located in the liver. For example, OCT and SDH.

ASPARTATE TRANSAMINASE (AST; EC 2.6.1.1) AND ALANINE TRANSAMINASE (ALT; EC 2.6.1.2)

Colorimetric Method

In the colorimetric method these enzymes, otherwise called glutamic oxaloacetic transaminase (GOT) and glutamic pyruvic transaminase (GPT), catalyze the transfer of α-amino groups from specific amino acids to α-ketoglutaric acid to yield glutamic acid and oxaloacetic acid or pyruvic acid. The keto acids formed are then determined colorimetrically. The method described here is of Reitman and Frankel (1957). For measuring these enzyme activities, the kits are available from commercial sources. The reaction for AST is

$$\text{Aspartate} + \alpha\text{-Ketoglutaric acid} \xrightarrow{\text{AST}} \text{Oxaloacetic acid} + \text{Glutamic acid} \quad (6)$$

The reaction for ALT is

$$\text{Alanine} + \alpha\text{-Ketoglutaric acid} \xrightarrow{\text{ALT}} \text{Pyruvic acid} + \text{Glutamic acid} \quad (7)$$

Sample Preparation

Draw the blood into a tube either containing anticoagulant (plasma) or one which does not contain an anticoagulant (serum). Care should be taken to avoid hemolysis. Centrifuge the blood at 700 χg for 10 min to separate plasma or serum from the packed cells or from the clot.

Reagents

1. *Phosphate buffer, pH 7.5 (0.2 M):* Dissolve 0.445 g of monobasic sodium phosphate and 4.51 g of dibasic sodium phosphate in distilled water and make up the volume to 200 ml.
2. D-*Aspartate (0.5 M):* Dissolve 6.655 g of DL-aspartic acid (FW 133.1) in phosphate buffer and make up the volume to 100 ml.
3. DL-*Alanine (0.5 M):* Dissolve 4.45 g of DL-alanine (FW 89.09) in phosphate buffer and make up the volume to 100 ml.
4. *α-Ketoglutaric acid (18 mM):* Dissolve 0.303 g of α-ketoglutaric acid, monosodium salt (FW 168.08) in phosphate buffer and make up the volume to 100 ml.
5. *2,4-Dinitrophenylhydrazine (2,4-DNPH:0.1%):* Dissolve 200 mg of 2,4-DNPH in 200 ml of 2 *N* HCl.
6. *Sodium hydroxide (0.4 N):* Dissolve 16 g of NaOH in distilled water and make up to a final volume of 1 liter.

Procedure

Pipette 0.1 ml of serum/plasma into a test tube. To this, add 0.4 ml substrate (DL-aspartate for AST and DL-alanine for ALT) and 0.1 ml of α-ketoglutaric acid to measure the respective aminotransferases. The reaction is carried out at 37°C for 30–60 min. After incubation, add 0.5 ml of 2,4-DNPH to arrest the reaction. After keeping the tubes for 20 min at room temperature, add 5 ml of NaOH and mix thoroughly. The color developed because of the resulting phenylhydrazones should be read after 5 min in a spectrophotometer at A_{505}. The color intensity is proportional to the transaminase activity.

Prepare a standard curve by taking different concentrations of sodium pyruvate (0.15 to 0.75 μmol/ml). Develop the color as mentioned above with 2,4-DNPH and NaOH. The sample values should be calculated from this standard curve. The activities can be represented as μmol glutamate/min/ml serum or plasma. This can be converted to International units (unit is defined as the amount of enzyme that converts 1 μmol of substrate per min) or Sigma-Frankel units (one unit will form 4.82×10^{-4} μmol glutamate per min).

UV Method

The UV method is discussed below. Karmen (1955) was first to demonstrate the kinetic assay for AST, and Wroblewski and LaDue (1955) were the first to describe ALT. There are two reactions involved in this UV assay.

The reactions for AST are

$$\text{L-Aspartate} + \text{2-Oxoglutarate} \xrightarrow{\text{AST}} \text{Oxaloacetate} + \text{L-Glutamate} \quad (8)$$

$$\text{Oxaloacetate} + \text{NADH} \xrightarrow{\text{MDH}} \text{L-Malate} + \text{NAD} \quad (9)$$

The reactions for ALT are

$$\text{L-Alanine} + \text{2-Oxoglutarate} \xrightarrow{\text{ALT}} \text{Pyruvate} + \text{L-Glutamate} \quad (10)$$

$$\text{Pyruvate} + \text{NADH} \xrightarrow{\text{LDH}} \text{Lactate} + \text{NAD} \quad (11)$$

In the first reactions [Eq. (8) and (10)], these enzymes catalyze the transfer of amino group from aspartate or alanine to 2-oxoglutarate, with the generation of glutamate and oxaloacetate or pyruvate. The oxaloacetate and pyruvate are reduced by the respective dehydrogenases [malate dehydrogenase and lactate dehydrogenase; Eqs. (9) and (11)] to malate and lactate. During this conversion, an equivalent amount of NADH is oxidized to NAD.

Sample Preparation

This is similar to the one reported in colorimetric determination of these enzyme activities. Care should be taken for any hemolysis.

Reagents

1. *Tris buffer, pH 7.4 (160 mM):* Dissolve 9.69 g of tris base (MW 121.1) in 450 ml of distilled water and adjust the pH to 7.4 with HCl and make up the volume to 500 ml.
2. L-*Aspartate (1 M):* Dissolve 13.31 g of L-aspartic acid (FW 133.1) in distilled water and make up the final volume to 100 ml.

3. L-*Alanine (2 M):* Dissolve 8.9 g of L-alanine (FW 89.09) in distilled water and make up the final volume to 100 ml.
4. *2-Oxoglutarate (120 mM):* Dissolve 2.02 g of 2-oxoglutaric acid, monosodium salt (FW 168.08) in distilled water and make up the final volume to 100 ml.
5. *MDH or LDH:* 40 units per ml, from a commercial source.
6. *NADH (1.5 mM):* Dissolve 100 mg of NADH in 10 ml of tris buffer. Prepare fresh.
7. *Ethylenediaminetetraacetic acid (EDTA: 50 mM):* Dissolve 1.861 g of EDTA, disodium salt (FW 372.2) in tris buffer and make up the final volume to 100 ml.

Procedure

Pipette 0.1 ml of the sample in semimicrocuvette. To this, add 0.6 ml of tris buffer, 0.1 ml L-aspartate (for AST)/L-alanine (for ALT), 0.1 ml 2-oxoglutarate, 0.05 ml MDH (for AST)/LDH (for ALT), 0.05 ml NADH and 0.1 ml EDTA. Monitor the reaction over a period of 10 min at a constant temperature (20–30°C) at A_{340} in a spectrophotometer.

The decrease in A_{340}, resulting from oxidation of NADH to form NAD, is proportional to the respective enzyme activity. The units of enzyme activity have been defined in the respective colorimetric procedure.

AST and ALT activities in serum are the most frequently measured indicators of hepatotoxicity. ALT is primarily localized in liver, whereas AST is present in a wide variety of tissues, including heart, skeletal muscle, kidney, and brain in addition to liver (Boyde and Latner, 1961). In a series of experiments, Molander et al. (1955) have used serum AST as a more sensitive index of hepatocellular injury then ChE or ALP measurements, in rats treated with CCl_4. Balazs et al. (1961) used serum ALT as a marker for hepatotoxicity caused by CCl_4, ethionine, thioacetamide, dimethylnitrosamine, or allyl alcohol. Serum ALT elevations were observed following acute treatment of all these agents, but with ethionine, the elevation was not pronounced. This finding is consistent with the known effects of ethionine. Ethionine does not cause extensive centrilobular necrosis; instead, it induces fatty infiltration. There are several reports that ALT is an insensitive measure of hepatic steatosis (Asada, 1958; Ghoshal et al., 1969). On the other hand, agents associated with severe necrosis produce dramatic elevation of serum ALT. Investigators have also observed an excellent correlation between elevations in serum ALT and hepatocellular necrosis produced by several chemicals including chloroform and acetaminophen (Plaa and Hewitt, 1982; Purushotham et al., 1988).

ISOCITRATE DEHYDROGENASE

Colorimetric Method

In the colorimetric method this enzyme (also known as ICD [$NADP^+$] and EC 1.1.1.42) converts isocitrate to α-ketoglutarate. Since this is primarily of liver origin (Hoffman, 1970), marked elevations in serum ICD are used as screening test for hepatotoxicants. ICD activity was first measured by Wolfson and Williams-Ashman (1957), and their procedure went through several modifications subsequently. The procedure described here is that of King (1967) and Bergmeyer (1974) with slight modifications. Like for the other enzymes, ICD assay kits are available from commercial sources.

Sample Preparation

This is similar to the one described for AST and ALT. Particular care should be taken to avoid any hemolysis, since ICD is found in high concentrations in erythrocytes.

Reagents

1. *Phosphate buffer, pH 7.5 (0.2 M):* Dissolve 2.78 g of monobasic sodium phosphate (FW 138) and 2.84 g of dibasic sodium phosphate (FW 142) in 100 ml of distilled water separately. Mix 16 ml of monobasic and 84 ml of dibasic solutions to get pH 7.5. Confirm the pH before use.
2. DL-*Isocitric acid (8 mM):* Dissolve 206 mg of DL-isocitric acid, trisodium salt (FW 258.07) in phosphate buffer and make up the final volume to 100 ml.
3. *Manganese chloride ($MnCl_2$ 10 mM):* Dissolve 125.8 mg of $MnCl_2$ (FW 125.8) in phosphate buffer and make up the final volume to 100 ml.
4. *NADP (0.75 mM):* Dissolve 6 mg in 1 ml distilled water.
5. *2,4-Dinitrophenylhydrazine (2,4-DNP; 0.02%):* Dissolve 20 mg of 2,4-DNP in 100 ml of 1 *N* HCl.
6. *5% EDTA:* Dissolve 5 g of EDTA, sodium salt in distilled water and the final volume is made up to 100 ml.
7. *Sodium hydroxide (0.5 N):* Dissolve 16 g of NaOH in distilled water and final volume is made up to 800 ml.

Procedure

The reaction mixture of 1 ml contains 0.5 ml isocitrate, 0.2 ml $MnCl_2$, 0.1 ml NADP, and 0.2 ml of serum. Start the reaction by keeping the tubes in water bath at 37°C.

$$\text{L-Isocitrate} + \text{NADP} \xrightarrow{\text{ICD}} \text{NADPH} + \alpha\text{-Ketoglutarate} + CO_2 \quad (12)$$

In this reaction [Eq. (12)], actual product is oxalosuccinate, which is then immediately decarboxylated to form α-ketoglutarate. ICD activity is proportional to the α-ketoglutarate produced. Blanks containing all the contents of reaction mixture except NADP, should be maintained. Following 30 min of incubation, add 1 ml of 2,4-DNP and 0.5 ml of EDTA. Mix the contents and incubate in water bath at 37°C for 30 min.

$$\alpha\text{-Ketoglutarate} + 2,4\text{-DNP} \longrightarrow \text{Phenylhydrazone (color)} \quad (13)$$

Following this incubation, add 4 ml of sodium hydroxide and mix thoroughly. The reddish brown color developed [Eq. (13)] should be read at $A_{390 \pm 10}$ in a spectrophotometer within 5 min after developing the color. The color intensity is proportional to the ICD activity.

Prepare a standard curve by taking different cocentrations of α-ketoglutarate (0.05 to 0.2 μmol/ml) and by developing the color as mentioned above starting from 2,4-DNP addition. The sample values should be calculated from this standard curve. The ICD activity can be represented as μmol α-ketoglutarate/min/ml serum.

UV Method

The UV method described here is similar to the method of Wolfson and Williams-Ashman (1957), as described by Tietz (1970). This kinetic procedure involves determining the increase in NADPH, since ICD catalyzes the oxidative decarboxylation of isocitric acid to α-ketoglutaric acid in the presence of Mn^{2+} [Eqs. (14) and (15)] and the hydrogen transfer coenzyme in this reaction is NADP.

$$\text{L-Isocitrate} + \text{NADP} \xrightarrow{\text{ICD}} \text{NADPH} + H^+ + \text{Oxalosuccinate} \quad (14)$$

$$\text{Oxalosuccinate} + H^+ \xrightarrow{\text{ICD}} \alpha\text{-Ketoglutarate} + CO_2 \quad (15)$$

It is not clear whether the oxidation and decarboxylation reactions occur simultaneously or in separate steps. Nevertheless, increase in NADPH is proportional to ICD activity.

Sample Preparation

Similar to the one reported for colorimetric method.

Reagents

1. *Tris buffer, pH 7.5 (0.1 M):* Dissolve 12.1 g of tris base in distilled water, adjust pH to 7.5 with HCl, and make up the volume to 1 liter.
2. DL-*Isocitric acid (0.1 M):* Dissolve 0.265 g of DL-isocitric acid, trisodium salt in 0.85% NaCl and make up the final volume to 10 ml.
3. *Mn^{2+} (0.015 M):* Dissolve 300 mg of $MnCl_2$ $4H_2O$ in 0.85% NaCl and make up the final volume to 100 ml.
4. *NADP (1 mM):* Dissolve 8.3 mg of NADP in 10 ml of 0.85% NaCl, to be prepared fresh every day.

Procedure

Add serum or plasma (up to 0.5 ml) to a cuvette. To this, add 1.5 ml tris buffer followed by 0.5 ml $MnCl_2$, 0.5 ml NADP, and 0.1 ml isocitrate. Monitor the reaction over a period of 10 min at 30°C or at a constant temperature at A_{340}. The increase in absorbance at A_{340}, because of NADPH formation from NADP, is proportional to the ICD activity. The unit of enzyme activity can be represented as nmol NADPH formed/min/ml serum or plasma.

ICD has received attention as a marker of hepatic disease. Elevations of this enzyme in serum or plasma is relatively specific for liver disease. This enzyme is present in the cytoplasm of hepatocytes and changes in serum ICD parallel those in transaminases in patients with hepatobiliary disease. However, ICD is a less sensitive marker than are the transaminases. During CCl_4 toxicity, ICD was less sensitive than SDH. Korsrud et al. (1973) reported elevations of serum enzymes during thioacetamide-, dimethylnitrosamine-, and diethanolamine-induced hepatotoxicity.

SORBITOL DEHYDROGENASE

We discuss this enzyme under the UV method. This enzyme (also known as SDH, L-Iditol dehydrogenase, and EC 1.1.1.14) was first described by Blakely in 1951, which catalyzes the reversible oxidation-reduction reaction, involving the interconversion of fructose and sorbitol. This enzyme is primarily localized in liver and very little is present in blood. Consequently, appearance of SDH in serum is a good indicator of hepatocellular injury (Gerlach, 1957; Secchi and Ghidonie, 1962). However, transaminases are preferred over SDH, since SDH levels remain elevated for much shorter periods of time than transaminases (Asada and Galambos, 1963; Wiesner et al., 1965). The method described here is that of Asada and Galambos (1963), and Wiesner et al. (1965).

Sample Preparation

This is similar to the one reported for AST and ALT. Since red blood cells do not contain SDH, slight hemolysis does not affect the test (Richterich, 1969).

1. *Tris buffer, pH 7.5 (0.07 M):* Dissolve 8.477 g of Trizma base in distilled water, adjust pH to 7.5 with HCl, and make up the volume to 1 liter.
2. D-*Fructose (4 M: MW 180):* Dissolve 72 g of D-fructose in tris buffer and make up the volume to 100 ml.
3. *NADH (0.55 mM: MW 709.4):* Dissolve 8 mg of NADH in 20 ml of tris buffer.

Procedure

Pipette 0.5 ml of serum into the cuvette. Start the reaction by the addition of 1.5 ml of tris buffer followed by 0.5 ml of D-fructose and 0.5 ml of NADH. Monitor the reaction over a period of 10 min at a constant temperature (20–30°C) at A_{340} in a spectrophotometer.

The reaction is

$$\text{D-Fructose} + \text{NADH} \xrightarrow{\text{SDH}} \text{D-Sorbitol} + \text{NAD} \qquad (16)$$

The decrease in absorbance at A_{340}, resulting from the oxidation of NADH to form NAD, is proportional to the SDH activity. The activity can be represented as μmol NADH oxidized/min/ml serum, or can be converted to the enzyme units. One sigma unit of SDH is 1 nmol of NADH oxidized per hour at 25°C under the assay conditions described. One International Unit of an enzyme is defined as that amount which will convert 1 μmol of substrate per minute under the specified conditions of procedure.

SDH has received the attention of scientists as a marker of hepatic injury, since SDH elevations are relatively specific for liver disease and changes in SDH levels in serum parallel those in transaminases (Asada and Galambos, 1963; Curtis et al., 1972). However, this enzyme is very unstable and care should be taken during analysis. In an early series of experiments, Korsrud et al. (1972) determined 9 serum enzymes during CCl_4 hepatotoxicity in an attempt to identify those enzymes that would respond quantitatively to varying doses of CCl_4 and would indicate even minimal liver damage. From these studies, SDH was found to be the most sensitive index of liver injury than ICD, fructose, 1,6-diphosphate aldolase, AST, and ALT. In another set of experiments, Korsrud et al. (1973) reported that SDH was not a preferentially sensitive index of liver injury in diethanolamine-treated rats, since five enzymes (SDH, ICD, ALD, ALT and AST) were equally responsive. It was later suggested that SDH was the best enzymic index of liver injury when minimal damage or minimal changes are being assessed. Kroner and Planker (1980) reported that CCl_4 and D-galactosamine elevated SDH levels significantly in female rats.

ORNITHINE CARBAMOYL-TRANSFERASE

These enzymes are also known as Carbamoylphosphate, L-ornithine carbamoyl-transferase, OCT, and EC 2.1.3.3. OCT catalyzes the transfer of the carbamyl group from carbamyl phosphate to ornithine to produce citrulline. This enzyme is found predominantly in the mitochondrial fraction of hepatocytes (Reichard, 1959, 1960; Curtis et al., 1972), and very minute amounts are detected in the serum. OCT is a more sensitive indicator of hepatotoxicity than transaminases, in humans as well as experimental animals (Reichard, 1959, 1962). The assay of this enzyme has not met with wider clinical acceptance because of its tedious procedure. The procedure is lengthy and cumbersome.

OCT activity can be measured directly by following the appearance of citrulline or indirectly by arsenolysis, in which the enzyme catalyzes the reverse reaction of citrulline to ammonia, carbon dioxide, and ornithine. In the forward reaction, citrulline is determined colorimetrically with diacetyl monoxime after destruction of serum urea with urease (Ceriotti and Gazzaniga, 1966, 1967; Vassef, 1978). In the reverse reaction, OCT activity can be determined by the production of $^{14}CO_2$ from (^{14}C-ureido) L-citrulline (Reichard, 1964; Korsrud et al., 1973; Drotman, 1975) or by the production of ammonia. The formation of ammonia can be analyzed by conversion to indophenol (Konttinen, 1967, 1968) or by a microdiffusion procedure in a Conway cell (Reichard, 1957; Reichard and Reichard, 1958; Demange et al., 1962).

Sample Preparation

Obtain serum or plasma from experimental animals as described under transaminases.

Radioisotopic Assay

The method described here is adapted from that of Reichard (1964) which is modified by Korsrud et al. (1973). The reaction is

$$\text{Citrulline} \xrightarrow{\text{OCT, arsenate}} CO_2 + NH_3 + \text{Ornithine} \qquad (17)$$

Reagents

1. *(^{14}C)-Citrulline-arsenate reagent:* Dissolve 1.75 g of L-citrulline and 15.6 g of dibasic sodium arsenate (Na_2AsSO_4, $7H_2O$) in distilled water. To this, add 45.45 μCi of (^{14}C-ureido)-L-citrulline, adjust pH to 7.1, and dilute to a final volume of 100 ml with distilled water. This solution contains 50 μmol of (^{14}C)-citrulline (0.004545 μCi/μmol; 10,000 dpm/μmol) in 0.5 ml of 0.5 *M* arsenate buffer, pH 7.1.
2. *Sulfuric acid (9 N):* Dilute 50 ml of concentrated H_2SO_4 to 100 ml with distilled water in a volumetric flask.
3. *Hyamine hydroxide (1 M):* This can be obtained commercially.
4. *Scintillation fluid:* Dissolve 4 g of Omnifluor (New England Nuclear), containing 98% 2,5-diphenyloxazole and 2% of *p*-bis (*O*-methylstyryl-benzene) in 1 liter of toluene. An appropriate commercially available scintillation cocktail can also be employed.

Procedure

Add serum or plasma (up to 0.5 ml) and 0.5 ml of ^{14}C-citrulline arsenate reagent to a 25-ml Erlenmeyer flask fitted with a polypropylene center well. Remove 20 ml of air from the flask by using a syringe and needle. Now, incubate the flask for 18 h at 37°C with gentle agitation. After incubation, stop the reaction by adding 0.5 ml of 9 *N* sulfuric acid with a 1.0-ml syringe fitted with a 20-guage needle. Following this, add 0.25 ml of hyamine hydroxide to the center well with a syringe and needle. Then, incubate the flask for an additional 60 min to trap the evolved $^{14}CO_2$. After this incubation, remove the stopper and center well from the flask. Wipe the center well free of condensation and place in a counting vial containing 15 ml of scintillation fluid. $^{14}CO_2$ radioactivity can be measured by liquid scintillation spectrometry. The formation of $^{14}CO_2$ from (^{14}C-ureido)-L-citrulline is linear over a period of 24 h. Blanks containing all components of the reaction mixture except plasma or serum should be maintained. Subtract the blank values from sample values during calculation. The activity can be expressed as nmol citrulline converted/min/ml plasma or serum.

Colorimetric Method

In the colorimetric method, citrulline is decomposed by arsenolysis catalyzed by OCT. The liberated ammonia is determined as indophenol after reaction with phenolnitroprusside and alkaline hypochlorite (Konttinen, 1967, 1968).

Reagents

1. *Citrulline-arsenate reagent:* Dissolve 3.5 g of DL-citrulline and 15.6

g of dibasic sodium arsenate in distilled water, adjust pH to 7.1, and make up the volume to 100 ml.

2. *Phenol-nitroprusside reagent:* Dissolve 10 g of phenol and 50 mg of sodium nitroprusside in distilled water adding to a final volume of 400 ml. This solution should be stored in amber bottle at 4°C.
3. *Alkaline hypochlorite:* Dissolve 5 g of NaOH in 200 ml of distilled water. When it is cooled down, add 6 ml of sodium hypochlorite (10–14% Cl), and bring the final volume to 400 ml. This solution should be stored in amber bottle at 4°C.
4. *Standard ammonia solution:* Dissolve 464 mg of ammonium sulfate, $(NH_4)_2SO_4$, in 600 ml of distilled water. To this, add 2 ml of concentrated sulfuric acid and bring the volume to 1 liter. Dilute this stock solution five times (1:4) with distilled water before use (1.4 μmol/ml).

Procedure

Pipette serum or plasma (up to 0.5 ml) into a test tube. To this, add an equivalent volume of citrulline-arsenate reagent and incubate for 24 h at 37°C. Proper blanks should be maintained for all the samples. Add citrulline-arsenate reagent to the blank tubes after incubation of serum for 24 h at 37°C. From both sample and blank tubes, transfer 0.05 ml of reaction mixture to a tube containing 2 ml of phenol-nitroprusside solution. Mix the contents. To each tube, add 2 ml of alkaline hypochlorite solution, mix thoroughly and incubate at 37°C for 15 min. Read the color developed at A_{630} in a spectrophotometer. The standard curve can be prepared by taking different concentrations of standard ammonia solution and the color can be developed in accordance to the above procedure described for samples. The unit of enzyme activity can be represented as μmol citrulline/min/ml plasma or serum.

OCT is predominently found in mitochondrial fraction of liver cells (Curtis et al., 1972; Reichard, 1959, 1960) and is markedly elevated in the serum during both acute and chronic liver disease (Musser et al., 1966; Reichard, 1962). OCT is a very sensitive marker for hepatotoxicity, is increased more than aminotransferases and follows a temporal phase similar to that of aminotransferases (Reichard, 1959; Tegeris et al., 1969). OCT was also reported to be as sensitive an index of liver injury as AST in cattle and sheep poisoned with CCl_4, dimidium bromide, or sporidesmin (Ford, 1965). OCT activity in serum is a useful monitor of liver injury caused by a variety of hepatotoxicants and a very good correlation between serum OCT levels and quantified histological changes was observed (Curtis et al., 1972; Drotman, 1975; Drotman and Lawhorn, 1978; Korsrud et al., 1973).

FRUCTOSE 1,6-DIPHOSPHATE ALDOLASE

Colorimetric Method

This enzyme is also known as F-1,6-ALD, D-fructose-1,6-diphosphate, D-glyceraldehyde-3-phosphate-lyase, and EC 4.1.2.13. Below, we discuss the colorimetric method. This enzyme (ALD) is found mainly in liver, occurs at low levels in the serum and erythrocytes of healthy individuals. The increase in this enzyme activity has been noted after low doses of CCl_4 challenge to laboratory animals (Korsrud et al., 1972). ALD belongs to the class of enzymes called lyases, which reversibly cleave substrates without hydrolysis into two units, one or both of which contain a double-bonded carbon atom. Aldolase catalyzes the splitting of D-fructose-1,6-diphosphate (FDP) to D-glyceraldehyde-3-phosphate (GAP) and dihydroxyacetone phosphate (DAP), as depicted in Eq. (18).

$$\text{FDP} \xrightarrow{\text{ALD}} \text{DAP} + \text{GAP} \tag{18}$$

The procedure described here is adopted from the one outlined by Pinto et al. (1969a).

Sample Preparation

Prepare serum or plasma as described earlier. ALD in serum is quite stable and slight hemolysis does not alter the enzyme activity significantly.

Reagents

1. *Collidine buffer, pH 7.4 (0.1 M):* Dissolve 3.03 g of collidine in a final volume of 250 ml distilled water and adjust pH to 7.4 with 2 *N* HCl.
2. *Hydrazine, pH 7.4 (0.56 M):* Dissolve 7.29 g of hydrazine ($NH_2NH_2H_2SO_4$) into a final volume of 100 ml distilled water and adjust the pH to 7.4 with 10% NaOH. Hydrazine traps the triose phosphates and prevents any change in the ratio between the two forms.
3. *Iodoacetate, pH 7.4 (2 mM):* Dissolve 42 mg of sodium iodoacetate in distilled water, adjust pH to 7.4 with NaOH, and make volume up to 100 ml. Iodoacetate prevents the catalysis of glyceraldehyde-3-phosphate.
4. *Fructose 1,6-diphosphate, pH 7.4 (0.06 M):* Dissolve 320 mg of fructose-1,6-diphosphate, tetrasodium salt in 10 ml of collidine buffer.
5. *10% Trichloroacetic acid (TCA):* Dissolve 10 g of TCA in distilled water and make up the volume to 100 ml.

6. *0.75% Sodium hydroxide:* Dissolve 0.75 g of NaOH in distilled water and make up the volume to 100 ml.
7. *2,4-Dinitrophenylhydrazine (0.1%):* Dissolve 200 mg of dinitrophenylhydrazine in 200 ml of warm 2 *N* HCl.

Procedure

Pipette 0.5 ml collidine buffer plus 0.125 ml each of water, hydrazine, and iodoacetate solutions into each test tube. Place the tubes in a water bath maintained at 37°C. After a few minutes, add 0.5 ml serum followed by 0.125 ml FDP solution. Incubate the tubes for 60 min at 37°C. After 60 min of incubation, stop the reaction by adding 1.5 ml of 10% TCA. Proper blanks should be maintained by adding FDP after stopping the reaction with TCA. Centrifuge the contents to remove the precipitated protein. To 0.5 ml of this supernatant, add 0.5 ml of 0.75% NaOH to permit phosphate hydrolysis to occur. After 15 min, add 0.5 ml 2,4-dinitrophenylhydrazine and keep the tubes at 37°C for 10 min. The reaction is

$$\text{GAP} + \text{DAP} + \text{NaOH} + \text{2,4-DNP} \longrightarrow \text{Hydrazones (color)} \qquad (19)$$

After 10 min, add 4 ml of 0.75% NaOH. Read the color developed owing to the formation of dinitrophenylhydrazones at A_{540} in a spectrophotometer within 10 min.

The standard curve should be prepared by taking different concentrations of dihydroxyacetone (0.1 to 1 μmol/ml) and the color can be developed accordingly with NaOH and 2,4-DNP. Calculate the sample values from the standard curve. The unit of enzyme activity can be represented as nmol/min/ml serum or plasma.

UV Method

In the UV method aldolase reaction is coupled with two other enzyme reactions [Eqs. (20)–(22)]. Triosephosphate isomerase (TIM) is added to ensure rapid conversion of all glyceraldehyde-3-phosphate (GAP) to dihydroxyacetone phosphate (DAP). Glycerol-3-phosphate dehydrogenase (GDH) is added to reduce DAP to glycerol-3-phosphate, with NADH acting as hydrogen donor (Pinto et al., 1969b).

$$\text{Fructose-1,6-diphosphate} \xrightarrow{\text{ALD}} \text{GAP} + \text{DAP} \qquad (20)$$

$$\text{GAP} \xrightarrow{\text{TIM}} \text{DAP} \qquad (21)$$

$$\text{DAP} + \text{NADH} \xrightarrow{\text{GDH}} \text{glycerol-3-phosphate} + \text{NAD} \qquad (22)$$

The method described here is similar to the one reported by Pinto et al. (1969b).

Sample Preparation

Similar to that reported in colorimetric method.

Reagents

1. *Tris/collidine buffer, pH 7.4 (0.1 M):* Dissolve 3.25 g of tris base or 3.03 g of collidine in distilled water, adjust pH to 7.4 with 2 *N* HCl, and bring the volume up to 250 ml.
2. *FDP (22 mM):* Dissolve 112 mg of FDP, sodium salt in a final volume of 10 ml distilled water. Prepare fresh every day.
3. *NADH (1.5 mM):* Dissolve 100 mg of NADH in a final volume of 10 ml tris/collidine buffer. Prepare fresh every day.
4. *GDH/TIM:* Use the ammonium sulfate suspension of these two enzymes at a concentration of 10 mg/ml.

Procedure

Pipette 0.1 ml of serum or plasma into a semimicrocuvette. To this, add 0.5 ml tris/collidine buffer followed by 0.1 ml NADH, 10 μl of GDH/TIM suspension, 0.1 ml FDP, and 0.2 ml distilled water. Monitor the reaction at 37°C over a period of 10 min at A_{340} in a spectrophotometer. The decrease in A_{340}, because of NADH conversion to NAD is proportional to ALD activity. In this reaction, 2 moles of NADH are oxidized per mol of FDP. Represent the unit of enzyme activity as μmol NADH oxidized/ml serum or plasma/min.

Korsrud et al. (1972) determined serum activity of nine hepatic enzymes and found that ALD was one of the less sensitive markers for CCl_4 hepatotoxicity. However, the dose of thioacetamide, dimethylnitrosamine, and diethanolamine required to elevate serum ALD was less than that needed to elevate serum OCT (Korsrud et al., 1973). On the other hand, ALD was equally sensitive like SDH, ICD, ALT, and AST, to diethanolamine hepatotoxicity (Korsrud et al., 1973).

MEASUREMENT OF HEPATOTOXIC END POINTS BY USING *IN VITRO* MODELS

Measurement of cell viability is an integral part in understanding the toxicity of a given chemical. The routine procedures are membrane function tests such as trypan blue exclusion or determination of enzyme leakage. Leakage of a wide variety of enzymes may be monitored. However, LDH is a convenient marker and most widely employed, because of the stability of this enzyme activity in the incubation or culture medium.

LACTATE DEHYDROGENASE

This enzyme is also known as LDH, L-Lactate; NAD Oxidoreductase, and EC 1.1.1.27. LDH, a cytosolic enzyme, catalyzes the reversible oxidation of lactate to pyruvate. The reaction equilibrium strongly favors the reduction of pyruvate to lactate. However, the direct reaction, employing L-lactate and NAD, offers advantages over the reverse reaction. The original method was described by Wacker et al. (1956) and Amador et al. (1963). The present method is of Richards et al. (1975) with slight modifications. For assaying LDH, the enzyme assay kits can be obtained from commercial sources.

Sample Preparation

For determining LDH leakage, LDH released into the medium and total LDH available in the cells need to be measured. For LDH in the medium, place samples of the incubate (0.5 ml each) in 1.8 ml microcentrifuge tubes containing 0.5 ml of dibutylphthalate and centrifuge for 30 s at 13,000 × g. The top layer should be analysed for LDH activity. For total LDH, use 50 μl of a sonicated sample of incubate in the assay.

Procedure

Place 50 μl of the sample in a semimicrocuvette. Start the reaction by the addition of 0.75 ml of 250 m*M* tris, pH 8.9 containing 65 m*M* L-lactate and 11 m*M* NAD^+. Monitor the reaction over a period of 10 min at a constant temperature (20–30°C) at A_{340} in a spectrophotometer.

The reaction is

$$\text{L-Lacate} + NAD^+ \xrightarrow{\text{LDH}} \text{Pyruvate} + \text{NADH} \qquad (23)$$

The increase in A_{340} [Eq. (23)], resulting from the utilization of NAD to form NADH is proportional to LDH activity. The enzyme activity is normally represented as units. One unit of LDH activity is defined as an increase in absorbance of 0.001/min/ml of medium (Wacker et al., 1956). LDH leakage can be expressed as % of total LDH in the cells released into the medium, and is a measure of cellular injury.

From all of the above procedures, one can determine whether the test chemical is hepatotoxic and one can locate the site of toxic injury as well as the extent of injury based on the particular enzyme elevations. Once it is established that the compound is hepatotoxic, the next logical approach is to investigate the mechanism of hepatotoxicity. The following methodology relates to a number of selected biochemical techniques useful in the investigation of the mechanisms of hepatotoxicity.

BIOCHEMICAL MARKERS OF XENOBIOTIC BIOACTIVATION SYSTEM AS THE TARGET OF TOXICITY

The prevailing theory for the mechanism of hepatocellular injury by CCl_4 and $CBrCl_3$ states that these compounds are bioactivated to the $^{\cdot}CCl_3$ free radical by cytochrome P450-mediated reactions (Sipes et al., 1977; Wolf et al., 1980; Cheeseman et al., 1985). The $^{\cdot}CCl_3$ free radicals initiate the autocatalytic lipid peroxidation by attacking the methyl bridges of unsaturated fatty-acid chains of microsomal lipids (Cheeseman et al., 1985). There is also evidence for the formation of trichlorocarbon peroxy radicals ($CCl_3OO\cdot$), which may be the ultimate moieties initiating cellular injury (Connor et al., 1986). The principal mechanism put forth for the potentiation of CCl_4 hepatotoxicity by phenobarbital (Frank et al., 1982), DDT (McLean and McLean, 1966), and other modulators (Hewitt et al., 1980; Harris and Anders, 1981) of toxicity is enhanced metabolic activation (Cheeseman et al., 1985) followed by increased lipid peroxidation. The peroxidative process initiated by the $\cdot CCl_3$ free radical is thought to result in early morphological alteration of ER, loss of catalytic activity, and content of cytochrome P450 metabolizing system in SER, loss of glucose-6-phosphatase activity, loss of protein synthesis, and eventually, through as yet unidentified pathways, to cell death (Recknagel and Glende, 1973).

Initiation of lipid peroxidation is not unique to CCl_4 or the liver. Lipid peroxidation *in vivo* has been suggested to be of basic importance in aging, in damage to cells by oxidant injury associated with air pollution, in some phases of atherosclerosis, in some forms of liver injury, and in oxygen toxicity (Recknagel et al., 1982). Although these mechanisms explain why cell death occurs in CCl_4 toxicity, the mechanism for the progression of liver injury upon the administration of a toxic dose of CCl_4 has been largely unexplored. Recent developments in this area (Mehendale, 1984; Mehendale, 1989a,b) permit one to address this aspect of the halomethane toxicology. Ordinarily, a low-nontoxic dose of CCl_4 causes a stimulation of hepatocellular regeneration (Mehendale, 1984) and tissue repair in response to the limited hepatocellular necrosis. This biological response to cellular injury results in overcoming the limited injury and hence, the lesion does not progress. Recent evidence (Mehendale and Klingensmith, 1988; Bell et al., 1988; Purushotham et al., 1988; Kodavanti et al., 1989a,b,c; Mehendale et al., 1989a,b) permits to propose that the progressive phase of CCl_4 hepatotoxicity is closely associated with the biological issue of whether sufficient level of relatively early hepatocellular regeneration is stimulated. This compelling evidence has resulted in a novel hypothesis concerning the mechanism of the progressive phase of halomethane hepatotoxicity (Mehendale, 1989a,b).

CYTOCHROME P450

Cytochrome P450, a hemoprotein found in the smooth endoplasmic reticulum (SER) and to some extent in mitochondria of liver cells, has been implicated as a terminal oxidase in the metabolism of drugs and other xenobiotics. The reduced form of the pigment combines with carbon monoxide to form a complex having an absorption maximum at 450 nm and a minimum at 405 nm. The widely accepted procedure for measuring cytochrome P450 levels is that of Omura and Sato (1964).

Reagents

1. *Phosphate buffer (0.05 M), pH 7.4:* The method of preparing phosphate buffer has been described in the earlier section (ICD: prepare 0.2 *M* buffer and dilute four times to get 0.05 *M* concentration).
2. Sodium dithionite ($Na_2S_2O_4$): Use powdered sodium dithionite as a pinch of few grains to reduce P450.CO complex.
3. *Carbon monoxide gas:* Obtain from local commercial source.

Sample Preparation

Prepare microsomes (SER) from 10% liver homogenate by a series of centrifugations and suspend the final pellet obtained after 105,000 × g centrifugation for 1 h in 0.05 *M* phosphate buffer (pH 7.6). The protein concentration should be approximately 2 mg/ml.

Procedure

Pipette 2 ml of sample (microsomes) into each of two matched cuvettes (reference and sample) with stoppers. Determine the baseline by using a spectrophotometer by scanning from 510 to 400 nm. Bubble carbon monoxide gently into the sample cuvette for 20 s. A pasteur pipet tapered down to a small capillary can be used for CO bubbling in a vacuum hood. Add a few milligrams (a small pinch) of solid sodium dithionite. Invert the cuvette by covering with a stopper and gas the sample cuvette again with carbon monoxide for an additional 20 s. Treat the reference cuvette only with a few milligrams of sodium dithionite. Mix well and record the spectrum from 510 to 400 nm. Calculate the quantity of cytochrome P450 from the optical density difference (450–480 nm) and the molar extinction coefficient of 91 $mM^{-1}cm^{-1}$.

NADPH-CYTOCHROME P450 REDUCTASE

NADPH-cytochrome P450 reductase is conveniently measured by its NADPH-cytochrome *c* reductase activity (Phillips and Langdon, 1962). The assay is an indirect measure of NADPH cytochrome P450 reductase activity; measurement of actual NADPH cytochrome P450 reductase activity requires anaerobic conditions and rapid reaction techniques (Peterson et al., 1978). A small peptide of the reductase is required for NADPH cytochrome P450 reductase activity but not cytochrome *c* reductase activity; if no proteolysis has occurred, the two activities are closely correlated (Vermilion and Coon, 1978).

Reagents

1. *0.5 mM cytochrome c in 10 mM potassium phosphate, pH 7.7:* Dissolve 65 mg of cytochrome *c* (MW approximately 13,000) in 8 ml of 10 m*M* phosphate buffer and make up the volume to 10 ml.
2. *0.3 M potassium phosphate buffer, pH 7.7:* Dissolve 4.08 g monobasic potassium phosphate (FW 130.09) and 6.85 g of dibasic potassium phosphate (FW 228.2) separately in a final volume of 100 ml each with distilled water. Mix mono- and dibasic solutions to get pH 7.7 by using a pH meter.
3. *10 mM NADPH (MW 833.4):* Dissolve 83.3 mg in a final volume of 10 ml phosphate buffer. Prepare fresh daily.

Sample Preparation

Prepare microsomes as described previously (look in cytochrome P450 assay).

Procedure

Into a semimicrocuvette, pipet 0.08 ml cytochrome *c*, 0.2 ml enzyme sample, and 0.71 ml of 0.3 *M* phosphate buffer. Mix the contents and preincubate at 30°C in a thermostatically controlled spectrophotometric cuvette holder. Start the reaction by adding 0.01 ml of NADPH and measure disappearance of NADPH at A_{550} for about 3 min. Activity can be calculated as follows:

$$\frac{A_{550}\text{min}^{-1}}{0.021 \times \text{mg protein}} = \text{nmol cytochrome } c \text{ reduced/min/mg protein}$$

AMINOPYRINE DEMETHYLASE

Many drugs are dealkylated by microsomal enzymes according to the following:

$$\underset{\displaystyle \text{H}}{\text{R–}\overset{}{\underset{|}{\text{N}}}\text{–CH}_3} \xrightarrow[\text{NADPH}+\text{O}_2]{\text{microsomes}} \text{R–NH}_2 + \text{HCHO} \qquad (24)$$

The rate of metabolism may be followed by measuring either the formation of the demethylated product or the quantity of formaldehyde formed, or both. Here, we have described both the procedures to measure formaldehyde as well as 4-aminoantipyrine during dealkylation of amionopyrine [Eq. (25)].

$$\text{Aminopyrine} \xrightarrow{\nearrow \text{HCHO}} \text{Monomethyl-4-aminoantipyrine} \xrightarrow{\nearrow \text{HCHO}} \text{4-aminoantipyrine} \qquad (25)$$

Theoretically, 2 moles of formaldehyde should be formed for each mole of 4-aminoantipyrine produced with aminopyrine as substrate. Aminopyrine is a model substrate to study *N*-dealkylation of drugs.

Sample Preparation

Prepare hepatic microsomes by homogenizing liver sample, followed by a series of centrifugations to remove cell debris and mitochondria. Take the pellet after centrifugation at 105,000 × g and suspend in phosphate buffer (pH 7.5).

Formaldehyde Determination

For the determination of formaldehyde, the procedure described here is that of Nash (1953). The Nash reaction has been widely used, since it is simple, fast, and accurate. Formaldehyde formed during incubation will be trapped as the semicarbazone and measured by colorimetric procedure, based on Hantzsch reaction (Nash, 1953; Cochin and Axelrod, 1959). The Hantzsch reaction requires a β-diketone (acetylacetone), an aldehyde (formaldehyde), and an amine (NH_3 from ammonium acetate). The resulting product is 3,5-diacetyl-1,4-dihydrolutidine, which can be measured at A_{415}.

Reagents

1. *180 mM sodium phosphate buffer, pH 7.5:* Dissolve 2.7 g monobasic salt in a final volume of 125 ml distilled water. Dissolve 6.39 g of dibasic salt in a final volume of 250 ml distilled water. Mix monobasic

and dibasic solutions appropriately to get pH 7.5 by using a pH meter.

2. *200 mM Aminopyrine:* Dissolve 462 mg of aminopyrine HCl (FW 231.3) in a final volume of 10 ml distilled water.
3. *200 mM Semicarbazide:* Dissolve 223 mg in a final volume of 10 ml with distilled water.
4. *200 mM $MgCl_2$:* Dissolve 2.033 g of $MgCl_2 \cdot 6H_2O$ (FW 203.3) in distilled water and make up the volume to 50 ml.
5. *100 mM Glucose-6-phosphate:* Dissolve 336 mg of glucose-6-phosphate, dipotassium salt (FW 336.3) in a final volume of 10 ml distilled water.
6. *Glucose-6-phosphate dehydrogenase (G-6-PD):* Prepare a solution containing 20 units per 1 ml.
7. *3.3 mM NADP:* Dissolve 24.5 mg of NADP (MW 743.4) in 8 ml of phosphate buffer and make up the volume to 10 ml.
8. *20% Trichloroacetic acid (TCA):* Dissolve 20 g of TCA in distilled water and make up the volume to 100 ml.
9. *NASH reagent (2 M ammonium acetate, 0.05 M glacial acetic acid, and 0.02 M acetylacetone):* Dissolve 15 g ammonium acetate in distilled water. To this, add 1.5 ml glacial acetic acid and 1 ml acetylacetone and make up the volume to 100 ml with distilled water.

Procedure

Add 0.5 ml phosphate buffer, 0.05 ml $MgCl_2$, 0.4 ml glucose-6-phosphate, 0.05 ml G-6-PD, 0.3 ml NADP, 0.1 ml semicarbazide, and 0.1 ml aminopyrine to the tubes. Maintain proper blanks by adding phosphate buffer instead of aminopyrine. Start the reaction by adding 0.6 ml of microsomes (0.4 to 0.8 mg protein). Incubate at 37°C for 30 min. Stop the reaction by adding 1 ml of 20% TCA. Centrifuge the contents and take 1 ml of supernatant for determining formaldehyde. Add 1 ml of Nash reagent to the above sample and incubate the tubes at 37°C for 40 min. Read the color at A_{415}. Prepare a standard graph using 0.05 to 0.5 μmols of formaldehyde and calculate the sample values from this. The unit of enzyme activity can be represented as μmol formaldehyde formed/mg protein/h.

4-Aminoantipyrine Determination

The analysis of 4-aminoantipyrine involves diazotization followed by coupling with α-naphthol to yield a colored product which can be measured directly at A_{540} after acidification, or the colored complex can be extracted into a small volume of isoamyl alcohol for higher sensitivity (Brodie and Axelrod, 1950). When measuring aminopyrine demethylase activity using this procedure, omit semicarbazide from the incubation medium.

Reagents

1. *0.2% Sodium nitrite:* Dissolve 0.2 g of sodium nitrite in distilled water and make up the volume to 100 ml.
2. *5% α-Naphthol in ethanol:* Add 5 g of α-naphthol to 90 ml ethanol and make up the volume to 100 ml.
3. *1% Ammonium sulfamate:* Dissolve 1 g in a final volume of 100 ml with distilled water.
4. *4 N NaOH:* Dissolve 16 g NaOH (MW 40) in distilled water and make up the volume to 100 ml.
5. *HCl:* Use concentrated HCl. Obtain commercially.
6. *Isoamyl alcohol:* Obtain commercially.

Procedure

Add 0.5 ml of sodium nitrite to 2.5 ml of supernatant after incubation. Cool for 10 min and add 0.5 ml ammonium sulfamate. Mix by shaking and wait for 5 min. Add 0.1 ml of ethanolic solution of resublimed α-naphthol followed by 1 ml of NaOH. Allow to stand for 10 min and add 0.5 ml HCl followed by 3.0 ml amyl alcohol. Shake well and centrifuge. Draw off amyl alcohol and read the color at A_{540} after 10 min. Maintain blanks without substrate and develop the color. Prepare a standard graph taking different concentrations of 4-aminoantipyrine (0.5 to 4 μg/ml) in 5% TCA. Calculate the sample values from the standard graph. The unit of enzyme activity can be represented as μg aminoantipyrine formed/mg protein/h.

ANILINE HYDROXYLASE

The rate of aniline metabolism can be determined by measuring the quantity of *p*-aminophenol formed:

$$\text{Aniline} \xrightarrow[\text{NADPH} + O_2]{\text{microsomes}} p\text{-aminophenol} \qquad (26)$$

The procedure described here is simple and sensitive (Brodie and Axelrod, 1948).

Reagents

1. *0.18 M Sodium phosphate buffer, pH 7.5:* Dissolve 5.4 g monobasic sodium phosphate and 6.39 g dibasic sodium phosphate in a final volume of 250 ml with distilled water. Mix mono- and dibasic solutions appropriately to get pH 7.5 by using a pH meter.
2. *0.1 M Aniline (MW 129.62):* Dissolve 129.6 mg in a final volume of 10 ml distilled water.

3. *0.2 M $MgCl_2$*: Dissolve 2.033 g of $MgCl_2 \cdot 6H_2O$ (FW 203.31) in a final volume of 50 ml with distilled water.
4. *0.1 M Glucose-6-phosphate (FW 336.3):* Dissolve 336 mg in 10 ml of phosphate buffer.
5. *Glucose-6-phosphate dehydrogenase (G-6-PD):* Prepare a solution containing 20 units in one ml.
6. *3 mM NADP:* Dissolve 24.5 mg of NADP (MW 743.4) in 10 ml of phosphate buffer.
7. *10% Na_2CO_3*: Dissolve 10 g in distilled water and make up the volume to 100 ml.
8. *2% Phenol in 0.2 N NaOH:* Prepare 0.2 *N* NaOH by dissolving 0.8 g in a final volume of 100 ml with distilled water. Take 2 ml phenol and add to 98 ml of the above NaOH solution.
9. *20% Trichloroacetic acid (TCA):* Dissolve 20 g of TCA in distilled water and make up the volume to 100 ml.

Sample Preparation

Prepare microsomes as described previously (look in aminopyrine demethylase assay).

Procedure

Add 0.5 ml phosphate buffer, 0.05 ml $MgCl_2$, 0.4 ml glucose-6-phosphate, 0.05 ml G-6-PD, 0.3 ml NADP, and 0.1 ml aniline to the tubes. Maintain blank tubes by adding phosphate buffer instead of aniline. Start the reaction with 0.6 ml microsomes. Incubate at 37°C for 20 min. Care should be taken regarding incubation time, since *p*-aminophenol is not stable in incubation medium. However, the disappearance of *p*-aminophenol is minimal in the presence of aniline and this can be further minimized by decreasing the incubation time. Stop the reaction by adding 1 ml of 20% TCA and centrifuge the contents. Take 1 ml of supernatant for determining *p*-aminophenol and add 0.5 ml of 10% Na_2CO_3, followed by 1 ml of 2% phenol. Read the blue color at A_{630} after 30 min. Prepare a standard graph using 0.05 to 0.4 μmoles of *p*-aminophenol and calculate the sample values from this. The unit of enzyme activity can be represented as μmol *p*-aminophenol formed/mg protein/h.

BENZO(*A*)PYRENE HYDROXYLASE (BPH)

The rate of benzo(*a*)pyrene (BP) metabolism can be determined by measuring the quantity of 3-hydroxybenzo(*a*)pyrene (3-HBP) formed below:

$$\text{BP} \longrightarrow \text{3-HBP} + \text{other hydroxy and quinone derivatives} \quad (27)$$

The procedure described here is very simple and sensitive (Nebert and Gelboin, 1968).

Reagents

1. *100 mM Tris-HCl buffer, pH 7.5:* Dissolve 6.055 g tris base (MW 121.1) in 450 ml distilled water, adjust the pH with HCl, and make up the volume to 500 ml.
2. *15 mM $MgCl_2$:* Dissolve 0.305 g of $MgCl_2 \cdot 6H_2O$ (FW 203.31) in a final volume of 100 ml with distilled water.
3. *14.4 mM NADPH:* Dissolve 120 mg NADPH (MW 833.4) in 10 ml of tris buffer.
4. *3.2 mM Benzo(a)pyrene:* Dissolve 20.2 mg benzo(*a*)pyrene (MW 252.3) in DMSO and make up the volume to 25 ml.
5. *Acetone:* Obtain commercially.
6. *Hexane:* Obtain commercially.
7. *1 N NaOH:* Dissolve 4 g in distilled water and make up the volume to 100 ml.
8. *0.1 N H_2SO_4:* Take 1 ml concentrated H_2SO_4 and add to 179 ml distilled water.

Sample Preparation

Prepare microsomes as described previously (look in aminopyrine demethylase assay).

Procedure

Add 0.5 ml tris buffer, 0.2 ml $MgCl_2$, 0.025 ml NADPH, and 0.025 ml benzo(*a*)pyrene to a test tube. Incubate the reaction mixture at 37°C. Start the reaction with 0.25 ml microsomal sample containing 0.5 to 0.8 mg protein. Incubate at 37°C for 30 min. Stop the reaction by adding 1 ml of cold acetone. Maintain proper blanks by adding acetone prior to incubation. After the addition of acetone, add 3.25 ml hexane and incubate the mixture with shaking at 37°C for 10 min. Take 1 ml of organic phase and extract with 3.0 ml of 1 *N* NaOH. Determine the concentration of extracted, hydroxylate benzo(*a*)pyrene in the alkali phase spectrofluorometrically with excitation at 396 nm and emission at 522 nm. Prepare standard graph by using quinine sulfate or 3-hydroxybenzo(*a*)pyrene. Prepare standard solutions in 0.1 *N* H_2SO_4. Calculate the sample values from the standard graph. One unit of enzyme activity can be defined as that amount catalyzing the formation of hydroxylated product causing fluorescence equivalent to that of 1 pmole of 3-hydroxybenzo(*a*)pyrene in a 30 min incubation at 37°C.

LIPID PEROXIDATION *IN VITRO*

Several procedures for the detection and quantitation of lipid peroxidation in tissue samples or whole animals have been developed (Recknagel et al., 1982). The reaction of malonaldehyde, a degradation product of peroxidized lipids, with thiobarbituric acid (TBA) to produce a TBA-malonaldehyde chromophore has been taken as an index of lipid peroxidation and the most widely used method for detecting lipid peroxidation *in vitro*. Because malonaldehyde is rapidly metabolized in whole animals (Plaa and Witchi, 1976), this procedure cannot be used for *in vivo* preparations. The procedure described here is similar to that described by Buege and Aust (1978).

Reagents

1. *TCA-TBA-HCl reagent:* Add 15 g of trichloroacetic acid and 0.375 g of thiobarbituric acid to 80 ml of 0.25 N HCl and make up the volume to 100 ml.

Sample Preparation

Prepare liver microsomes as described previously (look in aminopyrine demethylase assay).

Procedure

Incubate microsomal fractions (2 mg protein/ml) with CCl_4 or the compound being tested as the substrate for 15 min at 37°C in Warburg flasks. Take a 0.25-ml aliquot and combine with 2 ml TCA-TBA-HCl reagent. Mix thoroughly and heat it for 15 min in a boiling water bath. After cooling, remove the flocculent precipitate by centrifugation at 1000 × g for 10 min. Measure the color of the sample at A_{535} against a blank that contained all the reagents minus the lipid. Also, prepare a standard curve using malondialdehyde (bis-dimethyl acetal), which can be obtained commercially.

LIPID PEROXIDATION *IN VIVO*

The determination of conjugated dienes in lipid extracts of hepatic tissue or its subcellular fractions is an accepted procedure for detecting lipid peroxidation *in vivo* (Davis and Mehendale, 1980). The ultraviolet difference spectra of peroxidized lipids show an absorption maximum at 233 nm with a secondary absorption maximum between 260 and 280 nm due to the presence of ketone dienes, commonly referred to as conjugated dienes. The appearance of conjugated dienes after treatment *in vivo* with toxicants is an unmistakable indication that lipid peroxidation has taken place.

Several other methods for the measurement of lipid peroxidation have been described. For example, the iodometric procedure of Bunyan et al. (1969) has been employed. A comparatively new approach is to measure the presence of fluorescent products. A variety of molecules that occur commonly in the tissue may react with malonaldehyde and yield characteristic fluorescent chromophores (Chio and Tappel, 1969). Malonaldehyde undergoes decomposition, and the decomposition products may also lead to fluorescent products when they react with proteins (Shin et al., 1972). Measurement of these fluorescent products seems to offer a workable way for detecting lipid peroxidation in biological systems and tissues (Rechnagel et al., 1982). Another interesting approach is the measurement of hydrocarbon gases. Gases such as ethane and pentane appear early in the course of autoxidation of edible fats (Frankel et al., 1961). Measurement of ethane and pentane has been employed for quantitation of lipid peroxidative process *in vitro* and *in vivo*. Ethane is the predominant gas produced during autoxidation of linolenic acid (Lieberman and Mapson, 1964; Riley et al., 1974) and pentane is the major gaseous hydrocarbon arising during thermal decomposition (Dillard et al., 1977) and iron-catalyzed decomposition of linoleic and arachidonic acid hydroperoxides (Dumelin and Tappel, 1977). Although the measurement of hydrocarbon gas production is an alternate procedure for the determination of lipid peroxidation, this technique cannot identify the tissue or subcellular organelle from which these substances arise. In addition, precautions must be taken to prevent or estimate the evolution of hydrocarbon gases by microorganisms in the gastrointestinal tract when *in vivo* studies are undertaken. Furthermore, putrefaction of dead cells and perhaps dying cells, or treatment-related enhanced apoptosis also contribute to the production of these gases; hence, this method inherently suffers from the inability to descriminate primary lipid peroxidation versus these other contributions, which can be substantial after treatment with xenobiotics.

The procedure described here for *in vivo* lipid peroxidation is a widely used and simple one (Klaassen and Plaa, 1969).

Reagents

1. *Chloroform-methanol (2:1 v/v):*
2. *Homogenizing buffer (0.32 M sucrose; 3 mM EDTA):* Dissolve 10.95 g of sucrose and 0.1 g of EDTA in distilled water and make up the volume to 100 ml.

Procedure

Homogenize the liver tissue in 2 volumes of homogenizing buffer. Then, extract the lipid from 0.5 ml aliquot of the homogenate into 9.5 ml of chloroform-methanol. Remove the tissue precipitate by filtration and bring up the volume to 10 ml with chloroform-methanol. Wash this with 2 ml saline (to remove nonlipid material), followed by two washes of each 0.5 ml with upper-phase solvent (upper phase of a mixture of 25 ml water and 95 ml chloroform-methanol). Add methanol to clear the samples and scan from 230 to 280 nm by using a spectrophotometer. Use the lower-phase solvent as the reference blank.

COVALENT BINDING OF REACTIVE METABOLITES TO MACROMOLECULES

Although the concept of lipid peroxidation is one of the truly important concepts of current experimental pathology and toxicology, it does not appear to serve as a universal mechanism of liver injury. A number of drugs and chemicals, e.g., acetaminophen (Gillette, 1975; Mitchell et al., 1976), furosemide (Gillette, 1975; Mitchell et al., 1976), 1,1-dichloroethylene (Jaegar et al., 1973), trichloroethylene (Allemand et al., 1978), bromobenzene (Reynolds, 1972; Gillette, 1975), and dimethylnitrosamine (Gillette, 1975), produce hepatic damage but do not appear to promote lipid peroxidation. Rather, these agents are thought to be converted to highly reactive, electrophilic metabolites by the hepatic MFO system. Following formation, the metabolites, which are considered to be the ultimate toxicants, can interact with hepatic constituents (e.g., protein, lipid) to form alkylated or arylated derivatives. Various investigators postulate that the binding of reactive metabolites to hepatic macromolecules can initiate cellular damage through processes as yet unidentified. The procedure described here is similar to the one employed by Davis and Mehendale (1980).

Reagents

1. *Radioactive tracer* (^{14}C): Obtain from commercial source.
2. *Ethanol, methanol or TCA:* Obtain from commercial source.
3. *1 N Sodium hydroxide:* Dissolve 1 g NaOH in 1 liter of distilled water.
4. *Chloroform-methanol (2:1 v/v):* Mix 2 volumes of chloroform to 1 volume of methanol.
5. *Liquid scintillation fluid:* Obtain from commercial source.

Procedure

Incubate the enzyme system (microsomes or liver homogenates) with radioactive tracer (the compound under investigation for covalent binding to macromolecules) for about 1 h.

For covalent binding to proteins, precipitate the enzyme system with organic solvent (ethanol or methanol, more than two volumes) or trichloroacetic acid (5% final concentration) after incubation. Collect the pellet after centrifugation at 10,000 $\times$ g for 15 min. The sensitivity of the residual substrate to the solvent must be considered as well as the solubility. The protein pellet should be washed by vigorous mixing or homogenization. The process of homogenization, centrifugation, and decantation of supernatant should be repeated several times until significant radioactivity no longer appears in the wash fractions. Usually, this takes 7–10 washes. As a practical approach when 2 to 3 consecutive washes show the same or declining radioactivity, which is not substantial, it is considered that further washing is no longer necessary. At that point, the protein samples should be dried by heating at 60°C for 2 h. The protein should be dissolved in 1 *N* NaOH (about 1–2 ml) for 1 h at 60°C. Insoluble material can be removed by centrifugation, and the protein in an aliquot of each sample should be then measured (Lowry et al., 1951), as the recovery is variable. A larger aliquot of each sample should be added to 10 ml of scintillation cocktail capable of holding water, and chemiluminescence is allowed to decay overnight at room temperature in the dark prior to counting. Express the results in terms of nanomoles of adduct per mg protein.

For covalent binding to lipids, extract the lipids from 0.5 ml of the enzyme system after incubation with the radioactive tracer. Proceed in a similar way as for the diene conjugation assay. After repeated washes, heat the chloroform-methanol extract at 80°C for 3 h (Maling et al., 1974). Wash this preparation with upper phase (mixture of 25 ml water and 95 ml chloroform-methanol) until no radioactivity can be detected in the water phase. At this point, dry the chloroform layer under nitrogen and suspend in a known volume of chloroform. An aliquot of each sample should be added to 10 ml of scintillation cocktail and the radioactivity can be measured in scintillation counter. Express the results in terms of nanomoles of adduct per mg lipid.

There are several excellent reviews regarding the formation and detoxification of reactive metabolites and their interaction with hepatic constituents (Mitchell and Jollow, 1975; Mazel and Pessayre, 1976; Gillette, 1977; Mitchell and Boyd, 1978; Sipes and Gandolfi, 1982; Mitchell et al., 1984; Cheesman et al., 1985). However, a detailed discussion will be beyond the scope of this chapter. One important aspect is that hepatotoxicity need not be correlated with pharmacokinetics of the parent substance or even its major metabolites, but may be correlated with the formation of quantitatively minor, highly reactive intermediates (Mitchell and Boyd, 1978). However, there are several factors such as glutathione, epoxide hydrolase, and glutathione-*s*-transferase that play a significant role in preventing covalent binding of reactive intermediates to some extent. When low doses of acetaminophen (Mitchell et al., 1976) or bromobenzene (Jollow et al., 1974) are administered, the toxic metabolite combines with glutathione, thereby preventing arylation of macromolecules and necrosis. As the dose is increased, availability of glutathione is decreased and a sharp increase in covalent binding occurs at a threshold dose, eventually leading to cell necrosis.

GLUCOSE-6-PHOSPHATASE

This enzyme is also known as G-6-Pase and EC 3.1.3.9. Glucose-6-phosphatase is associated with the smooth endoplasmic reticulum, and depression of its activity appears to specifically reflect injury to this organelle (Recknagel and Glende, 1973). The functional integrity of the enzyme is dependent on the presence of phospholipid. Peroxidative decomposition of microsomal lipids, as occurs with CCl_4, results in a significant loss of G-6-Pase activity (Klingensmith and Mehendale, 1983a,b). The method to determine G-6-Pase activity is based on the hydrolysis of glucose-6-phosphate to inorganic phosphate by the enzyme source (microsomes or liver homogenate) (Lowery et al., 1981).

Reagents

1. *60 mM Histidine-Imidazole buffer (pH 6.8):* Dissolve 0.817 g imidazole and 1.862 g histidine in distilled water and bring to a final volume of 200 ml.
2. *100 mM Glucose-6-phosphate:* Dissolve 336 mg of glucose-6-phosphate (MW 336.3) in 10 ml distilled water.
3. *1 M KCl (FW 74.55):* Dissolve 7.455 g in distilled water and bring to a final volume of 100 ml.
4. *50% Trichloroacetic acid (TCA):* Dissolve 50 g TCA in distilled water and make up the volume to 100 ml.

Sample Preparation

Prepare microsomes or liver homogenates as mentioned earlier (look in aminopyrine demethylase assay).

Procedure

Incubate 1 ml reaction mixture containing 0.5 ml of histidine buffer, 0.1 ml KCl, 0.1 ml glucose-6-phosphate, and 0.3 ml enzyme source, at 37°C for 15–30 min after due standardization. Stop the reaction by adding 0.1 ml of TCA. Centrifuge the tubes and estimate inorganic phosphate in the supernatant by the method of Carter and Karl (1982). The enzyme activity can be expressed as μmol P_i/mg protein/h.

Depression of G-6-Pase activity has been used to document microsomal damage produced by addition of agents to *in vitro* incubation systems. Several investigators (Glende et al., 1976; Benedetti et al., 1977) have used this enzyme to demonstrate that the key event in CCl_4-induced alteration of microsomal enzyme activity is lipid peroxidation and not covalent binding of CCl_4 derived free radicals to microsomal lipids. Klingensmith and Mehendale (1983a) have used this enzyme to compare the CCl_4-induced liver injury in rats following pretreatment with 10 ppm chlordecone, 10 ppm mirex, or 225 ppm phenobarbital. Grice (1972) suggested that alterations in G-6-Pase activity might serve as an indicator of incipient liver damage that might occur in advance of histologically detectable organ damage. However, there are other reports indicating that G-6-Pase is not the most sensitive test for detecting minimal damage. Klaassen and Plaa (1969) found depression of G-6-Pase activity only at 0.3 ml CCl_4/kg or greater dose, whereas hepatic triglyceride accumulation and BSP retention (Klaassen and Plaa, 1968) occurs at doses below 0.3 ml CCl_4/kg. Besides, histomorphological changes are observed in rats treated with 0.1 ml CCl_4/kg (Lockard et al., 1983a,b). Regardless of the relative sensitivity of this enzyme, it can be used as a diagnostic tool to quantify the effects designed to increase or decrease the toxicity of known hepatotoxicants in experimental toxicology.

MARKERS FOR THE SUPPRESSION OR STIMULATION OF CELL REGENERATION IN HEPATOTOXICITY

This is an important parameter that has not received sufficient attention of toxicologists. This has been neglected considering this event as a temporal one. Since replacement of dead cells and tissue repair must occur after cellular necrosis by stimulated cell division, the importance of hepatocellular regeneration cannot be overemphasized. Involvement of cell regeneration phenomenon in lethal interaction of chemicals has been recently reported (Lockard et al., 1983a,b; Kodavanti et al., 1989a,b,c). They have observed stimulation of hepatocellular regeneration, which occurs within 6 h after the administration of a low dose of CCl_4 alone, normally allows the animals to recover from the limited injury. Such a stimulation of hepatocellular regeneration is not expressed in animals pretreated with 10 ppm dietary chlordecone indicating that the suppressed hepatocellular regeneration might be responsible for the dramatic amplification of what is normally recoverable injury (Mehendale, 1984). Recent studies designed to investigate the validity of this hypothesis using partial hepatectomy as an ideal model to stimulate hepatocellular regeneration have provided supporting evidence (Bell et al., 188; Kodavanti et al., 1989a,b,c). In these studies, prior exposure to chlordecone did not interfere with the hepatocellular regeneration stimulated by partial hepatectomy. When CCl_4 was administered to these animals at the time of maximal regeneration stimulated by partial hepatectomy, the animals were significantly protected from the hepatotoxic and lethal effects of CD + CCl_4. When CCl_4 was administered 7 days after partial hepatectomy, when the hepatocellular regeneration had waned, the protection was also abolished (Bell et al., 1988; Kodavanti et al., 1989c). These studies indicate that potentiated hepatotoxicity of CCl_4 enters the progressive phase of toxicity due to lack of cell division and repair mechanisms. Thus, tissue repair mechanisms at the cellular level, primarily by way of cell division, are the determinants of whether the animals recover from toxicity or enter the progressive and irreversible phase, culminating in death. Here, some of the key parameters by which one can investigate whether regeneration and repair phenomena are involved in hepatotoxicity have been described.

IN VIVO ^{3}H-THYMIDINE INCORPORATION INTO HEPATOCELLULAR NUCLEAR DNA

Biochemically, this is a very ideal parameter to assess the cell division phenomenon. This parameter followed by autoradiographic examination of liver sections for ^{3}H-thymidine label, which will be evident as silver grains, has been successfully used for studying cell regeneration involvement in hepatotoxicity caused by large doses of CCl_4 (Kodavanti et al., 1989a) or smaller doses of CCl_4 in combination with other interactants like chlordecone (Kodavanti et al., 1989b; 1989c). The procedure described here is essentially that of Chang and Looney (1965).

Reagents

1. ^{3}H-Thymidine: 50 μCi/0.5 ml.
2. *2.2 M Sucrose (MW 342.3):* Dissolve 753 g in 500 ml of distilled water and make up the volume to 1000 ml.
3. *0.25 M Sucrose (MW 342.3):* Dissolve 42.8 g in distilled water and make up the volume to 500 ml.
4. *0.6 N Perchloric acid (PCA):* Add 8.61 ml of 70% PCA to 91.39 ml of distilled water.
5. *0.2 N PCA:* Add 8.61 ml of 70% PCA to 291.39 ml of distilled water.
6. *0.5 N PCA:* Add 7.17 ml of 70% PCA to 92.83 ml of distilled water.
7. *Diphenylamine:* Dissolve 1.5 g of diphenylamine in 100 ml of glacial acetic acid. To this, add 1.5 ml of concentrated sulfuric acid.
8. *Acetaldehyde (16 mg/ml):* Add 0.406 ml acetaldehyde to 19.594 ml distilled water.
9. *Diphenylamine Reagent:* On the day of experiment, mix 0.1 ml of acetaldehyde with 20 ml of diphenylamine.

Animal Treatment

Inject the animals with the ^{3}H-thymidine (50 μCi/300 g rat) intraperitoneally 2 h before sacrificing. This time period falls within the linearity for ^{3}H-thymidine incorporation (Chang and Looney, 1965: Kodavanti et al., 1989a,b.)

Isolation of Nuclei and DNA

Remove the livers and homogenize in 20 volumes of 2.2 *M* sucrose (*d* = 1.28). Centrifuge the homogenates at 40,000 × g for 1 h in an ultracentrifuge. The intact cells, connective tissue, mitochondria, and erythrocytes migrate to the top, and all the nuclei in the homogenate sediment forming a thin pellet. Wash the nuclei twice with 0.25 *M* sucrose solution and suspend them in 0.25 *M* sucrose (Chauveau et al., 1956).

Add 1 ml nuclei fraction to 0.5 ml of 0.6 *N* PCA, mix slowly, and allow to stand on ice for 10 min. Centrifuge the aliquot at 10,000 × g for 20 min and discard the supernatant. Wash the pellet twice with 0.2 *N* PCA, suspend 6 ml of 0.5 *N* PCA and heat for 20 min at 70–80°C. Centrifuge the suspension at 10,000 × g for 20 min and use the supernatant for DNA assay as well as for ^{3}H-thymidine label in that fraction.

Measure the DNA content in triplicate with the diphenylamine reaction as described by Burton (1956). To 1 ml of the above supernatant, add 2 ml of diphenylamine reagent, and incubate for 16–18 h at room temperature in dark. Read at A_{600}.

Radiometry for ^{3}H-thymidine label

Add two 200 μl aliquots of the above supernatant to 10 ml of Flo-Scint-TM II (Radiomatic Instruments and Chemical Co., Inc., Tampa, FL) or any liquid scintillation cocktail and quantitate the radioactivity using scintillation counter. ^{3}H-Thymidine incorporation data can be represented as cpm/mg DNA.

AUTORADIOGRAPHY OF LIVER SECTIONS

It would be more confirmative if *in vivo* ^{3}H-thymidine incorporation is followed by autoradiographic examination of liver sections for ^{3}H-thymidine label. The most useful isotope for autoradiography is tritium, an unstable isotope of hydrogen which emits weak beta radiation. The half-life of tritium is 12.5 years and the mean beta-particle energy is 0.018 MeV. In biological materials, the approximate range of the tritium beta particle is around 1 μm although 1 or 2% of the beta particles do have ranges up to 3 μm. The principle in autoradiography is that the label in the samples emits radiation. When the emulsion is exposed to this radiation, it gives spots in the form of black grains. These are called silver grains. The number of silver grains in a cell and the number of cells containing these silver grains will give an index of cell division. The procedure described here is liquid emulsion dipping method outlined by Gude (1968). Critical care should be taken by not exposing the reagents or slides to light during critical steps. We have indicated this in the procedure.

Reagents

1. *Subbing solution:* Dissolve 0.1 g gelatin and 10 mg chrome alum (chromic potassium sulfate dodecahydrate) in a final volume of 100 ml distilled water.
2. *Liquid emulsion:* Prepare this solution in light-tight dark room. Use Kodak NTB-2 nuclear track emulsion. It is solid at room temperature and becomes liquid when warmed to 45°C. For tritium autoradiography, use the emulsion diluted with an equal part of distilled water. This diluted emulsion produces a thin, even coating on the slide. Thicker coating merely increases the amount of emulsion available for the production of silver grains due to extraneous causes (background). Thicker emulsion also decreases the microscopic resolution of the preparation. NTB-2 can be purchased in 4-oz bottles and stored in the refrigerator. Heat each new batch to 45°C for 20 min and then mix it with 112 ml of distilled water, mixing carefully to avoid producing bubbles. Then, distribute the diluted emulsion into nylon scintillation vials in aliquots of about 10 ml. Each scintillation vial holds enough emulsion to dip about 30 slides and should be used only once. The vials of diluted emulsion should be stored in light-tight containers in the refrigerator and can be kept for many months. Vial caps with cork inserts should be avoided since these can give off organic compounds which may produce background grains in the emulsion during storage.
3. *Kodak D-19 developer:* Obtain this from Eastman Kodak Company. Dissolve 595 g in 3.8 liters of distilled water.
4. *Kodak rapid fixer:* Obtain this from Eastman Kodak Company. The package contains solution A (boric acid) and solution B (sulfuric acid). Add 473 ml of solution A to 950 ml distilled water. To this, add 52 ml of solution B with rapid agitation and make up the volume to 1.9 liters with distilled water.
5. *Giemsa stain:* Giemsa stock solution may be bought commercially or prepared from the dry powder. To use the dry powder, add 0.5 g of the stain to 33 ml glycerol and hold at 60°C for 2 h. Add 33 ml methanol and stir well and filter through millipore. This stock solution can be kept for many months.
6. *Hematoxylin stain:* Dissolve 50 g of ammonium or potassium alum in 1 liter of distilled water. Dissolve 1 g hematoxylin crystals in this solution. Then, add 0.2 g sodium iodate, 1 g citric acid, and 50 g chloral hydrate. Stir well until all components are in solution. The final color of the stain is reddish-violet and this solution can be kept for months.
7. *Alcoholic eosin (1%):* Prepare eosin stock solution by dissolving 1 g in 20 ml distilled water and then add, 80 ml 95% alcohol. For working

eosin solution, mix 1 part of eosin stock solution with 3 parts of 80% alcohol, just before use. Add 0.5 ml glacial acetic acid to each 100 ml stain and stir well.

Procedure

Inject the rats with ^{3}H-thymidine (50 μCi/rat) 2 h before killing. This 2-h time period falls within the linearity of ^{3}H-thymidine incorporation and also maximum incorporation can be observed in accord with previously reported findings (Chang and Looney, 1965; Kodavanati et al., 1989a,b,c). Remove the livers, slice them and fix in 10% formalin. After fixation, process the liver slices for embedding in paraffin. Prepare sections of 5 μm thickness, place them on subbed slide and dewax by dipping in xylene.

Subbing Slides (Carry Out Under Daylight). Carefully clean the slides and dip in "subbing solution." Allow them to dry before being used for cytological preparations. The subbing is designed to ensure that the autoradiographic emulsion will adhere to the slide and not slip during the development and fixing procedures. With well cleaned slides, film slippage is usually low regardless of whether the slides are subbed. However, if there is no technical reason for not using subbed slides, it is safer to use them.

Dipping Slides (Carry Out Under a Safelight, But Preferably in Dark). The dark room used for autoradiography should be completely light-tight and have a double light-tight door. The autoradiographic emulsion used in these experiments requires a Kodak Wratten Series II (red) safelight lit by a bulb of 25 W or less. The darkroom should also be equipped with a constant temperature water bath set to hold 45 ± 1°C, a glass or lucite dipping chamber slightly larger than the microscope slide, a slide rack, and a nonsparking table top fan. The safe light should be kept as far from the working area as possible. Care in arranging the working area makes it possible to carry out the operations with little or no light.

If one suspects that the emulsion may have picked up some background grains before use, it is wise to dip a test slide, dry it for 20 min, and then develop it before dipping the experimental slides. It is possible to remove background grains from the emulsion by treating the coated slides with H_2O_2 vapor before beginning the autoradiography. One should however check that H_2O_2 does not affect the specimen used in a particular experiment.

For dipping slides, immerse the dipping chamber in the 45°C water bath. Warm the vial of emulsion to 45°C for 10 min and then pour into the chamber, taking care that no bubbles are formed. Dip the slides, withdraw slowly, drain, and place vertically in rack. When all slides have been dipped, turn on the fan and dry the slides in the dark for about an hour. Dried slides are stored in light-tight black plastic boxes with a small container of granular calcium carbonate (Drierite) included to maintain dryness.

Storage. Store the black boxes of slides in a light-proof container in a 4°C cold-room. Estimate the shortest time which might be necessary for exposure and develop one slide after this time. If no grains are visible, wait twice the length of time and develop a second slide. Increase the length of time between each test slide until enough grains are visible to allow calculation of the proper time. Then, develop the rest of the slides.

Development (Carry Out Under a Safe Light, But Preferably in Dark). Although the exact development temperature is not critical, it should be about 18–20°C. It is important that slides and all developing solutions should be at the same temperature, so that the film of emulsion will not be affected. Process the slides by dipping the slides serially as follows:

Kodak D-19 developer: 2 1/2 min
H_2O stop bath: 30 s
Kodak rapid fixer (undiluted): 3 min
Distilled H_2O: 5×5 minutes
Agitate the slides gently from time to time in each solution.

Staining Autoradiographs. There are several ways of staining preparations through autoradiographic emulsion. Here, we have described the procedures for Giemsa stain as well as Hematoxylin and Eosin.

Giemsa Staining. Add about 1 part of Giemsa stock solution to 20 parts of 0.01 *M* phosphate buffer, pH 6.8, and immediately add the wet slides. The staining time depends on the preparation and on the batch of Giemsa stain and is usually somewhere between 2 min and 45 min. A metallic film forms rapidly over the surface of the Giemsa-buffer solution and will adhere to the slide if the slide is pulled out of the solution. For this reason, at the end of the staining period the stain is washed off the slides by pouring distilled water into the staining dish. Then air dry the slides, cover with a coverglass, and mount in Permount.

Hematoxylin and Eosin Staining. Keep the slides in hematoxylin for 15 min. After this, wash the slides in running tap water for 20 min. Then, counterstain with eosin from 15 s to 2 min depending on the age of the eosin and the depth of the counterstain desired. For even staining results, dip slides several times before allowing them to set in the eosin for the desired time. Dehydrate in 95% and absolute alcohols, two changes of 2 min each or until excess eosin is removed. Clear in xylene, two changes of 2 min each and mount in Permount.

Quantitation of Label. Estimate number of silver grains per cell and the number of hepatic nuclei bearing the label in 15–20 nonoverlapping areas of the stained liver sections. Total number of cells taken into account in each animal should exceed 1000. The unit for labeling index, can be expressed as (number of labeled cells/number of total cells) × 100.

IN VIVO UNSCHEDULED DNA SYNTHESIS AND S-PHASE SYNTHESIS

Measurement of *in vivo* unscheduled DNA synthesis (UDS) and S-phase synthesis (SPS) can be used as an indicator of cell proliferation induced by toxicants (Mirsalis et al., 1982; Mirsalis, 1982; Mirsalis and Butterworth, 1982) or hepatocyte proliferation in the absence of necrosis (Loury et al., 1987). Proliferation by either mechanism has been associated with an increased incidence of hepatocellular adenomas in rodents (Mirsalis et al., 1985; Loury et al., 1987).

Procedure

Hepatocyte Isolation and Culture. Prepare primary hepatocyte cultures from rats or mice treated with various hepatotoxicants. Perfuse livers *in situ* for approximately 2 min with a solution of 0.5 m*M* EGTA in Hanks balanced salt solution without Ca^{2+} or Mg^{2+} followed by 12 min in a 37°C solution of type 1 collagenase (100 units/ml) in Williams medium E (Gibco, Santa Clara, CA).

Obtain a single-cell suspension of hepatocytes and collect all the cells by centrifugation at 50 × g for 5 min. Resuspend in cold medium and filter through sterile gauze. Determine the viability of cells by trypan blue exclusion method.

Seeding of Hepatocytes. Seed approximately 6×10^5 cells into each well of a Falcon (Oxnard, CA) 9.6 cm^2, 6-well culture plate. Each well should contain a 25-mm Thermanox cover slip (Biolabs, Northbrook, IL) in Williams medium E supplemented with 2 m*M* glutamine, 50 μg/ml gentamycin sulfate, and 10% fetal bovine serum. After 1.5 to 2.0 h incubation in a humidified atmosphere at 37°C and 5% CO_2, wash the cultures to remove nonviable cells. Perform further washes and subsequent culturing using serum free Williams' E medium.

Cell Culture Fixation, and Staining. Incubate cultures in Williams' E containing 10 μCi/ml ^3H-(methyl)-thymidine (specific activity, approximately 80 Ci/mmol) for 4 h at 37°C, 5% CO_2, followed by 14 to 18 h in Williams' E containing 0.25 m*M* unlabeled thymidine. Then, wash the cultures twice with Williams' E, followed by 10 min in 1% sodium citrate to swell cells, fix in 1:3 glacial acetic acid:ethanol for 30 min, and wash three to six times with deionized water. Mount with dried coverslips to glass slides using Permount.

The other procedures such as dipping the slides, storage, development, and staining are similar to that and clearly mentioned in "Autoradiography of Liver sections."

Measurement of UDS. Quantitative autoradiographic grain counting can be accomplished as described by Mitchell and Mirsalis (1984). Code the slides prior to scoring to preclude scorer bias. Select an area of a slide randomly and count 50 morphologically unaltered cells. The highest of two nuclear-size areas over the cytoplasma and adjacent to the nucleus should be subtracted from the nuclear count to determine the net grains/nucleus (NG). Determine the percentage of cells undergoing repair (% IR), which is the unit of expression for UDS as the percent of those cells exhibiting 5 or more NG. Score at least three slides for each animal or concentration for a total of at least 150 cells per animal.

Measurement of S-phase Synthesis. The number of cells in *S*-phase can be easily distinguished in autoradiographic preparations. Code the slides prior to scoring to preclude scorer bias. Count the number of *S*-phase cells and non-*S*-phase cells beginning with a randomly selected patch of a slide. Score from 1000 to 2000 cells from each of three slides per animal for a total of 3000 to 6000 cells per animal. SPS responses less than 0.5% could be considered negative. Responses of 0.5 to 1.0% could be considered to be equivocal. Responses greater than 1.0% (when concurrent controls were <0.2%) are considered positive. The unit of expression is again % cells undergoing repair in *S*-phase (%*S*).

Mirsalis and his associates (Mirsalis, 1982; Mirsalis et al., 1982, 1985, 1989; Mitchell and Mirsalis, 1984) have done considerable amount of work using *in vivo* UDS and SPS on several toxicants in an attempt to study the genotoxic hepatocarcinogens and also as an indicator of hepatocelullar proliferation. They have screened 24 chemicals using Fischer-344 rats and B6C3F1 mice following *in vivo* treatment. 11-Aminoundecanoic acid, benzyl acetate, bis(2-chloro-1-methylethyl)ether (BCMEE), C.I. Solvent Yellow 14, cinnamaldehyde, cinnamyl anthranilate, dichloromethane, dichlorvos, glutaraldehyde, 4-4′-methylenedianiline (MDA), 4-nitrotoluene, 4,4′-oxydianiline, a polybrominated biphenyl mixture (PBB), reserpine, 1,1,2,2-tetrachloroethane, 1,1,2-trichloroethane, trichloroethylene, and 2,6-xylidine all failed to induce UDS in rats and/or mice. Dinitrotoluene and Michler's ketone induced positive UDS response in rat while *N*-nitrosodiethanolamaine and selenium sulfide induced equivocal UDS results in mouse and rat, respectively. BCMEE, bromoform, chloroform, PBB, 1,1,2-trichloroethane, and trichloroethylene were all potent inducers of SPS in mouse liver while C.I. Solvent Yellow

14, and 1,1,2,2-tetrachloroethane yielded equivocal SPS results in rat and mouse, respectively. Although these studies indicate that most of the test compounds do not induce UDS in the liver, the significant *S*-phase response induced by many of these compounds, especially the halogenated solvents, may be an important event in their hepatocarcinogenicity.

MARKERS OF COMPROMISED ENERGY STATUS IN HEPATOTOXICITY

Zimmerman and Ishak (1987) classified hepatotoxins into direct and indirect toxins. Direct toxins injure hepatocyte and its organelles by a direct physicochemical effect. In this category, tissue injury leads to metabolic defects. However, indirect hepatotoxins produce hepatic injury by interfering with a specific metabolic pathway. In this category, metabolic injury leads to tissue injury. Although Zimmerman and his associates classified hepatotoxins as direct and indirect ones, this distinction is not very clear since hepatotoxicity of these toxins enters the progressive phase only because of metabolic defects. Hence, more attention should be drawn to explore how cellular, biochemical, as well as metabolic homeostasis mechanisms might be altered to precipitate toxicity. It has been established that potentiated CCl_4 hepatotoxicity by chlordecone dramatically decreases cellular ATP levels as well as glycogen before cell necrosis was evident (Lockard et al., 1983a,b; Mehendale, 1984; Bell and Mehendale, 1987; Kodavanti et al., 1988, 1990). Since energy status was compromised before cell necrosis, this phenomenon together with other events could lead to cell damage. Here, we have described the key parameters indicative of hepatic energy status during toxic injury.

GLYCOGEN

Glycogen is the major source of energy for the liver and glucose is stored in the form of glycogen in liver tissue. Depletion of glycogen clearly indicates perturbed energy status in that tissue. The procedure described here is derived from the description of Carrol et al. (1956).

Sample Preparation

Prepare 10% homogenates of liver in 5% TCA and centrifuge at 6,000 × g for 15 min. Take the supernatant and determine glycogen.

Reagents

1. *5% Trichloroacetic acid (TCA):* Dissolve 10 g of TCA in distilled water and make up the volume to 200 ml.

2. *Ethanol:* 95% ethanol.
3. *Anthrone reagent:* Dissolve 0.4 g of anthrone in 200 ml of H_2SO_4.

Procedure

To one volume of the sample, add five volumes of 95% ethanol and allow to stand over night at 5° C. After precipitation is complete, centrifuge at 6000 × g for 15 min. Decant the supernatant and dissolve the residue in 1 ml distilled water. Add 5 ml of anthrone reagent to this by constant mixing. Keep the tubes in boiling water bath for 15 min. After cooling, read the color at A_{620} in a spectrophotometer. Prepare proper blanks. Derive standard graph by taking 10–100 μg glucose and calculate the sample values from the standard graph. The data can be presented as μg/g liver.

ADENOSINE TRIPHOSPHATE (ATP)

Maintenance of ATP levels is a very important factor in the cellular homeostasis. Under low energy levels, the animals will tend to be more sensitive to hepatotoxicants. Besides, several cellular processes including cell regeneration need ATP, for example: energy equivalent to 20 ATP molecules is required to move the chromatids to the poles (Nichlas, 1965). Several hepatotoxins including CCl_4 and ethionine decrease cellular ATP levels (Zimmerman and Ishak, 1987; Kodavanti et al., 1988, 1990). The procedure described here is obtained from Lamprecht and Trautschold (1974).

Sample Processing

Freeze-clamp the liver and homogenize 1 g liver tissue immediately in 4 ml of 0.75 *M* PCA. Take 1 ml of the homogenate and centrifuge at 6000 × g for 30 min. Take the supernatant and adjust the pH to 7.5 with 1 *N* NaOH. Bring the volume to 2.5 ml with distilled water.

Reagents

1. *Buffer, 50 mM tris, 50 mM KCl, 0.2 mM EDTA, 6.7 mM $MgCl_2$:* Dissolve 1.211 g of Tris-base (FW 121.1), 0.745 g of KCl (FW 74.55), 0.0149 g of EDTA (FW 372.2), and 0.272 g of $MgCl_2$ (FW 203.3) in distilled water, adjust pH to 7.5, and the final volume is made up to 200 ml.
2. *0.1 M NADP (FW 810.4):* Dissolve 81 mg in 1 ml of distilled water.
3. *1 M glucose (FW 180.2):* Dissolve 1.8 g in distilled water to a final volume of 10 ml.
4. *Hexokinase:* Obtain commercially, 10 mg/ml.
5. *G-6-PDH:* Obtain commercially, 2 units/5 μl.

Procedure

Take 1 ml of the above sample and add 1 ml of buffer, 0.015 ml of NADP, 0.03 ml of glucose, and 0.94 ml of H_2O. Take the reading at 340 nm. This could be considered as blank. Then, add 10 μl of hexokinase and 5 μl of G-6-PDH to the above, incubate at 37°C, and take the time scan at 340 nm for about 10 min. The principle involved in this method is

$$\text{Glucose} + \text{ATP} \xrightarrow{\text{Hexokinase}} \text{Glucose-6-phosphate} + \text{ADP} \quad (28)$$

$$\text{G-6-P} + \text{NADP} \xrightarrow{\text{G-6-PDH}} \text{Phosphogluconate} + \text{NADPH} \quad (29)$$

Make sure that the reaction is complete (linearity) and take the reading. This is considered as sample reading. The difference between sample reading and blank gives the real value for calculating ATP in the sample. For standard curve, take 0 to 500 nmol of ATP (about 5 to 6 concentrations) and measure absorbance in the same way as done for the samples. Calculate the ATP values in sample from the standard graph and express as μmol/g liver.

MAGNESIUM-DEPENDENT ADENOSINE TRIPHOSPHATASE (Mg^{2+}-ATPase)

Magnesium-dependent ATPase is a key enzyme in synthesis of ATP molecules and is involved in the terminal step of oxidative phosphorylation. Alterations in this enzyme activity together with ATP levels are helpful measures to assess perturbations in ATP turnover under the influence of toxicants. Kodavanti et al. (1988, 1990) recently investigated the potential inhibition of hepatomitochondrial Mg^{2+}-ATPase during xenobiotic-potentiated CCl_4 hepatotoxicity. The procedure described here is a coupled enzymatic method and is based on the method described by Fritz and Hamrick (1966).

Sample Preparation

Prepare mitochondria following a series of centrifugations as outlined by Green et al. (1957). Homogenize the liver in cold sucrose solution and centrifuge at 750 × g for 10 min to remove nuclei and cell debris. Take the supernatant and centrifuge at 9000 × g for 20 min. Suspend the resulting pellet in sucrose solution and centrifuge again at 9000 × g for 20 min. Suspend the pellet containing the mitochondria in sucrose solution. Mg^{2+}-ATPase in this fraction is stable when quick frozen in liquid nitrogen and stored at −80°C.

Reagents

1. *200 mM Imidazole (FW 68.1) buffer:* Dissolve 13.62 g in 800 ml distilled water, adjust the pH to 7.5 with HCl, and make up the volume to 1 liter.
2. *Salts (1.7 M NaCl, 0.34 M KCl, 0.085 M $MgCl_2$:* Dissolve 24.85 g NaCl (FW 58.45), 6.35 g KCl (FW 74.56), and 4.32 g $MgCl_2$ (FW 203.3) in imidazole buffer to a final volume of 250 ml.
3. *Pyruvate kinase (PK)-LDH mixture:* Dilute the stock solution obtained from commercial source to 900 units of PK and 1200 units of LDH per 1 ml with buffer.
4. *6 mM NADH (MW 709.4):* Dissolve 42.5 mg in 10 ml buffer.
5. *15 mM Ouabain (FW 728.6):* Dissolve 1.09 g of ouabain in warm distilled water to a final volume of 100 ml.
6. *100 mM ATP.2.5 H_2*O (FW 551.1): Dissolve 595 mg in 10 ml distilled water.
7. *15 mM Phosphoenolpyruvate (MW 234.1):* Dissolve 35.1 mg in 10 ml buffer.
8. *Sucrose solution (0.32 M sucrose, 1 mM EDTA, 10 mM imidazole):* Dissolve 109.5 g sucrose, 0.681 g imidazole, and 0.336 g of EDTA in 800 ml of distilled water, adjust the pH to 7.5 with HCl, and make up the volume to 1 liter with distilled water.

Procedure

Pipette 2.165 ml imidazole buffer, 0.175 ml salts, 0.01 ml PK-LDH, 0.2 ml ouabain, 0.1 ml phosphoenolpyruvate, 0.15 ml ATP, and 0.1 ml NADH into a cuvette. Add 0.1 ml of mitochondrial fraction equivalent to a protein content of 40–60 μg. Measure the absorbance changes in reaction mixture at 340 nm by using a Gilford spectrophotometer with a temperature-controlled chamber at 37°C. The enzyme reaction is

$$\text{ATP} \xrightarrow{\text{Mg}^{2+}\text{-ATPase}} \text{ADP} + \text{P}_i \tag{30}$$

$$\text{Phosphoenolpyruvate} + \text{ADP} \xrightarrow{\text{PK}} \text{Pyruvate} + \text{ATP} \tag{31}$$

$$\text{Pyruvate} + \text{NADH} \xrightarrow{\text{LDH}} \text{Lactate} + \text{NAD} \tag{32}$$

The decrease in NADH concentration is proportional to the enzyme activity. The change in A_{340} over a period of 10 min can be used in calculating the specific activity. The enzyme activity is expressed as μmol P_i/mg protein/h. The Mg^{2+}-ATPase can be delineated into oligomycin-sensitive and -insensitive forms by including 5 μM oligomycin in the reaction mixture.

ALTERED CALCIUM HOMEOSTASIS IN HEPATOTOXICITY

The intracellular distribution of calcium exerts a profound influence on cell metabolism, motility, and division (Joseph et al., 1983; Rasmussen and Barrett, 1984). Thus, intracellular calcium homeostasis is of importance to cell viability. The distribution of Ca^{2+} within hepatocytes is complex and involves binding to cell macromolecules (e.g., glycoproteins, proteins, and phospholipids) and compartmentation within subcellular organelles. The critically important cellular calcium pool for regulation of intracellular events is the free Ca^{2+} in the cytosol; the concentration of this pool is approximately 0.1 μM. The cytosolic-free Ca^{2+} concentration is regulated by energy-requiring transport systems located in the plasma membrane, endoplasmic reticulum, and mitochondrion. An ATP-dependent Ca^{2+} pump in the plasma membrane extrudes Ca^{2+} from the cytoplasm. In addition, the membrane of the endoplasmic reticulum and the inner mitochondrial membrane have leak systems that permit active extrusion of Ca^{2+} from the cytosol and a passive leak of Ca^{2+} back into the cytosol. About 65 to 80% of the total cell calcium may be contained within the mitochondria; this pool of calcium appears to serve as the primary buffer for the cytoplasmic free-Ca^{2+} pool (Joseph et al., 1983). As cytosolic free-Ca^{2+}concentration increases, mitochondria accumulate calcium to maintain a normal concentration of free-Ca^{2+} in the cytosol. Although the endoplasmic reticulum transporter has a relatively low capacity to accumulate calcium, especially when compared to mitochondria, sequestration of calcium by the endoplasmic reticulum does play a role in controlling cytosolic free-Ca^{2+} concentration. Of importance, in terms of the role of cell calcium in the pathogenesis of cell injury, is the fact that substantial calcium is contained in the mitochondrial and endoplasmic reticular pools; the redistribution of calcium from these pools into the cytosol can elevate the cytosolic free-Ca^{2+} concentration to the point where initiation of cell injury may occur (Carafoli and Tiozzo, 1968; Kroner and Planker, 1980). Alternatively, of even greater importance is the potentially much greater influx of external Ca^{2+} from outside the cells, since the concentration gradient is greater outside by 5–10 10^3-fold.

The role of Ca^{2+} as a mediator of toxicant-induced cell death has been a subject of interest for more than three decades. To some extent, it is clear that Ca^{2+} does play a significant role in toxic cell injury. Reynolds and his associates (Thiers et al., 1960; Reynolds, 1963, 1964) found that CCl_4 produces a biphasic increase in hepatocellular Ca^{2+}. A similar biphasic pattern was also observed in chlordecone-potentiated CCl_4 hepatotoxicity (Agarwal and Mehendal, 1984a,b). In these studies, the association between increased hepatic Ca^{2+} content and appearance of necrosis indicate that Ca^{2+} influx into hepatocyte initiates a series of cytotoxic events that results in cell death.

In vitro studies using isolated hepatocytes indicated that CCl_4 (Casini and Farber, 1981) as well as other toxicants such as galactosamine (El-Mofty et al., 1975; Schanne et al., 1980) increase hepatic Ca^{2+} content and the toxicity is dependent on Ca^{2+} concentration of the incubation medium. Chlorpromazine, which inhibits Ca^{2+} flux across membranes, reduced galactosamine-induced cell death (Schanne et al., 1980). However, later reports are controversial and indicate that toxic injury to isolated hepatocytes is not dependent on extracellular Ca^{2+} (Smith et al., 1981; Acosta and Sorenson, 1983; Farris and Read, 1985; Smith and Sandy, 1985). Likewise, the membrane-active toxicants like A23187 (a Ca^{2+} ionophore), melittin, phalloidin, ethylmethanesulfonate, and silica, produced cytotoxicity in the absence of extracellular Ca^{2+} (Schanne et al., 1979). Fariss et al. (1985) proposed that discrepancy between these studies is due to the presence or absence of vitamin E in the incubation medium. Though these observations suggested that toxicant-induced hepatocyte death is not mediated by an influx of extracellular Ca^{2+}, this phenomenon might not exist at all situations. A typical situation of such type of phenomenon is potentiated CCl_4 toxicity by chlordecone. Agarwal and Mehendale (1984 a,b, 1986) have observed a steady increase of Ca^{2+} levels in cytosolic, mitochondrial, as well as microsomal fractions with time. Under these conditions, extracellular Ca^{2+} influx appears to play a predominant role. Of particular interest in this regard are other studies (Agarwal and Mehendale, 1986) indicating that mitochondrial, microsomal, or cytosolic Ca^{2+} levels in liver cells obtained after *in vivo* treatment, did not decrease at any time point during chlordecone-potentiated CCl_4 toxicity.

As mentioned before, studies performed under conditions where extracellular Ca^{2+} was decreased much below physiological conditions are difficult to interpret for two distinctly separate problems. First, these are totally unphysiological conditions. Hence, the hepatocytes may not be expected to function in any normal physiological fashion. Second, absence of extracellular Ca^{2+} (a completely nonphysiological condition) is likely to trigger a cascade of events intracellularly, primarily leading to dislodgement of internal stores of Ca^{2+} intended for mobilization to extracellular space in order to ''achieve'' a physiological condition.

Further studies on these lines have indicated that transient increases in cytosolic pool is mainly due to release of Ca^2 from intracellular stores like microsomes and mitochondria. Studies have also indicated that Ca^{2+} pumping activity was inhibited within minutes after CCl_4 administration (Moore et al., 1976; Moore, 1980; Moore and Ray, 1983). In addition, Bellomo et al. (1983) and Nicotera et al. (1985) demonstrated that menadione and *t*-butylhydroperoxide inhibited plasma membrane Ca^{2+}-ATPase and this inhibition of Ca^{2+} extrusion across plasma membrane may be critical to buffer the increased cytosolic Ca^{2+} when the intracellular organelles fail to sequester the excess Ca^{2+}. In total, all these studies come to a conclusion that toxicant-induced increase in the cytosolic free-Ca^{2+} appear to initiate cytotoxic events that eventually result in death of hepatocyte.

Elucidation of the role of calcium in liver injury is complicated by the technical difficulties involved in the determination of calcium distribution within the hepatocyte. It seems clear that simple measurement of total cellular calcium content is an inadequate approach. Compartmentalization of cellular Ca^{2+} can be estimated in hepatocytes with the metallochrome indicator arsenazo III in isolated hepatocytes *in vitro* (Murphy et al., 1980; Bellomo et al., 1982) or fluorescent Ca^{2+} indicators such as Quin-2, Fura-2, or Indo-3 in isolated hepatocytes (Thomas et al., 1984). However, these techniques are complicated and may be invalid for use with some toxicants. An alternative method for estimating increases in cytosolic free-Ca^{2+} concentration involves the determination of phosphorylase *a* activity (Bellomo et al., 1982). Here, we have described some of the procedures related to the measurements of free-Ca^{2+} and other Ca^{2+}-regulated biochemical events.

PHOSPHORYLASE *a*

For the last few years, phosphorylase *a* activity has been successfully used to indicate cytosolic free Ca^{2+} levels. Phosphorylase *a* is a dimer, involved in glycogenolysis, and is regulated both by cAMP and Ca^{2+}. Hence, it becomes necessary to ensure that cAMP is not increased when phosphorylase *a* is used to indicate free-Ca^{2+}. Measurement of phosphorylase *a* activity can be accomplished by determining the incorporation of the glucosyl moiety of glucose-1-phosphate into glycogen as revealed by release of inorganic phosphate (Hue et al., 1975).

Sample Preparation

Homogenize the liver in 10 volumes of ice-cold homogenizing buffer and use for enzyme assay or obtain cytosolic fraction by a series of centrifugations. Take the supernatant after centrifugation at 105,000 × g for 1 h and use for enzyme assay.

Reagents

1. *Homogenizing buffer (100 mM NaF; 20 mM EDTA; 0.5% glycogen; 50 mM glycylglycine), pH 7.4:* Dissolve 0.42 g sodium fluoride (NaF, MW 42), 0.672 g EDTA (MW 336.2), 0.5 g glycogen, and 0.66 g glycylglycine (FW 132.1) in 80 ml distilled water, adjust the pH to 7.4 and make up the volume to 100 ml.
2. *Incubation buffer (100 mM glucose-1-phosphate; 2% glycogen; 0.3 M NaF, 1 mM caffeine), pH 6.1:* Dissolve 376 mg glucose-1-phosphate (FW 376.16), 200 mg glycogen, 126 mg NaF, and 1.94 mg

caffeine in 8 ml distilled water, adjust the pH to 6.1, and make up the volume to 10 ml. Presence of caffeine is intended to inhibit phosphorylase b.

3. *1.2 M Trichloroacetic acid (TCA):* Dissolve 19.61 g of TCA in distilled water and make up the volume to 100 ml.

Procedure

Incubate 100 μl of cytosolic fraction or liver homogenate with an equal volume of incubation buffer for 30 to 60 min at 37°C. Stop the reaction with 0.5 ml of 1.2 *M* ice-cold TCA followed by 3.6 ml water. Centrifuge and determine inorganic phosphate in the supernatant (Carter and Karl, 1982). Enzyme activity can be expressed as μmol P_i/mg protein/h.

CYCLIC ADENOSINE MONOPHOSPHATE (cAMP)

The procedure described here is essentially taken from the instruction manual supplied by NEN for measuring cAMP by radioimmunoassay. We have described nonacetylated procedure for cAMP (^{125}I) radioimmunoassay, since we could measure cAMP in liver tissue employing this procedure quite satisfactorily. The principle involved in this procedure is common like the other radioimmunoassays; there will be competition between a fixed radioactive and an unknown nonradioactive antigens for a fixed number of antibody-binding sites. Decreased radioactivity in antigen-antibody complex indicates the presence of greater nonradioactive antigen, i.e., cAMP.

Sample Preparation

Homogenize the frozen tissue sample at 4°C with 6% TCA to make a 1 ml 10% (w/v) homogenate. To determine the recovery of cyclic AMP during extraction, add to each sample extract approximately 4000 cpm of ^3H-cyclic AMP as a marker. At a specific activity of approximately 37 Ci/mmol and at a 50% counting efficiency, this represents approximately 0.1 pmol (100 fmol) of cyclic AMP. This amount must be taken into account when calculating the cyclic AMP content of the tissue as determined by the radioimmunoassay. Centrifuge TCA extracts at 2500 × g at 4°C for 15 min. Collect the supernatant and extract 4 times with 5 volumes of water-saturated ether. Discard the ether phase. This will remove TCA from the solution. Place samples in a water bath at 70–80°C and evaporate to dryness under a stream of air. Dissolve the residue in assay buffer (volume depends on the amount of cAMP in the sample) and use 100 μl directly in the radioimmunoassay.

Reagents

The following reagents will be supplied in cAMP radioimmunoassay kit.

A. *cAMP Sodium acetate buffer:* One vial of concentrated buffer is supplied. Dilute to 500 ml with distilled water. The final solution will contain sodium acetate buffer, pH 6.2, and a stabilizer. The diluted buffer is stable for at least two months when stored at 2–8°C.

B. *cAMP Standard:* One vial of lyophilized standard is supplied. Reconstitute the contents with exactly 2 ml of distilled water. The reconstituted solution will contain cAMP at a concentration of 5000 pmol/ml, sodium acetate buffer, 0.1% sodium azide and an inert ingredient. the cAMP standard has been calibrated spectrophotometrically using the molar absorption coefficient, $\epsilon = 14.6 \times 10^3$ 1 mol^{-1} at 259 nm, pH 6.9. The reconstituted standard is stable for at least two months when stored at 2 to 8°C.

C. *cAMP Antiserum complex:* One vial of lyophilized, pretreated, first and second antibody is supplied. Reconstitute with 21 ml of distilled water. The resulting solution will contain the prereacted antibody complex, 0.005% thimerosal and an inert ingredient, in sodium phosphate buffer, pH 6.0. The reconstituted antiserum complex is stable for at least two months when stored at 2 to 8°C.

D. *cAMP* (^{125}I) Tracer (succinyl cAMP tyrosine methyl ester (^{125}I): Two vials of concentrated tracer are supplied. Each vial contains approximately 1.5 μCi on calibration date in one ml of a 1:1 *n*-propanol:water solution. Use one vial at a time as directed. Add 5 ml of distilled water to each vial as required. The concentrate and diluted tracer are stable for at least 2 months when stored at 2 to 8°C.

E. *cAMP Carrier serum:* Two vials of lyophilized carrier serum are supplied. Use one vial at a time. Reconstitute the contents of each vial with exactly 6 ml of distilled water. (One vial of reconstituted carrier serum is sufficient for 100 tubes). The resulting solution will contain carrier serum, 0.1% sodium azide, a stabilizer, and an inert ingredient in sodium acetate buffer, pH 6.2. Store at 2 to 8°C.

F. *cAMP Precipitator:* Two vials containing 50 ml are supplied ready to use. Each vial contains a precipitation enhancer and sodium azide in sodium acetate buffer. Store at 2 to 8°C. Shake well before using.

G. *Working tracer:* Prepare working tracer solution by adding one volume of diluted cAMP (^{125}I)-tracer to one volume of the reconstituted cAMP carrier serum. Do not store and reuse this solution.

Procedure

Add 100 μl of each standard solution (0 to 50 pmol/ml) or sample to the appropriate tubes. Maintain blank tubes by adding 200 μl of assay buffer. Add 100 μl of the working tracer solution to all tubes. Add 100 μl of antiserum complex to all tubes, except blank tubes. Mix all tubes, cover and incubate overnight (16 to 18 h) at 2 to 8°C. After incubation, add 0.5 ml of cold cAMP precipitator to all the other tubes. Mix well with a vortex mixer and centrifuge

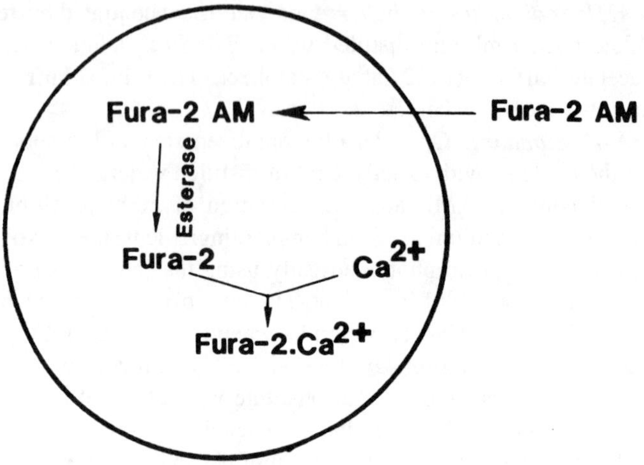

Figure 1. Mechanism for the measurement of Fura-2 fluorescence in a hepatocyte.

in cold for 15 min at approximately 1200 × g. Decant by gently inverting all tubes once, preferably at the same time, discarding the supernatant into a radioactive waste container. Keeping the tubes inverted, place them on absorbent paper for blotting. Allow tubes to drain for 20 to 30 s. Count all tubes in a gamma counter optimized for (^{125}I). At the usual counting efficiency of 50 to 70%, a counting time of 1 min should be sufficient. The sample values can be obtained from standard graph and expressed as pmol/g liver.

FREE CALCIUM MEASUREMENT USING FURA-2

Currently, the most popular method for measuring free-Ca^{2+} in mammalian cells is to monitor the fluorescence of Ca^{2+} indicators like Quin-2 or Fura-2 (Tsien, 1980, 1983; Tsien et al., 1982; Grynkiewicz et al., 1985). These calcium indicators like Fura-2 are loaded into intact cells by incubating with a membrane-permeant acetoxymethyl ester derivative (Fura-2 AM). Cytosolic esterases split off the ester groups and leave the membrane-impermeant Fura-2 trapped in cytosol. This Fura-2 binds to Ca^{2+} present in cytosol and gives fluorescence (Figure 1). The procedure described here is similar to that of Cooper et al. (1985).

Reagents

1. *Fura-2 AM (1 mM):* Dissolve 1 mg completely in 1 ml volatile solvent such as acetone or chloroform and then transfer 50 μl aliquots to 20 microfuge tubes. Evaporate all the volatile solvent and store desiccated at −20°C or lower. This sample should be reconstituted as needed using 50 μl DMSO.

2. *Krebs-Ringer bicarbonate buffer containing 1.3 mM $CaCl_2$* and 15 mM glucose: Dissolve 8.3 g NaCl (142 m*M*, MW 58.4), 0.35 g KCl (4.8 m*M*, MW 74.5), 0.14 g KH_2PO_4 (1.05 m*M*, MW 136.1), 0.41 g $MgSO_4 \cdot 7H_2O$ (1.65 m*M*, MW 246), 4.2 $NaHCO_3$ (50 m*M*, MW 84), 0.19 g $CaCl_2 \cdot 2H_2O$ (1.3 m*M*, FW 147), and 2.7 g glucose (15 m*M*, MW 180.2) in 900 ml distilled water, adjust the pH to 7.4, and make up the volume to 1 liter.
3. *Hanks buffer containing 1 mM $CaCl_2$ and 1 mM $MgCl_2$*: Dissolve 8 g NaCl (137 m*M*, MW 58.4), 0.4 g KCl (5.4 m*M*, FW 74.6), 0.2 g $MgSO_4 \cdot 7H_2O$ (0.825 m*M*, MW 246), 0.045 g Na_2HPO_4 (0.32 m *M*, FW 142), 0.06 g KH_2PO_4 (0.44 m*M*, FW 136.1), 0.11 g $CaCl_2 \cdot 2H_2O$ (1 m*M*, FW 110.9), and 0.2 g $MgCl_2 \cdot 6H_2O$ (1 m*M*, FW 203.3) in 900 ml distilled water, adjust the pH to 7.4 and make up the volume to 1 liter.
4. *Digitonin (2 mM, MW 1229.34):* Dissolve 0.245 g of digitonin in distilled water and make up the volume to 100 ml.
5. *40 mM EGTA in 25 mM Tris, pH 8.3:* Dissolve 0.303 g tris base (MW 121.1) in 90 ml distilled water, adjust the pH to 8.3, and make up the volume to 100 ml. Dissolve 1.52 g EGTA (MW 380.4) in 100 ml of the above tris buffer.

Isolation of Hepatocytes and Fura-2 AM Loading

Isolate the hepatocytes by using the standard collagenase perfusion method (Moldeus et al., 1978). Wash the cells three times and resuspend in Krebs-Ringer bicarbonate buffer containing 1.3 m*M* $CaCl_2$, 15 m*M* glucose, and 2% dialyzed bovine serum albumin (BSA) at a concentration of 10 mg, dry weight/ml. After 10 min preincubation at 37°C under an atmosphere of O_2:CO_2 (95:5%), add Fura-2 AM at a final concentration of 1 to 5μM. After 30 min of incubation, wash the cells twice to remove extracellular Fura-2 AM and then resuspend at a concentration of 3 to 4 mg dry weight/ml in a modified Hank's buffer containing 1 m*M* $CaCl_2$, 1 m*M* $MgCl_2$, and 0.2% dialyzed BSA.

Fluorescence Measurement

For determination of Fura-2 fluorescence, incubate 2 ml of cells for 10 min at 37°C before transferring to a cuvette in Fluorolog (Spex Industries, NJ). Maintain the cells at 37°C, constantly stirred and gassed continuously over the surface with O_2:CO_2 (19:1). Determine the fluorescence at an excitation wavelength of 340 nm and emission wavelength of 505 nm. For studying the effect of a toxicant, add that compound to the cells after the fluorescence trace had settled and the fluorescence (F) was determined. On completion of the fluorescence measurements, permeabilize the cells with 20 μl of digitonin (20 μ*M* final concentration) to determine the maximum fluorescence of the Ca^{2+} saturated Fura-2 (F_{max}) and then 100 μl EGTA (2 m *M* final concentration) to determine the minimum fluorescence of Fura-2 in the absence of free-Ca^{2+} (F_{min}). Control incubations should be carried with nonloaded cells to determine the contribution to the fluorescence made by changes in redox state of the nicotinamide nucleotides on treatment of the cells and Fura-2 fluorescence should be corrected for this. The (Ca^{2+}) before and after toxicant addition can be calculated based on the equation outlined by Tsien et al. (1982).

$$(Ca^{2+})_i = K_D(F - F_{min}) / (F_{max} - F)$$

CALMODULIN (CaM) BY RADIOIMMUNOASSAY

Calmodulin (CaM) is now well accepted as the major calcium-binding protein, acts as an intracellular receptor for Ca^{2+}, and regulates a broad spectrum of fundamental cellular processes (Mikkelsen and Wallach, 1974; Miller et al., 1975; DeJersey and Martin, 1980; Means et al., 1982; Johnson and Mills, 1986). CaM is a heat-stable protein with low molecular weight, and it is well accepted that 4 mols of Ca^{2+} bind to each molecule of CaM with dissociation constants in the micromolar range (Means et al., 1982; Wallace et al., 1982; Johnson and Mills, 1986). Neither ionic Ca^{2+} nor CaM alone are biologically active in cells. When Ca^{2+} binds to CaM, the hydrophobicity of CaM increases and Ca^{2+}-CaM complex regulates several enzyme systems by direct interaction with proteins that control a large number of cellular activities including cell division (Welsh et al., 1978; Boynton et al., 1980; Chafouleas et al., 1981, 1982; Cheung, 1984). Recently, Orrenius and his colleagues have investigated the involvement of Ca^{2+} and CaM in activation of cytosolic proteases and phospholipases which destroy the cytoskeletal proteins (Nicotera et al., 1986a,b, 1989).

The procedure for the determination of CaM levels in nuclear, cytosol, microsomal, and mitochondrial fractions or whole liver homogenate is a radioimmunoassay method described by Chafouleas et al. (1979). The radioimmunoassay kits can be obtained from commercial sources (New England Nuclear, NEN). The procedure described here is essentially taken from the Instruction Manual supplied by NEN. The principle involved in this procedure is similar like the other radioimmunoassays. There will be competition between a fixed radioactive and an unknown nonradioactive antigens for a fixed number of antibody-binding sites. Decreased radioactivity in antigen-antibody complex indicates the presence of greater nonradioactive antigen, i.e., CaM.

Sample Preparation

After isolation, homogenize frozen liver samples in 10 m*M* imidazole buffer (pH 7.5) containing 1 m*M* 2-mercaptoethanol, 10 m*M* $MgCl_2$, and 0.15 *M* NaCl and obtain different subcellular fractions by differential centrifugation or use the whole homogenate. Heat each fraction for 5 min at 95°C and centrifuge at 5000 × g for 15 min. Use the clear supernatant for assay.

Reagents

The following reagents will be supplied in CaM radioimmunoassay kit.

A. *Calmodulum assay buffer:* One vial of assay buffer containing 50 ml of 0.125 *M* borate (pH 8.4) with 0.075 *M* NaCl, 0.2% BSA, 1 m*M* EGTA, and 0.1% sodium azide is supplied. No further dilution is necessary.

B. *Calmodulin* (^{125}I) *tracer:* One vial of lyophilized tracer containing 2 μCi of calmodulin (^{125}I) is supplied. Reconstitute by adding 11 ml of calmodulin assay buffer. Mix well and allow the vial to stand 10 min at room temperature to ensure complete dissolution.

C. *Calmodulin antibody:* One vial of lyophilized antibody is supplied. This antibody was raised in sheep and purified by affinity chromatography. Reconstitute the antibody by adding 11 ml of calmodulin assay buffer. Mix well and allow the vial to stand 10 min at room temperature to ensure complete dissolution.

D. *Calmodulin second antibody diluent:* One vial of 55 ml diluent is supplied. Use the entire amount to reconstitute the calmodulin second antibody.

E. *Calmodulin second antibody:* One vial of lyophilized second antibody is supplied. Reconstitute with second antibody diluent.

Procedure

Add 0.1 ml of each standard solution (0.31 to 20 ng) or sample to appropriate tubes. Maintain blank tubes by adding 0.2 ml assay buffer. Add 0.1 ml of tracer (^{125}I) to all the tubes. Then, add 0.1 ml of calmodulin antibody to all the tubes, except blank tubes. Mix well, cover and incubate all the tubes overnight (16 to 24 h) at 4°C. After incubation, place the tubes in ice bath and add 0.5 ml of camodulin second antibody. Incubate all these tubes for another 30 min at 4°C and then, centrifuge at 5000 × g for 15 min. Decant by gently inverting all tubes once, preferably at the same time, discarding the supernatant into a radioactive waste container. Keeping the tubes inverted, place them on absorbent paper for blotting. Count all tubes in a gamma counter optimized for (^{125}I). At the usual counting efficiency of 50 to 70%, a counting time of 1 min should be sufficient. The sample values can be obtained from standard graph and reported as ng/mg protein. Protein content in the sample can be determined using folin-phenol reagent (Lowry et al., 1951).

CALCIUM-ACTIVATED NEUTRAL PROTEASE

Calcium-activated protease (EC. 3.4.22.17) was initially described by Krebs and co-workers in the 1960s (Meyer et al., 1966; Huston and Krebs, 1968). Liver parenchymal cells, like other mammalian cells, contain both lysosomal and nonlysosomal pathways for the degradation of intracellular proteins (Knowles and Ballard, 1976; Neff et al., 1979; Amenta and Brocher, 1981; Bigelow et al., 1981; DeMartino, 1982). These endoproteases can activate many regulatory enzymes such as phosphorylase-*b*-kinase (Meyer et al., 1966), phosphorylase phosphatase (Mellgren et al., 1979), and the proenzyme of a cyclic nucleotide-independent protein kinase (Inoue et al., 1977) by a limited proteolysis. In subcellular localization studies by Reville et al. (1976), calcium-activated neutral proteases (CANP) were found predominently in soluble fraction (cytosol), and small amounts found in mitochondria-microsomal fractions were considered to be due to contamination by cytosol. However, Beer et al. (1982) reported two CANP similar to that of cytosol in purified mitochondria. The major CANP activity is half-maximally activated with 25 μM Ca^{2+} (high-affinity form; CANP-1) and the other is half-maximally activated with 750 μM Ca^{2+} (low-affinity form; CANP-II). These proteases can be measured by determining the hydrolysis of ^{14}C-radiolabeled casein. The hydrolysis of casein has been previously used to measure CANP (Beer et al., 1982).

Sample Preparation

Either cytosol or mitochondria, obtained through the standard differential centrifugation procedures can be used for protease assay.

Reagents

1. *Tris-HCl buffer (pH 7.4;200 mM):* Dissolve 2.422 g of Tris-base (MW 121.1) in distilled water, adjust pH to 7.4 with HCl, and make up the volume to 100 ml.
2. *Mercaptoethanol (60 mM):* Dissolve 0.042 ml in distilled water and make up the volume to 10 ml.
3. Casein (80 μg/10μl; 40,000 cpm): Dissolve 8 mg in 1 ml of distilled water.
4. *Casein (20 mg/ml):* Dissolve 0.2 g in distilled water and make up the volume to 10 ml.
5. *10% Trichloroacetic acid (TCA):* Dissolve 10 g of TCA in a final volume of 100 ml with distilled water.
6. *$CaCl_2$, H_2O (1 mM; FW 147):* Dissolve 14.7 mg in a final volume of 100 ml with distilled water.

7. *$CaCl_2$, H_2O (25 mM; FW 147):* Dissolve 0.367 g in a final volume of 100 ml with distilled water.
8. *Ethyleneglycol-bis-(aminoethyl ether)-tetraacetic acid (EGTA; 100 mM; MW 380.4):* Dissolve 3.8 g in a final volume of 100 ml with tris buffer.
9. *Leupeptin (1 mg/ml):* Dissolve 10 mg in 10 ml of distilled water.

Procedure

The reaction mixture of 100 μl should contain 35 μl of enzyme sample, 25 μl tris buffer, 10 μl mercaptoethanol, 10 μl casein (^{14}C), 10 μl $CaCl_2$ (1 m*M* for CANP-I and 25 m*M* for CANP-II), and 10 μl of distilled water. Initiate the reaction with addition of $CaCl_2$ and incubate the reaction mixture for 30 min in a shaking water bath at 25°C. Control reactions contained 10 μl EGTA and 10 μl leupeptin instead of $CaCl_2$ and water. Terminate the reaction by adding 150 μl ice-cold 10% TCA and add 50 μl of casein (20 mg/ml) to aid in protein precipitation. Maintain duplicate reaction mixtures which were not incubated at 25°C, precipitate them immediately with TCA, and use them to determine background levels. Keep all the tubes on ice for 10 min and centrifuge at 7,700 × g for 10 min. Precipitation of protein with TCA at a final concentration of 5% for 10 min proved optimal for separating precipitable from nonprecipitable (^{14}C) peptides. After centrifugation, take 0.1 to 0.2 ml of supernatant (nonprecipitated ^{14}C peptides) and add 10 ml of Flo-Scint-TMII (Radiomatic Instruments and Chemical Co., Inc., Tampa, FL) or any liquid scintillation cocktail and quantitate the radioactivity by using a scintillation counter. To differentiate calcium-activated protease, subtract the value obtained in the presence of leupeptin and EGTA from the value obtained in the absence of leupeptin and EGTA. The unit of protease activity can be defined as that amount of protease which releases 100 cpm of ^{14}C casein after 30 min under standard assay conditions (DeMartino and Croall, 1982).

There are several reports concerning the association between perturbations in calcium homeostasis and cellular necrosis under several toxic phenomena including hypoxic or ischemic conditions (Brattin et al., 1984; Robinson et al., 1985; Johnson et al., 1987). One implication of these findings is that increased cytosolic calcium levels may result in activation of Ca^{2+}-dependent degradative enzymes such as proteases, phospholipases, and endonucleases, which in turn may cause cell damage and death. Stimulation of nonlysosomal proteolytic systems have been reported under cystamine-induced hepatotoxicity (Nicotera et al., 1986a,b). Several studies are needed to make a general statement regarding the role of proteases in cytoskeletal destruction leading to cell necrosis under toxic conditions.

MARKERS FOR THE ROLE OF GLUTATHIONE TURNOVER IN HEPATOTOXICITY

Glutathione (GSH) is a ubiquitous tripeptide, is present in large concentrations in the liver, and plays a major role in the elimination of a large number of nucleophilic exogenous toxicants. Thus, it is important in detoxification of several chemicals. Toxicant by itself could be nucleophilic or converted to nucleophiles by microsomal enzymes. During toxicity phase, GSH levels are drastically decreased and, most often, enzymes responsible in such processes are also affected. Glutathione metabolism is also important in quenching reactive intermediates and radical species generated by oxidative toxicity. The enzymes involved in the GSH-mediated removal of oxyradicals, and peroxides include GSH-peroxidase, catalase, and superoxide dismutase. The ultimate result of oxidative stress is the formation of glutathione disulfide (GSSG) at the expense of reducing equivalents like NADPH (Figure 2). GSSG-reductase is very necessary in maintaining cellular homeostasis of reduced GSH. This section is devoted to the methods usually employed to measure GSH, other thiols, disulfides, and related enzymes.

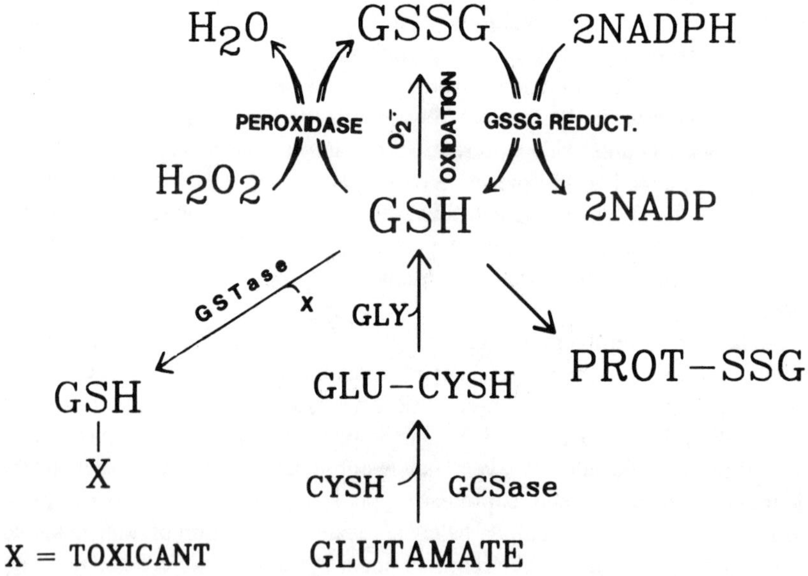

Figure 2. GSH Metabolism in liver.

GLUTATHIONE (GSH), RELATED THIOLS, AND DISULFIDES

Several methods are available to measure GSH, thiols, and disulfides. GSH constitutes a major portion of soluble thiols in the cell or tissue and, therefore, the usual method employed for GSH measurement relates to measurements of total thiols in the soluble fraction of liver homogenate. This method uses 5′,5′-dithiobisnitrobenzoic acid reduction. The color developed in the presence of reducing agent (thiols) is measured spectrophotometrically. The disadvantages of this method are that (1) it overestimates the level of GSH, (2) it cannot differentiate various thiols and their disulfides, and (3) it is relatively insensitive. A method which is very sensitive and reliable is the high performance liquid chromatographic (HPLC) separation and quantitation of all of the thiol components. Usually UV, electrochemical detector, or dual Hg/Au electrodes are used. Reed et al. (1980) have described the HPLC method which relies on derivatized GSH and disulfides for the detection of the thiol components. This method is very sensitive and has been widely used. Other methods are available with slight modifications, related to the specific needs or type of tissues involved. Here we have described the method of Reed et al. (1980).

In this method, the analysis is based upon the well-known reaction of iodoacetic acid with thiols to form *s*-carboxymethyl derivatives (Hirs, 1967) followed by chromophore derivatization of amino groups with Sanger's reagent (1-fluoro-2,4-dinitrobenzene; Sanger, 1967). These derivatives are rapidly separated by HPLC to permit nanomole level analysis of GSH, GSSG and related amino acids.

Reagents

1. *35% Perchloric acid (PCA):* Mix equal volumes of 70% PCA and distilled water.
2. *Iodoacetic acid:* Dissolve 16 mg in 1 ml distilled water.
3. *Sanger's reagent:* Add 1.5 ml of 1-fluoro-2,4-dinitrobenzene in 98.5 ml absolute ethanol.

Sample Preparation

Weigh 1 g of liver in a beaker. Add immediately 0.1 ml of 35% PCA directly on the liver which was kept on ice. Mince the liver with scissors. Add a few milliliters of distilled water to facilitate homogenization by using a tissue homogenizer (Brinkman Polytron). Rinse the Polytron and bring the volume to 20 ml with distilled water. Mix the homogenate thoroughly and centrifuge at 5000 × g for 10 min. Take 1 ml supernatant; neutralize by adding excess sodium bicarbonate powder until the effervescence subsides.

Immediately after this, add 0.1 ml of iodoacetic acid solution, mix well, and keep in dark for 1 h at room temperature. After 1 h, add 0.5 ml of Sanger's reagent and 0.5 ml absolute ethanol. Leave in dark for a minimum of 4 h (or overnight) at room temperature. Store at $-20°C$ until HPLC analysis (storage for about 2 weeks does not cause any change in GSH, GSSG, or related thiols).

HPLC Analysis

Column: Spherisorb, NH_2, 5 μm, 25 cm.

Solvent A: 4:1, methanol:water (v/v).

Solvent B: Mix 272 g of sodium acetate trihydrate, 122 ml distilled water, and 378 ml glacial acetic acid. Add 200 ml of the resulting solution to 800 ml of solvent A.

The gradient for HPLC analysis depends on the ion-exchange capacity of the column. Typically an isocratic period of 10 min should be followed by a gradient started at 5 to 25% and programmed linearly to 95% solvent B over a 20–40 min period. Any standard HPLC unit with gradient master and UV detector with integrator/plotter can be used. After setting up all the parameters for HPLC such as flow rate (1 ml/min), gradient (solvent A at 100% and solvent B at 0), and detector wavelength (252 nm), samples can be injected. Depending on concentration of GSH, other thiols, and disulfides, inject 20–100 μl of desired sample. Immediately 5 min after injection of sample, start the gradient with steady increase of B to 95% in 20 min. Calculation factors can be computed in integrator, or the concentration can be calculated from the area of a particular peak.

Prepare a standard curve with various concentrations of derivatized thiols and disulfides to identify unknown concentrations in the tissue. GSH concentration in liver is in order of 5 m*M* and varies with animal species.

Changes in GSH and related disulfide levels can be elicited by several toxic stimuli to cells. These effects can be conveniently studied using isolated cells or organs and are good models to study several electrophilic or radical-producing chemicals. Menadione is one of such compounds that has been studied for its ability to oxidize GSH in isolated liver perfusion experiments. The changes in GSH are accompanied by other cytotoxic changes such as altered Ca^{2+} transport in hepatocytes (Mehendale et al., 1985). The phase II conjugation system has been reported to be induced by phenobarbital (Utley and Mehendale, 1989). Some studies, specifically involving GSH metabolism in isolated cells or organ perfusions, where serum or albumin are supplemented in the medium, are complicated because of the ability of GSH to bind to albumin and ultimately get oxidized (Joshi et al., 1986, 1987, 1988). Therefore, special care should be taken while interpreting the results. Next, we discuss the enzymes of glutathione metabolism.

GLUTAMYLCYSTEINE SYNTHETASE

γ-Glutamylcysteine Synthetase (γ-GCSase) catalyzes the first step in the synthesis of GSH (Meister, 1974). This enzyme couples glutamate and cysteine moieties forming glutamylcysteine. This is an energy dependent reaction and requires ATP. The procedure described here is that of Sekura and Meister (1977). The reaction is as follows:

$$\text{Glutamate} + \text{L-Aminobutyrate} \xrightarrow[\text{ATP} \curvearrowright \text{ADP} + P_i]{} \text{Glutamyl-aminobutyrate} \quad (33)$$

In this procedure, P_i liberated from ATP during formation of glutamyl-aminobutyrate from glutamate and L-aminobutyrate is measured. The release of P_i is directly proportional to the glutamyl-aminobutyrate formed. Measure P_i as an end point to determine γ-GCSase.

Reagents

1. *0.2 M Na-glutamate (MW 169.1):* Dissolve 3.38 g in distilled water and make up the volume to 100 ml.
2. *0.2 M* L-Aminobutyrate (MW 103.1): Dissolve 2.06 g in distilled water and make up the volume to 100 ml.
3. *0.2 M* $MgCl_2 ,6H_2O$ (FW 203.3): Dissolve 4.07 g in distilled water and make up the volume to 100 ml.
4. *50 mM ATP, disodium salt (FW 551.2):* Dissolve 0.275 g in distilled water and make up the final volume to 10 ml.
5. *0.2 M Tris buffer, pH 8.2:* Dissolve 2.42 g tris base (MW 121.1) in distilled water, adjust pH to 8.2 with HCl, and make up the volume to 100 ml.
6. *40 mM EDTA, disodium salt (FW 336.2):* Dissolve 1.34 g in a final volume of 100 ml with 0.2 *M* tris buffer.
7. *0.4% Bovine serum albumin:* Dissolve 4 mg in 10 ml distilled water.
8. *50% Trichloroacetic acid (TCA):* Dissolve 50 g in distilled water and make up the volume to 100 ml.
9. *0.1 M Tris buffer, pH 7.4:* Dissolve 1.211 g tris base (MW 121.1) in distilled water, adjust pH to 7.4 with HCl, and make up the volume to 100 ml.

Sample Preparation

Homogenize liver samples in 0.1 *M* tris buffer, pH 7.4, and centrifuge at 10,000 × g for 15 min. Take the postmitochondrial supernatant and use it for assay.

Procedure

Add 0.05 ml glutamate, 0.05 ml aminobutyrate, 0.1 ml $MgCl_2$, 0.1 ml ATP, 0.05 ml EDTA, 0.05 ml albumin, and 0.5 ml 0.2 *M* tris buffer to a test tube. Preincubate the reaction mixture for 2 min and start the reaction by adding 0.1 ml enzyme sample. Maintain appropriate blanks in which enzyme sample or substrate are individually omitted. Incubate at 37°C for 15 to 30 min and stop the reaction by adding 0.1 ml 50% TCA. Centrifuge and take the supernatant for P_i assay.

Determine P_i by using malachite green (Carter and Karl, 1982), described earlier in section 3.2. The unit of enzyme activity is represented as μmol P_i/mg protein/h.

Cells maintain relatively high concentrations of GSH by GSSG-reductase system. Cells also have the ability to maintain normal levels of GSH by the synthetic processes. The key enzyme involved in GSH synthesis is γ-GCSase. Altered levels of γ-GCSase have been associated with the toxicity of several chemicals (Richman et al., 1973; Joshi et al., 1988).

GLUTATHIONE DISULFIDE REDUCTASE

This is also known as GSSG-reductase and EC 1.6.4.2. GSSG-reductase is important in maintaining cellular homeostasis of reduced glutathione. GSSG, which is a toxic metabolite, is usually generated intracellularly as a result of oxidative changes in the cell. One very efficient mechanism to rid GSSG is its rereduction in the presence of NADPH. This pathway is very efficient in the cell and the reaction is catalyzed by GSSG-reductase. A simple spectrophotometric method which follows GSH generation from GSSG in the presence of EDTA and reduced NADP (NADPH) can reliably give a good estimate of activity because GSSG-reductase has high substrate specificity, and thus other disulfides are less likely to be cleaved by this enzyme. Rate of NADPH oxidation directly reflects the enzyme activity. The method described here is of Carlberg and Mannervik (1975).

Sample Preparation

Homogenize 1 g liver in 10 ml of 50 m*M* potassium phosphate buffer, pH 7 containing 1 m*M* EDTA, and 1% triton x-100 by using an ultra turrax and glass homogenizer. Centrifuge the homogenate at 105,000 × g for 30 min. Take the supernatant and use it for assay.

Reagents

1. *0.2 M Sodium phosphate buffer, pH 7.6:* Dissolve 2.78 g monobasic sodium phosphate (FW 138) and 2.84 g dibasic sodium phosphate (FW 142) separately in a final volume of 100 ml with distilled water.

Mix both the solutions appropriately to get pH 7.6 by using a pH meter.

2. *10 mM GSSG:* Dissolve 61.3 mg GSSG (MW 612.6) in 10 ml phosphate buffer.
3. *1 mM NADPH (MW 833.4):* Dissolve 8.334 mg in 10 ml phosphate buffer. Prepare fresh daily.
4. *5 mM EDTA:* Dissolve 0.168 g EDTA, disodium salt (FW 336.2) in phosphate buffer and make up the volume to 100 ml.

Procedure

Add 0.5 ml sodium phosphate buffer, 0.1 ml GSSG, 0.1 ml NADPH, and 0.18 ml distilled water to semimicrocuvette. Mix and place in a thermostatic spectrophotometer cuvette chamber. Add 20 μl of enzyme preparation, mix, and record NADPH oxidation or decrease in absorbance over 10 min at A_{340}. Appropriate blanks should be run simultaneously in the absence of enzyme. The oxidation of 1 μmol NADPH per mg protein per min under the specified conditions can be used as a unit of GSSG-reductase activity.

GLUTATHIONE-*S*-TRANSFERASES

These are also known as GSTases and EC.2.5.1.18. GSTases appear to perform several detoxification functions. This group of enzymes are analogous to serum albumin in the broad spectrum of compounds that serve as ligands, but differ significantly in that the transferases have an additional and specific site for GSH. They appear to catalyze all reactions when GSH acts as a nucleophile, providing that the electrophilic substrate is bound to the enzyme (Jakoby and Habig, 1980).

The method for measuring GSTase activity usually involves simple spectrophotometric assay unless there is a special interest in isoenzyme pattern. The basic principle behind spectrophotometric method is that electrophilic substrates will bind to GSH with the help of enzyme and form a colored GSH-substrate complex. This has a specific absorption maximum. Several substrates are available for GSTases; however, two substrates are widely used (1-chloro-2,4-dinitrobenzene; 1,2-dichloro-4-nitrobenzene) for spectrophotometric measurement of activity. The method described here is that of Habig et al. (1974), which uses 1-chloro-2,4-dinitrobenzene as substrate.

Reagents

1. *0.2 M Potassium phosphate buffer, pH 7.2:* Dissolve 2.722 g potassium phosphate monobasic (FW 136.09) and 4.564 g potassium phosphate dibasic (FW 228.2) separately in a final volume of 100 ml each with distilled water. Mix mono- and dibasic solutions to get pH 7.2 by using a pH meter.

2. *0.1 M 1-Chloro-2,4-dinitrobenzene (MW 202.55):* Dissolve 202 mg in a final volume of 10 ml with ethanol.
3. *10 mM Glutathione, reduced (MW 307.3):* Dissolve 30.7 mg in 10 ml of phosphate buffer.

Sample Preparation

Homogenize tissue in phosphate buffered saline (pH 7.2) using Polytron (Brinkman Instruments, Westbury, NY), and prepare cytosolic, mitochondrial and microsomal fractions after a series of centrifugations. Mitochondrial and microsomal pellets can be disrupted by sonication and the contents should be diluted with phosphate buffered saline (pH 7.2).

Procedure

Add 0.5 ml potassium phosphate buffer, 0.1 ml GSH, 0.01 ml 1-chloro-2,4-dinitrobenzene as a substrate, and 0.39 ml distilled water in a semimicrocuvette and mix the contents. Place the cuvette in a temperature-controlled chamber in the spectrophotometer. Add 50 μl of enzyme source and record the increase in A_{340} over a period of 10 min. The appropriate blanks without enzyme should be scanned simultaneously. For the calculation purpose, use the absorption coefficient of 9.6 mM^{-1} cm^{-1}.

The formation of GSH conjugates because of interaction of GSH with the xenobiotics is regulated by GSTase. The ultimate carcinogens, benzo-(*a*)pyrene-7,8-diol-9,10-epoxides, and aflatoxin B_1-8,9-epoxide are substrates for GSTases (Jernstrom et al., 1982, 1985). The role of GSTases in detoxification has been reviewed extensively (Jakoby et al., 1976; Ketterer et al., 1984).

GLUTATHIONE PEROXIDASE

This is also known as glutathione-H_2O_2-oxidoreductase and EC.1.11.1.9. Hydrogen peroxides and other organic hydroperoxides are generated in the cell as a result of endogenous metabolic processes and oxidative stress caused by toxic insult. The reactive species like H_2O_2 are very unstable and can cause tissue damage by oxidation. Glutathione-peroxidase (GSH-peroxidase) and catalase are very active in removing this reactive species. Se-dependent GSH-peroxidase catalyzes the breakdown of inorganic peroxides; however, Se-independent peroxidase catalyzes the breakdown of organic peroxides (Lawrence and Burk, 1976). Spectrophotometric method of glutathione-peroxidase described here is of Lawrence and Burk (1976).

Reagents

1. *0.5 M Potassium phosphate buffer, pH 7.0:* Dissolve 6.8 g potassium

phosphate monobasic (FW 136.09) and 11.41 g potassium phosphate dibasic (FW 228.2) separately in a final volume of 100 ml each with distilled water. Mix mono- and dibasic solutions appropriately to get to pH 7 by using a pH meter.

2. *10 mM Ethylenediamine tetracetic acid (EDTA):* Dissolve 0.366 g of EDTA, disodium salt (FW 336.2) in phosphate buffer and make up the volume to 100 ml.
3. *10 mM Sodium azide (MW 65.01):* Dissolve 65 mg in 100 ml of distilled water.
4. *GSSG-reductase:* 500 units per 2.8 ml.
5. *10 mM GSH (307.3):* Dissolve 30.7 mg in 10 ml phosphate buffer.
6. *2 mM NADPH (MW 833.4):* Dissolve 16.7 mg in 10 ml phosphate buffer. Prepare fresh daily.
7. *15 mM Cumene (isopropylbenzene) hydroperoxide (MW 155.2):* Add 29 μl of 80% solution to 9.971 ml phosphate buffer.
8. *2.5 mM H_2O_2 (FW 34.02):* Add 2.83 μl of 30% solution to phosphate buffer and make up the volume to 10 ml.
9. *0.25 M Sucrose (MW 342.3):* Dissolve 8.557 g in distilled water and make up the volume to 100 ml.
10. *1.14 M NaCl (MW 58.5):* Dissolve 6.67 g in distilled water and make up the volume to 100 ml.

Sample Preparation

Perfuse wash livers with 1.14 *M* NaCl and homogenize in 0.25 *M* sucrose. Prepare the supernatant fraction by centrifugation at 105,000 × g for 1 h and use it for assay.

Procedure

Add in a semimicrocuvette 0.1 ml of each potassium phosphate buffer, EDTA, NaCl, GSH, NADPH, and H_2O_2 or cumene hydroperoxide. To this, add 5 to 6 μl of GSSG-reductase and 0.344 ml distilled water. Place the cuvette in a spectrophotometer cuvette holder and add 50 μl of enzyme preparation. Record disappearance of NADPH over 10 min at A_{340}. For the reference or blank cuvette, add distilled water instead of enzyme preparation. Enzyme activity can be calculated from the slope of the lines and can be represented as μmol NADPH oxidized/mg protein/min.

GSH-peroxidase is involved in hydroperoxide metabolism. Thus, GSH-peroxidase plays an important role in the metabolism of xenobiotics and/or in the metabolism of peroxyradicals produced by xenobiotics (Wendel et al., 1979; Wendel and Feuerstein, 1981).

CATALASE

This is also known as hydrogen peroxide, hydrogen peroxide oxidoreductase, and EC 1.11.1.6. Catalase acts on hydrogen peroxides, generated either via metabolism of endogenous substances or via metabolism of exogenous compounds. Since it removes reactive hydrogen peroxide from the cell, it is important in detoxification mechanism. Several methods have been employed for measuring catalase activity. However, most of them have some drawbacks in the accuracy of measuring the catalysis of hydrogen peroxide. The major factor in variability is the rapidity of its action on H_2O_2. The method described here is a simple spectrophotometric method and yet is fairly sensitive and specific (Beers and Sizer, 1952) and directly measures the decrease in H_2O_2. Using digital spectrophotometer with built in kinetic programs and temperature control system, it is possible to accurately measure the catalysis of H_2O_2 by catalase.

Sample Preparation

Homogenize the liver in 0.05 *M* phosphate buffer (pH 7) and subject the homogenates to differential centrifugation. Obtain the supernatant after centrifugation at 105,000 × g and use it for assay.

Reagents

1. *Potasssium phosphate buffer, pH 7:* Dissolve 3.26 g of KH_2PO_4 and 5.93 g of K_2HPO_4 in distilled water, check the pH for 7, and make up the volume to 1000 ml.
2. *19 mM H_2O_2:* Add 0.216 ml of H_2O_2 (30% solution) to 99.784 ml of the above phosphate buffer.

Procedure

To 1 ml quartz microcuvette, add 0.15 ml of the H_2O_2-phosphate buffer. Add 25 μl H_2O to the reference cuvette and add 25 μl of 105,000 × g supernatant to the sample cuvette. Make sure that the reaction is linear with protein and time of reaction. Monitor the rate of H_2O_2 depletion in the sample cuvette at 240 nm by using a Gilford Response spectrophotometer. Using the extinction coefficient of 43.6 $M^{-1}cm^{-1}$, express the results as μmol H_2O_2 consumed/mg protein/min.

Radicals which are generated from metabolism of toxicants could induce destruction of several macromolecules in cells. As a defense against such insults, cells possess basically three enzymes (SOD, catalase, and GSH-peroxidase) in scavenging free radicals. Catalase is very active in removing peroxyradicals and has been able to protect cells from toxic injury (Ito et al., 1983).

SUPEROXIDE DISMUTASE (SOD)

Superoxides are both oxidants and reductants. Hence, these can modify a variety of biologically important molecules (Fridovich, 1974). SOD catalyzes the dismutation of the superoxide radical into H_2O_2 (Fridovich, 1974). Three distinct types of SOD have been described, which catalyze the same reaction with comparable efficiency (McCord et al., 1977). Biological protection against O_2 is afforded by the SODs which often have a highly localized distribution. Enzymatic assay for SOD is performed by using the xanthine-xanthine oxidase assay (Beauchamp and Fridovich, 1971).

Sample Preparation

Homogenize fresh liver tissue in 4 volumes of 50 m*M* potassium phosphate buffer (pH 7.8) by a motor-driven Teflon-pestle homogenizer, followed by sonication at maximum output of the Biosonic IV (Tekmar Inst.,). Sonicate the homogenates in 1 min bursts for a total of 5 min on ice. For mitochondrial SOD determinations, centrifuge the homogenate at 12,000 × g for 10 min in a refrigerated centrifuge and suspend mitochondrial pellet in phosphate buffer. Freeze and thaw once. This is followed by 4 × 15 s sonication on ice. Suspend the fraction at a protein concentration of 10 mg/ml.

Reagents

1. *100 mM Potassium phosphate buffer (pH 7.8):* Dissolve 1.36 g potassium phosphate monobasic (FW 136.09) and 2.282 g potassium phosphate dibasic (FW 228.2) separately in a final volume of 100 ml each with distilled water. Mix mono- and dibasic solutions appropriately to get pH 7.8 by using a pH meter.
2. *10 mM diethylenetriaminepentaacetic acid (MW 393.35):* Dissolve 39.3 mg in 10 ml phosphate buffer.
3. *Catalase:* Prepare a solution containing 1 unit catalase in 20 μl with distilled water.
4. *0.56 mM Nitroblue tetrazolium (NBT):* Dissolve 4.86 mg of nitroblue tetrazolium chloride monohydrate (FW 867.8) in 10 ml phosphate buffer.
5. *1 mM Xanthine (FW 152.11):* Dissolve 15.2 mg in 100 ml phosphate buffer.
6. *Xanthine oxidase:* Prepare a solution containing 0.05 units per ml.

Procedure

Add to a semimicrocuvette 0.45 ml phosphate buffer, 0.1 ml dinitroblue tetrazolium, 0.1 ml xanthine, 20 μl xanthine oxidase, 20 μl catalase, and 0.1 ml enzyme preparation. Follow nitroblue tetrazolium reduction at A_{560} in a recording spectrophotometer. Reaction mixture without enzyme should be used as a blank. The initial rate of reduction of NBT is diminished in the presence of SOD because the enzyme scavenges the superoxide radical. One unit of enzyme activity is defined as the amount of enzyme that decreases the initial rate to 50% of its maximal value for the particular tissue being assayed.

ACKNOWLEDGMENTS

This effort was supported in part by funds from United States Environmental Protection Agency grants nos. R-811072 and CR-811053, Department of the Air Force grant no. AFOSR-88-0009, and by the Training Program in Environmental Toxicology funded by the National Institute of Environmental Health Sciences, ES-07045. HMM is the recipient of The 1988 Burroughs Wellcome Toxicology Scholar Award. Excellent assistance in the preparation of this chapter by Dr. Urmila Kodavanti is greatly appreciated.

REFERENCES

Acosta D and Sorenson EMB: Role of calcium in cytotoxic injury of cultured hepatocytes. *Ann NY Acad Sci 1983; 407:78.*

Agarwal AK and Mehendale HM: Excessive hepatic accumulation of intracellular Ca^{2+} in chlordecone potentiated CCl_4 toxicity. *Toxicology* 1984a; 30:17.

Agarwal AK and Mehendale HM: CCl_4-induced alterations in Ca^{2+}-homeostasis in chlordecone and phenobarbital pretreated animals. *Life Sci* 1984b; 34:141.

Agarwal AK and Mehendale HM: Effect of chlordecone on carbon tetrachloride induced increase in calcium uptake in isolated perfused rat liver. *Toxicol Appl Pharmacol* 1986; 83:342.

Allemand HD, Pessayre D, Descatoire V, Degott C, Feldman G, and Benhamou J-P: Metabolic activation of trichloroethylene into a chemically reactive metabolite toxic to rat liver. *J Pharmacol Exp Ther* 1978; 204:714.

Amador E, Dorfman LE, and Wacker WEC: Serum lactic dehydrogenase: and analytical assessment of current assays. *Clin Chem* 1963; 9:391.

Amenta JS and Brocher SC: Mechanisms of protein turn over in cultured cells. *Life Sci* 1981; 28:1195.

Arkesteijn CLM: A kinetic method for 5′-nucleotidase using stabilized glutamate dehydrogenase. *J Clin Chem Clin Biochem* 1976; 14:155.

Arnonson NN Jr and Touster O: Isolation of rat liver plasma membrane fragments in isotonic sucrose. *Methods Enzymol* 1974; 31:90.

Asada M: Transaminase activity in liver damage. 1. Study on experimental liver damage. *Med J Osaka Univ* 1958; 9:45.

Asada M and Galambos JT: Sorbitol dehydrogenase and hepatocellular injury: an experimental and clinical study. *Gastroenterology* 1963; 44:578.

Balazs R, Murry RK, McLaughlan JM, and Grice HC: Hepatic tests in toxicity studies on rats. *Toxicol Appl Pharmacol* 1961; 3:71.

Beauchamp C and Fridovich I: Superoxide dismutase: improved assays and an assay applicable to acrylamide gels. *Anal Biochem* 1971; 44:276.

Beer DG, Hjelle JJ, Peterson DR, and Malkinson AM: Calcium activated proteolytic activity in rat liver mitochondria. *Biochem Biophys Res Comm* 1982; 109:1276.

Beers RF and Sizer IW: A spectrophotometric method for measuring the breakdown of hydrogen peroxide by catalase. *J Biol Chem* 1952; 195:133.

Belfield A and Goldberg DM: Normal ranges and diagnostic value of serum 5′-nucleotidase and alkaline phosphatase activities in infancy. *Arch Dis Child* 1971; 46:842.

Bell AN and Mehendale HM: Comparative changes in hepatic DNA, RNA, protein, lipid and glycogen induced by a subtoxic dose of CCl_4 in chlordecone, mirex and phenobarbital pretreated rats. *Toxicol Lett* 1987; 35:191.

Bell AN, Young RA, Lockard VG, and Mehendale HM: Protection of chlordecone potentiated CCl_4 hepatotoxicity and lethality by partial hepatectomy. *Arch Toxicol* 1988; 61:392.

Bellomo G, Jewell SA, Thor H, and Orrenius S: Regulation of intracellular calcium compartmentation: studies with isolated hepatocytes and *t*-butylhydroperoxide. *Proc Natl Acad Sci USA* 1982; 79:6842.

Bellomo G, Mirabelli F, Richelmi P, and Orenius S: Critical role of sulfhydryl groups in ATP-dependent Ca^{2+} sequestration by the plasma membrane fraction from rat liver. *FEBS Lett* 1983; 163:136.

Benedetti A, Casini AF, Feralli M, and Comporti M: Studies on the relationships between carbon tetrachloride-induced alterations of liver microsomal lipids and impairment of glucose-6-phosphatase activity. *Exp Mol Pathol* 1977; 27:309.

Bergmeyer HU: *Methods of Enzymatic Analysis,* vol II, ed 2, New York, Academic Press, 1974.

Bessey OA, Lowry OH, and Brock MJ: A method for the rapid determination of alkaline phosphatase with five cubic millimeters of serum. *J Biol Chem* 1946; 164:321.

Bigelow S, Hough R, and Rechsteiner M: The selective degradation of injected proteins occurs principally in the cytosol rather than in lysosomes. *Cell* 1981; 25:83.

Blakley RL: The metabolism and anti-ketogenic effects of sorbitol. Sorbitol dehydrogenase. *Biochem J* 1951; 49:257.

Boyde TRC and Latner AL: Starch gel electrophoresis of transaminase in human tissue extracts and serum. *Biochem J* 1961; 82:52.

Boynton AL, Whitfield JF, and MacManus JP: Calmodulin stimulates DNA synthesis by rat liver cells. *Biochem Biophys Res Comm* 1980; 95:745.

Brattin WJ, Pencil SD, Waller RL, Glende EA, and Recknagel RO: Assessment of the role of calcium ion in halocarbon hepatotoxicity. *Environ Hlth Perspect* 1984; 57:321.

Bridges JW, Benford DJ, and Hubbard SA: Mechanisms of toxic injury. *Ann NY Acad Sci* 1983; 407:42.

Brodie BB and Axelrod J: The estimation of acetanilide and its metabolic products, aniline, *N*-acetyl-*p*-aminophenol, and *p*-aminophenol (free and total conjugated) in biological fluids and tissues. *J Pharmacol Exp Ther* 1948; 94:22.

Brodie BB and Axelrod J: The fate of aminopyrine (pyramidon) in man and methods for the estimation of aminopyrine and its metabolites in biological material. *J Pharmacol Exp Ther* 1950; 99:171.

Buege JA and Aust SD: Microsomal lipid peroxidation. *Methods Enzymol* 1978; 52;302.

Bunyan J, Murrell EA, Green J, and Diplock AT: On the existence and significance of lipid peroxides in vitamin E deficient animals. *Br J Nutr* 1969; 21:475.

Burton K: A study of the conditions and mechanism of the diphenylamine reaction for the colorimetric estimation of DNA. *Biochem J* 1956; 62:315.

Cameron R, Kellen J, Kolin A, et al: γ-Glutamyl-transferase in putative premalignant liver cell preparations during hepatocarcinogenesis. *Cancer Res* 1978; 38:823.

Carafoli E and Tiozzo R: A study of energy-linked calcium transport in liver mitochondria during CCl_4 intoxication. *Exp Mol Pathol* 1968; 9:131.

Carlberg I and Mannervik B: Purification and characterization of the flavoenzyme glutathione reductase from rat liver. *J Biol Chem* 1975; 250:5475.

Carrol NV, Longley RW, and Roe JH: The determination of glycogen in liver and muscle by use of anthrone reagent. *J Biol Chem* 1956; 220:583.

Carter SG and Karl DW: Inorganic phosphate assay with malachite green: an improvement and evaluation. *J Biochem Biophys Methods* 1982; 7:7.

Casini AF and Farber JL: Dependence of the carbon tetrachloride-induced death of cultured hepatocytes on the extracellular calcium concentration. *Am J Pathol* 1981; 105:138.

Ceriotti G and Gazzaniga A: A sensitive method for serum ornithine carbamyltransferase determination. *Clin Chim Acta* 1966; 14:57.

Ceriotti G and Gazzaniga A: Accelerated micro and ultramicro procedure for ornithine corbamyltransferase (OCT) determination. *Clin Chim Acta* 1967; 16:436.

Chafouleas JG, Bolton WE, Hidaka H, Boyde AE, and Means AR: Calmodulin and cell cycle: involvement in regulation of cell-cycle progression. *Cell* 1982; 28:41.

Chafouleas JG, Dedman JR, Munjall RP, and Means AP: Calmodulin: development and application of a sensitive radioimmunoassay. *J Biol Chem* 1979; 254:10262.

Chafouleas JG, Pardue RL, Brinkley BR, Dedman JR, and Means AR: Regulation of intracellular levels of calmodulin and tubulin in normal and transformed cells. *Proc Natl Acad Sci USA* 1981; 78:996.

Chang LO and Looney WB: A biochemical and autoradiographic study of the *in vivo* utilization of tritiated thymidine in regenerating rat liver. *Cancer Res* 1965; 25:1817.

Chauveau J, Moule Y, and Rouiller CH: Isolation of pure and unaltered liver nuclei morphology and biochemical composition. *Exp Cell Res* 1956; 11:317.

Cheeseman KH, Albano EF, Tomasi A, and Slater TF: Biochemical studies on the metabolic activation of halogenated alkanes. *Envir Hlth Perspec* 1985; 64:85.

Cheung WY: Calmodulin: its potential role in cell proliferation and heavy metal toxicity. *Federation Proc* 1984; 43:2995.

Chio KS and Tappel AL: Synthesis and characterization of the fluorescent products derived from malanaldehyde and amino acids. *Biochemistry* 1969; 8:2821.

Cochin J and Axelrod J: Biochemical and pharmacological changes in the rat following chronic administration of morphine, nalorphine, and normorphine. *J Pharmacol Exp Ther* 1959; 125:105.

Connor HD, Thurman RG, Galiz MD, and Mason RP: The formation of a novel free radical metabolite from CCl_4 in the perfused rat liver and *in vivo*. *J Biol Chem* 1986; 261:4542.

Cooper RH, Coll KE, and Williamson JR: Differential effects of phorbol ester on phenylephrine and vasopressin-induced Ca^{2+} mobilization in isolated hepatocytes. *J Biol Chem* 1985; 260:3281.

Curtis SJ, Moritz M, and Snodgrass PJ: Serum enzymes derived from liver cell fractions. 1. The response to carbon tetrachloride intoxication in rats. *Gastroenterology* 1972; 62:84.

Davis ME and Mehendale HM: Functional and biochemical correlates of chlordecone exposure and its enhancement of CCl_4 hepatotoxicity. *Toxicology* 1980; 15:91.

DeJersey J and Martin RB: Lanthanide probes in biological systems: the calcium binding site of pancreatic elastage as studied by terbium fluorescence. *Biochemistry* 1980; 19:1127.

Demange G, Ballan P, Moretti GF, Staeffen J, Catanzano G, and Lafaix JM: Dosage de l'ornithine carbamyl transferase du serum: une methods simple st sensible. *Presse Med* 1962; 70:1032.

DeMartino GN: Cytoplasmic proteases of rat liver parenchymal cells. *Biochem Biophys Res Comm* 1982; 108:1325.

DeMartino GN and Croal DE: Calcium-dependent proteases in neuroblastoma cells. *J Neurochem* 1982; 38:1642.

Dillard CJ, Dumelin EE, and Tappel AL: Effect of dietary vitamin E and expiration of pentane and ethane by the rat. *Lipids* 1977; 12:109.

Drotman RB: A study of kinetic parameters for the use of serum ornithine carbamyl transferase as an index of liver damage. *Fd Cosmet Toxicol* 1975; 13:649.

Drotman RB and Lawhorn GT: Serum enzymes as indicators of chemically induced liver damage. *Drug Chem Toxicol* 1978; 1:163.

Dumelin EE and Tappel AL: Hydrocarbon gases produced during *in vitro* peroxidation of polyunsaturated fatty acids and decomposition of performed hydroperoxides. *Lipids* 1977; 12:894.

El-Mofty SK, Scrulton MC, Serroni A, Nicolini C, and Farber JL: Early reversible plasma membrane injury in galactosamine-induced liver cell death. *Am J Pathol* 1975; 79:579.

Farris MW, Pascoe GA, and Reed DJ: Vitamin E reversal of the effect of extracellular calcium on chemically induced toxicity in hepatocytes. *Science* 1985; 227:751.

Farris MW and Reed DJ: Mechanisms of chemical induced toxicity. II. Role of extracellular calcium. *Toxicol Appl Pharmacol* 1985; 79:296.

Fiske CH and Subbarow Y: The colorimetric determination of phosphorus. *J Biol Chem* 1925; 66:375.

Ford EJH: Changes in the activity of ornithine carbamyl trasferase (OCT) in the serum of cattle and sheep with hepatic lesions. *J Comp Pathol* 1965; 75:299.

Frank H, Haussman HJ, and Remmer H: Metabolic activation of carbon tetrachloride: induction of cytochrome P450 with phenobarbital or 3-methylcholanthrene and its effects on covalent binding. *Chem Biol Interact* 1982; 40:193.

Frankel EN, Nowakowska J, and Evans CD: Formation of methyl azelaldehydate on autoxidation of lipids. *J Am Oil Chem Soc* 1961; 38:161.

Fridovich I: Superoxide dismutases. *Adv Enzymol Relat Areas Mol Biol* 1974; 41:35.

Fritz PJ and Hamrick ME: Enzymatic analysis of adenosine triphosphatase. *Enzymol Acta Biocat* 1966; 30:57.

Gerlach U: Patholgischer ubertritt von sorbitdehydrogenase ins Blut bei leberekrankungen. *Klin worchenschr* 1957; 35:1144.

Ghoshal AK, Porta EA, and Hartroft WS: The role of lipoperoxidation in the pathogenesis of fatty livers induced by phosphorous poisoning in rats. *Am J Pathol* 1969; 54:275.

Gillette JR: Mechanism of hepatic necrosis induced by halogenated aromatic hydrocarbons, in Keppler D (ed): *The Pathogenesis and Mechanisms of Liver Cell Necrosis*. Lancaster, UK, MTP Press, 1975, pp 239–254.

Gillette JR: Kinetics of reactive metabolites and covalent binding *in vivo* and *in vitro*, in Jollow DJ, Kocsis JJ, Snyder R, and Vanio H (eds): *Biological Reactive Intermediates*. New York, Plenum Press, 1977, pp 25–41.

Glende EA, Jr, Hruszkewycz AM, and Recknagel RO: Critical role of lipid peroxidation in carbon tetrachloride induced loss of aminopyrine demethylase, cytochrome P-450 and glucose-6-phosphatase. *Biochem Pharmacol* 1976; 25:2163.

Goldberg DM: 5′-Nucleotidase: recent advances in cell biology, methodology and clinical significance. *Digestion* 1973; 8:87.

Goldberg DM and Martin JV: Role of γ-glutamyl transpeptidase activity in the diagnosis of hepatobiliary disease. *Digestion* 1975; 12:232.

Green DE, Lester RL, and Ziegler DM: Studies on the mechanism of oxidative phosphorylation: 1. Preparation and properties of a phosphorylating electron transfer particle from beef heart mitochondria. *Biochim Biophys Acta* 1957; 23:516.

Grice HC: The changing role of pathology in modern safety evaluation. *CRC Crit Rev Toxicol* 1972; 1:119.

Grynkiewicz G, Poenie M, and Tsien RY: A new generation of Ca^{2+} indicators with greatly improved fluorescence properties. *J Biol Chem* 1985; 260:3440.

Gude WD (ed): *Autoradiographic Techniques,* New Jersey, Prentice-Hall, 1968.

Gutman AD: Serum alkaline phosphatase activity in diseases of the skeletal and hepatobiliary systems. A consideration of the current status. *Am J Med* 1959; 27:875.

Habig WH, Pabst MJ, and Jakoby WB: Glutathione *S*-transferases. The first enzymatic step in mercapturic acid formation. *J Biol Chem* 1974; 249:7130.

Hagerstrand I: Distribution of alkaline phosphatase activity in healthy and diseased human liver tissue. *Acta Pathol Microbiol Scand A* 1975; 83:519.

Harris RN and Anders MW: 2-Propanol treatment induces selectively the metabolism of carbon tetrachloride to phosgene. *Drug Metab Dispos* 1981; 9:551.

Hewitt WR, Miyajima H, Cote MG, and Plaa GL: Modification of haloalkane induced hepatotoxicity by exogenous ketones and metabolic ketosis. *Fed Proc* 1980; 39:3118.

Hill PG and Sammons HG: An assessment of 5′-nucleotidase as a liver function test. *Quart J Med* 1967; 36:457.

Hirs CHW (ed): *Methods in Enzymology,* vol 11, New York, Academic Press, 1967, pp 199–203.

Hoffman WS: *The Biochemistry of Clinical Medicine, 4th ed year book, Chicago, 1970.*

Hue L, Bontemps F, and Hers HG: The effect of glucose and of potassium ions on the interconversion of the two forms of glycogen phosphorylase and of glycogen synthetase in isolated rat liver preparations. *Biochem J* 1975; 152:105.

Huston RB and Krebs EG: Activation of skeletal muscle phosphorylase kinase by Ca^{2+} II. Identification of the kinase activating factor as a proteolytic enzyme. *Biochemistry* 1968; 7:2116.

Inoue M, Kishimoto A, Takai Y, and Nishizuka Y: Studies on a cyclic nucleotide-independent protein kinase and its proenzyme in mammalian tissues II. Proenzyme and its activation by calcium dependent protease from rat brain. *J Biol Chem* 1977; 252:7610.

Ito S, Inoue S, and Fujita K: The mechanism of toxicity of 5-*S*-cysteinyldopa to tumour cells. Hydrogen peroxide as a mediator of cytotoxicity. *Biochem Pharmacol* 1983; 32:2079.

Jakoby WB and Habig WH: Glutathione transferases, in Jakoby WB (ed): *Enzymatic Basis of Detoxication,* vol II. Academic Press, New York, 1980, pp 69.

Jakoby WB, Ketley JN, and Habig WH: in Arias IM and Jakoby WB (eds): *Glutathione Metabolism and Function,* New York, Raven Press, 1976, p 213.

Jaegar RJ, Trabulus MJ, and Murphy SD: Biochemical effects of 1,1-dichloroethylene in rats: dissociation of its hepatotoxicity from a lipoperoxidative mechanism. *Toxicol Appl Pharmacol* 1973; 24:457.

Jernstrom B, Babson JR, Moldeus P, Holmgren A, and Reed DJ: Glutathione conjugation and DNA-binding of (±)-*trans*-7,8-dihydroxy-7,8-dihydrobenzo(*a*)pyrene and (±)-7β,8α-dihydroxy-9α,10α-epoxy-7,8,9,10-tetrahydrobenzo(*a*)pyrene in isolated rat hepatocytes. *Carcinogenesis* 1982; 3:861.

Jernstrom B, Martinez M, Meyer DJ, and Ketterer B: Glutathione conjugation of the carcinogenic and mutagenic electrophile (±)-7β, 8α-dihydroxy-9-α,10α-oxy-7,8,9,10-tetrahydrobenzo(a)pyrene catalyzed by purified rat liver glutathione transferases. *Carcinogenesis* 1985; 6:85.

Johnson JD, Conroy WG, and Isom GE: Alteration of cytosolic calcium levels in PC 12 cells by potassium cyanide. *Toxicol Appl Pharmacol* 1987; 88:217.

Johnson JD and Mills JS: Calmodulin. *Med Res Rev* 1986; 6:341.

Jollow DJ, Mitchell JR, Zampaglione N, and Gillette JR: Bromobenzene-induced liver necrosis: protective role of glutathione and evidence for 3,4-bromobenzene oxide as the hepatotoxic metabolite. *Pharmacology* 1974; 11:151.

Joseph SK, Coll KE, Cooper RH, Marks JS, and Williamson JR: Mechanisms underlying calcium homeostasis in isolated hepatocytes. *J Biol Chem* 1983; 258:731.

Joshi UM, Dumas M, and Mehendale HM: Glutathione turnover in perfused rabbit lung. Effect of external glutathione. *Biochem Pharmacol* 1986; 35:3409.

Joshi UM, Prasada Rao KS, and Mehendale HM: Glutathione status in constituted physiological fluids containing albumin. *Int J Biochem* 1987; 19:1029.

Joshi UM, Kodavanti PRS, and Mehendale HM: Glutatione metabolism and utilization of external thiols by cigarette smoke-challenged, isolated rat and rabbit lungs. *Toxicol Appl Pharmacol* 1988; 96:324.

Karmen A: A note on the spectrophotometric assay of glutamic-oxalacetic transaminase in human blood serum. *J Clin Invest* 1955; 34:131.

Ketterer B, Meyer D, Taylor J, and Burchell B: Glutathione transferases and their role in xenobiotic metabolism. *Trends Pharmacol Sci* 1984; 5:319.

Kilhoffer M, Demaille JG, and Gerard D: Terbium as luminescent probe of calmodulin calcium-binding sites. Domains I and II contain the high-affinity site. *FEBS Lett* 1980; 116:269.

Kim NK, Yasmineh WG, Freier EF, et al: Value of alkaline phosphatase 5′-nucleotidase, γ-glutamyl transferase and glutamate dehydrogenase activity measurements (single and combined) in serum is diagnosis of metastasis to the liver. *Clin Chem* 1977; 23:2034.

King J: Thesis, Inst Med Lab Tech London, 1967.

Klaassen CD and Plaa GL: Effect of carbon tetrachloride on the metabolism, storage and excretion of sulfobromophthalein. *Toxicol Appl Pharmacol* 1968; 12:132.

Klaassen CD and Plaa GL: Comparison of the biochemical alterations elicited in livers from rats treated with carbon tetrachloride, chloroform, 1,1,2-trichloroethane and 1,1,1-trichloroethane. *Biochem Pharmacol* 1969; 18:2019.

Klingensmith JS and Mehendale HM: Destruction of hepatic mixed-function oxidase parameters by CCl_4 in rats following acute treatment with chlordecone, mirex, and phenobarbital. *Life Sci* 1983a; 33:2339.

Klingensmith JS and Mehendale HM: Hepatic microsomal metabolism of CCl_4 after pretreatment with chlordecone, mirex, or phenobarbital in male rats. *Drug Metab Dispos* 1983b; 11:329.

Knowles SE and Ballard FJ: Selective control of the degradation of normal and aberrent proteins in Reuber H35 hepatoma cells. *Biochem J* 1976; 156:609.

Kodavanti PRS, Joshi UM, and Mehendale HM: Hepatic energy status during CCl_4 toxicity in rats pretreated with chlordecone, mirex and phenobarbital. *The Toxicologist* 1988; 8:66.

Kodavanti PRS, Joshi UM, Young RA, Meydrech EF, and Mehendale HM: Protection of hepatotoxic and lethal effects of CCl_4 by partial hepatectomy. *Toxicol Pathol* 1989a; 17:494.

Kodavanti PRS, Joshi UM, Young RA, Bell AN, and Mehendale HM: Role of hepatocellular regeneration in chlordecone potentiated hepatotoxicity of carbon tetrachloride. *Arch Toxicol* 1989b, 63:367.

Kodavanti PRS, Joshi UM, Lockard VG, and Mehendale HM: Chlordecone (Kepone^R) potentiated carbon tetrachloride hepatotoxicity in partially hepatectomized rats—a histomorphometric study. *J Appl Toxicol* 1989c; 9:367

Kodavanti PRS, Kodavanti VP, and Mehendale HM: Altered hepatic energy status in chlordecone (keponel®)-potentiated cd_4 hepatotoxicity. *Biochem Pharmacol* 1990; 40:859.

Konttinen A: Simple method for determination of ornithine transcarbamylase activity in serum. *Clin Chim Acta* 1967; 18:147.

Konttinen A: A further simplified method of ornithine carbamolytransferase measurement. *Clin Chim Acta* 1968; 21:29.

Korsrud GO, Grice HG, Goodman TK, Knipfel JE, and McLaughlan JM: Sensitivity of several serum enzymes for the detection of thioacetamide-, dimethylnitroamine- and diethanolamine-induced liver damage in rats. *Toxicol Appl Pharmacol* 1973; 26:299.

Korsrud GO, Grice HC, and McLaughlan JM: Sensitivity of several serum enzymes in detecting carbon tetrachloride-induced liver damage in rats. *Toxicol Appl Pharmacol* 1972; 22:474.

Kroner H and Planker M: The role of calcium in liver cell damage. Comparative studies with carbon tetrachloride and D-galactosamine. *Path Res Pract* 1980; 169:298.

Lamprecht W and Trautschold I: Determination with hexokinase and glucose-6-phosphate dehydrogenase, in Bergmeyer HU (ed): *Methods of Enzymatic Analysis,* New York, Verlag Chemie Weinheim Academic Press, 1974, vol 4, p 2101.

Lawrence RA and Burk RF: Glutathione peroxidase activity in selenium deficient rat liver. *Biochem Biophys Res Comm* 1976; 71:952.

Lieberman M and Mapson LW: Genesis and biogenesis of ethylene. *Nature* 1964; 204:343.

Lockard VG, Mehendale HM, and O'Neal RM: Chlordecone-induced potentiation of carbon tetrachloride hepatotoxicity: a light and electron microscopic study. *Exp Mol Pathol* 1983a; 39:230.

Lockard VG, Mehendale HM, and O'Neal RM: Chlordecone-induced potentiation of carbon tetrachloride hepatotoxicity: a morphometric and biochemical study. *Exp Mol Pathol* 1983b; 39:246.

Loury DJ, Goldsworthy TL, and Butterworth BE: The value of measuring cell replication as a predictive index of tissue specific tumorigenic potential, in Butterworth BE and Slaga T (eds): *Banbury Report 25: Non Genotoxic Mechanisms of Carcinogenesis.* New York: Cold Spring Harbor Laboratory Press, 1987, p 119.

Lowery R, Glende EA Jr, and Recknagel RO: Destruction of liver microsomal calcium pump activity by carbon tetrachloride and bromotrichloromethane. *Biochem Pharmacol* 1981; 30:135.

Lowry OH and Lopez JA: The determination of inorganic phosphate in the presence of labile phosphate esters. *J Biol Chem* 1946; 162:421.

Lowry OH, Rosebrough NJ, Farr AL, and Randall RJ: Protein measurement with Folin-phenol reagent. *J Biol Chem* 1951; 193:265.

Maling HM, Eichelbaum FM, Saul W, Sipes IG, Brown EAB, and Gillette JR: Nature of the protection against carbon tetrachloride-induced hepatotoxicity produced by pretreatment with dibenamine [*N*-(2-chloroethyl) dibenzylamine]. *Biochem Pharmacol* 1974; 23:1479.

Mancini RE and Kocsis JJ: Dimethylsulfoxide increases the lethality of CCl_4 in rats but decreases its hepatotoxicity. *Toxicol Appl Pharmacol* 1974; 27:206.

Mazel P and Pessayre D: Significance of metabolite mediated toxicities in the safety evaluation of drugs and chemicals, in Mehlman MA, Shapiro RE, and Blumenthal H (eds): *Advances in Modern Toxicology,* vol 1, pt 1: *New Concepts in Safety Evaluation.* New York, Hemisphere, 1976, p 307.

McCord JM, Crapo JD, and Fridovich I: Superoxide dismutase assay: a review of methodology, in Michelson AM, McCord JM, and Fridovich I (eds): *Superoxide and Superoxide Dismutases,* New York, Academic Press, 1977, pp 11–17.

McLean AEK and McLean EK: The effect of diet and 1,1,1-trichloro-2,2-bis (*p*-chlorophenyl) ethane (DDT) on microsomal hydroxylating enzymes and on sensitivity of rats to carbon tetrachloride poisoning. *Biochem J* 1966; 100:564.

Means AR, Tash JS, and Chafouleas JG: Physiological implications of the presence, distribution and regulation of calmodulin in eukaryotic cells. *Physiol Rev* 1982; 62:1.

Mehendale HM: Potentiation of halomethane hepatotoxicity: chlordecone and carbon tetrachloride. *Fundam Appl Toxicol* 1984; 4:295.

Mehendale HM: Mechanism of the lethal interaction of chlordecone and CCl_4 at nontoxic doses. *Toxicol Lett* 1989a; 49:215.

Mehendale HM: Amplification of hepatotoxicity and lethality of CCl_4 and $CHCl_3$ by chlordecone. *Rev Biochem Toxicol* 1989b; 10:91.

Mehendale HM and Klingensmith JS: *In vivo* metabolism of CCl_4 by rats pretreated with chlordecone, mirex, or phenobarbital. *Toxicol Appl Pharmacol* 1988; 93:247.

Mehendale HM, Purushotham KR, and Lockard VG: The time-course of liver injury and ^{3}H-thymidine incorporation in chlordecone-potentiated $CHCl_3$ hepatotoxicity. *Exp Mol Pathol* 1989; 51:31.

Mehendale HM, Svensson SA, Baldi C, and Orrenius S: Accumulation of Ca^{2+} induced by cytotoxic levels of menadione in the isolated, perfused rat liver. *Eur J Biochem* 1985; 149:201.

Meister A: *The Enzymes,* ed 3, vol 10. New York, Academic Press, 1974, pp 671.

Mellgren RL, Aylward JH, Killilea SD, and Lee EYC: The activation and dissociation of a native high molecular weight form of rabbit skeletal muscle phosphorylase phosphatase by endogenous Ca^{2+} dependent proteases. *J Biol Chem* 1979; 254:648.

Meyer WL, Fisher EH, and Krebs EG: Activation of skeletal muscle phosphorylase b kinase by Ca^{2+}. *Biochemistry* 1966; 3:1033.

Mikkelsen RB and Wallach DFH: High affinity calcium binding sites on erythrocyte membrane proteins. Use of lanthanides as fluorescent probes. *Biochim Biophys Acta* 1974; 363:211.

Miller TL, Nelson DJ, Brittain HG, Richardson FS, and Martin RB: Calcium binding sites of rabbit troponin and carp parvalbumin. *FEBS Lett* 1975; 58:262.

Mirsalis JC: Use of *in vivo* DNA repair assay as an indicator of genotoxic exposure, in Bridges BA, Butterworth BE, and Weinstein IB (eds): *Banbury Report 13: Indicators of Genotoxic Exposure.* New York, Cold Spring Harbor Laboratory Press, 1982.

Mirsalis JC and Butterworth BE: Induction of unscheduled DNA synthesis in rat hepatocytes following *in vivo* treatment with dinitrotoluene. *Carcinogenesis* 1982; 3:241.

Mirsalis JC, Tyson CK, and Butterworth BE: Detection of genotoxic carcinogens in the *in vivo—in vitro* hepatocyte DNA repair assay. *Environ Mutagen* 1982; 4:553.

Mirsalis JC, Tyson CK, Loh EN, Steinmetz KL, Bakke JP, Hamilton CM, Spak K, and Spalding JW: Induction of hepatic cell proliferation and unscheduled DNA synthesis in mouse hepatocytes following *in vivo* treatment. *Carcinogenesis* 1985; 6:1521.

Mirsalis JC, Tyson CK, Steinmetz KL, Loh EK, Hamilton CM, Bakke JP, and Spalding JD: Measurement of unscheduled DNA synthesis and *S*-phase synthesis in rodent hepatocytes following *in vivo* treatment: testing of 24 compounds. *Environ Molec Mutagen* 1989; 14:155.

Mitchell JR and Boyd MR: Dose thresholds, host susceptibility and pharmacokinetic considerations in the evaluation of toxicity from chemically reactive metabolites, in Plaa GL and Duncan WA (eds): *Proceedings of the First International Congress on Toxicology,* New York, Academic Press, 1978, p 169.

Mitchell JR and Jollow DJ: Metabolic activation of drugs to toxic substances. *Gastroenterology* 1975; 68:392.

Mitchell JR, Nelson SD, Thorgeirsson SS, McMurty RJ, and Dybing E: Metabolic activation: biochemical basis for many drug-induced liver injuries. *Prog Liver Dis* 1976; 5:259.

Mitchell AD and Mirsalis JC: Unscheduled DNA synthesis as an indicator of genotoxic exposure, in Ansari AA and de Serres F (eds): *Single Cell Mutation Monitoring System: Methodologies and Applications.* New York, Plenum Press, 1984, pp 165–216.

Mitchell JR, Smith CV, Lauterburg BH, Hughes H, Corcoran GB, and Horning EC: Reactive metabolites and the pathophysiology of acute lethal cell injury, in Mitchell JR and Horning MG (eds): *Drug Metabolism and Drug Toxicity.* New York, Raven Press, 1984, p 301.

Molander DW, Wroblewski F, and LaDue JS: Serum glutamic oxalacetic transaminase as an index of hepatocellular injury. *J Lab Clin Med* 1955; 46:831.

Moldeus P, Hogberg J, and Orrenius S: Isolation and use of liver cells, in Fleischer S and Parker L (eds): *Methods in Enzymology,* vol LII. New York, Academic Press, 1978, pp 60–71.

Moore L: Inhibition of liver microsome calcium pump by *in vivo* administration of CCl_4, $CHCl_3$ and 1,1-dichloroethylene (vinylidene chloride). *Biochem Pharmacol* 1980; 29:2505.

Moore L, Davenport GR, and Landon EJ: Calcium uptake of a rat liver microsomal subcellular fraction in response to in vivo administration of carbon tetrachloride. *J Biol Chem* 1976; 251:1197.

Moore L and Ray P: Enhanced inhibition of hepatic microsomal calcium pump activity by CCl_4 treatment of isopropanol-pretreated rats. *Toxicol Appl Pharmacol* 1983; 71:54.

Murphy E, Coll K, Rich TL, and Williamson JR: Hormonal effects on calcium homeostasis in isolated hepatocytes. *J Biol Chem* 1980; 255:6600.

Musser AW, Ortigoza C, Vazquez M, and Riddick J: Correlation of serum enzymes and morphology alterations of the liver with special reference to serum guanase and ornithine carbamyl transferase. *Am J Clin Pathol* 1966; 46:82.

Naftalin L, Child VJ, and Morley DA: Observations on the site of origin of serum γ-glutamyl transpeptidase. *Clin Chim Acta* 1969; 26:297.

Nash T: The colorimetric estimation of formaldehyde by means of the Hantzsch reaction. *Biochem J* 1953; 55:416.

Nebert DW and Gelboin HL: Substrate-inducible microsomal aryl hydroxylase in mammalian cell culture. 1. Assay and properties of induced enzyme. *J Biol Chem* 1968, 243:6242.

Neff NT, DeMartino GN, and Goldberg AL: The effect of protease inhibitors and decreased temperature on the degradation of different classes of proteins in cultured hepatocytes. *J Cell Physiol* 1979; 101:439.

Nichlas RB: Chromosome velocity during mitosis as a function of chromosome size and position. *J Cell Biol* 1965; 25:119.

Nicotera P, Moore M, Mirabelli F, Bellomo G, and Orrenius S: Inhibition of hepatocyte plasma membrane Ca^{2+}-ATPase activity by menadione metabolism and its restoration by thiols. *FEBS Lett* 1985, 181:149.

Nicotera P, Hartzell P, Davis G, and Orrenius S: The formation of plasma membrane blebs in hepatocytes exposed to agents that increase cytosolic Ca^{2+} is mediated by the activation of nonlysosomal proteolytic system. *FEBS Lett* 1986a; 209:139.

Nicotera P, Hartzell P, Baldi C, Svensson SA, Bellomo G, and Orrenius S: Cystamine induces toxicity in hepatocytes through the elevation of cytosolic Ca^{2+} and the stimulation of a nonlysosomal proteolytic system. *J Biol Chem* 1986b; 261:14628.

Nicotera P, Thor H, and Orrenius S: Cytosolic free Ca^{2+} and cell killing in hepatoma 1c1c7 cells exposed to chemical anoxia. *FASEB J* 1989; 3:59.

Omura T and Sato R: The carbon monoxide-binding pigment of liver microsomes. 1. Evidence for its hemoprotein nature. *J Biol Chem* 1964; 239:2370.

Orlowski M and Meister A: γ-Glutamyl-*p*-nitroanilide: a new convenient substrate for the determination and study of L- and D-γ-glutamyl-transpeptidase activities. *Biochim Biophys Acta* 1963; 73:679.

Peterson JA, Ebel RE, and O'Keefe DH: Dual-wavelength stopped-flow spectrophotometric measurement of NADPH-cytochrome P-450 reductase. *Methods Enzymol* 1978; 52:221.

Phillips AH and Langdon RG: Hepatic triphosphopyridine nucleotide-cytochrome *c* reductase: isolation, characterization and kinetic studies. *J Biol Chem* 1962; 237:2652.

Pinto PVC, Kaplan A, and Van Dreal PA: Aldolase. II. Spectrophotometric determination using an ultraviolet procedure. *Clin Chem* 1969b; 15:349.

Pinto PVC, Van Dreal PA, and Kaplan A: Aldolase: I. Colorimetric determination. *Clin Chem* 1969a; 15:339.

Plaa GL and Hewitt WR: Quantitative evaluation of indices of hepatotoxicity, in Plaa GL and Hewitt WR (eds): *Toxicology of the Liver*. New York, Raven Press, 1982, p 103.

Plaa GL and Witschi H: Chemicals, drugs and lipid peroxidation. *Annu Rev Pharm Toxicol* 1976; 16:125.

Popper H and Schaffner F: Drug-induced hepatic injury. *Ann Intern Med* 1959; 51:1230.

Purushotham KR, Lockard VG, and Mehendale HM: Amplification of chloroform hepatotoxicity and lethality by dietary chlordecone in mice. *Toxicol Pathol* 1988; 16:27.

Rasmussen H and Barrett PG: Calcium messenger system: an integrated view. *Physiol Rev* 1984; 64:938.

Recknagel RO and Glende EA, Jr: Carbon tetrachloride hepatotoxicity: an example of lethal cleavage. *CRC Crit Rev Toxicol* 1973; 2:263.

Recknagel RO, Glende EA, Jr, Waller RL, and Lowrey K: Lipid peroxidation: biochemistry, measurement and significance in liver cell injury, in Plaa GL and Hewitt WR (eds): *Toxicology of the Liver*. New York, Raven Press, 1982.

Reed DJ, Babson JR, Beatty PW, Brodie AE, Ellis WW, and Potter DW: High-performance liquid chromatography analysis of nanomole levels of glutathione, glutathone disulfide, and related thiols and disulfides. *Anal Biochem* 1980; 106:55.

Reichard H: Determination of ornithine carbamyl transferase with micro-diffusion technique. *Scand J Clin Lab Invest* 1957; 9:311.

Reichard H: Ornithine carbamoyl transferase in dog serum on intravenous injection of enzyme, choledochus ligation and carbon tetrachloride poisoning. *J Lab Clin Med* 1959; 53:417.

Reichard H: Ornithine carbamoyl transferase activity in human tissue homogenates. *J Lab Clin Med* 1960; 56:218.

Reichard H: Studies on ornithine carbamoyl transferase activity in blood and serum. *Acta Med Scand* 1962; 172 (suppl 390):1.

Reichard H: Determination of ornithine carbamoyl transferase in serum: a rapid method. *J Lab Clin Med* 1964; 63:1061.

Reichard H and Reichard P: Determination of ornithine carbamyl transferase in serum. *J Lab Clin Med* 1958; 52:709.

Reitman S and Frankel S: A colorimetric method for the determination of serum glutamic oxalocetic and glutamic pyruvic transaminases. *Am J Clin Pathol* 1957, 28:56.

Reville WJ, Goll DE, Stromer MH, Robson RM, and Dayton WR: A Ca^{2+}-activated protease possibly involved in myofibrillar protein turnover. *J Cell Biol* 1976; 70:1.

Reynolds ES: Liver parenchymal cell injury. I. Initial alterations of the cell following poisoning with carbon tetrachloride. *J Cell Biol* 1963, 19:139.

Reynolds ES: Liver parenchymal cell injury. II. Cytochemical events concerned with mitochondrial dysfunction following poisoning with carbon tetrachloride. *Lab Invest* 1964; 13:1457.

Reynolds ES: Comparison of early injury to liver endoplasmic reticulum by halomethanes, hexachloroethane, benzene, toluene, bromobenzene, ethionine, thioacetamide and dimethylnitrosamine. *Biochem Pharmacol* 1972; 21:2555.

Richards AH, Lubinski RM, and Vanderlinde RE: Studies on the kinetic assay of lactate dehydrogenase activity. *Clin Chem* 1975; 21:1018.

Richman PG, Orlowski M, and Meister A: Inhibition of γ-glutamylcysteine synthetase by L-methionine-*S*-sulfoximine. *J Biol Chem* 1973; 248:6684.

Richterich R: *Clinical Chemistry Theory and Practice.* New York, Academic Press, 1969, p 433.

Riley CA, Cohen G, and Lieberman M: Ethane evolution: a new index of lipid peroxidation. *Science* 1974; 183:208.

Robinson CP, Baskin SI, Visnich N, and Franz DR: The effects of cyanide and its interactions with norepinephrine on isolated aorta strips from the rabbit, dog and ferret. *Toxicology* 1985; 35:59.

Ross MH, Fly JO, and Archer AG: Alkaline phosphatase activity and pH optima. *J Biol Chem* 1951; 192:561.

Sanger F: Oxidation of insulin by performic acid. *Nature (London)* 1967; 160:295.

Schanne FAX, Kane AB, Young EE, and Farber JL: Calcium dependence of toxic cell death: a common final pathway. *Science* 1979; 206:700.

Schanne FAX, Pfau RG, and Farber JL: Galactosamine-induced cell death in primary cultures of rat hepatocytes. *Am J Pathol* 1980; 100:25.

Schaffner F and Raisfeld IH: Drugs and the liver: a review of metabolism and adverse reactions. *Adv Intern Med* 1969; 15:221.

Secchi GC and Ghidonie A: Interasse clinico della determinazione dell' attivita' sorbitoldeidrogenasica plasmatica in patologia epatica. *Enzymol Biol Clin* 1962; 2:99.

Seitanidis B and Moss DW: Serum alkaline phosphatase and 5′-nucleotidase levels during normal pregnancy. *Clin Chim Acta* 1969; 25:183.

Sekura R and Meister A: γ-Glutamylcysteine synthetase. Further purification, ''half of the sites'' reactivity, subunits, and specificity. *J Biol Chem* 1977; 252:2599.

Shin BC, Huggins JW, and Caraway KL: Effects of pH, concentration and aging on the malonaldehyde reaction with proteins. *Lipids* 1972; 7:229.

Sipes IG and Gandolfi AJ: Bioactivation of aliphatic organohalogens: formation, detection and relevance, in Plaa GL and Hewitt WR (eds): *Toxicology of the Liver.* New York, Raven Press, 1982, p 181.

Sipes IG, Krishna G, and Gillette JR: Bioactivation of carbon tetrachloride, chloroform, and bromotrichloromethane: role of cytochrome P-450. *Life Sci* 1977, 20:1541.

Smith MR and Sandy MS: Role of extracellular Ca^{2+} in toxic injury: comparative studies with the isolated perfused rat liver and isolated hepatocytes. *Toxicol Appl Pharmacol* 1985; 81:213.

Smith MT, Thor H, and Orrenius S: Toxic injury to isolated hepatocytes is not dependent on extracellular calcium. *Science* 1981; 213:1257.

Song CS, Kappas A, and Bodansky O: 5′-Nucleotidase of plasma membranes of the rat liver: studies on subcellular distribution. *Ann NY Acad Sci* 1969; 166:565.

Szasz G: A kinetic photometric method for serum γ-glutamyltranspeptidase. *Clin Chem* 1969; 15:124.

Tegeris AS, Smalley HE, Jr, Earl FL, and Curtis JL: Ornithine carbamyl transferase as a liver function test: comparative studies in dog, swine and man. *Toxicol Appl Pharmacol* 1969; 14:54.

Thiers RE, Reynolds ES, and Vallee BL: The effect of carbon tetrachloride poisoning on subcellular metal distribution in rat liver. *J Biol Chem* 1960; 235:2130.

Thomas AP, Alexander J, and Williamson JR: Relationship between inositol polyphosphate production and the increase of cytosolic free Ca^{2+} induced by vasopressin in isolated hepatocytes. *J Biol Chem* 1984; 259:5574.

Tietz NW: *Fundamentals of Clinical Chemistry*. Philadelphia, WB Saunders, 1970, pp 449–453.

Tietz NW, Burtis CA, Duncan P, Ervin K, Petitclerc CJ, Rinker AD, Shirey D, and Zygowicz ER: A reference method for measurement of alkaline phosphatase activity in human serum. Study group on alkaline phosphatase. *Clin Chem* 1983; 29:751.

Tsien RY: New calcium indicators and buffers with high selectivity against magnesium and protons: design, synthesis, and properties of prototype structures. *Biochemistry* 1980; 19:2396.

Tsien RY: Intracellular measurements of ion activities. *Annu Rev Biophys Bioeng* 1983; 12:91.

Tsien RY, Pozzan T, and Rink TJ: Calcium homeostasis in intact lymphocytes: Cytoplasmic free calcium monitored with a new, intracellularly trapped fluorescent indicator. *J Cell Biol* 1982; 94:325.

Tsien RY, Pozzan T, and Rink TJ: Measuring and manipulating cytosolic Ca^{2+} with trapped indicators. *Trends Biochem Sci* 1984; 9:263.

Utley WS and Mehendale HM: Phenobarbital-induced cytosolic cytoprotective mechanisms that offset increases in NADPH cytochrome P-450 reductase activity in menadione mediated cytotoxicity. *Toxicol Appl Pharmacol* 1989; 99:323.

Van der Slik WI, Persijin JP, Englesman E, et al: Serum 5′-nucleotidase. *Clin Biochem* 1970; 3:59.

Vassef AA: Direct micromethod for colorimetry of serum ornithine carbamoyl transferase activity, with use of linear standard curve. *Clin Chem* 1978; 24:101.

Vermilion JL and Coon MJ: Purified liver microsomal NADPH-cytochrome P-450 reductase: spectral characterization of oxidation-reduction states. *J Biol Chem* 1978; 253:2694.

Wacker WEC, Ulmer DD, and Vallec BL: Metalloenzymes and myocardial infarction. *N Engl J Med* 1956; 255:449.

Wallace RW, Tallant EA, Dockter ME, and Cheung WY: Calcium binding domains of calmodulin: sequence of fill as determined with terbium luminescence. *J Biol Chem* 1982; 257:1845.

Welsh MJ, Dedman JR, Brinkley BR, and Means AR: Calcium dependent regulator protein: localization in mitotic apparatus of eukaryotic cells. *Proc Natl Acad Sci. USA* 1978; 75:1867.

Wendel A and Feuerstein S: Drug-induced lipid peroxidation in mice. Modulation by monooxygenase activity, glutathione and selenium status. *Biochem Pharmacol* 1981; 30:2513.

Wendel A, Feuerstein S, and Konz KH: Acute paracetamol intoxication of starved mice leads to lipid peroxidation *in vivo. Biochem Pharmacol* 1979; 28:2051.

Wendel A, Feuerstein S, and Konz KH: Acute paracetamol intoxication of starved mice leads to lipid peroxidation *in vivo. Biochem Pharmacol* 1979; 28:2051.

Whitfield JB, Pounder RE, Neale G, et al: Serum γ-glutamyl transpeptidase activity in liver disease. *Gut* 1972; 13:702.

Wiesner IS, Rawnsley HM, Brooks FP, and Senior JR: Sorbitol dehydrogenase in the diagnosis of liver disease. *Am J Dig Dis* 1965; 10:147.

Wolf CR, Harrelson NG, Nastainczyk JWM, Philpot RM, Kalyanaraman B, and Mason RP: Metabolism of CCl_4 in hepatic microsomes and reconstituted monooxygenase systems and its relationship to lipid peroxidation. *Mol Pharmacol* 1980; 18:553.

Wolfson SK and Williams-Ashman HG: Isocitric and 6-phosphogluconic dehydrogenases in human blood serum. *Proc Soc Exp Biol Med* 1957; 96:231.

Wroblewski F and LaDue JS: SGOT activity as an index to liver cell injury. *Ann Intern Med* 1955; 43:345.

Zien M and Discombe G: Serum γ-glutamyl transpeptidase as a diagnostic aid. *Lancet* 1970; 2:748.

Zimmerman HJ: Serum enzyme measurement in experimental hepatotoxicity. *Isr J Med Sci* 1974; 10:328.

Zimmerman HJ: Experimental hepatotoxicity, in Born GVR, Eichler O, Farah A, Herkin H, and Welch AO (eds): *Handbook of Experimental Pharmacology,* vol XVI/5. Berlin, Springer-Verlag, 1976, pp 1–120.

Zimmerman HJ: *Hepatotoxicity: the Adverse Affects of Drugs and Other Chemicals on the Liver.* New York, Appleton-Century-Crofts, 1978.

Zimmerman HJ and Ishak KG: Hepatic injury due to drugs and toxins, in MacSweet RNM, Anthony PP, and Schever PJ (eds): *Pathology of Liver,* ed 2. New York, Churchill Livingstone, 1987, p 503.

Zimmerman HJ and Seeff LB: Enzymes in hepatic disease, in Coodley EL (ed): *Diagnostic Enzymology.* Philadelphia, Lea and Febiger, 1970, pp 1–38.

Zygowiez ER, Sunderman FW, Horak E, and Dooley JF: Inhibition by concanavalin A as the basis for a specific assay of serum 5′-nucleotidase activity. *Clin Chem* 1977; 23:2311.

Chapter

8

Biochemical Aspects of Fatty Liver

Mario U. Dianzani
Department of Experimental Medicine and Oncology
Section of General Pathology
University of Torino
Torino, Italy

Fatty liver is one of the most common and specific responses of the liver to injury. Morphologists use this name to indicate a condition characterized by the appearance of droplets of fat inside hepatocytes. In fact, fat is normally present in hepatocytes, but cannot be shown by the usual light microscopic methods. According to biochemists, fatty liver is a pathological increase in fat content in hepatocytes. Synonyms are liver steatosis, fat infiltration, or degeneration. The simple terms used, however, had different meaning for the old biochemists. In fact, the term "infiltration" was used for cases where the accumulation of fat in hepatocytes was from outside, whereas "degeneration" referred to a local net synthesis of fat. This distinction cannot be retained today, since it is now known that in every case fat, in the form of precursors, always comes from outside.

Fat is an ester or an amide of a fatty acid. So, the definition given above refers in principle not only to the cases where the accumulation occurs in the form of triglicerides (TG), but even to those cases where the chemical nature of the accumulating fat is different. At present, the term "fatty liver" is, however, mostly restricted to TG accumulation, whereas other names (phospholipidosis, gangliosidosis, cerebrosidosis, etc.) are used to indicate other types of accumulations.

TG accumulation inside hepatocytes starts with the formation of small droplets surrounded by a membrane, originating from Golgi cisternae. Over time, several small droplets fuse to form a single large drop pushing the cytoplasm and the nucleus at the periphery of the cell. Cell size is then increased; thus the term "ballooning" is frequently used. Notwithstanding the increase in cell size, the mass of the cytoplasm is generally decreased at this stage. In fact, the cytoplasm is very frequently reduced to a small layer enveloping the fatty drop, just sited under the plasma membrane. The size of the entire liver is also strongly increased and the color of the organ is yellowish and pale, due to the color of fat and also to the decrease in the amount of blood. In fact, the enlargement of the hepatocytes decreases the space for blood in capillaries. When considering the pathogenesis of fatty livers, therefore, one has to remember that fatty livers are at the same time ischemic.

According to the intralobular site of the onset of fat deposition, two main types of fatty liver are usually considered: centrolobular and peripheral.

In the first case, the cells surrounding the centrolobular vein are involved, whereas in the second infiltration starts from the periportal spaces. Most toxic fatty livers belong to the first type, probably due to the fact that the poison is metabolized to a more toxic form by the cytochrome P450-dependent microsomal monooxygenase system. Cytochrome P450 has been found to be more concentrated in the centrolobular than in the peripheral cells (Gooding et al., 1978). Even some types of nutritional fatty liver, however, start as centrolobular forms; for instance, choline-deficient fatty liver. The reason for this is unknown, and it is also unknown why most fatty livers related to a block in protein synthesis start as peripheral forms.

It has been stated earlier that small fat droplets tend to fuse and that a unique large drop is the final result of fusion. There is, however, an exception to this. It is represented by Wolman disease, an inborn lysosomal disease characterized by the congenital lack of acid lipase. In this case, TG accumulate inside lysosomes and fusion does not occur, the cell acquiring the aspect of a foamy cell with the nucleus occupying a central position.

In order to understand the pathogenesis of different forms of fatty liver, a short description of the TG metabolism in this organ is needed.

SHORT SURVEY ON TRIGLYCERIDE METABOLISM IN LIVER CELLS

Fat is absorbed from the intestine in the form of free fatty acids, which are set free from TG by the intestinal lipases. Enterocytes quickly retransform free fatty acids (FFA) into TG or phospholipids. Only FFA with a number of carbon atoms lower than 10 (Isselbacher, 1965) are not included in lipid molecules, and can reach the liver directly by the porta vein. Inside enterocytes, TG and phospholipids are combined with apoproteins to form chylomicrons (Isselbacher and Budz, 1963; Lopez, 1971; Scanu, 1972). These are released to the lymph and then reach the blood at the level of the left succlavia vein. In the blood, chylomicrons undergo remodelation and exchanges of both their protein and lipid moieties, with final decrease in their size (Scanu, 1972). In fasting subjects, chylomicrons are absent and the only forms of circulating lipoproteins are those secreted by the liver. Lipoprotein lipase, an enzyme produced by endothelial cells (Lopez, 1971; Blanchette-Mackie and Scow, 1971) releases FFA from TG still bound to the lipoproteins micelle. It has been shown that this enzyme is activated by apo-C, the major lipoprotein component of very low-density lipoproteins (VLDL), and especially by its subfraction apo-C_2. Apo-C is present in trace amounts, even in chylomicrons. FFA are now ready to cross cell membranes, including those of hepatocytes. Most of them, however, reach adipocytes, where they become quickly reincluded in TG molecules.

Further use of TG requires hydrolysis by the local lipase. The activity of this enzyme depends upon the intracellular concentration of cAMP, which results from the balance between its synthesis by adenylate cyclase and its destruction by a phosphodiesterase. Synthesis of the first enzyme is under the control of several hormones (catecholamines, STH, ACTH, glucagon, and cortisol) and is also stimulated by prostaglandins (Maickel et al., 1965; Polonovski, 1976). Phosphodiesterase activity, on the contrary, is inhibited by methylxanthines, including caffeine and theophyllin (Butcher and Sutherland, 1962; Klainer et al., 1962; Murad et al., 1962; Sutherland et al., 1962; Triner and Nahas, 1965). FFA released from TG are set free to the blood, where they become bound to the albumins. In this way, they can reach peripheral cells, including liver cells. After crossing plasma membranes, FFA are then bound to transport proteins; especially to ligandin (proteins Y) and to proteins Z (Arias et al., 1976). The bond with proteins Z is more specific, but proteins Y prevail quantitatively.

FFA coming from outside, however, do not represent the only components of the intracellular free fatty acids pool. In fact, a certain number of them are synthesized inside hepatocytes from acetyl-CoA. The synthesis proceeds by the reverse pathway usually leading to beta oxidation. Synthesis requires, however, a high ratio of reduced to oxidized pyridine nucleotides (Rebouças and Isselbacher, 1964) as well as energy in the form of ATP. FFA are now available to be either reincluded in lipid molecules or oxidized to produce energy. Oxidation occurs mostly in mitochondria, but a certain amount of fatty acids, especially the hydroxy-long-chained fatty acids, can be oxidized in peroxisomes.

In any case, oxidation occurs after preliminary activation of FFA to acyl-CoA derivatives. As acyl-CoAs cannot cross the mitochondrial membrane, they must be first transformed in acyl-carnitines, which are able to penetrate across these membranes. Inside mitochondria, acyl-carnitines react with CoA to reform acyl-CoAs, whereas carnitine can return to the cytoplasm to renew its shuttle work. Acyl-CoAs are now ready to be oxidized on the internal mitochondrial membrane. The oxidizing system is composed by several enzymes. The first attack is produced by acyl-CoA dehydrogenases, which contain FAD. Resulting enoyl-CoAs are then hydrated by specific enoyl-CoA hydratase, producing β-hydroxyacyl-CoAs.

These are finally cleaved by keto-thiolases in the presence of free CoA-SH to form acetyl-CoA and an acyl-CoA containing two carbon atoms less than that entering the cycle. It is clear from this short description that several co-factors (such as ATP, CoA, carnitine, FAD, and NAD) are needed in the cycle. Moreover, its functioning requires the integrity of the mitochondrial structure, as well as a continuous supply of ATP through oxidative phosphorylation.

The main metabolic destiny of FFA inside hepatocytes is, however, represented by their inclusion in lipid molecules. These may remain inside the cell as structural components, or may be ''secreted'' into the blood plasma as lipoproteins. Structural lipids are mostly concentrated within membranes. Both in these subcellular structures and in lipoprotein micelles, the contact between hydrophobic components, as TG, cholesterol, and cholesterol esters, and hydrophilic ones, as proteins, is usually mediated by phospholipids, which display amphiphilic properties. In lecithins, the phosphorylcholine moiety represent the hydrophobic moiety, i.e., the portion containing the two fatty acid residues contacts TG. It is to be remembered, however, that in certain instances the contact TG with TG can be done even through the methyl side chains of certain aminoacids, which are hydrophobic.

The fate of lipids incorporated into membranes is mostly represented by their hydrolysis inside secondary lysosomes, after formation of autophagic vacuoles and their fusion with primary lysosomes. A minor portion of lipids can exchange, however, with other intracellular lipid-containing structures, thus escaping or delaying their intralysosomal destiny. We have no idea about the quantitative relevance of these aspects, as well as of the conditions regulating the formation of secondary lysosomes. The hypothesis that a stimulus for autophagy is represented by some lipoperoxidative events taking place in membranes still needs further proof.

If true, this hypothesis would, however, attribute some relevance to lipid peroxidation as a physiological mechanism participating in cell catabolism. There are several indications for the possible relevance of lipid peroxidation even in other sections of cell physiology, and especially in the control of cell multiplication, but again evidence is lacking. The fact that 4-hydroxy-nonenal, one of the major products of lipid peroxidation, is able to influence the cell cycle at concentrations of 10^{-6} *M* that can be reached inside the cells (Esterbauer et al., 1982; Poli et al., 1985a) cannot be retained as a definite proof.

The same can be said about several other actions of 4-HNE, such as the inhibition of ornithine decarboxylase and of *S*-adenosyl-methionine decarboxylase (Dianzani, 1979b; Dianzani, 1982), inhibition of both DNA and RNA polymerases (Dianzani, 1982; Schauenstein et al., 1977) or block of the expression the oncogene *c*-myc in an erythroleukemic cell strain (Fazio et al., 1987). Even the fact that lipid peroxidation is very low in embryonic and regenerating tissues (Cheeseman et al., 1985, 1986, 1987) as well as in tumor cells (Ahmed and Slater, 1981; Bartoli and Galeotti, 1979; Dianzani et al., 1982, 1984; Dianzani and Rossi, 1981; Fonnesu et al., 1966; Ugazio et al., 1967) may also be an interesting coincidence.

Another possibility regarding arachidonic acid, as well as other polyunsaturated fatty acids contained within cell membranes, is represented by their transformation into prostaglandins or leukotrienes. Again, we do not have any idea about the quantitative relevance of this aspect.

The main pathway for TG discharge from liver cells is represented by lipoprotein secretion. TG are mostly transported by VLDL, whose main protein is apo-C, whereas apo-B and apo-E are minor components. The different portions of lipoproteins are synthesized in different parts of the cell. In fact, apolipoproteins are produced, as all other cytoplasmic proteins, at the level of the rough endoplasmic reticulum (RER), whereas phospholipids are synthesized either in the smooth endoplasmic reticulum (SER) or in the cytosol. The primary assembly of lipoprotein portions occurs at the level of the SER, but their final arrangement and maturation takes place only in the Golgi apparatus. Gradient fractionation of the Golgi vesicles have allowed the separation of three different subfractions, usually referred to as F_1 (secretory), F_2 (intermediate), and F_3 (formative subfraction). F_1 displays a very high galactosyl-transferase activity, whereas F_2 content of this enzyme is about one-half of that found in F_1 and F_3 is only one-fifth. After injection of ^{14}C palmitate, the labeling starts on the F_3 and then reaches the other two fractions.

One hour after the injection, the extent of labeling in F_1 is three times, and that in F_2 is two times higher than that in F_3. This shows that lipid-containing material first appears in F_3, which for this reason is referred to as the ''formative'' subfraction, and proceeds toward the other two subfractions, becoming accumulated in F_1, where it undergoes its maximal galactosylation and probably also galactosamination. For these reasons F_1 is considered as the secretive and maturative side.

The mature micelles now leave the Golgi region and move toward the plasma membrane, probably due to depolymerization of tubulin, the main constituent of microtubules. The micelles, at this moment, are still enveloped by a dependance of the Golgi sacculi, which fuses with the plasma membrane after crossing of the cytoplasm.

The connection between the lipid-containing vesicles, or liposomes, (Baglio and Farber, 1965) and the plasma membrane is done by microtubules.

Their main protein, tubulin, is a macromolecule resulting from polymerization of several subunits. These are heterodimers formed by two different chains, usually referred to as alpha and beta chains. The alpha chains contain a site that is able to bind to colchicine, as well as another to bind to podophyllotoxin. It is connected to the beta chain by a GDP molecule.

The beta chain contains a cAMP-phosphorylable site, as well as a site that is able to bind to vinblastin or vincristin. It is bound to a GTP molecule, which represents the connecting point with the alpha chain of the next subunit. It has been well established that both phosphorylation of the peculiar site in the beta chain, and binding with colchicine, podophyllotoxin, vinblastin, or vincristin prevent the polymerization of subunits, as well as the depolymerization of the well-formed macromolecule. Therefore, treatment with cAMP or with substances able to increase its intracellular level, or also with colchicine or other inhibitors, block the microtubules from functioning (Gabriel et al., 1977; Wilson and Leduc, 1963).

It is noteworthy that in isolated hepatocytes, treatment with colchicine results in TG accumulation (Gravela et al., 1977). This general description of TG metabolism was necessary to have a clear overview of the possible causes for fat accumulation within hepatocytes. We now understand that fat increase in liver cells is due to an increase of TG precursors from outside or to some internal derangement of hepatocyte metabolism. Increased arrival from outside may occur in the form of chylomicrons or in the form of FFA bound to albumin. In both cases, the hepatocyte is not primarily involved and behaves as an innocent bystander of events born in other districts. In the second case, however, the liver cell is primarily affected, but the mechanisms leading to fat accumulation are numerous.

In fact, TG content may increase from (1) an increased local synthesis of fatty acids from acyl-CoA, (2) a decreased oxidation, sparing for TG synthesis extra amounts of FFA, (3) a decline in lysosomal acid lipase action, or (4) a block in TG secretion in the form of lipoproteins. Further differentiation in the last group can be made for the same reasons leading to decreased secretion.

These may reside (4a) in a defect in the assembly due either to deficient synthesis in protein or phospholipids components, or (4b) in changes in their main structure, resulting in the formation of a nonsecretable micelle, or (4c) in defects of the Golgi apparatus affecting formation or maturation, or even (4d) in blocks occurring in the intracellular traffic of micelles, preventing their movements toward the plasma membrane. Finally, another hypothetical reason might be (4e) a defect in crossing the plasma membrane (Figure 1).

In the description of the several different forms of fatty liver, this general scheme will be as follows. It is, however, important to show immediately that several different mechanisms can coexist and collaborate in increasing fat. This usually occurs especially in the late phases of steatogenic treatments,

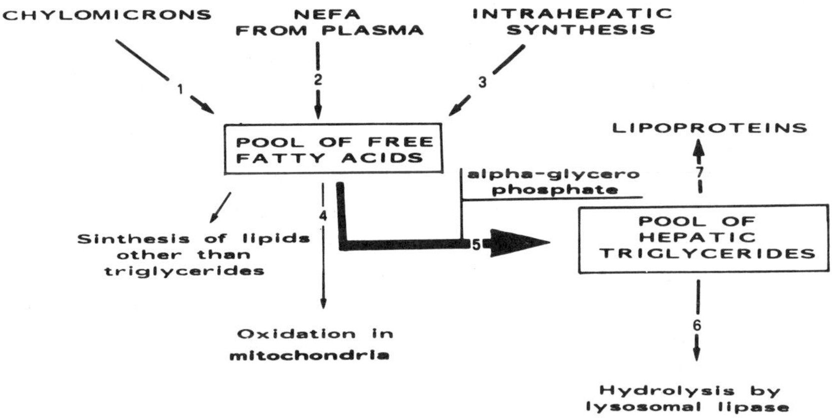

Figure 1. The main pathways for triglyceride metabolism in liver cells.

where functional derangement may affect several cell structures at the same time. For this reason, the position in the scheme of each single type of fatty liver refers to the onset of fat accumulation.

FATTY LIVERS DUE TO INCREASED ARRIVAL OF PRECURSORS FFA FROM OUTSIDE

It was reported earlier that fatty acids are transported by the blood to liver cells bound either to chylomicrons or to albumin. As only FFA can cross the plasma membrane, chylomicron-bound TG must undergo hydrolysis through lipoprotein lipase action before becoming able to enter the hepatocyte. Inside these cells, FFA must then be resynthesized to TG to produce fat accumulation. Resynthesis is done by a synthetase present in the cytosol. The net amount of TG being actually present inside the cell represents, therefore, the balance between synthesis and loss of such substances.

The rate of synthesis is only limited by the availability of precursors, and especially of FFA, as glycerol is abundantly produced in cytosol. The availability of FFA depends upon the balance between the sum of FFA from the blood plus FFA synthesized inside the cell from acyl-CoAs and the amount of FAA being oxidized in mitochondria or peroxisomes or used for the synthesis of lipids other than TG. Thus, when TG loss in the form of secreted lipoproteins is not impaired and FFA oxidation is normal, an excess amount of FFA from the blood can provoke fat accumulation. When oxygen supply to the liver is in some way restricted, even a moderate increase of FFA by the blood may be sufficient.

When increased FFA blood supply is the only cause for fat accumulation, hepatocyte function is preserved, at least at the beginning of the process.

After long-term accumulation, however, some damage may appear. There are mainly two ways for such type of damage. The first mechanism is represented by the increase in cell size, caused by the presence of the large fat droplets, which are added to the preexisting cytoplasmic material.

Enlargement of the cells provokes mutual compression, with definite decrease in the absorbing surface, as well as restriction of capillaries. Thus, cell nutrition may become so deficient as to decrease the original cell mass, and atrophy of the cytoplasm occurs. The second way for damage to occur is by the possible onset of lipid peroxidation. This results from the large increase in peroxidizable substrates (unsaturated fatty acids), which are not accompanied by a corresponding increase in the concentration of antioxidants, especially in vitamin E. This provokes a derangement in the peroxidative balance (Hartroft and Porta, 1967).

Such a situation is very frequent in fatty livers following long-term feeding on high-fat diets, which are rarely well balanced for the relative contents of peroxidation substrates (polyunsaturated fatty acids) and antioxidants, and is definitely expected to occur in fatty livers following fat mobilization from adipocytes. In fact, in this last case, the activation of lipase sets free from adipocytes FFA, and not corresponding amounts of vitamin E or of other antioxidants.

Fatty Liver Due to Increased FFA Attached to Lipoproteins

Chylomicrons are a source for liver FFA within a few hours after the meal. When they are present in the blood, the plasma becomes milky, and is clarified over time under the action of lipoprotein lipase. If fatty meals are repeated and abundant, fatty liver due to excess FAA from the blood may develop. It is noteworthy that in hyperchylomicronemia, a congenital disease, there is no fatty liver. In fact, this disease is due to the congenital deficiency of lipoproteinlipase, and no FFA are released in such instances.

Some return to the liver of lipoproteins can occur even hours after a meal, mostly in the form of VLDL in the rat and of LDL in human. The amount of VLDL returning to the rat liver is, however, restricted, as most of the VLDL is used by peripheral tissues. In any case, FAA is able to cross the liver's plasma membrane, and possibly a certain amount of apoprotein is also captured. In any case of LDL, what returns to the liver is especially cholesterol, which is oxidized in mitochondria to form cholic acid.

The capture by hepatocytes of LDL components requires the presence of apolipoprotein E. In any case, the amount of FFA or of other components of these lipoproteins taken up by the liver is not sufficient to produce accumulation of fat.

Fatty Liver Due to Increases of Free Fatty Acids Bound to Albumin

This type of fatty liver presumes increased mobilization of FFA from adipocyte TG, and this requires activation of the local lipase. We have seen that this may depend on activation of adenylate cyclase synthesis or decrease of cAMP phosphodiesterase.

The first type of fatty liver is relatively frequent in the practical clinics, due to the high frequency of situations, including stress or long-term cortisol therapy. Experimentally, fatty livers have been described after repeated injections of catecholamines (epinephrine: Lombardi, 1966; norepinephrine: Feigelson et al., 1961), cortisol (Hill and Droke, 1963; Hill et al., 1965; Felt et al., 1962) or ACTH (Paoletti et al., 1961).

The type of fatty liver related to cAMP phosphodiesterase inhibition is mostly encountered after treatment with methylxanthines, as caffeine or theophylline (Estler, 1973; Estler and Ammon, 1966). It is conceivable that it may represent a cause for clinical fatty liver in humans consuming excessive amounts of coffee or tea.

FATTY LIVERS DUE TO INTRAHEPATIC CAUSES

Fatty Livers Due to Increased Local Synthesis of Free Fatty Acids from Acetyl-CoA

The increase in the local synthesis of FFA from acetyl-CoA is considered mainly in four forms of fatty livers: (1) in that following a long-term supply of a diet deficient in polyunsaturated (essential) fatty acids (EFA); (2) in that following treatment with barbiturates; (3) in that provoked by ethanol treatment; and (4) in that following treatment with isopropanol.

Fatty Liver After Feeding on an EFA-Deficient Diet

Feeding rats or mice on an EFA-deficient diet results in fat accumulation in the liver. The onset is evident 3–4 days after beginning the treatment, and the cells first involved are those at the periphery of the lobule. Maximal accumulation occurs after 3–4 weeks of continuous feeding on the diet.

Severe morphological derangements have been described in mitochondria, which become larger and swollen, but display a higher density in their matrix, which is different from other types of swelling (Hayashida and Portman, 1960; Johnson, 1963; Klein and Johnson, 1954; Levin et al., 1957; Smith and De Luca, 1964; Wilson and Leduc, 1963). The cristae, moreover, tend to occupy the center of the particle, in a peculiar parallel arrangement.

The endoplasmic reticulum is apparently normal, whereas an increase in the number of microbodies (peroxisomes) has been described. Lysosomes are usually preserved.

It is noteworthy that these damages are consistent only after 2–3 weeks feeding on the diet. Under the biochemical point of view, mitochondria display higher oxidation, but there is an uncoupling of oxidative phosphorylation (Klein and Johnson, 1954). The fact that even these types of damage occur late disproves the theory that they are important in explaining the onset in fat accumulation, which is already evident after a few days. Moreover, it has been seen (Poli et al., 1981b) that the extent of fat accumulation is higher after a few weeks than after 3–4 months treatment, when mitochondrial damage is maximal. Thus, mitochondrial damage may collaborate in maintaining high-fat levels in late stages, but cannot explain the onset.

Some authors have proposed decreased lipoprotein secretion as the main mechanisms for fat accumulation (Fukazawa et al., 1971). The derangement would result from the changes in phospholipid composition. In fact, phospholipids are essential components of lipoprotein micelles and a change in their composition might justify a decrease in their function. The decrease in lipoprotein secretion, however, has been described only after 3–4 months of feeding on the diet, and does not work well for the first 2 months (Poli et al., 1981b). So, even this mechanism is not relevant to explain the onset in fat accumulation, but only in the maintenance of high-fat levels in late stages.

A net increase in FFA synthesis from acetyl-CoA has been described to take place very early in EFA deficiency, coincidently with the onset in fat accumulation (Allmann et al., 1963; Allmann and Gibson, 1962, 1965; Demeyer et al., 1974). As EFA display an inhibitory action at the genetic level on the synthesis of new fatty acids, it has been proposed that their removal from the diet is responsible for the increase in synthesis found early in EFA deficiency.

This hypothesis is substantiated by the study of the changes in fatty acid composition in animals on this diet. Arachidonic and linoleic acids decrease quickly, especially at membrane level (Burdino et al., 1976; Hayashida and Portman, 1960), whereas oleic and palmitoleic acids do increase (Burdino et al., 1976; Morton and Horner, 1961). Moreover, eicosatrienoic acid, usually not represented in normal rat tissue, does appear in consistent amounts in both its isomeric forms (Burdino et al., 1976). This acid is synthesized from oleic acid by a metabolic sequence strongly resembling that usually employed for the conversion of linoleic to arachidonic acids (Fulco and Mead, 1959). Eicosatrienoic acid rapidly substitutes archidonic acid in membranes (Aes-Jørgensen, 1961). It has been claimed that its production occurs from the adaptive synthesis of an enzymatic system usually repressed by linoleic acid. A similar effect might work even at the level of the new synthesis of long-chained saturated acids.

The change in phospholipid composition in membranes might explain their specific damage in rats fed the EFA-deficient diet, whereas the late decrease in lipoprotein secretion might have a similar explanation as it regards lipoprotein composition, as well as the decrease of HDL with respect to the other lipoproteins (Sano and Privett, 1980). The susceptibility to lipid peroxidation of the liver from rats fed the EFA-deficient diet is not decreased with respect to normal liver. Even the stimulation by CCl_4 is not modified. This means that eicosatrienoic acid is a good substrate for lipid peroxidation (Poli et al., 1981b).

Barbiturate-Induced Fatty Liver

Treatment with phenobarbital or with other barbiturates (5 mg/100 b.w.) in the rat three times of six to eight intervals (Ugazio et al., 1971) provokes fat accumulation in the liver (Chalvardjian, 1970; Salvador et al., 1970); this is more evident in male than in female rats. It has been claimed that the mechanism consists in stimulation of fatty acid synthesis from acetyl-CoA (Tuma et al., 1974).

Other mechanisms, as increased supply of FFA by the blood, block in lipoprotein secretion or decreased oxidation, have been shown not to be involved. Increase in FFA synthesis might be related to the net increase in membrane lipids, as related to the well-known inducing effect of phenobarbital on the smooth endoplasmic reticulum (SER).

Another possibility, however, might be an increase in the ratio of reduced to oxidized forms of pyridine nucleotides following the rapid oxidation of barbiturates in SER chain. In any case, fatty liver is transient and TG accumulation not very high. Increase in lipid peroxidation has been described even in this type of fatty liver (Hahn et al., 1976, 1977), but the causes are not yet understood. A possible reason might be the relative increase in peroxidizable substrates, as related to the synthesis of new membranes and not accompanied by a correspondent increase in natural antioxidants, such as vitamin E.

Another possibility might be even an increase in production of superoxide anion and of hydrogen peroxide due to the lack of balance between phenobarbital dehydrogenation and cytochrome P450 oxidation. In such conditions, electrons might bypass from $FADH_2$ to oxygen, so forming $O_2^{\cdot-}$. Superoxide anion would take up hydrogen to form hydrogen peroxide. $OH^{\cdot}$ free-radical formation would then occur in the presence of iron, either by a Fenton reaction or by a Haber-Weiss reaction. Oxygen free radicals would then start lipid peroxidation.

Ethanol-Induced Fatty Liver

Ethanol administration produces several types of liver damage. The typical damage after acute administration of alcohol is fatty liver, whereas necrosis, as shown both by morphological examination and by measurement of plasma transaminase, is absent. Fatty liver occurs even after chronic treatment, both in rats and in chronic drinkers, but necrosis is always present in this case. Moreover, the presence of Mallory bodies (alcoholic hyaline) is a rule in these conditions, and inflammation, probably started by necrosis, is very frequent. In a certain number of affected humans, as well as in baboons (Lieber, 1981, 1983), cirrhosis or cancer develop. This never happens, however, in the rat.

Acute fatty liver following oral treatment by 6–8 g ethanol/kg b.w. in the rat, is usually centrolobular. Time course of fat accumulation is shown in Figure 2. TG accumulation starts at 5–6 h after poisoning and reaches the top between 6 and 12 h, normal values being seen again after 24 h. Male rats are usually more responsive than females, and fasted rats more than fed ones. Accumulation of TG in hepatocytes has been seen even with isolated cells, as incubated in the presence of ethanol.

Several reasons have been reported by different authors to be responsible for this accumulation. They include increased FFA supply by the blood, possibly due to the toxic stress (Bode et al., 1970; Brodie et al., 1961; Dich et al., 1983; Estler, 1973; Martin and Garner, 1977) The question, in this case, is if increased supply can entirely explain the accumulation, or if other mechanisms are also needed. Hypophysectomy or adrenectomy, which strongly decrease fat mobilization from adipocytes, are able to prevent ethanol-induced fat accumulation in the liver (Mallov and Witt, 1966).

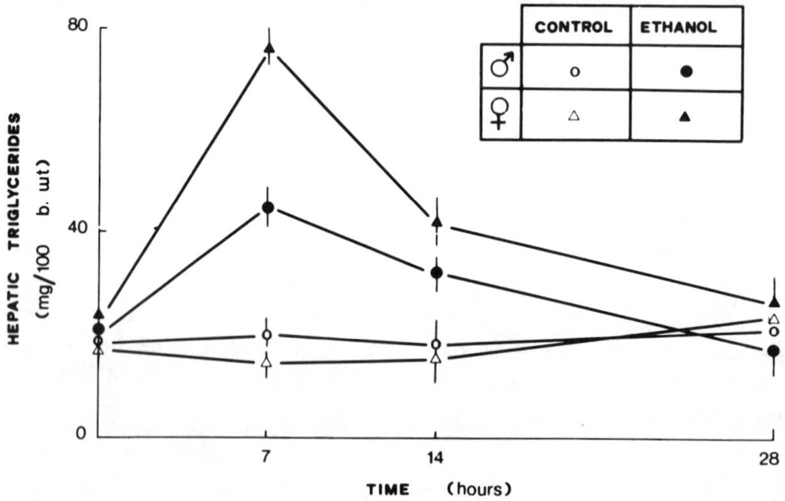

Figure 2. Time course of triglyceride accumulation in the liver of rats after acute ethanol (7 g/kg b.w.t.).

Poggi and Di Luzio, however, were unable to show an increased supply of labeled ^{14}C palmitate by the blood to the liver when ethanol was given to rats previously transplanted with adipose lobules containing ^{14}C-labeled TG (Poggi and Di Luzio, 1964). Moreover, adrenalectomy and hypophysectomy are protective even in the case of CCl_4- or of white phosphorus-induced fatty livers, where primary damage to hepatocytes is not doubtful. Blood supply of FFA would therefore be only permissive for fat accumulation, whose real cause would consist in liver cell derangement.

A lot of intrahepatic causes have been claimed to work in producing fat accumulation after ethanol. Several authors consider the block of fatty acids oxidation by mitochondria as the major cause (Di Luzio, 1966a; Di Luzio and Costales, 1964; Di Luzio and Hartman, 1967). Actually, mitochondria become swollen after ethanol and show a net decrease in their ability to oxidize fatty acids (Bernstein and Penniall, 1978; Blomstrand et al., 1973; Di Luzio, 1966b; Di Luzio and Costales, 1964; Di Luzio and Hartman, 1967; Koch et al., 1977, 1978; Lieber and Schmid, 1961; Lindros, 1970; Rebouças and Isselbacher, 1964; Wahid et al., 1980; Wooles, 1966; Wooles and Weymouth, 1968; Zakim and Green, 1968), whereas ATP concentration in the whole liver decreases (Gordon, 1972, 1977; Hyams and Isselbacher, 1964; Thayer, 1984; Walker and Gordon, 1970).

Mitochondrial damage, however, occurs rather late after acute ethanol, and the possibility that it can entirely explain the onset of TG accumulation receives from this little evidence. It seems, however, possible that the block in oxidation may contribute to explain fat accumulation in late stages of acute poisoning, and also that occurring after chronic treatment (Dianzani, 1985).

Other authors have described a decline in lipoprotein secretion after acute ethanol *in vivo* (Dajani and Kouyoumjian, 1967; Koga and Hirayama, 1968; Madsen, 1969; Mookerjea and Chow, 1969) or in perfusion experiments (Isselbacher, 1965; Schapiro et al., 1964, 1965), whereas others found unchanged values (Beard et al., 1965; Kiessling and Tobe, 1964; Lieber, 1981, 1983). The authors finding a decreased secretion think that the main reason for this is a block in apolipoprotein synthesis. This might be the consequence of the high production of acetaldehyde coming from ethanol oxidation. Aldehydes inhibit protein synthesis (Dianzani, 1979b, 1982), but acetaldehyde is active, under this respect, only at rather high concentration (Dianzani, 1978, 1979a, 1985).

Perin et al. (1974), in any case, reported effects of acetaldehyde *in vitro* on protein synthesis that might be consistent with this hypothesis. Acetaldehyde is the main product of ethanol catabolism, and is surely more toxic on the cells than ethanol itself. Another possibility suggested by the experiments of Gabriel et al. (1977) was that acetaldehyde and possibly other aldehydes derived from lipid peroxidation might act as inhibitors of the secretion itself. In fact, the aldehydes block tubulin *in vitro*. Baraona et al.

(1977) added further experimental evidence to this point, whereas Berman et al. (1983) did not find any morphological equivalent of this damage. Mailliard et al. (1984) found a defect in the assembly of glycoproteins in the Golgi apparatus.

Lieber (1981, 1983) thinks that the decrease in lipoprotein secretion after ethanol is only relative to the excess of TG produced due to the high ex novo synthesis of FFA in liver cell. In other words, the lipoprotein secretion would be either normal or slighty increased, and so unable to compensate for the strong increase in TG production.

Increased TG production has been proven by the experience of Rebouças and Isselbacher (1964) and of several other authors (Gordon, 1972, 1977; Lieber, 1981, 1983; Lieber and Schmid, 1961). The main reason for this increase is the ratio between the reduced and the oxidized forms of pyridine nucleotides (Beard et al., 1965; Lieber, 1983; Lieber et al., 1963; Lieber and Schmid, 1961; Van Waes and Lieber, 1980; Veech et al., 1972) that would be a consequence of the lack of balance between ethanol dehydrogenation and available oxygen. It is noteworthy, however, that even this explanation is not accepted by some workers. For instance, Slater et al. (1964a,b) demonstrated that the increase in the ratio occurred very early after ethanol, but returned to normal also very early, and was practically normal when fat accumulation began. Moreover, no fatty liver is observed after injection of sorbitol, which also produces an early increase in the ratio.

Ethanol is oxidized to acetaldehyde by four different pathways. One of them is cytosolic and accounts for about 80–90% ethanol oxidation in normal conditions. The involved enzyme is the NAD-dependent ethanol dehydrogenase. Three other pathways are related to the SER metabolic chain. Normally they are used at a minor rate, but become more important quantitatively after chronic treatment. In fact, microsomal ethanol oxidizing system (MEOS) is inducible by ethanol itself.

The first microsomal pathway is the usual monooxygenase system, using cytochrome P450, or possibly a molecular variant of this, to reduce oxygen. The existence of a specific system for ethanol has been debated, but at present there is a general agreement about its presence and its relevance, which becomes consistent in chronically treated animals or patients. The discovery of an animal species lacking ethanol dehydrogenase (the deer mouse or Peromyscus maniculatus), but nevertheless susceptible to cell damage by ethanol has brought definite evidence for the importance of the MEOS system in ethanol pathology (Lieber, 1983).

The second microsomal pathway represents only a bypass from the first one. In fact, flavoprotein can release one electron to oxygen, thus forming the superoxide anion $O_2^{\bar{\cdot}}$. This can act as a further hydrogen acceptor, thus producing hydrogen peroxide, H_2O_2. In the presence of catalase, mostly present in peroxisomes, H_2O_2 is decomposed to water and oxygen, which can oxidize ethanol to acetaldehyde. The production of the superoxide anion can occur independently upon the nature of the attacked substrate, the only condition being the presence of excess substrate with reference to the amount of oxygen available. Therefore, $O_2^{\cdot -}$ formed during the oxidation of whatsoever substrate can oxidize ethanol. The demonstration of this possibility has been for a long time the main argument of the opponents to the theory of the existence of MEOS as a direct ethanol-oxidizing enzymatic chain.

In the presence of iron, $O_2^{\cdot -}$ and H_2O_2 can generate OH^-, as well as $OH^\cdot$ free radicals. This can occur either by Fenton or by Haber-Weiss reaction:

1. Fenton reaction:

$$H_2O_2 + Fe^{2+} \longrightarrow OH^\cdot + OH^- + Fe^{3+}$$

2. Haber-Weiss reaction:

$$H_2O_2 + O_2^{\cdot -} + Fe^{3+} \longrightarrow OH^\cdot + OH^- + Fe^{2+}$$

$OH^\cdot$ is extremely reactive and short-lived. In the presence of ethanol, this becomes oxidized to acetaldehyde, with possible intermediate formation of the ethanol free radical (Albano et al., 1986, 1987, 1988; Cederbaum and Dicker, 1983; Cederbaum et al., 1979; Dianzani, 1985, 1987; Ekström et al., 1986).

It is well known that all the three microsomal pathways increase in activity after long-term ethanol administration, or after previous induction with other substrates, as phenobarbital. The pathways involving free-radical production, i.e., the second and third microsomal pathway as described, are counteracted by antioxidants, by free-radical scavengers, as well as by enzymes protecting against free-radical formation. One of these is superoxide dismutase, which is able to form H_2O_2 and oxygen from two molecules of $O_2^{\cdot -}$, according to the following reaction:

$$O_2^{\cdot -} + O_2^{\cdot -} \overset{2H^+}{\rightleftarrows} H_2O_2 + O_2$$

Superoxide dismutase is mainly present in the cytosol, so it can work on $O_2^{\cdot -}$ escaping from microsomal membranes (or being present at their surface). It is, however, difficult to act on the substrate when this is deeply sited within the membrane. Moreover, $O_2^{\cdot -}$ is much more short lived in lipids than in the aqueous phase, so it seems possible that the amounts of this substance escaping from the lipid phase of the membranes toward the cytoplasm are rather restricted.

We have already mentioned catalase among the enzymes that protect against H_2O_2-induced damage. It has been stated, however, that catalase is mostly contained in peroxisomes, and so it is not expected to be protective toward hydrogen peroxide produced in membranes. Another enzyme attacking H_2O_2 is peroxidase. Among the different peroxidases, glutathione peroxidases seem to be especially adapt at protecting the cells, as they can act on H_2O_2 in the presence of G-SH, which is first converted to the thiol free radical G-$S^{\cdot}$ and then to G-S-S-G. A peculiar glutathione peroxidase is able, moreover, to act even on lipoperoxides. So, the intracellular level of GSH can be retained as important in the protection from ethanol-induced damage.

Acetaldehyde is considered by several authors as the most toxic ethanol catabolite. It has been proven to be able to do damage to mitochondria, to microsomes (Lieber, 1983), to protein synthesis, to tubulin (Baraona et al., 1977; Gabriel et al., 1977), as well as to other cell functions. As an aldehyde, acetaldehyde can react both with $-NH_2$ and with $-SH$ groups (Kitson, 1985), forming in the first case Schiff's bases and in the second thiazolidine β-carboxylic acids. So far, acetaldehyde has been found to bind covalently to proteins (Donohue et al., 1983). In these ways, it is considered as a potential poison of a number of cell functions. For this reason, some authors prefer speaking in terms of aldehydism instead of alcoholism (Von Wartburg and Bühler, 1984).

As other aldehydes, acetaldehyde is, however, furtherly metabolized by normal tissues. In the presence of oxygen, aldehyde dehydrogenases are mostly concerned. They exist in the tissues mainly in two forms: a form provided with low K_m and high affinity for the substrate and another one provided with high K_m and low affinity (Dianzani, 1982). These forms are represented differently in the different cell organelles. The internal mitochondrial membrane mostly contains the high-affinity enzyme, whereas the cytosol mostly contains the low-affinity one. So, higher concentrations of aldehyde are needed to be attacked in the cytosol than in mitochondria. Aldehyde dehydrogenases are flavoproteins; the possibility exists, therefore, for free-radical production during substrate oxidation, mostly in form of $O_2^{\cdot -}$. This may become a further mechanism for cell damage.

When added to microsomes or to isolated hepatocytes, acetaldehyde stimulates lipid peroxidation more than ethanol itself (Müller and Sies, 1982; Rossi et al., 1986; Stege, 1982).

Another debated point for ethanol toxicity is in fact represented by the increase in lipid peroxidation seen by several authors after ethanol administration *in vivo,* or also after incubation of homogenates or cells with this substance *in vitro*.

The first observation by Di Luzio et al. (Di Luzio, 1966b; Di Luzio and Costales, 1964; Di Luzio and Hartman, 1967), that ethanol-induced fatty liver is partially prevented by prior administration of antioxidants, led Comporti et al. (1967) to study lipid peroxidation in the homogenates or microsomes of rats treated *in vivo* with ethanol or added with ethanol *in vitro*. The authors used the estimation of accumulated MDA by a thiobarbituric acid test as a measure of lipid peroxidation (Eber et al., 1962; Poli et al., 1981a). Several much more toxic aldehydes, both saturated and unsaturated, are produced. The most toxic group is that of 4-hydroxy-2,3-*trans*-unsaturated aldehydes, among which 4-hydroxy-nonenal (HNE) is the most represented (Benedetti et al., 1980, 1982; Eber et al., 1962; Esterbauer et al., 1982). The effects of this aldehyde on several enzymatic activities and cell function have been extensively investigated (Dianzani, 1979b, 1982; Schauenstein et al., 1977). Moreover, it is now known that the kinetics of lipid peroxidation may be influenced by several conditions and by different stimulators (Eber et al., 1962; Esterbauer et al., 1982). In addition, we are aware that metabolism of the formed aldehydes occurs at rather higher rate, but with differences in the specific rates for single aldehydes. So, the amounts of MDA or of other aldehydes we can find in the incubation media at given times after starting lipid peroxidation *in vitro* do not correspond to the net amount so far produced, but only to the balance between production and differential removal. So, an increase in TBA-reacting material or of other aldehydes can occur even in conditions of decreased lipid peroxidation, when the removal is in some way decreased more than lipid peroxidation itself. In any case, studying the accumulation of aldehydes in incubation experiments can indicate only the susceptibility of the tissue to lipid peroxidation, but in no case is a definite statement of increased lipid peroxidation occurring in the tissue before the preparation of the homogenates. Increased susceptibility to lipid peroxidation can be produced by a decrease in natural antioxidants, as we know that such a decrease shortens the lag phase before the onset of lipid peroxidation by homogenates.

These considerations may lead one to understand why Hashimoto and Recknagel (1968), who estimated lipid peroxidation by measuring the appearance of the diene conjugation band at 233 nm in the microsomal phospholipids, found "no chemical evidence" for increased lipid peroxidation after acute ethanol. Torrielli et al. (1978) were able to confirm both increased MDA production and the absence of increase in the diene conjugation band. These allowed the conclusion that what was really increased after ethanol was not lipid peroxidation, but susceptibility to lipid peroxidation, possibly due to a decrease in antioxidants. Several authors have reported a decrease of GSH after either acute or chronic ethanol (Fernandez and Videla, 1981; Fernandez et al., 1982; Shaw et al., 1983; Videla et al., 1982; Videla and Valenzuela, 1982; Vina et al., 1980). Other authors have reported an increased production of MDA after chronic ethanol (MacDonald, 1973; Reitz, 1975; Shaw et al., 1981, 1983; Suematsu et al., 1981; Takada et al., 1970; Videla et al., 1982; Videla and Valenzuela, 1982; Vina et al., 1980) and also the appearance of the diene conjugation band (Shaw et al., 1983). In acute poisoning, Corongiu et al. (1983) have reported a late appearance of the diene conjugation band in microsomal phospholipids by using second-derivative spectrometry, i.e., a method more sensitive than that used by Hashimoto and Recknagel (1968). The reason for this delayed appearance of the band might be related to a difference in the kinetics of lipid peroxidation as induced by ethanol.

The aldehyde pattern, as shown recently (Dianzani, 1985, 1987), is, however, not very different from that seen in the cases of CCl_4- or ADP/iron-induced lipid peroxidation. Rossi et al. (1986) studied in a parallel way the production of TBA-reacting substances and GSH decline in homogenates. They showed that the two curves have opposite direction, but very similar aspect. Moreover, total glutathione content does not change, as all GSH is converted to G-S-S-G. This rules out the idea that a direct reaction of acetaldehyde with GSH is responsible for the disappearance of the last substance. Rossi et al. (1986) saw also that acetaldehyde itself stimulates MDA production, being more active than ethanol. This may be interpreted as due to free-radical production (probably $O_2^{\cdot -}$) during the oxidation of acetaldehyde, which is accomplished by flavoenzymes, as aldehyde oxidase and xanthin oxidase.

The hypothesis has been, therefore, advanced that increased lipid peroxidation after ethanol and acetaldehyde were related to a radical reaction accompanying oxidation of these substances.

These experiments were extended by Albano et al. (1986, 1987, 1988; Dianzani, 1985), who were able to detect, by using *N*-oxy-pyridine-tetramethylbutylnitrone as a spin trap, an ethanol-derived free radical in the presence of liver microsomes. This free radical was surely hydroxyethyl free radical as shown by experiments done with ethanol containing ^{13}C. The presence of the ^{13}C isotope instead of the usual ^{12}C gives typical deformation of the signal, which is useful for recognizing the presence of the isotope in it. The signal production was much higher in animals previously treated repeatedly with ethanol, i.e., with specifically induced MEOS, or with phenobarbital. It was also strongly stimulated by iron and inhibited by about 85% by desferroxamine, a chelator of iron. Inhibitors of the electron flow along the microsomal chain, as SKF 525 A or *p*-chloro-mercuribenzoate, strongly inhibit the signal formation, whereas carbon monoxide, which blocks the chain at the level of cytochrome P_{450}, slightly intensifies it.

As in the experiments of Rossi et al. (1986), regarding MDA production, the addition to the system of catalase inhibitor strongly intensified it. This means that the formation of hydrogen peroxide is a step in the formation of the signal. Moreover, the scavengers of the $OH^{\cdot}$ free radical, as mannitol or benzoate, strongly decreased the formation of the signal. The interpretation of these results is that the hydroxyethyl free radical is formed by direct action of $OH^{\cdot}$ on ethanol in a nonenzymic way. $OH^{\cdot}$ is formed in the system by a Fenton's or a Haber-Weiss reaction in the presence of iron, as described above. The possibility that a certain amount of the hydroxyethyl free radical is formed at the level of ethanol dehydrogenation by the NADP-dependent enzyme remains, however, open. Hydroxyethyl free radical is probably less short-lived than $OH^{\cdot}$, but direct data on the half-life are still missing. If its half-life is really longer than that of $OH^{\cdot}$, free-radical formation may be even interpreted as a protective reaction, and ethanol as an $OH^{\cdot}$ scavenger.

These results represent the final demonstration that ethanol stimulates lipid peroxidation, its action being, however, mediated through the production of oxygen free radicals, as $OH^{\cdot}$ and $O_2^{\cdot}$, and of hydroxyethyl free radical. These are the most probable promoters of lipid peroxidation.

The problem remains, however, as far as it regards the meaning of increased lipid peroxidation in the pathogenesis of cell damage. The opinion of the author is that lipid peroxidation cannot be retained as responsible for fat accumulation, at least in acute fatty liver. This opinion derived from the demonstration that even in the case of CCl_4-induced fatty liver, fat accumulation is not dependent on lipid peroxidation (Dianzani, 1987; Dianzani and Poli, 1985; Poli et al., 1981a, 1985b, 1987) as the block of lipoprotein secretion, which is the most important and early mechanism for this accumulation depend on covalent binding of trichloromethyl free radicals to lipids and proteins, and not on lipid peroxidation. Moreover, ADP-iron produces a more intense stimulation of lipid peroxidation than ethanol without producing fat accumulation to a considerable extent. It has only to be remembered that after acute ethanol most of this substance is oxidized by the cytosolic ethanol dehydrogenase, i.e., by a pathway not involving the production of free radicals. The constant, but never complete, protection afforded by both liposoluble (Di Luzio, 1966b; Di Luzio and Costales, 1964; Di Luzio and Hartman, 1967; Torrielli et al., 1973, 1978) and hydrosoluble antioxidants (Torrielli et al., 1973, 1978), as GSH and other SH-containing substances, might be explained by secondary events, as the prevention of the formation of aldehydes affecting lipoprotein secretion or protein synthesis in late stages of poisoning. Under this point of view HNE is much more powerful than acetaldehyde, so it is expected to play some role over time.

Chronic ethanol treatment has been especially studied in rats and in baboons (Lieber, 1981, 1983). The attempt was to try to reproduce the aspect usually found in heavy-drinker humans. The baboon is the only known laboratory animal able to develop cirrhosis after chronic ethanol, whereas only fatty liver, necrosis, hepatitis, and production of hyaline Mallory bodies occur in rats. Fatty liver results in this case from the sum of several different causes. First, those causes responsible for acute fatty liver can play a role even in the chronic case. Loss of oxidation, block of protein synthesis, and decreased lipoprotein secretion are also effective in this case. In fact, mitochondrial damage is more pronounced than after acute treatment, and the same occurs in the case of protein synthesis.

An additional cause for membrane damage is, however, represented by an increased need in choline, which was demonstrated several years ago by the group of Takeuchi in Japan (1968, 1969). Deficiency in choline might provoke the substitution of phosphatidylcholine with phosphatidylserine or phosphatidylethanolamine both in membranes and in lipoprotein phospholipids. This would more easily produce membrane damage, as well as a decreased secretion of lipoproteins. In fact, the substitutes are less polar than phosphatidylcholine and less effective in connecting the hydrophobic with the hydrophilic portions of the macromolecules.

In CCl_4-induced microsomal damage (Benedetti et al., 1982; Comporti, 1985) it has been observed that the phosphatidylethanolamine and phosphatidylserine portions of membranes are damaged by lipid peroxidation earlier than the phosphatidylcholine portions. Thus, it seems possible that the introduction in phospholipids of increased amounts of such components may facilitate the onset of lipid peroxidation even in the case of ethanol poisoning.

Furthermore, additional reasons for chronic liver damage after ethanol may be attributed to stomach and intestine damage, which can modify the absorption of nutrients. Moreover, the fact that a portion of the caloric intake is covered by ethanol may decrease the intake of other necessary nutrients. Thus, protein, fat, selenium, and vitamin deficiencies may develop. Increased iron absorption (Mazzanti et al., 1987; Nordmann et al., 1987) may also contribute. In any case, a change in the ionic pumps has been hypothesized as a possible further mechanism of damage. Acetaldehyde (Gonzales-Calvin et al., 1983) and HNE (Paradisi et al., 1982, 1985) are able to inhibit plasma-membrane ATPase, as well as adenylate cyclase (Dianzani, 1982; Dianzani and Poli, 1982, Dianzani et al., 1990). This enzyme, on the contrary, is stimulated by HNE at concentrations lower than 10^{-6} *M* after short-term (5-min) incubation. Stimulation of adenylate cyclase has been reported even after ethanol *in vivo* (Whetton et al., 1983).

The increase in lipid peroxidation after chronic ethanol has been well substantiated by several authors. Some (Reitz, 1975) detected increased production of MDA by the homogenates, whereas others preferred to measure pentane or ethane expiration by treated animals (Müller and Sies, 1982) or patients (Moscarella et al., 1984). Continuous flow of toxic aldehydes and continuous destruction of membranes may be the causes for cell death, which are frequent after chronic treatment, being different from the acute one. In the case of CCl_4, it has been established that acute cell death can be prevented by prior administration of vitamin E or of promethazine, which block lipid peroxidation but not covalent binding (Meldolesi, 1968; Poli et al., 1985b, 1987). We have for the moment no proof that this happens even in the case of ethanol treatment. In fact, the experiment is complicated here by the fact that cell death does not occur after acute ethanol, whereas the chronic model introduced a number of different variables.

It has been shown that the 4-hydroxy-2,3-*trans*-en-al series, and especially HNE and 4-hydroxy-octenal (HOE), are chemoattractants for polynucleated neutrophyls (Curzio and Dianzani, 1983; Curzio et al., 1982, 1986, 1987). Since these aldehydes, as produced during lipid peroxidation, can escape from still-living cells, they can play a role for white cell recruitment before cell death. So, phagocytes may bring to the field an additional source of free radicals, which may be important for hepatitis production. Preliminary experiments have shown that in exsudates high amounts of HNE and other aldehydes are contained. Moreover, the plasma of alcoholic patients contains an increased amount of MDA and HNE bound to the proteins. The bond between HNE and serum albumin, as studied by Curzio et al. (1986, 1987), is reversible. This bond may represent a mechanism for release of the aldehydes in sites remote from those where they were produced.

Chronic ethanol feeding produce the appearance of the alcoholic hyaline (Mallory bodies) in liver cells. The substance results from the assembly of intermediate filaments (Denk and Eckersdorfer, 1977; Denk et al., 1976; Franke et al., 1979; Okamura et al., 1975), but the cause for this is unknown. The possibility that aldehydes may play a role cannot be discarded and represents a useful tool for future experiments.

Fatty Liver Due to Isopropanol

A single dose of isopropanol (6 g/kg b.w. by stomach tube) provokes TG accumulation in rat liver, starting as early as 4 h after treatment (Nordmann et al., 1973). Increased level of nonesterified fatty acids from the blood has been proven (Nordmann et al., 1973). The major reason for an increase in TG is, however, an increase in the synthesis of new FFA from acetyl-CoA, occurring probably for the same reason previously discussed in the case of ethanol. Like ethanol, isopropanol is an inducer of the microsomal oxidizing chain, and this might explain an increased need for structural lipids (Nordmann et al., 1973). Moreover, isopropanol impairs lipoprotein secretion, whereas fatty acid oxidation in mitochondria is not modified. Albano et al. (1986, 1987, 1988) have found that isopropanol, as well as several other alcohols, are metabolized by liver microsomes with production of a free radical. This has been trapped in the same way as the hydroxyethyl free radical. Therefore, it seems probable that a radical mechanism works even in the case of isopropanol.

FATTY LIVER DUE TO DECREASED FATTY ACID OXIDATION

Historically, this cause for fatty liver was firstly proposed to explain TG accumulation following treatments with CCl_4, white phosphorus, ethionine and ethanol treatments, as well as choline-deficiency, where heavy damage to mitochondria had been described (Dianzani, 1954, 1957). In the most cases, however, TG accumulation started earlier than any demonstrable loss of fatty acid oxidation (Dianzani and Marinari, 1961). So, it was concluded that mitochondrial damage can only cooperate with other causes in maintaining a high TG level, but cannot be retained as responsible for its onset (Dianzani, 1978, 1979a). The problem remains, however, regarding the possible existence of other forms of fatty livers where the onset itself can be explained by decreased oxidation of fatty acids.

In principle, decreased oxidation may be related to one or more of the following mechanisms: (1) decreased oxygen supply, (2) decreased substrate availability, (3) decreased rate of oxidation due to enzymatic defects.

The first mechanism is fully realized in hypoxia. Among the different forms of hypoxia are especially the hypoxic, anemic and the stagnant ones. Hypoxic hypoxia, which is related to a decrease of oxygen supply, might be responsible for TG accumulation in liver cells sometimes found in men who died from a restriction of air. Even humans with heavy forms of tuberculosis may sometimes present fatty liver presumably related with such a cause. Anemic hypoxia has been retained as responsible for fatty livers of humans with severe anemic syndromes, and especially with pernicious anemia. A form of stagnant hypoxia frequently accompanied by fatty liver is present in chronic liver stasis. Morbid anatomists usually refer to this type of liver damage as *nutmeg liver*. In fact, the region surrounding the centrolobular vein is atrophic due to strong vein dilatation, whereas the periphery of the lobule is filled with fat. Some fat may appear even in liver cells of animals receiving small nonlethal amounts of cyanide or those breathing carbon monoxide.

Decreased substrate supply to mitochondria may occur when the activation reaction of fatty acids, i.e. the Lippmann reaction (R–COOH + ATP + CoA–SH → R CO–S–CoA + AMP + PP), is in some way impaired or when the reaction of acyl–CoA–SH with carnitine is not working. We now know that in the dog pantothenic acid deficiency is characterized by fatty liver (Fidanza et al., 1967). Pantothenic acid is used for the synthesis of CoA, which is needed for fatty acid activation needed for acyl-CoA oxidation in mitochondria. Other animal species, however, do not present fatty livers after pantothenic acid deficiency.

Fatty acid activation requires energy in the form of ATP. Lack of ATP may perhaps explain the transient increase in TG seen in the liver after giving dinitrophenol to rats (Dianzani and Scuro, 1956), and may collaborate to maintain high-TG values in the late stages of fatty livers provoked by treatments that damage mitochondria, such as those with CCl_4, white phosphorus, ethionine, as well as choline deficiency.

Decreased rate of oxidation may be the cause also of fatty livers seen in vitamin PP or in vitamin B_{12} deficiencies (Di Luzio and Costales, 1964; Tandler et al., 1968). TG accumulation in these cases, however, is never heavy. Special attention should be given to fatty liver seen in rats after repeated intraperitoneal injection of 4-pentenoic acid (80 mg/kg b.w., 6–8 times during a fasting period of 48 h). It has been shown that this substance specifically blocks acyl-CoA.

Histologically, the liver cells are filled up with very small vesicles containing fat. This picture is very similar to that described in the Reye syndrome, a disease of childhood characterized by microvesicular fatty infiltration of the liver, as well as of other organs including the brain. The reason for this syndrome is thought to be mitochondrial damage, for reasons not yet understood (Thayer, 1984).

The hypothesis has also been advanced that fatty liver following arsenite poisoning is related to a block in fatty acid oxidation. Intraperitoneal injection of 25 mg/kg b.w. As_2O_3 provokes within half an hour a strong decrease in oxidations in the liver (Webb, 1966). According to this author, the damage would follow the interaction of As^{3+} with the –SH groups of lipoic acid, which is involved in fatty acid oxidation. Dihydrolipoate dehydrogenase has been found to be inhibited. In principle, As^{3+} might even act on other –SH groups, as those required for protein synthesis. Thus, fat accumulation might be dependent on a block in lipoprotein synthesis, at least in the late stages of poisoning.

FATTY LIVER DUE TO DECREASED TRIGLYCERIDE HYDROLYSIS OR RELEASE

It has been already reported that TG are removed from the liver mainly in two ways: hydrolysis inside lysosomes and "secretion" in the form of lipoproteins. Both a decreased hydrolysis and a block in secretion may, therefore, provoke fat accumulation. Decreased hydrolysis occurs only in the rare Wolman's disease, related to deficiency of acid lipase. Block in secretion, however, is by far the most frequent cause for fatty liver.

Fatty Livers Due to Block in the Synthesis of the Phospholipid Moiety of the Lipoprotein Molecule

Phospholipids in lipoproteins are mostly represented by lecithins and cephalins, as well as by inositolphosphate derivatives. Lecithins are esters of glycerol, where two of the hydroxyls are esterified by fatty acids, whereas the third one is bound to phosphorylcholine. In cephalins, phosphorylcholine is substituted by either phosphorylserine or by phosphorylethanolamine. Lecithins, due to the presence of choline, are much more polar than cephalins. Choline deficiency is one of the most studied forms of experimental fatty livers. It can obtained by two main methods: long-term feeding on a diet devoid of choline (Best and Hershey, 1932; Best and Channon, 1935; Best et al., 1936, 1949), or by force-feeding of young Sprague-Dawley rats on a diet containing a high level of fat and at the same time deficient in choline (Lombardi, 1966; Lombardi et al., 1964, 1966). These rats usually double their weight in a week, when they are about 150 g, and have, therefore, higher needs for choline than most animals; moreover, the high-fat intake increases the need for choline.

By the first method, fat accumulation starts 3–4 days after the beginning of the experiment. By the short-term treatment, the onset of fat accumulation occurs about 15 h after force-feeding by stomach tube. The shorter time required to see TG accumulation is one of the reasons for the preference most recent authors had for the second method.

Choline-deficiency fatty liver is centrolobular. In long-term treatments, severe mitochondrial damage is seen 3–4 days after the start. For this reason, the hypothesis was advanced that the main mechanism for fat accumulation was damage to oxidation in mitochondria (Chen et al., 1972; Dianzani, 1954, 1957). Lombardi et al. (1964, 1966; Lombardi, 1966; Lombardi and Recknagel, 1962), however, showed by using the short-term treatment that mitochondrial damage is still missing at 15 h, when TG accumulation is already evident. As lecithins are substituted for very early on by cephalins in phospholipids from both membranes and lipoproteins, the hypothesis has been advanced that the real cause for fatty liver is the lower polarity of the new phospholipids, with consequential partial loss of the function of phospholipids in connecting hydrophobic (TG) and hydrophilic (mainly proteins) portions of the lipoprotein micelles. Total protein synthesis, as well as lipoprotein secretion, are not decreased (Gravela et al., 1971, 1974; Lombardi et al., 1964, 1966), but secreted lipoproteins contain less TG.

Lipid peroxidation was found to be increased in long-term choline deficiency by Ugazio et al. (1967; Ugazio, 1971), who were able to detect an increase in lipid firmly bound to proteins (lipofuscins). Recently, Perera et al. (1985), as well as Shinozuka et al. (1987), have been able to find increased MDA production by homogenates of rat liver after long-term treatment. An increase in the diene conjugation band is also present in liver fat (Perera et al., 1985; Shinozuka et al., 1987), as well as in microsomal phospholipids (Hong Tan et al., 1987). Lombardi (1987) refers, however, the diene conjugation band not to choline deficiency, but to the nature of the diet used, which would contain peroxidized lipids with high diene conjugation band by itself.

Regarding the possible reason for increased lipid peroxidation, Ugazio et al. (1967; 1971) put forward the hypothesis that it might be related to the increase in the peroxidizable substrate (i.e., unsaturated fatty acids) without a corresponding increase in the diet of liposoluble antioxidants, such as vitamin E. This would produce a derangement in the lipoperoxidative balance. According to Benedetti et al. (1977, 1980), in CCl_4 poisoning lipid peroxidation would start from the phosphatidylcholine and phosphatidylserine portions of membrane phospholipids. So, the possibility arises that the substitution of phosphatidylcholine with these substances might represent another reason for increased susceptibility to lipid peroxidation in choline deficiency. The extent and quality of damage due to increased lipid peroxidation in this condition is, however, difficult to establish. Glucose-6-phosphatase, the marker enzyme of the SER, which is quickly inhibited after CCl_4, and whose loss is surely dependent upon lipid peroxidation, is not modified in choline deficiency (Koga and Kirayama, 1968). This seems to show that in any case the sites affected firstly by lipid peroxidation are different in choline deficiency and in CCl_4 treatment.

Possibly related to deficient synthesis of methyl groups or to their transport is fatty liver in rats fed a diet deficient in vitamin B_{12} (Bennett et al., 1981). One cannot exclude, however, that this deficiency is the result of blocking protein or nucleic acid synthesis. Deficiency in methyl groups probably takes place even in fatty livers due to treatment with excess nicotinamide, a substance that is eliminated by urine in the form of *N*-methyl derivatives. Their synthesis is accomplished in the liver, which becomes depleted in free methyl groups. The endogenous synthesis of choline becomes lower, and cephalins substitute for lecithins in both membranes and lipoproteins.

Fatty Liver Due to Impairment in Protein Synthesis

Several types of fatty livers, both dietetic and toxic, belong to this section. In fact, apolipoprotein defects can be explained either by decreased synthesis or by structural or conformational changes. Decreased synthesis can be the consequence of aminoacids restriction in the diet or of damage in the proteosynthetic mechanism.

Fatty Liver Due to Protein-Deficient Diets. The importance of the protein content and quality in the diet, in order to maintain a normal level in liver TG, was first shown by Best et al. (1936, 1949; Best and Hershey, 1932; Best and Channon, 1935), who showed that in choline deficiency the concomitance of a decreased protein supply did increase fat accumulation. It quickly became clear that the total amount of protein in the diet is important in order to prevent fat accumulation.

In rats, diets containing less than 12–15% casein are steatogenic, the maximal effect being reached with 3% casein. In order to have a pure protein-deficiency diet, one has to add to the protein-free or protein-low diet a sufficient amount of substances containing labile methyl groups, as well as –SH groups. With this type of diet, fat accumulation starts in the peripheral regions of the lobule, whereas in choline or methyl groups deficiency it is centrolobular (Best et al., 1949).

The morphological changes include mitochondrial swelling, degranulation of the RER, as well as dilatation of the cisternae of the SER (McLean and McLean, 1966, 1969). Glycogen content usually increases, probably due to the early decrease in the activity of glucose-6-phosphatase (Garlick et al., 1975). mRNA for albumin synthesis is also quickly decreased (Pani et al., 1975, 1976). Protein catabolism in lysosomes increases strongly.

It is difficult to find a unique cause for fatty liver. Mitochondrial damage may be retained as responsible for decreased fatty acid oxidation (Tulpule, 1955), but probably, as in other forms of fatty livers, this damage works only in late stages. The main cause for fat accumulation is probably represented by a blockage in protein synthesis, provoking a defect in lipoprotein assembly and therefore a decrease in lipoprotein secretion.

A clinical expression of this type of fatty liver is represented by the so-called kwashiorkhor, which exist in two different forms, usually referred to as ''simple'' and ''marasmatic'' (Golden et al., 1985). In the first form, which is also less severe, the most important liver lesion is fat infiltration, related to lack of amino acids in the diet.

Administration of milk proteins is usually followed by full recovery. In the marasmatic form several other parenchymal changes take place until death. According to Golden et al. (1985), marasmatic kwashiorkhor would be complicated by several other defects in nutrition, especially deficiency in vitamin E and of other antioxidants, both organic and inorganic, such as selenium and zinc.

Moreover, very often there is an increase in iron absorption. This would produce an increase in lipid peroxidation, provoking the bulk of cell damage. Lack in dietary amino acids, as well as in antioxidants, would prevent the formation of protecting enzymes, as superoxide dismutase, and glutathione peroxidases.

Fatty Liver Due to Diets Unbalanced in Amino Acids. Fatty livers are often encountered after subjects are fed diets containing the required amount of proteins, depending upon the quality of such proteins. Substitution of casein with vegetable proteins or with blood fibrin or albumin provokes TG accumulation in the liver. Some of such fatty livers may be explained by the peculiar composition in amino acids of the given proteins, which would contain too high amounts of other amino acids. A diet deficient in threonine or of tryptophane has been really proven to produce fatty liver (Sidransky and Verney, 1964).

Force-feeding with threonine-deficient diet produce fatty liver in rats receiving a normal amount of other amino acids (Sidransky et al., 1965, 1974, 1977). Protein synthesis in these conditions is not decreased when totally considered. Only the synthesis of apolipoproteins, which contain abundant threonine, is decreased (Nimni and Bavetta, 1961; Sidransky and Verney, 1964; Sidransky et al., 1965, 1974, 1977; Waterlow, 1961).

Lack of single amino acids, however, does not explain all the forms of fatty liver of this type. Excess of single amino acids in diets containing all others in the required amount can produce fatty liver, indeed. This is the case, for instance, with an excess methionine, or excess tryptophane (Harbitz et al., 1984). In rats given a diet containing 6% fibrin, the administration of leucine, isoleucine, valine, hystidine, methionine, and phenylalanine severely increases fatty infiltration (Benton et al., 1955).

Generally, the best prevention of this type of fatty liver is when the rats are given complete amino acid mixtures. The reasons why an excess of amino acids in the diet can produce fatty liver probably depend on the presence on the cell surface of receptors for different classes of amino acids, and not for single ones (Christensen, 1964; Christensen and Lang, 1965). When an amino acid belonging to a given class is in excess, it becomes preferentially fixed to the receptor, and therefore internalized more frequently, at the expense of the other members of the same class.

Thus, an intracellular defect in these amino acids will take place.

Fatty Liver Due to Damage of the Protein Synthetic System

The system for protein synthesis may become damaged at different levels: at that of DNA transcription, or at that of the synthesis of mRNA, or at that of the function of ribosomes. Moreover, this can be affected at several sites, as aminoacyl-tRNA fixation, translation, and initiation.

Fatty Liver Due to Damage of the Transcriptive Function of DNA. Several inhibitors of protein synthesis work by preventing mRNA production due to a change in the template function of DNA. Some of these inhibitors are steatogenic, others are not. For instance, actinomycin D, one of the most powerful substances of this group, is very poorly steatogenic (Sarma et al., 1969; Reid et al., 1970, 1971).

The reason for this is that the action of actinomycin D is not restricted to liver cells, but is extended to most cells of the organism. This substance is also able to block protein synthesis in adipocytes, thus preventing the formation of active adenylate cyclase. There is decreased release of FFA from adipocytes into the blood. These are known to display a permissive action of fat accumulation, so that this does not occur, notwithstanding the high level of protein inhibition (Dianzani and Gravela, 1975).

On the contrary, aflatoxins are powerful steatogenic substances. They are inactive outside the liver, as they become activated to hydroxy derivatives by the hydroxylating systems of SER only in liver cells (Purchase and Steyn, 1969; Saunders et al., 1972; Schabort and Steyn, 1969). This steatosis is peripheral and cell death is frequent. Over time, hepatomas are frequently seen (Carnagan, 1967; Lancaster et al., 1961; Vesselinovich et al., 1972; Wogan, 1966, 1968). All the poison injected in the rat is found in the liver within 1 h (Butler and Clifford, 1965). The first observed change is the block in pyrimidine incorporation into RNA, especially at the nuclear level (Sporn et al., 1966).

Block of protein synthesis (Dajani and Kouyoumijan, 1967), as well as dissociation in polyribosomal profiles (Roy, 1968; Sidransky et al., 1977), follow in short term. TG accumulation occurs at about 6 h after treatment. As the block in mRNA synthesis takes place within an hour, this means that 6 h is the time needed for exhaustion of the function of preexisting apolipoproteins.

Both *in vitro* and *in vivo,* aflatoxin B_1, the most powerful compound of the group, binds covalently to DNA. The OH– group in activated aflatoxin and the N7 position of guanine in DNA are involved in the bond.

According to several authors, further hydroxylation of aflatoxin would be needed to form the bond, which would involve a 2,3-epoxy group. In fact, 2,3-dihydro-3-hydroxy-*N*-guanyl-aflatoxin B_1 has been found in the urine of treated rats (Autrup et al., 1979; Croy and Wogan, 1981; Daoud and Irving, 1977; Kwiram et al., 1978; Lutz et al., 1973, 1980), as well as in the reaction mixture.

According to other authors, another epoxy derivative would be 8,9-epoxy-aflatoxin B_1. This compound has also been isolated from the reaction mixture, in the presence of liver microsomes. The block in mRNA synthesis would be a consequence of the deformation of the DNA template due to the covalent bond of the substances with DNA itself.

An additional minor mechanism for blockage in mRNA synthesis could be, however, the direct inhibition of the specific RNA polymerase (Barker et al., 1963; Moulé and Frayssinet, 1972; Yu, 1977). Also, microsomal RNA would be affected. In later stages, mitochondrial damage also seems to collaborate in maintaining high-fat levels. It is noteworthy that even DNA synthesis in such particles is affected rather early (Friedman and Wogan, 1970), as well as the intramitochondrial protein synthesis.

Nirajan and Avadhani (1980) have supposed a direct activation of aflatoxin inside mitochondria. Further damage to mitochondria consists in swelling and loss in oxidation.

Glutathione has a protective action towards aflatoxins. In fact, it binds to aflatoxin, thus forming the adduct 2,3-dihydro-2(*S*-glutathionyl)-3-hydroxy-aflatoxin B_1, which does not bind to DNA and is eliminated by the kidney as such (Ch'ih et al., 1983; Degen and Neuman, 1978; O'Brien et al., 1983).

Possibly, fatty liver observed after treatment with cancerogenic azodyes also belongs to this group. It is well known that azodyes are activated by SER monoxygenase system and that they react with DNA either through their free radical or through an epoxy group resulting from metabolism in the SER.

Fatty Liver Due to Alkylating Agents. Several alkylating agents are steatogenic. They are able to alkylate both nucleic acids and protein, so their mechanism of action is often composite. As in the case of aflatoxins, one of the targets is DNA, so fatty liver may result from inhibition of mRNA synthesis due to a change in template DNA. Other targets are, however, mRNA, tRNA, and rRNA, whose functions may also become impaired. As in the case of aflatoxin, the change in DNA may also produce mutations, so these substances are not only steatogenic, but carcinogenic.

One of the most powerful alkylating agents is ethionine, an amino acid. It binds to liver ATP shortly after injection with production of the relatively stable adenosyl-*S*-ethionine (Farber and Corban, 1958; Farber and Popper, 1950). ATP becomes insufficient for both amino acid and fatty acid activation. According to Farber's group, this is the most important reason to explain fatty liver and succeeds in increasing fat 5–6 h after injection, especially in female rats. Additional causes for fat accumulation are, however, ethylation of DNA, mRNA, rRNA, and tRNA, as effected by adenosyl-*S*-ethionine.

Dimethylnitrosamine and diethylnitrosamine are also powerful alkylating, carcinogenic, and steatogenic substances. Fatty liver is evident in this case 5–6 h after treatment with 30–50 mg/kg b.w.t. Both substances are activated by the enzymes of the SER (Brouwers and Emmelot, 1960; Kawanishi et al., 1985; Magee, 1971). Induction of such enzyme by previous treatments strongly increases toxicity (Garner and McLean, 1969; Seawright and McLean, 1967), whereas previous feeding a protein-deficient diet results both in decrease in cytochrome P450 and in toxicity (McLean and McLean, 1969; Menzel et al., 1982).

Even CCl_4-induced destruction of the enzymatic chain strongly decreases toxicity (Pound and Lawson, 1975a,b,c). The same occurs after giving silver nitrate (Pani et al., 1976) or dimethylnitrosamine itself (Pound, 1975; Pound and Lawson, 1975a,c). The active product is thought to be diazomethane, which would be formed from dimethylnitrosamine by oxidative demethylation (Czigan et al., 1973; Miller and Miller, 1965). In DNA, purines are especially affected. 3-Methyl-guanine, 7-methyl-guanine, 9-methyl-guanine, 3-methyl-adenine, and 1-methyl-adenine have been isolated from DNA-treated animals. DNA alkylation produces both mutation and block of transcription (Daoud and Irving, 1977; Herron and Shank, 1980; Pegg, 1978; Pegg and Hui, 1978; Umbenhauer and Pegg, 1981; Ying et al., 1980). The synthesis of mRNA is clearly inhibited, but mRNA is also itself alkylated. Moreover, dissociation of polyribosomal profiles has been seen before a demonstrable block in the synthesis of mRNA, suggesting an additional alkylation of rRNA or tRNA (Gravela et al., 1974; Sidransky et al., 1974, 1977; Villa-Trevino, 1967; Wooles, 1966, 1968). Increase in MDA production has been shown after dimethylnitrosamine by Slater (1972). Moreover, increased production of $O_2^{\cdot-}$ has been reported by Sealy et al. (1978). Partial protection by the antioxidant diethyldithiocarbamate has also been described (Ying et al., 1980).

The cause for onset in lipid peroxidation is not known, but it seems probable that free radicals are produced in the SER during the metabolism reactions in which dimethylnitrosamine is involved. The relevance of increased lipid peroxidation with respect to fat accumulation has not been established as yet.

Fatty Liver Due to Block in mRNA Synthesis Not Related to Changes in Template DNA. mRNA synthesis can be blocked by direct action on mRNA polymerase, as well as due to a decrease in the available substrates. The first mechanism works in animal treated by amanitins, one of the types of poisons of the mushroom *Amanita phalloides*.

In fact, amanitins, and especially α-amanitin, are specific inhibitors of RNA polymerase B (Buku et al., 1971; Lindell et al., 1970; Montanaro et al., 1971, 1973a,b; Montecuccoli et al., 1973). Shortly after the administration of amanitins, disaggregation of the nucleolar matter can be shown morphologically (Fiume and Laschi, 1965; Marinozzi and Fiume, 1971; Petrov and Sekeris, 1971). Inhibition of RNA polymerase B takes place within 5 min, whereas total RNA is strongly decreased after 1 h.

Fat accumulation starts several hours after treatment, being more evident at 24 h. It has been found that the penetration of amanitin inside hepatocytes is mediated by the same transport system allowing the penetration in the cell of biliary salts. This system works through receptor polypeptides sited on the sinusoidal portion of the hepatocyte membrane. Two of the involved polypeptides, molecular weighing, respectively, 48,000 and 54,000, have been characterized (Kroncke et al., 1986).

It is questionable why in most types of fatty livers related to the block in protein synthesis the onset in fat accumulation occurs between 5 and 6 h after treatment, while in the case of amanitins it can be seen only after 24 h. Since the block in mRNA synthesis occurs within 5 min and that of protein synthesis is present after 10 min, it is clear that the lag time before fat accumulation is not used for poison activation. Possible speculation is that some additional events, possibly related to the block in the synthesis of mRNA, may expand the half-life of preexisting lipoproteins. This is, however, only matter for future work.

An example of block of mRNA synthesis related to lack of substrates is that occurring after treatment with D-galactosamine. This substance has been extensively used by Keppler et al. (1968; Keppler, 1976; Keppler and Decker, 1969), at the dose of 200 mg/kg b.w.t. in the rat. Morphologically, dilatation of the cisternae of the SER, degranulation of the RER, decline in glycogen granules, and appearance of a number of autophagic vacuoles take place within a few minutes. At 2 h after injection, severe damage affects nucleoli, which undergo a fibrillar disintegration (Scharnbeck et al., 1972). At 3 h palmitoyl-CoA oxidation in mitochondria strongly decreases (Mangheney-Andreani et al., 1982). 5′-Nucleotidase, a marker of plasma membranes, decreases strongly at 2–3 h (El-Mofty et al., 1975).

Biochemically, the main damage is the profound depletion in UTP, which becomes trapped in the formation of uridindiphosphogalactosamine (Keppler, 1976; Keppler and Decker, 1969; Keppler et al., 1968; Konishi et al., 1974; Meszaaros et al., 1974; Rasenack et al., 1980). Intense galactosamination of the membrane structures is thought to be responsible for loss in the ionic pumps. The impairment in the calcium pump, with consequent increase in the intracellular calcium, is considered to be responsible for cell death. Treatments of rats by uridine is protective toward all these types of damage (Shinozuka et al., 1973).

Feeding on a diet rich in the pyrimidine orotic acid (1–1.5%) produces fatty liver in the rat (Sidransky et al., 1965; Standerfer and Handler, 1955). The same result is seen even after endovenous continuous infusion (Feigelson et al., 1961). Female rats are more susceptible than the males. Fat accumulation is typically periportal (Handschumacher et al., 1960; Sidransky et al., 1965) and starts 2–3 days after feeding the diet (Novikoff et al., 1966).

A block in lipoprotein secretion is the main cause for TG accumulation (Windmüller, 1964). In fact, a decrease in VLDL apolipoproteins has been shown. It is noteworthy that HDL apolipoprotein and total protein synthesis are normal or slightly increased (Novikoff et al., 1966; Waterlow, 1961). According to Novikoff et al. (1966), the block in synthesis would effect specifically VLDL apolipoproteins, possibly due to a diversion of protein synthesis toward synthesis of specific proteins. This would be a consequence of excess pyrimidine in the diet. Administration of excess uracyl or excess thymine, however, does not produce similar results (Standerfer and Handler, 1955).

Administration of adenine is able to prevent orotic acid-induced fatty liver. This might mean that orotic acid increases the request for adenine, either due to involvement of adenine in side reactions or due to need for balance among the bases. Adenine is able to even prevent ethionine-induced fatty liver, but in this case the most probable explanation is that this purine allows the restoration of normal ATP pool. In fact, as it was described above, ethionine sequesters ATP in the formation of adenosyl-*S* ethionine.

Even in orotic acid-induced fatty liver an increase in MDA production has been reported (Kinsella, 1967a; Torrielli et al., 1971). According to Kinsella (1967a) this would be the consequence of oxido-reduction reactions involving orotic acid, which is converted to dihydro-orotic acid by a dehydrogenase in the presence of NADH + H^+. No diene conjugation band is, however, produced in microsomal phospholipids (Torrielli et al., 1971), and no relationship seems to exist between increased MDA production and fat accumulation. The fact that SER has been reported to be damaged (Novikoff et al., 1966) might introduce the possibility that metabolism of orotic acid in the SER monooxygenase system could involve free-radical production. This hypothesis, however, needs to be tested.

Fatty Liver Due to Damage in Amino-Acid Activation. Amino-acid activation requires ATP to form aminoacyl-AMP. The last substance will then react with the specific tRNA to form aminoacyl-tRNA. A lack in ATP or a decrease in the activity of the enzymes involved may decrease the formation of aminoacyl-tRNAs, with the consequent decline in protein synthesis.

The most known example of fatty liver due to ATP sequestration is that of ethionine, as reported above. It has been said, however, that this is not the unique mechanism for ethionine-induced fatty liver. In fact, adenosyl-*S*-ethionine can ethylate DNA and RNA, thus inhibiting protein synthesis in other ways. With ethionine, fatty liver appears 5–6 h after treatment, female rats being more sensitive than males.

In any case, a block in lipoprotein secretion takes place before the onset of TG accumulation (Harris and Robinson, 1961; Lombardi and Recknagel, 1962; Lombardi et al., 1964, 1966). Further reasons for fat accumulation working in later stages are damage to mitochondria (Hutterer, 1963) and the consequential further decrease in ATP concentration. Recently it has been reported that plasma-membrane adenylate cyclase activity strongly increases after ethionine, as after CCl_4 or ethanol. Thus, cAMP content does increase (Wong and Kisilevsky, 1984).

No reports refer to damages in aminoacyl-tRNA synthesis as to possible causes for fatty liver. Alkylation of RNA certainly occurs after treatment with alkylating agents. Even this mechanism, therefore, may collaborate to increase TG concentration in the liver after treatment with such substances.

Fatty Liver Due to Impaired Fixation of Aminoacyl-tRNAs to the Large Ribosomal Subunit. The most important poison of aminoacyl-tRNA fixation to the large ribosomal subunit is puromycin, which is thought to compete with aminoacyl-tRNA for the ribosomal site due to its conformational similarity with this substance (Nathans and Lipmann, 1960, 1961; Ying et al., 1980). Puromycin produces fat accumulation in the liver starting about 16 h after administration. A block in protein synthesis and in lipoprotein secretion is evident as early as 8 h after treatment (Rao et al., 1970).

Another group of substances preventing the fixation of certain aminoacyl-tRNAs to the ribosomal subunit is that of tetracyclines. These must be used at very high concentration, however, to produce fatty liver in the rat. It has been reported that tetracyclines, and especially oleandomycin, specifically block the fixation of phenylalanyl-tRNA to its receptor (Igarashi and Kaji, 1970; Last, 1969).

Fatty Liver Due to the Inhibitors of the Translation. This group includes several substances. The extent of fatty accumulation is, however, not always corresponding to their activity as inhibitors of translation. In fact, extrahepatic effects, and especially those on adipocyte adenylate cyclase, are very often present, with consequent decrease in FFA supply to hepatocytes by the blood.

Cycloheximide and emetine, an antimalaric drug, are the most studied examples. Emetine is much more steatogenic than cycloheximide (Bar-on et al., 1972, 1973; Dianzani and Gravela, 1975; Gravela et al., 1971, 1974; Verbin et al., 1969, 1971). Both substances also display an extrahepatic action, and produce a decrease in FFA concentration in the blood practically at the same extent. So, in order to explain the difference in fat accumulation, one must hypothesize about some additional mechanism taking place in the liver. The extent in the block in lipoprotein secretion is practically identical in both cases (Dianzani and Gravela, 1975). Both cycloheximide and emetine block translation and produce ''freezing'' of polyribosomal profiles (Gravela et al., 1971, 1974; Dianzani and Gravela, 1975; Siegel and Sisler, 1963) usually produced by CCl_4 or white phosphorus.

Other inhibitors of translation are diphtheria toxin, as well as a group of substances produced by plants, such as abrin, ricin, modeccin, viscumin, and volkensin.

Diphtheria toxin is moderately steatogenic for rabbit liver, fat accumulation being demonstrable 16 h after treatment, coincidently with the block in lipoprotein secretion (Casu et al., 1966). Ricin and abrin are poorly steatogenic and display mainly extrahepatic actions (Montanaro et al., 1973a,b; Olsnes and Pihl, 1973; Sperti et al., 1971, 1973).

In the liver, their primary target seems to be the Kupffer cell. In hepatocytes, fat increases at 6 h, coincidently with the block in protein synthesis. Dissociation of polyribosomal profiles is evident as early as 3 h after treatment.

Modeccin, a peptide from *Adenia (Modecca) digitata,* a Passifloracea from South Africa, as a difference from ricin and abrin, can easily cross the hepatocyte plasma membrane (Barbieri et al., 1982, 1984). It blocks protein synthesis and provokes damage to mitochondria and to the endoplasmic reticulum.

Very similar is the behavior of volkensin, another peptide from *Adenia volkensia,* a plant from Kenya. Fatty liver and cell death are always present. Viscumin, a protein from *Viscum album,* is also steatogenic and produces cell death in the liver (Olsnes et al., 1982).

Fatty Liver Due to Inhibition of the Initiation. The most typical inhibitors of the initiation reaction, such as pyrocathecol violet and aurintricarboxylic acid, cannot cross the plasma membrane of the liver cells, so they can act only *in vitro* after its disruption. For this reason they are not steatogenic (Dianzani, 1975, 1978, 1979a).

CCl_4 and white phosphorus are powerful steatogenic and necrogenic agents. Among other actions, they display an inhibitory action on protein synthesis, mostly affecting the initiation reaction. In fact, these substances are powerful dissociators of the polyribosomal profiles when used alone, but do not dissociate when the rats are pretreated with either cycloheximide or emetine, i.e., with poisons ''freezing'' translation (Diazani, 1975, 1978, 1979a; Dianzani and Gravela, 1975; Gravela, 1973; Smuckler and Koplitz, 1974). This result is only explained if CCl_4 and white phosphorus do not act on translation, but on reattachment of ribosomes to mRNA (initiation).

Nothing is known about the involved mechanism, but one can remember that pretreatment with GSH affords some protection. As SH groups are necessary for the initiation reaction, one possibility is that the inhibition is in some way related to oxidation of these groups. Of course, other explanations (for instance, haloalkylation of important protein or co-factors in the case of CCl_4) are possible.

It is noteworthy that neither promethazine nor vitamin E, which prevent the onset of lipid peroxidation after CCl_4, are protective toward the loss in protein synthesis produced by CCl_4. This shows that haloalkylation, and not lipid peroxidation, are primarily involved in the inhibition mechanism (Poli et al., 1982, 1985b, 1987).

According to a recent report by Clawson et al. (1987), CCl_4 would inhibit protein synthesis by damaging 2-*O*-ribose methylation in rRNA. Maximal inhibition (60%) was found at 1.5 h after giving the poison, the normal values being restored with 4–5 h. According to Clawson, such times would fit with those needed for a block in protein synthesis after CCl_4. Dissociation of polyribosomal profiles, however, occurs a few minutes after CCl_4 (Gravela, 1973; Dianzani, 1975, 1978, 1979a; Dianzani and Gravela, 1975), and the block in protein synthesis has been also reported to occur very early (Gravela, 1973; Smuckler and Koplitz, 1974; Dianzani, 1975, 1978, 1979a).

Carbon Tetrachloride–Induced Fatty Liver

CCl_4 is a powerful steatogenic substance, used for decades in laboratories, especially for its high reproducibility of results. It also produces a certain degree of cell death, occurring in some cells within 1 h after treatment. Repeated injections in the rat are cyrrhogenic and to a certain extent carcinogenic.

CCl_4 is the model for a class of halogenated alkanes and alkenes. Several of such compounds, such as CCl_4 itself, are extensively used in the solvent industry. Others, such as chloroform, have been used in anesthesiology in the past, and others, such as halothane, are still being used for this purpose. Other haloalkanes, such as 1,2-dibromoethane, are used in agriculture. Therefore, the knowledge of the mechanism of action of such substances has social relevance.

Animals treated with either CCl_4 or with other substances of the group in sufficient amounts usually present first of all a toxic shock. This is characterized by activation of adipocyte lipase by the adenylcyclase mechanism, and consequently by increased FFA supply to hepatocytes. This explains why fat accumulation in the liver is to some extent prevented by previous adrenalectomy or hypophysectomy, as well as by spinal cord transection (Calvert and Brody, 1961; Butler, 1961; Larson and Plaa, 1965). A first interpretation of CCl_4-induced fatty liver has been, therefore, related to fat mobilization following shock.

Further experiments, and definitely those done on isolated hepatocytes (Gravela et al., 1977), have shown, however, that the main mechanism for fatty liver occurs inside liver cells. Increased FFA supply by the blood is, therefore, only permissive for fatty liver and may collaborate to produce a high TG level in the liver *in vivo*.

The toxic action of haloalkanes is related to their metabolism taking place in the smooth endoplasmic reticulum of liver cell. In this cell site, CCl_4 is homolytically cleaved, giving rise to $CCl_3^{\cdot}$ free radical and to a chlorine ion. The free radical has been clearly demonstrated by spin trapping with phenylbutylnitrone both after pulse radiolysis and in biological specimens, either *in vitro* or *in vivo* (Tomasi et al., 1983). There is, however, evidence that in the presence of oxygen, a second more reactive free radical, i.e., $CCl_3O_2^{\cdot}$, is also formed.

The intensity of metabolism clearly explains the differences in toxicity or different haloalkanes and haloalkenes. CCl_4 is less toxic than $CBrCl_3$ and more toxic than $CFCl_3$ (Slater, 1972, 1982). Cottrell (1958) showed that the difference in toxicity corresponds with a difference in the activation energy of these compounds. That having the lowest activation energy is also the most toxic ($CBrCl_3$).

The intensity of metabolism of a definite substance depends, however, also on another factor. We know that toxicity and metabolism of CCl_4 are very low in newborn rats, where the monooxygenase system of the SER is not fully working as yet (Dawkins, 1963). Toxicity and metabolism are also low in regenerating liver (Cheeseman, 1987). Destruction of the efficiency of the chain by pretreatment with inhibitors as SKF 525A (Glende and Recknagel, 1970; Seawright and McLean, 1967; Ugazio et al., 1967), or with a small dose of CCl_4 itself (Guzelian and Swisher, 1979; Pound, 1975; Ugazio et al., 1973), or with lead nitrate (Pani et al., 1975) also decreases CCl_4 toxicity.

Atrophy of the endoplasmic reticulum provoked by a protein-deficient diet produces the same effect (McLean and McLean, 1966, 1969). A decreased toxicity even follows the contemporary treatment with CCl_4 and of other compounds metabolized by the same metabolic chain of the endoplasmic reticulum, such as phenobarbital or propylgallate (Torrielli and Slater, 1971; Torrielli and Ugazio, 1970; Ugazio and Torrielli, 1969).

During the last few years, the question asked is whether CCl_4 may be attacked in cellular sites other than SER. Data for a metabolism of CCl_4 inside mitochondria have been described by Brattin et al. (1984).

Albano et al. (1985) found an early involvement of the mitochondrial calcium pump with discharge in the cytosol of intramitochondrial calcium. As this phenomenon is not prevented by antioxidants, such as promethazine or vitamin E, it is clearly independent upon lipid peroxidation. Moreover, the authors found haloalkylation of mitochondrial lipids and proteins occurring in coincidence with damage to the calcium pump. Finally, Tomasi et al. (1987) have succeeded in trapping a $CCl_3^{\cdot}$ free-radical signal in mitochondria.

Other discussions regards the nuclear envelope. As this is a dependence of the SER, there is no difficulty in accepting that it can be able to attack CCl_4. Levy and Brabec (1984) have given substantial evidence for this. Moreover, haloalkylation of nuclear nucleic acids and protein has been reported (Diaz Gomez and Castro, 1980).

The problem of CCl_4 metabolism in plasma membranes has received, on the contrary, only negative results.

Recently, however, Ungemach (1985) reported the presence of increased diene conjugation and increased MDA production by plasma membranes from hepatocytes incubated in the presence of ADP-iron or CCl_3Br. These changes were associated with a decrease in thymidine transport, as well as an increased permeability to K^+. The analysis of membrane phospholipids revealed a loss in polyunsaturated fatty acids and a three- to fourfold increase in lysophosphatides. This last phenomenon was thought to be related to the activation of plasma-membrane phospholipase A_2, probably mediated by the decrease in membrane fluidity following lipid peroxidation (Paradisi et al., 1988; Verheij et al., 1981).

Phospholipase A_2 would remove especially oxidized fatty acids from phospholipids, thus releasing hydroperoxides for glutathione peroxidase action (Van Kujik et al., 1987). Recent experiments by the group of Slater (LePage et al., 1988), failed to show any presence of the CCl_4-metabolizing chain in purified plasma membranes. More recently, Loeper et al. (1990) produced however evidence for the presence of small amounts of cytochrome P450.

We found no evidence for increased lipid peroxidation in hepatocyte plasma membranes after incubation with ADP-iron (Poli et al., 1987). The use of phospholipase inhibitors did not significantly modify CCl_4- or ADP/iron-induced irreversible cell damage.

Plasma-membrane adenylate cyclase, however, clearly increases 30 min after CCl_4 administration into rats (Mourelle and Rubalcava, 1981; Paradisi and Dianzani, 1979; Paradisi et al., 1982, 1985). Moreover, 4-HNE, a product of lipid peroxidation, increases adenylate cyclase of purified plasma membranes after short-term incubation (the first 5 min) at low concentration (10^{-6}

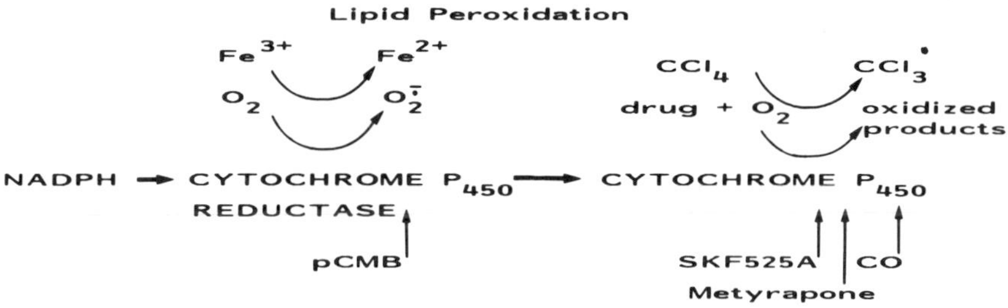

Figure 3. Generation of oxygen radical species and of $CCl_3^{\cdot}$ in the chain of the smooth endoplasmic reticulum.

M), but produces inhibition for a longer time of incubation and for higher concentrations (Paradisi et al., 1982; Dianzani et al., 1990).

In rats given CCl_4, cAMP concentration in the liver is increased as early as 30 min after treatment (Paradisi and Dianzani, 1979). Structural changes in plasma membranes are shown by freezing-fracture experiments (James et al., 1982, 1986; Paradisi, Losa et al., unpublished results). Calcium ATPase is not significantly affected by CCl_4, as well as by low HNE concentration (Parola et al., 1990), calmodulin is also not affected (Barrera et al., 1987). 5′-Nucleotidase and Na^+-K^+-ATPase are also affected by relatively high HNE concentrations (Paradisi et al., 1982, 1985).

Recent unpublished experiments by our group point to some effect of CCl_4 on plasma-membrane phosphoinositol metabolism, but it is still difficult to explain how this occurs. In conclusion, the problem of CCl_4-action on plasma membranes is still open.

The major site for CCl_4 metabolism and activation remains in any case the drug metabolizing in the SER. Most free-radical production is thought to occur at the level of cytochrome P450 (Castro et al., 1973, 1978), but the long-debated possibility of a further production at the level of the flavoprotein is still tenable (Slater, 1972, 1982) (Figure 3).

$CCl_3^{\cdot}$ free radical has been clearly identified by spin-trapping both in the homogenate or microsomes from the poisoned rat and in microsomes or hepatocytes *in vitro* (Connor et al., 1986; Davies and Slater, 1986; Tomasi et al., 1980). The trapped signal surely regards CCl_4, as was demonstrated in experiments where $^{13}CCl_4$ was used instead of $^{12}CCl_4$. The presence in $^{13}CCl_4$ of one extra nucleon in the carbon atom produces perturbation in its magnetic properties, producing a typical double-peak signal. This coincides with the usual signal displayed by $^{12}CCl_4$ (Tomasi et al., 1980).

In the presence of oxygen, $CCl_3^{\cdot}$ is quickly converted to CCl_3O_2- (trichlormethylperoxy) free radical (Forni et al., 1983). This is much more reactive than $CCl_3^{\cdot}$ and has not been identified by spin-trapping as yet. In fact, the $CCl_3^{\cdot}$ signal decreases in intensity in the presence of oxygen. *In vitro* $CCl_3O_2^{\cdot}$ reacts with polyunsaturated fatty acids much more quickly than

$CCl_3^{\cdot}$ does, the reaction being prevented by the presence of the antioxidant promethazine. The reaction of $CCl_3^{\cdot}$ however, is very poorly inhibited by promethazine.

A third type of CCl_4-derived free radical has been reported recently: $C^{\cdot}OCl$ is formed from $CCl_3O_2^{\cdot}$. It would be quickly transformed into carbonic acid. In fact, the formation of CO_2- from it has been described (Connor et al., 1986).

$CCl_3^{\cdot}$ and $CCl_3O_2^{\cdot}$, the two major free radicals, produce in the cell structures they contact two types of reactions: haloalkylation and hydrogen abstraction.

By the first reaction, trichloromethyl group becomes covalently bound to the affected molecule. This reaction has been proven to occur with lipids, proteins, and nucleic acids or purine and pyrimidine bases. The second reaction will produce $CHCl_3$ from the free radical. Chloroform is then further metabolized by the same drug-metabolizing system, with formation of other free radicals. This occurs, however, at a minor rate due to the higher activation energy of $CHCl_3$ with respect to CCl_4. The molecule undergoing hydrogen abstractions is very often transformed into a free radical. This happens usually with polyunsaturated fatty acids contained in the membranes, which are very prone to dehydrogenation.

This starts the process of lipid peroxidation. The onset of this process consists in the homolytical cleavage of a covalent bond between a hydrogen atom and carbon atom in a fatty-acid chain. This fatty-acid free radical then fixes oxygen, forming lipoperoxide free radicals. The reaction continues with the breaking of the chain and the formation of more simple derivatives. Among them, substances containing carbonyls are the most important ones (Esterbauer et al., 1982; Poli et al., 1985a). A number of aldehydes and ketones are formed during peroxidation *in vitro* of microsomal or isolated hepatocytes lipids. About 25 have been identified by mass spectroscopy in the unpolar or less polar subfraction.

Among them, some are saturated (hexanal, heptanal, nonanal, decanal, propional, etc.), while others are unsaturated (hexanal, nonenal, decenal, etc.). A special class is formed by the 4-hydroxy-2,3-*trans*-unsaturated aldehydes, the most represented of which are 4-hydroxy-nonenal, first identified by Benedetti et al. (1982), 4-hydroxy-hexenal, 4-hydroxy-undecenal, and 4-hydroxy-octenal, in magnitude order. Adihydroxy-2,3-*trans*-unsaturated aldehyde has been also identified as 4,5-dihydroxydecenal.

The aldehydes produced during lipid peroxidation display a rather high toxicity, surely much more intense than that of MDA, which is usually measured as a marker of lipid peroxidation. MDA is practically nontoxic, whereas both saturated and unsaturated aldehydes described above display a clear toxicity.

The most toxic compounds are represented, however, by the aldehydes of the 4-hydroxy-unsaturated series, and especially HNE, which is abundantly produced. These aldehydes can escape from still-living cells, can bind to albumin, and can be released in remote sites, thus spreading far from the production sites lipid peroxidation toxicity. The kinetics of CCl_4-stimulated lipid peroxidation is definitely different from that of one stimulated by ADP-iron (Figure 4).

With CCl_4, relatively less MDA and more unsaturated aldehydes are produced than with ADP-iron. Moreover, the total amount of aldehydes formed per gram substrate is also much higher in the presence of ADP-iron. This means either that after CCl_4 a considerable amount of aldehydes produced are not recovered, due to secondary reactions (possibly metabolism), or that a certain portion of polyunsaturated fatty acids follows after CCl_4 a pathway different from that of aldehyde production.

There is even the possibility that haloalkylation occurs at the level of the fatty acid itself; this reaction might prevent lipid peroxidation, or might give rise to haloalkylated aldehydes not detectable in the usual aldehyde pool.

It seems possible that $CCl_3^{\cdot}$ and $CCl_3O_2^{\cdot}$ have different responsibilities in toxicity. In fact, promethazine or previous enrichment with vitamin E completely prevent lipid peroxidation as well as certain types of cell damage, whereas they are completely unable to prevent haloalkylation, as well as

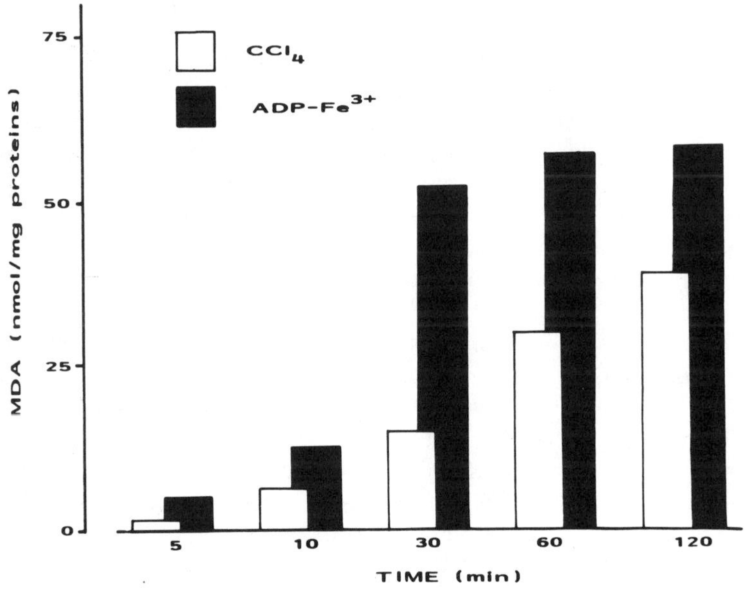

Figure 4. Stimulation of malonaldehyde production by isolated hepatocytes after incubation either with CCl_4 (I29 μM) or with ADP-iron.

certain other types of cell damage (Dianzani, 1987; Dianzani and Poli, 1985; Poli et al., 1981a, 1982, 1985b).

It was stated earlier that promethazine is able to block the reaction *in vitro* of $CCl_3O_2^{\cdot}$ with polyunsaturated fatty acids, but is unable to prevent the binding of $CCl_3^{\cdot}$ to other organic reactants. It seems very probable that lipid peroxidation mostly depends upon $CCl_3O_2^{\cdot}$ free radical, whereas $CCl_3^{\cdot}$ is mostly responsible for haloalkylation (Forni et al., 1983).

Table 8–1 gives a summary of the enzymes and functions that are inhibited by CCl_4, but which are protected by previous treatment with promethazine or of vitamin E.

The inactivation of the metabolic chain of the SER by CCl_4 is unaffected by these antioxidants (Poli et al., 1981a), so the inactivation is probably due to haloalkylation. Haloalkylation of cytochrome P450 has been demonstrated as a major factor in the inactivation of the chain and at the same time a proof of the prevalent formation of the $CCl_3^{\cdot}$ free radical at this site (Cheeseman, 1987; De Groot and Haas, 1981; Fernandez et al., 1982). The damage of the calcium pump in mitochondria is also dependent upon haloalkylation and is not prevented by vitamin E, as reported above.

The block in protein synthesis and the related dissociation in polyribosomal profiles are also dependent upon haloalkylation. It is noteworthy that also lipoprotein secretion depends upon this mechanisms. The onset of fat accumulation seems, therefore, to depend mainly upon haloalkylation and not upon lipid peroxidation.

Lipid peroxidation, however, is responsible for the inactivation of glucose-6-phosphatase, another microsomal marker, and of galactosyltransferase, an enzyme of the Golgi vesicles. Cells treated with promethazine or with vitamin E do not release in the incubation medium lactate dehydrogenase. Leakage from the cells of this enzyme is usually considered as the most suitable marker for irreversible lesion, i.e., for cell death. Cell death, at least in the case of the acute CCl_4 poisoning, seems to be related to lipid peroxidation and not to haloalkylation. This does not exclude, of course, other mechanisms from collaborating in the production of cell death in later stages of poisoning.

In the same way, we cannot exclude that some of the damage that begins with haloalkylation will be caused in later stages even by lipid peroxidation.

As an example, it has been shown that peroxidized lipids interact with cytochrome P450 with consequent inactivation of this haemoprotein. Antioxidants are able to prevent partially the loss of cytochrome P450 produced in isolated hepatocytes by ADP-iron. Moreover, the treatment of hepatocytes with 2.5×10^{-4} HNE lowers by 47% cytochrome P450 levels. In the case of microsomes, 4×10^{-4} HNE is needed to inactivate cytochrome P450 by 50%.

As far as the mechanism for the onset of fatty infiltration is concerned,

TABLE 8–1.
CCl_4 Poisoning of Rat Hepatocytes in Single-Cell Suspension. Effect of Promethazine Addition to the Incubation Medium or of Animal Preloading with Vitamin E*

	CCl_4-Induced Inhibition	Protection Exerted by Promethazine or Vitamin E
Glucose-6-phosphatase activity	35–40%	High or complete
Cytochrome P450 content	30–35%	Very low or absent
Aminopyrene demethylase	60–80%	Very low or absent
Protein synthesis	40–50%	Absent
Protein, lipoprotein secretion	65–80%	Absent
Plasmamembrane integrity	Time dependent	High or complete

*Cells have been incubated at 37°C for 60 min (where not otherwise indicated) with 129 μM CCl_4 in the presence or in the absence of 50–100 μM promethazine. In the case of vitamin E, cells were isolated from rats which received i.p. 100 mg/kg b.w.t. fifteen hours before sacrifice.

it is now clear that CCl_4 works in several ways. Historically, the first advanced hypothesis was that fat accumulation was related mostly to the impairment of fatty acid oxidation in mitochondria, due to the severe damage affecting these particles shortly after poisoning. In fact, mitochondria are swollen and show uncoupling of oxidative phosphorylation so that a severe loss in ATP takes place. This may produce fat accumulation in late stages of poisoning, but surely does not contribute to the onset. In fact, loss in fatty acid oxidation in mitochondria is present only 12–16 h after poisoning with a single dose, notwithstanding the fact that mitochondrial changes are morphologically evident as early as 30 min after treatment.

Figure 5 shows the chronology of the events after CCl_4. It is clear that the damage to SER occurs very early and that the block in protein synthesis is already demonstrable 5–10 min after poisoning, practically simultaneous with the first appearance of fat. According to Recknagel et al. (1960) an early cause for fat accumulation is the block in lipoprotein secretion, possibly consequent to the block in protein synthesis. This has been shown even with isolated hepatocytes (Poli et al., 1985a,b, 1987).

An overview of the chronology of the first increase of TG in different types of fatty liver (Table 8–2) shows that in most cases where the cause for the damage is a block in protein synthesis TG started to increase 5–6 h after the beginning of the experiment. This delay has been interpreted as related to the time needed for the consumption of preexisting apolipoprotein molecules (Dianzani, 1979b, 1982; Dianzani and Gravela, 1975).

Thus, there is a great contrast between the time for onset after CCl_4 or CCl_3Br (30–60 min) and that seen with most protein synthesis inhibitors. This means that CCl_4 does not cause fat accumulation in the liver due to a block in protein synthesis. Pretreatment of rats with vitamin E, or incubation of hepatocytes in a medium containing promethazine, completely abolishes aldehyde formation without affecting haloalkylation and without preventing the block protein synthesis and lipoprotein secretion, as well as in fat accumulation.

This clearly shows that haloalkylation is the primary event in producing TG accumulation. Very early after CCl_4, lipoproteins accumulating in the Golgi subfractions become haloalkylated (Poli et al., 1987), both in the lipid and in the protein moieties. The hepatocytes continue, however, to secrete lipoproteins to a certain extent for some time.

The secreted lipoproteins are less haloalkylated in the protein moiety than those remaining inside the cells. This may mean that haloalkylation of apolipoprotein prevents secretion. Lack of protection toward loss of lipoprotein secretion even occurs in hepatocytes enriched with vitamin E. Both enrichment with vitamin E and treatment with promethazine prevent, however, cell death, as shown by both prevention of cell staining and of lactate dehydrogenase leakage.

CHRONOLOGY OF CCl_4-INDUCED LIVER INJURY

WITHIN 5 MIN.	DETECTION OF CONJUGATED DIENES IN LIVER MICROSOMAL LIPIDS.
	COVALENT BINDING OF $CCl_3^{\cdot}$ TO CELLULAR LIPIDS AND PROTEINS.
AFTER 5 MIN.	TRYGLICERIDES ACCUMULATION INTO THE SECRETORY FRACTION OF GOLGI APPARATUS VESICLES.
AFTER 10-15 MIN.	DISSOCIATION OF POLIRIBOSOMAL PROFILES.
	DECLINE IN PROTEIN SYNTHESIS.
	RELEASE OF Ca^{2+} BY MITOCHONDRIA IN ISOLATED HEPATOCYTES.
AFTER 15 MIN.	INHIBITION OF MICROSOMAL GLUCOSE-6-PHOSPHATASE ACTIVITY.
	INHIBITION OF GALACTOSYL TRANSFERASE IN GOLGI APPARATUS.
	TRYGLICERIDES ACCUMULATION INTO THE FORMATIVE FRACTION OF GOLGI APPARATUS VESICLES.
AFTER 10-30 MIN.	MORPHOLOGICAL ABERRATIONS OF THE ENDOPLASMIC RETICULUM.
AFTER 30 MIN.	RELEASE OF Ca^{2+} BY SMOOTH ENDOPLASMIC RETICULUM VESICLES.
AFTER 30-40 MIN.	APPEARANCE OF MITOCHONDRIAL SWELLING.
AFTER 60 MIN.	DECLINE OF THE HEPATIC MIXED FUNCTION OXIDASE SYSTEM.
AFTER 60-90 MIN.	UNCOUPLING OF MITOCHONDRIAL OXIDATIVE PHOSPHORYLATION.
AFTER 120 MIN.	INCREASE IN ACTIVITY OF LYSOSOMAL ENZYMES.
AFTER 12 HOURS	BLOCK OF MITOCHONDRIAL FATTY ACID OXIDATION.
AFTER 12-24 HOURS	PEAK OF LIVER STEATOSIS AFTER CCl_4 ADMINISTRATION.
WITHIN 24-144 HOURS	REPARATIVE PHASE.

LIVER NECROSIS, THAT MAY BE ALREADY EVIDENT AFTER 1-2 HOURS, IS CLEAR AFTER 6 HOURS, REACHING A MAXIMUM AFTER 24 HOURS.

Figure 5. Chronology of liver changes after intragastric administration of CCl_4 in the rat (100–250 μl/100 g b.w.t.).

It is still unclear why haloalkylation of the apoliproteins prevents secretion. The first damage might affect either the assembly of lipoproteins, or the transit of the well-formed micelles through the Golgi apparatus, or possibly maturation by galactosylation. It was reported earlier, however, that the last reaction is damaged by CCl_4, protection being afforded by promethazine or vitamin E.

TABLE 8–2.
Time Required for the Appearance of Fatty Liver Following Various Treatments with Hepatotoxic Agents

Treatments	Given Dose (for 100 g b.w.t.)	Latency Time
CCl_4	0.26–2.58 m*M*	1
$CBrCl_3$	0.26 m*M*	1
CS_2 (males)	2.58 m*M*	12
CS_2 (females)	2.58 m*M*	6
Yellow phosphorus	1–5 mg	6
Ethionine (females)	100 mg	7
Dimethylnitrosamine	20 mg	5–6
Cycloeximide	0.1 mg	5–6
Aflatoxin	0.1 mg	6
Emetine	2 mg	3
Ethanol	800 mg	6
α-Amanitine	0.5 μg	24
Diet coline-deficient	—	12

Galactosylation block depends upon lipid peroxidation, but its prevention does not prevent block in secretion. The accumulation of ^{14}C palmitate in the Golgi subfraction is significantly increased as early as 15 min after poisoning in the Golgi ''formative'' subfraction (F_3), but already at 5 min in the secretory pole (F_1). This means that the inhibition in the transit occurs at least as early as the impairment in protein synthesis.

Even these changes are not prevented by vitamin E enrichment, notwithstanding that vitamin E content is increased by eight- to ninefold. The fact that ^{14}C-palmitate accumulation occurs first at the secretive then at the formative pole of the Golgi apparatus means that 'secretion' of lipoproteins is more damaged than their assembly. This fits with the previously shown fact that poisoned cells are still able to ''secrete'' haloalkylated lipoproteins to a certain extent.

The reason for haloalkylation in the Golgi apparatus are not understood as yet. In fact, there is the possibility that the components of the lipoprotein micelles are haloalkylated prior to entering the Golgi structures. We cannot exclude, however, that haloalkylation occurs inside them. In this case, we would have to accept the idea that Golgi membranes contain a certain amount of the metabolic chain. There are already some positive results favoring this interpretation, but clear evidence is still lacking.

The reported experiments clearly separate the causes for the onset in the increase in TG and those for acute cell death. It is noteworthy that this is prevented by promethazine or by vitamin E loading even in the case of hepatocyte incubation in the presence of either paracethamol or 1,2-dibromoethane (Albano et al., 1983, 1986; Tomasi et al., 1983), two substances able to decrease intracellular GSH content and at the same time to stimulate lipid peroxidation.

In conclusion, TG accumulation after CCl_4 begins with haloalkylation-dependent block in the exit of lipoprotein micelles from the Golgi apparatus. Continuous FFA supply to the liver from the blood is, however, permissive for accumulation. Over time, lipid peroxidation may contribute to accumulation, especially due to the interaction of aldehydes (especially HNE) with tubulin (Gabriel et al., 1977). Tubulin is involved in the transport of the micelles from Golgi apparatus to the plasma membrane. Mitochondrial damage, as well as the inhibition of protein synthesis, contribute to TG accumulation in later stages.

White Phosphorus–Induced Fatty Liver

White phosphorus is a powerful steatogenic and necrogenic agent, being different from the two other allotropic states of the same element, i.e., respectively, red and black phosphorus. This is the unique case of a substance whose toxicity depends upon its physical state, and not upon chemical composition.

White phosphorus is usually given in oil solution, either by stomach tube or, especially in the old literature, subcutaneously. TG accumulation in the liver starts at about 5 h after treatment, as in the case of other substances inhibiting protein synthesis.

In this case, however, infiltration starts centrolobular. Protein synthesis after phosphorus is impaired at 1–2 h, prior than fat accumulation (Dianzani, 1975, 1978, 1979a; Seakins and Robinson, 1964). Dissociation of polyribosomal profiles is very similar to that seen after CCl_4 (Dianzani, 1972; Pani et al., 1976; Torrielli et al., 1973); as in the case of CCl_4, it is prevented by pretreatment with either cycloeximide or emetine. This result also shows that white phosphorus acts at the level of the initiation reaction.

This mechanism fits with the rather early block in lipoprotein secretion, as described by Lombardi and Recknagel (1962), and by Seakins and Robinson (1964). Mitochondrial damage is very strong and early. It may contribute to maintain high TG levels in later stages (Dianzani, 1972, 1975, 1978, 1979a). Lipid peroxidation has been claimed to play some role in the pathogenesis of the damage (Ghoshal et al., 1971). In fact, an increase in MDA production by homogenates has been found a few hours after poisoning. This finding, however, is still controversial (Dianzani, 1972; Pani et al., 1976) and in any case not as intense as after CCl_4. Moreover, glucose-6-phosphatase, which is affected by CCl_4 early on remains active. As cell death appears rather early, there is the problem in determining whether increased lipid peroxidation is a cause or a consequence of this phenomenon.

The fate of white phosphorus inside the liver cell is not known. After injection with ^{32}P, this becomes accumulated in microsomes and in the cytosol. ^{32}P, however, is nontoxic, so there is little concern about white phosphorous toxicity. Pretreatment with phenobarbital, which induces the SER drug-metabolizing system, does not increase white phosphorous toxicity (Pani et al., 1976). It is interesting, however, that oxygenated derivatives of phosphorous, such as phosphonates and phosphinites, have been reported to react with cytochrome P450, producing its inactivation (Dahl and Hodgson, 1978).

The technical difficulties encountered when using white phosphorus have lessened interest for this poison during the past years. This is a pity, since the model of a substance indebted for its toxicity to its physical state, and not to its chemical composition, is unique.

Other Types of Fatty Liver

Other types of fatty liver cannot be classified as yet due to insufficient knowledge.

One of the most interesting cases is that of fatty liver following administration of the so-called rare earths, lantanium, cerium, praseodimium, neodimium, and samarium, which are usually given by endovenous injection (2 mg/kg b.w. in the rat). Fat accumulations belong to the centrolobular type and are evident a few hours after injection. Recovery of normal TG values is reached usually 2–3 days after injection (Snyder et al., 1960a, b).

Shortly after intravenous injection an increase in blood FFA has been observed, probably due to the stress. Coincidently with the onset of the accumulation of TG, there is a decrease in lipoprotein secretion (Lombardi and Recknagel, 1962), accompanied by a severe damage of the smooth endoplasmic reticulum, with dilation of the cisternae and degranulation (Magnusson, 1963). Apparently this form of fatty liver belongs to the group of those related to inhibition of protein synthesis, but the evidence for this is far from convincing. First of all, we do not know at which stage protein synthesis is inhibited.

Moreover, Glenn et al. (1963) reported a decreased incorporation of palmitate in phosphatidylethanolamine, as well as an increased incorporation of it in cardiolipin and in phosphatidylinositol, thus suggesting a disorder in phospholipids synthesis. In later stages, damage to mitochondria is also evident (Neubert and Hoffmeister, 1960; Snyder et al., 1960b). Mitochondrial damage may contribute to maintain high TG levels in the liver in late stages of poisoning. As in the case of other poisons, interest about the mechanism of action of the rare earths has diminished recently.

Sporidesmin-induced fatty liver has been studied by several authors in the past. It provokes a peripheral fat infiltration (Webb, 1966), related with a loss in protein synthesis and with a block in lipoprotein secretion (Peters and Smith, 1964; Slater and Greenbaum, 1965). The dose rate is 1 mg/kg b.w. orally. Heavy damage to lysosomes, mitochondria, and endoplasmic reticulum, as well as to microvilli, leading to cholostasis and necrosis, has been reported (Eber et al., 1962).

It is thought that the toxic action of sporidesmin is related to its autoxidation. This would produce the formation of the free radical O_2-, which would represent the damaging agent (Munday, 1982). To my knowledge, however, there is no report of increased lipid peroxidation as yet.

Carbon disulfide has been especially studied in Great Britain by the group of De Matteis. It provokes in the rat a centrolobular fatty liver at the dose of 250 μl/100 g b.w. (Ugazio et al., 1971). Fat infiltration is evident after 5–6 h, as in the cases of other inhibitors of protein synthesis (Bond and De Matteis, 1969). This has been found to be inhibited, whereas dissociation of polyribosomal profiles occurs.

The action of CS_2 would be a consequence of the strong decrease of cytochrome P450 action this substance produces (Bond and De Matteis, 1969). In fact, CS_2 toxicity increases after induction of the enzymatic system by phenobarbital (De Lamatre et al., 1983). These facts favor the hypothesis that toxicity is related to the metabolism of the poison in the SER. Here, CS_2 might undergo oxidation to the intermediate compound S–CO–S, which would then be cleaved homolitically to give carbonyl sulfide (S–CO) and a single sulfur atom. This is highly reactive. After fixation to the structures of the SER, it could start the damage leading to severe fatty infiltration and necrosis (Dalvi et al., 1974).

Thioacetamide is another poison producing centrolobular fat infiltration and cell death (Barker et al., 1963; Thoënes and Bannasch, 1962). It provokes heavy nucleolar damage, as well as impairment in both RNA synthesis and release to the cytoplasm (Smuckler and Koplitz, 1974). The large (50 S) ribosomal subunits have been found to be intensively phosphorylated, but the mechanism for this is still not understood (Gressner and Greiling, 1978). Mitochondrial damage has been also described (Müller and Dargel, 1984).

Paracetamol at high dosage produces a heavy fatty liver of the centrolobular type, as well as necrosis (Portmann et al., 1975). GSH level declines strongly both in the liver of poisoned animals and in hepatocytes studied *in vitro* after paracethamol addition (Albano et al., 1985).

This decline has been considered as a possible cause for cell death, as well as for the intense stimulation of lipid peroxidation (Chu and Bhakthan, 1978; Crowe et al., 1979; de Vries, 1981; Fiaccavento et al., 1980a,b; Reid et al., 1970; Reiter and Wendel, 1983; Suarez et al., 1972; Younes and Seegers, 1985; Younes et al., 1986). It is thought that the real toxic agent is not paracetamol itself, but a metabolic product. This would be a free radical in nature, possibly semiquinone imine (Ross et al., 1984). This free radical would sequester GSH and this might be the major reason for GSH sudden decline. Moreover, the free radical becomes covalently bound to proteins

(Devalia and McLean, 1983; Hoffmann et al., 1985). Paracetamol free radical has been demonstrated *in vitro,* after oxidation of the drug in the presence of H_2O_2 and of peroxidase, but has not been found *in vivo* as yet. An intermediate thyil free radical has been, however, demonstrated after interaction with GSH (Ross et al., 1984) (Figure 6).

As in the case of CCl_4, paracetamol, therefore, produces at least two types of mechanisms for early damage: covalent binding and lipid peroxidation. Antioxidants are able to prevent both lipid peroxidation and cell death in isolated hepatocytes, but are unable to prevent covalent binding, decline in GSH, and damage to the endoplasmic reticulum (Poli et al., 1987). The same things occur with another haloalkane, 1,2-dibromoethane (Tomasi et al., 1983). There is considerable evidence that in several cases acute cell death produced by poisons stimulating lipid peroxidation is related to this last event.

Considerable interest has been given during the last years to halogenated arenes, and especially to bromobenzene, a substance producing, at high dosage, a heavy steatonecrosis in the mouse. A few hours after bromobenzene, lipid peroxidation increases enormously, much more than after other known poisons (Casini et al., 1985; Comporti, 1985). The most probable cause for increase in lipid peroxidation is, according to the Comporti group, the profound GSH decline following treatment. This is probably due to bromobenzene metabolism in the smooth endoplasmic reticulum, probably generating free radicals that become early covalently bound to GSH and to proteins. Also, oxygen free-radical species would be produced (Khanna and Smith, 1985; Trush et al., 1982). According to Comporti, covalent binding to proteins or lipids would be not involved in the production of cell death (Casini et al., 1985; Comporti, 1985), whereas lipid peroxidation, following GSH depletion, would be directly involved. In fact, Throlox C, a vitamin E derivative, affords

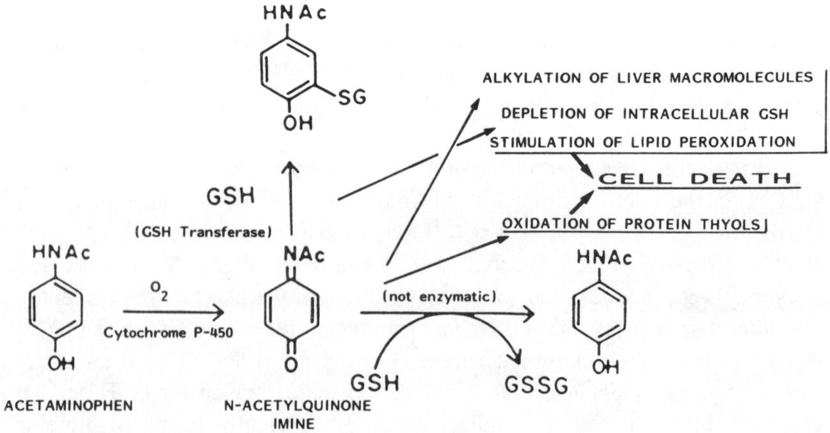

Figure 6. Metabolism of paracetamol in liver cells.

protection against cell death, whereas prior treatment of mice with a diet containing high amounts of iron strongly increases the damage (Khanna and Smith, 1985).

Unfortunately, high dosages, close to the lethal one, are required to get these results, and it is quite impossible to obtain the data by using the model of isolated hepatocytes, due to the very low solubility in water of bromobenzene.

The pathogenesis of fatty infiltration has not been studied as yet.

REFERENCES

Aaes-Jørgensen E: Essential fatty acids. *Physiol Rev* 1961; 41:1–51.

Adamstone FB and Spector H: Tryptophan deficiency in the rat. Histologic changes induced by forced feeding of an acid-hydrolized casein diet. *Arch Pathol* 1950; 49:173–184.

Ahmed SM and Slater TF: Lipid peroxidation in a microsomal fractions obtained from some rat and mouse tumours, in Slater TF and Garner A (eds): *Recent Advances in Lipid Peroxidation and Tissue Injury*. London, Brunel University Press, 1981, pp 177–195.

Albano E, Bellomo G, Carini R, Biasi F, Poli G, and Dianzani MU: Mechanisms responsible for carbon tetrachloride-induced perturbation of mitochondrial calcium homeostasis. *FEBS* 1985; 192:184–188.

Albano E, Poli G, Chiarpotto E, Biasi F, and Dianzani MU: Paracethamol-stimulated lipid peroxidation in isolated rat and mouse hepatocytes. *Chem Biol Interac* 1983; 47:249–263.

Albano E, Rundgren M, Harvison PJ, Nelson SD, and Moldeus P: Mechanisms of *N*-acetyl-*p*-benzoquinone imine cytotoxicity. *Mol Pharm* 1985; 28:306–311.

Albano E, Tomasi A, Goria-Gatti L, Carini R, Vannini V, and Dianzani MU: Free radical metabolism of ethanol, in Rice-Evans C (ed): *Free Radicals, Cell Damage and Disease*. London, Richelieu Press, 1986, pp 117–126.

Albano E, Tomasi A, Goria Gatti L, and Dianzani MU: Spin trapping of free radical species produced during the microsomal metabolism of ethanol. *Chem Biol Interac* 1988; 65:223–236.

Albano E, Tomasi A, Goria Gatti L, Poli G, Vannini V, and Dianzani MU: Free radical metabolism of alcohols by rat liver microsomes. *Free Rad Res Comms* 1987; 3:243–249.

Allmann DW and Gibson DM: Fatty acids synthesis from malonyl CoA in linoleate deficiency. *Fed Proc* 1962; 21:288.

Allmann DW and Gibson DM: Fatty acid synthesis during early linoleic acid deficiency in the mouse. *J Lipid Res* 1965; 6:51–62.

Allmann DW, Den H, Rahman MS, and Gibson DM: Palmitate synthesis in linoleic acid deficiency. *Fed Proc* 1963; 22:362.

Arias IM, Fleischner G, Listowsky I, Kamisaka K, Mishkin S, and Gatmaitan Z: On the structure and function of ligandin and Z protein, in Taylor W (ed): *The Hepatobiliary System,* vol 7. New York, Plenum Press, 1976, pp 81–107.

Autrup H, Essigman JM, Croy RG, Trump BF, Wogan GN, and Harris CC: Metabolism of aflatoxin B_1 and identification of the major aflatoxin B_1-DNA adducts formed in cultured human bronchus and colon. *Cancer Res* 1979; 39:694–698.

Baglio C and Farber E: Ultrastructural consequences of biochemical lesions in the liver induced by ethionine. *Fed Proc* 1965; 24:556.

Baraona E, Leo MA, Borowsky SA, and Lieber CS: Pathogenesis of alcohol-induced accumulation of protein in the liver. *J Clin Invest* 1977; 60:546–554.

Barbieri L, Aron GM, Irvin JD, and Stirpe F: Purification and partial characterization of another form of the antiviral protein from the seeds of *Phytolacca americana* L. (pokeweed). *Biochem J* 1982; 203:55–59.

Barbieri L, Falasca AI, and Stirpe F: Volkensin, the toxin of *Adenia volkensii* (kilyambiti plant). *FEBS Lett* 1984; 171:277–279.

Barker EA, Smuckler EA, and Benditt EP: Effects of thioacetamide and yellow phosphorus poisoning on protein synthesis in vivo. *Lab Invest* 1963; 12:955–960.

Bar-on H, Kook AI, Stein O, and Stein Y: Assembly and secretion of very low density lipoproteins by rat liver following inhibition of protein synthesis with cycloheximide. *Biochim Biophys Acta* 1973; 306:106–114.

Bar-on H, Stein O, and Stein Y: Multiple effects of cycloheximide on the metabolism of triglycerides in the liver of male and female rats. *Biochim Biophys Acta* 1972; 270:444–452.

Barrera G, Parola M, Paradisi L, Albin L, and Dianzani MU: Subcellular calmodulin distribution in rat liver after CCl_4 poisoning. *Cell Biochem Function* 1987; 5:129–133.

Bartoli GM and Galeotti T: Growth-related lipid peroxidation in tumour microsomal membranes and mitochondria. *Biochim Biophys Acta* 1979; 574:537–541.

Beard JD, Barboriak JJ, and Hamilton LH: Plasma lipids of dogs during and after chronic ethanol administration. *Fed Proc* 1965; 24:277.

Benedetti A, Casini AF, Ferrali M, and Comporti M: Early alterations induced by carbon tetrachloride in the lipids of the membranes of the endoplasmic reticulum of the liver cell. II. Distribution of the alterations in the various lipid fractions. *Chem Biol Interac* 1977; 17:167–183.

Benedetti A, Comporti M, and Esterbauer H: Identification of 4-hydroxynonenal as a cytotoxic product originating from the peroxidation of liver microsomal lipids. *Biochim Biophys Acta* 1980; 620:281–296.

Benedetti A, Esterbauer H, Ferrali M, Fulceri R, and Comporti M: Evidence for aldehydes bound to liver microsomal protein following CCl_4 or $BrCCl_3$ poisoning. *Biochim Biophys Acta* 1982; 711:345–356.

Bennett MA, Hellerman J, and Donnelly AJ: Influence of vitamin B_{12} on liver disease due to 'labile methyl'-free diet. *Cancer Res* 1955; 15:398–402.

Bennett RA, Hessigmann JM, and Wogan GN: Excretion of aflatoxin-guanine adduct in the urine of aflatoxin B_1-treated rats. *Cancer Res* 1981; 41:650–654.

Benton DA, Harper AE, Winje ME, and Elvehjem CA: Compounds other than threonine that affect fat deposition in the livers of rats fed low protein diets. *J Biol Chem* 1955; 214:677–684.

Berman WJ, Gil J, Jennett RB, Tuma D, Sorrell MF, and Rubin E: Ethanol, hepatocellular organelles and microtubules. A morphometric study in vivo and in vitro. *Lab Invest* 1983; 48:760–767.

Bernstein JD and Penniall R: Effects of chronic ethanol treatment upon rat liver mitochondria. *Biochem Pharmacol* 1978; 27:2337–2342.

Best CH and Channon HJ: CCCXIV. The action of choline and other substances in the prevention and cure of fatty livers. *Biochem J* 1935; 29:2651–2658.

Best CH, Grant R, and Ridout JH: The 'lipotropic' effect of dietary protein. *J Physiol (London)* 1936; 86:337–342.

Best CH, Hartroft WS, Lucas CC, and Ridout JH: Liver damage produced by feeding alcohol or sugar and its prevention by choline. *Br Med J* 1949; 2:1101–1106.

Best CH and Hershey JM: Further observations on the effects of some components of crude lecithine on depancreatized animals. *J Physiol (London)* 1932; 75:49–55.

Blanchette-Mackie EJ and Scow RO: Sites of lipoprotein lipase activity in adipose tissue perfused with chylomicrons. Electron microscope cytochemical study. *J Cell Biol* 1971; 51:1–25.

Blomstrand R, Kager L, and Lantto O: Studies on the ethanol-induced decrease of fatty acid oxidation in rat and human liver slices. *Life Sci* 1973; 13:1131–1141.

Bode C, Stahler E, Kono H, and Goebell H: Effects of ethanol on free coenzyme A, free carnitine and their fatty acid esters in rat liver. *Biochim Biophys Acta* 1970; 210:448–455.

Bond EJ and De Matteis F: Biochemical changes in rat liver after administration of carbon disulphide, with particular reference to microsomal changes. *Biochem Pharmacol* 1969; 18:2531–2549.

Brattin WJ, Pencil SD, Waller RL, Glende EA, and Recknagel RO: Assessment of the role of calcium ion in halocarbon hepatotoxicity. *Environm Health Perspectives* 1984; 57:321–323.

Brault D, Neta P, and Patterson LK: The lipid peroxidation model for halogenated hydrocarbon toxicity. Kinetics of peroxyl radical processes involving fatty acid and Fe(III) porphyrins. *Chem Biol Interac* 1985; 54:289–297.

Brodie BB, Butler WM, Horning MG, Maickel RP, and Maling HM: Alcohol-induced triglyceride deposition in liver through derangement of fat transport. *Amer J Clin Nutr* 1961; 9:432–435.

Brouwers JAJ and Emmelot P: Microsomal *N*-demethylation and the effect of the hepatic carcinogen dimethylnitrosamine on amino acid incorporation into the proteins of rat livers and hepatomas. *Exp Cell Res* 1960; 19:467–474.

Buku A, Campadelli-Fiume G, Fiume L, and Wieland Th: Inhibitory effect of naturally occurring and chemically modified amanitoxins on RNA polymerase of rat liver nuclei. *FEBS Lett* 1971; 14:42–44.

Burdino E, Danni O, Dianzani MU, Milillo PA, Poli G, Sena LM, Torrielli MV, and Ugazio G: Halogenoalkane toxicity in different experimental conditions, in Dianzani MU, Ugazio G, and Sena LM (eds): *Recent Advances in Biochemical Pathology: Toxic Liver Injury.* Torino, Minerva Medica, 1976, pp 51–64.

Butcher RW and Sutherland EW: Adenosine 3′,5′-phosphate in biological materials. *J Biol Chem* 1962; 237:1244–1250.

Butler TC: Reduction of carbon tetrachloride in vivo and reduction of carbon tetrachloride and chloroform in vitro by tissues and tissue constituents. *J Pharmacol Exp Ther* 1961; 134:311–319.

Butler WH and Clifford JI: Extraction of aflatoxin from rat liver. *Nature* 1965; 206:1045.

Calvert DN and Brody TM: The effects of thyroid function upon carbon tetrachloride hepatotoxicity. *J Pharmacol Exp Ther* 1961; 134:304–310.

Carnaghan RBA: Hepatic tumours and other chronic liver changes in rats following a single oral administration of aflatoxin. *Br J Cancer* 1967; 21:811–814.

Casini AF, Pompella A, and Comporti M: Liver glutathione depletion induced by bromobenzene, iodobenzene, and diethylmaleate poisoning and its relation to lipid peroxidation and necrosis. *Am J Pathol* 1985; 118:225–237.

Castro JA, De Ferreyra EC, De Castro CR, Diaz Gomez MI, D'Acosta N, and De Fenos OM: Studies on the mechanism of cystamine prevention of several liver structural and biochemical alterations caused by carbon tetrachloride. *Toxicol Appl Pharmacol* 1973; 24:1–19.

Castro JA, De Ferreyra EC, De Castro CR, De Fenos OM, Gram TE, Reagan RL, and Guarino AM: Mechanism of chloramphenicol prevention of carbon tetrachloride-induced liver damage. *Exp Mol Pathol* 1978; 28:395–405.

Casu A, Monacelli R, and Pala V: Triglyceride behaviour in liver and serum of rabbits treated with diphteria toxin. *Experientia* 1966; 22:544–545.

Cederbaum AI and Dicker E: Inhibition of microsomal oxidation of alcohols and of hydroxyl-radical-scavenging agents by the iron-chelating agent desferroxiamine. *Biochem J* 1983; 210:107–113.

Cederbaum AI, Miwa G, Cohen G, and Lu AYH: Production of hydroxyl radicals and their role in the oxidation of ethanol by a reconstituted microsomal system containing cytochrome P-450 purified from phenobarbital-treated rats. *Biochem Biophys Res Comm* 1979; 91:747–754.

Chalvardjian A: Mode of action of choline. V. Sequential changes in hepatic and serum lipids of choline-deficient rats. *Canad J Biochem* 1970; 48:1234–1240.

Channon HJ and Wilkinson H: XLII. Protein and the dietary production of fatty livers. *Biochem J* 1935; 29:350–356.

Cheeseman KH: Personal communication, 1987.

Cheeseman KH, Collins M, Maddix S, Milia A, Proudfoot K, Slater TF, Burton GW, Webb A, and Ingold KU: Lipid peroxidation in regenerating rat liver. *FEBS Lett* 1986; 209:191–196.

Cheeseman KH, Emery S, Maddix J, Slater TF, Burton GW, and Ingold KH: *Biochem. J.* 1988; 250:247–252.

Cheeseman KH, Maddix S, Milia A, Chiarpotto E, Biasi F, Albano E, and Poli G: Iron-induced lipid peroxidation and the destruction of cytochrome P-450, in Poli G, Cheeseman KH, Dianzani MU, and Slater TF (eds): *Free Radicals in Liver Injury*. Oxford, IRL Press, 1985, pp 77–80.

Chen WJ, Chi EY, and Smuckler EA: Carbon tetrachloride-induced changes in mixed function oxidases and microsomal cytochromes in the rat lung. *Lab Invest* 1977; 36:388–394.

Chen SH, Estes LW, and Lombardi B: Lecithin depletion in hepatic microsomal membranes of rats fed on a choline-deficient diet. *Exp Mol Pathol* 1972; 17:176–186.

Ch'ih JJ, Lin T, and Devlin TM: Activation and deactivation of aflatoxin B_1 in isolated rat hepatocytes. *Biochem Biophys Res Comm* 1983; 110:668–674.

Chu S and Bhakthan NMG: Experimental acetaminophen-induced hepatic necrosis. Biochemical and electron microscopic study of cystamine protection. *Lab Invest* 1978; 39:193–203.

Christensen HN: A transport system serving for mono- and diamino acids. *Proc Natl Acad Sci USA* 1964; 51:337–344.

Christensen HN and Lang M: An amino acid transport system of unassigned function in the Ehrlich ascites tumor cell. *J Biol Chem* 1965; 240:3601–3608.

Clawson GA, MacDonald JR, and Woo CH: Early hypomethylation of 2′-*O*-ribose moieties in hepatocyte cytoplasmic ribosomal RNA underlies the protein synthetic defect produced by CCl_4. *J Cell Biol* 1987; 105:705–711.

Comporti M: Biology of disease. Lipid peroxidation and cellular damage in toxic liver injury. *Lab Invest* 1985; 53:599–623.

Comporti M, Benedetti M, and Chieli E: Studies on in vitro peroxidation of liver lipids in ethanol-treated rats. *Lipids* 1973; 8:498–502.

Comporti M, Hartman A, and Di Luzio NR: Effect on in vivo and in vitro ethanol administration in liver lipid peroxidation. *Lab Invest* 1967; 16:616–624.

Connor HD, Thurman RG, Galizi MD, and Mason RP: The formation of a novel free radical metabolite from CCl_4 in the perfused rat liver and in vivo. *J Biol Chem* 1986; 261:4542–4548.

Corongiu FP, Dore M, Vargiolu S, Montaldo C,and Congiu L: Protection by lead nitrate on the 'in vitro' lipid peroxidation of rat liver. *Proc. Intern. Conf. of Industrial Environmental Xenobiotics,* Prague, 13–15 September, 1977, pp 287–289.

Corongiu FP, Lai M, and Milia A: Carbon-tetrachloride, bromotrichloromethane and ethanol acute intoxication. *Biochem J* 1983; 212:625–631.

Cottrell TC: *The Strength of Chemical Bonds,* London, Butterworth, 1958.

Cottrell RC, Lake BG, Phillips JC, and Gangolli SD: The hepatic metabolism of ^{15}N labelled dimethylnitrosamine in the rat. *Biochem Pharmacol* 1977; 26:809–813.

Craddock VM: Effect of a single treatment with the alkylating carcinogens dimethylnitrosamine, diethylnitrosamine and methyl methanesulfonate, on liver regenerating after partial hepatectomy. I. Test for induction of liver carcinomas. *Chem Biol Interac* 1975a; 10:313–321.

Craddock VM: Effect of a single treatment with alkylating carcinogens dimethylnitrosamine, diethylnitrosamine and methyl methanesulfonate, on liver regenerating after partial hepatectomy. II. Alkylation of DNA and inhibition of DNA replication. *Chem Biol Interac* 1975b; 10:323–332.

Crowe CA, Yong AC, Calder IC, Ham KN, and Tange JD: The nephrotoxicity of *p*-aminophenol. I. The effect on microsomal cytochromes, glutathione and covalent binding in kidney and liver. *Chem Biol Interac* 1979; 27:235–243.

Croy RG and Wogan GN: Temporal patterns of covalent DNA adducts in rat liver after single doses of aflatoxin B_1. *Cancer Res* 1981; 41:197–203.

Curzio M and Dianzani MU: Effects of hydroxyalkenals on neutrophil migration. *Atti Acc Sci Torino* 1983; 117:245–250.

Curzio M, Di Mauro C, Esterbauer H, and Dianzani MU: Chemotactic activity of aldehydes. Structural requirements. Role in inflammatory process. *Biomed Pharmacother* 1987; 41:304–314.

Curzio M, Esterbauer H, Di Mauro C, Cecchini G, and Dianzani MU: Chemotactic activity of the lipid peroxidation product 4-hydroxy-nonenal and homologous aldehydes. *Biol Chem Hoppe Seyler* 1986; 367:321–329.

Curzio M, Torrielli MV, Giroud JP, Esterbauer H, and Dianzani MU: Neutrophil chemotactic responses to aldehydes. *Res Commun Chem Pathol Pharm* 1982; 36:463–476.

Czygan P, Greim H, Garro AJ, Hutterer F, Schaffner F, Popper H, Rosenthal O, and Cooper DY: Microsomal metabolism of dimethylnitrosamine and the cytochrome P-450 dependency of its activation to a mutagen. *Cancer Res* 1973; 33:2983–2986.

Dahl HR and Hodgson E: Complexes of trivalent oxygenated phosphorus compound with cytochrome P-450 and cytochrome P-420: the origin of double Soret spectra. *Chem Biol Interac* 1978; 21:137–154.

Dajani RM and Kouyomyian C: A probable direct role of ethanol in the pathogenesis of fat infiltration in the rat. *J Nutr* 1967; 91:535–539.

Dalvi RR, Poore RE, and Neal RA: Studies of the metabolism of carbon disulphide by rat liver microsomes. *Life Sci* 1974; 14:1785–1796.

D'Andrea AD and Haseltine WA: Modification of DNA by aflatoxin B_1 creates alkali-labile lesions in DNA at positions of guanine and adenine. *Proc Natl Acad Sci USA* 1978; 75:4120–4124.

Daoud AH and Irving CC: Methylation of DNA in rat liver and intestine by dimethylnitrosamine and *N*-methylnitrosourea. *Chem Biol Interac* 1977; 16:135–143.

Davies MJ and Slater TF: Electron spin resonance spin trapping studies on the photolytic generation of halocarbon radicals. *Chem Biol Interac* 1986; 58:137–147.

Dawkins MJR: Carbon tetrachloride poisoning in liver of the new-born rat. *J Pathol Bacteriol* 1963; 85:189–196.

Degen GH and Neumann HG: The major metabolite of aflatoxin B_1 in the rat is a glutathione conjugate. *Chem Biol Interac* 1978; 22:239–255.

de Groot H and Haas W: Self-catalized O_2-independent inactivation of NADPH- or dithionite-reduced microsomal cytochrome P-450 by carbon tetrachloride. *Biochem Pharmacol* 1981; 30:2343–2347.

de Lamatre JG, Hoffmeier CA, Lacko AG, and Roheim PS: Distribution of apolipoprotein A-IV between the lipoprotein and the lipoprotein-free fractions of rat plasma: possible role of lecithin: cholesterol acyltransferase. *J Lipid Res* 1983; 24:1578–1585.

Demeyer DI, Tan WC, and Privett OS: Effect of essential fatty acid deficiency on lipid metabolism in isolated fat cells of epididymal fat pads of rats. *Lipids* 1974; 9:1–7.

Denk H and Eckersdorfer R: Colchicine-induced Mallory body formation in the mouse. *Lab Invest* 1977; 36:563–565.

Denk H, Eckersdorfer R, Gschnait F, Konrad K, and Wolff K: Experimental induction of hepatocellular hyalin (Mallory bodies) in mice by griseofulvin treatment. 1. Light microscopic observations. *Lab Invest* 1976; 35:377–382.

Devalia JL and McLean AEM: Covalent binding and the mechanism of paracetamol toxicity. *Biochem Pharmacol* 1983; 32:2602–2603.

de Vries J: Hepatotoxic metabolic activation of paracetamol and its derivatives phenacetin and benorilate: oxygenation or electron transfer? *Biochem Pharmacol* 1981; 30:399–402.

Dianzani MU: Uncoupling of oxidative phosphorylation in mitochondria from fatty livers. *Biochem Biophys Acta* 1954: 14:514–532.

Dianzani MU: The content of adenosine polyphosphates in fatty livers. *Biochem J* 1957; 65:116–124.

Dianzani MU: Liver steatosis induced by white phosphorus. *Morgagni* 1972; 5:1–24.
Dianzani MU: Toxic liver injury by protein synthesis inhibitors, in Popper H and Schaffner F (eds): *Progress in Liver Disease.* New York, Grune and Stratton, 1975, pp 232–245.
Dianzani MU: Biochemical aspect of fatty liver, in Slater TF (ed): *Biochemical Mechanisms of Liver Injury.* New York, Academic Press, 1978, pp 145–195.
Dianzani MU: Reactions of the liver to injury: fatty liver, in Farber E and Fisher M (eds): *Toxic Injury of the Liver.* New York, Marcel Decker, 1979a, pp 281–331.
Dianzani MU: Biological activity of methyl glyoxal and related aldehydes in submolecular pathology and cancer, in *Ciba Foundation Symposium* 67. Amsterdam, Elsevier, 1979b, pp 245–265.
Dianzani MU: Biochemical effects of saturated and unsaturated aldehydes, in McBrien DCH and Slater TF (eds): *Free Radicals Lipid Peroxidation and Cancer.* London, Academic Press, 1982, pp 129–158.
Dianzani MU: Lipoperoxidation in ethanol poisoning: a critical reconsideration. *Alcohol and alcoholism* 1985; 20:161–173.
Dianzani MU: The role of free radicals in liver damage. *Proc Nutr Soc* 1987; 46:43–52.
Dianzani MU and Gravela E: Inhibition of protein synthesis in carbon-tetrachloride-induced liver injury, in Keppler D (ed): *Pathogenesis and Mechanisms of Liver Cell Necrosis.* Lancaster, MTP, 1975, pp 225–228.
Dianzani MU and Marinari UM: The octanoate oxidation by mitochondria from fatty livers. *Biochim Biophys Acta* 1961; 48:552–561.
Dianzani MU and Poli G: Lipid peroxidation and haloalkylation in CCl_4-induced liver injury, in Poli G, Cheeseman KH, Dianzani MU and Slater TF (eds): *Free Radicals in Liver Injury.* Oxford, IRL Press, 1985, pp 149–158.
Dianzani MU and Rossi MA: Lipid peroxidation in tumours, in Pani P, Feo F, and Columbano A (eds): *Recent Trends in Chemical Carcinogenesis.* Cagliari, ESA, 1981, pp 243–257.
Dianzani MU and Scuro S: The effects of some inhibitors of oxidative phosphorylation on the morphology and enzymic activities of mitochondria. *Biochem J* 1956; 62:205–215.
Dianzani MU, Gabriel L, Gravela E, and Paradisi L: Interference of CCl_4 metabolites with subcellular structures, in Dianzani MU, Ugazio G, and Sena LM (eds): *Recent Advances in Biochemical Pathology. Toxic Liver Injury.* Torino, Minerva Medica, 1976, pp 309–319.
Dianzani MU, Canuto RA, Garcea R, Biocca ME, Pascale R, Pirisi L, and Feo F: The subcellular distribution and properties of aldehyde dehydrogenase of hepatoma AH-130, in *Internal Workshop on Membranes in Tumor Growth,* Rome, 1982, pp 541–548.
Dianzani MU, Canuto RA, Rossi MA, Poli G, Garcea R, Biocca ME, Cecchini G, Biasi F, Ferro M, and Biasi AM: Further experiments on lipid peroxidation in transplanted and experimental hepatomas. *Toxicol Pathol* 1984; 12:189–199.
Dianzani MU, Paradisi L, Parola M, Barrera G, and Rossi MA, in Craster A, de Paulet et al. (eds): *Free Radicals, Lipoproteins and Membrane Lipids.* Plenum Press, New York, 1990, pp 171–181.
Diaz Gomez MI and Castro JA: Nuclear activation of carbon tetrachloride and chloroform. *Res Commun Chem Pathol Pharm* 1980; 27:191–194.

Dich J, Bro B, Grunnet N, Jensen F, and Kondrup J: Accumulation of triacylglycerol in cultured rat hepatocytes is increased by ethanol and by insulin and dexamethasone. *Biochem J* 1983; 212:617–623.

Di Luzio NR: Prevention of the acute ethanol-induced fatty liver by the simultaneous administration of antioxidants. *Life Sci* 1964; 3:113–120.

Di Luzio NR: Influence of intravenously administered hexahydrocoenzyme Q_4 on liver injury. *Life Sci* 1966a; 5:1467–1478.

Di Luzio NR: A mechanism of the acute ethanol-induced fatty liver and the modification of liver injury by antioxidants. *Lab Invest* 1966b; 15:50–63.

Di Luzio NR and Costales F: Inhibition of the carbon tetrachloride induced fatty liver by antioxidants. *Fed Proc* 1964; 23:520.

Di Luzio NR and Costales F: Inhibition of the ethanol and carbon tetrachloride-induced fatty liver by antioxidants. *Exp Mol Pathol* 1965; 4:141–154.

Di Luzio NR and Hartman AD: Role of lipid peroxidation in the pathogenesis of the ethanol-induced fatty liver. *Fed Proc* 1967; 26:1436–1442.

Donohue TM, Tuma DJ, and Sorrell MF: Binding of metabolically derived acetaldehyde to hepatic proteins in vitro. *Lab Invest* 1983; 49:226–229.

Eber L, Fitzgerald PJ, and Herman L: Liver cell changes following ethionine. *Fed Proc* 1962; 21:303.

Ehrenreich JH, Bergeron JJM, Siekievitz P, and Palade GE: Golgi fractions prepared from rat liver homogenates. 1. Isolation procedure and morphological characterization. *J Cell Biol* 1973; 59:45–72.

Ekström G, Cronholm T, and Ingelman-Sundberg M: Hydroxyl-radical production and ethanol oxidation by liver microsomes isolated from ethanol-treated rats. *Biochem J* 1986; 233:755–761.

El-Mofty SK, Scrutton MC, Serroni A, Nicolini C, and Farber JL: Early, reversible plasma membrane injury in galactosamine-induced liver cell death. *Amer J Pathol* 1975; 79:579–596.

Esterbauer H, Cheeseman KH, Dianzani MU, Poli G, and Slater TF: Separation and characterization in the aldehydic products of lipid peroxidation stimulated by ADP-Fe^{2+} in rat liver microsomes. *Biochem J* 1982; 208:129–140.

Estler CJ: Comparison of hepatic triglyceride content and hepatic lipid secretion after various doses of ethanol. *Biochem Pharmacol* 1973; 24:1871–1873.

Estler CJ and Ammon HPT: Deposition of fat in the liver following administration of caffeine. *Experientia* 1966; 22:589–590.

Farber E and Corban MS: Sex difference in ethionine inhibition of hepatic protein synthesis. *J Biol Chem* 1958; 233:625–630.

Farber E and Popper H: Production of acute pancreatitis with ethionine and its prevention by methionine. *Proc Soc Exp Biol Med* 1950; 74:838–840.

Fazio V, Barrera G, Farace MG, Frati L, Manzari V, and Dianzani MU: HNE, a product of cellular lipid peroxidation, markedly affects *c*-myc and globin gene in K562 cell line, in *Third Annual Meeting on Oncogenes,* abstracts. Frederick, MD, 1987, pp 155.

Feigelson EB, Pfaff WW, Karmen A, and Steinberg D: The role of plasma fatty acids in development of fatty liver. *J Clin Invest* 1961; 40:2171–2179.

Felt V, Röhling S, Vohnout S, and Reichl D: On the mechanism of action of glucocorticoids on phospholipid metabolism in rabbits. *J Endocrinol* 1962; 24:309–314.

Fernandez V and Videla LA: Effect of acute and chronic ethanol ingestion on the content of reduced glutathione of various tissues of the rat. *Experientia* 1981; 37:392–394.

Fernandez G, Villaruel MC, De Toranzo EGD, and Castro JA: Covalent binding of carbon tetrachloride metabolites to the heme moiety of cytochrome P-450 and its degradation products. *Res Commun Chem Pathol Pharm* 1982; 35:283–290.

Fidanza A, Cavina G, Moretti G, and Mollica A: Fatty acid metabolism: liver lipids of pantothenic acid-deficient rats treated with omega-methylpantothenic acid. II. Gas-chromatographic analysis of lipid fraction. *Chem Abstr* 1967; 67:88691.

Fischer G, Domingo M, Lodder D, Katz N, Reinacher M, and Eigenbrodt E: *Virchows Arch* 1987; B53:359–364.

Fiume L and Laschi R: Lesioni ultrastrutturali prodotte nelle cellule parenchimali epatiche dalla falloidina e dalla alfa-amanitina. *Lo Sperimentale* 1965; 115:288–297.

Fiaccavento C, Drago A, Pagliuca C, and Torrielli MV: Osservazioni sperimentali sul danno epatico conseguente alla somministrazione di paracetamolo e tetracloruro di carbonio. *Boll Soc Ital Biol Sper* 1980a; 56:912–918.

Fiaccavento C, Rossi MA, and Torrielli MV: Relationship between GSH depletion and "in vitro" lipoperoxidative processes in liver of rats intoxicated with paracetamol. *Boll Soc Ital Biol Sper* 1980b; 56:42–48.

Fonnesu A, Del Monte U, and Olivotto M: Consumo di ossigeno e perossidazione lipidica delle frazioni citoplasmatiche extramitocondriali di fegato e di epatoma-ascite di Yoshida. *Lo Sperimentale* 1966; 116:353–372.

Forni LG, Packer JE, Slater TF, and Wilson RL: Reaction of the trichloromethyl and halothane-derived peroxy radicals with unsaturated fatty acids: a pulse radiolysis study. *Chem Biol Interac* 1983; 45:171–177.

Fox TR, Schumann AM, and Watanabe PG: *Arch Toxicol Suppl* 1987; 10:217–227.

Fox TR and Watanabe PG: *Science* 1985; 228:596–597.

Franke WW, Denk H, Schmid E, Osborn M, and Weber K: Ultrastructural, biochemical, and immunologic characterization of Mallory bodies in livers of griseofulvin-treated mice. Fimbriated rods of filaments containing prekeratin-like polypeptides. *Lab Invest* 1979; 40:207–220.

Friedman MA and Wogan GN: Liver nuclear RNA metabolism in rats treated with aflatoxin B_1. *Life Sci* 1970; 9:741–747.

Friedrich-Freksa SH, Gossner W, and Borner P: *Z Krebsforsch* 1969; 72:226–239.

Fukazawa T, Privett OS, and Takahashi Y: Effect of EFA deficiency on lipid transport from liver. *Lipids* 1971; 6:388–393.

Fulco AJ and Mead JF: Metabolism of essential fatty acids. VIII. Origin of 5,8,11-eicosatrienoic acid in the fat-deficient rat. *J Biol Chem* 1959; 234:1411–1416.

Furukawa K, Numoto S, Furuya K, Furukawa N, and Williams GM: *Cancer Res* 1985; 45:5011–5019.

Furuya K, Maeura Y, and Williams GM: *Toxicol Pathol* 1984; 12:136–142.

Gabriel L, Burdino E, Torrielli MV, and Ugazio G: Interference of DPPD with hepatic drug metabolizing system. *Pharm Res Comm* 1974; 6:127–134.

Gabriel L, Bonelli G, and Dianzani MU: Inhibition of colchicine binding to rat liver tubulin by aldehydes and by linoleic acid hydroperoxide. *Chem Biol Interac* 1977; 19:101–109.

Galand P, Jacobovitz D,and Alexandre K: *Int J Cancer* 1988; 41:155–161.

Garlick PJ, Millward DJ, James WPT, and Waterlow JL: The effect of protein deprivation and starvation on the rate of protein synthesis in tissues of the rat. *Biochim Biophys Acta* 1975; 414:71–84.

Garner RC and McLean AEM: Increased susceptibility to carbon tetrachloride poisoning in the rat after pretreatment with oral phenobarbitone. *Biochem Pharmacol* 1969; 18:645–650.

Gerber MA and Thung SN: *Am J Pathol* 1980; 98:395–400.

Germain L, Goyette R, and Marceau N: *Cancer Res* 1985; 45:673–681.

Ghadially FN (ed): *Ultrastructural Pathology of the Cell.* London and Boston, Butterworths, 1975.

Gellert J, Alderman J, and Lieber CS: Interaction between ethanol metabolism and mixed-function oxidation in alcohol dehydrogenase positive and negative deermice. *Biochem Pharmacol* 1986; 35:1037–1041.

Ghoshal AK and Farber E: Mechanisms of initiation in the carcinogenic process induced by dietary imbalance. *Models and Mechanisms in Chemical Carcinogenesis, IV Sardinian International Meeting,* Alghero, 23–27 October, 1987, abstract, p 125.

Ghoshal AK, Porta EA, and Hartroft WS: Isotopic studies on the absorption and tissue distribution of white phosphorus in rats. *Exp Mol Pathol* 1971; 14:212–219.

Ghoshal AK, Rushmore TH, and Farber E: Initiation of carcinogenesis by a dietary deficiency of choline in the absence of added carcinogens. *Cancer Lett* 1987; 36:289–296.

Glende EA and Recknagel RO: Microsomal electron transport and in vitro CCl_4-induced lipoperoxidation. *Fed Proc* 1970; 29:755.

Glenn JL, Opalka E, and Tischer K: The incorporation of labeled palmitic acid into the phospholipids of normal and fatty livers. *J Biol Chem* 1963; 238:1249–1254.

Gonda HM, Judah DJ, Arora HL, Neal GE, and Jones G: *Cancer Res* 1976; 36:2399–2407.

Gonda TJ, Sheiness DK, and Bishop JM: *Mol Cell Biol* 1982; 2:617–624.

Golden MHN, Golden BE, and Bennet FI: Relationships of trace elements deficiences to malnutrition, in Chandra RK (ed): *Nutrition and Children,* (Nestle' Nutrition, Vevey). New York, Raven Press, 1985, pp 185–207.

Gonzalez-Calvin JL, Saunders JB, and Williams R: Effects of ethanol and acetaldehyde on hepatic plasma membrane ATPase. *Biochem Pharmacol* 1983; 32:1723–1728.

Gooding PE, Chayen J, Sawyer B, and Slater TF: Cytochrome P-450 distribution in rat liver and the effect of sodium phenobarbitone administration. *Chem Biol Interac* 1978; 20:299–310.

Gordon ER: Effect of an intoxicating dose of ethanol on lipid metabolism in an isolated perfused rat liver. *Biochem Pharmacol* 1972; 21:2991–3004.

Gordon ER: ATP metabolism in an ethanol-induced fatty liver. *Biochem Pharmacol* 1977; 26:1229–1234.

Gossner VW and Friedrich-Freksa H: *Z Naturforsch* 1964; 19b:862–863.

Goyette M, Petropoulos CJ, Shank PR, and Fausto N: *Science* 1983; 219:510–512.

Gravela E: Evidence for a reduced active life-span of messenger RNA in liver of rats poisoned with carbon tetrachloride. *Exp Mol Pathol* 1973; 19:79–93.

Gravela E and Dianzani MU: Studies on the mechanism of CCl_4-induced polyribosomal damage. *FEBS Lett* 1970; 9:93–96.

Gravela E, Feo F, Canuto RA, Garcea R, and Gabriel L: Functional and structural alterations of liver ergastoplasmic membranes during DL-ethionine hepatocarcinogenesis. *Cancer Res* 1975; 35:3041–3047.

Gravela E, Pani P, Ferrari A, and Mazzarino C: Effects of the CCl_4-cycloheximide interaction on protein synthesis and lipid metabolism in rat liver. *Biochem Pharmacol* 1971; 20:3423–3430.

Gravela E, Pani P, and Bertone G: Changes in hepatotoxicity by dimethylnitrosamine in relation to modifications of the drug metabolizing enzyme system. *Ital J Biochem* 1974; 23:402–411.

Gravela E, Poli G, Albano E, and Dianzani MU: Studies on fatty liver with isolated hepatocytes. I. The action of colchicine, phalloidin, cytochalasin B, and cycloheximide on protein and triglyceride synthesis and secretion. *Exp Mol Pathol* 1977; 27:339–352.

Gressner AM and Greiling H: The state of ribosomal protein phosphorylation during thioacetamide-induced liver injury. *Exp Mol Pathol* 1978; 28:39–47.

Grollman AP: Inhibitors of protein biosynthesis. V. Effects of emetine on protein and nucleic acid biosynthesis in HeLa cells. *J Biol Chem* 1968; 243:4089–4094.

Grollman AP and Huang MT: Inhibitors of protein synthesis in eukaryotes: tools in cell research. *Fed Proc* 1973; 32:1673–1678.

Grunnet N, Kondrup J, and Dich J: Effect of ethanol on lipid metabolism in cultured hepatocytes. *Biochem J* 1985; 228:673–681.

Guzelian PS and Swisher RW: Degradation of cytochrome P-450 haem by carbon tetrachloride and 2-allyl-2-isopropylacetamide in rat liver in vivo and in vitro. Involvement of non-carbon monoxide-forming mechanisms. *Biochem J* 1979; 184:481–489.

Hahn HKJ, Barak AJ, Tuma DJ, and Sorrell MF: Phenobarbital-induced enzymatic and non-enzymatic lipid peroxidation in rat liver microsomes. *Biochem Pharmacol* 1977; 26:164–165.

Hahn HKJ, Tuma DJ, Barak AJ, and Sorrell MF: Effect of phenobarbital on lipid peroxidation in the liver. *Biochem Pharmacol* 1976; 25:769–772.

Handschumacher RE, Creasey WA, Jaffe JJ, Pasternak CA, and Hankin L: Biochemical and nutritional studies on the induction of fatty liver by dietary orotic acid. *Proc Natl Acad Sci* 1960; 46:178–186.

Harbitz I, Wallin B, Hauge JG, and Morland J: Effect of ethanol metabolism on initiation of protein synthesis in rat hepatocytes. *Biochem Pharmacol* 1984; 33:3465–3470.

Harris PM and Robinson DS: Ethionine administration in the rat. I. Effects on the liver and plasma lipids and on the disposal of dietary fat. *Biochem J* 1961; 80:352–360.

Hartroft WS and Porta EA: Present knowledge of ceroid pigment, in *Present Knowledge in Nutrition,* ed 3, The Nutrition Foundation, 1967, pp 28–31.

Hashimoto S and Recknagel, RO: No chemical evidence of hepatic lipid peroxidation in acute ethanol toxicity. *Exp Mol Pathol* 1968; 8:225–242.

Hayashida T and Portman OW: Swelling of liver mitochondria from rats fed diets deficient in essential fatty acids. *Proc Soc Exp Biol* 1960; 103:656–659.

Herron DC and Shank RC: Methylated purines in human liver DNA after probable dimethylnitrosamine poisoning. *Cancer Res* 1980; 40:3116–3117.

Hill RB and Droke WA: Production of fatty liver in the rat by cortisone. *Proc Soc Exp Biol* 1963; 114:766–769.

Hill RB, Droke WE, and Hays AP: Hepatic lipid metabolism in the cortisone-treated rat. *Exp Mol Pathol* 1965; 4:320–327.

Hoffmann KJ, Streeter AJ, Axworthy DB, and Baillie TA: Structural characterization of the major covalent adduct formed in vitro between acetaminophen and bovine serum albumin. *Chem Biol Interac* 1985; 53:155–172.

Hong Tan K, Meyer DJ, and Ketterer B: Lipid peroxidation in choline-methionine deficiency. *Free Rad Res Comms* 1987; 3:273–278.

Hutterer F: Role of impaired transmethylation of epinephrine in ethionine-induced hepatic fat accumulation. *Exp Mol Pathol* 1963; 2:541–548.

Hyams DE and Isselbacher KJ: Prevention of fatty liver by administration of adenosine triphosphate. *Nature* 1964; 204:1196–1197.

Igarashi K and Kaji A: Relationship between sites 1, 2 and acceptor, donor sites for the binding of aminoacyl tRNA to ribosomes. *Eur J Biochem* 1970; 14:41–46.

Ingall A, Lott KAK, Slater TF, Finch S, and Stier A: Metabolic activation of carbon tetrachloride to a free-radical product: studies using a spin trap. *Biochem Soc Trans* 1978; 6:962–964.

Isselbacher KJ: Metabolism and transport of lipid by intestinal mucosa. *Fed Proc* 1965; 24:16–22.

Isselbacher KJ and Budz DM: Anaerobic production of alpha-glycerolphosphate by cell-free preparations of rat hepatomas and the effect of potassium cyanide. *Nature* 1963; 200:364–365.

Isselbacher KJ and Greenberger NJ: Metabolic effects of alcohol on the liver. *New Engl J Med* 1964; 270:402–410.

James JL, Friend DS, MacDonald JR, and Smuckler EA: Alterations in hepatocyte plasma membrane in carbon tetrachloride poisoning. Freeze-fracture analysis of gap junction and electron spin resonance analysis of lipid fluidity. *Lab Invest* 1986; 54:268–274.

James JL, Moody DE, Chan CH, and Smuckler EA: The phospholipids of the hepatic endoplasmic reticulum. *Biochem J* 1982; 206:203–210.

Johnson RM: Swelling studies on liver mitochondria from essential fatty acid deficient rats. *Exp Cell Res* 1963; 32:118–129.

Kawanishi T, Ohno Y, Takahashi A, Takanaka A, Kasuya Y,and Omori Y: Relation between hepatic microsomal metabolism of *N*-nitrosamines and cytochrome P-450 species. *Biochem Pharmacol* 1985; 34:919–924.

Keppler D: in Dianzani MU, Ugazio G,and Sena LM (eds): *Recent Advances in Biochemical Pathology: Toxic Liver Injury*. Torino, Minerva Medica, 1976, p 132.

Keppler D and Decker K: Studies on the mechanism of galactosamine hepatitis: accumulation of galactosamine-1-phosphate and its inhibition of UDP-glucose pyrophosphorylase. *Eur J Biochem* 1969; 10:219–225.

Keppler D, Lesch R, Reutter W, and Decker K: Experimental hepatitis induced by D-galactosamine. *Exp Mol Pathol* 1968; 9:279–290.

Keppler DOR, Rudigier JFM, Bischoff E, and Decker KFA: The trapping of uridine phosphates by D-galactosamine, D-glucosamine, and 2-deoxy-D-galactose. A study on the mechanism of galactosamine hepatitis. *Eur J Biochem* 1970; 17:246–253.

Khanna RN and Smith AG: Fate of hexachlorobenzene in C57BL/10 mice with iron overload. *Biochem Pharmacol* 1985; 34:4157–4162.

Kiessling KH and Tobe U: Degeneration of liver mitochondria in rats after prolonged alcohol consumption. *Exp Cell Res* 1964; 33:350–354.

Kinsella JE: Increased lipoperoxide content of orotic acid-induced fatty liver. *Biochim Biophys Acta* 1967a; 137:205–207.

Kinsella JE: Protein and lipoperoxide levels in orotic acid-induced fatty liver. *Canad J Biochem* 1967b; 45:1206–1211.

Kitson TM: High concentration of aldehydes slow the reaction of cytoplasmic aldehyde dehydrogenase with thiol-group modifiers. *Biochem J* 1985; 228:765–767.

Klainer LM, Chi YM, Freidberg SL, Rall TW, and Sutherland EW: Adenyl cyclase. IV. The effects of neurohormones on the formation of adenosine 3′,5′-phosphate by preparations from brain and other tissues. *J Biol Chem* 1962; 237:1239–1243.

Klein PD and Johnson RM: Phosphorus metabolism in unsaturated fatty acid-deficient rats. *J Biol Chem* 1954; 211:103–110.

Koch OR, Boveris A, Sirotzky de Favelukes S, Schwarcz de Tarlovsky M, and Stoppani AOM: Biochemical lesions of liver mitochondria from rats after chronic alcohol consumption. *Exp Mol Pathol* 1977; 27:213–220.

Koch OR, Roatta de Conti LL, Bolanos LP, and Stoppani AOM: Ultrastructural and biochemical aspects of liver mitochondria during recovery from ethanol-induced alteration. *Am J Pathol* 1978; 90:325–337.

Koga S and Hirayama C: Disturbed release of lipoprotein from ethanol-induced fatty liver. *Experientia* 1968; 24:438–439.

Konishi Y, Shinozuka H, and Farber JL: The inhibition of rat nuclear ribonucleic acid synthesis by galactosamine and its reversal by uridine. *Lab Invest* 1974; 30:751–756.

Kroncke KD, Fricker G, Meier PJ, Gerok W, Wieland T, and Kurz J: alpha-Amanitin uptake into hepatocytes. Identification of hepatic membrane transport systems used by amatoxins. *J Biol Chem* 1986; 261:12562–12567.

Kwiram AL, Liu YC, Farquhar MN, and Smuckler EA: Binding of aflatoxin to DNA as studied by optical detection of magnetic resonance. *Biochem Biophys Res Commun* 1978; 29:1354–1359.

Laconi E, Rao PM, Rajalakshmi S, and Sarma DSR: Possible mechanisms of liver tumor promotion in the rat as revealed by orotic acid model, in *Models and Mechanisms in Chemical Carcinogenesis, IV Sardinian International Meeting,* Alghero, 23–27 October, 1987, abstract, p 57.

Lancaster MC, Jenkins FP, and McL Philp J: Toxicity associated with certain samples of groundnuts. *Nature* 1961; 192:1095–1096.

Larson RE and Plaa GL: A correlation of the effect of cervical cordotomy hypothermia, and catecholamines on carbon tetrachloride-induced hepatic necrosis. *J Pharmacol Exp Ther* 1965; 147:103–111.

Last JA: Studies on the binding of tetracycline to ribosomes. *Biochim Biophys Acta* 1969; 195:506–514.

LePage RN, Cheeseman KH, Osman N, and Slater TF: *Cell Biochem. Function,* 1988; 6:87–99.

Levin E, Johnson RM, and Albert S: Mitochondrial changes associated with essential fatty acid deficiency in rats. *J Biol Chem* 1957; 228:15–21.

Levy GN and Brabec MJ: Binding of carbon tetrachloride metabolites to rat hepatic mitochondrial DNA. *Tox Lett* 1984; 22:229–234.

Lieber CS: Metabolic effects of ethanol on the liver and other digestive organs. *Clin Gastroenterol* 1981; 10:315–342.

Lieber CS: Ethanol metabolism and toxicity, in Hodgson E, Bend JR, and Philpot RM (eds): *Reviews in Biochemical Toxicology*. New York, Elsevier Biomedical, 1983, pp 267–311.

Lieber CS and Schmid R: The effect of ethanol on fatty acid metabolism: stimulation of hepatic fatty acid synthesis in vitro. *J Clin Invest* 1961; 40:394–399.

Lieber CS, Jones DP, Mendelson JM, and DeCarli LM: Fatty liver, hyperlipemia and hyperuricemia produced by prolonged alcohol consumption despite adequate dietary intake. *Trans Ass Amer Physic* 1963; 76:289–301.

Lin JY, Pao CC, Ju ST, and Tunk TC: Polyribosome disaggregation in rat liver following administration of the phytotoxic proteins, abrin and ricin. *Cancer Res* 1972; 32:943–947.

Lindell TJ, Weinberg F, Morris PW, Roeder RG, and Rutter WJ: Specific inhibition of nuclear RNA polymerase II by alpha-amanitin. *Science* 1970; 170:447–449.

Lindros KO: Suppression by ethanol of acetate and hexanoate oxidation studied by non-recirculating liver perfusion. *Biochem Biophys Res Comm* 1970; 41:635–641.

Loeper J, Descatoire V, Maurice M, Beaune P, Feldmann G, Larrey D, and Pessayre D: *Hepatology*, 1990; 11:850–858.

Lombardi B: Fatty liver. Considerations on the pathogenesis of fatty liver. *Lab Invest* 1966; 15:1–20.

Lombardi B: Choline-devoid diet and hepatocancerogenesis in the rat, in *Models and Mechanisms in Chemical Carcinogenesis, IV Sardinian International Meeting,* Alghero, 23–27 October, 1987, abstract, p 119.

Lombardi B and Recknagel RO: Interference with secretion of triglycerides by the liver as a common factor in toxic liver injury. With some observations on choline deficiency fatty liver. *Amer J Pathol* 1962; 40:571–586.

Lombardi B, Ugazio G, and Raick AN: Rate of accumulation and source of triglycerides in the liver of choline deficient rats. *Fed Proc* 1964; 23:126.

Lombardi B, Ugazio G, and Raick AN: Choline-deficiency fatty liver: relation of plasma phospholipids to liver triglycerides. *Amer J Physiol* 1966; 210:31–36.

Lopez A: Serum lipid transport systems: recent advances. *Lipids* 1971; 6:369–377.

Lusty CJ: Lipoyl dehydrogenase from beef liver mitochondria. *J Biol Chem* 1963; 238:3443–3452.

Lutz LM, Glende EA, and Recknagel RO: Protection by diethyldithio-carbamate against carbon tetrachloride lethality in rats and against carbon tetrachloride-induced lipid peroxidation in vitro. *Biochem Pharmacol* 1973; 22:1729–1734.

Lutz WK, Jaggi W, Lüthy J, Sagelsdorff P, and Schlatter C: In vivo covalent binding of aflatoxin B_1 and aflatoxin M_1 to liver DNA of rat, mouse and pig. *Chem Biol Interac* 1980; 32:249–256.

MacDonald CM: The effects of ethanol on hepatic lipid peroxidation and on the activities of glutathione reductase and peroxidase. *FEBS Lett* 1973; 35:227–230.

Madsen NP: Reduced serum very low-density lipoprotein levels after acute ethanol administration. *Biochem Pharmacol* 1969; 18:261–262.

Magee PN: The possible role of nucleic acid methylasis in the induction of cancer. *Cancer Res* 1971; 31:599–604.

Magnusson G: The behavior of certain Lanthanons in rats. Studies on the excretion, subcellular liver distribution, and toxicology after intravenous administration. *Acta Pharmacol Toxicol* 1963; vol 20, suppl 3.

Maickel RP, Davies JI, and Weiss B: Cyclic 3′,5′-AMP: an intermediate in the activation of the adipose tissue lypolitic activity. *Fed Proc* 1965; 24:299.

Mailliard ME, Sorrell MF, Valentine GD, and Tuma DJ: Impaired plasma membrane glycoprotein assembly in the liver following acute ethanol administration. *Biochem Biophys Res Commun* 1984; 123:951–958.

Malloy S and Witt P: Effect of stress and tranquilization on plasma free fatty acid levels in the rat. *J Pharmacol Exp Ther* 1961; 132:126–130.

Mangeney-Andreani M, Sire O, Montagne-Clavel J, Nordmann R, and Nordmann J: Inhibitory effect of D-galactosamine administration on fatty acid oxidation in rat hepatocytes. *FEBS Lett* 1982; 145:267–270.

Marinozzi V and Fiume L: Effects of alpha-amanitin on mouse and rat liver cell nuclei. *Exp Cell Res* 1971; 67:311–322.

Martin CN and Garner RC: Aflatoxin B-oxide generated by chemical or enzymic oxidation of aflatoxin B_1 causes guanine substitution in nucleic acids. *Nature* 1977; 267:863–865.

Mazzanti R, Srai KS, Debnam ES, Boss AM, and Gentilini P: The effect of chronic ethanol consumption on iron absorption in rats. *Alcohol and Alcoholism* 1987; 22:47–52.

McCay PB, Lai EK, Poyer JL, DuBose CM, and Janzen EG: Oxygen- and carbon-centered free radical formation during carbon tetrachloride metabolism. Observation of lipid radicals in vivo and in vitro. *J Biol Chem* 1984; 259:2135–2143.

McLean AEM and McLean EK: The effect of diet and 1,1,1-trichloro-2,2-bis-(*p*-chlorophenyl)ethane (DDT) on microsomal hydroxylating enzymes and on sensitivity of rats to carbon tetrachloride poisoning. *Biochem J* 1966; 100:564–571.

McLean AEM and McLean EK: Diet and toxicity. *Brit Med Bull* 1969; 25:278–280.

Meldolesi J: Protective effects of alpha-tocopherol on the hepatotoxicity of carbon tetrachloride: an electron microscope study. *Exp Mol Pathol* 1968; 9:141–147.

Menzel HJ, Kladetzky RG, and Assmann G: One-step screening method for the polymorphism of apolipoproteins A-I, A-II, and A-IV. *J Lipid Res* 1982; 23:915–922.

Meszaros K, Antoni F, Mandl J, Hrabak A, and Garzo T: Effects of D-galactosamine on nucleotide metabolism and on microsomal membranes in mouse liver. *FEBS Lett* 1974; 44:141–145.

Miller JA and Miller EC: Metabolism of drugs in relation to carcinogenogenicity. *Ann NY Acad Sci* 1965; 123:125–140.

Montanaro N, Novello F, and Stirpe F: Effects of alpha-amanitin on ribonucleic acid polymerase II of rat brain nuclei and on retention of avoidance conditioning. *Biochem J* 1971; 125:1087–1090.

Montanaro L, Novello F, and Stirpe F: Inhibition of ribonucleic acid and of protein synthesis in the organs of rats and mice poisoned with alpha-amanitin. *Biochim Biophys Acta* 1973a; 319:188–198.

Montanaro L, Sperti S, and Stirpe F: Inhibition by ricin of protein synthesis in vitro. Ribosomes as the target of the toxin. *Biochem J* 1973b; 136:677–687.

Montecuccoli G, Novello F, and Stirpe F: Effect of alpha-amanitin poisoning on the synthesis of deoxyribonucleic acid and of protein in regenerating rat liver. *Biochim Biophys Acta* 1973; 319:199–208.

Mookerjea S and Chow A: Impairment of glycoprotein synthesis in acute ethanol intoxication in rats. *Biochim Biophys Acta* 1969; 184:83–92.

Mookerjea S and Sadhu DP: Metabolic depression by arsenite and malonate and cholesterologenesis. *Arch Biochem Biophys* 1955; 58:232–236.

Morton RA and Horner AA: The liver-lipid constituents of male and female rats. 1. Effects of the fat-deficiency syndrome. *Biochem J* 1961; 79:631–635.

Moscarella S, Laffi G, Coletta D, Arena V, Cappellini AP, and Gentilini P: in Gentilini P and Dianzani MU (eds): *Frontiers of Gastrointestinal Research, Liver Cirrhosis,* vol 8. P Rozen, Tel Aviv, Karger, Basel, 1984, p 208.

Moule' Y and Frayssinet C: Enzymic conversion of aflatoxin B_1 to a derivative inhibition in vitro transcription. *FEBS Lett* 1972; 25:52–56.

Mourelle M and Rubalcava B: Regeneration of the liver after carbon tetrachloride. Differences in adenylate cyclase and pancreatic hormone receptors. *J Biol Chem* 1981; 256:1656–1660.

Müller B and Dargel R: Structural and functional impairment of mitochondria from rat livers chronically injured by thioacetamide. *Acta Pharmacol Toxicol* 1984; 55:126–132.

Müller A and Sies H: Role of alcohol dehydrogenase activity and of acetaldehyde in ethanol-induced ethane and pentane production by isolated perfused rat liver. *Biochem J* 1982; 206:153–156.

Munday R: Studies on the mechanism of toxicity of the mycotoxin, sporidesmin. I. Generation of superoxide radical by sporidesmin. *Chem Biol Interac* 1982; 41:361–374.

Murad F, Chy YM, Rall TW, and Sutherland EW: Adenyl cyclase. III. The effect of catecholamines and choline esters on the formation of adenosine-3′-5′-phosphate by preparations from cardiac muscle and liver. *J Biol Chem* 1962; 237:1253–1238.

Nanni G, Pronzato MA, Averame MM, Gambella GR, Cottalasso D, and Marinari UM: Influence of acute ethanol intoxication on rat liver Golgi apparatus glycosylation activities. *FEBS Lett* 1978; 93:242–246.

Nathans D and Lipmann F: Amino acid transfer from sRNA to microsome. II. Isolation of a heat-labile factor from liver supernatant. *Biochim Biophys Acta* 1960; 43:126–128.

Nathans D and Lipmann F: Amino acid transfer from aminoacyl-ribonucleic acids to protein on ribosomes of *Escherichia coli. Proc Natl Acad Sci* 1961; 47:497–504.

Neubert D and Hoffmeister I: Intracellülare lokalisation von fehsubstanzen bei experimenteller leberverfehung. *Naunyn-Schmiedeberg's Arch Exp Pathol Pharmarkol* 1960; 237:519–537.

Nimni ME and Bavetta LA: Dietary composition and tissue protein synthesis. I. Effect of tryptophan deficiency. *Proc Soc Exp Biol Med* 1961; 108:38–43.

Niranjan BJ and Avadhani NG: Activation of aflatoxin B_1 by a mono-oxygenase system localized in rat liver mitochondria. *J Biol Chem* 1980; 255:6575–6578.

Nordmann R, Giudicelli Y, Beaugé F, Clément M, Ribiére C, Rouach H, and Nordmann J: Studies on the mechanisms involved in the isopropanol-induced fatty liver. *Biochim Biophys Acta* 1973; 326:1–11.

Nordmann R, Ribière C, and Rouach H: Involvement of iron and iron-catalyzed free radical production in ethanol metabolism and toxicity. *Enzyme* 1987; 37:57–69.

Novello F, Fiume L, and Stirpe F: Inhibition by alpha-amanitin of ribonucleic acid polymerase solubilized from rat liver nuclei. *Biochem J* 1970; 116:177–180.

Novikoff AB, Roheim PS, and Quintana N: Changes in rat liver cells induced by orotic acid feeding. *Lab Invest* 1966; 15:27–49.

O'Brien K, Moss E, Judah D, and Neal G: Metabolic basis of the species difference to aflatoxin B_1 induced hepatotoxicity. *Biochem Biophys Res Commun* 1983; 114:813–821.

Okamura K, Harwood TR, and Yokoo H: Isolation and electrophoretic study on Mallory bodies from the livers of alcoholic cirrhosis. *Lab Invest* 1975; 33:193–199.

Olsnes S and Pihl H: Isolation and properties of Abrin: a toxic protein inhibiting protein synthesis. Evidence for different biological functions of its two constituent-peptide chains. *Eur J Biochem* 1973; 35:179–185.

Olsnes S, Stirpe F, Sandvig K, and Pihl A: Isolation and characterization of Viscumin, a toxic lectin from *Viscum album* L. (Mistletoe). *J Biol Chem* 1982; 257:13263–13270.

Packer TF, Willson RL, Bahnemann D, and Asmus KD: *J Chem Soc Perkins Trans* 1980; 11:296.

Pain VM, Clemens MJ, and Garlick PJ: The effect of dietary protein deficiency on albumin synthesis and on the concentration of active albumin messenger ribonucleic acid in rat liver. *Biochem J* 1978; 172:129–135.

Pani P, Corongiu FP, Sanna A, and Congiu L: Protection by lead nitrate against carbon tetrachloride hepatotoxicity. *Drug Metab Disposition* 1975; 3:148–154.

Pani P, Columbano A, Dessi' S, Porcu M, and Congiu L: Influence of lead nitrate on dimethylnitrosamine intoxication. *Chem Biol Interac* 1976; 15:107–116.

Paoletti R, Smith RL, Maickel RP, and Brodie BB: Identification and physiological role of norepinephrine in adipose tissue. *Biochem Biophys Res Comm* 1961; 5:424–429.

Paradisi L and Dianzani MU: Cyclic nucleotides levels in the liver of rats treated with CCl_4. *Chem Biol Interac* 1979; 26:1–9.

Paradisi L, Losa GA, Cervo A, Barrera G, Parola M, and Dianzani MU: *Hepatology,* 1988; 4530.

Paradisi L, Parola M, Ribotta PB, Carasso MC, Bosia B, Amoroso L, and Dianzani MU: Effetti della somministrazione di tetracloruro di carbonio sulla attivita' della adenilato ciclasi e 3′,5′-fosfodiesterasi in plasmamembrane di fegato di ratto. *Atti Acc Sci Torino* 1982; 116:395–400.

Paradisi L, Panagini C, Parola M, Barrera G, and Dianzani MU: Effects of 4-hydroxynonenal on adenylate cyclase and 5′-nucleotidase activities in rat liver plasma membranes. *Chem Biol Interac* 1985; 53:209–217.

Parodi S, Balbi C, Taningher M, Pala M, Marchesini F, Bordone R, Sarma DSR, and Santi L: 1% Orotic acid in the diet, as a promoter of rat liver carcinogenesis, is inducing DNA fragmentation or changes in chromatin conformation? In *Models and Mechanisms in Chemical Carcinogenesis, IV Sardinian International Meeting,* Alghero, 23–27 October, 1978, abstract, pp 48–49.

Parola M, Albano E, Autelli R, Barrera G, Biocca ME, Paradisi L, and Dianzani MU: *Chem Biol Interact* 1990; 73:103–119.

Pegg AE: Effect of pretreatment with other dialkylnitrosamines on excision from hepatic DNA of O^6-methylguanine produced by dimethylnitrosamine. *Chem Biol Interac* 1978; 22:109–116.

Pegg AE and Hui G: Removal of methylated purines from rat liver DNA after administration of dimethylnitrosamine. *Cancer Res* 1978; 38:2011–2017.

Perera MIR, Demetris AJ, Katyal SL, and Shinozuka H: Lipid peroxidation of liver microsome membranes induced by choline-deficient diets and its relationship to the diet-induced promotion of the induction of gamma-glutamyltranspeptidase-positive foci. *Cancer Res* 1985; 45:2533–2538.

Perin A, Scalabrino G, Sessa A, and Arnaboldi A: In vitro inhibition of protein synthesis in rat liver as a consequence of ethanol metabolism. *Biochim Biophys Acta* 1974; 366:101–108.

Peters JA and Smith LM: The composition of the liver lipids of sheep and the effect of early sporidesmin poisoning. *Biochem J* 1964; 92:379–385.

Petrov P and Sekeris CE: Early action of alpha-amanitin on extranucleolar ribonucleoproteins, as revealed by electron microscopic observations. *Exp Cell Res* 1971; 69:393–401.

Poggi M and Di Luzio NR: The role of liver and adipose tissue in the pathogenesis of the ethanol-induced fatty liver. *J Lipid Res* 1964; 5:437–445.

Poli G, Albano E, Biasi F, Cecchini G, Carini R, Bellomo G, and Dianzani MU: Lipid peroxidation stimulated by carbon tetrachloride or iron and hepatocyte death: protective effect of vitamin E, in Poli G, Cheeseman KH, Dianzani MU, and Slater TF (eds): *Free Radicals in Liver Injury*. Oxford, IRL Press, 1985b, pp 207–215.

Poli G, Albano E, and Dianzani MU: The role of lipid peroxidation in liver damage. *Chem Phys Lipids* 1987; 45:117–142.

Poli G, Cheeseman KH, Slater TF, and Dianzani MU: The role of lipid peroxidation in CCl_4-induced damage to liver microsomal enzymes: comparative studies in vitro using microsomes and isolated liver cells. *Chem Biol Interac* 1981a; 37:13–24.

Poli G, Chiarpotto E, Biasi F, Pavia R, Albano E, and Dianzani MU: Enzymatic impairment induced by biological aldehydes in intact rat liver cells. *Res Commun Chem Pathol Pharm* 1982; 38:71–76.

Poli G, Dianzani MU, Cheeseman KH, Slater TF, Lang J, and Esterbauer H: Separation and characterization of the aldehydic products of lipid peroxidation stimulated by carbon tetrachloride or ADP-iron in isolated rat hepatocytes and rat liver microsomal suspensions. *Biochem J* 1985a; 227:629–638.

Poli G, Longhin PP, Paradisi L, and Dianzani MU: EFA-deficient rats and their responsiveness to CCl_4 poisoning. *Atti Acc Sci Torino* 1981b; 115:81–89.

Polonovski J: Quelques aspects du métabolisme des lipoprotéines. *Biochimie* 1976; 58:971–980.

Portmann B, Talbot IC, Day DW, Davidson AR, Murray-Lyon IM, and Williams R: Histopathological changes in the liver following a paracetamol overdose: correlation with clinical and biochemical parameters. *J Pathol* 1975; 117:169–181.

Pound AW: The effect of a dose of dimethylnitrosamine on the toxicity of a subsequent dose and on the toxicity of carbon tetrachloride in mice. *Br J Exp Pathol* 1975; 56:271–275.

Pound AW and Lawson TA: Partial hepatectomy and toxicity of dimethylnitrosamine and carbon tetrachloride, in relation to the carcinogenic action of dimethylnitrosamine. *Br J Cancer* 1975a; 32:596–603.

Pound AW and Lawson TA: Protection by carbon tetrachloride against the toxic effects of dimethylnitrosamine in mice. *Br J Exp Pathol* 1975b; 56:77–82.

Pound AW and Lawson TA: Reduction of carbon tetrachloride toxicity by prior administration of single small dose in mice and rats. *Br J Exp Pathol* 1975c; 56:172–179.

Purchase IFH and Steyn M: The metabolism of aflatoxin B_1 in rats. *Br J Cancer* 1969; 23:800–805.

Rao KS, Glende EA, and Recknagel RO: Effect of drug pretreatment on carbon tetrachloride-induced lipid peroxidation in rat liver microsomal lipids. *Exp Mol Pathol* 1970; 12:324–331.

Rasenack J, Koch HK, Nowack J, Lesch R, and Decker K: Hepatotoxicity of D-galactosamine in the isolated perfused rat liver. *Exp Mol Pathol* 1980; 32:264–275.

Rebouças G and Isselbacher KJ: Studies on the pathogenesis of the ethanol-induced fatty liver. II. Effect of ethanol on palmitate-^{14}C-metabolism by the isolated perfused rat liver. *J Clin Invest* 1964; 43:1338–1342.

Recknagel RO and Lombardi B: Studies of biochemical changes in subcellular particles of rat liver and their relationship to a new hypothesis regarding the pathogenesis of carbon tetrachloride fat accumulation. *J Biol Chem* 1961; 236:564–569.

Recknagel RO, Lombardi B, and Schotz MC: A new insight into pathogenesis of carbon tetrachloride fat infiltration. *Proc Soc Exp Biol Med* 1960; 104:608–610.

Reid IM, Sarma DSR, and Sidransky H: Actinomycin D and hepatic polyribosomal disaggregation. Evidence for different half-lives of mRNA of free and membrane-bound polyribosomes. *Lab Invest* 1971; 25:141–148.

Reid IM, Shinozuka H, and Sidransky H: Polyribosomal disaggregation induced by puromycin and its reversal with time. An ultrastructural study of mouse liver. *Lab Invest* 1970; 23:119–127.

Reiter R and Wendel A: Drug-induced lipid peroxidation in mice. IV. In vitro hydrocarbon evolution, reduction of oxygen and covalent binding of acetaminophen. *Biochem Pharmacol* 1983; 32:665–670.

Reitz RC: A possible mechanism for the peroxidation of lipids due to chronic ethanol ingestion. *Biochim Biophys Acta* 1975; 380:145–154.

Ross D, Albano E, Nilsson U, and Moldeus P: Thiyl radicals-formation during peroxidase-catalyzed metabolism of acetaminophen in the presence of thiols. *Biochem Biophys Res Commun* 1984; 125:109–115.

Rossi MA, Giordano M, and Dianzani MU: Effetto "in vitro" dell'etanolo e dell'acetaldeide sulla perossidazione lipidica e sul contenuto di glutatione di omogenati di fegato. *Boll Soc Ital Biol Sper* 1986; 62:1–6.

Roy AK: Effects of aflatoxin B_1 on polysomal profiles and RNA synthesis in rat liver. *Biochim Biophys Acta* 1968; 169:206–211.

Salvador RA, Atkins C, Haber S, and Conney AH: Changes in the serum concentration of cholesterol, triglycerides and phospholipids in the mouse and rat after administration of either chlorcyclizine or phenobarbital. *Biochem Pharmacol* 1970; 19:1463–1469.

Sano M and Privett OS: Effects of an essential fatty acid deficiency on serum lipoproteins in the rat. *Lipids* 1980; 15:337–344.

Sarma DSR, Reid IM, and Sidransky H: The selective effect of actinomycin D on free polyribosomes of mouse liver. *Biochem Biophys Res Commun* 1969; 36:582–588.

Saunders FC, Barker EA, and Smuckler EA: Selective inhibition of nucleoplasmic rat liver DNA-dependent RNA polymerase by aflatoxin B_1. *Cancer Res* 1972; 32:2487–2494.

Scanu AM: Structural studies on serum lipoproteins. *Biochim Biophys Acta* 1972; 265:471–508.

Schabort JC and Steyn M: Substrate and phenobarbital inducible aflatoxin-4-hydroxylation and aflatoxin metabolism by rat liver microsomes. *Biochem Pharmacol* 1969; 18:2241–2252.

Schapiro RH, Drummey GD, Shimizu Y, and Isselbacher KJ: Studies on the pathogenesis of the ethanol-induced fatty liver. II. Effect of ethanol on palmitate-I-^{14}C metabolism by the isolated perfused rat liver. *J Clin Invest* 1964; 43:1338–1343.

Schapiro RH, Scheig RL, Drummey GD, Mendelson JH, and Isselbacher KJ: Effect of prolonged ethanol ingestion on the transport and metabolism of lipids in man. *N Engl J Med* 1965; 272:610–616.

Scharnbeck H, Schaffner F, Keppler D, and Decker K: Ultrastructural studies on the effect of choline orotate on galactosamine induced hepatic injury in rats. *Exp Mol Pathol* 1972; 16:33–46.

Schauenstein E, Esterbauer H, and Zollner H: *Aldehydes in Biological Systems*. London, Pion Ltd., 1977, pp 1–205.

Seakins A and Robinson DS: Changes associated with the production of fatty livers by white phosphorus and by ethanol in the rat. *Biochem J* 1964; 92:308–312.

Sealy RC, Swartz HM, and Olive PL: Electron spin resonance-spin trapping. Detection of superoxide formation during aerobic microsomal reduction of nitro-compounds. *Biochem Biophys Res Commun* 1978; 82:680–684.

Seawright AA and McLean AEM: The effect of diet on carbon tetrachloride metabolism. *Biochem J* 1967; 105:1055–1061.

Shaw S, Jayatilleke, Ross WA, Gordon ER, and Lieber CS: Ethanol-induced lipid peroxidation: potentiation by long-term alcohol feeding and attenuation by methionine. *J Lab Clin Med* 1981; 98:415–424.

Shaw S, Rubin KD, and Lieber CS: Depressed hepatic glutathione and increased diene conjugates in alcohol liver disease. *Dig Dis Science* 1983; 28:585–589.

Shigeta Y, Nomura F, Iida S, Leo MA, Felder MR, and Lieber CS: Ethanol metabolism in vivo in the microsomal ethanol-oxidizing system in deermice lacking alcohol dehydrogenase (ADH). *Biochem Pharmacol* 1984; 33:807–814.

Shinozuka H, Gupta C, Hattori A, Betschart JM, and Virji MA: Choline-deficiency, lipid peroxidation, liver cell surface receptor alteration and liver tumor promotion, in *Models and Mechanisms in Chemical Carcinogenesis, IV Sardinian International Meeting*, Alghero, 23–27 October, 1987, abstract, p 124.

Shinozuka H, Martin JT, and Farber JL: The induction of fibrillar nucleoli in rat liver cells by D-galactosamine and their subsequent reformation into normal nucleoli. *J Ultrastruct Res* 1973; 44:279–292.

Sidransky H and Farber E: Chemical pathology of acute amino acid deficiencies. II. Biochemical changes in rats fed threonine- or methionine-devoid diets. *Arch Pathol* 1958b; 66:135–149.

Sidransky H and Verney E: Chemical pathology of acute amino acid deficiencies. VII. Morphologic and biochemical changes in young rats force-fed arginine-, leucine-, isoleucine-, or phenilalanine-devoid diets. *Arch Pathol* 1964; 78:134–148.

Sidransky H, Verney E, and Wagle DS: Effect of dietary orotic acid on liver glycogen. *Proc Soc Exp Biol Med* 1965; 120:408–412.
Sidransky H, Verney E, and Murty CN: Studies on mRNA of the livers of rats force-fed and threonine-devoid or complete diet. *J Nutr* 1974; 104:726–732.
Sidransky H, Verney E, Murty CN, Sarma DSR, and Reid M: Effect of aflatoxin B_1 on hepatic polyribosomes and protein synthesis in the rat. *Chem Biol Interac* 1977; 18:69–82.
Siegel MR and Sisler HD: Inhibition of protein synthesis in vitro by cycloheximide. *Nature* 1963; 200:675–676.
Singer SJ and Nicholson GL: The fluid mosaic model of the structure of cell membranes. *Science* 1972; 175:720–731.
Slater TF: *Free Radical Mechanisms in Tissue Injury*. London, Pion Ltd., 1972, pp 1–283.
Slater TF: Activation of carbon tetrachloride: chemical principles and biological significance, in McBrien DCH and Slater TF (eds): *Free Radical, Lipid Peroxidation and Cancer*. London, Academic Press, 1982, pp 243–274.
Slater TF and Greenbaum AL: Changes in lysosomal enzymes in acute experimental liver injury. *Biochem J* 1965; 96:484–491.
Slater TF, Sträuli UD, and Sawyer BC: Changes in liver nucleotide concentrations in experimental liver injury. I. Carbon tetrachloride poisoning. *Biochem J* 1964a; 93:260–266.
Slater TF, Sawyer BC, and Sträuli UD: Changes in liver nucleotide concentrations in experimental liver injury. II. Acute ethanol poisoning. *Biochem J* 1964b; 93:267–270.
Smith JA and De Luca HF: Structural changes in isolated liver mitochondria of rats during essential fatty acids deficiency. *J Cell Biol* 1964; 21:15–26.
Smith LC, Pownall HJ, and Gotto AM: The plasma lipoproteins: structure and metabolism. *Annu Rev Biochem* 1978; 47:751–777.
Smuckler EA, Iseri OA, and Benditt EP: An intracellular defect in protein synthesis induced by carbon tetrachloride. *J Exp Med* 1962; 116:55–71.
Smuckler EA and Koplitz M: Thioacetamide-induced alterations in nuclear RNA transport. *Cancer Res* 1974; 34:827–838.
Snyder F, Baker F, Rafter J, and Kyker GC: Octanoate oxidation in the cerium-induced fatty liver. *Biochim Biophys Acta* 1960a; 43:554–555.
Snyder F, Cress EA, and Kyker GC: Rare-earth fatty liver. *Nature* 1960b; 185:480–481.
Sperti S, Montanaro L, and Mattioli A: Studies on diphtheria toxin. The effect of GTP on the toxin-dependent adenosine diphosphate ribosylation of rat liver aminoacyl transferase II. *Chem Biol Interac* 1971; 3:141–148.
Sperti S, Montanaro L, Mattioli A, and Stirpe F: Inhibition by ricin of protein synthesis in vitro: 60S ribosomal subunit as the target of the toxin. *Biochem J* 1973; 136:813–815.
Sporn MB, Wesley Dingman C, Longacre Phelps H, and Wogan GN: Aflatoxin B_1: binding to DNA in vitro and alteration of RNA metabolism in vivo. *Science* 1966; 151:1539–1541.
Standerfer SB and Handler P: Fatty liver induced by orotic acid feeding. *Proc Soc Exp Biol Med* 1955; 90:270–271.

Stege TE: Acetaldehyde-induced lipid peroxidation in isolated hepatocytes. *Res Commun Chem Pathol Pharm* 1982; 36:287–297.

Suarez KA, Carlson GP, Fuller GC, and Fausto N: Differential acute effects of phenobarbital and 3-methylcolantrene pretreatment on carbon tetrachloride-induced hepatotoxicity in rats. *Toxicol Appl Pharmacol* 1972; 23:171–.

Suematsu T, Matsumura T, Sato N, Miyamoto T, Ooka T, Kamada T, and Abe H: Lipid peroxidation in alcoholic liver disease in humans, *Alcoholism: Clin Exp Res* 1981; 5:427–430.

Sutherland EW, Rall TW, and Menon T: Adenyl cyclase. I. Distribution, preparation, and properties. *J Biol Chem* 1962; 237:1220–1227.

Swenson DH, Lin JK, Miller EC, and Miller JA: Aflatoxin B_1-2,3-oxide as a probable intermediate in the covalent binding of aflatoxin B_1 and B_2 to rat liver DNA and ribosomal RNA in vivo. *Cancer Res* 1977; 37:172–181.

Takeuchi J, Takada A, Ebata K, Sawae G, and Okumura Y: Effect of alcohol on the liver of rats. I. Effect of a single intoxicating dose of alcohol on the livers of rats fed a choline-deficient diet or a commercial ration. *Lab Invest* 1968; 19:211–217.

Takeuchi J, Takada A, Kanayama R, Ohara N, and Okumura Y: Effect of alcohol on the liver of rats. II. Factors contributing to elevations of plasma transaminase activities and hepatic cell necrosis following a single administration of alcohol in rats. *Lab Invest* 1969; 21:398–405.

Takada A, Ikegami F, Okumura Y, Hasumura Y, Kanayama R, and Takeuchi J: Effect of alcohol on the liver of rats. III. The role of lipid peroxidation and sulphydryl compounds in ethanol-induced liver injury. *Lab Invest* 1970; 23:421–428.

Tandler B, Erlandson RA, and Wynder EL: Riboflavin and mouse hepatic cell structure and function. I. Ultrastructural alterations in simple deficiency. *Am J Pathol* 1968; 52:69–96.

Thayer WS: Inhibition of mitochondrial fatty acid oxidation in pentenoic acid-induced fatty liver. A possible model for Reye's syndrome. *Biochem Pharmacol* 1984; 33:1187–1194.

Thayer WS and Rubin E: Effects of chronic ethanol intoxication on oxidative phosphorylation in rat liver submitochondrial particles. *J Biol Chem* 1979; 254:7717–7723.

Thöenes W and Bannasch P: Electronen- und lichtmikroskopische untersuchungen am cytoplasma der leberzellen nach akuter und chronischer thioacetamid-vergiftung. *Virchows Arch Pathol Anat* 1962; 335:556–583.

Tomasi A, Albano E, Banni S, Botti B, Corongin F, Dessí MA, Iannone A, Vannini V, and Dianzani MU: *Biochem J,* 1987; 246:313–317.

Tomasi A, Albano E, Dianzani MU, Slater TF, and Vannini V: Metabolic activation of 1,2-dibromoethane to a free radical intermediate by rat liver microsomes and isolated hepatocytes. *FEBS Lett* 1983; 160:191–194.

Tomasi A, Albano E, Lott KAK, and Slater TF: Spin trapping of free radical products of CCl_4 activation using pulse radiolysis and high energy radiation procedures. *FEBS Lett* 1980; 122:303–306.

Torrielli MV and Ugazio G: Effect of DPPD on the orotic acid-induced fatty liver in the rat. *Life Sci* 1970; 9:1–7.

Torrielli MV and Slater TF: Inhibition of NADPH-cytochrome *c* reductase by propyl gallate. *Biochem Pharmacol* 1971; 20:2027–2032.

Torrielli MV, Dianzani MU, and Ugazio G: Behaviour of lipoperoxidation in rat liver during orotic acid treatment. *Life Sci* 1971; 10:99–111.

Torrielli MV, Pani P, Gabriel L, and Gravela E: The pathophysiological significance of the hepatic DMES in the liver damage induced by CCl_4 and white phosphorus. *Proc Eur Soc Study Drug Toxicity* 1973; 15:294–300.

Torrielli MV, Gabriel L, and Dianzani MU: Ethanol-induced hepatotoxicity: experimental observations on the role of lipid peroxidation. *J Pathol* 1978; 126:11–25.

Triner L and Nahas GG: Acidosis: effect on lipolytic activity of norepinephrine in isolated fat cells. *Science* 1965; 150:1725–1726.

Trush MA, Mimnaugh EG, and Gram TE: Activation of pharmacologic agents to radical intermediates. Implication for the role of free radicals in drug action and toxicity. *Biochem Pharmacol* 1982; 31:3335–3346.

Tsokos-Kuhn JO, Smith CV, Mitchell JR, Tate CA, and Entman ML: Evidence for increased membrane permeability of plasmalemmal vesicles from livers of phenobarbital-induced CCl_4 intoxicated rats. *Mol Pharmacol* 1986; 30:444–451.

Tulpule PJ and Williams JN: Study of the role of essential fatty acids in liver metabolism. *J Biol Chem* 1955; 217:229–234.

Tuma DJ, Sorrell MF, and Barak AJ: Hepatic and serum lipid patterns during development of phenobarbital induced fatty livers in rats. *Proc Soc Exp Biol Med* 1974; 146:953–956.

Ugazio G: Comportamento della lipoproteina nella steatosi epatica. *Quaderni Sclavo Diagnostica* 1971; 7:944–1003.

Ugazio G and Torrielli MV: Effect of propyl gallate on carbon tetrachloride induced fatty liver. *Biochem Pharmacol* 1969; 18:2271–2274.

Ugazio G, Gabriel L, and Burdino E: Osservazioni sperimentali sui lipidi accumulati nel fegato di ratti alimentati con dieta colino-priva. *Lo Sperimentale* 1967; 117:1–17.

Ugazio G, Congiu M, Masala BL, and Serra MU: Steatosi epatica da solfuro di carbonio. *Studi Sassaresi* 1971; 49:213–222.

Ugazio G, Koch RR, and Recknagel RO: Reversibility of liver damage in rats rendered resistant to carbon tetrachloride by prior carbon tetrachloride administration: bearing on the lipoperoxidation hypothesis. *Exp Mol Pathol* 1973; 18:281–289.

Umbenhauer DR and Pegg AE: Metabolism of dimethylnitrosamine and subsequent removal of O^6-methylguanine from DNA by isolated rat hepatocytes. *Chem Biol Interac* 1981; 33:229–238.

Ungemach FR: Plasma membrane damage of hepatocytes following lipid peroxidation: involvement of phospholipase A_2, in Poli G, Cheeseman KH, Dianzani MU, and Slater TF (eds): *Free Radicals in Liver Injury*. Oxford, IRL Press, 1985, pp 127–134.

van Kuijk FJGM, Sevanian A, Handelman GJ, and Dratz HA: A new role for phospholipase A_2: protection of membranes from lipid peroxidation damage. *TIBS* 1987; 12:31–34.

Van Waes L and Lieber CS: Alcohol liver injury, in Farber E and Fisher MM (eds): *Toxic Injury of the Liver*. New York, Marcel Decker, 1980, pp 629–653.

Vasudevan S, Laconi E, Rao PM, Rajalakshmi S, and Sarma DSR: Can metabolic disturbances generate tumor promoters? In *Models and Mechanisms in Chemical Carcinogenesis, IV Sardinian International Meeting,* Alghero, 23–27 October, 1987, abstract, p 67.

Veech RL, Guynn R, and Veloso D: The time-course of the effects of ethanol on the redox and phosphorylation states of rat liver. *Biochem J* 1972; 127:387–397.

Verbin RS, Goldblatt PJ, and Farber E: The biochemical pathology of inhibition of protein synthesis in vivo. The effects of cycloheximide on hepatic parenchymal cell ultrastructure. *Lab Invest* 1969; 20:529–536.

Verbin RS, Longnecker DS, Liang H, and Farber E: Some observations on the acute histopathologic effects of cycloheximide in vivo. *Amer J Pathol* 1971; 62:111–126.

Vereij HM, Slotboom AJ, and de Haas GH: Structure and functions of phospholipase A_2. *Rev Physiol Biochem Pharmacol* 1981; 91:91–203.

Vesselinovitch SD, Mihailovich N, Wogan GN, Lombard LS, and Rao KWN: Aflatoxin B_1, a hepatocarcinogen in the infant mouse. *Cancer Res* 1972; 32:2289–2291.

Videla LA and Valenzuela A: Alcohol ingestion, liver glutathione and lipoperoxidation: metabolic interrelations and pathological implications. *Life Sci* 1982; 31:2395–2407.

Videla LA, Fernandez V, De Marinis A, Fernandez N, and Valenzuela A: Lipoperoxidation pressure and glutathione status following acetaldehyde and aliphatic alcohol pretreatments in the rat. *Biochem Biophys Res Comm* 1982; 104:965–970.

Villa-Trevino SI: A possible mechanism of inhibition of protein synthesis by dimethylnitrosamine. *Biochem J* 1967; 105:625–631.

Vina J, Estrela JM, Guerri C, and Romero FJ: Effect of ethanol on glutathione concentration in isolated hepatocytes. *Biochem J* 1980; 188:549–552.

Von Euler LH and Windmüller HG: Fatty liver in the rat after intravenous infusion of orotic acid. *Proc Soc Exp Biol Med* 1967; 125:1251–1254.

Von Wartburg JP and Bühler R: Biology of disease. Alcoholism and aldehydism: new biochemical concepts. *Lab Invest* 1984; 50:5–15.

Wahid S, Khanna JM, Carmichael FJ, and Israel Y: Mitochondrial functions following chronic ethanol treatment: effect of diet. *Res Commun Chem Pathol Pharm* 1980; 30:477–491.

Walker JEC and Gordon ER: Biochemical aspects associated with an ethanol-induced fatty liver. *Biochem J* 1970; 119:511–516.

Waterlow JC: Oxidative phosphorylation in the livers of normal and malnourished human infants. *Proc Roy Soc B (London)* 1961; 155:96–114.

Webb JL: *Enzymes and Metabolic Inhibitors,* vol 2. New York, Academic Press, 1966, pp 615–656.

Wendel A and Hallbach J: Quantitative assessment of the binding of acetaminophen metabolites to mouse liver microsomal phospholipid. *Biochem Pharmacol* 1986; 35:385–389.

Whetton AD, Needham L, Dodd NJF, Heyworth CM, and Houslay MD: Forskolin and ethanol both perturb the structure of liver plasma membranes and activate adenylate cyclase activity. *Biochem Pharmacol* 1983; 32:1601–1608.

Wilson JW and Leduc EH: Mitochondrial changes in the liver of essential fatty acid-deficient mice. *J Cell Biol* 1963; 16:281–296.

Windmüller HG: An orotic acid-induced, adenine-reversed inhibition of hepatic lipoprotein secretion in the rat. *J Biol Chem* 1964; 239:530–537.

Wogan GN: Chemical nature and biological effects of the aflatoxins. *Bact Rev* 1966; 30:460–470.

Wogan GN: Biochemical responses to aflatoxins. *Cancer Res* 1968; 28:2282–2287.

Wong KH and Kisilevsky R: Hepatic adenylate cyclase and phosphodiesterase activity during acute ethionine intoxication. *Exp Mol Pathol* 1984; 40:122–134.

Wooles WR: Depressed hepatic fatty acid oxidation as a factor in the etiology of acute ethanol-induced fatty liver. *Life Sci* 1966; 5:267–276.

Wooles WR and Weymouth RJ: Prevention of the ethanol-induced fatty liver by chlorcyclizine-induced maintainance of hepatic lipid oxidation. *Lab Invest* 1968; 18:709–714.

Yang CS, Strickhart FS, and Woo GK: Inhibition of the mono-oxigenase system by butylated hydroxyanisole and butylated hydroxitoluene. *Life Sci* 1975; 15:1497–1505.

Yarmolinsky MB and de la Haba GL: Inhibition by puromycin of amino acid incorporation into protein. *Proc Natl Acad Sci USA* 1959; 45:1721–1729.

Ying TS, Sarma DSR, and Farber E: The sequential analysis of liver cell necrosis. Inhibition of diethylnitrosamine- and dimethylnitrosamine-induced acute liver cell death by posttreatment with diethyldithiocarbamate. *Amer J Pathol* 1980; 99:159–174.

Younes M and Siegers CP: The role of iron in the paracetamol- and CCl_4-induced lipid peroxidation and hepatotoxicity. *Chem Biol Interac* 1985; 55:327–334.

Younes M, Cornelius S, and Siegers CP: Ferrous ion supported in vivo lipid peroxidation induced by paracetamol. Its relation to hepatotoxicity. *Res Commun Chem Pathol Pharm* 1986; 51:89–99.

Young DL, Powell G, and McMillan WO: Phenobarbital-induced alterations in phosphatidylcholine and triglyceride synthesis in hepatic endoplasmic reticulum. *J Lipid Res* 1971; 12:1–8.

Yu FL: Mechanism of aflatoxin B_1 inhibition of rat hepatic nuclear RNA synthesis. *J Biol Chem* 1977; 252:3245–3251.

Zakim D and Green J: Relation of ethanol inhibition of hepatic fatty acid oxidation to ethanol-induced fatty liver. *Proc Soc Exp Biol Med* 1968; 127:138–142.

Chapter

9

Free Radical Damage and Lipid Peroxidation

Richard O. Recknagel and Eric A. Glende, Jr.
Department of Physiology and Biophysics
School of Medicine
Case Western Reserve University
Cleveland, Ohio

Robert S. Britton
Department of Internal Medicine
Division of Gastroenterology
St. Louis University Medical Center
St. Louis, Missouri

INTRODUCTION

In the study of hepatotoxicity there are a number of curious paradoxes. For example, iron is present in a number of enzymes such as cytochrome P450, the cytochromes, and cytochrome oxidase, to name a few. Iron is an indispensible component of the diet, and without it life as we know it is impossible. Yet, under certain circumstances iron can itself act as a hepatotoxin, and it plays a catalytic role in the hepatotoxicity of a number of xenobiotics. The same is true for molecular oxygen. Except for certain anaerobic microorganisms, atmospheric molecular oxygen (dioxygen) is indispensible for life. Yet, under certain circumstances oxygen can be dramatically toxic. Another case in point involves a kind of dual role for cytochrome P450. This enzyme is a key component of the mixed-function oxidase system of the hepatocellular endoplasmic reticulum (ER). As such it serves an indispensible function in a wide variety of reactions concerned with the detoxification of xenobiotics, i.e., of compounds that are "strangers to life." Yet there are a large number of xenobiotics that are not toxic in and of themselves, but whose toxicological potentialities are elicited when they are activated (converted metabolically) to toxic products by the cytochrome P450 mixed-function oxidase system. The mechanisms through which iron, oxygen, and certain hepatotoxins effect tissue damage frequently involve formation of intermediate free-radical species and the initiation of lipid peroxidation. The liver cell guards against free-radical damage and lipid peroxidation by protective mechanisms. Unchecked iron toxicity is obviated by sequestration of iron in ferritin and hemosiderin. The potential toxicity of the partial reduction products of oxygen is minimized by the catalytic action of superoxide dismutase (SOD), the glutathione (GSH) peroxidases, and catalase. Lipid peroxidation is contained by antioxidant mechanisms involving vitamin E, GSH, GSH peroxidases, and phospholipase A_2. Membrane damage ensues when these protective mechanisms are overwhelmed.

The aim of this chapter is to describe essentials of the chemistry of lipid peroxidation and the role of iron in initiation and propagation of lipid peroxidation. The major cellular antioxidant mechanisms will be presented. How lipid peroxidation damages membranes will be discussed, as well as possible mechanisms through which localized membrane damage may initiate cascades of secondary events leading to pathological involvement of parts of the cell not involved in the initial localized lipid peroxidation. Although no attempt will be made to cover the entire field of free-radical-dependent hepatic cell injury, selected aspects of a number of instances will be discussed. This will provide an opportunity to show how the general principles of free-radical-dependent lipid peroxidation apply to concrete cases.

A very large literature bears on the subject of this chapter. For example, a relevant and excellent review (Horton and Fairhurst, 1987) cites 686 literature references. It seems obvious to the authors that citation of the complete relevant literature would be out of place in this book. To keep bibliographic citations to a reasonable minimum, rather than concerning ourselves with matters of priority, we have favored citing references that in our judgement will provide interested readers with convenient entrees into the literature.

GENERAL COMMENTS ON XENOBIOTIC CELL INJURY

The liver is subject to direct injurious attack by a wide variety of so-called primary hepatotoxins (Zimmerman, 1978) including alcohols; quinones; aliphatic and aromatic halogenated hydrocarbons; sulfur compounds; complex organic compounds of biological origin, e.g., phalloidin and aflatoxin; heavy metals; rare earths; elemental phosphorus; and many others. To our knowledge, in no case has a complete description been given for the biochemical and cell physiological mechanisms set into motion by the appearance of any primary hepatotoxin in the liver cell, nor for any case has a complete description been given for the full set of causes ultimately responsible for final cell death. To provide such a complete description turns out to be an exceedingly difficult and complex task. Within the context of the study of basic mechanisms, lipid peroxidation has attracted a great deal of attention. This is due in part to the wide range of xenobiotics known to cause peroxidative decomposition of liver lipids (Table 9–1, and see Kappus, 1987) and to the fact that lipid peroxidation is so destructive for biological membranes.

Another reason for the great interest in lipid peroxidation is that for any particular xenobiotic agent, once it can be established that lipoperoxidation is involved, the investigator is provided with a range of new opportunities for further analysis of the overall problem. Indeed, in some cases study of the associated lipid peroxidation and its sequelae has provided an exceptionally deep penetration into our understanding of possible mechanisms linking initial appearance of the toxigenic agent in the liver cell and the ultimate pathological consequences. It should be added, however, that in some cases of xenobiotic toxicity where lipid peroxidation is involved (e.g., CCl_4, $CBrCl_3$, bromobenzene), the ultimate cell injury appears to depend in a complex way on a combination of lipoperoxidative effects and covalent binding of xenobiotic metabolites to membrane lipids and proteins.

DESTRUCTIVE EFFECTS OF LIPID PEROXIDATION

Unchecked peroxidative decomposition of membrane structural lipids is catastrophic for living systems. There are numerous examples. To name a few, in experimental vitamin E deficiency (Bunyan et al., 1960) peroxidative decomposition of red cell lipids occurs which leads to increased red cell permeability and hemolysis. Lipid peroxidation results in mitochondrial swelling and disintegration (Hoffsten et al., 1962). Disintegration of lysosomes has been correlated with peroxidative decomposition of lysosomal lipids (Wills and Wilkinson, 1966). Peroxidative decomposition of rat liver microsomal lipids results in loss of associated glucose-6-phosphatase and cytochrome P450 (Recknagel and Glende, 1973), loss of UDP-glucuronyltransferase, and loss of latency of nucleoside diphosphatase (de Groot et al., 1985). Loss of the capacity of liver microsomes to sequester Ca^{2+} ions as a result of microsomal lipid peroxidation is particularly significant (see below and Table 9–2; note the rapidity with which Ca^{2+} sequestration capacity collapses with onset of lipid peroxidation).

It is obvious from these examples that lipid peroxidation leads to loss of normal structure and function of biomembranes. The membrane degeneration

TABLE 9–1. Xenobiotics Causing Hepatic Lipid Peroxidation

Toxic Agent	Animal	Method of Detection	References
Halogenated Hydrocarbons			
CCl_4	Rat, human	Conjugated dienes	Recknagel and Ghoshal (1966); Hashimoto et al. (1968)
$CBrCl_3$	Rat	Conjugated dienes	Koch et al. (1974)
$CHCl_3$	PB* induced rat	Conjugated dienes	Brown (1972)
Halothane	Rat hepatocytes, PB induced rat	Conjugated dienes, MDA evolution, chemiluminescence	Tomasi et al. (1983); Brown (1972)
Bromobenzene	Mouse	Conjugated dienes, carbonyls	Casini et al. (1985)
Iodobenzene	Mouse	Conjugated dienes, carbonyls	Casini et al. (1985)
1,1,2,2-Tetra-chloroethane	Mouse	Conjugated dienes	Tomokuni (1970)
1,2-dibromoethane	Rat, rat hepatocytes	Conjugated dienes MDA evolution	Nachtomi and Alumot (1972); Albano et al. (1984)
Alcohols			
Ethanol	Baboon	Conjugated dienes	Shaw et al. (1981)
Ethanol	Human	Conjugated dienes	Shaw et al. (1983)
Ethanol†	Rat	Conjugated dienes	Sippel (1983)
Isopropanol	Rat	Conjugated dienes	Videla et al. (1982)
Propanol	Rat	Conjugated dienes	Videla et al. (1982)
Allyl alcohol	Mouse	Ethane evolution, liver MDA	Jaeschke et al. (1987)
Hydroperoxides			
Cumene hydroperoxide	Rat hepatocytes	MDA evolution	Sies and Summer (1975)
t-Butyl hydroperoxide	Rat and mouse hepatocytes	MDA evolution	Rush et al. (1985)

Redox Cycling Compounds			
Paraquat	Rat hepatocytes	MDA evolution	Watanabe et al. (1986)
Adriamycin	Rat	Fluorescent products	Thayer (1988)
Other Compounds			
Acetaldehyde	Rat	Conjugated dienes	Videla et al. (1982)
Acrylonitrile	Rat	Conjugated dienes	Silver and Szabo (1982)
Cadmium	Rat hepatocytes	MDA evolution	Stacy et al. (1980)
Cocaine	Mouse	Conjugated dienes	Kloss et al. (1983)
Endotoxin	Mouse	Ethane evolution, liver MDA	Peavy and Fairchild (1986)
Ethoxycoumarin	Rat hepatocytes	MDA evolution	Gerson et al. (1986)
Paracetamol	Mouse	Ethane evolution, liver MDA	Wendel and Feuerstein (1981)
2-(Ethylmercurimercapto)-benzoic acid	Rat hepatocytes	MDA evolution	Anundi et al. (1979)
P-Phenylenediamine	Rat hepatocytes	MDA evolution	Anundi et al. (1979)
Iodoacetamide	Rat hepatocytes	MDA evolution	Anundi et al. (1979)
Chloramine-T	Rat hepatocytes	MDA evolution	Anundi et al. (1979)
Chloroacetamide	Rat hepatocytes	MDA, ethane evolution	Anundi et al. (1979, 1980)
Diethylmaleate	Rat hepatocytes, mouse	MDA, conjugated dienes, carbonyl functions	Anundi et al. (1979); Casini et al. (1985)
Sodium vanadate	Rat hepatocytes	MDA evolution	Stacey and Klaassen (1981)

*PB, phenobarbital.
[†] Not confirmed in some laboratories.

TABLE 9–2.
Rapid Loss of Microsomal Ca^{2+} Sequestering Capacity as a Consequence of Lipid Peroxidation*

Length of Incubation (min)	MDA Production (μg/mg protein)	Microsomal Ca^{2+} Uptake (% of control)
0.5	0.07 ± 0.04	95.3 ± 3.5
1.0	0.13 ± 0.02	86.7 ± 6.5
1.5	0.19 ± 0.07	58.0 ± 24
2.0	0.41 ± 0.24	17.0 ± 14
3.0	1.15 ± 0.61	5.8 ± 6.2
4.0	1.86 ± 0.35	0.8 ± 0.3

*Conditions: rat liver microsomes incubated at 37° in the presence of 40 μM $FeSO_4$ and a NADPH-generating system (Reprinted with permission from *Biochemical Pharmacology,* Vol 32, Waller, Robert L., Glende, Eric A. Jr., and Recknagel, Richard O.: *Carbon Tetrachloride and Bromotrichloromethane Toxicity.* 1983, Pergamon Press.)

depends, at least in part, on radical-radical interactions that are themselves part of the process of termination of the lipid peroxidation (see below).

ESSENTIALS OF THE CHEMISTRY OF LIPID PEROXIDATION

The structural phospholipids of cell membranes (plasma membrane, ER, etc.) have a high density of polyenoic fatty acyl side chains, predominantly at the number 2 position of the glycerol carbon chain. The detailed chemistry of lipid peroxidation is very complex (Porter, 1984; Kappus, 1985; Simic and Taylor, 1987; Horton and Fairhurst, 1987). It is useful to divide the overall process into the three stages of initiation, propagation, and termination. Briefly, the relatively low bond dissociation energy of 75 kcal/mole for abstraction of a bis-allelic hydrogen atom of a polyenoic fatty acid (Egger and Cocks, 1973) renders these hydrogen atoms particularly susceptible to attack by free radicals (Figure 1A). Such an attack constitutes the initiation step.

Carbon-centered radicals can initiate the process, e.g., the trichloromethyl free radical ($\cdot CCl_3$), as in carbon tetrachloride hepatotoxicity. Initiation of lipid peroxidation is also associated with generation of oxygen-centered radicals. With the important exception of its participation in radical reactions, molecular O_2 is relatively unreactive. All kinds of flammable and burnable materials, including gun powder, gasoline, paper, matches, cloth, wood, paint, sugars, fats, and proteins, are in continuous contact with O_2 yet no

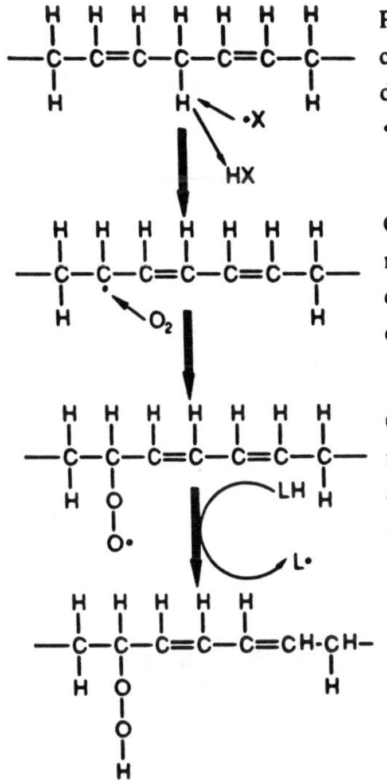

Figure 1. Initiation of lipid peroxidation. On the naturally occurring polyenoic fatty acid (A), the bis-allelic hydrogen atoms are on the methylene carbon atoms between the two double bonds. The carbon chain pictured should be imagined to be linked at one end via an ester bond to the β-carbon of the glycerol moiety of a membrane phospholipid.

combustion takes place. Combustion will take place when the temperature is elevated. Molecular O_2 is a diradical; it has two unpaired electrons. The unpaired electrons have parallel spins and for reaction (other than radical reactions) to take place there has to be a spin inversion, which requires energy input. This takes place when the temperature is raised. Despite the fact that O_2 is relatively inert, it is the major terminal oxidant for biological oxidations, as a result of enzymic catalysis. A large fraction of the O_2 consumed in normal respiration, of the order of 90% or more, undergoes tetravalent reduction in the mitochondrial cytochrome-cytochrome oxidase chain to form H_2O. The remaining 10% is consumed by reactions in the ER, in the cytoplasm, in other parts of the cell, and in mitochondrial O_2 consumption not involving cytochrome oxidase. The overall reaction catalyzed by the mitochondrial respiratory chain is

$$4H^+ + 4e^- + O_2 \longrightarrow 2H_2O$$

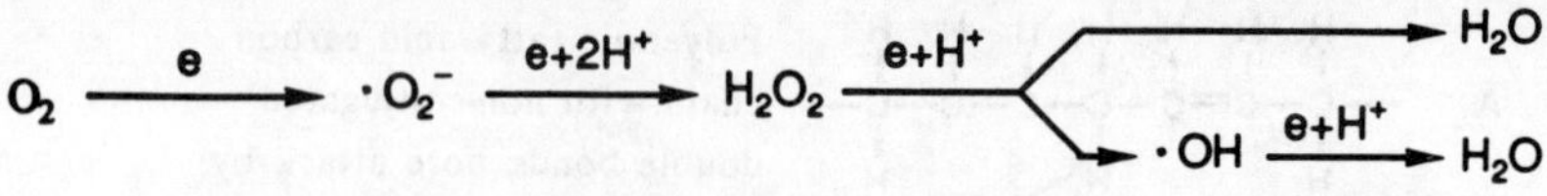

Figure 2. Sequential one-electron reduction of O_2 to water.

If this overall reaction is written as a series of one-electron steps we obtain the sequence as given in Figure 2.

The intermediate reduction products superoxide ($\cdot O_2^-$), hydrogen peroxide, and hydroxyl radical ($\cdot OH$) are responsible (under certain conditions) for the toxicity of O_2. They participate in the initiation of lipid peroxidation. See Halliwell and Gutteridge (1984, 1985a) for reviews. The exact mechanisms involved are not known with certainty. Superoxide probably does not initiate lipid peroxidation; it is more active as a reductant than as an oxidant. However, $\cdot O_2^-$ is in equilibrium with the perhydroxyl radical, $\cdot OOH$:

$$\cdot O_2^- + H^+ \longrightarrow \cdot OOH, pK_a \text{ ca. } 4.8$$

The acid conditions of ischemia and anoxia would favor appearance of $\cdot OOH$, which is much more reactive than $\cdot O_2^-$.

It is now widely recognized that ionic iron (Fe^{2+}, Fe^{3+}) and low-molecular-weight iron chelates can catalyze the initiation of lipid peroxidation. Lipid peroxidation can be completely prevented, or instantly stopped, by chelation of Fe^{3+} with desferrioxamine or other iron chelators (α,α'-dipyridyl, EDTA). Further, reduced iron plays a key role as indicated by the absolute requirement for reducing equivalents, usually present as NADPH or a NADPH generating system (Kornbrust and Mavis, 1980). The precise role played by iron in the initiation step has been widely debated. The hydroxyl radical, a much stronger oxidant than $\cdot O_2^-$, is generated in the so-called iron-catalyzed or superoxide-driven Haber-Weiss reaction (Figure 3).

For occurrence of a toxicologically meaningful rate of generation of $\cdot OH$ via the iron-catalyzed Haber-Weiss reaction, elevated rates of generation of $\cdot O_2^-$ and H_2O_2 would be needed. Elevated rates of $\cdot O_2^-$ occur in the redox cycling associated with the metabolism of a number of drugs and xenobiotics, and, with an elevated rate of $\cdot O_2^-$ generation an elevated rate of generation of

$$\cdot O_2^- + Fe^{3+} \longrightarrow O_2 + Fe^{2+} \quad \text{(reduction of } Fe^{3+} \text{ by } \cdot O_2^-)$$

$$Fe^{2+} + H_2O_2 \longrightarrow Fe^{3+} + OH^- + \cdot OH \quad \text{(Fenton reaction)}$$

$$\text{net reaction: } \cdot O_2^- + H_2O_2 \longrightarrow O_2 + OH^- + \cdot OH \quad \text{(Haber- Weiss reaction)}$$

Figure 3. Iron catalyzed Haber-Weiss reaction.

H_2O_2 would also occur via dismutation of $\cdot O_2^-$ catalyzed by superoxide dismutase (SOD):

$$\cdot O_2^- + \cdot O_2^- + 2H^+ \longrightarrow O_2 + H_2O_2$$

The rate of the dismutation reaction catalyzed by SOD is extremely fast and the steady-state concentration of $\cdot O_2^-$ is kept very low. Hydrogen peroxide, generated by the dismutation reaction, is reduced to water very effectively by cellular GSH-peroxidases (see below). Thus, SOD and the GSH-peroxidases are key enzymic components in the complex of cellular antioxidant mechanisms.

Whether ·OH participates directly in initiation of lipid peroxidation, i.e., by direct hydrogen abstraction at the methylene carbon of polyenoic fatty acyl side chains of membrane phospholipids, has not been settled definitively. The extreme reactivity of ·OH makes it unlikely that it could migrate from sites of origin in the cytosol to polyenoic fatty acids in membrane lipids. This difficulty has led to the suggestion (Halliwell and Gutteridge, 1985b) that iron in close proximity to lipids of organelle membranes could catalyze "site-selective" generation of ·OH, with subsequent initiation of lipid peroxidation. Direct participation by ·OH has been questioned (Aust et al., 1985) in favor of initiation by an oxidant requiring both Fe^{3+} and Fe^{2+}, as well as O_2. Since optimization of initiation of lipid peroxidation occurs at a 1:1 ratio of Fe^{3+} to Fe^{2+}, it has been postulated that the active oxidant is a Fe^{3+}-O_2-Fe^{2+} complex, but its precise nature has not been established (Minotti and Aust, 1987).

Once initiated, the lipid peroxidation is propagated. In Figure 1, note that after attack by the radical ·X, the products of the reaction are HX and a lipid free radical. Resonance of the free-radical electron results in shift of a double bond and generation of relatively stable lipid conjugated dienes (Figure 1B). Detection of the latter, which absorb intensely at 233 nm, has been widely used to determine whether lipid peroxidation has occurred *in vivo* or *in vitro* (Recknagel and Glende, 1984). Carbon-centered radicals (·L) react very rapidly with molecular oxygen to form oxygen-centered lipid peroxyl radicals (LOO·, Figure 1c). The latter, after abstraction of a labile hydrogen from a neighboring polyenoic fatty acid, form corresponding lipid hydroperoxides (LOOH, Figure 1d). Endoperoxides are also formed (Figure 4). These processes alone allow for a continuing linear chain reaction, since one new carbon-centered lipid radical appears for every lipid peroxyl radical formed (Figure 5).

Clearly, the linear progagation of the chain reaction depends on reaction of O_2 with the carbon-centered lipid radical ·L and the further reaction of the oxygen-centered lipid peroxyl radical (LOO·) with another polyenoic fatty acid in the hydrophobic core of the affected membrane.

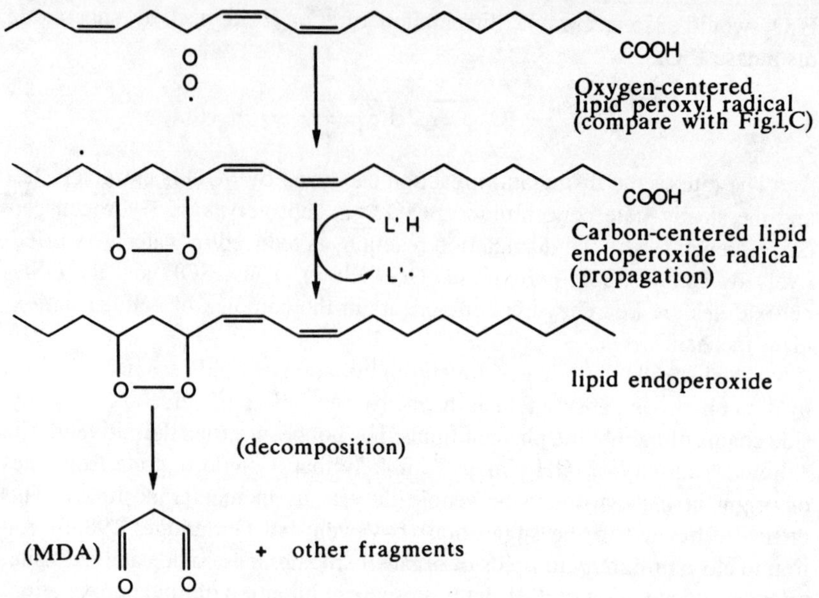

Figure 4. Formation of lipid endoperoxide and decomposition to yield malonic dialdehyde (MDA).

An important feature of the overall process is the occurrence of reductive decomposition of LOOH by Fe^{2+} to yield lipid alkoxyl radicals (LO·), and oxidative decomposition by Fe^{3+} to yield lipid peroxyl radicals (LOO·). Secondary generation of alkoxyl radicals probably predominates, since reaction of Fe^{2+} with lipid hydroperoxides is much faster than Fe^{3+} reactions (Simic and Taylor, 1987). The new free radicals that appear (Figure 6) can initiate new chains of lipid hydroperoxide formation.

Thus, once lipid peroxidation is initiated, the process can be propagated via autocatalysis, which is dependent not only on oxygen, but on metal catalyzed decomposition of transiently appearing lipid hydroperoxides. Even-

$$L\cdot + O_2 \longrightarrow LOO\cdot$$

$$LOO\cdot + L'H \longrightarrow LOOH + L'\cdot$$

$$L'\cdot + O_2 \longrightarrow L'OO\cdot$$

$$L'OO\cdot + L''H \longrightarrow L'OOH + L''\cdot \text{ ;etc.}$$

Figure 5. Linear chain propagation of lipid peroxidation.

$$LOOH + Fe^{2+} \longrightarrow LO\cdot + Fe^{3+} + OH^-$$

$$LOOH + Fe^{3+} \longrightarrow LOO\cdot + Fe^{2+} + H^+$$

Figure 6. Two modes of iron-catalyzed decomposition of a lipid hydroperoxide.

tual decomposition of the peroxidized fatty acids gives rise to a variety of stable end products. Endoperoxide decomposition yields malonic dialdehyde (Figure 4), easily detected in the widely used thiobarbituric acid (TBA) test (Recknagel et al., 1982). A variety of other carbonyl compounds also appear (Esterbauer, 1985), some of which are toxic (Benedetti et al., 1980, and see below). Peroxidative decomposition of lipids also yields ethane, from omega-3 fatty acids, e.g., linolenic acid; and pentane, from omega-6 fatty acids, e.g., linoleic acid (Sevanian and Hochstein, 1985; Horton and Fairhurst, 1987). Gas chromatographic detection of these short-chain hydrocarbons in exhaled air of experimental animals has proven to be a powerful noninvasive technique for monitoring lipid peroxidation *in vivo* (Wendel, 1987).

In living systems, propagation of lipid peroxidation is terminated through a variety of mechanisms all of which have in common the elimination of free radicals from the peroxidizing system. Termination occurs through end-stage condensation (i.e., radical-radical interaction) of a variety of oxygen-centered and carbon-centered radicals generated in the overall process (Simic and Taylor, 1987). These reactions would result in cross-linking of fatty acyl side chains of neighboring phospholipids. For example, in CCl_4-initiated microsomal lipid peroxidation *in vivo* and *in vitro,* it has been shown that grossly abnormal lipids of high molecular mass appear, evidently due to cross-linking (Link et al., 1984). In CCl_4-dependent lipid peroxidation, abnormal branched-chain fatty acids occur as a result of end-condensation of trichloromethyl radicals with lipid radicals (Gordis, 1969; Link et al., 1984). Lipid radicals can also abstract hydrogen from protein functional groups (Wolff et al., 1984) which can result in lipid-protein cross-linking. Thus, lipid radical-radical interactions, and lipid-protein radical-radical interactions curtail the autocatalytic propagation of the lipid peroxidation. In a biological setting, however, it is precisely these cross-linking reactions that no doubt contribute significantly to the membrane deterioration. Furthermore, the bifunctional aldehyde MDA can react with amine groups to form a Schiff base (Figure 7). Formation of such Schiff bases can inactivate enzymes. The products are high-molecular-weight fluorescent aggregates known as lipofuscin, ceroid, or age pigments. MDA cross-linking can also occur with the amino groups of

$$\underset{}{RNH_2} + R'NH_2 + \underset{\text{(MDA)}}{O{=}CH-CH_2-CH{=}O} \longrightarrow \underset{\text{(Schiff base)}}{R{-}N{=}CH-CH{=}CH-NH-R'} + 2H_2O$$

Figure 7. Reaction of MDA with amine groups to form a Schiff base.

phosphatidylethanolamine and phosphatidylserine. Cross-linking of lipid peroxidation products also occurs with DNA. Fluorescence of the cross-linked adducts is the basis for a method used to detect lipid peroxidation (Dillard and Tappel, 1984). Lipoperoxidative membrane breakdown also involves attack on membrane protein SH groups by 4-hydroxyalkenals (see below).

PROTECTION AGAINST LIPOPEROXIDATION

Since unchecked lipid peroxidation is so destructive for biomembranes, it is not surprising that mechanisms have evolved to protect against oxidative deterioration of membrane lipids. As discussed immediately above, one aspect of the termination of the lipoperoxidative chain reaction is radical-radical end-condensations. This leads to abnormal lipid-lipid, lipid-protein, and protein-protein configurations that are detrimental for the membrane involved. If the toxicological insult to the cell is not lethal, sooner or later the damaged membrane will be repaired. In addition, cells in general and liver cells in particular are equipped with elaborate mechanisms guarding against lipid peroxidation. With regard to initiation, the cell judiciously maintains very low concentrations of catalytically active iron, and, as already mentioned, concentrations of $\cdot O_2^-$ and H_2O_2 are kept very low via catalytic action of SOD and the GSH-peroxidases. Once initiated, other mechanisms curtail propagation. It has been known for many years that lipid peroxidation in a liver homogenate will start as soon as the supply of endogenous GSH is exhausted, and that addition of GSH promptly stops further peroxidation (Christopherson, 1966, 1968). Similarly (see Horton and Fairhurst, 1987, for further literature citations) during incubations of rat liver microsomes, usually with added ADP-Fe^{3+} and a source of NADPH, the attendant lipid peroxidation can be markedly inhibited if the cytosolic extract is added to the microsomal incubation. These inhibitory effects depend in a complex way on vitamin E, on GSH, on cytosolic and microsomal glutathione peroxidases, and on other enzymes.

In playing its role as a critically important antioxidant, vitamin E (α-tocopherol) donates a hydrogen atom from the chromanol ring hydroxyl group to the lipid radical, generating a tocopheroxyl radical. In rat liver microsomes the ratio of polyunsaturated fatty acids to vitamin E is about 1000:1. The rate of reaction of vitamin E with lipid hydroperoxyl radicals, however, is much faster (ca. 10^4) than the rate of reaction of lipid hydroperoxyl radicals with neighboring polyunsaturated fatty acids; hence vitamin E acts as an effective "chain breaker," curtailing progression of the lipid peroxidation. Because of its initially low concentration, in serving as a "lipid radical scavenger," the concentration of vitamin E in the membrane would rapidly become vanishingly small in the absence of a mechanism to reduce the tocopheroxyl radical back to tocopherol. This problem has been extensively studied. In chemical systems ascorbic acid can reduce the tocopheroxyl radical, but whether such reduction is significant in biological systems has not been settled definitively (McCay, 1985).

Whether water-soluble ascorbic acid in the cytosol would have access to the lipid-soluble tocopheroxyl radical in the membrane of the ER seems questionable, at least in the absence of enzymic facilitation. The situation with respect to a role for GSH is more clear, although far from settled in all respects. It was reported in 1982 (Reddy et al., 1982) that GSH addition markedly inhibited lipid peroxidation of isolated rat liver microsomes. It was suggested that the GSH effect was due to regeneration of vitamin E, mediated by a microsomally associated factor. Subsequent work (Haenen and Bast, 1983; Burk, 1983) showed clearly that the onset of lipid peroxidation was delayed by GSH in a concentration-dependent manner; the ultimate rate and extent of the lipid peroxidation, however, were not affected. The GSH-dependent factor was heat labile and specific for GSH; cysteine, mercapto-ethanol, dithiothreitol, propylthiouracil, and GSSG were inactive. This and related work (McCay et al., 1986) has led to the view that the tocopheroxyl radical can be reduced back to tocopherol by a microsomal enzyme that utilizes reducing equivalents from GSH. In this situation, α-tocopherol acts as a shuttle in the membrane transferring a hydrogen atom from GSH to the lipid radical. The radical reductase activity sustains the vitamin E content of the membrane at the expense of oxidation of GSH to GSSG. In the cytoplasm, GSSG is reduced back to GSH by NADPH-dependent glutathione reductase, which maintains the ratio of GSH to GSSG at about 100:1. Through this mechanism, vital protection against deleterious lipid peroxidation in the ER is linked to the considerable reductive potential of the cytoplasm, i.e., NADPH availability. If the supply of NADPH becomes limited, glutathione disulfide (GSSG) tends to accumulate and it will be transported out of the cell (Sies and Akerboom, 1984, and see below).

If α-tocopherol donates a hydrogen atom to a lipid hydroperoxyl radical (LOO·), a lipid hydroperoxide (LOOH) would appear. This would break the lipoperoxidative chain reaction. However, the danger of alkoxyl radical (LO·) generation from decomposition of LOOH functions catalyzed by Fe^{2+} would remain. Based on the work of Ursini and colleagues (see Ursini and Bindoli, 1987, for a review), it is now known that rat liver microsomes contain a phospholipid hydroperoxide glutathione peroxidase. This selenium-containing enzyme utilizes GSH for the reduction of phospholipid hydroperoxides to the corresponding alcohols. The classical cytoplasmic glutathione peroxidases will not reduce the lipid hydroperoxide function if the lipid hydroperoxide is present in a phospholipid; prior action of phospholipase A_2 is required (see below). The phospholipid hydroperoxide glutathione peroxidase of Ursini and co-workers reduces phospholipid hydroperoxides directly, without intervention of phospholipase A_2. It is presumed that this mechanism, in conjunction with the chain-breaking property of vitamin E, would normally protect the ER from excessive lipoperoxidative degeneration.

As is well-known, the polyenoic fatty acids (18:3, 20:4, 22:6) most susceptible to peroxidation are predominantly bound by ester linkage to the number 2 carbon of the glycerol moiety of phospholipids. Lipid hydroperoxides in this position, in addition to their reduction to alcohols by the phospholipid hydroperoxide glutathione peroxidase, can be eliminated through combined action of phospholipase A_2 and soluble GSH-dependent peroxidases. Phospholipase A_2 cleaves phospholipid ester bonds in the number 2 position. Normally the enzyme is relatively inactive. Through mechanisms that are not fully known, when the normal configurational structure of a membrane is altered as a result of insertion of oxygen functions, as in lipoperoxidation, the affected membrane site becomes accessible to phospholipase A_2, and this enzyme preferentially hydrolyzes peroxidized fatty acid esters in phospholipid membranes relative to rates of hydrolysis of nonperoxidized fatty acid esters (Sevanian and Kim, 1985). The hepatocellular cytoplasm contains a selenium-dependent glutathione peroxidase which reduces free fatty acid hydroperoxides to the corresponding fatty acid alcohol. The active site of the enzyme is reduced selenium (see Simic, 1981, for literature citations) which reduces the lipid hydroperoxide to the lipid alcohol:

$$\text{Enz-SeH} + \text{ROOH} \longrightarrow \text{Enz-SeOH} + \text{ROH}$$

The next step specifically requires GSH:

$$\text{Enz-SeOH} + \text{GSH} \longrightarrow \text{Enz-SeSG} + \text{H}_2\text{O}$$

Reaction with another molecule of GSH reconstitutes the enzyme:

$$\text{Enz-SeSG} + \text{GSH} \longrightarrow \text{Enz-SeH} + \text{GSSG}$$

The net reaction is

$$\text{ROOH} + 2\text{GSH} \longrightarrow \text{ROH} + \text{GSSG} + \text{H}_2\text{O}$$

This seleno-glutathione peroxidase will also reduce H_2O_2 to H_2O; note that GSSG is a product of the reaction.

The hepatocellular cytoplasm also contains a number of glutathione *S*-transferase isozymes, some of which have glutathione peroxidase-like activity (Horton and Fairhurst, 1987). These enzymes do not contain selenium. The reaction, quite different from that catalyzed by selenium glutathione peroxidase, involves attack by the glutathione thiolate anion (GS^-) on electrophilic oxygen of the hydroperoxide, resulting in reduction of the hydroperoxide to the corresponding alcohol and generation of unstable glutathione sulfenic acid (Prohaska, 1980).

$$ROOH + GSH \longrightarrow [GSOH] + ROH$$

The sulfenic acid then reacts non-enzymically with another molecule of GSH.

$$[GSOH] + GSH \longrightarrow GSSG + H_2O$$

The net reaction is the same as for the selenium glutathione peroxidase. Cytosolic NADPH-dependent glutathione reductase reduces GSSG back to GSH.

Neither the cytosolic selenium dependent glutathione peroxidase nor the peroxidase-like glutathione *S*-transferases can reduce lipid hydroperoxide functions when the peroxidized fatty acyl carbon chain is present in a phospholipid. It has been conclusively demonstrated (Tan et al., 1984; Sevanian et al., 1983; Sevanian and Kim, 1985, van Kuijk et al., 1985, 1987) that the peroxidized fatty acid must first be cleaved from the parent membrane phospholipid before the hydroperoxide function can be reduced to the alcohol by soluble GSH-dependent peroxidases. From the point of view of protection against lipoperoxidative membrane damage, this system theoretically could function as a terminator of the process by straightforward removal of lipid fatty acid hydroperoxides anticipatory to their reduction to fatty alcohols. Such a function would require rapid activation of phospholipase A_2. In the case of the action of CCl_4 and $CBrCl_3$ on rat hepatocytes (see below), phospholipase A_2 is not activated until well after initiation of lipid peroxidation. The relatively late onset of phospholipase A_2 activation suggests that a major role for phospholipase A_2 and the soluble glutathione peroxidases may be in membrane repair. Once cleaved from peroxidized phospholipids, any oxidative threat posed by appearance of free fatty acid hydroperoxides would quickly be eliminated by reduction of the fatty acid hydroperoxides to the corresponding alcohols, mediated by the soluble glutathione peroxidases. Repair of the lysophospholipid remaining in the membrane would be achieved by reacylation with an appropriate long-chain fatty acyl coenzyme A derivative. Currently, this is an active area of research and final answers are not available. Nevertheless, it can be said that the recent findings with respect to the link between GSH and vitamin E, and the link between phospholipase A_2 and the GSH-peroxidases aid greatly in understanding much earlier work on the consequences of vitamin E and selenium-deficient diets. Space limitations do not permit a survey of this interesting subject.

OXIDATIVE STRESS

Oxygen is necessary for life. Yet, as has been mentioned above, it can participate in abnormal and highly toxic metabolic sequences. All aerobic cells face the constant danger of O_2-dependent injury. During the long course of evolution the antioxidant mechanisms described above have evolved to guard cells against their built-in prooxidant tendencies. The resulting dynamic balance between prooxidant tendencies and antioxidant protection is a fundamental characteristic of aerobic life. This realization has emerged from work of the past 10 to 15 years on widely diverse subjects including the tissue injury occasioned by ischemia followed by reflow, inflammation, drug and xenobiotic toxicity, the biochemistry of GSH, vitamin E and selenium deficiency, aging, carcinogenesis, and many others (Sies, 1985). The term oxidative stress or oxidant stress has been defined as a condition in which the prooxidant-antioxidant balance is disturbed in favor of prooxidation over antioxidation. For example, in the case of CCl_4 hepatotoxicity, although reductive dehalogenation is the initial toxigenic reaction (see below), the rate of generation of carbon-centered ($\cdot CCl_3$) and oxygen-centered ($\cdot OOCCl_3$) radicals overwhelms the protective mechanisms. The resulting O_2-dependent lipid peroxidation constitutes the oxidative stress. Conversely, oxidative stress can involve situations in which antioxidant protective mechanisms have been weakened, as in vitamin E and selenium deficiency. As a corollary to the central role played by GSH in cellular antioxidant defense mechanisms, any decrease in the supply of GSH constitutes an oxidative stress, as this term is being used here. A decrease in the supply of GSH can come about through conjugation reactions, e.g., with iodoacetamide, diethyl maleate, 1-chloro-2,4-dinitrobenzene, iodomethane, or by conjugation with metabolic derivatives of bromobenzene, allyl alcohol, acetaminophen, and other drugs and xenobiotics. GSH can also be depleted by oxidation to GSSG, e.g., by diazene reagents such as azoester and diamide, by organic hydroperoxides such as *t*-butyl hydroperoxide, and by H_2O_2, as in redox cycling (see below). The supply of GSH can also be reduced by inhibiting NADPH-dependent GSH reductase with 1,3-bis(2-chloroethyl)-1-nitrosourea (BCNU) (Babson and Reed, 1978; Meredith and Reed, 1983). These conditions can impose or contribute to an oxidative stress. This does not mean that any decrease in the supply of GSH will inevitably have pathological consequences. Hepatic levels of GSH fall during fasting, but they return to normal on refeeding, without harming the liver. When, however, a fall in the level of GSH is linked to or accompanied by other circumstances, such as appearance in the cell of carbon-centered radicals, or oxygen-centered radicals combined with iron mobilization, the fall in GSH can contribute significantly to a shift in the prooxidant-antioxidant balance in favor of the former. In many, but not in all cases, the oxidative stress occasioned by GSH depletion will involve lipid peroxidation.

A complete discussion of all situations involving GSH depletion is not possible within the confines of this chapter. We have preferred to stress some of the main themes as illustrated by selected examples.

Some authors have restricted use of the term oxidative stress to the more limited context of oxidative conditions that shift the GSH-GSSG balance toward increased GSSG formation. This more limited use of the term can lead to confusion. Hepatic GSSG is preferentially released across the canalicular membrane into the nascent bile via an active transport mechanism that also transports glutathione-*S*-conjugates (Akerboom et al., 1982). The intrahepatocellular rate of oxidation of GSH to GSSG, and the rate of release of GSSG into the bile, are greatly increased by organic hydroperoxides (*t*-butyl hydroperoxide), and by redox cycling xenobiotics such as nitrofurantoin (Akerboom et al., 1982) and diquat (Lauterburg et al., 1984). Thus, the particular type of oxidative stress imposed by agents that oxidize GSH to GSSG can be detected by an increase in the rate of excretion of GSSG into the bile. BCNU, by inhibiting reduction of GSSG back to GSH, also leads to increased biliary excretion of GSSG (Smith and Mitchell, 1985). An increase in plasma concentration of GSSG has also been used as an index of "oxidative stress" (Adams et al., 1983). However, if the term oxidative stress is confined strictly to those situations in which oxidative conditions lead to an increased rate of oxidation of GSH to GSSG, and consequently to increased concentrations of GSSG in bile and plasma, then CCl_4 hepatotoxicity, which does not significantly affect the GSH-GSSG steady state, would not be considered as imposing an oxidative stress. With oxidative stress defined more broadly as any condition in which prooxidant tendencies outweigh antioxidant protection, then CCl_4 hepatotoxicity clearly does involve oxidative stress. The different uses of the term have led to exactly opposite conclusions as to whether acetaminophen hepatotoxicity involves oxidative stress (see below). In our view, the broader definition of the term, as given by Sies (1985), is more useful.

GLUTATHIONE DEPLETION AND OXIDATIVE STRESS

As can be seen from the discussion immediately above, the subject of glutathione depletion and its consequences constitutes an important subset of the larger and more general subject of oxidative stress. Glutathione is L-γ-glutamyl-L-cysteinyl-glycine. In rat liver the cytosolic concentration of GSH is about 5 m*M*. The 2GSH $\leftrightarrow$ GSSG steady state, catalyzed by NADPH-dependent glutathione reductase, is kept far to the left; the GSSG concentration is only about 1–3% of the concentration of GSH. The cytosolic pool of GSH represents about 85% of total hepatocellular GSH; mitochondrial GSH accounts for the remaining 15%. The cytosolic pool of GSH turns over rapidly with a $t_{1/2}$ of about 2 h. In contrast, the distinct mitochondrial pool turns over much more slowly, with a $t_{1/2}$ of about 30 h (Meredith and Reed, 1982). Given the rapid rate of breakdown and resynthesis of the cytosolic pool, it is understandable that availability of dietary cysteine would be necessary for maintenance of normal GSH levels in the cytosol. For example, in rats which feed diurnally, hepatic GSH reaches a peak level at about 0600 h after which it declines steadily to roughly half-maximum by 1800 hours; it recovers on refeeding (Isaacs and Binkley, 1977). GSH content of mouse liver cells can be reduced about 50% by starvation, and feeding a pure sucrose diet for 1 or 2 days is even more effective (Wendel and Feuerstein, 1981). This level of GSH depletion does not, in and of itself, pose a serious oxidative stress. However, in experimental animals (usually rats or mice) with liver GSH depleted by fasting or by sucrose-feeding certain xenobiotics will produce a severe oxidative stress. The aryl halides bromobenzene and iodobenzene have been particularly well studied in this regard (Casini et al., 1985; Comporti, 1985, 1987). Bromobenzene undergoes cytochrome P450-dependent conversion to 3,4-bromobenzene epoxide. Covalent binding of the epoxide to cellular macromolecules has long been thought to be the basic mechanism of bromobenzene-induced centrilobular necrosis. A major route of disposal of the epoxide, however, involves reaction with GSH, catalyzed by glutathione-*S*-transferases. The resulting glutathionyl conjugate is ultimately excreted as a mercapturic acid. As a consequence of the conjugation reaction, in glucose-fed mice a severe depletion of liver GSH results. Liver necrosis is evident in about one-half of the animals at about 16 to 20 h after bromobenzene intoxication, when hepatic GSH levels are depressed about 90%. Iodobenzene, somewhat more toxic than bromobenzene, produces similar effects. In the animals exhibiting hepatic necrosis, lipid peroxidation and necrosis occur at the same low-threshold value of liver GSH content, i.e., at about 10% of control values. In the animals not exhibiting necrosis, GSH levels were somewhat less severely depressed and there was little or no lipid peroxidation. Significantly, the level of covalent binding of bromobenzene metabolites was more or less uniform in all the intoxicated mice whether necrosis was evident or not. Thus, in these experiments necrosis correlated with lipid peroxidation and not with covalent binding of bromobenzene metabolites. Furthermore, when Trolox C, a water-soluble lower homologue of vitamin E, was administered at 9 and 13 h after bromobenzene intoxication, none of the animals exhibited either signs of hepatic necrosis or hepatic lipid peroxidation, whereas covalent binding of bromobenzene metabolites to liver protein was not depressed relative to bromobenzene-intoxicated control animals not given Trolox C. These results point to oxidative stress as a significant component in aryl halide hepatotoxicity.

Allyl alcohol liver injury is characterized by periportal necrosis. The alcohol is metabolized by alcohol dehydrogenase to acrolein, a highly reactive electrophile that combines with the nucleophilic SH group of GSH in a reaction catalyzed by glutathione-*S*-transferases. A dramatic decline in liver cell GSH results (Ohno et al., 1985; Jaeschke et al., 1987). With isolated hepatocytes treated with acrolein, a similar dramatic fall in GSH occurs, and in this system the release of hepatocyte cytosolic enzymes (i.e., cell necrosis) correlated time-wise with onset of lipid peroxidation (Zitting and Heinonen, 1980). When mice on a 48-h sucrose diet were given allyl alcohol, hepatic GSH was almost totally lost within 15 min. Massive lipid peroxidation ensued, as evidenced by an enormous increase in ethane evolution, which correlated with severe hepatocellular necrosis (Jaeschke et al., 1987). Clearly a sucrose diet followed by allyl alcohol administration imposed a lethal oxidative stress on the liver.

For the toxigenic aryl halides and for allyl alcohol it is doubtful whether the GSH depletion alone is responsible for the onset of the massive lipid peroxidation. A number of compounds, e.g., diethyl maleate (DEM), 1-chloro-2,4-dinitrobenzene, iodomethane, and phorone, are conjugated directly with GSH without any requirement for prior metabolic activation. The conjugation reactions are catalyzed by glutathione-*S*-transferases and result in depletion of GSH stores in the liver. Administration of DEM or phorone to sucrose-fed mice markedly reduced liver GSH, yet there was no detectable liver cell necrosis and lipid peroxidation was either absent (DEM) or minimal (phorone). Further, acrolein itself does not initiate lipid peroxidation. These results have led to speculation that in addition to the oxidative stress imposed by lowered levels of GSH, generation of oxygen-centered radicals may be involved in allyl alcohol toxicity (Jaeschke et al., 1987). For the toxigenic aryl halides (Comporti, 1987) and for allyl alcohol (Jaeschke et al., 1987), onset of the lipid peroxidation evidently requires the presence of catalytically active iron, since in both of these cases lipid peroxidation and hepatic necrosis were essentially completely prevented by desferrioxamine. To our knowledge, the mechanisms have not yet been clarified as to how iron-dependent initiation and/or propagation of lipid peroxidation become involved in the metabolism of the aryl halides and allyl alcohol.

The analgesic and antipyretic drug acetaminophen (paracetamol, *N*-acetyl-*p*-aminophenol, APAP) is hepatotoxic in large doses. Horton and Fairhurst (1987) may be consulted for a succinct review. Briefly, the metabolic pathways to formation of glucuronide and sulfate conjugates have high affinity, but they are saturable. At high doses increasing amounts of acetaminophen are metabolized via a cytochrome P450-dependent pathway (Mitchell et al., 1973a) to an active metabolite believed to be *N*-acetyl-*p*-benzoquinoneimine (NAPQI). The greater sensitivity of mouse and hamster liver cells to acetaminophen relative to the rat and human has been ascribed to differences in the rate of formation of NAPQI (Tee et al., 1987). NAPQI is conjugated with GSH in a reaction catalyzed by glutathione-*S*-transferases. When the hepatocellular GSH level falls to 20 to 30% of normal, the active metabolite covalently binds to protein SH groups. In the early 1970s it was shown that covalent binding of acetaminophen metabolites correlated rather closely with cellular necrosis (Jollow et al., 1973; Potter et al., 1973; Mitchell et al., 1973b). This led to the view that the liver cell necrosis of acetaminophen toxicity was fundamentally a question of covalent binding of the active metabolite. This covalent binding hypothesis has been questioned in a number of reports. For example, when glutathione reductase of cultured rat liver hepatocytes was inhibited with BCNU, acetaminophen toxicity was strikingly potentiated without any increase in covalent binding of [^{3}H]-acetaminophen metabolites (Gerson et al., 1985). In the same report it was shown that acetaminophen toxicity could be abolished by pretreatment of the hepatocytes with the iron chelator desferrioxamine and restored by subsequent addition of $FeCl_3$, without any detectable alterations in the degree of covalent binding. Clearly, in these experiments, acetaminophen toxicity was completely dissociated from covalent binding of [^{3}H]-acetaminophen metabolites. These experiments do not rule out the possibility that covalent binding of acetaminophen metabolites may play some role in acetaminophen toxicity *in vivo*. However, they do weaken covalent binding as a unitary hypothesis for acetaminophen toxicity and they point toward oxidative stress as an alternate mechanism. There has been some confusion on the latter point. In experiments on rats, Smith and Mitchell (1985) could find no evidence that biliary efflux of GSSG was increased as a consequence of acetaminophen toxicity and concluded that ". . . a role for oxidant stress mechanisms . . . in acetaminophen-induced hepatotoxicity is unsubstantiated and unlikely." Within the framework of GSH biochemistry, if the term oxidant stress is limited to conditions that shift the GSH-GSSG steady state toward increased GSSG, then this conclusion is justified. Nevertheless, an impressive body of work has established that a component of acetaminophen toxicity does involve an oxidative stress, as that term has been broadly defined. For example, acetaminophen intoxication leads to appearance of MDA in rat liver and lipid conjugated dienes in hamster liver (see Horton and Fairhurst, 1987, for references). In starved mice liver GSH is reduced to roughly 10% of control levels about 2 h after administration of hepatotoxic doses of acetaminophen at which time ethane evolution is greatly increased (Wendel et al., 1979). In the same report it was noted that administration of a large dose of diethyl maleate to starved mice also drove hepatic GSH content down about 90%, but there was no significant lipid peroxidation. Thus, acetaminophen toxicity appears to depend not only on the reduction of GSH stores, which compromises the GSH-dependent antioxidant mechanisms, but also on active initiation of lipid peroxidation. The latter appears to be due to a pathway of NAPQI formation that involves cytochrome P450-dependent one-electron oxidation of acetaminophen to the semiquinone free radical (De Vries, 1981;

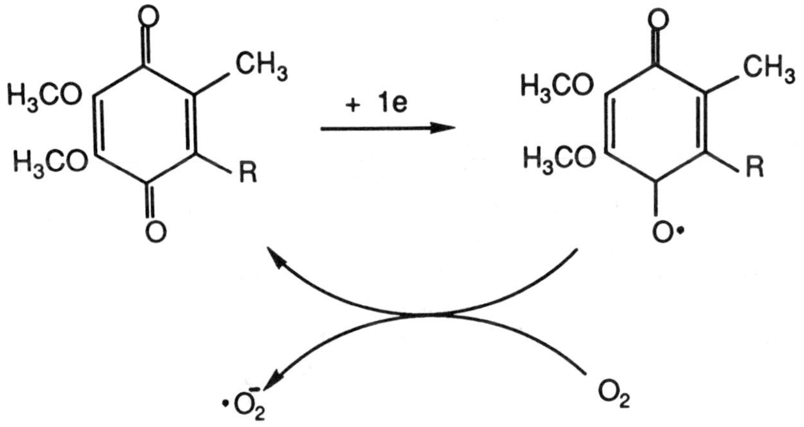

Figure 8. Generation of superoxide ($\cdot O_2^-$) by redox cycling of ubiquinone-semiquinone. For coenzyme Q_{10}, R is $(CH_2CH{=}CCH_3{-}CH_2)_{10}H$.

Rosen et al., 1983). According to this view, the semiquinone free radical would then reduce O_2 to superoxide anion free radical ($\cdot O_2^-$) which, via dismutation catalyzed by SOD would generate H_2O_2. In the presence of catalytically active iron, via generation of hydroxyl radical or other active oxygen species, lipid peroxidation could be initiated.

REDOX CYCLING

Normally at least 98% of oxygen consumption by the mitochondria involves terminal reduction of O_2 to H_2O by cytochrome oxidase. Cytochrome oxidase releases no detectable oxygen radicals into free solution. Evidently, during the course of evolution this enzyme evolved so as to keep all radical intermediates of oxygen reduction firmly bound to the protein. However, other components of the electron transport chain leak a few electrons onto O_2 while passing the great bulk of the electron flow on to the next component of the chain. The electron leak generates $\cdot O_2^-$. For example, whereas two electrons are transferred from oxidizable substrates to the respiratory chain, ubiquinone (coenzyme Q) is reduced in a one-electron transfer. A very large fraction of the ubisemiquinone free radical generated by the one electron reduction is reoxidized by the cytochrome *b* complex, but a small fraction can also reduce O_2 to $\cdot O_2^-$ (Fisher, 1988). The cyclic one-electron reduction of ubiquinone followed by reoxidation by O_2 to yield $\cdot O_2^-$ can be considered as the prototype of quinone redox cycling (Figure 8).

Normally, the steady-state concentrations of $\cdot O_2^-$ and H_2O_2 in the cytosol are kept very low by cytosolic CuZnSOD and cytosolic GSH peroxidases and catalase. CuZnSOD in the space between the inner and outer membranes of the mitochondria, and MnSOD in the mitochondrial matrix, with associated GSH peroxidases, maintain very low concentrations of the aggressively reactive O_2 species in these organelles. However, toxigenic levels of $\cdot O_2^-$ and H_2O_2 can be generated by the cyclic reduction and reoxidation of a number of quinones including menadione, and adriamycin (Smith et al., 1985). Quinone redox cycling is catalyzed by flavoenzymes including NADPH-cytochrome P450 reductase, NADPH-cytochrome b_5 reductase and NADH-ubiquinone oxidoreductase. A variety of other drugs and xenobiotics can be metabolically reduced and subsequently reoxidized by O_2 to yield $\cdot O_2^-$. These redox-cycling compounds include nitrofurans, the herbicides paraquat and diquat, the antibiotic cephaloridine, certain antimalarial drugs, mitomycin C, β-lapachone, alloxan, dialuric acid, and 6-hydroxydopamine. Space does not permit a detailed discussion. Horton and Fairhurst (1987) and Sies (1985) should be consulted for details and extensive literature citations. The essential point is that the cyclic reduction and autoxidation (i.e., direct oxidation by O_2) of free-radical intermediates generates $\cdot O_2^-$. Through dismutation catalyzed by SOD, substantial quantities of H_2O_2 can be generated, which in turn through the iron-catalyzed Haber-Weiss reaction will generate $\cdot OH$. Also, reduction of H_2O_2 by GSH peroxidases will lower cellular levels of GSH, and, in some cases (e.g., menadione) direct reaction of the drug with GSH will also lower cellular GSH levels (Smith et al., 1985). The oxidation of GSH to GSSG, and the appearance of active species of partially reduced O_2 associated with the metabolism of these redox-cycling drugs and xenobiotics can impose a severe oxidative stress. In many, but not all cases, the oxidative stress will be associated with initiation of lipid peroxidation.

It is particularly interesting to note that for isolated hepatocytes treated with adriamycin (Meredith and Reed, 1983) substantial lipid peroxidation and leakage of lactic dehydrogenase occurred only when adriamycin treatment was combined with BCNU. The combined treatment drove the mitochondrial pool of GSH down to less than 10% of control levels within about 3 h, at which time lipid peroxidation and enzyme leakage were substantial. Neither adriamycin alone nor BCNU alone could reduce the mitochondrial pool to such a low level, and treatment with either agent alone was much less toxic and produced little or no lipid peroxidation. Evidently, in this case, maintenance of an adequate supply of mitochondrial GSH is crucial in protection against the oxidative stress. Depletion of mitochondrial GSH is also critical for expression of ethacrynic acid hepatotoxicity (Meredith and Reed, 1982).

Menadione hepatotoxicity is a well-studied example of toxigenic oxidative stress that does not appear to be accompanied by lipid peroxidation. In this case an elevated concentration of cytosolic Ca^{2+} resulting from an alteration in normal protein thiol homeostasis is thought to be the key to the toxicity (Bellomo and Orrenius, 1985; Orrenius and Bellomo, 1986). The elevated levels of cytosolic Ca^{2+} activate cell proteases and plasma membrane phospholipases. In the case of cystamine-induced toxicity of rat hepatocytes, activation of a nonlysomal proteolytic system correlated with the toxicity, rather than activation of phospholipases (Nicotera et al., 1986).

EXPERIMENTAL IRON OVERLOAD

The two major causes of iron overload in humans are increased iron absorption (as occurs in hereditary hemochromatosis) and parenteral administration of iron (e.g., transfusions as a source of iron in patients with refractory anemias). Since the body lacks any effective means to excrete excess iron, when the iron burden exceeds the body's capacity for safe storage the result is widespread damage to the liver, heart, and endocrine glands. Hepatotoxicity is the most common finding in patients with iron overload. The liver is the major recipient of excess iron, and massive deposition of iron in hepatocytes can lead to the eventual development of fibrosis and cirrhosis. Clinical evidence suggests that the amount of hepatic iron is a major determinant of liver injury. Fibrosis or cirrhosis usually do not occur unless hepatic iron levels exceed a threshold concentration of about 4,000 μg Fe per gram wet weight of liver (normal: 200 μg/g). Conversely, removal of excess iron by phlebotomy or chelation produces clinical improvement. For a review of the causes and consequences of iron overload, see Gordeuk et al. (1987).

Because of the powerful pro-oxidant action of iron salts *in vitro,* it has long been suspected that lipid peroxidation may play a role in the injurious effects of iron overload. It is now clear that iron overload in experimental animals can result in lipid peroxidation *in vivo,* as evidenced by increased alkane exhalation, elevated tissue MDA levels, and conjugated dienes in tissue lipids (Britton et al., 1987). In experimental iron overload, lipid peroxidation does not occur until iron concentrations exceed a threshold level. There is a paucity of data regarding lipid peroxidation in patients with iron overload.

Bacon and his associates have developed a model of dietary iron overload in the rat, in which finely divided particles of elemental iron are added to the diet (Bacon et al., 1983). This model mimics the periportal distribution of hepatic iron seen in hereditary hemochromatosis in humans, it achieves comparable hepatic iron levels to those seen in patients with iron overload, and hepatic fibrosis eventually results. Using this model and measuring conjugated dienes to localize the sites of peroxidation, these workers have demonstrated that lipid peroxidation occurs *in vivo* in both hepatic mitochondria and endoplasmic reticulum, and that the former have a lower-threshold iron concentration for peroxidation (approximately 1,000 and 2,000 μg/g wet wt, respectively). Additionally, the lipid peroxidation in liver mitochondria and microsomes is associated with functional abnormalities in these organelles, including impaired mitochondrial oxidative metabolism and decreased cytochrome P450 content in the microsome fraction. Although this finding does not prove direct causality, it seems most likely that lipid peroxidation is responsible since similar functional deficits are produced by iron-induced lipid peroxidation in these organelles *in vitro*. Increased lysosomal fragility has been detected in homogenates of liver samples from both iron-loaded human patients and experimental animals (Peters and Seymour, 1976; Peters et al., 1985); however, the existence of *in vivo* lipid peroxidation in lysosomes from iron-loaded livers has not yet been assessed.

The form of intracellular iron responsible for initiating hepatic lipid peroxidation in iron overload is not known. Presumably, with progressively increasing iron deposition, the capacity of the hepatocyte to maintain iron in storage forms is exceeded, with a resultant increase in the putative intracellular transit pool of iron. It is widely accepted that cells contain iron in low-molecular-weight form(s), perhaps chelated to amino acids, citrate, or nucleotides (Jacobs, 1977; Bacon and Tavill, 1984). Recent evidence from experiments utilizing rats with dietary iron overload supports the concept that iron overload results in an increase in an hepatic cytosolic pool of low-molecular-weight iron which is catalytically active in stimulating lipid peroxidation (Britton et al., 1987).

CARBON TETRACHLORIDE AND BROMOTRICHLOROMETHANE

The hepatotoxicity of halogenated hydrocarbons in its general aspects will be discussed elsewhere in this book. Comments here will be confined to CCl_4 and $CBrCl_3$. CCl_4 and its closely related congener $CBrCl_3$ are classic examples of toxigenic halogenomethanes. $CBrCl_3$ is considerably more toxic than CCl_4, a fact that depends on a lower bond dissociation energy for the C–Br bond relative to the C–Cl bond (Koch et al., 1974). In the case of CCl_4, it is well known that its toxicity depends on reductive dehalogenation by a specific cytochrome P450 isozyme (see Recknagel and Glende, 1973; Recknagel et al., 1977, Brattin et al., 1985, and Recknagel et al., 1989, for reviews on CCl_4 hepatotoxicity). Recently it has been shown that isolated rat liver mitochondria also carry out reductive dehalogenation of CCl_4 (Tomasi et al., 1987). The one-electron reduction generates trichloromethyl radical:

$$CCl_4 + e^- \longrightarrow \cdot CCl_3 + Cl^-$$

That $\cdot CCl_3$ radicals are generated from CCl_4 metabolism both *in vivo* and *in vitro* has been conclusively demonstrated by use of the spin trap phenyl-*t*-butylnitrone. Further reaction of $\cdot CCl_3$ radicals with molecular O_2 to form trichloromethyl peroxyl radicals ($\cdot OOCCl_3$) occurs very rapidly. In the ER, presumably near to the critical cytochrome P450 isozyme, the free radicals $\cdot CCl_3$ and $\cdot OOCCl_3$ attack methylene carbons (see $\cdot X$ in Figure 1) of polyenoic fatty acids of the membrane phospholipids, thus initiating onset of lipid peroxidation. Covalent binding of CCl_4 cleavage products to microsomal lipids and proteins also occurs. As a result of the combined effects of lipid peroxidation and covalent binding of CCl_4 metabolic products, concomitant prompt onset of structural and functional derangeament of the ER ensues. Cytochrome P450 activity, and *pari passu,* mixed-function oxidase activity are lost rapidly. Interestingly, microsomal NADPH-cytochrome P450 reductase activity is essentially unaffected. A decline of hepatic protein synthesis and glucose-6-phosphatase activity are early manifestations. A particularly interesting effect is rapid collapse of the capacity of the ER to sequester Ca^{2+} ions (Lowrey et al., 1981a). In contrast to the rapid collapse of structure and function in the ER, the mitochondrial elements of the cell are remarkably resistant both structurally and functionally. Mitochondrial resistance is also reflected in considerably less mitochondrial lipid peroxidation and covalent binding of CCl_4 cleavage products relative to the ER (Rao and Recknagel, 1968, 1969). Since $\cdot CCl_3$ and $\cdot OOCCl_3$ can act as initiators of lipid peroxidation, Fe^{2+} does not appear to be required at this stage. In fact, it has been shown experimentally that CCl_4 can initiate lipid peroxidation in NADPH-supplemented rat liver microsomes *in vitro* in the absence of any metals chelateable by high concentrations of EDTA. However, CCl_4 hepatotoxicity *in vivo* does depend to a certain extent on iron availability since it has been shown (Younes and Siegers, 1985) that treatment of rats with the iron chelator desferrioxamine significantly attenuated CCl_4-dependent lipid peroxidation and liver injury. This case illustrates a probable role for iron in the propagation stage of the lipid peroxidation (Figure 6). It has been noted by a number of investigators that there is little or no change in hepatocellular GSH content associated with CCl_4 hepatotoxicity. CCl_4 and $CBrCl_3$ belong to a class of radical-generating xenobiotics whose hepatotoxicity does not depend on exhaustion of GSH-dependent protective mechanisms.

An interesting feature of CCl_4 hepatotoxicity involves competition with oxygen. It has been shown conclusively that molecular O_2 competes with CCl_4 for available cytochrome P450 (de Groot and Noll, 1986). Because of this competition, the metabolism and toxicity of CCl_4, *in vitro,* are maximal at low PO_2's in thc neighborhood of 5–10 mmHg. Since the PO_2 is known to be low in the centrilobular regions of the liver, this circumstance explains, at least in part, the predominantly centrilobular disposition of CCl_4 liver injury in the whole animal. Availability of the "oxystat" system used in this work has led (deGroot et al., 1988) to the demonstration that CCl_4-dependent death of hepatocytes *in vitro* clearly depends on the induced lipid peroxidation and not on covalent binding of CCl_4-derived metabolic cleavage products.

SECONDARY MECHANISMS: CALCIUM HOMEOSTASIS

Lipid peroxidation and covalent binding of CCl_4-cleavage products appear to be confined initially to loci in the ER in and around cytochrome P450 and, to a lesser extent, in the mitochondrial elements of the cell (Tomasi et al., 1987). A problem arises as to the mechanism or mechanisms that link these early events of CCl_4 metabolism to pathological degeneration in other cell organelles and in particular in the plasma membrane. This is the so-called problem of "pathological action at a distance" (Recknagel, 1983). Free-radical products of CCl_4 metabolism and free-radical intermediates arising during microsomal or mitochondrial lipid peroxidation are probably too reactive and too short-lived to move from their loci of origin through the cytosol to other parts of the cell. This fundamental problem, not unique to the case of CCl_4 hepatotoxicity, gave rise to the idea that nonradical toxicological second messengers probably appear during CCl_4 metabolism and consequent lipoperoxidation. Work in this area has been dominated by two ideas. One is that nonradical products of intracellular membrane lipid peroxidation, in particular 4-hydroxyalkenals, could act as the putative toxicological second messengers (Benedetti et al., 1980; Comporti, 1985, Esterbauer, 1985). The 4-hydroxyalkenals, in particular 4-hydroxynonenal, are quite toxic. They form so-called Michael addition compounds by conjugation with protein SH groups. There appears to be little doubt that the 4-hydroxyalkenals contribute to breakdown of the ER in CCl_4 and bromobenzene toxicity. These compounds, however, are rapidly detoxified in the cell cytoplasm by conjugation with GSH, catalyzed by glutathione *S*-transferase (Alin et al., 1985), and by reduction to corresponding alcohols, catalyzed by alcohol dehydrogenase (Esterbauer, 1985). For these reasons it has been questioned (Recknagel, 1983; Recknagel and Glende, 1988b) whether the 4-hydroxyalkenals could play a significant role as toxicological second messengers capable of traversing the GSH-rich hepatocellular cytoplasm.

Another possibility, viz., that eventual plasma membrane breakdown in CCl_4 toxicity might involve a disturbance of hepatocellular Ca^{2+} homeostasis emerged from work (Moore et al., 1976; Lowrey et al., 1981a, 1981b) which showed that the Ca^{2+} sequestering capacity of the ER was very sensitive to CCl_4 metabolism *in vivo* and *in vitro* (also, see Table 9–2). This work prompted the suggestion (Recknagel, 1983) that CCl_4 metabolism at the level of the ER might result in a protracted and unphysiological rise in the concentration of Ca^{2+} in the cytoplasm with toxicological consequences. The fact that CCl_4 added to isolated hepatocytes causes glycogen mobilization (Dolak et al., 1985) and conversion of glycogen phosphorylase to the active *a* form (Long and Moore, 1986) supported the idea, since these effects depend on a rise of cytosolic Ca^{2+} concentration. Further, in experiments based on use of the

Ca^{2+}-fluorescent probe Fura-2, it has been shown that isolated hepatocytes experience a small but significant rise in cytosolic Ca^{2+} when exposed to CCl_4 at concentrations sufficiently low (<1.0 m*M*) to not cause direct permeabilization of the cell plasma membrane (Dolak et al., 1988). The rise in cytosolic Ca^{2+} may be a key link in a chain of causality leading from metabolism of CCl_4 and $CBrCl_3$ to ultimate cellular necrosis. Otherwise latent phospholipase A_2 is known to be activated by Ca^{2+} ions (van den Bosch, 1980; Chien et al., 1980) and hydrolysis of a relatively small fraction of membrane phospholipids has profound degenerative consequences for the biological membrane involved (see Dolak et al., 1985). Thus, a complete theory for CCl_4- and $CBrCl_3$-dependent cellular necrosis can be seen to emerge, leading from CCl_4 activation, through lipid peroxidation and covalent binding of cleavage products, to a rise in cytosolic Ca^{2+} concentration and finally to activation of latent phospholipase A_2 (Figure 9, also see Recknagel and Glende, 1988a,b, and Recknagel et al., 1989).

For isolated hepatocytes exposed to low concentrations of CCl_4, phos-

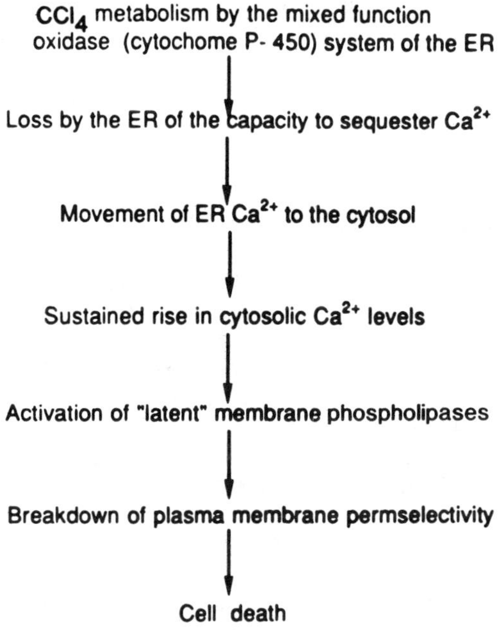

Figure 9. The Ca^{2+} hypothesis for CCl_4 hepatotoxicity.

pholipase A_2 activation just begins at about 15 min and becomes significant by 30 min (Glende and Pushpendran, 1986). It has also been reported that phospholipase C is activated in CCl_4 hepatotoxicity (Schwertz and Lamb, 1982; Lamb and Schwertz, 1982; Lamb et al., 1984). Further, in a study of effects of $CBrCl_3$ on isolated rat hepatocytes (Ungemach, 1985, 1987) it was shown that plasma membrane function deteriorated about 60 min after the time of maximum lipid peroxidation, which occurred at 30 min. At the time of collapse of plasma membrane function, plasma membrane phosphatidylcholine and phosphatidylethanolamine were decreased and their respective lysophosphatides were increased. This effect on plasma membrane phospholipids was ascribed to activation of phospholipase A_2 consequent on a disturbance in Ca^{2+} homeostasis. Although these experimental results are consistent with the calcium hypothesis for the toxicity of CCl_4 and $CBrCl_3$ (Figure 9), the hypothesis is far from proven. Currently, it is a guide for further work.

ACKNOWLEDGMENTS

R.S. Britton was the recipient of a Medical Research Council of Canada Fellowship. Original work reported herein was supported by research grant no. RO1-ES 01821 (E.A. Glende, Jr., Principal Investigator) from the National Institutes of Health, U.S.P.H.S.

REFERENCES

Adams JD, Jr., Lauterburg BH, and Mitchell JR: Plasma glutathione and glutathione disulfide in the rat: regulation and response to oxidative stress. *J Pharmacol Exp Ther* 1983; 227:749–754.

Akerboom TPM, Bilzer M, and Sies H: Competition between transport of glutathione disulfide (GSSG) and glutathione-*S*-conjugates from perfused rat liver into bile. *FEBS Lett* 1982; 140:73–76.

Albano E, Poli G, Tomasi A, Bini A, Vannini V, and Dianzani MU: Toxicity of 1,2-dibromoethane in isolated hepatocytes: role of lipid peroxidation. *Chem Biol Interac* 1984; 50:255–265.

Alin P, Danielson UH, and Mannervik B: 4-Hydroxyalk-2-enals are substrates for glutathione transferase. *FEBS Lett* 1985; 179:267–270.

Anundi I, Hogberg J, and Stead AH: Glutathione depletion in isolated hepatocytes: its relation to lipid peroxidation and cell damage. *Acta Pharmacol Toxicol* 1979; 45:45–51.

Anundi I, Rajs J, and Hogberg J: Chloroacetamide hepatotoxicity: hydropic degeneration and lipid peroxidation. *Toxicol Appl Pharmacol* 1980; 55:273–280.

Aust SD, Morehouse LA, and Thomas CE: Role of metals in oxygen radical reactions. *J Free Rad Biol Med* 1985; 1:3–25.

Babson JR and Reed DJ: Inactivation of glutathione reductase by 2-chloroethyl nitrosourea-derived isocyanates. *Biochem Biophys Res Commun* 1978; 83:754–762.

Bacon BR and Tavill AS: Role of the liver in normal iron metabolism. *Semin Liver Dis* 1984; 4:181–192.

Bacon BR, Tavill AS, Brittenham GH, Park CH, and Recknagel RO: Hepatic lipid peroxidation *in vivo* in rats with chronic iron overload. *J Clin Invest* 1983; 71:429–439.

Bellomo G and Orrenius S: Altered thiol and calcium homeostasis in oxidative hepatocellular injury. *Hepatology* 1985; 5:876–882.

Benedetti A, Comporti M, and Esterbauer H: Identification of 4-hydroxynonenal as a cytotoxic product originating from the peroxidation of liver microsomal lipids. *Biochim Biophys Acta* 1980; 620:281–296.

Brattin WJ, Glende EA, Jr., and Recknagel RO: Pathological mechanisms in carbon tetrachloride hepatotoxicity. *J Free Rad Biol Med* 1985; 1:27–38.

Britton RS, Bacon BR, and Recknagel RO: Lipid peroxidation and associated organelle dysfunction in iron overload. *Chem Phys Lipids* 1987; 45:207–239.

Brown BR, Jr: Hepatic microsomal lipoperoxidation and inhalation anesthetics. *Anesthesiology* 1972; 36:458–465.

Bunyan J, Green J, Edwin EE, and Diplock AT: Studies on vitamin E. 5. Lipid peroxidation in dialuric acid-induced hemolysis of vitamin E-deficient erythrocytes. *Biochem J* 1960; 76:47–51.

Burk RF: Glutathione-dependent protection by rat liver microsomal protein against lipid peroxidation. *Biochim Biophys Acta* 1983; 757:21–28.

Casini AF, Pompella A, and Comporti M: Liver glutathione depletion induced by bromobenzene, iodobenzene, and diethylmaleate poisoning, and its relation to lipid peroxidation and necrosis. *Am J Pathol* 1985; 118:225–237.

Chien KC, Sherman SC, Mittnacht S, Jr., and Farber JL: Membrane structure and function subsequent to calcium activation of an endogenous phospholipase. *Arch Biochem Biophys* 1980; 205:614–622.

Christophersen BO: Oxidation of reduced glutathione by subcellular fractions of rat liver. *Biochem J* 1966; 100:95–101.

Christophersen BO: The inhibitory effect of reduced glutathione on the lipid peroxidation of the microsomal fraction and mitochondria. *Biochem J* 1968; 106:515–522.

Comporti M: Biology of disease: lipid peroxidation and cellular damage in toxic liver injury. *Lab Invest* 1985; 53:599–623.

Comporti M: Glutathione depleting agents and lipid peroxidation. *Chem Phys Lipids* 1987; 45:143–169.

de Groot H, Littauer A, Hugo-Wissemann D, Wissemann P, and Noll T: Lipid peroxidation and cell viability in isolated hepatocytes in a redesigned oxystat system: evaluation of the hypothesis that lipid peroxidation, preferentially induced at low oxygen partial pressures, is decisive for CCl_4 liver cell injury. *Arch Biochem Biophys* 1988; 264:591–599.

de Groot H and Noll T: The crucial role of low steady state oxygen partial pressures in haloalkane free-radical mediated lipid peroxidation: possible implications in haloalkane liver injury. *Biochem Pharmacol* 1986; 35:15–19.

de Groot H, Noll T, and Tolle T: Loss of latent activity of liver microsomal membrane enzymes evoked by lipid peroxidation. Studies of nucleoside diphosphatase, glucose-6-phosphatase, and UDP glucuronyltransferase. *Biochim Biophys Acta* 1985; 815:91–96.

De Vries J: Hepatotoxic metabolic activation of paracetamol and its derivatives phenacetin and benorilate: Oxygenation or electron transfer? *Biochem Pharmacol* 1981; 30:399–402.

Dillard CJ and Tappel AL: Fluorescent damage products of lipid peroxidation, in Packer L (ed): *Methods in Enzymology,* vol. 105, New York, Academic Press, 1984, pp 337–341.

Dolak JA, Glende EA, Jr., and Recknagel RO: The significance of glycogen mobilization and phospholipase A_2 activation in carbon tetrachloride liver cell injury, in Poli G, Cheeseman KH, Dianzani MU, and Slater TF (eds): *Free Radicals in Liver Injury*. Oxford, IRL Press Ltd., 1985, pp 117–125.

Dolak JA, Waller RL, Glende EA, Jr., and Recknagel RO: Liver cell calcium homeostasis in carbon tetrachloride liver cell injury: new data with Fura2. *J Biochem Toxicol* 1988; 3:329–342.

Egger KW and Cocks AT: Homopolar and heteropolar bond dissociation energies and heats of formation of radicals and ions in the gas phase I. Data on organic molecules. *Helv Chim Acta* 1973; 56:1516–1536.

Esterbauer H: Free radicals and lipid peroxidation, in Poli G, Cheeseman KH, Dianzani MU, and Slater TF (eds): *Free Radicals in Liver Injury*. Oxford, IRL Press Ltd., 1985, pp 29–47.

Fisher AB: Intracellular production of oxygen-derived free radicals, in Halliwell B (ed): *Oxygen Radicals and Tissue Injury* (Proceedings of an Upjohn Symposium). Bethesda, Federation of American Societies for Experimental Biology, 1988, pp 34–39.

Gerson RJ, Casini A, Gilfor D, Serroni A, and Farber JL: Oxygen-mediated cell injury in the killing of cultured hepatocytes by acetaminophen. *Biochem Biophys Res Commun* 1985; 126:1129–1137.

Gerson RJ, Serroni A, Gilfor D, Ellen JM, and Farber JL: Killing of cultured hepatocytes by the mixed-function oxidation of ethoxycoumarin. *Biochem Pharmacol* 1986; 35:4311–4319.

Glende EA, Jr and Pushpendran CK: Activation of phospholipase A_2 by carbon tetrachloride in isolated rat hepatocytes. *Biochem Pharmacol* 1986; 35:3301–3307.

Gordeuk VR, Bacon BR, and Brittenham GM: Iron overload: causes and consequences. *Annu Rev Nutr* 1987; 7:485–508.

Gordis E: Lipid metabolites of carbon tetrachloride. *J Clin Invest* 1969; 48:203–209.

Haenen GRMM and Bast A: Protection against lipid peroxidation by a microsomal glutathione-dependent labile factor. *FEBS Lett* 1983; 159:24–28.

Halliwell B and Gutteridge JMC: Oxygen toxicity, oxygen radicals, transition metals and disease. *Biochem J* 1984; 219:1–14.

Halliwell B and Gutteridge JMC: *Free Radicals in Biology and Medicine*. Oxford, Clarendon Press, 1985a.

Halliwell B and Gutteridge JMC: The importance of free radicals and catalytic metal ions in human disease. *Mol Aspects Med* 1985b; 8:89–193.

Hashimoto S, Glende EA, Jr, and Recknagel RO: Hepatic lipid peroxidation in acute fatal human carbon tetrachloride poisoning. *N Eng J Med* 1968; 279:1082–1085.

Hoffsten PE, Hunter FE, Jr, Gebicki JM, and Weinstein J: Formation of "lipid peroxide" under conditions which lead to swelling and lysis of rat liver mitochondria. *Biochem Biophys Res Commun* 1962; 7:276–280.

Horton AA and Fairhurst S: Lipid peroxidation and mechanisms of toxicity. *CRC Crit Rev Toxicol* 1987; 18:27–79.

Isaacs J and Binkley F: Glutathione dependent control of protein disulfide content by subcellular fractions of hepatic tissue. *Biochim Biophys Acta* 1977; 497:192–204.

Jacobs A: Low molecular weight intracellular iron transport compounds. *Blood* 1977; 50:433–439.

Jaeschke H, Kleinwaechter C, and Wendel A: The role of acrolein in allyl alcohol-induced lipid peroxidation and liver cell damage in mice. *Biochem Pharmacol* 1987; 36:51–57.

Jollow DJ, Mitchell JR, Potter WZ, Davis DC, Gillette JR, and Brodie BB: Acetaminophen-induced hepatic necrosis. II. Role of covalent binding *in vivo*. *J Pharmacol Exp Ther* 1973; 187:195–202.

Kappus H: Lipid peroxidation mechanisms, analysis, enzymology and biological relevance, Sies, H (ed): in *Oxidative Stress*. New York, Academic Press, pp 273–310, 1985.

Kappus H: A survey of chemicals inducing lipid peroxidation. *Chem Phys Lipids* 1987; 45:105–115.

Kloss MW, Rosen GM, and Rauckman EJ: Evidence of enhanced in vivo lipid peroxidation after acute cocaine administration. *Toxicol Lett* 1983; 15:65–70.

Koch RR, Glende EA, Jr, and Recknagel RO: Hepatotoxicity of bromotrichloromethane: bond dissociation energy and lipoperoxidation. *Biochem Pharmacol* 1974; 23:2907–2915.

Kornbrust DJ and Mavis RD: Microsomal lipid peroxidation. I. Characterization of the role of iron and NADPH. *Molec Pharmacol* 1980; 17:400–407.

Lamb RG, McCue SB, Taylor DR, and McGuffin MA: The role of phospholipid metabolism in bromobenzene- and carbon tetrachloride hepatic injury. *Toxicol Appl Pharmacol* 1984; 75:510–520.

Lamb RG and Schwertz DW: The effects of bromobenzene and carbon tetrachloride exposure *in vitro* on phospholipase C activity of rat liver cells. *Toxicol Appl Pharmacol* 1982; 63:216–229.

Lauterburg BH, Smith CV, Hughes H, and Mitchell JR: Biliary excretion of glutathione and glutathione disulfide in the rat: regulation and response to oxidative stress. *J Clin Invest* 1984; 73:124–133.

Link B, Durk H, Thiel D, and Frank H: Binding of trichloromethyl radicals to lipids of the hepatic endoplasmic reticulum during tetrachloromethane metabolism. *Biochem J* 1984; 223:577–586.

Long RM and Moore L: Elevated cytosolic calicum in rat hepatocytes exposed to carbon tetrachloride. *J Pharmacol Exptl Therap* 1986; 238:186–191.

Lowrey K, Glende EA, Jr, and Recknagel RO: Rapid depression of rat liver microsomal calcium pump activity after administration of carbon tetrachloride or bromotrichloromethane and lack of effect after ethanol. *Toxicol Appl Pharmacol* 1981a; 59:389–394.

Lowrey K, Glende EA, Jr, and Recknagel RO: Destruction of liver microsomal calcium pump activity by carbon tetrachloride and bromotrichloromethane. *Biochem Pharmacol* 1981b; 30:135–140.

McCay PB: Vitamin E: Interactions with free radicals and ascorbate. *Annu Rev Nutr* 1985; 5:323–340.

McCay PB, Lai EK, Powell SR, and Breuggemann G: Vitamin E functions as an electron shuttle for glutathione dependent "free radical reductase" activity in biological membranes. *Federat Proc* 1986; 45:451.

Meredith MJ and Reed DJ: Status of the mitochondrial pool of glutathione in the isolated hepatocyte. *J Biol Chem* 1982; 257:3747–3753.

Meredith MJ and Reed DJ: Depletion *in vitro* of mitochondrial glutathione in rat hepatocytes and enhancement of lipid peroxidation by adriamycin and 1,3-bis(2-chloroethyl)-1-nitrosourea (BCNU). *Biochem Pharmacol* 1983; 32:1383–1388.

Minotti G and Aust SD: The requirement for iron (III) in the initiation of lipid peroxidation by iron (II) and hydrogen peroxide. *J Biol Chem* 1987; 262:1098–1104.

Mitchell JR, Jollow DJ, Potter WZ, Davis DC, Gillette JR, and Brodie BB: Acetaminophen-induced hepatic necrosis. I. Role of drug metabolism. *J Pharmacol Exp Ther* 1973a; 187:211–217.

Mitchell JR, Jollow DJ, Potter WZ, Gillette JR, and Brodie BB: Acetaminophen–induced hepatic necrosis. IV. Protective role of glutathione. *J Pharmacol Exp Ther* 1973b; 187:211–217.

Moore L, Davenport GR, and Landon EJ: Calcium uptake of a rat liver microsomal subcellular fraction in response to *in vivo* administration of carbon tetrachloride. *J Biol Chem* 1976; 251:1197–1201.

Nachtomi E and Alumot E: Comparison of ethylene dibromide and carbon tetrachloride toxicity in rats and chicks: blood and liver levels; lipid peroxidation. *Exp Molec Path* 1972; 16:71–78.

Nicotera P, Hartzell P, Davis G, and Orrenius S: The formation of plasma membrane blebs in hepatocytes exposed to agents that increase cytosolic Ca^{2+} is mediated by the activation of a nonlysosomal proteolytic system. *FEBS Lett* 1986; 209:139–144.

Ohno Y, Ormstad K, Ross D, and Orrenius S: Mechanism of allyl alcohol toxicity and protective effects of low molecular weight thiols studied with isolated rat hepatocytes. *Toxicol Appl Pharmacol* 1985; 78:169–179.

Orrenius S and Bellomo G: Toxicological implications of perturbation of Ca^{2+} homeostasis in hepatocytes, in Cheung WY (ed): *Calcium and Cell Function VI*. New York, Academic Press, 1986, pp 186–208.

Peavy DL and Fairchild EJ II: Evidence for lipid peroxidation in endotoxin-poisoned mice. *Infect Immun* 1986; 52:613–616.

Peters TJ, O'Connell MJ, and Ward RJ: Role of free-radical mediated lipid peroxidation in the pathogenesis of hepatic damage by lysosomal disruption, in Poli G, Cheeseman KH, Dianzani MU, and Slater TF (eds): *Free Radicals in Liver Injury*. Oxford, IRL Press Ltd., 1985, pp 107–115.

Peters TJ and Seymour CA: Organelle pathology in primary and secondary hemochromatosis with special reference to lysosomal changes. *Clin Sci Mol Med* 1976; 50:75–78.

Porter NA: Chemistry of lipid peroxidation, in Packer L (ed): *Methods in Enzymology,* vol 105. New York, Academic Press, 1984; pp 273–282.

Potter WZ, Davis DC, Mitchell JR, Jollow DJ, Gillette JR, and Brodie BB: Acetaminophen-induced hepatic necrosis. III. Cytochrome P-450 mediated covalent binding *in vitro. J Pharmacol Exp Ther* 1973; 187:203–210.

Prohaska JR: The glutathione peroxidase activity of glutathione *S*-transferases. *Biochim Biophys Acta* 1980; 611:87–98.

Rao KS and Recknagel RO: Early onset of lipoperoxidation in rat liver after carbon tetrachloride administration. *Exp Molec Pathol* 1968; 9:271–278.

Rao KS and Recknagel RO: Early incorporation of carbon-labeled carbon tetrachloride into rat liver particulate lipids and proteins. *Exp Molec Pathol* 1969; 10:219–228.

Rechnagel RO: Minireview: a new direction in the study of carbon tetrachloride hepatotoxicity. *Life Sci* 1983; 33:401–408.

Recknagel RO and Ghoshal AK: Lipoperoxidation as a vector in carbon tetrachloride hepatotoxicity. *Lab Invest* 1966; 15:132–148.

Recknagel RO and Glende EA, Jr: Carbon tetrachloride hepatotoxicity: an example of lethal cleavage. *CRC Crit Rev Toxicol* 1973; 2:263–297.

Recknagel RO and Glende EA, Jr: Spectrophotometric detection of lipid conjugated dienes, in Packer L (ed): *Methods in Enzymology,* vol 105. New York, Academic Press, 1984; pp 331–337.

Recknagel RO and Glende EA, Jr: Effect of carbon tetrachloride on liver cell calcium homeostasis, in Heilmann C (ed): *Calcium-Dependent Processes in the Liver.* London, MTP Press, Ltd, 1988a, pp 229–236.

Recknagel RO and Glende EA, Jr: The carbon tetrachloride hepatotoxicity model: Free radicals and calcium homeostasis, in Miguel T, Quintanilha T, and Weber H (eds): *Handbook of Free Radicals and Antioxidants in Biomedicine,* vol III. Boca Raton, CRC Press, 1989; pp 3–16.

Recknagel RO, Glende EA, Jr, Dolak JA, and Waller RL: Mechanisms of carbon tetrachloride toxicity. *Pharmacol Ther* 1989; 43:139–154.

Recknagel RO, Glende EA, Jr, and Hruszkewycz AM: Chemical mechanisms in carbon tetrachloride toxicity, in Pryor WA (ed): *Free Radicals in Biology,* vol III. New York, Academic Press, 1977; pp 97–132.

Recknagel RO, Glende EA, Jr, Waller RL, and Lowrey K: Lipid peroxidation: biochemistry, measurement, and significance in liver cell injury, in Plaa G and Hewitt WR (eds): *Toxicology of the Liver.* New York, Raven Press, 1982, pp 213–242.

Reddy CC, Scholz RW, Thomas CE, and Massaro EJ: Vitamin E dependent reduced glutathione inhibition of rat liver microsomal lipid peroxidation. *Life Sci* 1982; 31:571–576.

Rosen GM, Singletary WV, Jr, Rauckman EJ, and Killenberg PG: Acetaminophen hepatotoxicity: an alternate mechanism. *Biochem Pharmacol* 1983; 32:2053–2059.

Rush GF, Gorski JR, Ripple MG, Sowinski J, Bugelski P, and Hewitt WR: Organic hydroperoxide-induced lipid peroxidation and cell death in isolated hepatocytes. *Toxicol Appl Pharmacol* 1985; 78:473–483.

Schwertz DW and Lamb RG: The influence of carbon tetrachloride metabolism on the carbon tetrachloride-induced activation of rat liver cell phospholipase *C* activity. *Toxicol Appl Pharmacol* 1982; 65:402–412.

Sevanian A and Hochstein P: Mechanisms and consequences of lipid peroxidation in biological systems. *Annu Rev Nutr* 1985; 5:365–390.

Sevanian A, Muakkassah-Kelly SF, and Montestruque S: The influence of phospholipase A_2 and glutathione peroxidase on the elimination of membrane lipid peroxides. *Arch Biochem Biophys* 1983; 223:441–452.

Sevanian A and Kim E: Phospholipase A_2 dependent release of fatty acids from peroxidized membranes. *J Free Rad Biol Med* 1985; 1:263–271.

Shaw S, Rubin KP, and Lieber CS: Depressed hepatic glutathione and increased diene conjugates in alcoholic liver disease. *Dig Dis Sci* 1983; 28:585–589.

Shaw S, Jayatilleke E, Ross WA, Gordon ER, and Lieber CS: Ethanol-induced lipid peroxidation: potentiation by long-term alcohol feeding and attenuation by methionine. *J Lab Clin Med* 1981; 98:417–424.

Sies H: Oxidative stress, in Sies H (ed). New York, Academic Press, 1985.

Sies H and Akerboom TPM: Glutathione disulfide (GSSG) efflux from cells and tissues, in Packer L (ed): *Methods in Enzymology,* vol 105. New York, Academic Press, 1984; pp 445–451.

Sies H and Summer K-H: Hydroperoxide-metabolizing systems in rat liver. *Eur J Biochem* 1975; 57:503–512.

Silver EH and Szabo S: Possible role of lipid peroxidation in the actions of acrylonitrile on the adrenals, liver and gastrointestinal tract. *Res Commun Chem Pathol Pharmacol* 1982; 36:33–43.

Simic MG: Free radical mechanisms in autoxidation. *J Chem Ed* 1981; 58:125–131.

Simic MG and Taylor KA: Free radical mechanisms of oxidative reactions, in St Angelo AJ and Bailey ME (eds): *Flavor Quality of Fresh Meats: Warmed-Over Flavor*. New York, Academic Press, 1987, pp 69–117.

Sippel HW: Effect of an acute dose of ethanol on lipid peroxidation and on the activity of microsomal glutathione *S*-transferase in rat liver. *Acta Pharmacol Toxicol* 1983; 53:135–140.

Smith MT, Evans CG, Thor H, and Orrenius S: Quinone-induced oxidative injury to cells and tissues, in Sies H (ed): *Oxidative Stress*. London, Academic Press, 1985, pp 91–113.

Smith CV and Mitchell JR: Acetaminophen hepatotoxicity *in vivo* is not accompanied by oxidant stress. *Biochem Biophys Res Commun* 1985; 133:329–336.

Stacey NH, Cantilena LR, Jr, and Klaassen CD: Cadmium toxicity and lipid peroxidation in isolated rat hepatocytes. *Toxicol Appl Pharmacol* 1980; 53:470–480.

Stacey NH and Klaassen CD: Inhibition of lipid peroxidation without prevention of cellular injury in isolated hepatocytes. *Toxicol Appl Pharmacol* 1981; 58:8–18.

Tan K Hong, Meyer DJ, Belin J, and Ketterer B: Inhibition of microsomal lipid peroxidation by glutathione and glutathione transferases B and AA. *Biochem J* 1984; 220:243–252.

Tee LBG, Davies DS, Seddon CE, and Boobis AR: Species differences in the hepatotoxicity of paracetamol are due to differences in the rate of conversion to its cytotoxic metabolite. *Biochem Pharmacol* 1987; 36:1041–1052.

Thayer WS: Evaluation of tissue indicators of oxidative stress in rats treated chronically with adriamycin. *Biochem Pharmacol* 1988; 37:2189–2194.

Tomasi A, Albano E, Banni S, Botti B, Corongiu F, Dessi MA, Iannone A, Vannini V, and Dianzani MU: Free-radical metabolism of carbon tetrachloride in rat liver mitochondria: a study of the mechanism of action. *Biochem J* 1987; 246:313–317.

Tomasi A, Billing S, Garner A, Slater TF, and Albano E: The metabolism of halothane by hepatocytes: a comparison between free radical spin trapping and lipid peroxidation in relation to cell damage. *Chem Biol Interac* 1983; 46:353–368.

Tomokuni K: Studies on hepatotoxicity induced by chlorinated hydrocarbons. II. Lipid metabolism and absorption spectrum of microsomal lipid in mice exposed to 1,1,2,2-tetrachloroethane. *Acta Medicinae Okayama* 1970; 24:315–322.

Ungemach FR: Plasma membrane damage of hepatocytes following lipid peroxidation: involvement of phospholipase A_2, in Poli G, Cheeseman KH, Dianzani MU, and Slater TF (eds): *Free Radicals in Liver Injury*. Oxford, IRL Press Ltd., 1985, pp 127–134.

Ungemach FR: Pathobiological mechanisms of hepatocellular damage following lipid peroxidation. *Chem Phys Lipids* 1987; 45:171–205.

Ursini F and Bindoli A: The role of selenium peroxidases in the protection against oxidative damage of membranes. *Chem Phys Lipids* 1987; 44:255–276.

van den Bosch H: Intracellular phospholipase *A*. *Biochim Biophys Acta* 1980; 604:191–246.

van Kuijk FJGM, Handelman GJ, and Dratz EA: Consecutive action of phospholipase A_2 and glutathione peroxidase is required for the reduction of phospholipid hydroperoxides and provides a convenient method to determine peroxide values in membranes. *J Free Rad Biol Med* 1985; 1:421–427.

van Kuijk FJGM, Sevanian A, Handelman GJ, and Dratz EA: A new role for phospholipase A_2: protection of membranes from lipid peroxidation. *Trends Biochem Sci* 1987; 12:31–34.

Videla LA, Fernandez V, de Marinis A, Fernandez N, and Valenzuela A: Liver lipoperoxidative pressure and glutathione status following acetaldehyde and aliphatic alcohol pretreatment in the rat. *Biochem Biophys Res Commun* 1982; 104:965–970.

Waller RL, Glende EA, Jr, and Recknagel RO: Carbon tetrachloride and bromotrichloromethane toxicity: dual role of covalent binding of metabolic cleavage products and lipid peroxidation in depression of microsomal calcium sequestration. *Biochem Pharmacol* 1983; 32:1613–1617.

Watanabe N, Shiki Y, Morisaki N, Saito Y, and Yoshida S: Cytotoxic effects of paraquat and inhibition of them by vitamin E. *Biochim Biophys Acta* 1986; 883:420–425.

Wendel A: Measurement of *in vivo* lipid peroxidation and toxicological significance. *Free Rad Biol Med* 1987; 3:355–358.

Wendel A and Feuerstein S: Drug-induced lipid peroxidation in mice. I. Modulation by monooxygenase activity, glutathione and selenium status. *Biochem Pharmacol* 1981; 30:2513–2520.

Wendel A, Feuerstein S, and Konz K-H: Acute paracetamol intoxication of starved mice leads to lipid peroxidation *in vivo*. *Biochem Pharmacol* 1979; 28:2051–2055.

Wills ED and Wilkinson AE: Release of enzymes from lysosomes by irradiation and the relation of lipid peroxide formation to enzyme release. *Biochem J* 1966; 99:657–666.

Wolff SP, Garner A, and Dean RT: Free radicals, lipids and protein degradation. *Trends Biochem Sci* 1984; 11:27–31.

Younes M and Siegers CP: The role of iron in the paracetamol- and CCl_4-induced lipid peroxidation and hepatotoxicity. *Chem Biol Interac* 1985; 55:327–334.

Zimmerman H: *Hepatotoxicity*. New York, Appleton-Century-Crofts, 1978.
Zitting Z and Heinonen T: Decrease of reduced glutathione in isolated rat hepatocytes caused by acrolein, acrylonitrile, and the thermal degradation products of styrene copolymers. *Toxicology* 1980; 17:331–341.

Chapter

10

Drug-Induced Abnormalities of Liver Heme Biosynthesis

Francesco De Matteis
MRC Toxicology Unit
Medical Research Council Laboratories
Woodmansterne Road
Carshalton
Surrey, England

INTRODUCTION

All animal cells possess the ability to synthesize heme, which they utilize as the prosthetic group of various hemoproteins, essential enzymes in many vital processes; among these processes are oxygen transport and utilization, electron transport, and decomposition or transfer of peroxides. In the liver most of the heme synthesized by the parenchymal cells is apparently destined for the cytochromes of the P450 group, a family of related but distinct hemoproteins under the control of separate genes. Several hepatic cytochromes P450 can be induced by many foreign chemicals of pharmacological and toxicological interest, and this probably results in an increased demand for heme in the hepatocyte. In addition, many cytochromes P450 have the important property of accepting drugs at their active site, so that these foreign chemicals can be metabolized or, in some cases, the oxygen bound to the heme prosthetic group of the cytochrome gives rise to partially reduced molecular oxygen species. Reactive species (derived from either the drug substrate or from molecular oxygen) can be produced in these processes, and these may lead to increased degradation of heme or to a defect of an intermediary enzyme of the heme biosynthetic sequence with pronounced alterations in the heme and porphyrin homeostasis of the cell. For all these reasons, cytochrome P450 represents the focal point of interaction between drugs and heme metabolism, and alterations of heme metabolism caused by drugs are often initiated at the level of this important hemoprotein. These drug-induced alterations have been studied most extensively and are more readily demonstrable in the liver than in most other tissues, probably because those forms of cytochrome P450 which are inducible by foreign compounds and can bind and metabolize them are particularly developed in this organ.

Before summarizing the most recent advances in the field of drug-induced abnormalities of liver heme metabolism, the pathway of heme biosynthesis and its regulation will be briefly outlined, so that the mechanism of action of the various drugs can be discussed more easily.

Heme Biosynthesis, Significance of 5-Aminolevulinate Synthetase

The pathway of heme biosynthesis is given in Figure 1. The first intermediate is 5-aminolevulinate (ALA), an aminoketone which results from the condensation of glycine with succinyl-CoA, carried out by the enzyme ALA-synthetase within the matrix compartment of the mitochondrion. Two molecules of ALA are then condensed with each other in the cytosol to the monopyrrole precursor, porphobilinogen (PBG). Four molecules of PBG then join together to form the symmnetrical, linear tetrapyrrole, hydroxymethylbilane. This intermediate undergoes cyclization under the influence of uroporphyrinogen III synthetase to the asymmetrical uroporphyrinogen (uro'gen) III, which is then decarboxylated stepwise by a cytoplasmic decarboxylase to produce coproporphyrinogen III. It should be noted that the real intermediates in the pathway are not uroporphyrin and coproporphyrin but their colorless reduced porphyrinogens (hexahydroporphyrins), in which the pyrrole rings are joined together by methylene bridges. The corresponding porphyrins, oxidative by-products of the pathway, cannot themselves be metabolized. This has led to the proposal (Heikel et al., 1958) that in certain porphyrias where uroporphyrin accumulates, the mechanism responsible may be accelerated oxidation of uroporphyrinogen, causing an ''oxidative escape'' of the intermediate from the metabolic pathway, a concept that will be discussed at some length later. Once coproporphyrinogen III is formed, it is taken up into the mitochondrion, where the remaining steps of heme biosynthesis take place. Coproporphyrinogen is first oxidized to protoporphyrinogen by coproporphyrinogen oxidase (this metabolic step involves oxidative decarboxylation of two propionic acid side chains to vinyl groups). Protoporphyrinogen is then oxidized to protoporphyrin by another oxidase and, finally, the last enzyme of heme biosynthesis, ferrochelatase, inserts ferrous iron into protoporphyrin to produce heme.

There are two main reasons why the first enzyme of the pathway, ALA-synthetase, is thought to play a key role in the regulation of heme biosynthesis.

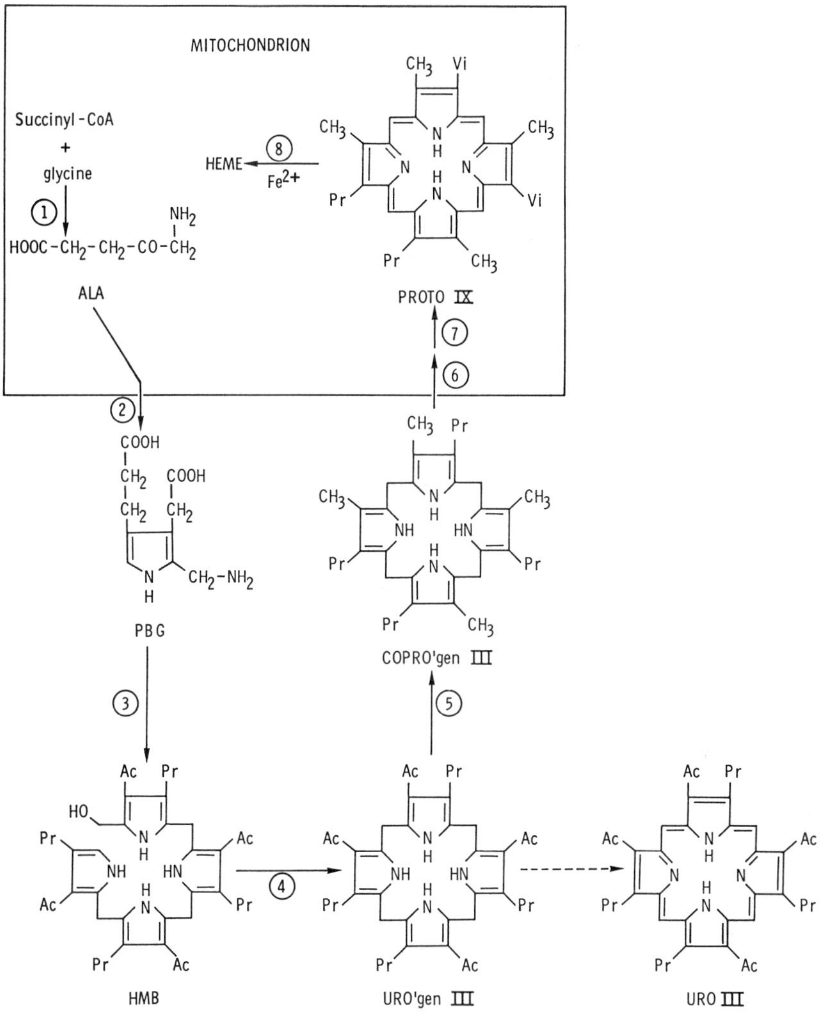

Figure 1. *The pathway of heme biosynthesis.* ALA: 5-aminolevulinate; PGB: porphobilinogen; HMB: hydroxymethylbilane; URO'gen: uroporphyrinogen; COPRO'gen: coproporphyrinogen; PROTO: protoporphyrin; URO: uroporphyrin. β-Substituents in pyrrole rings of tetrapyrroles are indicated thus, Ac: CH_2COOH; Pr: CH_2CH_2 COOH; Vi: $CH{=}CH_2$. Reactions are catalysed by (1) ALA-synthetase; (2) ALA-dehydratase; (3) PBG deaminase; (4) uroporphyrinogen III synthetase; (5) uroporphyrinogen decarboxylase; (6) coproporphyrinogen oxidase; (7) protoporphyrinogen oxidase; and (8) ferrochelatase. Also shown (dotted arrow) is oxidation of uroporphyrinogen to uroporphyrin, resulting in oxidative escape of uro'gen from the pathway.

First, its activity appears to be rate-limiting in the overall pathway, as best illustrated by the observation that an increased supply of ALA will readily increase the liver synthesis of porphyrins and heme *in vivo* and *in vitro,* while an increased supply of the precursors of ALA will produce little or no increased synthesis. Thus, at least under normal conditions, the rate of heme synthesis appears to be determined mostly by the supply of ALA.

The second reason why ALA-synthetase is considered important in the regulation of the pathway is that this enzyme is the site where heme, the end product, exercises a negative-feedback control on its own synthesis, a concept originally proposed for microbial systems (Burnham and Lascelles, 1963) and then applied to the liver of chicks, rodents (Granick, 1966), and humans (Meyer and Schmid, 1977).

A pool of cellular free heme (or regulatory heme) has been proposed to be involved by many authors, but although several lines of indirect evidence for the existence of this regulatory pool have been obtained in various laboratories (Druyan and Kelly, 1972; De Matteis, 1978; Grandchamp et al., 1981), its precise identity and intracellular localization have not yet been defined. It could be free heme or a pool in rapid equilibrium with free heme. There is evidence that the utilization of heme for the synthesis of liver hemoprotein—such as cytochromes of the P450 group—is not an irreversible process but that, to some extent, exchange between free heme and hemoprotein heme (as well as among different hemoprotein pools) may occur *in vivo* (Bock and Siekevitz, 1970; Sadano and Omura, 1985). Certain cytosolic proteins exhibit high affinity for heme and may have a role in transporting heme from the site of synthesis to those of utilization and degradation (Senjo et al., 1985; Vander Jagt et al., 1985; Vincent and Muller-Eberhard, 1985; and references therein), assisting at the same time the process of equilibration of heme across different intracellular pools. Therefore the heme status of the hepatocytes should be considered to be in a dynamic flux at all times, with the apoproteins of the various liver hemoproteins, the regulatory systems for ALA-S, and the pathways of heme degradation (to either bile pigments or abnormal degradation products) all competing with each other for the same pool of regulatory heme with rapid turnover (Figure 2). Direct measurements of the concentration of liver free heme are not yet possible, but some information about this concentration can be gained indirectly from the activity of heme-dependent enzymes, such as catalase and tryptophan pyrrolase, since the heme necessary for their synthesis (and activity) would be expected to be drawn from the intermediary "regulatory" pool. Reference to the use of these enzymes to gain information on the cellular heme status will be found in the following: Badawy et al. (1986), Bissell and Hammaker (1977), De Matteis (1978), and Sweeney (1976).

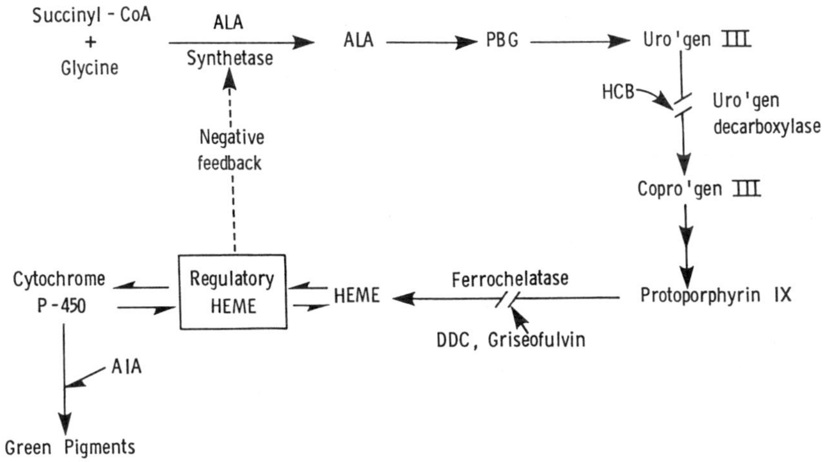

Figure 2. *The regulation of the pathway by heme and site of interaction of the three classes of drugs considered here.* AIA: 2-allyl-2-isopropylacetamide; DDC: 3,5-diethoxycarbonyl-1,4-dihydrocollidine; HCB: hexachlorobenzene. The mechanisms involved in the negative feedback of heme are discussed in the text. (Reproduced with permission.)

Role of Heme in the Regulation of the Pathway

The most important regulatory mechanism exercised by heme on ALA-synthetase involves repression of the synthesis of the mRNA for this enzyme, and drugs which stimulate the biosynthesis of ALA-synthetase do so by accelerating the transcription and increasing the amount of the mRNA for ALA-synthetase (Whiting, 1976; Yamamoto et al., 1982; Ades et al., 1987; Maguire et al., 1986). Although the effect of heme on ALA-synthetase has been studied mostly in hepatocytes or animals treated with inducers of the liver enzyme, where the ability of heme to prevent induction of ALA-synthetase has been documented, more recently, sensitive techniques for measuring ALA-synthetase activity have become available and an effect of heme on the basal, uninduced ALA-synthetase activity has also been demonstrated. The liver enzyme was shown to be markedly depressed by administration of either heme itself or ALA, the precursor most likely acting by producing heme intracellularly. Similar effects could be demonstrated in the brain, provided that ALA was given as the methyl ester, a form capable of crossing the blood-brain barrier, so as to generate excess heme within the brain tissue (De Matteis et al., 1981a). Corresponding changes in the basal level of ALA-synthetase mRNA have since been demonstrated after giving either heme or its precursor (Srivastava et al., 1988) compatible with heme acting, at least in part, at the transcriptional level in both tissues. These findings are in agreement with the original proposal by Granick (1966) that heme acts as as a repressor of the transcription of the ALA-synthetase gene and that drugs activate the transcription by interfering with the specific repressor function of heme.

An additional mechanism suggested for the heme control of ALA-synthetase biosynthesis is at the stage of the transfer of newly formed ALA-synthetase into the matrix compartment of the mitochondrion. ALA-synthetase is first made as a large precursor and can be isolated from the cytosol in this form. It then undergoes a process of maturation and translocation into the matrix compartment of the mitochondrion (the site where it is enzymatically active), and these stages can be apparently inhibited by excess heme as, after a heme load, accumulation of the cytosolic form of ALA-synthetase can be demonstrated at the expense of the mitochondrial physiologically active form (Yamauchi et al., 1980). There is discrepancy of view on the physiological significance of this heme effect: most authors consider the effect of heme on transfer and processing of ALA-synthetase less important as a regulatory mechanism than the heme-dependent repression of ALA-synthetase gene transcription; however, Hamilton et al. (1988) have recently proposed that in chicken embryo hepatocytes heme regulates ALA-synthetase primarily at the stage of translocation of newly synthesized enzyme into the mitochondrion.

Another mechanism by which heme may also affect ALA-synthetase is by modifying directly the activity of the enzyme. A direct inhibition of rat liver ALA-synthetase by heme has been documented with partially purified preparations of the enzyme (Scholnick et al., 1972; Whiting and Eliott, 1972), and the close association of the heme-synthesizing enzyme (the ferrochelatase) and ALA-synthetase within the mitochondrion may be suited to a direct control of the synthetase by the concentration of heme at the very site of heme synthesis. This may explain the observation that either heme or ALA given in large doses causes a decline of ALA-synthetase activity which is faster and more pronounced than that obtained with inhibitors of protein synthesis (De Matteis et al., 1981a). Therefore, although most authors consider the most significant regulatory effect of heme to be at the level of the transcription of the ALA-synthetase gene, nevertheless the possibility of a direct inhibitory effect of heme on ALA-synthetase cannot be excluded, perhaps as an additional "fine-tuning" mechanism in the end-product regulation of the pathway.

Drug-Induced Hepatic Porphyrias: Disorders of the Regulation of Liver Heme Biosynthesis

Under normal conditions the biosynthesis of ALA is tightly coupled to formation of heme, as ALA and other intermediates proceed unhindered along the biosynthetic chain with little accumulation. There are conditions, however, some produced by drugs, where this tight control breaks down and far more porphyrins and earlier precursors are formed than are turned into heme, so that the intermediate metabolites accumulate in excess. These conditions, known as porphyrias, should be regarded as disorders of the regulation of heme biosynthesis.

Granick (1986) first showed that drugs which induce accumulation of porphyrins in the liver of the intact animal or in hepatocytes cultured *in vitro* also stimulate markedly the *de novo* synthesis of the first enzyme of the pathway, ALA-synthetase, thus resulting in excess formation of ALA. He suggested that drugs and heme might compete for a binding site on a hypothetical heme-aporepressor protein so that in presence of a sufficiently high concentration of the drug inducer, heme would be displaced from the repressor, thus inactivating it and allowing transcription of the ALA-synthetase gene. Recent work has laid more emphasis on mechanisms which involve the size of the regulatory heme pool. Accordingly increased transcription may result from the ability of the ALA-synthetase-inducing drugs to cause depletion of the regulatory pool by one or more of the following mechanisms (De Matteis, 1978; Marks et al., 1988): (1) accelerated destruction of heme, (2) inhibition of heme biosynthesis, and (3) increased rate of combination of heme with apoproteins for synthesis or "reconstitution" of hemoproteins. Although all three mechanisms will result in stimulation of ALA-synthetase and in accumulation of the intermediates of the pathway, it is only where stimulation of ALA-synthetase is complicated by inhibition of heme biosynthesis (mechanism 2) that the accumulation of the intermediate of the pathway will be very pronounced—particularly of the intermediate which precedes immediately the enzymatic block. Since different enzymes in the biosynthetic chain are inhibited (or become limiting) after administration of different toxic agents, the profile of metabolites excreted in excess or accumulating in the tissues will also be different, so that distinct metabolic syndromes can be recognized.

Table 10–1 summarizes the various drug-induced abnormalities to be considered in this chapter and gives for each of them the salient feature of the biochemical picture, the main molecular lesion, and also the mechanism which is thought to be responsible for accumulation of porphyrins. In the discussion which follows, the main emphasis will be placed on the mechanisms by which different classes of foreign chemicals disturb the regulation of liver heme biosynthesis and induce hepatic porphyrias, paying particular attention to their interaction with cytochrome P450 as the initial step in the pathogenesis of the metabolic disorders.

This is a very complex field of study, which has attracted renewed attention in recent years, but many of the underlying mechanisms are not completely understood. Some of the mechanisms discussed in this chapter are, therefore, still hypothetical, but will nevertheless be considered, as they provide a convenient framework for interpreting existing data and also a suitable starting point for future experiments.

TABLE 10–1.
Drug-Induced Disorders of Porphyrin Metabolism to be Considered in This Chapter*

Drug	Salient Feature of Porphyric Disturbance	Main Molecular Lesion	Mechanism of Accumulation of Intermediates
2-Allyl-2-isopropyl-acetamide	ALA and PBG in liver and urine. Also, a variety of porphyrins	Alkylation of cytochrome P450 heme leading to depletion of cellular heme	Increased supply of precursors exceeds the capacity of intermediary enzymes of the pathway
3,5-Diethoxycarbonyl-1,4-dihydrocollidine	Marked accumulation of protoporphyrin	Conversion of cytochrome P450 heme into N-methyl-protoporphyrin, a targeted inhibitor of ferrochelatase	Increased synthesis, combined with decreased utilization of protoporphyrin
Hexachlorobenzene	Marked accumulation of uroporphyrin and other highly carboxylated porphyrins	Still hypothetical. An uncoupled cytochrome P450 may be responsible for increased oxidation of uroporphyrinogen III.	Oxidative escape of uroporphyrinogen, combined with impaired enzymatic decarboxylation

*The most widely studied drug is given for each disorder, though several other chemicals have been shown to be similarly effective (see text).

UNSATURATED DRUGS AS SUICIDE SUBSTRATES OF CYTOCHROME P450. DISORDER OF PORPHYRIN METABOLISM ASSOCIATED WITH INCREASED LIVER HEME TURNOVER

Following a single report (Duesberg, 1932) of a human patient who developed porphyria after treatment with Sedormid [*N*-(2-isopropylpent-4-enoyl)urea], this drug was shown to increase liver production of porphyrins in the experimental animal (Schmid and Schwartz, 1952). In the course of these studies Schwartz and Ikeda (1955) reported on certain unusual green porphyrins accumulating in the liver of rats and rabbits made porphyric with Sedormid. This important finding, unexplained for many years, was to prove fundamental to our understanding of the mode of action of Sedormid and many other, similarly acting, unsaturated compounds. The nonhypnotic congener, 2-allyl-2-isopropylacetamide (Figure 3a), subsequently used in preference to Sedormid in experimental studies (Goldberg and Rimington, 1955) was shown to be a powerful inducer of ALA-synthetase in the liver of rats (Marver et al., 1966). A selective destruction of liver cytochrome P450 pre-

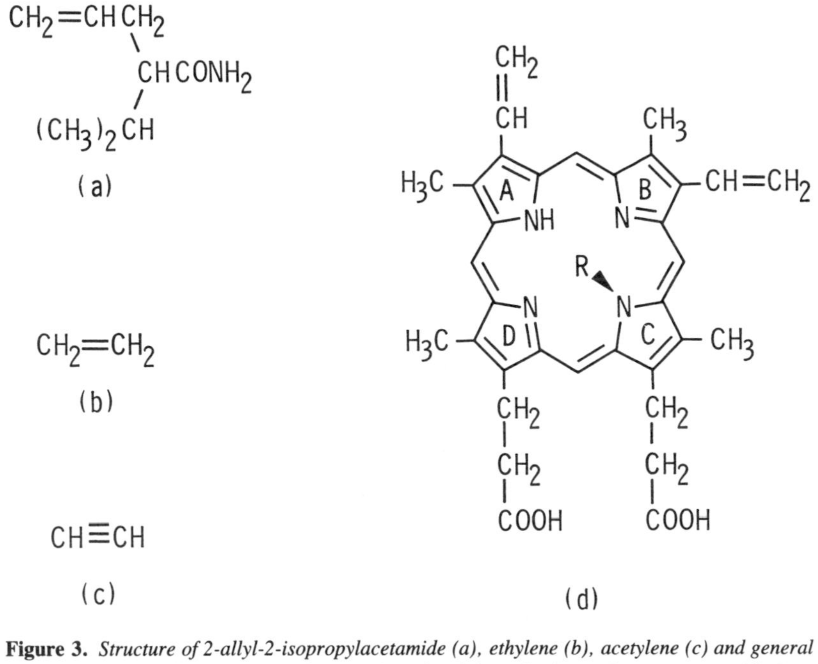

Figure 3. *Structure of 2-allyl-2-isopropylacetamide (a), ethylene (b), acetylene (c) and general formula for the N-alkyl protophorphyrins (d) isolated from the liver after treatment with these drugs.* Four structural isomers are possible depending on which pyrrole nitrogen is alkylated, the isomer shown being N_c, where ring C is *N*-alkylated. The alkyl group present on the pyrrole nitrogen, as well as the isomeric type of the porphyrin vary according to the drug (see text).

ceded and accompanied the induction of ALA-synthetase after a single dose of 2-allyl-2-isopropylacetamide to rats. Also, the heme prosthetic group of the cytochrome was shown to be converted in the process into unidentified products or green pigments (De Matteis, 1970), most probably the green porphyrins reported by Schwartz in similarly treated rabbits. The destruction of the heme moiety of cytochrome P450 required the active participation of the cytochrome in its drug-metabolizing role, as conversion of the unsaturated drug into a reactive derivative was necessary for heme destruction, an example of self-catalyzed or suicidal inactivation reaction (De Matteis, 1970; Levin et al., 1973), in favor of which more direct and conclusive evidence has recently been obtained (Loosemore et al., 1981; Ortiz de Montellano and Mico, 1981). Double-labeling experiments of heme, using [^{3}H] and [^{14}C]-ALA, to label liver heme pools with different turnover rates showed drug-dependent degradation not only for microsomal heme, but also for heme from other cellular compartments including the cytosol, and rapidly turning over heme was preferentially affected in all fractions of the liver hemogenate. Heme with regulatory function might be similarly modified so as to lose its activity as a feedback regulator of ALA-synthetase, thus explaining the increased induction of the enzyme (De Matteis, 1970, 1973).

N-Alkylation of Prosthetic Heme, a Mechanism for Depletion of Regulator Heme

2-Allyl-2-isopropylacetamide is not the only unsaturated drug capable of converting the heme of cytochrome P450 into green pigments. Several other drugs containing a side chain with a terminal double bond have been shown to be similarly effective (including allyl-barbiturates used therapeutically in humans, Levin et al., 1973) and, in addition, drugs containing an unsaturated acetylenic triple bond, among these the ethynyl-substituted steroids of the contraceptive pill (White and Muller-Eberhard, 1977; White, 1978, 1981). The simplest unsaturated compounds of both the alkene and alkyne series, ethylene and acetylene themselves (Figure 3b and c), were also active. A list of the unsaturated compounds capable of destroying cytochrome P450 and a more detailed coverage of the metabolism of these compounds and mechanisms of heme destruction can be found in Ortiz de Montellano and Correia (1983), White (1984), and De Matteis (1987), where additional references to the work summarized below are also given.

As metabolic activation of the unsaturated grouping was required for heme destruction, the possibility that a reactive metabolite of the unsaturated drugs, once produced by cytochrome P450, might then attack the heme prosthetic group had been considered (De Matteis, 1970). However, direct evidence of covalent interaction between the drug metabolite and heme itself has only been produced more recently (Ortiz de Montellano et al., 1978, 1979), by showing that the green pigments isolated after treatment with unsaturated drugs of both the alkene and alkyne series contained radioactivity from the unsaturated drug bound to the tetrapyrrole system in a 1:1 stoichiometry. The alkylated hemes contain iron in their native form, but readily lose iron on exposure to strong acids during the workup procedure, so that the green pigments are usually isolated as the corresponding porphyrins. These have been identified as *N*-alkylated porphyrins from their electronic spectra, titration behavior with acids and reactivity with metals (De Matteis and Cantoni, 1979) indicating that the site of covalent interaction of the drug metabolite is one of the four pyrrole nitrogens of the porphyrin system (Figure 3d).

Further characterization of the *N*-alkylated porphyrins has provided information on the nature of the alkylating metabolite, and also contributed to our understanding of the mechanism by which unsaturated drugs disturb the regulation of liver heme biosynthesis. Three different stages can be distinguished in these further studies and will now be discussed in turn. (1) The structure of some typical heme adducts has been elucidated, and this has provided evidence for a suicidal mechanism involving addition of oxygen to the unsaturated bond. (2) The *N*-alkylated pigments produced by 2-allyl-2-isopropylacetamide, ethylene, and other unsaturated drugs have been shown not to inhibit the enzyme ferrochelatase, thus providing an explanation for the modest accumulation of protoporphyrin seen in this type of hepatic porphyria. (3) Finally the alkylated heme products have been found to exhibit chiral properties, and study of these properties has provided further evidence for a role of apocytochrome P450 in alkylation of exchangeable heme pools and consequent depletion of regulatory heme.

Role of a Monooxygenated Intermediate in Heme Alkylation

Early attempts to determine the structure of the 2-allyl-2-isopropylacetamide green pigment by mass spectrometry were frustrated by secondary modification of the drug-derived alkyl group attached to the porphyrin system. Therefore, in further work, the simplest olefine, ethylene, was chosen as the suicide substrate, as ethylene was at least as effective as 2-allyl-2-isopropylacetamide at producing green pigments, yet lacked the complicating structural features (e.g., hetero atom substitution) which would lead to secondary modification of the adduct. The ethylene green pigment was identified as *N*-hydroxyethyl protoporphyrin, suggesting the involvement of a reactive monooxygenated intermediate in heme alkylation: not only were the electronic absorption spectrum and mass spectrometry data compatible with this structure, but the chromatographic behavior of the pigment could be modified by acetic anhydride confirming the presence of a free hydroxyl group (De Matteis et al., 1980b). *N*-(2-Oxopropyl)-protoporphyrin was isolated after propyne (Ortiz de Montellano and Kunze, 1981), suggesting a reactive monooxygenated intermediate also with drugs of the alkyne series. Monooxygenation of unsaturated, double and triple bonds between two carbons is known to give

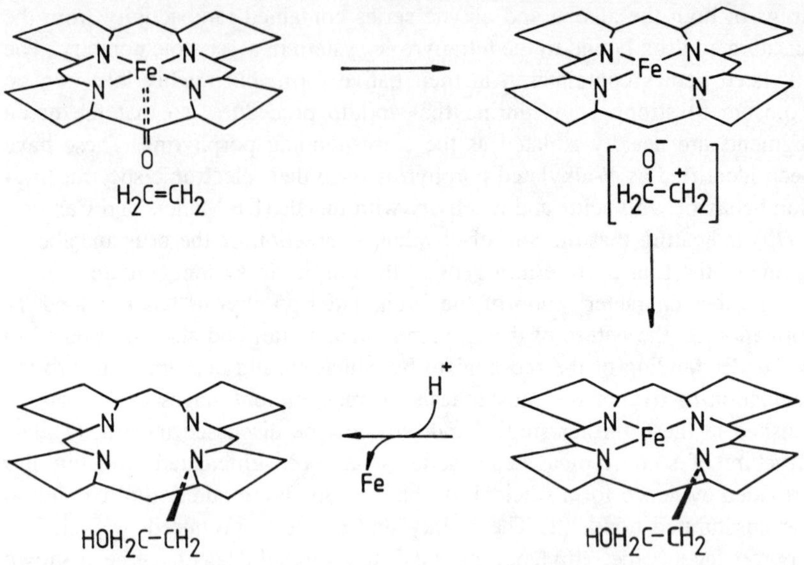

Figure 4. *Possible reactions involved in activation of ethylene by cytochrome P450 and consequent alkylation of one of its pyrrole nitrogens.* (1) Ethylene is bound close to the active oxygen species in the cytochrome P450 active center; (2) a reactive intermediate, such as that proposed by Ortiz de Montellano et al. (1983), is formed; (3) this alkylates one of the pyrrole nitrogens, giving rise to an alkylated heme; (4) the latter then loses iron on exposure to strong acids to generate the corresponding porphyrin. (Reproduced with permission.)

rise to reactive cyclic oxides, *oxiranes* and *oxirenes,* respectively; however, the involvement of an oxirane (or epoxide) intermediate in the alkylation of heme by olefines has been excluded by Ortiz de Montellano et al. (1983), who have proposed instead an acyclic monooxygenated intermediate which can proceed independently to either heme alkylation or to epoxide formation. The main stages of cytochrome P450 heme alkylation so far discussed are schematically summarized in Figure 4, where the native, iron-containing green pigment has been shown as bearing the alkyl group on one pyrrole nitrogen, to account for the selectivity of pyrrole nitrogen alkylation, briefly referred to below. The exact structure of the iron-containing alkylated pigment is not yet known.

Structural Isomerism and Chirality of the *N*-Alkylated Porphyrins

Extensive and elegant NMR investigations by Ortiz de Montellano and collaborators have also established that the four pyrrole nitrogens of the heme prosthetic group of cytochrome P450 are not randomly alkylated. Instead, different nitrogens are specifically substituted, depending the the unsaturated drug under study; for example, the nitrogen of pyrrole ring D (see Figure 3d for the nomenclature of the pyrrole rings) after ethylene (Ortiz de Montellano et al., 1981a) and the nitrogen of either ring C or D in the case of 2-allyl-2-isopropylacetamide (Ortiz de Montellano et al., 1984). The differential ring specificity of the various unsaturated drugs tested may have a bearing on the biochemical picture of the hepatic porphyria caused by 2-allyl-2-isopropylacetamide and other olefins, where a marked stimulation of ALA-synthetase is seen, but the accumulation of protoporphyrin is relatively modest. These drugs do not inhibit the enzyme ferrochelatase in the liver of treated animals, and the *N*-alkylated porphyrins isolated from the liver after their administration (in clear contrast to those produced by treatment with dihydropyridines, see later) are not inhibitory towards ferrochelatase *in vitro* (De Matteis et al., 1980a). An explanation for the lack of ferrochelatase inhibitory activity lies in the size of *N*-alkyl group, relatively large with olefines, and also in its location on propionate-substituted pyrrole rings (i.e., ring C or D), as these steric factors are known to reduce the affinity of an *N*-alkylated protoporphyrin for ferrochelatase (a point to be discussed in more detail later). With 2-allyl-2-isopropylacetamide and ethylene, accumulation of the intermediates is seen without loss of the intermediary enzymes, probably because the supply of heme precursors exceeds the capacity of the subsequent enzymes of the pathway, even though the activities of these enzymes are all normal. As most of the newly formed precursors are turned into heme, the increased ALA-synthetase activity should be viewed as a secondary response to the accelerated heme turnover, intended to replenish the heme pools depleted by the unsaturated drugs. Before discussing the role of cytochrome P450 in channelling newly synthesized heme into the wasteful pathway of drug-dependent alkylation, it is necessary to discuss briefly the chiral properties of the *N*-alkylated porphyrins due to the stereospecific orientation of heme in cytochrome P450.

In protoporphyrin IX the two faces of the porphyrin plane are not superimposable because of the asymmetric arrangement of the vinyl and propionate side chains. An *N*-alkylated protoporphyrin derivative may therefore exist in two chiral configurations, depending on whether the *N*-alkyl group lies above or below the porphyrin plane. Synthetic *N*-alkylated protoporphyrins are optically inactive, as they are the racemic mixtures of the two enantiomorphs. In contrast, the green pigments isolated from the liver of animals treated with propyne, ethylene, 2-allyl-2-isopropylacetamide, and secobarbitone all exhibited chiral properties, as shown by a circular dichroism spectrum. These findings indicated that alkylation of the tetrapyrrole macrocycle is stereospecific in cytochrome P450 and take place preferentially—if not exclusively—from one side of the porphyrin plane, most probably the oxygen-binding side of heme (Figure 4), as this side will be facing the drug to be metabolized and will therefore be directly exposed to the resulting electrophilic species. In recent work use has been made of the chiral properties of the green pigments to test the hypothesis that the apoprotein of cytochrome P450 may act as a catalytic center for conversion of exchangeable pools of heme into *N*-alkylated porphyrins in the liver. It will be recalled that the drug-dependent heme destruction is not confined to the heme of cytochrome P450, but has been shown to concern prelabeled heme (particularly rapidly turning over heme) from different fractions of the liver homogenate (De Matteis, 1973). Exogenous heme added to the microsomal incubation (Unseld and De Matteis, 1978) or injected into treated rats *in vivo* (Correia et al., 1979) is also utilized in the *N*-alkylation pathway. Most of the green pigments, once formed, diffuse away from the microsomal membranes (De Matteis, 1973), leaving behind a heme-free apoprotein which, at least to some extent, is capable of accepting new heme for reconstitution of cytochrome P450 (Sardana et al., 1976; Correia et al., 1979; Ortiz de Montellano and Correia, 1983). The hypothesis has therefore been put forward (De Matteis, 1978; Correia et al., 1979) that the apocytochrome P450 may accept exchangeable pools of liver heme for *N*-alkylation, leading to a state of regulatory heme depletion and to activation of ALA synthetase.

Isolated hepatocytes incubated in presence of 2-allyl-2-isopropylacetamide readily utilize exogenous heme for conversion into *N*-alkylated porphyrins. In order to distinguish the contribution of endogenous and exogenous heme pools to the production of *N*-alkylated porphyrins, a protoheme analogue, mesoheme, was added to the hepatocyte suspension together with 2-allyl-2-isopropylacetamide (De Matteis et al., 1986). The *N*-alkylated mesoporphyrin thus produced could be separated by high-pressure liquid chromatography (HPLC) from the corresponding *N*-alkylated protoporphyrin originating from the cellular, endogenous protoheme; both porphyrins were found to exhibit chiral properties, indicating that the *N*-alkylation reaction must have taken place on an enzymic template in the case of both endogenous and exogenous heme. This finding is therefore entirely compatible with the exchangeable heme being taken up by the apoproteins of cytochrome P450 before being *N*-alkylated.

Summary of Mechanisms Involved

In conclusion, 2-allyl-2-isopropylacetamide and other unsaturated suicide substrates of cytochrome P450 lead to a perturbation of liver heme homeostasis and to hepatic porphyria by a multistage mechanism centered on the apoprotein of cytochrome P450. The following successive stages are illustrated in Figure 5 (proceeding from the top of the figure in a clockwise direction). (1) An unsaturated drug, for example ethylene, binds the active site of cytochrome P450; (2) the heme moiety binds and activates molecular oxygen, and ethylene is then metabolized to a reactive monooxygenated derivative; (3) the heme prosthetic group is *N*-alkylated to produce the corresponding adduct or green pigment which then leaves the apoprotein site in the microsomal membranes; (4) fresh heme is taken up by the free apoprotein to reconstitute a functionally active cytochrome P450, ready to initiate a new cycle and to undergo renewed degradation. As a result of these cyclic events, the concentration of cellular free (regulatory) heme will decrease, leading to a secondary stimulation of ALA synthetase, a response mostly destined to sustain the accelerated heme turnover but also responsible for accumulation of heme precursors, as the intermediary enzymes of the pathway become limiting.

It will be noted from the model of Figure 5 that the crucial steps essential for drug-dependent depletion of exchangeable heme are (1) that the heme-free apoproteins should be capable of accepting new heme for reconstitution, and (2) that the reconstituted cytochrome P450 should be functionally competent, as to initiate a new cycle of suicidal catalysis. If either one of these requirements is not met, one might anticipate the suicide substrate to cause extensive loss of cytochrome P450 but relatively little depletion of cellular free heme, hence no pronounced effect on the activity of ALA synthetase. Recent work has shown that after treatment with 2-allyl-2-isopropylacetamide

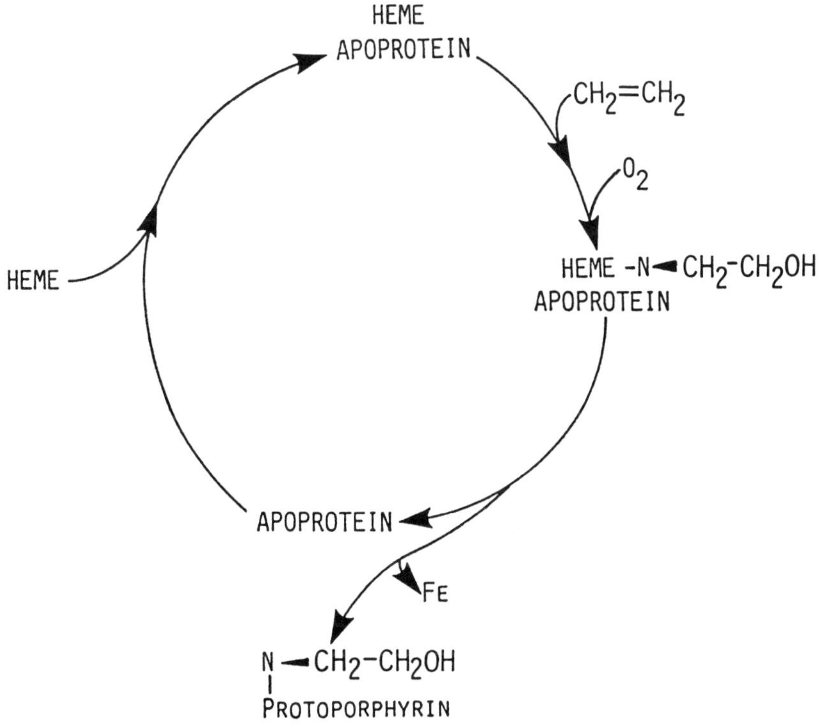

Figure 5. *The apoprotein of cytochrome P450 as a catalytic center for conversion of exchangeable pools of heme to N-alkylated porphyrins (or green pigments).* Ethylene or another unsaturated drug is metabolized to a reactive oxygenated derivative which alkylates one pyrrole nitrogen of the heme prosthetic group. The alkylated heme leaves the apoprotein site, fresh heme is taken up to undergo degradation, and depletion of regulatory heme results. (Reproduced in a modified form, with permission.)

and other suicide substrates of cytochrome P450, the product of heme degradation is in part retained by the apoprotein in a covalently bound form (Davies et al., 1986). This second fate of heme may be favored (as compared to the pathway leading to isolation of *N*-alkylation products) with suicide substrates such as carbon tetrachloride (Davies et al., 1985) or, in the case of 2-allyl-2-isopropylacetamide, with certain inducible forms of cytochrome P450 which lose heme but cannot accept new heme for reconstitution (Bornheim et al., 1987). Should this particular fate of heme clearly predominate in the cell after administration of a suicide substrate of cytochrome P450, one would expect the stimulation of ALA-synthetase (and secondary accumulation of porphyrins) to be absent or very slight. Experimental verification of this will provide further evidence for the validity of the model described in Figure 5.

DRUG-INDUCED PROTOPORPHYRIAS. DISORDERS OF PORPHYRIN METABOLISM ASSOCIATED WITH INHIBITION OF FERROCHELATASE

In contrast to the disorders of liver heme metabolism caused by 2-allyl-2-isopropylacetamide and related compounds, in the porphyrias caused by 3,5-diethoxycarbonyl-1,4-dihydrocollidine (DDC) and Griseofulvin (Figure 6a and b) the biochemical picture is clearly dominated by the accumulation of protoporphyrin. Both drugs cause inhibition of liver ferrochelatase (Onisawa and Labbe, 1963; Lockhead et al., 1967), and the block in heme biosynthesis at the level of this enzyme (Figure 2) explains why protoporphyrin accumulates so markedly. A stimulation of ALA synthetase is seen also with these drugs, again reflecting a depletion of the regulatory pool, due in this case to the pronounced inhibition of ferrochelatase and consequent block in heme biosynthesis, although accelerated destruction of heme also plays a part, as discussed later. The increased supply of ALA thus resulting, a compensatory response intended to overcome the enzymatic block, is an important contributing factor to the accumulation of protoporphyrin (De Matteis, 1978). This can be so pronounced, particularly in mice, that both erythrocytes and blood plasma become intensely fluorescent: solid casts of protoporphyrin pigment can be demonstrated in the liver and pigmented stones in the gall bladder, leading to such complications as jaundice, photosensitivity, and biliary cirrhosis after prolonged intoxication. All these symptoms are found in human erythropoietic protoporphyria, an inherited condition where a defect of ferrochelatase is present as a genetic trait (Elder, 1987) and where protoporphyrin again accumulates in vast excess.

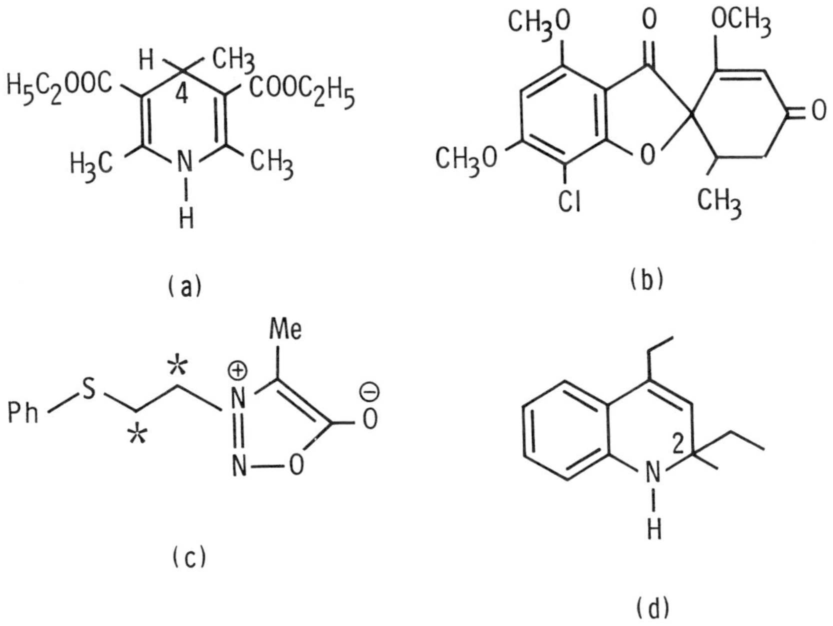

Figure 6. *Structure of Protoporphyria-inducing drugs.* (a) 3,5-diethoxycarbonyl-1,4-dihydrocollidine; (b) Griseofulvin; (c) 3-[2-(2,4,6-trimethylphenyl)-thioethyl]-4-methyl sydnone; (d) 2,4-diethyl-2-methyl-1,2-dihydroquinoline. On oxidation of (a), (b), and (c) by cytochrome P450, a fragment of the drug is liberated in a reactive form (see text) and alkylates the heme prosthetic group, giving rise to a *N*-alkylprotoporphyrin of the general formula given in Figure 3d.

Role of *N*-Alkyl Protoporphyrin Inhibitors of Ferrochelatase

The mechanism of induction of protoporphyria by DDC has been largely elucidated in recent years. Although studies carried out with griseofulvin have been less informative, there are important similarities between these two drugs in their mechanism of action, as in both cases an inhibitor of ferrochelatase can be isolated from the liver of treated animals. With DDC, where this has been investigated extensively, cytochrome P450 has been shown to be involved in initiating drug-induced protoporphyria by producing an inhibitor of ferrochelatase, suffering at the same time suicidal inactivation. The most important steps in the elucidation of the mechanism of action of DDC have been recently reviewed (De Matteis et al., 1987) and can be summarized as follows.

A powerful inhibitor of ferrochelatase was first isolated from the liver of animals treated with DDC by Tephly et al. (1979) and shown to be a modified porphyrin originating from prelabeled liver heme. Mice, a species more sensitive than rats to the porphyrogenic properties of DDC, also produced more inhibitor, while the aromatic nonporphyrogenic analogue of DDC,

3,5-diethoxycarbonylcollidine, did not cause accumulation of the inhibitor in the liver. The unusual porphyrin had the spectral characteristics of a *N*-alkylated protoporphyrin and was identified as *N*-methyl protoporphyrin: that is, protoporphyrin IX bearing a methyl group on one of its pyrrole nitrogen atoms (Figure 3d) (De Matteis et al., 1980b; Ortiz de Montellano et al., 1981b; Tephly et al., 1981). Further investigations showed that the methyl group present on the pyrrole nitrogen of *N*-methyl protoporphyrin was in fact the 4-methyl group of DDC: when mice were given either deuterated 4-CD_3-DDC or the 4-ethyl analogue of DDC, *N*-CD_3 protoporphyrin and *N*-ethyl protoporphyrin respectively, were isolated from their livers (De Matteis et al., 1981b), showing conclusively that the *N*-alkyl group originated from the drug. These observations are in agreement with the finding (Cole et al., 1981) that the inhibition of ferrochelatase caused by treatment of chicken hepatocyte cultures with different dihydropyridines depends on the presence and also on the nature of their 4-alkyl group.

Involvement of Cytochrome P450

Rapid loss of liver cytochrome P450 had been reported after administration of DDC to mice (Wada et al., 1968) and conversion of prelabeled microsomal heme into unidentified products (Abbritti and De Matteis, 1973). These two effects are reminiscent of those described in the previous section for 2-allyl-2-isopropylacetamide and other unsaturated suicide substrates of cytochrome P450, which also give rise to *N*-alkylated porphyrins by covalent attachment of a drug metabolite to the heme prosthetic group of the cytochrome. There is now good evidence that implicates cytochrome P450 in a suicidal inactivation reaction also with the dihydropyridines and with two additional classes of protoporphyria-inducing drugs more recently described. When isolated rat hepatocytes were incubated with DDC and several of its analogues, accumulation of *N*-alkyl protoporphyrins was found to be related to loss of cytochrome P450. The analogue lacking a 4-alkyl group (4-desmethyl DDC) was inactive in both respects; on increasing the size of the 4-substituent of the dihydropyridine from a methyl to an ethyl or propyl residue, the production of *N*-alkylated porphyrins increased and a corresponding greater loss of cytochrome P450 was observed. However, with the 4-isopropyl analogue an even greater loss of cytochrome P450 was observed but no alkylated product could be demonstrated, neither could inhibition of ferrochelatase be detected after treatment with this drug *in vivo*. It was, therefore, concluded that a dihydropyridine can donate its 4-alkyl substituent to nucleophilic sites in the molecule of the prosthetic heme of cytochrome P450; however, for a *N*-alkylated porphyrin to accumulate, the 4-alkyl group must possess special characteristics of size and shape, possibly because bulky residues will be sterically hindered in their access to the pyrrole nitrogen atoms and will attack instead more accessible sites in the molecule of heme (De Matteis et al.,

1987). Whether this second pathway of heme loss (not involving *N*-alkylated products) is responsible for the apoprotein-bound heme fragments which have been described more recently (Correia et al., 1987) remains to be determined.

The molecular mechanism suggested for the transfer by cytochrome P450 of the 4-alkyl group of DDC and other dihydropyridines to one pyrrole nitrogen of its heme prosthetic group (De Matteis et al., 1981b; Ortiz de Montellano and Correia, 1983; Marks et al., 1985) is based on that proposed by Loev and Snader (1965) for the oxidative aromatization of dihydropyridines in chemical systems. An initial oxidative attack at the dihydropyridine nitrogen can lead to aromatization by one of two competing pathways: (1) loss of the 4-alkyl group, or (2) loss of the 4-hydrogen atom, the former pathway predominating with leaving 4-alkyl residues of greater stability and under the influence of steric factors, such as the presence of bulky substituents in the 3 and 5 positions of the dihydropyridines, as discussed (De Matteis et al., 1987). In both the chemical and biological system the oxidative reaction involves electron transfer, without oxygen insertion, so in this case the cytochrome appears to operate as an oxidase rather than in its more usual role as a monooxygenease. Augusto et al. (1982) have proposed that with cytochrome P450 the first step is a one-electron oxidation of the dihydropyridine to its radical cation derivative, leading to elimination of the 4-alkyl group as an alkyl radical.

An *N*-alkylated porphyrin with inhibitory activity on ferrochelatase has also been isolated from the liver of mice treated with griseofulvin. Although this inhibitor porphyrin has so far been studied less extensively than that produced by DDC treatment, nevertheless its spectral characteristics are indistinguishable from those of *N*-methyl protoporphyrin. The HPLC properties of its four isomers are also identical to those of *N*-methyl protoporphyrin, and the four isomers are present in very similar proportions to those seen after DDC (De Matteis et al., 1987). Therefore, the *N*-methyl group of the griseofulvin porphyrin could also originate from the drug by a cytochrome P450-dependent mechanism, but this has not yet been established. A similar *N*-alkylated pigment with inhibitor properties towards ferrochelatase can be extracted in trace amounts from the liver of control mice that have received no drug treatment (Tephly et al., 1979, 1981), so the possibility that griseofulvin may accelerate an alternate pathway of production of *N*-methyl protoporphyrin is less likely but cannot yet be excluded.

More recently, metabolic transfer of a small fragment of a foreign chemical to the heme of cytochrome P450 has also been described for two additional classes of compounds. Here, too, inhibition of ferrochelatase and protoporphyria were observed after treatment and a *N*-alkylated porphyrin inhibitory towards ferrochelatase could be demonstrated. 3-[2-(2,4,6-Trimethylphenyl)-thioethyl]-4-methyl sydnone is one such suicide substrate of cytochrome P450 and *N*-vinyl protoporphyrin the corresponding inhibitor of ferrochelatase (Ortiz de Montellano et al., 1986), the alkylating vinyl fragment being derived from the two carbon atoms in the thioethyl residue of the sydnone (these two carbons are marked by an asterisk in Figure 6c). The other compound shown to be active in inducing protoporphyria and inhibiting ferrochelatase is 2,4-diethyl-2-methyl-1,2-dihydroquinoline (Figure 6d). The dihydroquinoline is oxidized by cytochrome P450 to the corresponding quinoline and in process releases the 2-ethyl and 2-methyl substituents in a reactive form, producing by alkylation of the prosthetic group a mixture of *N*-methyl and *N*-ethyl protoporphyrins (Powles et al., 1988).

It can, therefore, be concluded that production of a *N*-alkyl porphyrin with inhibitory properties towards ferrochelatase is of more common occurrence than originally suspected. Following the discovery of *N*-methyl protoporphyrin after DDC treatment, similar porphyrins also inhibitory towards ferrochelatase have been isolated after treatment with other dihydropyridines, griseofulvin, sydnones, and dihydroquinolines. In most cases metabolic transfer by cytochrome P450 of a small fragment of the foreign chemical to one pyrrole nitrogen of heme has been demonstrated, a fragment small enough to allow interaction of the modified porphyrin with the active site of ferrochelatase, as will now be discussed.

Mechanism of Inhibition of Ferrochelatase

In order to inhibit ferrochelatase, *N*-methyl protoporphyrin must bind the porphyrin-binding site of the enzyme. This is shown by the following findings: (1) when the porphyrin substrate was added before the inhibitor, significant protection of the enzyme was obtained, indicating that inhibitor and substrate compete for the same binding site. In contrast, no protection was afforded by preincubating the enzyme with its metal substrate (De Matteis et al., 1980a). (2) Only *N*-alkylated porphyrins with two carboxylic acid side chains can inhibit ferrochelatase, and these porphyrins are all substrates of ferrochelatase in their native state: that is, when they are not *N*-alkylated. In addition, the two propionic acid side chains of *N*-methyl protoporphyrin must be free for inhibition to occur, as the corresponding methyl ester derivatives are inactive (De Matteis et al., 1985). (3) Early studies using concentrations of porphyrin substrate up to 36 μM failed to detect any change in K_m value for the inhibited enzyme (De Matteis et al., 1980a; Houghton et al., 1982), suggesting that *N*-methyl protoporphyrin acted either as an irreversible inhibitor or as a tight-binding inhibitor. In more recent experiments Dailey and Fleming (1983) have used very much higher concentrations of porphyrin substrate and obtained evidence for a high affinity, tight-binding mechanism; the inhibition was found to be competitive with respect to the porphyrin substrate, with a K_i value of approximately 7 nM.

Differences in inhibitory activity between different isomers of the same *N*-alkylated porphyrin have been reported, particularly when zinc is chelated in the center of the tetrapyrrole or, alternatively, when the *N*-alkyl group is large in size. In both cases, the N_C and N_D isomers, where one propionic acid-substituted ring is *N*-alkylated, are less inhibitory than the corresponding N_A and N_B isomers, in which vinyl-substituted rings are *N*-alkylated (see Figure 3d for nomenclature of each pyrrole ring). These steric factors may explain why the *N*-alkylated protoporphyrins produced by ethylene or 2-allyl-2-isopropylacetamide do not inhibit ferrochelatase, as in this case the *N*-alkyl group is relatively large and is also present on propionic acid-substituted rings. Two different explanations have been suggested for the loss of activity seen in the N_C and N_D isomers when the *N*-alkyl group increases in size. First, it is possible that in the N_C and N_D isomers, where a propionic acid-substituted ring is *N*-alkylated, the alignment of the two propionic acid side chains, with respect to each other, may be sufficiently altered to reduce the affinity of the inhibitor for ferrochelatase (De Matteis et al., 1985). An alternate explanation is favored by Ortiz de Montellano et al. (1986). These authors also propose that the carboxylic acid residues of the porphyrin are involved in binding, by interacting with positively charged amino acid residues at the active center. They place these amino acid residues at the bottom of a pocket or groove. However, the region of the active site where ring C and D are contained is sterically constrained and will, therefore, accommodate an *N*-methyl group, but no larger *N*-alkyl groups. In contrast, rings A and B are contained in a relatively open region of the active site that will accommodate alkyl groups larger than methyl attached to rings A and B (see also McCluskey et al., 1988).

Differences in inhibitory activity between *N*-alkylated deuteroporphyrin, mesoporphyrin, and protoporphyrin have also been reported, particularly for the N_C and N_D isomers, suggesting that the vinyl groups of protoporphyrin and, to a lesser extent, the ethyl groups of mesoporphyrin may also be involved in binding (De Matteis et al., 1985). Therefore, the porphyrin-binding site at the active center (shared by the porphyrin substrate and the *N*-alkylated inhibitor) may include, in addition to appropriately spaced positive charges for binding of the two propionate, an area of interaction for the vinyl side chains of ring A and B of protoporphyrin. There is as yet no satisfactory explanation as to why *N*-methyl protoporphyrin should exhibit such a great affinity for ferrochelatase. It is difficult to visualize why a porphyrin-binding site intended for a planar porphyrin substrate should bind with such avidity a porphyrin analogue which is distorted and bears a bulky substituent on a pyrrole nitrogen, unless distortion of the substrate was a geometric feature of its transition state and the *N*-alkyl porphyrin a transition-state analogue.* Alternatively, since on attempted insertion of a divalent metal into *N*-alkyl porphyrins a charged species is formed, this may contribute to the stability of the enzyme-inhibitor complex.

*See note added in proof.

Main Stages in the Induction of Protoporphyria by Drugs

Figure 7 summarizes the sequence of events leading to formation of *N*-methyl protoporphyrin and inhibition of ferrochelatase after treatment with DDC. A powerful antimetabolite, *N*-methylprotoporphyrin, originates in the endoplasmic reticulum from metabolic transfer of the 4-methyl group of DDC to a pyrrole nitrogen of the heme prosthetic group of cytochrome P450. The *N*-alkylated porphyrin, a substrate analogue targeted to the enzyme ferrochelatase by the tetrapyrrolic structure of protoporphyrin, migrates within the cell and reaches the inner membrane of the mitochondria, where its presence leads to enzyme inhibition. Two remarkable features of this process are the cooperation of a foreign chemical (DDC) and a biological component (heme) in generating a targeted enzyme inhibitor, and also the need for *N*-methylprotoporphyrin to travel from the site of production to the mitochondrion, where it acts as an antimetabolite. The situation is quite similar with other protoporphyria-inducing agents except that, as already discussed, a

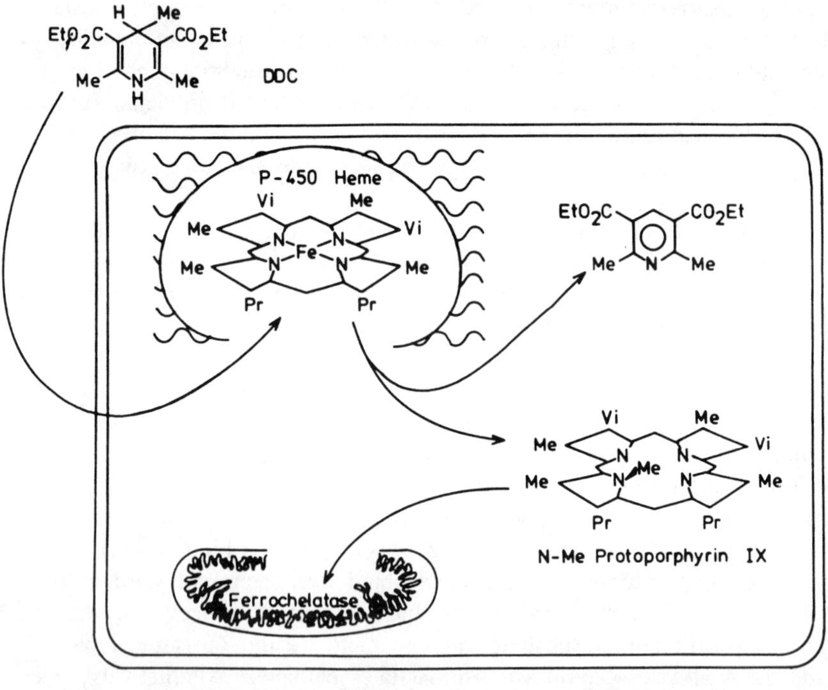

Figure 7. *Mechanism of inhibition of liver ferrochelatase by 3,4-diethoxycarbonyl-1,4-dihydrocollidine (DDC).* The drug first interacts with cytochrome P450 (in the endoplasmic reticulum of the hepatocytes), and during oxidative metabolism its 4-methyl group is transferred to one of the pyrrole nitrogens of the heme prosthetic group. The *N*-methyl protoporphyrin then migrates to the mitochondrion to inhibit ferrochelatase. (Reproduced with permission.)

different inhibitory porphyrin may be produced or—in the case of griseofulvin—the mechanism of production of the inhibitor has not yet been worked out.

A compensatory stimulation of ALA-synthetase follows and this, in combination with the inhibition of ferrochelatase, is responsible for the marked accumulation of protoporphyrin. Direct confirmation of the role of *N*-methyl protoporphyrin in the induction of protoporphyria has been obtained in mice and chick embryos (De Matteis and Marks, 1983) by treating them with an *N*-methylated dicarboxylic porphyrin and demonstrating loss of ferrochelatase activity as well as stimulation of ALA-synthetase and accumulation of protoporphyrin.

DRUG-INDUCED UROPORPHYRIAS. DISORDERS OF PORPHYRIN METABOLISM ASSOCIATED WITH INHIBITION OF UROPORPHYRINOGEN DECARBOXYLATION

In hepatic uroporphyria a block in heme metabolism also exists but at a different stage in the biosynthetic sequence, the stage of decarboxylation of uroporphyrinogen (uro'gen) III. In most cases a defect of the relevant enzyme, uro'gen decarboxylase, can be detected, thus accounting for the block in heme biosynthesis and characteristic accumulation of uroporphyrin. However, since the uro'gen intermediate has to be kept in the reduced state (i.e., as a hexahydroporphyrin) in order to be decarboxylated, a block in its metabolism can also be achieved by its oxidation to the corresponding porphyrin which cannot be further metabolized. Such a concept of an oxidative escape of the intermediate from the pathway of heme biosynthesis was first postulated by Heikel et al. (1958), but evidence in its favor has only been obtained more recently. Whether due to inhibition of the decarboxylase or to oxidative escape of the uro'gen intermediate, the block in heme biosynthesis is accompanied by a stimulation of ALA-S, a compensatory response analogous to that encountered in protoporphyria, again resulting from reduction of the heme feedback mechanism and here too contributing to a more severe porphyria.

The causative role of toxic factors, especially chronic alcoholism in human uroporphyria, had been suspected for many years, but the first unequivocal evidence that drugs could be responsible for such condition was the outbreak of human toxic porphyria in Turkey over 30 years ago. In 1956 many cases of cutaneous porphyria were reported with intense photosensitivity, pronounced skin pigmentation and hypertrichosis, liver enlargement, and marked uroporphyrinuria. Several thousand individuals of both sexes, all age groups, and three different ethnic groups were affected (Cam and Nigogosyan, 1963). Poisoning by hexachlorobenzene (HCB) (Figure 8a), a fungicidal agent

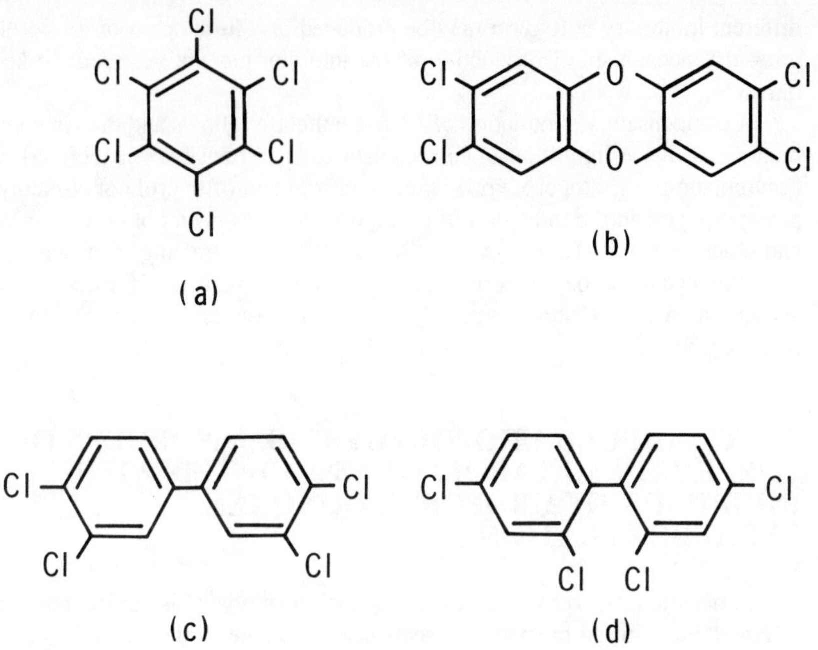

Figure 8. *Structure of Uroporphyria-inducing drugs.* (a) Hexachlorobenzene; (b) 2,3,7,8-tetrachlorodibenzo-*p*-dioxin; (c) 3,4,3′,4′-tetrachlorobiphenyl; (d) 2,4,2′,4′-tetrachlorobiphenyl. Note that in (d) presence of two chlorines in the 2 position sterically hinders a planar configuration, thus reducing markedly the activity of this biphenyl (see text).

introduced in bread made from dressed wheat intended for sowing, was suspected as the cause of these disturbances. Animal experiments soon confirmed that HCB could cause a condition of hepatic porphyria when fed to rats and rabbits (Ockner and Schmid, 1961; De Matteis et al., 1961). In both humans and experimental animals the picture of the metabolic disorder, characterized by the preponderance of uroporphyria and other highly carboxylated porphyrins, suggested an impairment of the uro'gen decarboxylation mechanism (De Matteis et al., 1961). Subsequent work has in fact demonstrated marked inhibition of liver uro'gen decarboxylase after treating animals with HCB. The time course of progressive enzyme inhibition in animals given HCB continuously correlates well with time-dependent increase in urinary excretion or liver concentrations of uroporphyrin (Elder, 1978).

Drug-Induced Uroporphyria as a Model for Human Sporadic Porphyria Cutanea Tarda

Several other polyhalogenated chemicals, among these polybrominated biphenyls (PBB) and 2,3,7,8-tetrachlorodibenzo-*p*-dioxin (TCDD) (Figure 8b), have since been shown to produce hepatic uroporphyria not only in experimental animals, but also in humans, as in the widespread PBB intoxication which affected both domestic animals and humans in Michigan and in people exposed to TCDD at work in factories in the United States and Czechoslovakia (Smith and De Matteis, 1980). In these cases of human toxic porphyria, whether due to HCB, TCDD, or other polyhalogenated compounds, the biochemical pictures and clinical manifestations are very similar to those of porphyria cutanea tarda (PCT), a heterogenous syndrome that comprises several different disorders (Sweeney, 1986). A rare familial variety of PCT has been described where a defect of uro'gen decarboxylase is inherited as an autosomal dominant characteristic and shows itself as low-enzyme activity not only in the liver but also in the erythrocytes and all other tissues so far investigated (Elder, 1987). In contrast, in the more common form of PCT, the so-called sporadic variety, which is accompanied by alcoholic liver damage, other forms of liver disease and almost invariably, by hepatic siderosis, the defect of uro'gen metabolism is confined to the liver (De Verneuil et al., 1978; Elder et al., 1978). The metabolic defect is present in the liver but not in the erythropoietic system also in porphyria induced by HCB (San Martin de Viale et al., 1976; Tiepermann et al., 1980) and other halogenated chemicals. In addition, the following important similarities between drug-induced uroporphyria and human sporadic PCT have been recognized: (1) in both conditions iron as well as estrogens will increase the severity of the metabolic disorder and of the associated defect of the uro'gen decarboxylase (Smith and De Matteis, 1980; Smith and Francis, 1981, 1983); (2) a genetic disposition of some kind has been suspected in both disorders, in order to explain why not all alcoholics with chronic liver damage will develop porphyria (Sweeney, 1986), and also to account for the finding that in Turkey, within the same family, only some members developed the condition (Dogramaci, 1964); finally, (3) impaired liver function and pathological changes in liver morphology are commonly encountered in both syndromes, ranging from degenerative and necrotic changes of the hepatocytes and fibrosis to increased incidence of hepatoma (Salata et al., 1985; Smith et al., 1989). In view of these widespread similarities, it is possible that the mechanism by which sporadic PCT and HCB porphyria are induced is also similar. However, this has not yet been established, as the nature of the genetic predisposition and the way in which it interacts with iron, liver damage, and environmental chemicals have still to be clarified. Possible ways in which genetic factors may be involved will be briefly considered below, after discussion of the more recent work on the mechanism of induction of uroporphyria by HCB, TCDD, and other drugs.

Mechanisms of Induction of Uroporphyria by Drugs

Sinclair and Granick (1974) were the first to reproduce the characteristic lesion of uroporphyria (a block in uro'gen metabolism with accumulation of uroporphyrin) in cultured chicken hepatocytes, by exposing them to one of several polychlorinated chemicals. In contrast to the intact rodent which will require prolonged and continuous treatment, from weeks to months, in order to develop uroporphyria, the chicken embryo hepatocytes in culture will respond markedly within a few hours and, largely because of this, have proved a very useful experimental system for investigating the mechanism of action of uroporphyria-inducing drugs.

It should be noted that the results obtained in this culture system have not always corresponded exactly to the findings in rodents. For example, the inhibition of uro'gen decarboxylase has been found to be less marked and more variable in the cell system (compare Lambrecht et al., 1988 with De Verneuil et al., 1983), in spite of the very pronounced accumulation of uroporphyrin. Also, a number of drugs usually thought to be inactive in rodents have been shown to be effective uroporphyria-inducing agents in cultures of chick embryo hepatocytes. These apparent anomalies are not yet completely understood but may still be compatible with a similar mechanism of action of the porphyria-inducing chemicals in the two systems, as discussed below.

Sinclair and Granick (1974) suggested that HCB and other halogenated chemicals were converted by drug-metabolizing systems into derivatives capable of inhibiting uro'gen decarboxylase, thus postulating that inhibitory drug metabolites were responsible for induction of uroporphyria. This hypothesis has received wide support until recently (it is still favored by some authors), as the following observations were compatible with it: (1) Pretreatment of the animals or of cultured hepatocytes with certain inducers of cytochrome P450 increased the degree of porphyria caused by a chlorinated chemical (Debets et al., 1980; Carpenter et al., 1984; De Verneuil et al., 1983; Wainstok de Calmanovici et al., 1984), while pretreatment with inhibitors of cytochrome P450 activity inhibited the development of porphyria (Sinclair and Granick, 1974; Debets et al., 1980; Sinclair et al., 1986; Smith et al., 1986). (2) The responsiveness to porphyria of C57BL/6 and DBA/2 mice and of crosses between these two strains could be correlated with inducibility of hepatic aryl hydrocarbon hydroxylase activity, i.e., the *Ah* phenotype (Jones and Sweeney, 1980), and more extensive work on different inbred strains of mice showed a similar, though only partial, correlation between the phenotype and susceptibility to drug-induced uroporphyria (Greig et al., 1984). (3) While TCDD or HCB were not themselves inhibitory to uro'gen decarboxylase, a powerful inhibitor could be isolated from the liver of animals treated with these drugs (Billi et al., 1986; Cantoni et al., 1984; Smith and Francis, 1987). (5) Finally, in microsomal incubation experiments, HCB was shown to give rise to reactive metabolites capable of interacting with proteins, and therefore potentially capable of modifying covalently uro'gen decarboxylase (Van Ommen et al., 1986).

The hypothesis of an inhibitory drug metabolite failed to account, however, for the ability of iron to exacerbate the metabolic condition, in favor of which further and more conclusive evidence had since been obtained (Sweeney et al., 1979; Smith and Francis, 1983). It was also difficult to visualize a reactive metabolite common to the various chemicals shown to be effective in the tissue-culture system (Ferioli et al., 1984) and more recently also in the intact rodent (Francis and Smith, 1987). The hypothesis also failed to explain why TCDD, a compound which is not appreciably metabolized (Rose et al., 1976) should be such a powerful uroporphyria-inducing agent both in culture and in the whole animal.

Possible Involvement of an Uncoupled Cytochrome P450 System

An alternate hypothesis has been put forward which still implicates the cytochrome P450 system in the induction of uroporphyria by drugs, but by an indirect, iron-mediated mechanism, a mechanism which also goes some way toward explaining the additional toxic liver changes frequently found in uroporphyria. The drugs responsible may interact with the NADPH-dependent reductase/cytochrome P450 system of the hepatocytes as poor substrates leading to production of oxygen species by an uncoupling mechanism and to associated mobilization of iron from ferritin in an active pro-oxidant form (De Matteis and Stonard, 1977). Uroporphyria may then result from some mechanism requiring interaction of reduced oxygen with nonheme iron, either through production of a long-lived inhibitor of uro'gen decarboxylase or by modification of the enzyme itself. A contributing factor may be direct oxidation of the substrate of the decarboxylase, uro'gen, as the oxidative derivative, uroporphyrin, cannot be metabolized further and would therefore accumulate (Ferioli et al., 1984).

Evidence in favor of an oxidative mechanism involving the NADPH-reductase/cytochrome P450 system has recently been obtained and the form(s) of cytochrome P450 induced by 3-methylcholanthrene and similarly acting inducers [cytochrome(s) P448] have been specifically implicated. Liver microsomes from rats treated with either HCB or 3-methylcholanthrene produce increased amounts of reactive oxygen species when incubated with NADPH *in vitro* (Urquhart and Elder, 1987). In additional experiments, liver microsomes from 3-methylcholanthrene-treated chick embryos were shown to catalyse the oxidation of porphyrinogens in presence of NADPH, when challenged with small amounts of a powerful uroporphyria-inducing agent (Sinclair et al., 1987). In the experiments of De Matteis et al. (1988) two tetrachlorobiphenyl isomers were tested, 3,4,3′,4′- and 2,4,2′,4′-tetrachlorobiphenyl (for short 3,4-TCB and 2,4-TCB, respectively, see Figure 8c and d). Only the first isomer (a planar biphenyl) was found to be a strong uroporphyria-inducing agent in cultures of chicken embryo hepatocytes, and only this isomer, when given to the chick embryo *in ovo,* proved an effective inducer of cytochrome(s) P448 and could also stimulate NADPH-dependent oxidation of uroporphyrinogen by microsomes *in vitro*. In contrast, 2,4-TCB (a non-planar isomer) was considerably less active in all these respects. Two different drug-mediated responses are apparently involved in the oxidation of porphyrinogens (see Figure 9): (1) first, induction of cytochrome(s) P448 has to take place; (2) then a planar polyhalogenated chemical can interact with the induced cytochrome so as to produce an oxidizing species. Action (1) above is also shared by β-naphthoflavone and, in the experiments of Sinclair et al. (1987), by 3-methylcholanthrene, probably because these inducers all stimulate the synthesis of a form of cytochrome which can recognize 3,4-TCB and bind it effectively. This type of oxidative response may also apply to degradation of unconjugated bilirubin as seen in the congenitally jaundiced Gunn rat, where again induction of cytochrome(s) P448 and subsequent interaction with a planar biphenyl are apparently involved (Trenti et al., 1989).

More recent work has been aimed at identifying the form of cytochrome which participates in drug-dependent porphyrinogen oxidation. β-Naphthoflavone, polyhalogenated biphenyls, and TCDD cause accumulation in rat liver microsomes of at least two distinct forms of cytochrome P450: cytochrome P450 IA1 and IA2, according to a recent nomenclature (Nebert et al., 1987). These two forms are also known in the rat as cytochrome P450c and P450d, respectively. There is evidence that cytochrome P450d has great affinity for either TCDD or polyhalogenated biphenyls, as 3 days after a single dose substantial binding of these inducers to cytochrome P450d (but not to

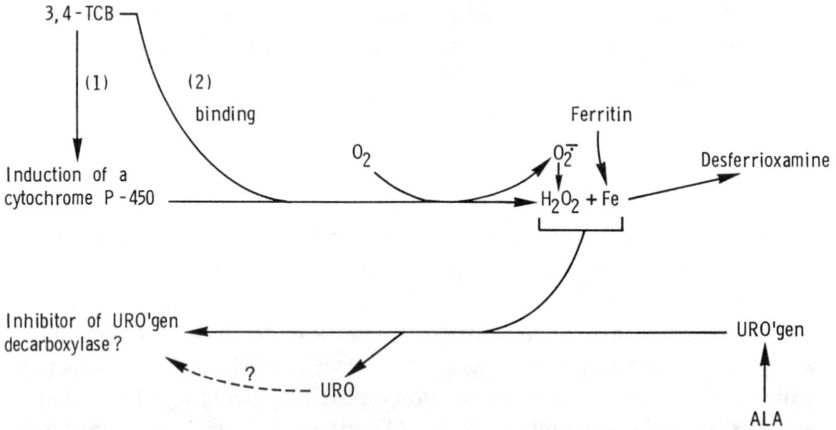

Figure 9. *Hypothetical mechanism for induction of uroporphyria by drugs.* The scheme attempts to bring together in a coherent mechanism the results so far obtained in various laboratories, but the role of the induced cytochrome and of iron in uroporphyrinogen oxidation and the mechanism of production of the inhibitor of the decarboxylase are still largely hypothetical and require further work. Also note that although hydrogen peroxide is involved in chemical and enzymic model systems, the oxidative species produced by microsomes has not yet been identified. The various steps illustrated are explained in the text.

cytochrome P450c) can be detected after immunoprecipitation of the cytochrome with the appropriate specific antibody (Voorman and Aust, 1987). Voorman and Aust (1987) also showed that several polybrominated biphenyls added *in vitro* to purified cytochrome P450d were powerful inhibitors of its estradiol 2-hydroxylase activity; the inhibitory activity decreased with increasing ortho substitution of the biphenyl, a finding reminiscent of the less-marked activity of 2,4-TCB (an ortho-substituted biphenyl) in promoting oxidation of porphyrinogens and in inducing uroporphyria (see above). In recent experiments (Jacobs et al., 1989) the porphyrinogen oxidation activity of rat microsomes obtained from liver and kidney after a variety of inducers has been compared to the amount of immunochemically detectable cytochrome P450c and P450d. The effect of antibodies selectively inhibitory to either cytochrome has been investigated, again with respect to the porphyrinogen oxidation activity of induced microsomes. Jacobs et al. (1989) have concluded from these studies that cytochrome P450d (rather than cytochrome P450c) is specifically involved in porphyrinogen oxidation. This may explain why in previous work (De Matteis et al., 1988) the extent of induction of 7-ethoxyresorufin deethylase activity (which is mostly dependent on cytochrome P450c) did not show a good correlation with either porphyrinogen oxidation or induction of uroporphyria, for example, in cultured rat hepatocytes, as these cells can synthesize cytochrome P450c in primary culture, but not, apparently, cytochrome P450d (Stewart et al., 1985).

Although the findings summarized above offer some support for the uncoupled cytochrome P450 mechanism originally proposed, the exact role of cytochrome P450 and of the bound inducer in porphyrinogen oxidation is still matter for speculation. On binding to the active site of cytochrome P450, the polyhalogenated inducer will probably elicit transition of the spin state of the heme iron from low to high spin and in this way increase its redox potential. This will facilitate reduction of heme leading to binding and reduction of oxygen. On account of the relative strength of the carbon–halogen bonds, oxygen insertion will be difficult and the bound chemical will, therefore, behave as a poor substrate. The oxygenated intermediate of the drug-cytochrome complex will instead break down, producing reactive oxygen species and regenerating the ferric cytochrome-drug complex, so as to initiate a new cycle. There is still uncertainty as to the nature of the oxidizing species involved in the microsomal reaction with both porphyrinogens and bilirubin. Also, it is not yet clear whether the cytochrome is the actual catalyst in the oxidation of these substrates; or whether, once the reactive oxygen species have been produced, they can diffuse away to interact with nonheme iron to generate a secondary (or proximal) oxidant. Such issues have a bearing on the role of iron in these processes, as will now be discussed.

Possible Mechanisms for the Exacerbation by Iron of Hepatic Uroporphyria

The ability of iron to worsen the course of human sporadic PCT and of the related uroporphyrias of the experimental animals has been alluded to in the previous section and is well documented (Sweeney, 1986; Taljaard et al., 1972; Smith and Francis, 1983). So is the protective effect of depleting the iron stores by low iron intake, phlebotomy, or treatment with desferrioxamine (DES) (Sweeney, 1986; Sweeney et al., 1979; Sinclair and Granick, 1974; Wainstock de Calmanovici et al., 1986). Iron stored as ferritin would be expected to be inactive, and mobilization of iron in an active form, complexed to an unknown endogenous ligand, has therefore been suggested to be required (De Matteis and Stonard, 1977; Rowley and Sweeney, 1984) before iron can potentiate the characteristic metabolic lesion of uroporphyria, the block in uroporphyrinogen metabolism. Iron is known to participate in peroxidative and free-radical reactions (Halliwell and Gutteridge, 1985; Aust et al., 1985), so its synergistic role in uroporphyria could result from its ability to interact with the oxidative species produced by cytochrome P450 (Ferioli et al., 1984). Accordingly, the catalytic activity of iron in assisting the oxidation of porphyrinogens was investigated in chemical and enzymatic systems to try and answer the following questions: (1) Could iron increase the oxidation of porphyrinogens to porphyrins, thus contributing to the severity of uroporphyria through greater oxidative escape of the intermediate, uro'gen? (2) Was iron involved in inhibition of uro'gen decarboxylase by generating an inhibitor of the enzyme, such as that isolated from porphyric livers?

The iron chelate of ethylenediaminetetraacetic acid (EDTA) and nitriloacetic acid were very active at promoting the hydrogen peroxide-dependent oxidation of porphyrinogen to porphyrins, and a similar role of iron was found for the NADPH-dependent oxidation of porphyrinogens by liver microsomes *in vitro*. In contrast, neither the iron chelate of desferrioxamine nor ferritin iron possessed prooxidant activity, but the latter could be mobilized in an active form by incubation with EDTA. Iron was also found to promote modification of the porphyrin pigment, leading to marked loss of its Soret absorbance, suggesting further (or alternative) oxidation of the porphyrin(ogen) to an open-chain degradation product. Oxidation of uro'gen to uroporphyrin and to the unidentified derivative(s) lacking Soret absorbance could under certain conditions be dissociated from each other, suggesting that they may be produced by two different mechanisms. Electron transfer (and hydrogen abstraction) may be involved in the conversion of uro'gen to uroporphyrin, whereas the degradation product may result from oxidative attack by hydroxyl radicals or, alternatively, by self-catalyzed ring opening of the corresponding heme derivative. The possibility that such a degradation product might act as a targeted inhibitor of uro'gen decarboxylase was considered, to account for the inhibitor of the enzyme which has been extracted from porphyric livers (De Matteis, 1988).

The existence of these two oxidative fates for the porphyrinogen [(1) oxidation to the corresponding porphyrin, and (2) conversion to derivative(s) which do not exhibit Soret absorbance] has also been reported by Francis and Smith (1988) on the basis of experiments carried out with the xanthine/xanthine oxidase system. In addition, when the oxidative derivatives of uro'gen which lacked Soret absorbance were tested *in vitro* for their effect on uro'gen decarboxylase, they were found to be significantly inhibitory (Francis and Smith, 1988). This was the first demonstration that an inhibitor of uro'gen decarboxylase could be produced by oxidation of uro'gen, lending support to the concept that iron may be implicated in converting the substrate into a targeted inhibitor of the enzyme. Elder et al. (1986) have suggested that uroporphyrin may chelate iron to produce the corresponding heme; this may then be taken up by the active center of uro'gen decarboxylase there to catalyze formation of hydroxyl radicals, causing site-selective enzyme inactivation. It is also possible, however, that an oxidative derivative of uroporphyrin-(ogen)—for example, an open-chain degradation product (whether produced from direct oxidation of the tetrapyrrole or of its iron-chelate)—may interact with the enzyme as a high-affinity inhibitor (De Matteis, 1988). If the interaction was reversible or at least thermo-labile, such a derivative could then be released, when the inhibited enzyme or other binding proteins of the liver cytosol are heat-denatured and unfolded, thus explaining why an active inhibitor of uro'gen decarboxylase can be extracted from porphyric livers (Cantoni et al., 1984) by heat treatment. It must be stressed that these possibilities are still very speculative, but elucidation of the structure of the inhibitor and study of the mechanism of enzyme inhibition should clarify these issues.

A similar mechanism can be postulated for the exacerbation by iron of human sporadic PCT, where no exposure to chlorinated chemicals is involved, but alcoholic liver damage is frequently found. Alcohol is known to stimulate free-radical processes (Reinke et al., 1987) and to induce a form of cytochrome P450 that is particularly prone to generate reduced oxygen species (Ingelman-Sundberg and Johansson, 1984), so this cytochrome could conceivably be involved in initiating an oxidative process in which iron may participate. An alternate explanation is that in this condition there may be increased production of reactive oxygen metabolites from sources other than cytochrome P450 and by a mechanism which is not directly dependent on exogenous chemicals. Recent work suggests that iron may produce toxic liver effects similar to the ones reported in drug-induced uroporphyria also in the absence of any drug treatment, presumably by interacting with oxygen metabolites produced by the hepatocytes during normal oxidative metabolism. For example, chicken hepatocytes in culture will accumulate significant amounts of uroporphyrin and other highly carboxylated porphyrins from added ALA in the absence of any exogenous drug, an effect which is increased by adding iron and markedly decreased by desferrioxamine (Ferioli et al., 1984; De Matteis et al., 1988).

In addition, mice of the C57BL10 strain, given a single injection of iron, will develop porphyria 6 months later (Smith et al., 1989) without any drug treatment. The porphyria takes longer to appear and is less marked than in mice given, in addition to iron, hexachlorobenzene, but has all the characteristic chemical and enzymatic features of drug-induced uroporphyria. In mice of this strain iron also potentiated the liver carcinogenicity of HCB, perhaps suggesting that iron in this strain can also activate an oxidative genotoxic mechanism (Smith et al., 1989). These findings provide support for the concept (De Matteis and Stonard, 1977) that chlorinated chemicals, alcoholic liver damage, and possibly genetic factors may all contribute to the production of uroporphyria and other toxic liver effects by sensitizing the liver to the prooxidant activity of iron, but iron itself (in conjunction with reactive oxygen metabolites) may be the real toxic agent responsible.

Uroporphyria could also be elicited in mice by combining 3-methylcholanthrene pretreatment with continuous administration of ALA (Urquhart et al., 1988), but neither 3-methylcholanthrene nor ALA were effective when given on their own. These authors also subscribe to an oxidative mechanism dependent on cytochrome P450 and explain their findings as follows. Activation of oxygen by the 3-methylcholanthrene–induced cytochrome P450 system produces a reactive species that reacts with a heme precursor, probably a porphyrinogen, to produce a site-directed irreversible inhibitor of uroporphyrinogen decarboxylase. Continuous administration of ALA increases the concentration of the heme precursor from which the inhibitor is derived and therefore results in a more rapid and pronounced porphyria.

Summary of Mechanisms Involved and Possible Role of Genetic Factors

The recent contributions from various laboratories can now be summarized in a hypothetical mechanism, centered on cytochrome P450 (Figure 9), the main features of which are still being investigated. A powerful uroporphyria-inducing chemical (for example, 3,4-TCB) first induces a special form of cytochrome P450, then binds it with high affinity, so as to facilitate reduction of molecular oxygen to an active form. This will then interact with iron (mobilized from ferritin), thus resulting in oxidation of uroporphyrinogen to uroporphyrin and also to a further unknown derivative, which may act as a targeted inhibitor of uro'gen decarboxylase. Since an induced cytochrome P450 system, nonheme iron, and uroporphyrinogen are all involved in producing the postulated inhibitor of uro'gen decarboxylase, this scheme explains not only the role of drugs in initiating the metabolic disorder, but also why excess iron and increased supply of ALA (the precursor of uro'gen) may under certain conditions potentiate the porphyria. Can a scheme of this type account for the mechanism of induction of sporadic PCT by chronic alcoholism and liver damage and can the role of the postulated genetic predisposition be visualized on this basis? It is not possible at the present state of knowledge to answer these questions with any degree of certainty. If a cytochrome P450-dependent mechanism similar to that discussed for drug-induced uroporphyria also applied to the human condition, then the predisposition of certain individuals could be explained on the basis of a genetic polymorphism in either amount or inducibility of the cytochrome responsible—for example, of the ethanol-inducible cytochrome P450. Alternatively, there may be a genetic difference in the ability to reduce and mobilize stored liver iron in the form required for the oxidation reaction: genetic difference in the ability to utilize iron for lipid peroxidation and other oxidative reactions of a potentially toxic nature are suggested by findings in mice (Cheeseman et al., 1985; Smith et al., 1986, 1989). A third possibility is that inherited difference may exist in the inducibility of ALA-S, thus producing in some people with chronic liver damage an increased supply of the heme precursor (from which the inhibitor of uro'gen decarboxylase originates), as in the ALA loading experiments of Urquhart et al. (1988) in mice. Genetic differences in ALA-S inducibility have been documented in mice (Hutton and Gross, 1970; Holley, 1987). Finally, inherited differences may exist in the ability to detoxify the reactive oxidative species involved (Rizzardini et al., 1988).

In conclusion, the metabolic defect of PCT may result, as in the experimental model of the disease produced in animals by polyhalogenated drugs, from an abnormal interaction of three normal body constituents: a porphyrin(ogen), oxygen (in a partially reduced form) and nonheme iron (mobilized from ferritin). The genetic predisposition to develop the disease may be due to one of the above-considered factors, and it is possible that different predisposing factors may be involved in different patients.

A final aspect of the metabolic disorder which has not yet been adequately explained is the long latency in the induction of porphyria in some species (including humans and experimental rodents). This may indicate the slow initiation of a chronically active process which, once set in motion, has some tendency to maintain itself. A small but significant accumulation of uroporphyrin is seen soon after starting treatment with HCB (Kennedy et al., 1986), but a much longer period is required for the uro'gen decarboxylase activity to decrease significantly and a marked porphyria to become established. It is not known whether the accumulating uroporphyrin is capable of participating in the reactions outlined in Figure 9, for example, by taking up iron (Elder et al., 1986) and undergoing self-catalyzed oxidation (De Matteis, 1988). If it does, it might provide an oxidative free-radical chain reaction of its own, capable of spreading and maintaining the toxic process and its manifestations. Perhaps accumulation of uroporphyrin up to a certain threshold level is needed before a vicious cycle leading to uro'gen decarboxylation inhibition and pronounced porphyria can be set in motion. A free-radical derivative of uroporphyrin has also been described and its toxicological significance has been suggested (Morehouse et al., 1987). The possibility that uroporphyrin itself may contribute to the liver toxicity should, therefore, be examined, also in view of the report that in other conditions where porphyrins accumulate in excess in the liver the incidence of hepatic cancer may be increased (Kauppinen and Mustajoki, 1988).

NOTE ADDED IN PROOF

Since this chapter was submitted to the Editors (March 1989), a paper has appeared describing a model for ferrochelatase (Dailey HA, Jones CS, and Karr SW: Interaction of free porphyrins and metalloporphyrins with mouse ferrochelatase. *Biochim Biophys Acta* 1989; 999:7–11). In this, Dailey et al. also propose that *N*-methyl protoporphyrin may inhibit ferrochelatase by acting as a stable transition state analog.

ACKNOWLEDGMENTS

I would like to thank Mrs. J. A. Bryant for much skill and patience in typing the manuscript. I am grateful to Dr. Silke Bernhard, Dahlem Workshop Reports editor, for permission to reproduce Figures 2, 5, and 7 from a previous paper of mine (Report of the Dahlem Workshop on Mechanisms of Cell Injury, pp 101–114, Wiley, Chichester, 1987), and MacMillan Press Ltd., London, for similar permission concerning Figure 4 (*Selectivity and Molecular Mechanisms of Toxicity,* pp 183–210, 1987).

ABBREVIATIONS

ALA:	5-aminolevulinate
AIA:	2-allyl-2-isopropylacetamide
DDC:	3,5-diethoxycarbonyl-1,4-dihydrocollidine
HCB:	hexachlorobenzene
PBG:	porphobilinogen
PCT:	porphyria cutanea tarda
3,4-TCB:	3,4,3′,4′-tetrachlorobiphenyl
2,4-TCB:	2,4,2′,4′-tetrachlorobiphenyl
TCDD:	2,3,7,8-tetrachlorodibenzo-*p*-dioxin
Uro'gen:	uroporphyrinogen

REFERENCES

Abbritti G and De Matteis F: Effect of 3,5-diethoxycarbonyl-1,4-dihydrocollidine on degradation of liver haem. *Enzyme* 1973; 16:196–202.

Ades IZ, Stevens TM, and Drew PD: Biogenesis of embryonic chick liver 5-aminolevulinate synthase: regulation of the level of mRNA by hemin. *Arch Biochem Biophys* 1987; 253:297–304.

Augusto O, Beilan HS, and Ortiz de Montellano PR: The catalytic mechanism of cytochrome P-450: spin-trapping evidence for one-electron substrate oxidation. *J Biol Chem* 1982; 257:11288–11295.

Aust SD, Morehouse LA, and Thomas CE: Role of metals in oxygen radicals reactions. *J Free Rad Biol Med* 1985; 1:3–25.

Badawy AA-B, Morgan CJ, and Davis NR: Tryptophan pyrrolase and the regulation of mammalian hepatic haem biosynthesis, in Nordmann Y (ed): *Porphyrins and Porphyrias. Colloque INSERM* 1986; 134:69–81.

Billi SC, Wainstok De Calmanovici R, and San Martin De Viale LC: Rat-liver porphyrinogen carboxy-lyase inhibition as a function of the degree of hexachlorobenzene-induced porphyria, in Morris CR and Cabral JRP (eds): *Hexachlorobenzene* (IARC Scientific Publication 77). Lyon, International Agency for Research on Cancer, 1986, pp 487–491.

Bissell DM and Hammaker LE: Effect of endotoxin on tryptophan pyrrolase and δ-aminolaevulinate synthase. *Biochem J* 1977; 166:301–304.

Bock KW and Siekevitz P: Turnover of heme and protein moieties of rat liver microsomal cytochrome b_5. *Biochem Biophys Res Commun* 1970; 41:374–380.

Bornheim LM, Underwood MC, Caldera P, Rettie AE, Trager WF, Wrighton SA, and Correia MA: Inactivation of multiple hepatic cytochrome P-450 isozymes in rats by allylisopropylacetamide: mechanistic implication. *Molec Pharmacol* 1987; 32:299–308.

Burnham BF and Lascelles J: Control of porphyrin biosynthesis through a negative-feedback mechanism. *Biochem J* 1963; 87:462–472.

Cam C and Nigogosyan G: Acquired toxic porphyria cutanea tarda due to hexachlorobenzene. *J Am Med Assoc* 1963; 183:88–91.

Cantoni L, Dal Fiume D, Rizzardini M, and Ruggieri R: *In vitro* inhibitory effect on porphyrinogen carboxylase of liver extracts from TCDD treated mice. *Toxicol Lett* 1984; 20:211–217.

Carpenter HM, Williams DE, Henderson MC, Bender RC, and Buhler DR: Hexachlorobenzene-induced porphyria in Japanese quail. Effect of pretreatment with phenobarbital or β-naphthoflavone. *Biochem Pharmacol* 1984; 33:3875–3881.

Cheeseman KH, Proudfoot KA, Maddix SP, Collins MM, Milia A, and Slater TF: Low rate of NADPH/ADP-iron dependent lipid peroxidation in hepatic microsomes of DBA/2 mice. *FEBS Lett* 1985; 184:343–346.

Cole SPC, Whitney RA, and Marks GS: Ferrochelatase-inhibitory and porphyrin-inducing properties of 3,5-diethoxycarbonyl-1,4-dihydro-2,4,6-trimethylpyridine and its analogues in chick embryo liver cells. *Molec Pharmacol* 1981; 20:395–403.

Correia MA, Decker C, Sugiyama K, Caldera P, Bornheim L, Wrighton SA, Rettie AE, and Trager WF: Degradation of rat hepatic cytochrome P-450 heme by 3,5-dicarbethoxy-2,6-dimethyl-4-ethyl-1,4-dihydropyridine to irreversibly bound protein adducts. *Arch Biochem Biophys* 1987; 258:436–451.

Correia MA, Farrell GC, Schmid R, Ortiz de Montellano PR, Yost GS, and Mico BA: Incorporation of exogenous heme into hepatic cytochrome P-450 *in vivo. J Biol Chem* 1979; 254:15–17.

Dailey HA and Fleming E: Bovine ferrochelatase: kinetic analysis of inhibition by *N*-methylprotoporphyrin, manganese and heme. *J Biol Chem* 1983; 258:11453–11459.

Davies HW, Britt SG, and Pohl LR: Inactivation of cytochrome P-450 by 2-isopropyl-4-pentenamide and other xenobiotics leads to heme-derived protein adducts. *Chem Biol Interact* 1986; 58:345–352.

Davies HW, Satoh H, Schulick RD, and Pohl LR: Immunochemical identification of an irreversible bound heme-derived adduct to cytochrome P-450 following CCl_4 treatment of rats. *Biochem Pharmacol* 1985; 34:3203–3206.

Debets F, Hamers W, and Strik J: Metabolism as a prerequisite for the porphyrinogenic action of polyhalogenated aromatics, with special reference to hexachlorobenzene and polybrominated biphenyls (Firemaster BP-6). *Int J Biochem* 1980; 12:1019–1025.

De Matteis F: Rapid loss of cytochrome P-450 and heme caused in the liver microsomes by the porphyrogenic agent 2-allyl-2-isopropylacetamide. *FEBS Lett* 1970; 6:343–345.

De Matteis F: Drug-induced destruction of cytochrome P-450. *Drug Metab Dispos* 1973; 1:267–272.

De Matteis F: Hepatic porphyria caused by 2-allyl-2-isopropyl-acetamide, 3,5-diethoxycarbonyl-1,4-dihydrocollidine, griseofulvin, in De Matteis F and Aldridge WN (eds): *Heme and Hemoproteins,* vol 44: *Handbook of Experimental Pharmacology,* Berlin, Springer Verlag, 1978, pp 129–155.

De Matteis F: Drugs as suicide substrates of cytochrome P-450, in De Matteis F and Lock EA (eds): *Selectivity and Molecular Mechanisms of Toxicity.* New York, Macmillan, 1987, pp 183–210.

De Matteis F: Role of iron in the hydrogen peroxide-dependent oxidation of hexahydroporphyrins (porphyrinogens): a possible mechanism for the exacerbation by iron of hepatic uroporphyria. *Mol Pharmacol* 1988; 33:463–469.

De Matteis F and Cantoni L: Alteration of the porphyrin nucleus of cytochrome P-450 caused in the liver by treatment with allyl-containing drugs. Is the modified porphyrin *N*-substituted? *Biochem J* 1979; 183:99–103.

De Matteis F, Gibbs AH, Farmer PB, and Lamb JH: Liver production of *N*-alkylated porphyrins caused in mice by treatment with substituted dihydropyridines: evidence that the alkyl group on the pyrrole nitrogen atom originates from the drug. *FEBS Lett* 1981b; 129:328–331.

De Matteis F, Gibbs AH, and Harvey C: Studies on the inhibition of ferrochelatase by *N*-alkylated dicarboxylic porphyrins: steric factors involved and evidence that the inhibition is reversible. *Biochem J* 1985; 226:537–544.

De Matteis F, Gibbs AH, and Holley AE: Occurrence and biological properties of *N*-methyl protoporphyrin. *Ann NY Acad Sci* 1987; 514:30–40.

De Matteis F, Gibbs AH, Jackson AH, and Weerasinghe S: Conversion of liver haem into *N*-substituted porphyrins or green pigments. Nature of the substituent at the pyrrole nitrogen atom. *FEBS Lett* 1980b; 119:109–112.

De Matteis F, Gibbs AH, and Tephly TR: Inhibition of protohaem ferro-lyase in experimental porphyria. Isolation and partial characterization of a modified porphyrin inhibitor. *Biochem J* 1980a; 188:145–152.

De Matteis F, Harvey C, and Martin SR: *N*-Alkylation of exogenous haem analogues caused by drugs in isolated hepatocytes. Structural isomerism and chirality of the resulting porphyrins. *Biochem J* 1986; 238:263–268.

De Matteis F, Harvey C, Reed C, and Hempenius R: Increased oxidation of uroporphyrinogen by an inducible liver microsomal system. Possible relevance to drug-induced uroporphyria. *Biochem J* 1988; 250:161–169.

De Matteis F and Marks GS: The effect of *N*-methylprotoporphyrin and succinylacetone on the regulation of heme biosynthesis in chicken hepatocytes in culture. *FEBS Lett* 1983; 159:127–131.

De Matteis F, Prior BE, and Rimington C: Nervous and biochemical disturbances following hexachlorobenzene intoxication. *Nature (London)* 1961; 191:363–366.

De Matteis F and Stonard M: Experimental porphyrias as models for human hepatic porphyrias. *Semin Hematol* 1977; 14:187–192.

De Matteis F, Zutterlund P, and Wetterberg L: Brain 5-aminolaevulinate synthase. Developmental aspects and evidence for regulatory role. *Biochem J* 1981a; 196:811–817.

De Verneuil H, Nordmann Y, Phung N, Grandchamp B, Aitken G, Grelier M, and Noire J: Familial and sporadic porphyria cutanea: two different diseases. *Int J Biochem* 1978; 9:927–931.

De Verneuil H, Sassa S, and Kappas A: Effects of polychlorinated biphenyl compounds, 2,3,7,8-tetrachlorodibenzo-*p*-dioxin and iron on hepatic uroporphyrinogen decarboxylase. *Biochem J* 1983; 214:145–151.

Dogramaci I: Porphyrias and porphyrin metabolism, with reference to porphyrias in childhood. *Adv Pediatr* 1964; 13:11–63.

Druyan R and Kelly A: The effect of exogenous δ-aminolaevulinate on rat liver heme and cytochromes. *Biochem J* 1972; 129:1095–1099.

Duesberg R: Toxische Porphyrie. *Münchener Med Wochensch* 1932; 79:1821–1822.

Elder GH: Porphyria caused by hexachlorobenzene and other polyhalogenated aromatic hydrocarbons, in De Matteis F and Aldridge WN (eds): *Heme and Hemoproteins,* vol 44; *Handbook of Experimental Pharmacology.* Berlin, Springer-Verlag, 1978, pp 157–200.

Elder GH: The porphyrias, in Holton JB (ed): *The Inherited Metabolic Diseases,* London, Churchill-Livingstone, 1987, pp 256–283.

Elder GH, Lee GB, and Tovey JA: Decreased activity of hepatic uroporphyrinogen decarboxylase in sporadic porphyria cutanea tarda. *N Eng J Med* 1978; 299:274–278.

Ferioli A, Harvey C, and De Matteis F: Drug-induced accumulation of uroporphyrin in chicken hepatocytes cultures. Structural requirements for the effect and role of exogenous iron. *Biochem J* 1984; 224:769–777.

Francis JE and Smith AG: Polycyclic aromatic hydrocarbons cause hepatic porphyria in iron-loaded C57 BL/10 mice: comparison of uroporphyrinogen decarboxylase inhibition with induction of alkoxyphenoxazone dealkylation. *Biochem Biophys Res Commun* 1987; 146:13–20.

Francis JE and Smith AG: Oxidation of uroporphyrinogen by hydroxyl radicals. Evidence for nonporphyrin products as potential inhibitors of uroporphyrinogen decarboxylase. *FEBS Lett* 1988; 233:311–314.

Goldberg A and Rimington C: Experimentally produced porphyria in animals. *Proc Roy Soc B* 1955; 143:257–280.

Grandchamp B, Bissell DM, Licko V, and Schmid R: Formation and disposition of newly synthesized heme in adult rat hepatocytes in primary culture. *J Biol Chem* 1981; 256:11677–11683.

Granick S: The induction *in vitro* of the synthesis of δ-aminolevulinic acid synthetase in chemical porphyria: a response to certain drugs, sex hormones, and foreign chemicals. *J Biol Chem* 1966; 241:1359–1375.

Greig JB, Francis JE, Kay SJE, Lovell DP, and Smith AG: Incomplete correlation of 2,3,7,8-tetrachlorodibenzo-*p*-dioxin hepatotoxicity with *Ah* phenotype in mice. *Toxicol Appl Pharmacol* 1984; 74:17–25.

Halliwell B and Gutteridge JMC: The importance of free radicals and catalytic metal ions in human diseases. *Mol Aspects Med* 1985; 8:89–193.

Hamilton JW, Bement WJ, Sinclair PR, Sinclair JF, and Wetterhahn KE: Expression of 5-aminolaevulinate synthase and cytochrome P-450 mRNAs in chicken embryo hepatocytes *in vivo* and in culture. Effect of porphyrinogenic drugs and haem. *Biochem J* 1988; 255:267–275.

Heikel T, Lockwood W, and Rimington C: Formation of non-enzymic haem. *Nature (London)* 1958; 182:313.

Holley AE: Studies on experimental hepatic porphyria. Ph.D. Thesis, University of Surrey, 1987.

Houghton JD, Honeybourne CL, Smith KM, Tabba HD, and Jones OTG: The use of *N*-methyl protoporphyrin dimethylester to inhibit ferrochelatase in *Rhodopseudomonas spheroides* and its effect in promoting biosynthesis of magnesium tetrapyrroles. *Biochem J* 1982; 208:479–486.

Hutton JJ and Gross SR: Chemical induction of hepatic porphyria in inbred strains of mice. *Arch Biochem Biophys* 1970; 141:284–292.

Ingelman-Sundberg M and Johansson I: Mechanisms of hydroxyl radical formation and ethanol oxidation by ethanol-inducible and other forms of rabbit liver-microsomal cytochrome P-450. *J Biol Chem* 1984; 259:6447–6458.

Jacobs JM, Sinclair PR, Bement WJ, Lambrecht RW, Sinclair JF, and Goldstein JA: Oxidation of uroporphyrinogen by methylcholanthrene-induced cytochrome P-450. Essential role of cytochrome P-450d. *Biochem J* 1989; 258:247–253.

Jones KG and Sweeney GD: Dependence of the porphyrogenic effect of 2,3,7,8-tetrachlorodibenzo(p)dioxin upon inheritance of aryl hydrocarbon hydroxylase responsiveness. *Toxicol Appl Pharmacol* 1980; 53:42–49.

Kauppinen R and Mustajoki P: Acute hepatic porphyria and hepatocellular carcinoma. *Brit J Cancer* 1988; 57:117–120.

Kennedy SW, Wigfield DC, and Fox GA: The delay in polyhalogenated aromatic hydrocarbon-induced porphyria: mechanistic reality or methodological artefact? *Toxicol Lett* 1986; 31:235–241.

Lambrecht RW, Sinclair PR, Bement WJ, Sinclair JF, Carpenter HM, Buhler DR, Urquhart AJ, and Elder GH: Hepatic uroporphyrin accumulation and uroporphyrinogen decarboxylase activity in cultured chick-embryo hepatocytes and in Japanese quail *(Coturnix coturnix Japonica)* and mice treated with polyhalogenated aromatic compounds. *Biochem J* 1988; 253:131–138.

Levin W, Jacobson M, Sernatinger E, and Kuntzman R: Breakdown of cytochrome P-450 heme by secobarbital and other allyl-containing barbiturates. *Drug Metab Dispos* 1973; 1:275–284.

Levitt M, Schacter BA, Zipursky A, and Israels LG: The non-erythropoietic component of early bilirubin. *J Clin Invest* 1968; 47:1281–1294.

Lockhead AC, Dagg JH, and Goldberg A: Experimental griseofulvin porphyria in adult and foetal mice. *Brit J Derm* 1967; 79:96–102.

Loev B and Snader KM: The Hantzsch reaction. I. Oxidative dealkylation of certain dihydropyridines. *J Org Chem* 1965; 30:1914–1916.

Loosemore MJ, Wogan GN, and Walsh C: Determination of partition ratios for allylisopropylacetamide during suicidal processing by a phenobarbitone-induced cytochrome P-450 isozyme from rat liver. *J Biol Chem* 1981; 256:8705–8712.

Maguire DJ, Day AR, Borthwick IA, Srivastava G, Wigley PL, May B, and Elliott WH: Nucleotide sequence of the chicken 5-aminolevulinate synthase gene. *Nucl Acids Res* 1986; 14:1379–1391.

Marks GS, Allen DT, Johnston CT, Sutherland EP, Nakatsu K, and Whitney RA: Suicidal destruction of cytochrome P-450 and reduction of ferrochelatase activity by 3,5-diethoxycarbonyl-1,4-dihydro-2,4,6-trimethylpyridine and its analogues in chick embryo liver cells. *Mol Pharmacol* 1985; 27:459–465.

Marks GS, McCluskey SA, Mackie JE, Riddick DS, and James CA: Disruption of hepatic heme biosynthesis after interaction of xenobiotics with cytochrome P-450. *FASEB J* 1988; 2:2774–2783.

Marver HS, Collins A, Tschudy DP, and Rechcigl M: δ-Aminolevulinic acid synthetase II. Induction in rat liver. *J Biol Chem* 1966; 241:4323–4329.

Meyer UA and Schmid R: The porphyrias, in Stanburg JB, Wyngaarden JB, and Frederickson DS (eds): *The Metabolic Basis of Inherited Disease,* Ed 4, Chap 50. New York, McGraw-Hill, 1977, pp 1166–1220.

McCluskey SA, Marks GS, Whitney RA, and Ortiz de Montellano PR: Differential inhibition of hepatic ferrochelatase by regioisomers of *N*-butyl-, *N*-pentyl, *N*-hexyl, and *N*-isopropyl protoporphyrin IX. *Mol Pharmacol* 1988; 34:80–86.

Morehouse KM, Moreno SNJ, and Mason RP: The one-electron reduction of uroporphyrin I by rat hepatic microsomes. *Arch Biochem Biophys* 1987; 257:276–284.

Nebert DW, Adesnik M, Coon MJ, Estabrook RW, Gonzalez FJ, Guengerich FP, Gunsalus IC, Johson EF, Kemper B, Levin W, Phillips IR, Sato R, and Waterman MR: The P-450 gene superfamily: recommended nomenclature. *DNA* 1987; 6:1–11.

Ockner RK and Schmid R: Acquired porphyria in man and rat due to hexachlorobenzene intoxication. *Nature (London)* 1961; 189:499.

Onisawa J and Labbe RF: Effects of diethyl-1,4-dihydro-2,4,6-trimethyl pyridine-3,5-dicarboxylate on the metabolism of porphyrins and iron. *J Biol Chem* 1963; 238:724–727.

Ortiz de Montellano PR, Mico BA, and Yost GS: Suicidal inactivation of cytochrome P-450. Formation of a heme-substrate covalent adduct. *Biochem Biophys Res Commun* 1978; 83:132–137.

Ortiz de Montellano PR, Kunze KL, Yost GS, and Mico BA: Self-catalyzed destruction of cytochrome P-450: covalent binding of ethynyl sterols to prosthetic heme. *Proc Natl Acad Sci USA* 1979; 76:746–749.

Ortiz de Montellano PR and Kunze KL: Cytochrome P-450 inactivation: structure of the prosthetic heme adduct with propyne. *Biochemistry* 1981; 20:7266–7271.

Ortiz de Montellano PR and Mico BA: Destruction of cytochrome P-450 by allylisopropylacetamide is a suicidal process. *Arch Biochem Biophys* 1981; 206:43–50.

Ortiz de Montellano PR, Beilan HS, Kunze KL, and Mico BA: Destruction of cytochrome P-450 by ethylene. Structure of the resulting prosthetic heme adduct. *J Biol Chem* 1981a; 256:4395–4399.

Ortiz de Montellano PR, Beilan HS, and Kunze KL: *N*-Methyl-protoporphyrin IX: chemical synthesis and identification as the green pigment produced by 3,5-diethoxycarbonyl-1,4-dihydrocollidine treatment. *Proc Natl Acad Sci USA* 1981b; 78:1490–1494.

Ortiz de Montellano PR and Correia MA: Suicidal destruction of cytochrome P-450 during oxidative drug metabolism. *Annu Rev Pharmacol Toxicol* 1983; 23:481–503.

Ortiz de Montellano PR, Costa AK, Grab A, Sutherland EP, and Marks GS: Cytochrome P-450 destruction and ferrochelatase inhibition, in Nordmann Y (ed): *Porphyrins and Porphyrias. Colloque INSERM* 1986; 134:109–117.

Ortiz de Montellano PR, Mangold BLK, Wheeler C, Kunze KL, and Riech NO: Stereochemistry of cytochrome P-450-catalyzed epoxidation and prosthetic heme alkylation. *J Biol Chem* 1983; 258:4208–4213.

Ortiz de Montellano PR, Stearns RA, and Langry KC: The allylisopropylacetamide and novonal prosthetic heme adducts. *Mol Pharmacol* 1984; 25:310–317.

Powles JE, Marks GS, Ortiz de Montellano PR, and Lukton D: The involvement of cytochrome P-450 in the porphyrinogenicity of 1,2-dihydro-2,4,6 trimethylquinoline (DTMQ) and 2,4-diethyl-1,2-dihydro-2-methylquinoline (DDMQ). *Con Fed Biol Soc Proc* 1988; 31:163.

Reinke LA, Lai EK, DuBose CM, and McCay PB: Reactive free radical generation *in vivo* in heart and liver of ethanol-fed rats: correlation with radical formation *in vitro*. *Proc Natl Acad Sci USA* 1987; 84:9223–9227.

Rizzardini M, Graziani A, Carugo C, and Cantoni L: Investigations of the role of free radical processes in hexachlorobenzene-induced porphyria in mice. *J Biochem Toxicol* 1988; 3:33–46.

Rose JQ, Ramsey JC, Wentzler TH, Hummel RA, and Gehring PJ: The fate of 2,3,7,8-tetrachlorodibenzo-*p*-dioxin following single and repeated oral doses to the rat. *Toxicol Appl Pharmacol* 1976; 36:209–226.

Rowley B and Sweeney GD: Release of ferrous iron from ferritin by liver microsomes: a possible role in the toxicity of 2,3,7,8-tetrachlorodibenzo-*p*-dioxin. *Can J Biochem Cell Biol* 1984; 62:1293–1300.

Sadano H and Omura T: Incorporation of heme to microsomal cytochrome P-450 in the absence of protein biosynthesis. *J Biochem* 1985; 98:1321–1331.

Salata H, Cortes JM, Enriquez de Salamanca R, Oliva H, Castro A, Kusak E, Carreno V, and Guio CH: Porphyria cutanea tarda and hepatocellular carcinoma. Frequency of occurrence and related factors. *J Hepatol* 1985; 1:477–487.

San Martin de Viale LC, del C Rios de Molina M, Wainstock de Calmanovici R, and Tomio JM: Experimental porphyria produced in rats by hexachlorobenzene, in Doss M (ed): *Porphyrins in Human Diseases*. Basel, Karger, 1976, pp 445–452.

Sardana MK, Rajamanickam C, and Padmanaban G: Differential role of heme in the synthesis of mitochondrial and microsomal hemoproteins, in Doss M (ed): *Porphyrins in Human Diseases*. Basel, Karger, 1976, pp 62–70.

Schmid R and Schwartz S: Experimental porphyria. III. Hepatic type produced by Sedormid. *Proc Soc Exp Biol Med* 1952; 81:685–689.

Schwartz S and Ikeda K: Studies of porphyrin synthesis and interconversion, with special reference to certain green porphyrins in animals with experimental hepatic porphyria, in Wolstenholme GEW (ed): *Ciba Foundation Symposium: Porphyrin Biosynthesis and Metabolism.* London, J. and A. Churchill Ltd., 1955, pp 209–226.

Scholnick PL, Hammaker LE, and Marver HS: Soluble δ-aminolevulinic acid synthetase of rat liver. II. Studies related to the mechanism of enzyme action and hemin inhibition. *J Biol Chem* 1972; 247:4132–4137.

Senjo M, Ishibashi T, and Imai Y: Purification and characterization of cytosolic liver protein facilitating heme transport into apocytochrome b_5 from mitochondria. *J Biol Chem* 1985; 260:9191–9196.

Sinclair PR, Bement WJ, Bonkovsky HL, Lambrecht RW, Frezza JE, Sinclair JF, Urquhart AJ, and Elder GH: Uroporphyrin accumulation produced by halogenated biphenyls in chick-embryo hepatocytes. *Biochem J* 1986; 237:63–71.

Sinclair PR and Granick S: Uroporphyrin formation induced by chlorinated hydrocarbons (lindane, polychlorinated biphenyls, tetrachlorodibenzo-*p*-dioxin). Requirements for endogenous iron, protein synthesis and drug-metabolizing activity. *Biochem Biophys Res Commun* 1974; 61:124–133.

Sinclair PR, Lambrecht R, and Sinclair J: Evidence for cytochrome P-450-mediated oxidation of uroporphyrinogen by cell-free liver extracts from chick embryos treated with 3-methylcholanthrene. *Biochem Biophys Res Commun* 1987; 146:1324–1329.

Smith AG and De Matteis F: Drugs and the hepatic porphyrias. *Clin Haemat* 1980; 9:399–425.

Smith AG and Francis JE: Increased inhibition of hepatic uroporphyrinogen decarboxylase by hexachlorobenzene in male rats given the oestrogenic drugs diethylstilboestrol and chlorotrianisene. *Biochem Pharmacol* 1981; 30:1849–1853.

Smith AG and Francis JE: Synergism of iron and hexachlorobenzene inhibits uroporphyrinogen decarboxylase in inbred mice. *Biochem J* 1983; 214:909–913.

Smith AG and Francis JE: Chemically-induced formation of an inhibitor of hepatic uroporphyrinogen decarboxylase in inbred mice with iron overload. *Biochem J* 1987; 246:221–226.

Smith AG, Cabral JRP, Carthew P, Francis JE, and Manson MM: Carcinogenicity of iron in conjunction with a chlorinated environmental chemical, hexachlorobenzene, in C57 BL/10 ScSn mice. *Int J Cancer* 1989; 43:492–496.

Smith AG, Francis JE, Kay SJE, Greig JB, and Stewart FP: Mechanistic studies of the inhibition of hepatic uroporphyrinogen decarboxylase in C57 BL/10 mice by iron-hexachlorobenzene synergism. *Biochem J* 1986; 238:871–878.

Srivastava G, Borthwick IB, Maguire DJ, Elferink CJ, Bawden MJ, Mercer JFB, and May BK: Regulation of 5-aminolevulinate synthase mRNA in different rat tissues. *J Biol Chem* 1988; 263:5202–5209.

Stewart AR, Wrighton SA, Pasco DS, Fagan JB, Li D, and Guzelian PS: Synthesis and degradation of 3-methylcholanthrene-inducible cytochromes P-450 and their mRNAs in primary monolayer cultures of adult rat hepatocytes. *Arch Biochem Biophys* 1985; 241:494–508.

Sweeney GD: Hepatic catalase activity during states of altered heme synthesis, in Doss M (ed): *Porphyrins in Human Diseases.* Basel, Karger, 1976, pp 53–61.

Sweeney GD: Porphyria cutanea tarda, or the uroporphyrinogen decarboxylase deficiency diseases. *Clin Biochem* 1986; 19:3–15.

Sweeney GD, Jones KG, Cole FM, Basford D, and Krestynski F: Iron deficiency prevents liver toxicity of 2,3,7,8-tetrachlorodibenzo-*p*-dioxin (TCDD). *Science* 1979; 204:332–335.

Taljaard JJF, Shanley BC, Deppe WM, and Joubert SM: Porphyrin metabolism in experimental hepatic siderosis in the rat. II. Combined effect of iron overload and hexachlorobenzene. *Brit J Haematol* 1972; 23:513–519.

Tephly TR, Coffman BL, Ingall G, Abou Zeit-Har MS, Goff HM, Tabba HD, and Smith KM: Identification of *N*-methyl protoporphyrin IX in livers of untreated mice and mice treated with 3,5-diethoxycarbonyl-1,4-dihydrocollidine: source of the methyl group. *Arch Biochem Biophys* 1981; 212:120–126.

Tephly TR, Gibbs AH, and De Matteis F: Studies on the mechanism of experimental porphyria produced by 3,5-diethoxycarbonyl-1,4-dihydrocollidine: role of a porphyrin-like inhibitor of protohaem ferro-lyase. *Biochem J* 1979; 180:241–244.

Tieperman R, Koss G, and Doss M: Uroporphyrinogen decarboxylase deficiency in experimental chronic hepatic porphyria. *Hoppe-Seyler's Z Physiol Chem* 1980; 361:1217–1222.

Trenti T, De Matteis F, Gibbs AH, and Greig J: Inducible bilirubin oxidase in the microsomal fraction of rat liver, in *Proceedings of the 2nd International Meeting on Free Radicals in Liver Injury*. Turin, June 9–11, 1989 (in press).

Unseld A and De Matteis F: Destruction of endogenous and exogenous haem by 2-allyl-2-isopropylacetamide: role of the liver cytochrome P-450 which is inducible by phenobarbitone. *Int J Biochem* 1978; 9:865–869.

Urquhart AJ and Elder GH: Hexachlorobenzene-induced oxygen activation by mouse liver microsomes: comparison with phenobarbitone and 20-methylcholanthrene. *Biochem Pharmacol* 1987; 36:3795–3796.

Urquhart AJ, Elder GH, Roberts AG, Lambrecht RW, Sinclair PR, Bement WJ, Gorman N, and Sinclair JA: Uroporphyria produced in mice by 20-methylcholanthrene and 5-aminolaevulinic acid. *Biochem J* 1988; 253:357–362.

Vander Jagt DL, Hunsaker LA, Garcia KB, and Royer RE: Isolation and characterization of the multiple glutathione *S*-transferases from human liver. Evidence for unique heme-binding sites. *J Biol Chem* 1985; 260:11603–11610.

Van Ommen B, Adang AEP, Brader L, Posthumus MA, Müller F, and Van Bladeren PJ: The microsomal metabolism of hexachlorobenzene. Origin of the covalent binding to protein. *Biochem Pharmacol* 1986; 35:3233–3238.

Vincent SH and Muller-Eberhard U: A protein of the Z class of liver cytosolic proteins in the rat that preferentially binds heme. *J Biol Chem* 1985; 260:14521–14528.

Voorman R and Aust SD: Specific binding of polyhalogenated aromatic hydrocarbon inducers of cytochrome P-450d to the cytochrome and inhibition of its estradiol 2-hydroxylase activity. *Toxicol Appl Pharmacol* 1987; 90:69–78.

Wada D, Yano Y, Urata G, and NaKao K: Behaviour of hepatic microsomal cytochromes after treatment of mice with drugs known to disturb porphyrin metabolism in liver. *Biochem Pharmacol* 1968; 17:595–603.

Wainstok de Calmanovici R, Billi SC, Aldonatti CA, and San Martin de Viale LC: Effect of desferrioxamine on the development of hexachlorobenzene-induced porphyria. *Biochem Pharmacol* 1986; 35:2399–2405.

Wainstok de Calmanovici R, del C Rios de Molina M, Taira de Yamasato MC, Tomio JM, and San Martin de Viale LC: Mechanism of hexachlorobenzene-induced porphyria in rats. Effect of phenobarbitone pretreatment. *Biochem J* 1984; 218:753–763.

White INH: Metabolic activation of acetylenic substituents to derivatives in the rat causing the loss of hepatic cytochrome P-450 and haem. *Biochem J* 1978; 174:853–861.

White INH: Destruction of liver haem by norethindrone. Conversion into green pigments. *Biochem J* 1981; 196:575–583.

White INH: Biliary excretion of green pigments produced by norethindrone in the rat. *Biochem Pharmacol* 1982; 31:1337–1342.

White INH: Suicidal destruction of cytochrome P-450 by ethynyl substituted compounds. *Pharmaceut Res* 1984; 141–188.

White INH and Muller-Eberhard U: Decreased liver cytochrome P-450 in rats caused by norethindrone or ethynyloestradiol. *Biochem J* 1977; 166:57–64.

Whiting MJ: Synthesis of δ-aminolaevulinate synthase by isolated liver polyribosomes. *Biochem J* 1976; 158:391–400.

Whiting MJ and Elliott WH: Purification and properties of solubilized mitochondrial δ-aminolevulinic acid synthetase and comparison with the cytosol enzyme. *J Biol Chem* 1972; 274:6818–6826.

Yamamoto M, Hayashi N, and Kikuchi GK: Evidence for the transcriptional inhibition by heme of the synthesis of δ-aminolevulinate synthase in rat liver. *Biochem Biophys Res Commun* 1982; 105:985–990.

Yamauchi K, Hayashi N, and Kikuchi G: Translocation of δ-aminolevulinate synthase from the cytosol to the mitochondria and its regulation by hemin in the rat liver. *J Biol Chem* 1980; 255:1746–1751.

Chapter

11

Alcohol-Induced Hepatotoxicity

Charles S. Lieber
Section of Liver Disease and Nutrition
Alcohol Research and Treatment Center
Bronx Veterans Affairs Medical Center and
Mt. Sinai School of Medicine
Bronx, New York

INTRODUCTION

There are up to 15 million alcoholics in the United States. In addition, approximately two-thirds of the adult population uses alcohol on occasion, and about 12% can be defined as ''heavy drinkers.'' Among the medical problems associated with alcoholism, hepatic disorders are at the forefront. The spectrum of alcoholic liver injury involves hepatic steatosis (fatty liver), early fibrosis (perivenular and perisinusoidal fibrosis), alcoholic hepatitis, and cirrhosis. Of all medical deaths attributable to alcoholism, 75 percent are the result of cirrhosis of the liver. In the United States, cirrhosis has overtaken diabetes as the fifth cause of death; in large urban areas, it has become the fourth cause of all deaths in the active group of 25 to 65 years (New York City Summary of Vital Statistics, 1986). Early recognition of the alcoholic nature of many patients' complaints may lead to intervention at the initial stages of the disease prior to the medical or social disintegration of the patient and prior to the development of irreversible cirrhosis.

PATHOLOGY OF THE VARIOUS STAGES OF ALCOHOLIC LIVER INJURY

Fatty Liver

Fat accumulation in the liver cells is the earliest and most common response to alcohol, and alcohol is its most common etiological factor.

Increased hepatic lipid content may be demonstrated by biochemical measurements before it becomes histologically apparent. Characteristic ultrastructural changes reveal enlarged and distorted mitochondria with shortened cristae containing crystalline inclusions. The endoplasmic reticulum shows vacuolar dilatation and proliferation. The clinical spectrum of alcoholic fatty liver may extend from silent nonsymptomatic hepatomegaly to severe hepatocellular failure with cholestasis and portal hypertension. Most of the patients with pure fatty liver are virtually asymptomatic. Hepatomegaly is the commonest clinical sign.

Early Fibrosis: Pericellular, Perisinusoidal Fibrosis and Perivenular (or Pericentral) Fibrosis or Sclerosis

Although it can occur anywhere in the hepatic lobule, the earliest deposition of fibrous tissue is generally seen around the central veins and venules (also now called terminal venules). Fibrosis or sclerosis around central veins has been described in alcoholic hepatitis *(vide infra)*. However, pericentral (or perivenular) fibrosis can be seen in the absence of widespread inflammation and necrosis, in association with what most pathologists would label as ''simple'' fatty liver (Van Waes and Lieber, 1977). Perivenular fibrosis is commonly associated with perisinusoidal and pericellular fibrosis.

An interesting clinical and morphologic pattern of acute alcoholic liver disease was described by Uchida and co-workers (1983). Histopathologically, it is characterized by striking microvesicular fatty change in the perivenular hepatocytes with little or no alcoholic hyalin. This entity was called alcoholic foamy degeneration (AFD) of the liver.

Alcoholic Hepatitis

This stage is characterized by the appearance of necrosis with an inflammatory reaction, including polymorphonuclear cells.

Histological characteristics of alcoholic hepatitis are ballooning and great disarray of hepatocytes. Mallory's alcoholic hyaline (irregular cytoplasmic bodies) can be considered as a diagnostic hallmark but is not always present.

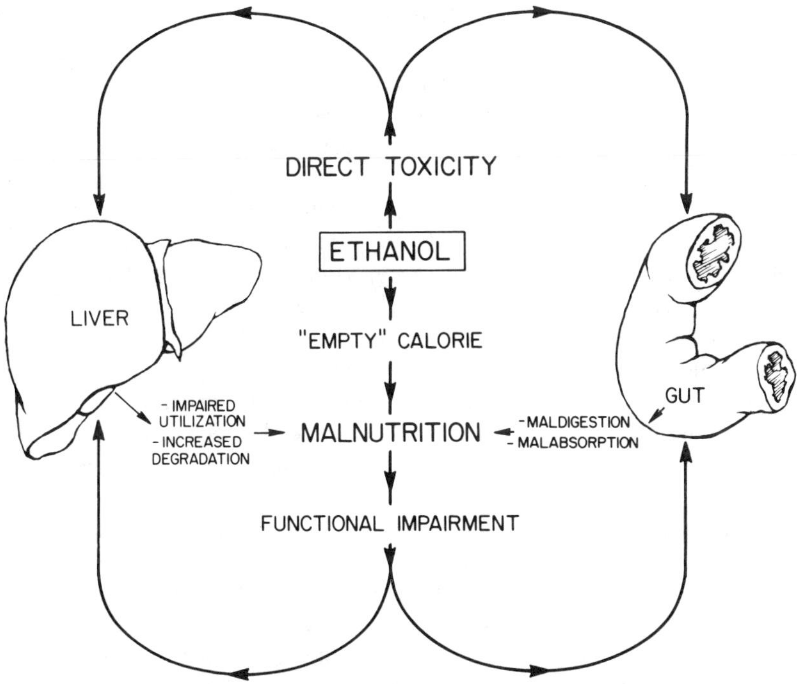

Figure 1. Interaction of direct toxicity of ethanol on liver and gut with malnutrition secondary to dietary deficiencies, maldigestion, and malabsorption.

Cirrhosis

Macroscopically the early cirrhotic liver is golden yellow with fine uniform nodules (from 1 to 5 mm) on the surface. Traditionally this type of cirrhosis is called micronodular or Laennec cirrhosis. Microscopically, scar tissue distorts the normal architecture of the liver by forming bands of connective tissue joining portal and central zones. In addition to fibrosis, various admixtures of fat, inflammatory reactions, and cholestasis may be seen.

PATHOGENESIS OF ALCOHOLIC LIVER INJURY

Nutritional Factors

Alcohol is both a food and a drug. Alcohol is rich in energy, and in many societies alcoholic beverages are considered part of the basic food supply. Alcohol is also consumed for its mood-altering effects and is thereby a psychoactive drug. Under both circumstances, a large intake of alcohol (alcohol and ethanol are used interchangeably in this review) may be the cause of primary malnutrition by displacing other nutrients in the diet because of the high-energy content of the alcoholic beverages *(vide infra)* (Figure 1) or because of associated socioeconomic and medical disorders. Until 2 decades ago, primary malnutrition (due to dietary deficiencies) was considered the main cause of liver disease in the alcoholic. As the overall nutrition of the population improved, more emphasis was placed on secondary malnutrition. Secondary malnutrition (Figure 1) may result from either maldigestion or malabsorption of nutrients caused by G.I. complications associated with alcoholism. Such primary and secondary malnutrition can affect virtually all nutrients, as described in detail before (Lieber, 1988a). Some metabolic nutritional effects of alcohol, such as hyperlipemia, hyperuricemia, and ketoacidosis, have also been discussed elsewhere (Lieber, 1982) and will not be reviewed here.

Biochemical, ultrastructural, and molecular aspects of the less well recognized nutritional disorders, such as alterations of nutrient activation, utilization, and degradation, which may also engender secondary malnutrition through altered bioavailability, have been summarized in a recent review (Lieber, 1988a).

Whereas traditionally, the disorders affecting the liver had been attributed exclusively to nutritional deficiencies accompanying alcoholism (Best et al., 1949), studies carried out over the last 3 decades indicate that in addition to the role of dietary deficiencies, alcohol per se can be incriminated as a direct etiological factor in the production of alcoholic liver disease. Indeed, it was demonstrated that even in the absence of dietary deficiencies, alcohol can lead to the development of fatty liver in humans (Lieber et al., 1963, 1965). An experimental model for this toxic effect was created by overcoming the natural aversion of the rodent for alcohol by incorporating the ethanol in a totally liquid diet (Lieber et al., 1963, 1965; DeCarli and Lieber, 1967; Lieber and DeCarli, 1982, 1986, 1989). Application of this liquid-diet feeding technique to the baboon resulted in the demonstration of a direct causal role of ethanol in the pathogenesis of liver cirrhosis (Lieber and DeCarli, 1974).

Hepatotoxic Effects of Ethanol and Its Metabolism

Only 2–10% of the ethanol absorbed is eliminated through the kidneys and lungs. The rest must be oxidized in the body, principally in the liver, which contains the bulk of the body's enzymes capable of ethanol oxidation. This relative organ specificity probably explains why ethanol oxidation produces striking metabolic imbalances in the liver. These effects are aggravated by the lack of feedback mechanism to adjust the rate of ethanol oxidation to the metabolic state of the hepatocyte, and the inability of ethanol, unlike other major sources of calories, to be stored or metabolized to a significant degree in peripheral tissues. When ethanol is present, it becomes the preferred fuel for the liver. By displacing up to 90% of all other substrates normally utilized by the liver (Lundquist and co-workers, 1962), ethanol literally takes over the intermediary metabolism of the liver. Ethanol metabolism in the liver results in the production of hydrogen and acetaldehyde (Figure 2). Each of these two products is directly responsible for a variety of metabolic alterations that play a role in the development of liver injury. A link between hepatotoxicity of ethanol and its metabolism could also explain some of the zonal changes in alcoholic liver disease *(vide infra)*.

Effects of Excessive Hepatic NADH Generation and Toxicity Associated with a "Hypermetabolic State"

Major metabolic consequences of hydrogen generation and the associated altered redox state (with a shift in the NADH/NAD ratios) have been discussed before (Lieber, 1982, 1988b). Regeneration of NAD from NADH ultimately requires oxygen. The observation of Trémolières and Carré (1961) showed

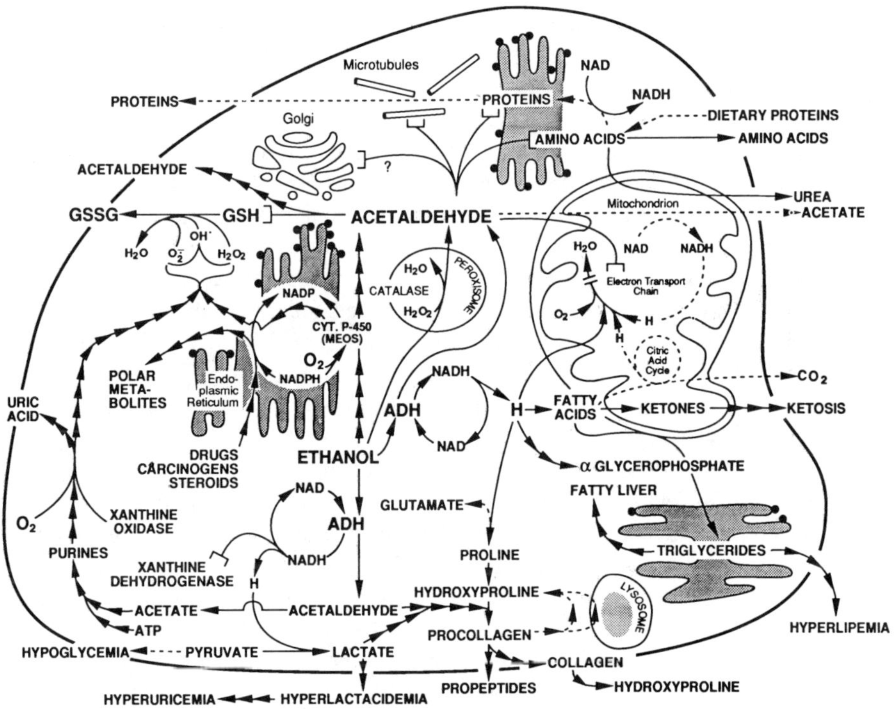

Figure 2. Oxidation of ethanol in the hepatocyte. Many disturbances in intermediary metabolism and toxic effects can be linked to (1) alcohol dehydrogenase (ADH), mediated generation of NADH, (2) the induction of microsomal enzymes, especially P450IIE1, and (3) acetaldehyde, the product of ethanol oxidation. Nicotinamide adenine dinucleotide (NAD); NADH, reduced NAD; GSH, reduced glutathione; GSSG, oxidized glutathione. The broken lines indicate pathways that are depressed by ethanol, whereas repeating arrows reflect stimulation or activation. The symbol ⊣ denotes interference or binding.

that oxygen consumption after an acute dose of alcohol increases more in alcoholics than in nonalcoholics. The cause of this is unknown. Two mechanisms have been postulated: (1) enhanced role of the induced MEOS (Pirola and Lieber, 1976) or (2) relative uncoupling of reoxidation of NADH in liver mitochondria (Israel and co-workers, 1975) possibly associated with increased catecholamine release or a hyperthyroidlike states. Some studies, however, have challenged the concept that chronic ethanol ingestion results in hyperthyroidism or liver hypermetabolism (Schaffer and co-workers, 1981; Teschke and co-workers, 1983). In any event, *in vivo* increased oxygen consumption can be compensated for by enhanced blood flow (Shaw and co-workers, 1977; Villeneuve and co-workers, 1981; Iturriaga and co-workers, 1980; Carmichael and co-workers, 1987). In baboons, defective O_2 utilization rather than lack of O_2 blood supply characterized liver injury produced by high concentrations of ethanol (Lieber et al., 1989). Indeed, in baboons fed control diets, hepatic vein catheterization revealed that a moderate ethanol dose resulted in increased splanchnic O_2 consumption, offset by an increase in flow sufficient to prevent lack of O_2 as judged from normal hepatic venous PO_2, tissue hemoglobin concentration (Hb), and O_2 saturation of Hb (SO_2). Flow-independent tissue O_2 consumption (VO_2) also increased. Hb, SO_2, and VO_2 were measured by hepatic surface reflectance spectrophotometry through peritoneoscopy. After high ethanol, VO_2 failed to increase (or even decreased), without any decrease in SO_2 or in hepatic venous PO_2, indicating impaired capacity of the hepatocytes to take up O_2, even in the presence of an ample O_2 supply.

A characteristic feature of the liver injury of the alcoholic is the predominance of the lesions in perivenular (also called centrilobular) zone or area 3 of the hepatic acinus (Edmondson and co-workers, 1967). Both in rats and baboons fed alcohol, and in alcoholic patients, the swelling of the hepatocytes is most conspicuous in the centrilobular (also now called perivenular) area. The mechanism for this zonal selectivity of the effects of ethanol remains unknown. Two distinct but not mutually exclusive hypotheses have been raised: the one claims that ethanol can produce hypoxic damage of perivenular hepatocytes, whereas the other postulates that conditions normally prevailing in perivenular zones enhance the metabolic toxicity of ethanol.

Zone 3, of course, is the one with the lowest oxygen tension, and lack of oxygen may contribute to liver damage. The hypoxia hypothesis originated from the observation that liver slices from rats fed alcohol chronically consume more oxygen than those of controls (Videla and Israel, 1970). It was then postulated that the increased consumption of oxygen would increase the gradient of oxygen tensions along the sinusoids to the extent of producing anoxic injury of perivenular hepatocytes (Israel and co-workers, 1975). Such a mechanism was illustrated experimentally when centrilobular liver necrosis was induced by hypoxia in chronic ethanol-fed rats (French and co-workers, 1984). Furthermore, both in alcoholics (Kessler and co-workers, 1954) and in animals fed alcohol chronically (Jauhonen and co-workers, 1982; Sato and co-workers, 1983), decreases in either hepatic venous oxygen saturation (Kessler and co-workers, 1954) or PO_2 (Jauhonen and co-workers, 1982) and in tissue oxygen tensions (Sato and co-workers, 1983) have been found during the withdrawal state. However, this decrease is within the range of values found in normal subjects. Moreover, the differences in hepatic oxygenation found during the withdrawal state disappeared (Jauhonen and co-workers, 1982; Shaw and co-workers, 1977) or decreased (Sato and co-workers, 1983) when alcohol was present in the blood.

An alternate and/or complementary hypothesis to explain the selective perivenular hepatotoxicity of ethanol postulates that the low oxygen tensions normally prevailing in perivenular zones could exaggerate the redox shift produced by ethanol (Jauhonen and co-workers, 1982). This was illustrated *in vivo* and in the isolated perfused liver, with varying the oxygen supply to reproduce the oxygen tensions prevailing *in vivo* along the sinusoid (Jauhonen and co-workers, 1985). Varying the oxygen tensions within the physiological range produced a redox gradient of both cytochrome oxidase and NAD^+ with a more reduced state at tensions normally prevailing in perivenular zones. The degree of reduction of cytochrome oxidase at these physiological oxygen tensions was not associated with impairment in the ability of the liver to consume oxygen and to produce ATP, suggesting a lack of cellular anoxia. Hepatic oxygen consumption was increased with 25 m*M* ethanol, but it had no direct effect on the state of reduction of cytochrome oxidase. The effects of ethanol and oxygen tensions on NADH fluorescence were additive, indicating that a greater redox shift should occur when ethanol is oxidized at oxygen tensions similar to those normally prevailing in perivenular zones than at those in periportal zones. This dependence of the ethanol-induced redox shift on oxygen tensions may contribute to the selective perivenular hepatotoxicity of alcohol (Jauhonen and co-workers, 1982).

Selective induction of the enzymes in this zone also contribute to the selective hepatotoxicity. The perivenular area is the zone with the greatest proliferation of the endoplasmic reticulum after chronic ethanol consumption (Iseri and co-workers, 1966) and where activity of the microsomal ethanol-oxidizing system (MEOS) was shown to be highest (Tsutsumi et al., 1989). Furthermore, the activity of ADH may also be highest in this zone, although this has been the subject of controversy, as discussed elsewhere (Lieber, 1988b). The preponderance of ethanol-induced lesions in the perivenular area could be explained by increased ethanol metabolism in this zone, if the lesions were secondary to the oxidation of ethanol and/or the metabolites formed *(vide infra)*.

Aggravation of Hepatotoxicity through Microsomal "Induction"

Chronic ethanol consumption results in proliferation of the membranes of the smooth endoplasmic reticulum, documented by subfractionation and chemical measurements in microsomes and by electron microscopy in animals (Iseri and co-workers, 1966) and humans (Lane and Lieber, 1966). This observation prompted the suggestion that liver microsomes could be the site for a distinct and adaptive system of ethanol oxidation. Such a system [microsomal ethanol-oxidizing system (MEOS)] has been demonstrated in liver microsomes *in vitro* (Lieber and DeCarli, 1968, 1970a). This system is distinct from alcohol dehydrogenase and catalase and is dependent on cytochrome P450 because (1) a P450 containing fraction can be isolated from liver microsomes that, although devoid of any alcohol dehydrogenase or catalase activity, oxidizes ethanol as well as higher aliphatic alcohols (e.g., butanol which is not a substrate for catalase) (Teschke et al., 1972; Mezey et al., 1973), (2) ethanol-oxidizing activity can be reconstituted *in vitro* with use of NADPH-cytochrome P450 reductase, phospholipid, and rat microsomal P450 (Ohnishi and Lieber, 1977; Miwa et al., 1978), and (3) a unique cytochrome P450 is induced in ethanol-treated rats (Ohnishi and Lieber, 1977). The ethanol-inducible form of P450 (LM3a), which has been purified from rabbit liver microsomes (Koop et al., 1982; Ingelman-Sundberg and Johansson, 1984), has a high capacity to metabolize not only ethanol but also 1-butanol, 1-pentanol, aniline (Morgan et al., 1982), acetaminophen (Morgan et al., 1982), carbon tetrachloride (Johansson and Ingelman-Sundberg, 1985), acetone (Koop and Casazza, 1985), *N*-nitrosodimethylamine (NDMA) (Yang et al., 1985), and benzene (Johansson and Ingelman-Sundberg, 1988). Similar results have been obtained with cytochrome P450j, a major hepatic isozyme purified from rats treated with isoniazid (Ryan et al., 1985, 1986). There is also evidence for the existence of an isozyme similar to P450j in humans (Wrighton et al., 1987; Song et al., 1986). Indeed, the purified human protein now has been obtained in a catalytically active form, with a high turnover rate for ethanol (Lasker et al., 1987a). Nebert et al. (1987) have proposed a new nomenclature in which the ethanol-inducible form is designated P450IIE1. Since other cytochrome P450 isozymes can also contribute to ethanol oxidation (Lasker et al., 1987b), the term microsomal ethanol-oxidizing system is being retained to refer to the overall capacity of microsomes to oxidize ethanol rather than to the fraction of the activity that is catalyzed solely by P450IIE1.

Inasmuch as compounds other than ethanol (e.g., acetone) can also serve as P450IIE1 inducers, and since the administration of ethanol (together with fat-containing diets), can be associated with ketonemia (Lefevre et al., 1970), the question has been raised whether the induction of P450IIE1 is mediated specifically by ethanol. Recent studies, however, have shown that the activity of P450IIE1 can be induced after short-term and relatively light consumption of ethanol, in the absence of increased acetonemia or hepatic steatosis (Lieber

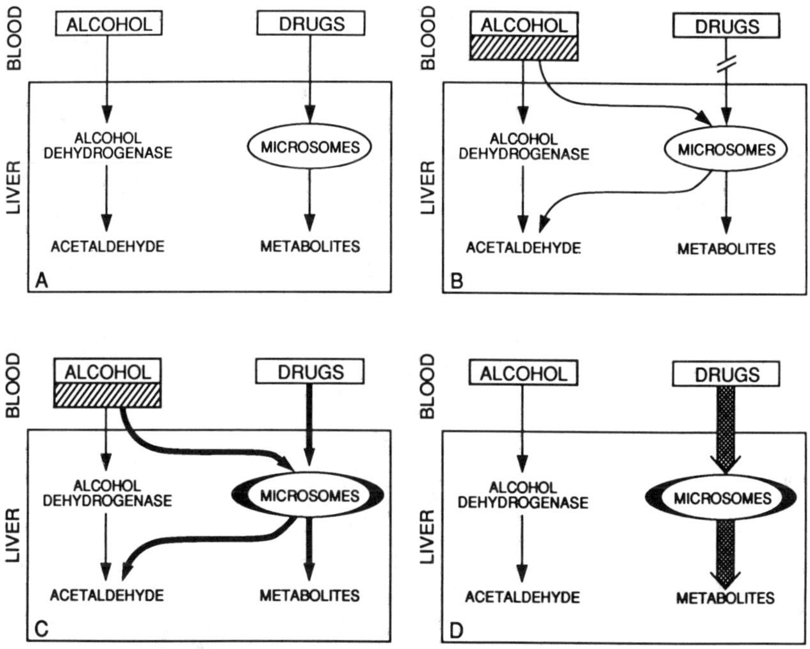

Figure 3. Hepatic ethanol-drug interactions involving the alcohol dehydrogenase pathway and liver microsomes. Alcohol is metabolized by alcohol dehydrogenase, and drugs by microsomes (A). Microsomal drug metabolism is inhibited in the presence of high concentrations of ethanol, in part through competition for a common microsomal detoxification process (B). Microsomal induction after long-term alcohol consumption contributes to accelerated ethanol metabolism at high blood ethanol levels (C). Increased drug metabolism and activation of xenobiotics (due to microsomal induction) persist after cessation of long-term alcohol consumption (D). Hatching indicates high blood alcohol levels. (From Lieber, 1988c.)

et al., 1988). Mechanisms involved in P450IIE1 regulation are under active investigation. Research with a P450IIE1 complementary DNA probe has indicated that the induction of P450IIE1 protein by ethanol-like agents may be regulated by posttranslational events (Song et al., 1986), whereas other studies have shown an increase in translatable P450IIE1 messenger RNA after treatment with ethanol (Kubota et al., 1988). The P450IIE1 gene was recently isolated, characterized, and localized to chromosome 7 in the rat (Umeno et al., 1988a) and 10 in humans (Umeno et al., 1988b).

Proliferation of the endoplasmic reticulum is thought to be "adaptive," since it is associated with enhanced activity of the microsomal enzymes involved in lipoprotein production and thus may contribute to the increased capacity of the liver to secrete fat, as lipoproteins, into the bloodstream. A concomitant increase in drug-metabolizing enzymes explains the enhanced metabolic tolerance of patients with alcoholism to a variety of drugs (Lieber, 1982) (Figure 3D). The activity of the MEOS is also increased in these patients, and this increase contributes to the accelerated rate of alcohol metabolism that occurs after chronic alcohol consumption (Figure 3C). Although these microsomal changes can be interpreted as ''adaptive'' alterations secondary to ''induction'' after chronic ingestion of alcohol, some injurious consequences may also ensue. Indeed, accelerated ethanol metabolism results in enhanced production of acetaldehyde and exacerbation of its various toxic manifestations *(vide infra),* including enhanced peroxidation. The latter may also be promoted more directly through enhanced free-radical formation by the induced microsomes. Indeed, it has been shown that long-term ethanol consumption is accompanied by increased formation of hydroxyl radicals (Krikun et al., 1984) and that P450IIE1 may be involved in this process in rabbits (Ingelman-Sundberg and Johansson, 1984). Other radicals, such as the hydroxyethyl radical (Reinke et al., 1987), are also produced in liver microsomes, and some of these may initiate lipid peroxidation *(vide infra).*

Increased microsomal activity may enhance the oxygen requirements, thereby aggravating whatever hypoxia may be present for unrelated reasons. Furthermore, some compounds acquire hepatotoxicity only after metabolism, or ''activation,'' by the enzymes of the endoplasmic reticulum. One such compound is carbon tetrachloride, the hepatotoxicity of which is greatly increased after chronic alcohol consumption, at least partly because of enhanced activation by microsomes (Hasumura and co-workers, 1974). It is likely that a similar mechanism increases susceptibility in these patients to the hepatotoxic actions of a variety of compounds in the environment (Lieber, 1988b), including anesthetic agents and many commonly used drugs, such as isoniazid and acetaminophen (Sato and co-workers, 1981). Micronutrients such as vitamins may also serve as substrates for the microsomes, and the ''induction'' of the microsomes may therefore alter vitamin requirements and even affect the integrity of the liver. Indeed, alcoholics have very low vitamin A levels in their livers (Leo and Lieber, 1982). In experimental animals, ethanol administration was shown to depress hepatic vitamin A levels, even when administered with an adequate diet (Sato and Lieber, 1981). When dietary vitamin A was virtually eliminated, the depletion rate of vitamin A from endogenous hepatic storage was two to three times faster in ethanol fed rats than in controls. Ethanol-induced vitamin-A depletion was severely exacerbated by concomitant drug administration (Leo et al., 1987) (Figure 4). In rats, severe vitamin A depletion was associated with the appearance of multivesicular lysosomes (Leo and co-workers, 1983). Such lesions were also commonly seen in alcoholic patients with low hepatic vitamin A levels. Vitamin A supplementation is sometimes used to correct the problems of night blindness and sexual dysfunctions of the alcoholic. Such therapy might also be useful with regard to the liver pathology. The therapeutic usage of vitamin A, however, is complicated by the fact that excessive amounts of vitamin A are known to be hepatotoxic and that the alcoholic has an enhanced susceptibility to this effect (Leo and co-workers, 1982; Leo and Lieber, 1988). In control rats, amounts of vitamin A equivalent to those commonly used for the treatment of the alcoholic were found to be without significant effects on the liver, but in animals chronically fed alcohol signs of toxicity developed, such as striking morphologic and functional alterations of the mitochondria (Leo and co-workers, 1982; Leo and Lieber, 1984). Enhanced toxicity was not associated with an increased vitamin A level in the liver; as a matter of fact, because (as mentioned before) alcohol administration tends to decrease vitamin A levels in the liver, even after vitamin A supplementation, alcohol-fed animals had vitamin A levels in the liver that were not higher than normal values, yet toxicity developed. One possible explanation, still to be proven, is that vitamin A toxicity may be mediated, at least in part, by the enhanced production of a toxic metabolite, as for the case of xenobiotic agents. Thus the toxicity could be related to the induction of the enzymes of the endoplasmic reticulum after chronic ethanol consumption.

Adverse Effects of Acetaldehyde

All known oxidative pathways of ethanol metabolism result in the production of acetaldehyde. Its metabolism and general effects have been re-

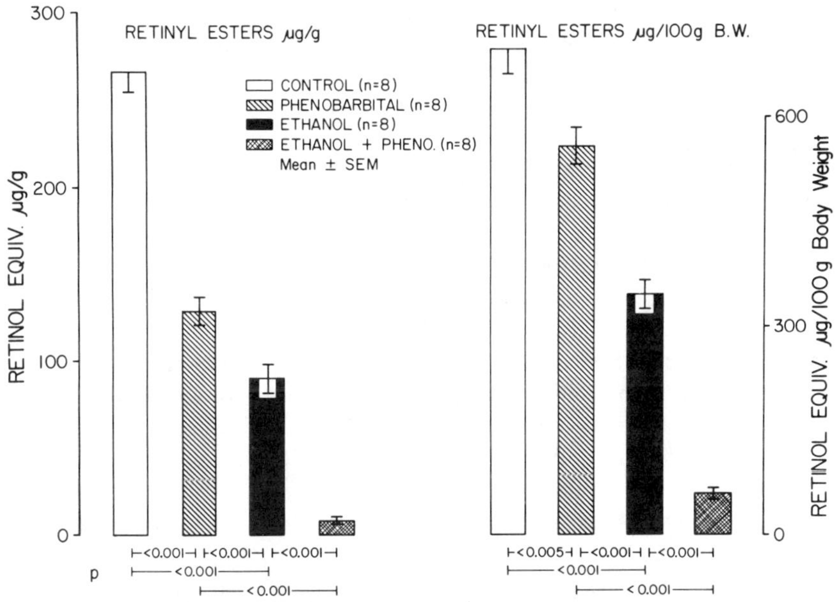

Figure 4. Effect of phenobarbital and/or ethanol on hepatic retinyl ester. Both drugs resulted in a significant depletion of hepatic retinyl ester. The combination of phenobarbital with ethanol produced a much more striking depletion, resulting in negligible hepatic retinyl ester levels. (From Leo et al., 1987.)

viewed elsewhere (Lieber, 1982, 1988d). Acetaldehyde is converted to acetate by acetaldehyde dehydrogenase, an enzyme with an interesting polymorphism and associated striking consequences in terms of ethanol intolerance and flushing, exhibited by subjects who harbor an inactive aldehyde dehydrogenase variant (Harada et al., 1980; Yoshida et al., 1984). These aspects will not be discussed in detail here; we shall focus instead on aspects of the toxicity of acetaldehyde most conspicuously related to liver injury.

Acetaldehyde: Protein Adduct Formation

Nomura and Lieber (1981) reported covalent binding of exogenously added acetaldehyde to proteins of liver microsomes and an even greater effect was found with acetaldehyde generated *in situ* (when ethanol was used). These results suggested a special capacity of "native" acetaldehyde (generated *in situ* in the membranes) to form covalent links with the associated proteins. Furthermore, this effect was significantly increased after chronic ethanol consumption, in parallel with the induction of MEOS activity. We have now found that this process occurs *in vivo* and that acetaldehyde selectively forms a stable adduct in rat liver microsomes with the ethanol-inducible cytochrome P450IIE1 (Behrens et al., 1988). In addition, acetaldehyde binds covalently to other hepatic macromolecules (Mauch et al., 1986; Lin et al., 1988), hepatic cell membranes (Barry et al., 1985, 1987) circulating proteins such as human serum albumin (Donohue et al., 1983), hemoglobin (Stevens et al., 1981), red blood cell membrane proteins (Gaines et al., 1977) and cytoskeletal proteins such as tubulin (Baraona and Lieber, 1983). The latter adduct formation may play a particularly significant role *(vide infra)*. It also is tempting to postulate that the acetaldehyde-protein adduct formation may be responsible, at least in part, for the appearance of antibodies against such adducts. Indeed, acetaldehyde adducts may serve as antigens generating an immune response in mice (Israel et al., 1986), and antibodies against acetaldehyde adducts are present in the serum of most alcoholics (Hoerner et al., 1986; Niemela et al., 1987). Hepatic injury may promote the release of significant amounts of acetaldehyde-altered proteins. This may be relevant to the clinical observation that, in addition to alcohol consumption, severity of liver disease may play a role in the appearance of circulating antibodies against acetaldehyde adducts, not only in alcoholics (Figure 5) but also in some nonalcoholics with liver injury (Figure 6). In turn, complement-binding acetaldehyde adduct-containing immune complexes may contribute to the perpetuation or exaggeration of liver disease. This may represent one of the immune mechanisms that could play a role in the pathogenesis of alcoholic liver injury.

Swelling of the Hepatocyte and Possible Relation to Necrosis

Two of the earliest and most conspicuous features of the hepatic damage

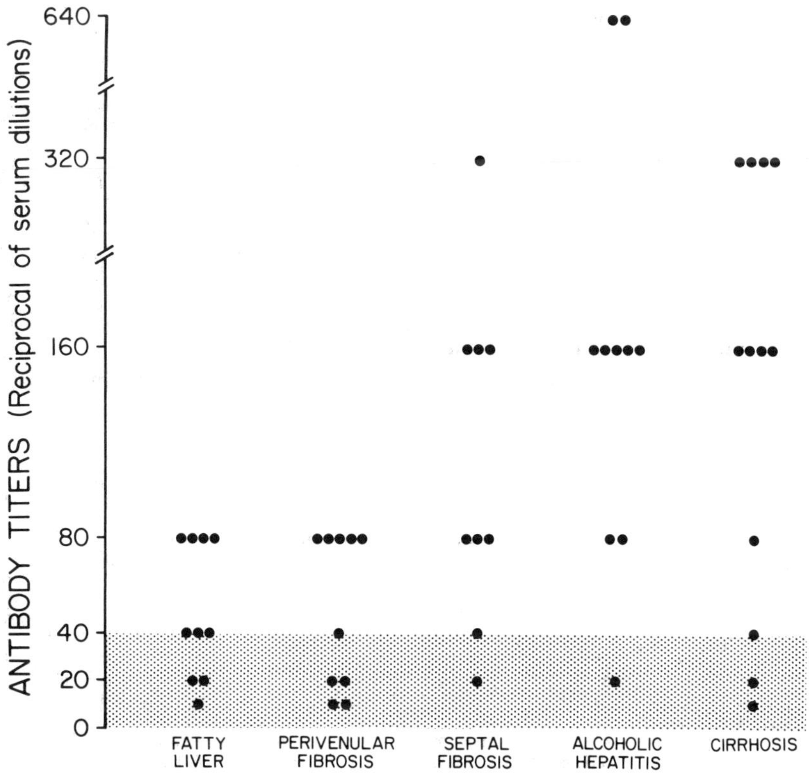

Figure 5. Serum antibody titers against acetaldehyde adducts in patients with alcoholic liver disease. The shaded area represents the range of antibody titers found in normal, nondrinking individuals. Antibody titers were significantly higher in patients with alcoholic hepatitis than in those with fatty liver ($p < 0.0005$) or perivenular fibrosis ($p < 0.02$), and higher in patients with cirrhosis or septal fibrosis than in those with fatty liver ($p < 0.02$) or perivenular fibrosis ($p < 0.05$) (analysis of variance with multiple comparisons). (From Hoerner et al., 1988.)

produced by alcohol are the deposition of fat and the enlargement of the liver. This hepatomegaly was traditionally attributed to the accumulation of lipids. However, in animals fed alcohol-containing diets, it was shown that lipids account for only half the increase in liver dry weight (Lieber and co-workers, 1965), while the other half is almost totally accounted for by an increase in proteins (Baraona and co-workers, 1975), possibly secondary to acetaldehyde induced impairment of microtubular-mediated protein secretion. However, the measured increase in export proteins such as albumin and transferrin accounted for only a small fraction of the total increase in cytosolic protein. This raised the possibility that other export or constituent proteins of the cytosol could also contribute to the increase. Indeed, Pignon and co-workers (1987) found that an increase in fatty acid-binding protein accounts for one-sixth to one-third of the increase in cytosolic protein induced by chronic ethanol feeding, thus becoming the largest known contributor to the ethanol-induced increase in these proteins. The results of Pignon and co-workers (1987) also showed a striking discrepancy between the alcohol-induced increase in esterified fatty acids in the liver (mainly as triglycerides) and the modest increase in nonesterified fatty acids. In addition to the ethanol-induced increase in the synthesizing activities (Joly and co-workers, 1974; M. J. Savolainen and co-workers, 1984), the increase of FABP could also favor the esterification of fatty acids, thereby playing a possible role in preventing potentially deleterious accumulation of nonesterified fatty acids (Brenner, 1984) and fatty acid–CoA esters (Powell and co-workers, 1986) secondary to the ethanol-induced inhibition of their oxidation. It is noteworthy that impairment of microtubules was also found to affect hepatocellular transport

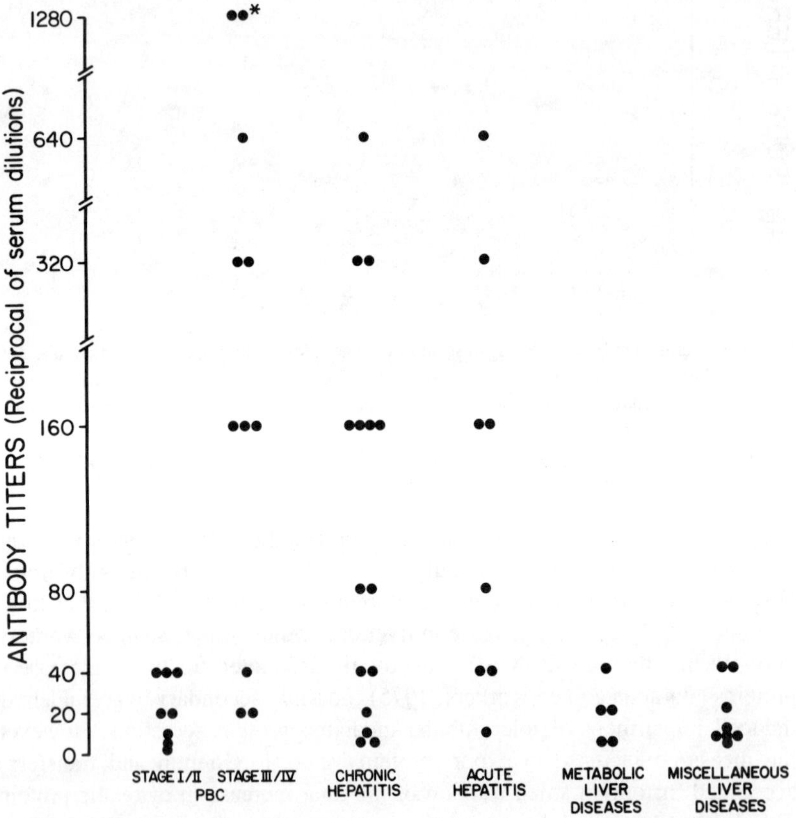

Figure 6. Serum antibodies against acetaldehyde adducts in patients with nonalcoholic liver diseases. Patients with primary biliary cirrhosis (PBC) at stages III and IV had significantly higher titers than those at stages I and II ($p < 0.005$, Mann-Whitney test). (From Hoerner et al., 1988.)

of biliary lipids (Gregory and co-workers, 1978) and be associated with engorgement of the Golgi (Matsuda and co-workers, 1979). The increase in hepatic protein observed after ethanol was not associated with changes in concentration (Baraona and co-workers, 1977), indicating that water was retained in proportion to the increase in protein. The mechanism of the water retention is not fully elucidated, but the rise in both protein and amino acids, plus a likely increase in associated small ions, could account osmotically for a large fraction of the water increase. The increases in lipid, protein, amino acid, water, and electrolytes result in increased size of the hepatocytes. In the rat, the swelling affects both perivenular and periportal hepatocytes. Similar changes have been observed in baboons fed 50% of energy as ethanol, although in this species the accumulation of fat greatly exceeds that of protein (M. J. Savolainen and co-workers, 1984) and the enlargement of the hepatocytes has, as in humans, a clear centrilobular distribution (Miyakawa and co-workers, 1985). The number of hepatocytes and the hepatic content of deoxyribonucleic acid (DNA) did not change after alcohol treatment, and thus the hepatomegaly is entirely accounted for by the increased cell volume (Baraona and co-workers, 1975, 1977). There is also an increase in the number of hepatic mesenchymal cells after ethanol feeding (Baraona and co-workers, 1975), but this increase does not significantly contribute to the hepatomegaly. The swelling of the hepatocytes after chronic alcohol administration was found to be associated with a reduction of the intercellular space and with portal hypertension (Orrego and co-workers, 1981). One suspects that ballooning and associated gross distortion of the volume of the hepatocytes may result in severe impairment of key cellular functions. In alcoholic liver disease some cells not uncommonly have a diameter which is increased 2–3 times, and thereby volume is increased tenfold or more. One may wonder to what extent this type of cellular disorganization, with protein retention and ballooning, may promote progression of the liver injury in the alcoholic. Indeed, there are causes of protein retention in the liver, such as α_1-antitrypsin deficiency, associated with progression to fibrosis and cirrhosis. By analogy, one can assume that protein retention may in some way also favor progression of liver disease in the alcoholic.

Structural and Functional Alterations of the Mitochondria and Plasma Membranes

As discussed before, studies with the electron microscope have revealed striking morphologic alterations, including swelling and abnormal cristae, in the liver mitochondria of patients with alcoholism. Controlled studies in animals and humans (Iseri and co-workers, 1966; Lane and Lieber, 1966; Rubin and Lieber, 1967; Lieber and Rubin, 1968) have shown that these changes are caused by alcohol itself, rather than other factors, such as a poor diet. These structural abnormalities are associated with functional impairments, especially decreased oxidation of fatty acids and of a variety of other substrates, including acetaldehyde (Hasumura and co-workers, 1976).

Mitochondria of alcohol fed animals have a reduction in cytochrome aa_3 and *b* content (Rubin and co-workers, 1970; Koch and co-workers, 1977). The respiratory capacity of the mitochondria was found to be depressed (Kiessling and Pilstrom, 1968; Rubin and co-workers, 1972; Gordon, 1973; Hasumura and co-workers, 1975) using pyruvate, succinate, and acetaldehyde as substrates. Oxidation of other substrates was also found to be reduced in mitochondria of ethanol-fed rats, except for α-glycerophosphate, the oxidation of which was reported by some to be increased (Kiessling, 1968) or unchanged (Pilstrom and Kiessling, 1972), whereas others found it to be decreased (Rubin and co-workers, 1972). Mg^{2+}-stimulated ATPase activity may be either decreased or normal, depending on experimental conditions (Hosein and co-workers, 1977). Use of free fatty acid devoid albumin *in vitro* and feeding of a high-protein low-carbohydrate diet *in vivo* was also found to prevent the ethanol-induced changes (Wahid and co-workers, 1980). These are, however, highly artificial conditions, and in human volunteers given ethanol, mitochondrial lesions developed even in the presence of a high-protein low-fat diet (Lieber and Rubin, 1968). Oxidative phosphorylation was found to be altered (Cederbaum and co-workers, 1974a). It is noteworthy that high concentrations of acetaldehyde, the product of ethanol metabolism, mimic the defects produced by chronic ethanol consumption on oxidative phosphorylation (Cederbaum and co-workers, 1974b). One may wonder to what extent chronic exposure to acetaldehyde is the cause for the defect observed after chronic ethanol consumption. Alterations of acetaldehyde metabolism associated with alcoholism have been discussed elsewhere (Lieber, 1988d). As has been pointed out, alcoholics may exhibit higher acetaldehyde levels than nonalcoholics for a given ethanol load and blood level. It is, therefore, reasonable to speculate that exposure to high acetaldehyde levels may in turn affect mitochondrial function and result in the vicious cycle depicted in Figure 7. In normal liver, higher acetaldehyde concentrations than those usually achieved after alcohol ingestion are required to produce toxicity. However, after chronic alcohol consumption, the liver mitochondria are unusually susceptible to the toxic effects of acetaldehyde, and a variety of important mitochondrial functions, such as fatty acid oxidation, are depressed, even in the presence of relatively low acetaldehyde concentrations (Matsuzaki and Lieber, 1977). It has been proposed that the functional changes in mitochondria may also be related to altered cellular membranes produced as a consequence of prolonged ethanol consumption (Waring and co-workers, 1981, 1982). Membrane binding of ethanol, anesthetics, and hydrophobic molecules in brain synaptosomes and liver mitochondria of rats is reduced after long-term consumption of ethanol. Membranes became resistant to structural disordering by ethanol and halothane. It has been suggested that the increased rigidity

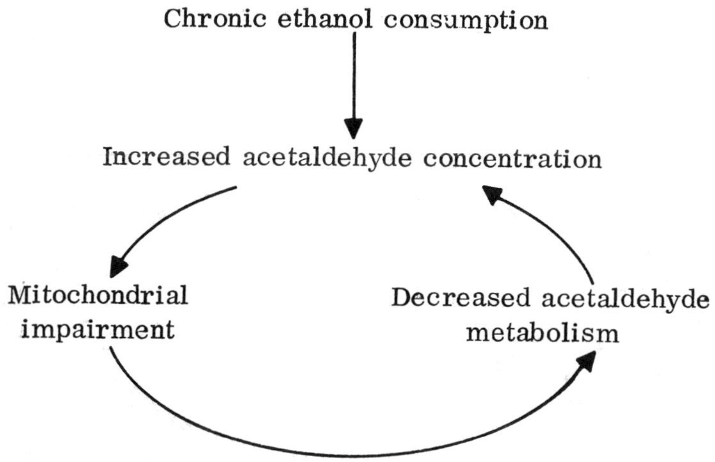

Figure 7. Possible relation between ethanol consumption, altered acetaldehyde levels, and mitochondrial impairment. (From Hasumura et al., 1975.)

impairs normal mitochondrial membrane function, although such a correlation has not always been verified (Gordon and co-workers, 1982).

Ethanol also strikingly alters other liver membranes, as reviewed by Sun and Sun (1985). Plasma membrane glycoprotein assembly in the liver following acute ethanol administration was significantly impaired (Mailliard and co-workers, 1984). Furthermore, hepatocytes isolated from ethanol-fed animals exhibit pronounced morphologic alterations of their plasma membranes by scanning electron microscopy (Figure 8) and a reduced content of alkaline phosphatase (despite an increase in total liver alkaline phosphatase) during incubation of the cells with 50 m*M* ethanol. Moreover, chronic ethanol feeding resulted in reduced recovery of alkaline phosphatase in hepatic plasma membranes. These findings suggest that the increased serum alkaline phosphatase levels observed in response to chronic ethanol feeding may be due, at least in part, to increased lability of this plasma membrane enzyme (Yamada and co-workers, 1985a). It is well known that ethanol also has striking effects on another plasma membrane enzyme, namely gamma-glutamyltranspeptidase (GGTP): GGTP activity is frequently elevated in alcoholics (Wietholz and Colombo, 1976; Horner and co-workers, 1979; Shaw and Lieber, 1980) and in rats given ethanol (Ishii and co-workers, 1978; Shaw and Lieber, 1980; Lahrichi and co-workers, 1982; Nishimura and Teschke, 1982; Yamada and co-workers, 1985b). It has now been found that factors other than a mere increase of liver GGTP may be responsible for the elevated serum levels of this enzyme seen after chronic ethanol consumption. Commercial pelleted diets as well as a reduction in dietary carbohydrates levels increased hepatic GGTP levels but an increase in serum levels was not observed (Yamada and co-workers, 1985b). Concomitant plasma membrane damage may be nec-

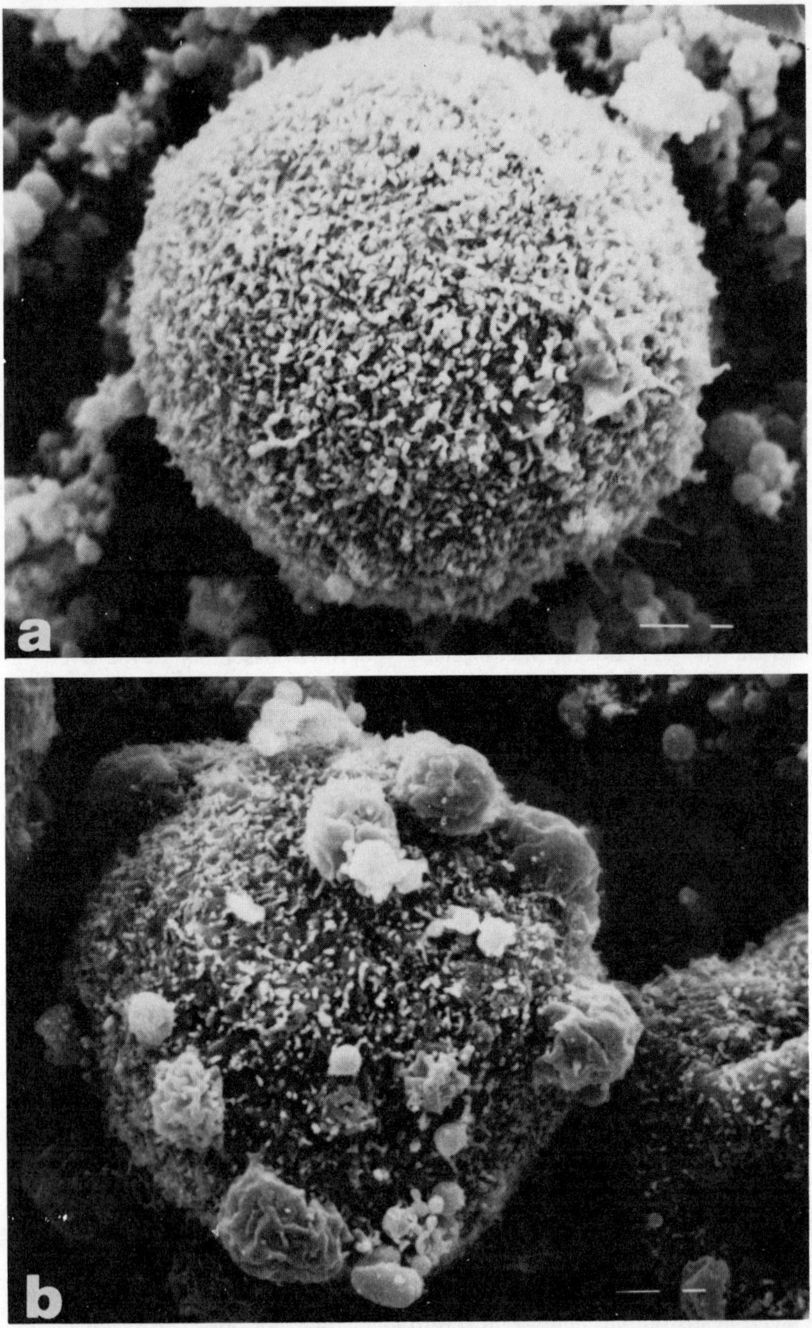

Figure 8. Scanning electron micrographs of isolated hepatocytes. (a) Typical hepatocyte from control rat showing abundant microvilli uniformly covering the cell surface. (b) Hepatocyte from ethanol-fed rat showing the appearance of blebs on the cell surface with a diminished number of microvilli. (a) × 6200; (b) × 6100. (From Yamada et al., 1985a.)

essary for an elevation of serum enzyme level to occur following chronic ethanol consumption (Yamada and co-workers, 1985a). ATPase is a third plasma membrane enzyme reported to be affected by ethanol. However, when the effects of ethanol and acetaldehyde on the activities of two hepatic plasma membrane ATPases—(Na^+, K^+) ATPase and $Mg2^+$ ATPase—were investigated over concentrations ranging from 8–90 m*M*, ethanol did not cause significant inhibition. Acetaldehyde produced noncompetitive inhibition of (Na^+, K^+) ATPase and $Mg2^+$ ATPase at concentrations of 6 and 56 m*M*, respectively, and 5′-nucleotidase activity was also inhibited at these concentrations. Because the inhibitory concentrations of both substances are higher than are usually found in alcoholic subjects or in experimental animals after alcohol feeding, it seems unlikely that direct suppression of ATPase activity by ethanol or acetaldehyde is responsible for the morphological abnormalities of alcohol-induced liver disease (Gonzalez-Calvin and co-workers, 1983). Similarly, the activation effects of ethanol on liver plasma membrane adenylate cyclase activity require unphysiologically high concentrations (Whetton and co-workers, 1983).

Various studies have demonstrated that the cell surface plays an important role in intracellular homeostasis through activities of membrane enzymes, hormone receptors, and transmembrane transport processes. Many of these dynamic features are influenced, in part, by the fluidity of the lipid bilayer because of its effect on the mobility and/or exposure of membrane proteins. There have been also several reports concerning the effect of ethanol on the fluidity of membranes. A currently held view is that *in vitro* ethanol renders membranes more fluid and that chronic ethanol administration alters membrane lipid composition, which results in less fluidity, considered as an adaptation to the acute fluidizing effect. Such an adaptation has been reported to occur in brain synaptosomes and erythrocytes (Chin and Goldstein, 1977; Rottenberg and co-workers, 1981), as well as in liver mitochondria (Waring and co-workers, 1981) *(vide supra)* and microsomes (Ponnappa and co-workers, 1982). However, when the effect of chronic ethanol consumption on liver plasma membranes (PM) was studied (Yamada and Lieber, 1984), it was found that contrasting with other membranes, chronic ethanol feeding resulted in an increase in hepatic plasma membrane fluidity as assessed by fluorescence anisotropy. Similar results were observed by Polokoff and co-workers (1985). This alteration was associated with an increase in the plasma membrane cholesterol ester content (Kim and co-workers, 1988). Thus, chronic ethanol feeding does not appear to result in a ‘homeoviscous adaptation’ of liver plasma membrane as reported for other membranes. Rather, an increase in fluidity has been observed. Such changes may relate to the specialized role of the liver in bile excretion, lipoprotein metabolism, and ethanol oxidation.

Promotion of Lipid Peroxidation; Interaction with Cysteine, Glutathione and Iron

L-Cysteine can complex with acetaldehyde and protects against death from acetaldehyde toxicity *in vivo* (Sprince and co-workers, 1974). Cysteine, *in vitro,* also afforded protection against the depression of CO_2 production from palmitate, octanoate, and ketoglutarate by acetaldehyde (Cederbaum and Rubin, 1976). Furthermore, cysteine is also one of the three amino acids which constitute glutathione. Binding of acetaldehyde with cysteine and/or glutathione may contribute to a depression of liver glutathione (Shaw and co-workers, 1981). Morton and Mitchell (1985) determined the turnover of GSH in individual animals by measuring the decrease in specific activity of GSH in bile over time after IV administration of [35S] cysteine. Rats fed ethanol chronically had significantly increased rates of GSH turnover. Consistent with the results of Vendemiale and co-workers (1984), the increase in turnover of GSH was not due to an increase in oxidation of GSH. There was, however, an association between GSH turnover and the activity of hepatic GGT in ethanol-fed rats. Contrasting with these effects of chronic ethanol consumption, acute ethanol administration inhibited GSH synthesis and produced an increased loss from the liver (Speisky and co-workers, 1985). GSH transferase activity (Kocak-Toker and co-workers, 1985) was decreased by acute ethanol administration, and GSH peroxidase after chronic treatment (Morton and Mitchell, 1985).

Glutathione offers one of the mechanisms for the scavenging of toxic free radicals. Although GSH depletion is not necessarily sufficient to cause lipid peroxidation, it is generally agreed upon that it may favor the peroxidation produced by other factors. A severe reduction in glutathione favors peroxidation (Wendel and co-workers, 1979), and the damage may possibly be compounded by the increased generation of active radicals by the ''induced'' microsomes following chronic ethanol consumption, as illustrated in Figure 2.

Enhanced lipid peroxidation, possibly mediated by acetaldehyde (DiLuzio and Stege, 1977), has been proposed as a mechanism for ethanol-induced fatty liver (DiLuzio and Hartman, 1967). The capacity of acetaldehyde to cause lipid peroxidation in the liver has been demonstrated in isolated perfused livers (Muller and Sies, 1982). It could promote lipid peroxidation indirectly, i.e., through depletion of glutathione (GSH). Theoretically, increased activity of microsomal NADPH oxidase following ethanol consumption (Lieber and DeCarli, 1970b; Reitz, 1975) could result in enhanced H_2O_2 and O_2^- production, thereby again favoring lipid peroxidation. It has also been claimed that the ethanol-inducible form of rabbit liver microsomal cytochromes P450 is associated with increased hydroxyl radical formation (Ingelman-Sundberg and Johansson, 1984). However, when the effect of chronic alcohol feeding on lipid peroxidation was studied in rat liver microsomes, a lack of relationship to hydroxyl radical generation was observed (Shaw and co-workers, 1984; Puntarulo and Cederbaum, 1988). Studies of Thomas and co-workers (1985) showed that ferritin can provide the iron necessary for initiation of lipid peroxidation. O_2^-, as generated by xanthine oxidase, reductively releases ferritin-bound iron. Once released, this iron can promote the peroxidation of phospholipid liposomes. It is noteworthy that in alcoholics the serum ferritin level was elevated in both groups immediately after a drinking bout, significantly more so in men with, than in those without, biochemical signs of liver injury (Välimäki and co-workers, 1983). The serum iron concentration was equally increased, but returned to normal during the first week of ethanol withdrawal. Experimentally, iron and ethanol treatments enhanced liver lipid peroxidation (MDA formation); an additive effect in lipid peroxidation was suggested to occur in this condition. In naive rats, very large amounts of ethanol (5–6 g/kg) are required to produce lipid peroxidation (DiLuzio and Hartman, 1967; MacDonald, 1973), whereas a smaller dose (3 g/kg) had no effect (Shaw and co-workers, 1981). By contrast, after chronic ethanol administration to the rat, even the smaller dose of ethanol administered acutely induced liver peroxidation, and this effect could be prevented, at least in part, by the administration of methionine, a precursor of glutathione (Shaw and co-workers, 1981). The ethanol-induced lipid peroxidation was even more striking in the baboon: administration of relatively small doses of ethanol (1–2 g/kg) after 5–6 h produced lipid peroxidation and GSH depletion. In the baboon chronically fed alcohol (50% of total calories for 1–4 years), alcoholic liver disease, including cirrhosis in some, developed and such animals exhibit evidence of enhanced hepatic lipid peroxidation and GSH depletion. These changes were observed following an overnight withdrawal from ethanol and were exacerbated by the readministration of ethanol. Evidence for GSH depletion and lipid peroxidation (enhanced diene conjugates) was found in liver biopsies of alcoholics who were withdrawn from alcohol (Shaw and co-workers, 1983). It is tempting to speculate that the propensity of the primates to develop more severe lesions than the rat after chronic ethanol consumption may in some way be related, at least in part, to the greater susceptibility to glutathione depletion and the initiation of lipid peroxidation. It is apparent, however, that glutathione depletion per se does not produce liver damage (Siegers and co-workers, 1977). As mentioned before, concomitant enhanced production of active radicals may be required, possibly resulting from the microsomal induction. Physiological substrates for xanthine oxidase, hypoxanthine and xanthine, as well as AMP, significantly increased in the liver after ethanol, together with an enhanced urinary output of allantoin (a final product of xanthine metabolism). Allopurinol pretreatment resulted in 90% inhibition of xanthine oxidase activity, and also significantly decreased ethanol-induced lipid peroxidation (Kato et al., 1990). This experimental evidence of a protective effect of allopurinol now awaits clinical confirmation.

Antioxidant protective mechanisms involve both enzymatic and nonenzymatic defense systems (Tribble et al., 1987). Impairments in such defense systems have been reported in alcoholics, including alterations of ascorbic acid (Bonjour, 1979), glutathione *(vide supra),* selenium (Tanner et al., 1986; Dworkin, 1985; Korpera et al., 1985), and vitamin E (Tanner et al., 1986; Yoshikawa et al., 1982; Bjorneboe et al., 1987). These changes could be due to direct effects of ethanol or to the malnutrition associated with alcoholism. Furthermore, these defense systems are mutually interrelated.

α-Tocopherol (αToc), the major antioxidant in the membrane, is viewed as the "last line" of defense against membrane lipid peroxidation (McCay, 1985; Niki, 1987). Bjorneboe et al. (1987) reported a reduced hepatic αToc content after chronic ethanol feeding in rats receiving adequate amounts of vitamin E as well as in alcoholics (Bjorneboe et al., 1988). Kawase et al. (1989) found that hepatic lipid peroxidation was significantly increased after chronic ethanol feeding in rats receiving a low vitamin E diet, indicating that dietary vitamin E is an important determinant of hepatic lipid peroxidation induced by chronic ethanol feeding. Both low dietary vitamin E and ethanol feeding significantly reduced hepatic αToc content, and the lowest hepatic αToc was found in rats receiving a combination of low vitamin E and ethanol. Furthermore, ethanol feeding caused a marked increase of hepatic α-tocopheryl quinone, a metabolite of αToc by free-radical reactions. Thus, the combination of ethanol with a low vitamin E intake results in a decrease of hepatic αToc content, which renders the liver more susceptible to free-radical attack.

Effects of Acetaldehyde on Free-Radical Generation and Enzyme Activities

Free-radical generation by neutrophils may provide a potential mechanism of cellular injury in acute alcoholic liver disease (William and Barry, 1987). This hypothesis is based on the observation that hepatocyte membranes altered by acetaldehyde stimulate neutrophils to produce superoxide anion. However, the concentration of acetaldehyde to which the liver membrane vesicles were exposed was unphysiologically high. In another study, an elevation of chemotactic factor inactivator was shown in alcoholic liver disease (Robbins et al., 1987). This may explain impaired neutrophil chemotaxis in patients with alcoholic hepatitis, possibly favoring both localized infections and overwhelming sepsis (by altering neutrophil influx to sites of inflammation). One mode of acetaldehyde toxicity may involve the interference with enzyme activities (Solomon, 1987), possibly secondary to binding with critical functional groups. Minute concentrations of acetaldehyde were found to impair the repair of alkylated nucleoproteins (Espina et al., 1988). Increased oxygen-radical generation by ethanol-induced microsomes may also be involved in enzyme inactivation (Dicker and Cederbaum, 1988).

Disulfiram, an inhibitor of acetaldehyde dehydrogenase, raises the acetaldehyde levels after drinking and thereby causes flushing and other adverse effects, useful to maintain abstinence. Moderate abnormalities of liver tests are occasionally encountered in patients undergoing treatment with this drug, but they may be more related to drinking than to the taking of disulfiram (Iber et al., 1987).

Role of Congeners

Majchrowicz and Mendelson (1970) showed that acetaldehyde levels in the blood of alcoholic patients were higher after bourbon ingestion than after grain ethyl alcohol and attributed this difference to the acetaldehyde content of the bourbon. However, Freund (1971) could find no measurable amounts of blood acetaldehyde after oral ingestion of an aqueous acetaldehyde solution ten times in excess of the concentration found in most bourbons. Since the amount of acetaldehyde in even the ''dirtiest'' of alcoholic beverages is about one-thousandth the amount produced by ethanol oxidation in the liver, it is probably not the acetaldehyde in the beverage that leads to higher blood acetaldehyde levels after ingestion, but more likely an effect of other congeners of the beverages. These congeners may somehow interfere with acetaldehyde metabolism, initiating a vicious cycle of mitochondrial impairment which further decreases acetaldehyde catabolism. For example, Rubenstein and co-workers (1975) found that pyrogallol (1 and 10 m*M* concentrations), a metabolic product of gallic acid found in tannins, inhibits rat liver aldehyde dehydrogenase activity *in vitro. In vivo,* Collins and co-workers (1974) have shown that pyrogallol (250 mg/kg, IP) increases acetaldehyde blood levels in rats when it is given 1 h before ethanol (3 gm/kg, IP). Higher aliphatic aldehydes were found to inhibit mitochondrial acetaldehyde metabolism (Hedlung and Kiessling, 1969). Thus elevated acetaldehyde levels following ingestion of alcoholic beverages, when compared with acetaldehyde levels following an ethanol-acetaldehyde mixture, suggest an effect of congeners on acetaldehyde metabolism. When taken in relatively moderate amounts for a short period of time, the congener content of alcoholic beverages did not appear to affect the degree of steatosis (DiLuzio, 1962), but the effects of higher amounts of congeners in conjunction with a greater alcohol intake over a longer period of time have not been extensively studied. Some evidence was presented that prolonged consumption of whiskey might exert more striking undesirable effects on the liver than pure ethanol (Jordo and Olsson, 1975) and that certain alcoholic beverages, particularly brandy, were more toxic to liver cell cultures than pure ethanol (Walker and co-workers, 1974).

One possible toxic congener is iron. Wine of certain vintages has been shown to contain large amounts of iron (Aron and co-workers, 1961; MacDonald, 1963; MacDonald and Baumslag, 1964), which was found to accumulate in liver and other organs of experimental animals given such wine over long periods of time (Aron and co-workers, 1961); this could also explain, at least in part, iron accumulation in various tissues (including the liver) of wine drinkers with cirrhosis of the liver (Aron and co-workers, 1961; MacDonald, 1963; MacDonald and Baumslag, 1964). By contrast, in non-wine-drinking alcoholics, hepatic iron excess is unusual (Barry, 1973; Jakobovits and co-workers, 1979), although increased iron absorption has been attributed to ethanol itself by some (Charlton and co-workers, 1964) but not others (Celada and co-workers, 1978). A unique situation is found in South African Blacks who may develop gross iron overload as a result of drinking home-brewed beer that contains large amounts of iron (Bothwell and co-workers, 1965).

Role of Alterations of The Immune System, Endotoxins, and Ethyl Esters

In the past few years, increasing evidence has suggested that lymphocytes are important in autoimmune tissue-damaging reactions. Accordingly, a cell-mediated mechanism has been implicated in the development of some chronic liver diseases. Several observations suggest that immunologic mechanisms are involved in the perpetuation of hepatic damage by alcohol. Alcoholic hyalin, an abnormal cytokeratin, possibly reflecting vitamin A deficiency *(vide infra),* has been postulated to trigger an immune response (French, 1981). Cytotoxic activity of mononuclear cells against autologous liver cells was demonstrated in baboons fed ethanol (Lue et al., 1981) and in alcoholics (Kakumu and Leevy, 1977; Poralla et al., 1984), with involvement of both T and non-T cells (Poralla et al., 1984). Alcohol abuse was also reported to lead to transient endotoxemia (Bode et al., 1987), which may contribute to the liver injury (Bhagwandeen et al., 1987). Although primary events have not been differentiated from changes secondary to liver injury, these studies justify further investigation of the role of immunologic factors in the development and progression of alcoholic liver disease.

Ethanol forms ethyl esters *in vivo*. Laposata and Lange (1986) found that concentrations of fatty acid ethyl esters in pancreas, liver, heart, and adipose tissue were significantly higher in acutely intoxicated subjects than in controls. Since this nonoxidative ethanol metabolism occurs in the organs most commonly injured by alcohol abuse, and since some of these organs lack an oxidative ethanol metabolizing system, it was postulated that fatty acid ethyl esters and their metabolism may play a role in the production of alcohol-induced injury.

Inhibition of Hepatic Regeneration

There is some debate whether ethanol interferes with hepatic regeneration. Whereas some found no effect on regeneration (Craig, 1975; Frank and co-workers, 1979), others (Leevy, 1963; Wands and co-workers, 1979) report inhibition of ^{3}H-thymidine incorporation. In liver biopsies, ethanol primarily suppresses mesenchymal cell replication (Leevy, 1966). In the partial hepatectomy model, ethanol decreases not only replication of hepatocytes, but also that of lipocytes (Tanaka et al., 1988). Ethanol inhibits the stimulation of ornithine decarboxylase activity resulting from hepatotoxicity (Pösö and Pösö, 1980a) and stabilizes this enzyme (Pösö and Pösö, 1980b). Under these conditions, it also inhibits the synthesis of RNA, probably at the transcriptional level (Pösö and Pösö, 1981). Others found no inhibition, by chronic ethanol feeding, of the induction of ornithine decarboxylase (after partial hepatectomy), but they did find significant inhibition of DNA synthesis (Diehl et al., 1988), an effect attenuated by malotilate (Takada et al., 1987). Ethanol inhibits the repair of DNA (Espina et al., 1988).

Disorders of Collagen Metabolism and Production of Cirrhosis

The present prevailing view is that alcoholic cirrhosis develops in response to alcoholic hepatitis. As discussed before, the latter is characterized by ballooning of the hepatocytes, extensive necrosis, and polymorphonuclear inflammation. In addition, there is increased deposition of collagen and glycosaminoglycans (Galambos and Shapiro, 1973). It is understandable that necrosis and inflammation may trigger the scarring process of cirrhosis, but one must question whether this is the only mechanism involved. In some populations in the United States (Nakano et al., 1982; Worner and Lieber, 1985) as well as in Japan (Takada et al., 1982; Minato et al., 1983; Hasumura et al., 1985) cirrhosis commonly develops in alcoholics without an apparent intermediate stage of florid alcoholic hepatitis. This observation raises the question of whether alcohol can promote development of cirrhosis without being preceded by alcoholic hepatitis. Indeed, in baboons fed alcohol, fatty liver developed, the hepatocytes increased in size, and there was some obvious ballooning. This ballooning was associated with some mononuclear inflammation, but very few of the polymorphonuclear cells so characteristic of human alcoholic hepatitis (Popper and Lieber, 1980). Although some clumping was apparent in the cytoplasm, by electron microscopy there was no alcoholic hyaline. Thus there was no florid picture of alcoholic hepatitis; yet, in one-third of the animals, typical cirrhosis developed (Figure 9). If one can extrapolate from these baboons to humans, it appears that full-blown alcoholic hepatitis may not be a necessary intermediate step in the development of alcoholic cirrhosis. Sequential clinical observations support this contention (Worner and Lieber, 1985) (Figure 10). This hypothesis in turn raises the question of which process, in the absence of alcoholic hepatitis, may initiate cirrhosis upon chronic consumption of ethanol. It is possible that cirrhosis develops in the presence of minimal inflammation and necrosis, which may

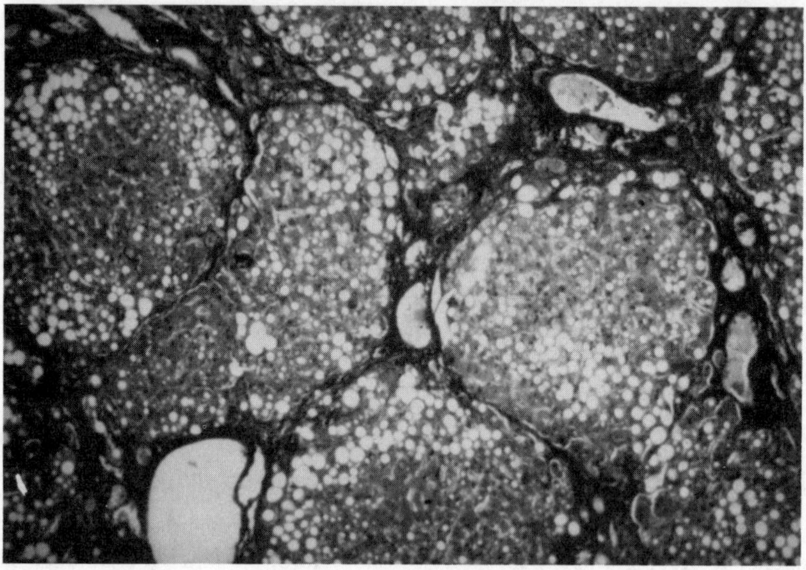

Figure 9. Cirrhosis in a baboon fed alcohol for four years. Fat is regularly distributed through nodules surrounded by connective tissue septa. Chromotropeaniline blue stain × 60. (From Lieber and DeCarli, 1974.)

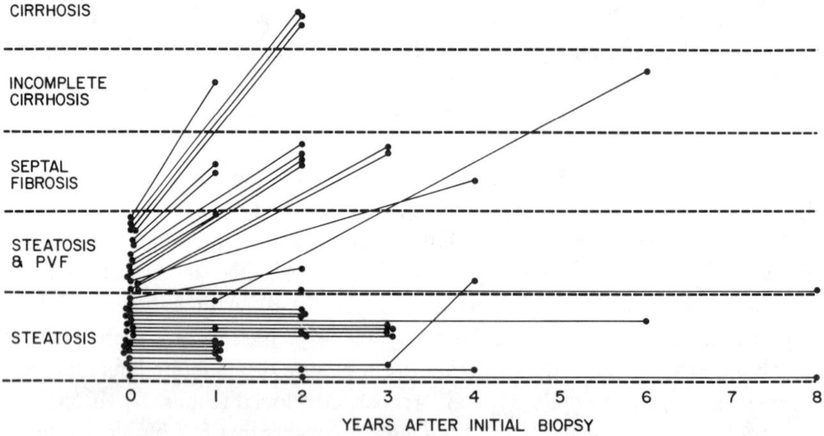

Figure 10. Progression of fibrosis in alcoholics without hepatitis followed up to 8 years after initial biopsy. Presence of perivenular fibrosis on the initial biopsy specimen is a harbinger of rapid development of fibrosis to more severe stages, including cirrhosis. (From Worner and Lieber, 1985.)

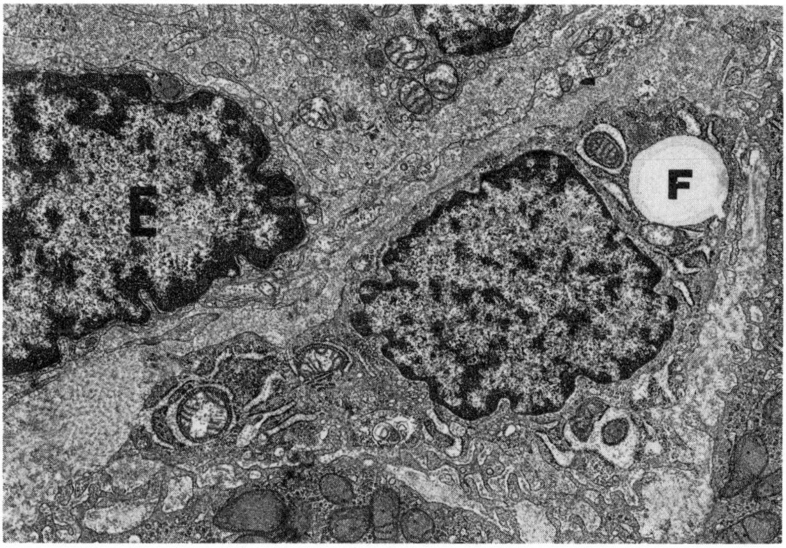

Figure 11. Transitional cell in the Disse space in a cirrhotic liver. The cell contains a small lipid droplet (F) with a total lipid volume density of less than 20% of the cell volume. The rER appears conspicuous. In the extracellular space, an abundant amount of collagen fibers is present. E, endothelial cell nucleus. × 7,000. (From Mak and Lieber, 1988.)

suffice to trigger the fibrosis (Lieber, 1977). Sequential biopsies in alcohol-fed baboons revealed that, in some animals, already at the early fatty liver stage an increased number of mesenchymal cells, particularly myofibroblasts, appear in the perivenular areas (Nakano and Lieber, 1982). Proliferation of these cells is eventually accompanied by deposition of abundant collagen bundles, first in the perivenular areas, leading to perivenular and pericellular fibrosis and ultimately to diffuse fibrosis and cirrhosis. Myofibroblasts have been isolated from surgical baboon liver biopsies by collagenase digestion and Percoll density gradient centrifugation (E. R. Savolainen and co-workers, 1984). By immunofluorescence, the cells synthesize collagen types I, III, and IV, and laminin. The importance of smooth muscle type cells was also shown by Voss and co-workers (1982), who cultured tissue specimens from human fibrotic liver. Myofibroblasts processes also extend into the Disse space. There, however, the most common mesenchymal cell is the lipocyte (also called fat-storing or Ito cell). Lipocytes are characterized electron microscopically by abundant lipid droplets (Mak and co-workers, 1984; Mak and Lieber, 1988), microfilament bundles, dense bodies, and pinocytic vesicles. After chronic alcohol consumption, about half the lipocytes are replaced by transitional cells between lipocytes and fibroblasts (Figure 11); these transitional cells were arbitrarily defined by a lipid droplet volume of less than 20%, measured by morphometry (Mak and co-workers, 1984; Mak and Lieber, 1988). The area of the rough endoplasmic reticulum in transitional cells is greater than in lipocytes. A similar increase in rough endoplasmic reticulum of lipocytes was observed in patients with fibrosis in the Disse space (Minato and co-workers, 1983). Transitional cells have abundant microfilaments, dense bodies, and pinocytic vesicles. They resemble myofibroblasts in the perivenular zones but they can be differentiated by the lack of surrounding basal laminae. Moreover, myofibroblasts typically have an indented nucleus. Transitional cells are surrounded by abundant collagen fibers, and can be seen in the perisinusoidal spaces throughout the lobule associated with a netlike fibrosis, sometimes linking up with the perivenular lesion. There was good correlation between hepatic fibrosis, the percent of transitional cells, and the area of their rough endoplasmic reticulum (Mak and co-workers, 1984). In addition to the extensive evidence concerning the role of myofibroblasts and Ito cells (with transitional cells) in the production of collagen, several studies reported evidence that hepatocytes can also be contributory. This was originally shown by Diegelmann and co-workers (1983) in hepatocytes in primary monolayer culture, and has been confirmed by Kamegaya and co-workers (1985) and by Clement and co-workers (1986).

The mechanisms of collagen degradation in the liver are complex. We have developed a quantitative assay for neutral collagenase activity in liver homogenate (Maruyama and co-workers, 1981), and there appears to be a paradoxical increase in the activity of this enzyme in animals fed ethanol, at least during the early stage of alcoholic liver injury (Maruyama and co-workers, 1982). At that stage, collagenolytic activity may be increased, together with increased collagen synthesis (Kato and co-workers, 1985). Subsequently, collagenase activity may decrease and contribute to the collagen accumulation (Maruyama and co-workers, 1982). Thus, cirrhosis might, in part, represent a relative failure of collagen degradation to keep pace with synthesis. In baboons given ethanol that developed significant fibrosis, molecular evaluation revealed that the RNA from these livers was more active *in vitro* protein synthesis, and the type I procollagen mRNA content was significantly higher (per liver RNA) as determined by hybridization analysis (Figure 12) (Zern et al., 1985).

Another possible mechanism whereby alcohol consumption may be linked to collagen formation is the increase in tissue lactate secondary to alcohol metabolism (Lieber and co-workers, 1971) (Figure 2). Elevated concentrations of lactate have been associated with increased peptidylproline hydroxylase activity both *in vitro* (Green and Goldberg, 1964) and *in vivo* (Lindy and co-workers, 1971). The hepatic free proline pool size, which has been incriminated in the regulation of collagen synthesis (Rojkind and DeLeon, 1970; Chvapil and Ryan, 1973), may be increased by ethanol (Häkkinen and Kulonen, 1975) and is expanded in human portal cirrhosis (Kershenobich and co-workers, 1970). In patients with alcoholic cirrhosis, increased serum-free proline and hydroxyproline have been reported (Mata and co-workers, 1975), but high hepatic concentration of free proline did not induce collagen synthesis, at least in rat liver (Forsander and co-workers, 1983). It has also been postulated that lactate may play a role (Kershenobich and co-workers, 1979) through inhibition of proline oxidase (Kowaloff and co-workers, 1977). This hypothesis illustrates once more the possible impact of ethanol-induced redox changes on intermediary metabolism, including the metabolism of collagen.

Associated Viral Disease

Alcoholism and viral hepatitis can occur in the same patients, but in many patients with alcoholic cirrhosis there is no evidence for antecedent viral hepatitis; similarly, in the baboons that developed alcoholic cirrhosis, no concomitant virus infection was detectable (Lieber, unpublished). Conversely, ethanol infusion did not appear to affect the course of hepatitis B virus infection in chimpanzees (Tabor and co-workers, 1978). Alcoholics have an increased

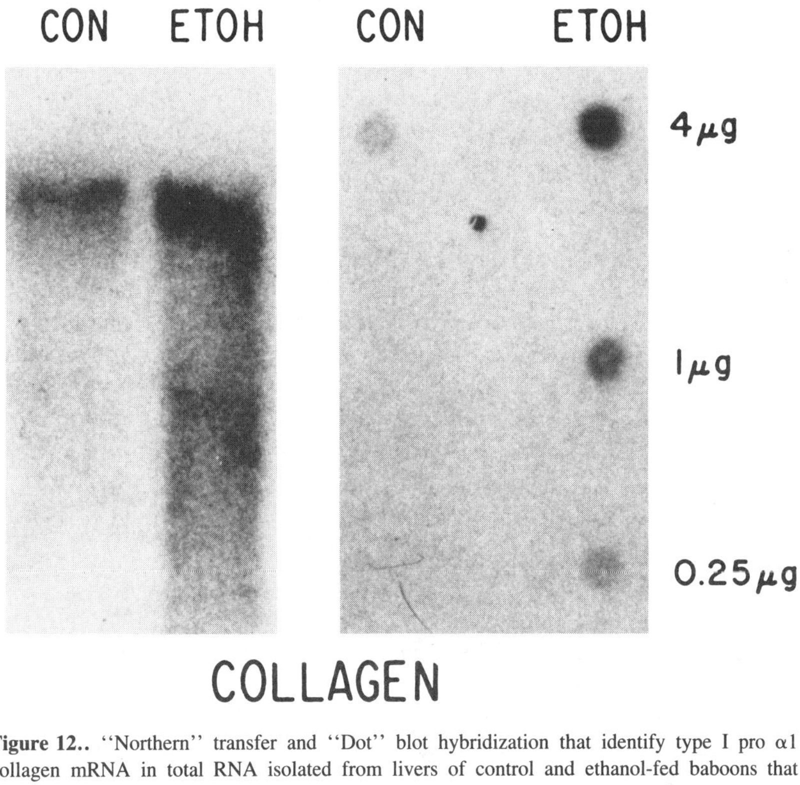

Figure 12.. "Northern" transfer and "Dot" blot hybridization that identify type I pro α1 collagen mRNA in total RNA isolated from livers of control and ethanol-fed baboons that developed fibrotic liver. Both hybridization techniques demonstrate more procollagen mRNA in the RNA from the ethanol-fed animals. (From Zern et al., 1985.)

incidence of viral hepatitis; there has been some question concerning the possible potentiation of alcohol-induced liver damage by the viral disease, but most observations do not substantiate such a mechanism. In particular, there is no evidence that the hepatitis B virus plays a role in alcoholic liver disease (Fong et al., 1988).

Summary

Oxidation of ethanol by alcohol dehydrogenase and the associated oxidation-reduction change produce a number of alterations including impairments in carbohydrate, lipid, and protein metabolism, and exacerbation of injury in the perivenular zones of the liver. A major role is also played by the cytochrome P450-dependent microsomal ethanol-oxidizing system (MEOS). The proposal, over 20 years ago, of this accessory but adaptive pathway of ethanol metabolism (Lieber and DeCarli, 1968) was met with more than a decade of skepticism, but it now has gained general acceptance, especially since the purification and characterization of the ethanol-inducible cytochrome P450 (P450IIE1) from the liver of animals and humans. P450IIE1 plays an important part in the metabolism of, and tolerance to, ethanol; it also has a unique capacity to activate xenobiotic agents. Indeed, microsomal induction provides an explanation for the increased susceptibility of heavy drinkers to the hepatotoxicity of industrial solvents, anesthetics, chemical carcinogens, commonly used drugs, and even over-the-counter analgesics and vitamins. The induction of this pathway also leads to increased production of acetaldehyde, a very reactive compound. Acetaldehyde, in turn, causes injury through the formation of adducts with proteins, resulting in antibody formation, enzyme inactivation, decreased DNA repair, and alterations in microtubules, mitochondria, and plasma membranes. Acetaldehyde also promotes glutathione depletion, toxicity mediated by free radicals, lipid peroxidation, and hepatic collagen synthesis. The earliest striking collagen deposition occurs around the terminal hepatic venules. This perivenular fibrosis has been identified as a precirrhotic marker of prognostic significance.

REFERENCES

Aron E, Paoletti C, Jobard P, and Gosse C: Recherches experimentales sur la cytosiderose au vin. *Arch Mal Appar Dig* 1961; 50:745–750.

Baraona E and Lieber CS: Effects of ethanol on hepatic protein metabolism, in *Alcohol and Protein Synthesis: Ethanol, Nucleic Acids and Protein Synthesis in the Brain and Other Organs* (NIAAA Research Monograph no 10, 75–95 DHHS Publ no (ADM) 83–1198), 1983.

Baraona E, Leo MA, Borowsky SA, and Lieber CS: Alcoholic hepatomegaly: accumulation of protein in the liver. *Science* 1975; 190:794–795.

Baraona E, Leo MA, Borowsky SA, and Lieber CS: Pathogenesis of alcohol-induced accumulation of protein in the liver. *J Clin Invest* 1977; 50:546–554.

Barry M: Iron and chronic liver disease. *JR Coll Phys London* 1973; 8:52–62.

Barry RE and McGivan JD: Acetaldehyde alone may initiate hepatocellular damage in acute alcoholic liver disease. *Gut* 1985; 26:1065–1069.

Barry RE, Williams AJK, and McGivan JD: The detection of acetaldehyde/liver plasma membrane protein adduct formed *in vivo* by alcohol feeding. *Liver* 1987; 7:364–368.

Behrens UJ, Hoerner M, Lasker JM, and Lieber CS: Formation of acetaldehyde adducts with ethanol-inducible P450IIE1 *in vivo*. *Biochem Biophys Res Commun* 1988; 154:584–590.

Best CH, Hartroft WS, Lucas CC, and Ridout JH: Liver damage produced by feeding alcohol or sugar and its prevention by choline. *J Brit Med* 1949; II:1001–1006.

Bhagwandeen BS, Apte M, Manwarring L, and Dickeson J: Endotoxin induced hepatic necrosis in rats on an alcohol diet. *J Pathol* 1987; 151:47–53.

Bjorneboe G-E A, Bjorneboe A, Hagen BF, Morland J, and Drevon CA: Reduced hepatic α-tocopherol content after long-term administration of ethanol to rats. *Biochim Biophys Acta* 1987; 918:236–241.

Bjorneboe G-E A, Johnsen J, Bjorneboe A, Marklund SL, Skylv N, Hoiseth A, Bache-Wiig J-E, Morland J, and Drevon CA: Some aspects of antioxidant status in blood from alcoholics. *Alcoholism: Clin Exp Res* 1988; 12:806–810.

Bode C, Kugler V, and Bode JC: Endotoxemia in patients with alcoholic and non-alcoholic cirrhosis and in subjects with no evidence of chronic liver disease following acute alcohol excess. *J Hepatol* 1987; 4:8–14.

Bonjour JP: Vitamins and alcoholism. *Int J Vit Nutr Res* 1979; 49:434–441.

Bothwell TH, Abrahams C, Bradlow BA, and Charlton RW: Idiopathic and banthu hemochromatosis. Comparative histological study. *Arch Path* 1965; 79:163–168.

Brenner RR: Effects of unsaturated acids on membrane structure and enzyme kinetics. *Prog Lipid Res* 1984; 23:69–96.

Carmichael FJ, Saldivia V, Israel Y, McKaigney JP, and Orrego H: Ethanol-induced increase in portal hepatic blood flow: interference by anesthetic agents. *Hepatology* 1987; 7:89–94.

Cederbaum AI, Lieber CS, and Rubin E: Effects of chronic ethanol treatment on mitochondrial functions. *Arch Biochem Biophys* 1974a; 165:560–569.

Cederbaum AI, Lieber CS, and Rubin E: The effect of acetaldehyde on mitochondrial function. *Arch Biochem Biophys* 1974b; 161:26–39.

Cederbaum AI and Rubin E: Protective effect of cysteine on the inhibition of mitochondrial function by acetaldehyde. *Biochem Pharmacol* 1976; 25:963–973.

Celada A, Rudolf H, and Donath A: Effect of a single ingestion of alcohol on iron absorption. *Am J Hematol* 1978; 5:225–237.

Charlton RW, Jacobs P, Sefiel H, and Bothwell TH: Effect of alcohol on iron absorption. *Brit Med J* 1964; 2:1427–1429.

Chin JH and Goldstein DB: Drug tolerance in biomembranes: a spin label study of the effect of ethanol. *Science* 1977; 196:684–685.

Chvapil M and Ryan JN: The pool of free proline in acute and chronic liver injury and its effect on the synthesis of collagen-globular protein. *Agents Actions* 1973; 3:38–44.

Clement B, Grimaud J-A, Campion J-P, Deugnier Y, and Guillouzo A: Cell types involved in collagen and fibronectin production in normal and fibrotic human liver. *Hepatology* 1986; 6:225–234.

Collins MA, Gordon R, Bidgeli MG, and Rubenstein JA: Pyrogallol potentiates acetaldehyde blood levels during ethanol oxidation in rats. *Chem Biol Interac* 1974; 8:127–130.

Craig J: Effects of ethanol and ethionine on DNA synthesis during experimental liver regeneration. *J Stud Alc* 1975; 36:148–157.

DeCarli LM and Lieber CS: Fatty liver in the rat after prolonged intake of ethanol with a nutritionally adequate new liquid diet. *J Nutr* 1967; 91:331–336.

Dicker E and Cederbaum A-I: Increased oxygen radical-dependent inactivation of metabolic enzymes by liver microsomes after chronic ethanol consumption. *FASEB J* 1988; 2:2901–2906.

Diegelmann RF, Guzelian PS, Gay R, and Gay S: Collagen formation by the hepatocyte in primary monolayer culture and *in vivo*. *Science* 1983; 219:1343–1345.

Diehl AM, Chacon M, and Wagner P: The effect of chronic ethanol feeding on ornithine decarboxylase activity and liver regeneration. *Hepatology* 1988; 8:237–242.

DiLuzio NR: Comparative study of the effect of alcoholic beverages on the development of the acute ethanol-induced fatty liver. *Quart J Stud Alcohol* 1962; 23:557.

DiLuzio NR and Hartman AD: Role of lipid peroxidation on the pathogenesis of the ethanol-induced fatty liver. *Fed Proc* 1967; 26:1436–1442.

DiLuzio NR and Stege TE: The role of ethanol metabolites in hepatic lipid peroxidation, in Fisher MM and Rankin JG (eds): *Alcohol and the Liver,* vol 3. New York, Plenum, 1977, pp 45–62.

Donohue TM, Jr, Tuma DJ, and Sorrell MF: Binding of metabolically derived acetaldehyde to hepatic proteins *in vitro*. *Lab Invest* 1983; 49:226–229.

Dworkin B, Rosenthal WS, Jankowski RH, et al: Low blood selenium levels in alcoholics with and without advanced liver disease. *Dig Dis Sci* 1985; 30:838–844.

Edmondson HA, Peters RL, Frankel HH, and Borowsky S: The early stage of liver injury in the alcoholic. *Medicine* 1967; 46:119–129.

Espina N, Lima V, Lieber CS, and Garro AJ: *In vitro* and *in vivo* inhibitory effect of ethanol and acetaldehyde on o^6 methylguanine transferase. *Carcinogenesis* 1988; 9:761–766.

Font TL, Govindarajan S, Valinluck B, and Redeker GA: Status of hepatitis B virus DNA in alcoholic disease: a study of a large urban population in the United States. *Hepatology* 1988; 8:1602–1604.

Forsander OA, Pikkarainen JAJ, Salaspuro M: A high hepatic concentration of free proline does not induce collagen synthesis in rat liver. *Hepato-Gastroenterology* 1983; 30:6–8.

Frank WO, Rayes AN, Washington A, and Holt PR: Effect of ethanol administration upon hepatic regeneration. *J Lab Clin Med* 1979; 93:402–413.

French SW: The mallory body: structure composition and pathogenesis. *Hepatology* 1981; 1:76–83.

French SW, Benson NC, and Sun PS: Centrilobular liver necrosis induced by hypoxia in chronic ethanol-fed rats. *Hepatology* 1984; 4:912–917.

Freund G: Alcohol, barbiturate, and bromide withdrawal syndrome in mice, in *Recent Advances in Studies of Alcoholism*. US Department of Health, Education and Welfare, Health Services and Mental Health Administration, National Institute of Mental Health, National Institute on Alcohol Abuse and Alcoholism. US Government Printing Office, Washington, DC, 1971, p 453.

Gaines KC, Salhany JM, Tuma DJ, and Sorrell MF: Reaction of acetaldehyde with human erythrocyte membrane proteins. *FEBS Lett* 1977; 75:115–117.

Galambos JT and Shapiro R: Natural history of alcoholic hepatitis. IV. Glycosaminoglycuronans and collagen in the hepatic connective tissue. *J Clin Invest* 1973; 52:2952–2962.

Gonzalez-Calvin JL, Saunders JB, and Williams R: Effects of ethanol and acetaldehyde on hepatic plasma membrane ATPases. *Biochem Pharmacol* 1983; 32:1723–1728.

Gordon ER: Mitochondrial functions in an ethanol-induced fatty liver. *J Biol Chem* 1973; 248:8271–8280.

Gordon ER, Rochman J, Aria M, and Lieber CS: Lack of correlation between hepatic mitochondrial membrane structure and functions in ethanol-fed rats. *Science* 1982; 216:1319–1321.

Green H and Goldberg B: Collagen and cell protein synthesis by an established mammalian fibroblast line. *Nature* 1964; 204:347–349.

Gregory DH, Vlahcevic ZR, Prugh MF, and Swell L: Mechanism of secretion of biliary lipids: role of a microtubular system in hepatocellular transport of biliary lipids in the rat. *Gastroenterology* 1978; 74:93–100.

Häkkinen H-M and Kulonen E: Effect of ethanol on the metabolism of alanine, glutamic acid, and proline in rat liver. *Biochem Pharmacol* 1975; 24:199–204.

Harada S, Misawa S, Agarwal DP, and Goedde HW: Liver alcohol and aldehyde dehydrogenase in the Japanese: isozyme variation and its possible role in alcohol intoxication. *Am J Hum Genet* 1980; 32:8–15.

Hasamura Y, Teschke R, and Lieber CS: Increased carbon tetrachloride hepatotoxicity, and its mechanism, after chronic ethanol consumption. *Gastroenterology* 1974; 66:415–422.

Hasumura Y, Teschke R, and Lieber CS: Acetaldehyde oxidation by hepatic mitochondria: decrease after chronic ethanol consumption. *Science* 1975; 189:727–729.

Hasumura Y, Minato Y, Nishimura M, Kaku Y, Tozuka S, Koyama W, Tanaka Y, and Takeuchi J: Hepatic fibrosis in alcoholics: morphologic characteristics, clinical diagnosis, and natural course. *Pathobiol Hepatic Fibrosis* 1985; 7:13–24.

Hedlung SG and Kiessling K-H: The physiological mechanism involved in hangover. I. The oxidation of some lower aliphatic fusel alcohols and aldehydes in rat liver and their effect on the mitochondrial oxidation of various substrates. *Acta Pharmacol Toxicol* 1969; 27:381–396.

Hoerner M, Behrens UJ, Worner T, and Lieber CS: Humoral immune response to acetaldehyde adducts in alcoholic patients. *Res Commun Chem Path Pharmacol* 1986; 54:3–12.

Hoerner M, Behrens UJ, Worner TM, Blacksberg I, Braly LF, Schaffner F, and Lieber CS: The role of alcoholism and liver disease in the appearance of serum antibodies against acetaldehyde adducts. *Hepatology* 1988; 8:569–574.

Horner F, Kellen JA, Kingston E, Maharaj N, and Malkin A: Dynamic changes of serum gamma glutamyl transferase in chronic alcoholism. *Enzyme* 1979; 24:217–223.

Hosein EA, Hofmann I, and Linder E: The influence of chronic ethanol feeding to rats on the integrity of liver mitochondrial membrane as assessed with the Mg^{2+}-stimulated ATPase enzyme. *Arch Biochem Biophys* 1977; 183:64–72.

Iber FL, Lee K, Lacoursiere R, and Fuller R: Liver toxicity encountered in the Veterans Administration trial of disulfiram in alcoholics. *Alcoholism: Clin Exp Res* 1987; 11:301–304.

Ingelmann-Sundberg M and Johansson I: Mechanisms of hydroxyl radical formation and ethanol oxidation by ethanol-inducible and other forms of rabbit liver microsomal cytochromes P-450. *J Biol Chem* 1984; 259:6447–6458.

Iseri OA, Lieber CS, and Gottlieb LS: The ultrastructure of fatty liver induced by prolonged ethanol ingestion. *Am J Path* 1966; 48:535–555.

Ishii H, Yasuraoka S, Shigeta Y, Takagi S, Kamiya T, Okuno F, Miyamoto K, Tsuchiya M: Hepatic and intestinal gamma-glutamyltranspeptidase activity: its activation by chronic ethanol administration. *Life Sci* 1978; 23:1393–1397.

Israel Y, Videla L, and Bernstein J: Liver hypermetabolic state after chronic ethanol consumption: hormonal interrelations and pathogenic implications. *Fed Proc* 1975; 34:2052–2059.

Israel Y, Hurwitz E, Niëmelä O, and Arnon R: Monoclonal and polyclonal antibodies against acetaldehyde-containing epitopes in acetaldehyde-protein adducts. *Proc Natl Acad Sci USA* 1986; 83:7923–7927.

Iturriaga H, Ugarte G, and Israel Y: Hepatic vein oxygenation, liver blood flow, and the rate of ethanol metabolism in recently abstinent alcoholic patients. *Eur J Clin Invest* 1980; 10:211–218.

Jakobovits AW, Morgan MY, and Sherlock S: Hepatic siderosis in alcoholics. *Dig Dis Sci* 1979; 24:305–310.

Jauhonen P, Baraona E, Lieber CS, and Hassinen IE: Dependence of ethanol-induced redox shift on hepatic oxygen tensions prevailing *in vivo*. *Alcohol* 1985; 2:163–167.

Jauhonen P, Baraona E, Miyakawa H, and Lieber CS: Mechanism for selective perivenular hepatotoxicity of ethanol. *Alcoholism: Clin Exp Res* 1982; 6:350–357.

Johannsson I and Ingelman-Sundberg M: Carbon tetrachloride-induced lipid peroxidation dependent on an ethanol-inducible form of rabbit liver microsomal cytochrome P-450. *FEBS Lett* 1985; 183:265–269.

Johannsson I and Ingelman-Sundberg M: Benzene metabolism by ethanol-, acetone-, and benzene-inducible cytochrome P-450 (IIE1) in rat and rabbit liver microsomes. *Cancer Res* 1988; 48:5387–5390.

Joly JG, Feinman L, Ishii H, and Lieber CS: Effect of chronic ethanol feeding on hepatic microsomal glycerophosphate acyltransferase activity. *J Lipid Res* 1974; 14:337–343.

Jordo L and Olsson R: Effect of long-term administration of different hard liquors and red wine on the rat liver. *Acta Pathol Microbiol Scand* 1975; 83:345–354.

Kakumu S and Leevy CM: Lymphocyte cytotoxicity in alcoholic hepatitis. *Gastroenterology* 1977; 72:594–597.

Kamegaya K, Okazaki I, Oda M, and Tsuchiya M: Some aspects of hepatic fibrogenesis in alcoholics—inductive effect of ethanol on liver cells to form collagen fibers, in Hirayama C and Kivirikko KI (eds): *Pathobiology of Hepatic Fibrosis*. New York, Elsevier, 1985, pp 25–34.

Kato S, Alderman J, Kawase T, Inatomi N, and Lieber CS: Role of xanthine oxidase in ethanol-induced lipid peroxidation. *Gastroenterology* 1990; 98:193–200.

Kato S, Murawaki Y, and Hirayama C: Effects of ethanol feeding on hepatic collagen synthesis and degradation in rats. *Res Comm Chem Path Pharmacol* 1985; 47:163–180.

Kawase T, Kato S, and Lieber CS: Lipid peroxidation and antioxidant defense systems in rat liver after chronic ethanol feeding. *Hepatology* 1989; 10:815–821.

Kershenobich D, Fierro FJ, and Rojkind M: The relationship between free pool of proline and collagen content in human liver cirrhosis. *J Clin Invest* 1970; 49:2246–2249.

Kershenobich D, Fierro FJ, and Rojkind M: The relationship between serum lactic acid and serum proline in alcoholic liver cirrhosis. *Gastroenterology* 1979; 77:A22 (abstract).

Kessler BJ, Liebler JB, Bronfin GJ, and Sass M: The hepatic blood flow and splanchnic oxygen consumption in alcoholic fatty liver. *J Clin Invest* 1954; 33:1338–1345.

Kiessling KH: Effect of ethanol on rat liver VI: A possible correlation between α-glycerophosphate oxidase activity and mitochondrial size in male and female rats fed ethanol. *Acta Pharmacol Toxicol* 1968; 26:246–252.

Kiessling KH and Pilstrom L: Effect of ethanol on rat liver. V. Morphological and functional changes after prolonged consumption of various alcoholic beverages. *J Stud Alcohol* 1968; 29:819–827.

Kim CI, Leo MA, Lowe N, and Lieber CS: Effects of vitamin A and ethanol on liver plasma membrane fluidity. *Hepatology* 1988; 8:735–741.

Kocak-Toker N, Uysal M, Aykac G, Sivas A, Yalcin S, and Oz H: Influence of acute ethanol administration on hepatic glutathione peroxidase and glutathione transferase activities in the rat. *Pharmacol Res Commun* 1985; 17:233–239.

Koch OR, Boveris A, Strotzky, de Favelukes S, Schwarcz de Tarlovsky M, and Stoppani AOM: Biochemical lesions of liver mitochondria from rats after chronic alcohol consumption. *Exp Mol Pathol* 1977; 27:213–220.

Koop DR and Casazza JP: Identification of ethanol-inducible P-450 isozyme 3a as the acetone and acetol monooxygenase of rabbit microsomes. *J Biol Chem* 1985; 260:13607–13612.

Koop DR, Morgan ET, Tarr GE, and Coon MJ: Purification and characterization of a unique isozyme of cytochrome P-450 from liver microsomes of ethanol-treated rabbits. *J Biol Chem* 1982; 257:8472–8480.

Korpera H, Kumpulainen J, Luoma PV, et al: Decreased serum selenium alcoholics as related to liver structure and function. *Am J Clin Nutr* 1985; 42:147–151.

Kowaloff EM, Phang JM, Granger AS, and Downing SJ: Regulation of proline oxidase activity by lactate. *Proc Natl Acad Sci* 1977; 74:5368–5371.

Krikun G, Lieber CS, and Cederbaum AI: Increased microsomal oxidation of ethanol by cytochrome P450 and hydroxyl radical-dependent pathways after chronic ethanol consumption. *Biochem Pharmacol* 1984; 33:3306–3309.

Kubota S, Lasker JM, and Lieber CS: Molecular regulation of ethanol-inducible cytochrome P450-IIE1 in hamsters. *Biochem Biophys Res Commun* 1988; 150:304–310.

Lahrichi M, Ratanasavanh D, Galteau M-M, and Diest G: Effect of chronic ethanol administration on gamma-glutamyltransferase activities in plasma and in hepatic plasma membranes of male and female rats. *Enzyme* 1982; 28:251–257.

Lane BP and Lieber CS: Ultrastructural alterations in human hepatocytes following ingestion of ethanol with adequate diets. *Am J Pathol* 1966; 49:593–603.

Laposata EA and Lange LG: Presence of nonoxidative ethanol metabolism in human organs commonly damaged by ethanol abuse. *Science* 1986; 231:497–499.

Lasker JM, Raucy J, Kubota S, Bloswick BP, Black M, and Lieber CS: Purification and characterization of human liver cytochrome P-450-ALC. *Biochem Biophys Res Commun* 1987a; 148:232–238.

Lasker JM, Tsutsumi M, Bloswick BP, and Lieber CS: Characterization of a benzoflavone (BF)-inducible hamster liver cytochrome P-450 isozyme catalytically similar to cytochrome P-450 ALC. *Hepatology* 1987b; 7:432 (abstract).

Leevy CM: *In vitro* studies of hepatic DNA synthesis in percutaneous liver biopsy specimens. *J Lab Clin Med* 1963; 61:761–779.

Leevy CM: Abnormalities of hepatic DNA synthesis in man. *Med (Baltimore)* 1966; 45:423–433.

Lefevre A, Adler H, and Lieber CS: Effect of ethanol on ketone metabolism. *J Clin Invest* 1970; 49:1775–1782.

Leo MA, Arai M, Sato M, and Lieber CS: Hepatotoxicity of moderate vitamin A supplementation in the rat. *Gastroenterology* 1982; 82:194–205.

Leo MA and Lieber CS: Hepatic vitamin A depletion in alcoholic liver injury. *N Eng J Med* 1982; 307:597–601.

Leo MA and Lieber CS: Normal testicular structure and reproductive function in deermice lacking retinol and alcohol dehydrogenase activity. *J Clin Invest* 1984; 73:593–596.

Leo MA and Lieber CS: Hypervitaminosis A: a liver lover's lament. *Hepatology* 1988; 8:412–417.

Leo MA, Lowe N, and Lieber CS: Potentiation of ethanol-induced hepatic vitamin A depletion by phenobarbital and butylated hydroxytoluene. *J Nutr* 1987; 117:70–76.

Leo MA, Sato M, and Lieber CS: Effect of hepatic vitamin A depletion on the liver in humans and rats. *Gastroenterology* 1983; 84:562–572.

Lieber CS: Fisher MM and Rankin JG (eds): *Pathogenesis of Alcoholic Liver Disease: An Overview.* New York, Plenum, 1977, pp 197–255.

Lieber CS: Metabolism and metabolic effects of alcohol. *Sem Hematol* 1980; 17:85–99.

Lieber CS: *Medical Disorders of Alcoholism: Pathogenesis and Treatment.* Philadelphia, Saunders, 1982.

Lieber CS: The influence of alcohol on nutritional status. *Nutr Rev* 1988a; 46:241–254.

Lieber CS: Metabolic effects of ethanol and its interaction with other drugs, hepatotoxic agents, vitamins and carcinogens (1988 update). *Sem Liver Dis* 1988b; 8:47–68.

Lieber CS: Biochemical and molecular basis for alcohol-induced injury to liver and other tissues. *N Eng J Med* 1988c; 319:1639–1650.

Lieber CS: Metabolic effects of acetaldehyde. *Biochem Soc Transac* 1988d; 16:241–247.

Lieber CS, Baraona E, Hernandez-Muñoz R, Kubota S, Sato N, Kawano S, Matsumura T, and Inatomi N: Impaired oxygen utilization: A new mechanism for the hepatotoxicity of ethanol in sub-human primates. *J Clin Invest* 1989; 83:1682–1690.

Lieber CS and DeCarli LM: Ethanol oxidation by hepatic microsomes: adaptive increase after ethanol feeding. *Science* 1968; 162:917–918.

Lieber CS and DeCarli LM: Hepatic microsomal ethanol oxidizing system: *in vitro* characteristics and adaptive properties *in vivo*. *J Biol Chem* 1970a; 245:2505–2512.

Lieber CS and DeCarli LM: Reduced nicotinamide-adenine dinucleotide phosphate oxidase: activity enhanced by ethanol consumption. *Science* 1970b; 170:78–80.

Lieber CS and DeCarli LM: An experimental model of alcohol feeding and liver injury in the baboon. *J Med Primatol* 1974; 3:153–163.

Lieber CS and DeCarli LM: The feeding of alcohol in liquid diets: two decades of application and 1982 update. *Alcoholism: Clin Exp Res* 1982; 6:523–531.

Lieber CS and DeCarli LM: The feeding of ethanol in liquid diets: 1986 update. *Alcoholism: Clin Exp Res* 1986; 10:550–553.

Lieber CS and DeCarli LM: Liquid diet technique of ethanol administration: 1989 update. *Alcohol Alcoholism* 1989; 24:197–211.

Lieber CS, Jones DP, Mendelson J, and DeCarli LM: Fatty liver, hyperlipemia and hyperuricemia produced by prolonged alcohol consumption, despite adequate dietary intake. *Trans Assoc Am Physicians* 1963; 76:289–300.

Lieber CS, Jones DP, and DeCarli LM: Effects of prolonged ethanol intake: production of fatty liver despite adequate diets. *J Clin Invest* 1965; 44:1009–1020.

Lieber CS, Lasker JM, DeCarli LM, Saeli J, and Wojtowicz T: Role of acetone, dietary fat and total energy intake in the induction of the hepatic microsomal ethanol oxidizing system. *J Pharmacol Exp Ther* 1988; 247:791–795.

Lieber CS and Rubin E: Alcohol fatty liver in man on a high protein and low fat diet. *Am J Med* 1968; 44:200–206.

Lieber CS, Rubin E, and DeCarli LM: Effects of ethanol on lipid, uric acid, intermediary and drug metabolism including the pathogenesis of the alcoholic fatty liver, in Kissin B and Begleiter H (eds): *The Biology of Alcoholism,* vol 1. New York, Plenum, 1971, pp 263–305.

Lin RS, Smith RS, and Lumeng L: Detection of a protein-acetaldehyde adduct in the liver of rats fed alcohol chronically. *J Clin Invest* 1988; 81:615–619.

Lindy S, Pedersen FB, Turto H, and Uitto J: Lactate, lactate dehydrogenase and protocollagen proline hydroxylase in rat skin autograft. *Hoppe-Seylers' Z Physiol Chem* 1971; 352:1113–1118.

Lue SL, Paronetto F, Lieber CS: Cytotoxicity of mononuclear cells and vulnerability of hepatocytes in alcoholic fatty liver of baboons. *Liver* 1981; 1:264–267.

Lundquist F, Tygstrup N, Winkler K, Mellemgaard K, and Munck-Petersen S: Ethanol metabolism and production of free acetate in the human liver. *J Clin Invest* 1962; 41:955–961.

MacDonald CM: The effect of ethanol on hepatic lipid peroxidation and on the activities of glutathione reductase and peroxidase. *FEBS Lett* 1973; 35:227–230.

MacDonald RA: Wines as a source of iron in hemochromatosis. *Nature (London)* 1963; 199:922.

MacDonald RA and Baumslag N: Iron in alcoholic beverages. Possible significance for hemochromatosis. *Am J Med Sci* 1964; 247:649–654.

Mailliard ME, Sorrell MF, Volentine GD, and Tuma DJ: Impaired plasma membrane glycoprotein assembly in the liver following acute ethanol administration. *Biochem Biophys Res Commun* 1984; 123:951–958.

Majchrowicz E and Mendelson JH: Blood concentrations of acetaldehyde and ethanol in chronic alcoholics. *Science* 1970; 168:1100–1102.

Mak KM and Lieber CS: Lipocytes and transitional cells in alcoholic liver disease: a morphometric study. *Hepatology* 1988; 8:1027–1033.
Mak KM, Leo MA, and Lieber CS: Alcoholic liver injury in baboons: transformation of lipocytes to transitional cells. *Gastroenterology* 1984; 87:188–200.
Maruyama K, Feinman L, Okazaki I, and Lieber CS: Direct measurement of neutral collagenase activity in liver homogenates from baboons and humans. *Biochim Biophys Acta* 1981; 658:124–131.
Maruyama K, Feinman L, Fainsilber Z, Nakano M, Okazaki I, and Lieber CS: Mammalian collagenase increases in early alcoholic liver disease and decreases with cirrhosis. *Life Sci* 1982; 30:1379–1384.
Mata JM, Kershenobich D, Villarreal E, and Rojkind M: Serum free proline and free hydroxyproline in patients with chronic liver diseases. *Gastroenterology* 1975; 68:1265–1269.
Matsuda Y, Baraona E, Salaspuro M, and Lieber CS: Effects of ethanol on liver microtubules and Golgi apparatus. *Lab Invest* 1979; 41:455–463.
Matsuzaki S and Lieber CS: Increased susceptibility of hepatic mitochondria to the toxicity of acetaldehyde after chronic ethanol consumption. *Biochem Biophys Res Commun* 1977; 75:1059–1065.
Mauch TJ, Donahue TM, Zetterman RK, Sorrell MF, and Tuma DJ: Covalent binding of acetaldehyde selectively inhibits the catalytic activity of lysine-dependent enzymes. *Hepatology* 1986; 6:263–269.
McCay PB: Vitamin E: interaction with free radicals and ascorbate. *Annu Rev Nutr* 1985; 5:323–340.
Mezey E, Potter JJ, and Reed WD: Ethanol oxidation by a component of liver microsomes rich in cytochrome P-450. *J Biol Chem* 1973; 248:1183–1187.
Minato Y, Hasumura Y, and Takeuchi J: The role of fat-storing cells in Disse space fibrogenesis in alcoholic liver disease. *Hepatology* 1983; 3:559–566.
Miwa GT, Levin W, Thomas PE, Lu AYH: The direct oxidation of ethanol by catalase- and alcohol dehydrogenase-free reconstituted system containing cytochrome P-450. *Arch Biochem Biophys* 1978; 187:464–475.
Miyakawa H, Iida S, Leo MA, Greenstein RJ, Zimmon DS, and Lieber CS: Pathogenesis of precirrhotic portal hypertension in alcohol-fed baboons. *Gastroenterology* 1985; 88:143–150.
Morgan ET, Koop DR, and Coon MJ: Catalytic activity of cytochrome P-450 isozyme 3a isolated from liver microsomes of ethanol-treated rabbits: oxidation of alcohols. *J Biol Chem* 1982; 257:13951–13957.
Morgan ET, Koop DR, and Coon MJ: Comparison of six rabbit liver cytochrome P-450 isozymes in formation of a reactive metabolite of acetaminophen. *Biochem Biophys Res Commun* 1983; 112:8–13.
Morton S and Mitchell MC: Effects of chronic ethanol feeding on glutathione turnover in the rat. *Biochem Pharmacol* 1985; 34:1559–1563.
Muller A and Sies H: Role of alcohol dehydrogenase activity and of acetaldehyde in ethanol-induced ethane and pentane production by isolated perfused rat liver. *Biochem J* 1982; 206:153–156.
Nakano M and Lieber CS: Ultrastructure of initial stages of perivenular fibrosis in alcohol-fed baboons. *Am J Path* 1982; 106:145–155.

Nakano M, Worner T, and Lieber CS: Perivenular fibrosis in alcoholic liver injury: ultrastructure of histologic progression. *Gastroenterology* 1982; 83:777–785.

Nebert DW, Adesnik M, Coon MJ, et al: The P-450 gene superfamily: recommended nomenclature. *DNA* 1987; 6:1–11.

New York City: Summary of vital statistics 1986. New York, Department of Health, Bureau of Health Statistics and Analysis.

Niki E: Interaction of ascorbate and α-tocopherol. *Ann NY Acad Sci* 1987; 493:186–199.

Niemela O, Klajner F, Orrego H, Vidins E, Blendis L, and Israel Y: Antibodies against acetaldehyde-modified protein epitopes in human alcoholics. *Hepatology* 1987; 7:1210–1214.

Nishimura M and Teschke R: Effect of chronic alcohol consumption on the activities of liver plasma membrane enzymes: gamma-glutamyltransferase, alkaline phosphatase and 5′-nucleotidase. *Biochem Pharmacol* 1982; 31:377–381.

Nomura F and Lieber CS: Binding of acetaldehyde to rat liver microsomes: enhancement after chronic alcohol consumption. *Biochem Biophys Res Commun* 1981; 100:131–137.

Ohnishi K and Lieber CS: Reconstitution of the microsomal ethanol-oxidizing system: qualitative and quantitative changes of cytochrome P-450 after chronic ethanol consumption. *J Biol Chem* 1977; 252:7124–7131.

Orrego H, Blendis LM, Crossley IR, Medline A, MacDonald A, Ritchie S, and Israel Y: Correlation of intrahepatic pressure with collagen in the Disse space and hepatomegaly in humans and in the rat. *Gastroenterology* 1981; 80:546–566.

Pignon J-P, Baraona W, Poynard T, Chaput J-C, and Lieber CS: Serum lipoproteins and alcoholic liver disease. *Gastroenterol Clin Biol* 1987; 11:460–472.

Pilstrom L and Kiessling KH: A possible localization of α-glycerophosphate dehydrogenase to the inner boundary membrane of mitochondria in livers from rats fed with ethanol. *Histochemistry* 1972; 32:329–334.

Pirola R and Lieber CS: Energy wastage in rats given drugs that induce microsomal enzymes. *J Nutr* 1976; 105:1544–1548.

Polokoff MA, Simon TJ, Ardon HR, Simon FR, and Iwahashi M: Chronic ethanol increases liver plasma membrane fluidity. *Biochemistry* 1985; 24:3114–4120.

Ponnappa BC, Waring A, Hoek JB, Rottenberg H, and Rubin E: Chronic ethanol ingestion increases calcium uptake and resistance to molecular disordering by ethanol in liver microsomes. *J Biol Chem* 1982; 257:10141–10146.

Popper H and Lieber CS: Histogenesis of alcoholic fibrosis and cirrhosis in the baboon. *Am J Pathol* 1980; 98:695–716.

Poralla T, Hutteroth TH, and Meyer Zum Buschenfelde KH: Cellular cytotoxicity against autologous hepatocytes in alcoholic liver disease. *Liver* 1984; 4:117–121.

Pösö AR and Pösö H: Inhibition of ornithine decarboxylase in regenerating rat liver by acute ethanol treatment. *Biochim Biophys Acta* 1980a; 606:338–346.

Pösö H and Pösö AR: Stabilization of tyrosine aminotransferase and ornithine decarboxylase in regenerating rat liver by ethanol treatment. *FEBS Lett* 1980b; 113:211–214.

Pösö H and Pösö AR: Inhibition of RNA and protein synthesis by ethanol in regenerating rat liver: evidence for transcriptional inhibition of protein synthesis. *Acta Pharmacol Toxicol* 1981; 49:125–129.

Powell GL, Tippett PS, Kiorpes TC, et al: Fatty acyl-CoA as an effector molecule in metabolism. *Fed Proc* 1986; 44:81 (abstract).

Puntarulo S and Cederbaum I: Effect of oxygen concentration on microsomal oxidation of ethanol and generation of oxygen radicals. *Biochem J* 1988; 251:787–794.

Reinke LA, Lai EK, DuBose CM, and McCay PB: Reactive free radical generation *in vivo* in heart and liver of ethanol-fed rats: correlation with radical formation *in vitro*. *Med Sci* 1987; 7:9223–9227.

Reitz RC: A possible mechanism for the peroxidation of lipids due to chronic ethanol ingestion. *Biochem Biophys Acta* 1975; 380:145–154.

Robbins RA, Zetterman RK, Kendall TJ, Gossman GL, Monsour HP, and Rennard SI: Elevation of chemotactic factor inactivator in alcoholic liver disease. *Hepatology* 1987; 7:872–877.

Rojkind M and DeLeon LD: Collagen biosynthesis in cirrhotic rat slices: a regulatory mechanism. *Biochem Biophys Acta* 1970; 217:512–522.

Rottenberg H, Waring A, Rubin E: Tolerance and cross-tolerance in chronic alcoholics: reduced membrane binding of ethanol and other drugs. *Science* 1981; 213:583–585.

Rubenstein JA, Collins MA, and Tabakoff B: Inhibition of liver aldehyde dehydrogenase by pyrogallol and related compounds. *Experientia* 1975; 31:414–415.

Rubin E and Lieber CS: Early fine structural changes in human liver induced by alcohol. *Gastroenterology* 1967; 52:1–13.

Rubin E, Beattie DS, and Lieber CS: Effects of ethanol on the biogenesis of mitochondrial membranes and associated mitochondrial functions. *Lab Invest* 1970; 23:620–627.

Rubin E, Beattie DS, Toth A, and Lieber CS: Structural and functional effects of ethanol on hepatic mitochondria. *Fed Proc* 1972; 31:131–140.

Ryan DE, Koop DR, Thomas PE, Coon MJ, and Levin W: Evidence that isoniazid and ethanol induce the same microsomal cytochrome P-450 in rat liver, an isozyme homologous to rabbit liver cytochrome P-450 isozyme 3a. *Arch Biochem Biophys* 1986; 246:633–644.

Ryan DE, Ramanathan L, Iida S, Thomas PE, Haniu M, Shively JE, Lieber CS, and Levin W: Characterization of a major form of rat hepatic microsomal cytochrome P-450 induced by isoniazid. *J Biol Chem* 1985; 260:6385–6393.

Sato N, Kamada T, Kawano S, Hayashi N, Kishida Y, Meren H, Yoshihara H, and Abe H: Effect of acute and chronic ethanol consumption on hepatic tissue oxygen tension in rats. *Pharmacol Biochem Behav* 1983; 18:443–447.

Sato M and Lieber CS: Hepatic vitamin A depletion after chronic ethanol consumption. *Gastroenterology* 1980; 79:1123 (abstract).

Sato C, Matsuda Y, and Lieber CS: Increased hepatotoxicity of acetaminophen after chronic ethanol consumption in the rat. *Gastroenterology* 1981; 80:140–148.

Savolainen MJ, Baraona E, Pikkarainen P, and Lieber CS: Hepatic triacylglycerol synthesizing activity during progression of alcoholic liver injury in the baboon. *J Lipid Res* 1984; 24:813–820.

Savolainen ER, Goldberg B, Leo MA, Velez M, and Lieber CS: Diagnostic value of serum procollagen peptide measurements in alcoholic liver disease. *Alcoholism: Clin Exp Res* 1984; 8:384–389.

Schaffer WT, Denckla WD, and Veech RL: Effect of chronic ethanol administration on O_2 consumption in whole body and perfused liver of the rat. *Alcoholism: Clin Exp Res* 1981; 5:192–197.

Shaw S, Heller EA, Friedman HS, Baraona E, and Lieber CS: Increased hepatic oxygenation following ethanol administration in the baboon. *Proc Soc Exp Biol Med* 1977; 156:509–513.

Shaw S, Jayatilleke E, and Lieber CS: Hepatic lipid peroxidation: potentiation by chronic alcohol feeding and attenuation by methionine. *J Lab Clin Med* 1981; 98:417–425.

Shaw S, Jayatilleke E, and Lieber CS: The effect of chronic alcohol feeding on lipid peroxidation in microsomes: lack of relationship to hydroxyl radical generation. *Biochem Biophys Res Commun* 1984; 118:233–238.

Shaw S and Lieber CS: Plasma amino acid abnormalities in the alcoholic: respective role of alcohol, nutrition and liver injury. *Gastroenterology* 1978; 74:677–682.

Shaw S and Lieber CS: Mechanism of increased gamma glutamyl transpeptidase after chronic alcohol consumption: hepatic microsomal induction rather than dietary imbalance. *Substance Alcohol Actions Misuse* 1980; 1:423–428.

Shaw S, Rubin KP, Lieber CS: Depressed hepatic glutathione and increased diene conjugates in alcoholic liver disease: evidence of lipid peroxidation. *Dig Dis Sci* 1983; 28:585–589.

Siegers C-P, Schütt A, and Strubelt O: Influence of some hepatotoxic agents on hepatic glutathione levels in mice. *Proc Eur Soc Toxicol* 1977; 18:160–162.

Solomon LR: Evidence for the generation of transaminase inhibitor(s) during ethanol metabolism by rat liver homogenates: a potential mechanism for alcohol toxicity. *Biochem Med Metab Biol* 1987; 38:9–18.

Song BJ, Gelboin HV, Park SS, Yang CS, and Gonzales FJ: Complementary DNA and protein sequences of ethanol-inducible rat and human cytochrome P-450s: transcriptional and post-transcriptional regulation of the rat enzyme. *J Biol Chem* 1986; 261:16689–16697.

Speisky H, MacDonald A, Giles G, Orrego H, and Israel Y: Increased loss and decreased synthesis of hepatic glutathione after acute ethanol administration. *Biochem J* 1985; 225:565–572.

Sprince H, Parker CM, Smith G, and Gonzales LJ: Protection against acetaldehyde toxicity in the rat by L-cysteine, thiamin and L-2-methylthiazolidene-4-carboxylic acid. *Agents Actions* 1974; 4:125–130.

Stevens VJ, Fantl WJ, Newman CB, Sims RV, Cerami A, and Peterson CM: Acetaldehyde adducts with hemoglobin. *J Clin Invest* 1981; 67:361–369.

Sun GY and Sun AY: Ethanol and membrane lipids. *Alcoholism: Clin Exp Res* 1985; 9:164–180.

Tabor E, Gerety RJ, Barker LF, Howard CR, and Zuckerman AJ: Effect of ethanol during hepatitis B virus infection in chimpanzees. *J Med Virol* 1978; 2:295–303.

Takada A, Nei J, Matsuda Y, and Kanayama R: Clinicopathological study of alcoholic fibrosis. *Am J Gastroenterol* 1982; 77:660–666.

Takada A, Nei J, Tamino H, and Takase S: Effects of malotilate on ethanol-inhibited hepatocyte regeneration in rats. *Hepatology* 1987; 5:336–343.

Tanaka Y, Funaki N, Mak KM, Kim CI and Lieber CS: Effects of ethanol and hepatic vitamin A on the proliferation of lipocytes in regenerating rat liver. *J Hepatology* 1991, (in press).

Tanner AR, Bantock I, Hinks L, Lloyd B, Turner NR, and Wright R: Depressed selenium and vitamin E levels in an alcoholic population. *Dig Dis Sci* 1986; 31:1307–1312.

Targan SR, Chow AW, and Guze LB: Role of anaerobic bacteria in spontaneous peritonitis of cirrhosis. Report of two cases and review of the literature. *Am J Med* 1977; 62:397–403.

Teschke R, Hasumura Y, Joly JG, and Lieber CS: Microsomal ethanol-oxidizing system (MEOS): purification and properties of a rat liver system free of catalase and alcohol dehydrogenase. *Biochem Biophys Res Commun* 1972; 49:1187–1193.

Teschke RE, Moreno F, Heinen E, Herrmann J, Kruskemper H-L, and Strohmeyer G: Hepatic thyroid hormone levels following chronic alcohol consumption: direct experimental evidence in rats against the existence of a hyperthyroid hepatic state. *Hepatology* 1983; 3:469–474.

Thomas CE, Morehouse LA, and Aust SD: Ferritin and superoxide-dependent lipid peroxidation. *J Biol Chem* 1985; 260:3275–3280.

Trémolières J and Carré L: Etudes sur les modalités d'oxydation de l'alcool chez l'homme normal et alcoolique. *Rev l'Alcohol* 1961; 7:202–227.

Tribble DL, Aw TY, and Jones DP: The pathophysiological significance of lipid peroxidation in oxidative cell injury. *Hepatology* 1987; 7:377–387.

Tsutsumi M, Lasker JM, Shimizu M, Rosman AS, and Lieber CS: The intralobular distribution of ethanol-inducible cytochrome P450IIE1 in rat and human liver. *Hepatology* 1989; 10:437–446.

Uchida T, Kao H, Quispe-Sjogren M, and Peters RL: Alcoholic foamy degeneration—a pattern of acute alcoholic injury of the liver. *Gastroenterology* 1983; 84:683–692.

Umeno M, McBride OW, Yang CS, Gelboin HV, and Gonzalez FJ: Human ethanol-inducible P450IIE1: complete gene sequence, promoter characterization, chromosome mapping, and cDNA-directed expression. *Biochemistry* 1988b; 27:9006–9013.

Umeno M, Song BJ, Kozak C, Gelboin HV, and Gonzalez FJ: The rat P450IIE1 gene: complete intron and exon sequence, chromosome mapping, and correlation of developmental expression with specific 5′ cytosine demethylation. *J Biol Chem* 1988a; 263:4956–4962.

Välimäki M, Härkönen M, and Ylikahri R: Serum ferritin and iron levels in chronic male alcoholics before and after ethanol withdrawal. *Alcohol Alcoholism* 1983; 18:255–260.

Van Waes L and Lieber CS: Early perivenular sclerosis in alcoholic fatty liver, an index of progressive liver injury. *Gastroenterology* 1977; 73:646–650.

Vendemiale G, Jayatilleke E, Shaw S, and Lieber CS: Depression of biliary glutathione excretion by chronic ethanol feeding in the rat. *Life Sci* 1984; 34:1065–1073.

Videla L and Israel Y: Factors that modify the metabolism of ethanol in rat liver and adaptive changes produced by its chronic administration. *Biochem J* 1970; 118:275–281.

Villeneuve JP, Pomier G, and Huet PM: Effect of ethanol on hepatic blood flow in unanesthetized dogs with chronic portal and hepatic vein catheterization. *Can J Physiol Pharmacol* 1981; 59:598–603.

Voss B, Rauterberg J, Pott G, Brehmer U, Allam S, Lehmann R, and Bassewitz DV: Nonparenchymal cells cultivated from explants of fibrotic liver resemble endothelial and smooth muscle cells from blood vessel walls. *Hepatology* 1982; 2:19–28.

Wahid S, Khanna JM, Carmichael FJ, and Israel Y: Mitochondrial function following chronic ethanol treatment: effect of diet. *Res Commun Chem Path Pharmac* 1980; 30:477–491

Walker F, Elmslie W, Fraser RA, Snape PE, and Watt GCM: Cytotoxic effect of alcohol on liver cells and fibroblast *in vitro*. *Scot Med J* 1974; 19:125–127.

Wands JR, Carter EA, Bucher NLR, and Isselbacher KJ: Inhibition of hepatic regeneration in rats by acute and chronic ethanol intoxication. *Gastroenterology* 1979; 77:528–531.

Waring AJ, Rottenberg H, Ohnishi T, and Rubin E: Membranes and phospholipids of liver mitochondria from chronic alcoholic rats are resistant to membrane disordering by alcohol. *Proc Natl Acad Sci USA* 1981; 78:2582–2586.

Waring AJ, Rottenberg H, Ohnishi T, and Rubin E: The effect of chronic ethanol consumption on temperature-dependent physical properties of liver mitochondrial membranes. *Arch Biochem Biophys* 1982; 216:51–61.

Wendel A, Fenerstein S, and Konz KH: Acute paracetamol intoxication of starved mice leads to lipid peroxidation *in vivo*. *Biochem Phys* 1979; 28:2051–2055.

Whetton AD, Needham L, Dodd NJF, Heyworth CM, and Houslay MD: Forskolin and ethanol both perturb the structure of liver plasma membranes and activate adenylate cyclase activity. *Biochem Pharmacol* 1983; 32:1601–1608.

Wietholz H and Colombo JP: Das Verhalten der gamma-glutamyltranspeptidae und anderer Leberenzyme in Plasma während der Alkohol-Entziehungskur. *Schweiz Med Wochen* 1976; 106:981–987.

William AJK and Barry RE: Free radical generation by neutrophils: a potential mechanism of cellular injury in acute alcoholic hepatitis. *Gut* 1987; 28:1157–1161.

Worner TM and Lieber CS: Perivenular fibrosis as precursor lesion of cirrhosis. *JAMA* 1985; 254:627–630.

Wrighton SA, Thomas PE, Ryan DE, and Levin W: Purification and characterization of ethanol-inducible human hepatic cytochrome P-450HLj. *Arch Biochem Biophys* 1987; 258:292–297.

Yamada S and Lieber CS: Decrease in microviscosity and cholesterol content of rat liver plasma membranes after chronic ethanol feeding. *J Clin Invest* 1984; 74:2278–2289.

Yamada S, Mak KM, and Lieber CS: Chronic ethanol consumption alters rat liver plasma membranes and potentiates release of alkaline phosphatase. *Gastroenterology* 1985a; 88:1799–1806.

Yamada S, Wilson JS, and Lieber CS: The effects of alcohol and diet on hepatic and serum gamma-glutamyltranspeptidase activities. *J Nutr* 1985b; 115:1285–1290.

Yang CS, Tu YY, Koop DR, and Coon MJ: Metabolism of nitrosamines by purified rabbit liver cytochrome P-450 isozymes. *Cancer Res* 1985; 45:1140–1145.

Yoshida A, Huang IY, and Ikawa M: Molecular abnormality of an inactive aldehyde dehydrogenase variant commonly found in Orientals. *Proc Natl Acad Sci USA* 1984; 81:258–261.

Yoshikawa Y, Takemura S, and Kondo M: α-Tocopherol level in liver diseases. *Acta Vitamnol Enzymol* 1982; 4:311–318.

Zern MA, Leo MA, Giambrone MA, and Lieber CS: Increased type I procollagen mRNA levels and *in vitro* protein synthesis in the baboon model of chronic alcoholic liver disease. *Gastroenterology* 1985; 89:1123–1131.

Chapter

12

Mechanisms of Cholestasis

Mary Vore
Department of Pharmacology
Graduate Center for Toxicology
University of Kentucky
Lexington, Kentucky

DEFINITION OF CHOLESTASIS

The term *cholestasis* is derived from the Greek work *chole,* meaning gall or bile, and means an arrest or stagnation in the flow of bile. The decrease in bile flow can be due to a decrease in the secretion of bile by the hepatocyte, termed intrahepatic cholestasis, or to the physical obstruction of the biliary tree, termed extrahepatic cholestasis. The term *intrahepatic cholestasis* was first used by Popper and Szanto (1956) and was based on a morphologic examination of human liver. These authors concluded that intrahepatic cholestasis was "not characterized by a definite morphologic substrate except for the presence of increased amounts of bile in liver and Kupffer cells as well as in concrements." They also noted that this picture was probably the result rather than the cause of intrahepatic cholestasis since a similar picture was seen with obstructive extrahepatic cholestasis and only rarely could the two be distinguished based on the morphologic data alone. Additional morphologic features of cholestasis, both intra- and extrahepatic, as seen by transmitting and scanning electron microscopy include the presence of biliary material in dilated bile canaliculi, a decreased number of canalicular microvilli, and distorted microvilli with a blunted or edematous appearance (Phillips et al., 1987). Hypertrophy of the Golgi apparatus, increased lysosomes, dilation and vesiculation of the rough endoplasmic reticulum, and hypertrophy of the smooth endoplasmic reticulum are also seen (Phillips et al., 1987). The bile canaliculi do not appear to rupture, but biliary contents are regurgitated across the cell via vesicles and are released at the sinusoidal surface by a reverse pinocytosis (Phillips et al., 1987).

The regurgitation of biliary contents into the blood appears to provide the basis for the biochemical changes seen. Bile acids and bilirubin are elevated in serum, giving rise to the cardinal clinical manifestations of cholestasis, itching, and jaundice, respectively. Serum cholesterol levels are also elevated, as are the levels of enzymes normally present in bile, including alkaline phosphatase, 5′-nucleotidase, gamma-glutamyl transpeptidase, and leucine aminopeptidase. Approximately 75% of patients with prolonged cholestasis, of extra- or intrahepatic origin, have alkaline phosphatase values elevated fourfold or more (Kaplan, 1987). The elevation of alkaline phosphatase occurs as the result of increased de novo synthesis of enzyme, apparently stimulated by the accumulated bile acids (Kaplan and Righetti, 1970; Hatoff and Hardison, 1980).

Cholestasis is thus a very broad term, which includes morphological, biochemical, and clinical features. It is necessarily imprecise, since many of these features depend on the duration and cause of the cholestasis. The physiological definition is therefore the most satisfying, since it is a functional one requiring only that bile flow is decreased. Direct determination of bile flow in humans is not practical, but is extremely useful in animal studies where a number of experimental models have been developed.

SECRETION OF BILE

As understanding of the functions of the hepatocyte developed, the focus on the mechanisms of cholestasis shifted from mechanical obstruction to pathologic changes in the hepatocyte itself. The recent appreciation of the hepatocyte as a secretory epithelial cell (Meier, 1988a) has provided the physiologic and biochemical foundations necessary for understanding the secretion of bile by the hepatocyte and, in time, for understanding the pathology in this process. A brief review of the mechanisms of bile secretion follows. The interested reader is referred to a number of chapters and reviews on this topic (Boyer, 1980; Blitzer and Boyer, 1982; Klaassen and Watkins, 1984; Coleman, 1987; Erlinger, 1987).

Morphology

A schematic view of the hepatocyte and the bile canaliculus is shown in Figure 1. Bile is secreted by the hepatocyte across the canalicular membrane and into the canaliculi. The canaliculus is approximately 1 μm in diameter and essentially represents the space between adjacent cells. There are no additional cells which line the canaliculus; rather the canalicular membrane

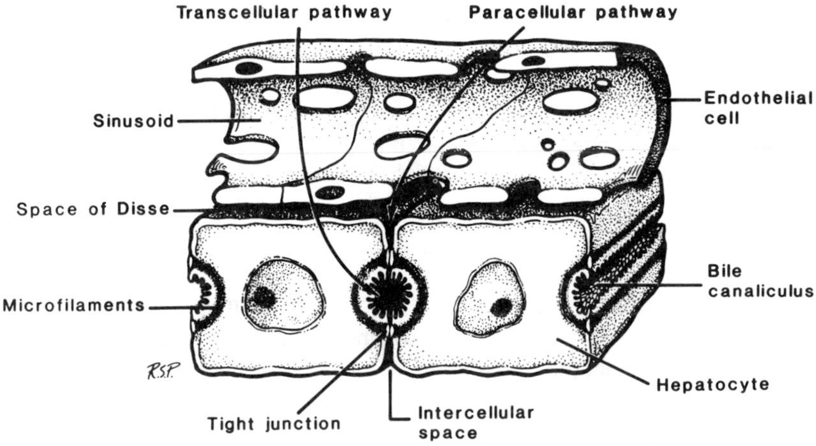

Figure 1. A schematic diagram of the hepatocyte and the bile canaliculus. Solutes may enter the canaliculus from the plasma either via the paracellular pathway (through the intercellular space and the tight junction) or via the transcellular pathway (across the basolateral and canalicular plasma membranes).

represents a specialized portion of the plasma membrane of the hepatocyte. The canalicular membrane is covered with microvilli which markedly expands its surface area. The canalicular lumen is separated from sinusoidal blood by intercellular junctions which are formed between adjacent cell membranes and extend the entire length of the canaliculi. They consist of (from the canaliculus out to the perisinusoidal space) a tight junction, an intermediate junction, and a desmosome. Intercellular gaps or gap junctions are also found as small patches and contain channels which provide a means of cell-to-cell communication. The permeability of the tight junctions is thought to be intermediate in nature, permitting fluid and electrolyte movement from the intercellular space across the junctional complex into the canalicular lumen via what is termed the paracellular pathway, described in greater detail below.

A zone of cytoplasm known as pericanalicular ectoplasm surrounds the canaliculus. It is devoid of organelles but has been shown to contain actin microfilaments, both thick (10 nm) and thin (5 nm), which form a web around the canaliculus. They are inserted into the junctional complex and extend into the canalicular microvilli. This contractile system of hepatocytes is thought to enhance bile flow by active contractions of the canaliculus and to provide "tone" to the canalicular wall (French and Davies, 1975; Phillips and Oda, 1974).

The canaliculus is closed at its origin near the central vein and forms a long, narrow channel which branches and joins with other canaliculi to form a three-dimensional network. Near the portal vein area, the canaliculi empty into the preductules (canal of Hering, cholangioles), the ductules, and finally into the bile ducts located in the portal tracts. The ductules are surrounded by cuboidal biliary epithelial cells that rest on a basement membrane (Phillips and Oda, 1974). Numerous microvilli project from the apical surface of the ductular cells into the lumen. Blood vessels, primarily arterial, also form a rich blood supply to the bile ducts. These observations have led to the suggestion that bile secreted at the level of the canaliculus is modified as it travels through the ductules (Erlinger, 1987).

Composition of Bile

Bile is composed primarily of water (97%), inorganic electrolytes (Na^+, K^+, Ca^{2+}, Mg^{2+}, Cl^-, $HCO3^-$) and organic solutes (bile acids, glutathione, cholesterol, phospholipids, and protein). The exact concentration of these varies among the species most commonly studied; the major components are detailed in Table 12–1. The electrolytes are generally present at concentrations similar to those in plasma with the major cation being Na^+; the bile acids, Cl^- and HCO_3^- serve as the balancing anions. The concentration of HCO_3^- is generally higher than that of plasma while that of chloride may be lower than plasma. This may be due to the presence of a chloride/bicarbonate exchange system located in the canalicular membrane as discussed below. The bile acids are usually present as the taurine and glycine conjugates with p*Ka* values of less than 2 and about 4 respectively, and are thus always ionized and exist in bile as bile salts. Taurocholate is the major bile acid in the rat, whereas in the human both taurine and glycine conjugates of the primary bile acids (cholate and chenodeoxycholate) and the secondary bile acids (primarily deoxycholate with small amounts of lithocholate) are present (Figure 2). Additional organic solutes include the bile pigments, small amounts of protein (primarily albumin derived from plasma) and other metabolites, primarily conjugates of endogenous hormones.

The osmolality of bile is generally equivalent to that of plasma and varies in parallel with the plasma. The bile salts are amphipathic molecules that form micelles at concentrations normally found in bile (Small, 1971), thus decreasing their osmotic activity. The inorganic electrolytes have been found to account for the total osmotic activity of bile, suggesting that the bile salts exist in their micellar form and have little or no osmotic activity (Wheeler and Ramos, 1960).

Bile Formation

The major force for the secretion of bile is generally accepted to be that generated by the active transport of osmotically active agents into a confined space and the resulting passive movement of water until osmotic equilibrium is reached (Diamond and Tormey, 1966; Oschman and Berridge, 1971). This concept is based on the initial observations of Schiff (1890) that feeding of

TABLE 12–1.
Bile Flow and Major Components of Bile

	Man	Rat	Dog	Rabbit
Flow (μl/min/kg)*	1.5–15.4	30–150	10	90
Na^+ (mEq/l)*	132–165	157–166	141–230	145–156
K^+ (mEq/l)*	4.2–5.6	5.8–6.4	4.5–11.9	3.6–6.7
Cl^- (mEq/l)*	96–126	96	31–107	77–99
HCO_3^- (mEq/l)*	17–55	24	14–61	40–63
Bile acids (mM)*	3–45	38†	16–187	6–24
Phospholipids (mg/liter)	1400–8100‡	412†	1620†	58†
Cholesterol (mg/liter)	970–3200‡	223†	250†	15†
Osmolarity (mOsmol/liter)†		332	293	312

Adapted from: Erlinger (1987)*, Klaassen and Watkins (1984)†, and Thureborn (1962)‡ bile acids to dogs with biliary fistula increased bile flow. The idea was substantiated by the more recent studies of Sperber (1959, 1965) who noted that secretion of any osmotically active compound into bile created an osmotic gradient and the subsequent movement of water and other diffusible solutes. Alternate mechanisms for the generation of bile flow include active transport of water, the vesicular transport of water by exocytosis, and the filtration of water by hydrostatic pressure. The active transport of water has never been demonstrated in a biological system and no evidence supports this hypothesis. The transport of proteins such as immunoglobulin A (IgA) has been shown to be mediated by exocytosis via vesicular pathways that involve the lysosomes and the Golgi (Delacroix et al., 1982). However, these vesicles are not present in sufficient quantities to account for the transport of bile water. The classic experiments of Brauer (Brauer et al., 1954; Brauer, 1965) demonstrated that bile production is independent of the perfusing pressure once a critical opening pressure is reached and the oxygen supply is not limiting. Studies in the isolated perfused liver have shown that bile is secreted at pressures which exceed the perfusion pressure; similar studies *in vivo* have shown that the bile secretory pressure exceeds the portal pressure (Richards, 1961). These data argue against hydrostatic pressure as a driving force for bile production. Thus, the transport of osmotically active organic and inorganic solutes into the bile canaliculi, followed by the passive movement of water, primarily via the paracellular pathway, is believed to provide the origin of water movement in the biliary tree. This canalicular bile is then modified as it passes through the ductules and ducts.

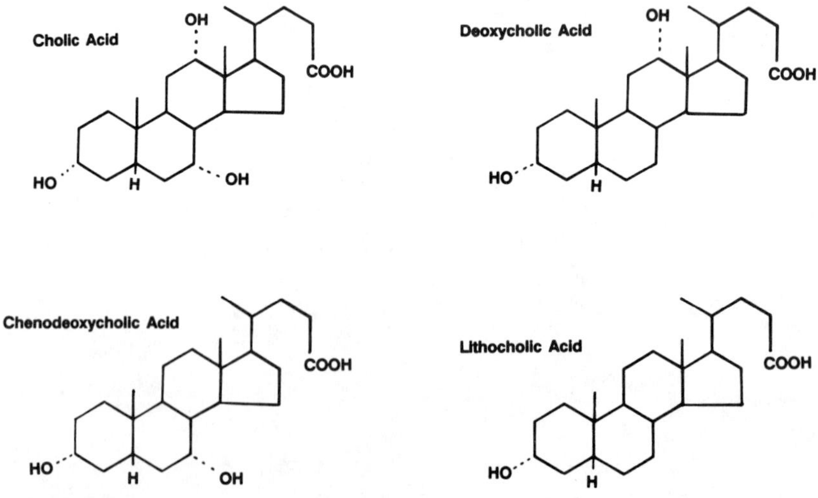

Figure 2. Structures of the primary and secondary bile acids.

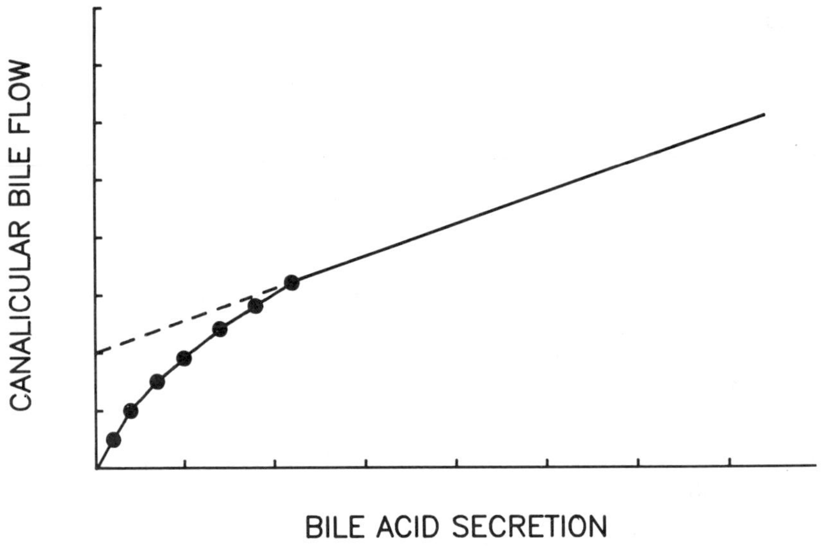

Figure 3. The relationship between bile acid secretion and canalicular bile flow. The solid line represents this relationship under normal conditions in which bile acids are infused to stimulate bile flow and bile acid secretion. The broken line represents a linear extrapolation of the solid line, yielding a positive intercept representing bile acid–independent bile flow. The dotted line represents the relationship where bile acid secretion rates are less than those seen under normal physiologic conditions and where bile acid secretion accounts for all of bile flow.

Canalicular bile can be estimated by measuring the biliary clearance of inert solutes which readily permeate the bile by simple, unrestricted diffusion and which are not reabsorbed or secreted by cells lining the ductules or ducts. Biliary clearance (C) is calculated as the product of bile flow (F) and the bile:plasma (B/P) concentration ratio of the solute at steady-state plasma concentrations: $C = F \times B/P$. Erythritol is the most commonly used solute which has been found to diffuse rapidly enough into bile in most species to reach equilibrium at high flow rates. Erythritol clearance has been shown to increase in parallel with increases in bile flow following infusion of bile acids. Since bile acids are secreted into the canaliculus by the hepatocyte, this is taken as evidence that erythritol clearance reflects the production of bile at its origin. Estimation of the clearance of less permeable solutes such as sucrose and inulin has been used as a tool to measure the permeability of the paracellular pathway, particularly in cholestasis (Forker, 1969; Layden et al., 1978; Krell et al., 1982; Jaeschke et al., 1983; Jaeschke et al., 1987). The theoretical basis and the limitations of this methodology have been reviewed by Forker (1968, 1970) and Wheeler et al. (1968).

The excellent correlation between the bile acid secretory rate in bile and bile flow over a broad range has led to the conclusion that bile acids are one of the osmotically active solutes that generate bile flow. This portion of bile flow is termed *bile acid–dependent flow* (Figure 3). As discussed above, bile acids exist in micellar form in bile and exert little osmotic activity. The bile flow attributed to the bile acids is therefore thought to be generated by the osmotic activity of the accompanying counterions (e.g., Na^+). At very low rates or in the absence of bile acid secretion, bile flow is not decreased proportionately. The portion of bile flow remaining in the absence of bile acid secretion is termed *bile acid–independent flow*.

Bile Acid–Dependent Flow

In view of the importance of bile acids in determining bile flow, much effort has been expended in understanding the mechanisms for their uptake by the hepatocyte, transport across the cell, and secretion into bile. The uptake of bile acids by the liver has been shown to be carrier-mediated and dependent on a Na^+ gradient. Saturable uptake of taurocholate has been demonstrated *in vivo* in the dog (Glasinovic, 1975a, 1975b), in the isolated perfused rat liver (Reichen and Paumgartner, 1975, 1976), in isolated hepatocytes (Anwer et al., 1976; Schwartz et al., 1975; Scharschmidt and Stephens, 1981), and in basolateral plasma membrane vesicles (Duffy et al., 1983; Inoue et al., 1982; Meier et al., 1984). These authors have also demonstrated that an out > in Na^+ gradient is required for uptake of taurocholate, and studies in plasma membrane vesicles have demonstrated a Na^+-dependent overshoot for taurocholate uptake. These data provide strong evidence that taurocholate uptake is a secondary active transport in which the energy for taurocholate uptake against a concentration gradient is derived from the transmembrane sodium gradient. The sodium gradient is maintained by Na^+,K^+-activated adenosinetriphosphatase (Na^+,K^+-ATPase), which is thought to function primarily in the basolateral domain of the plasma membrane and extrude sodium in exchange for potassium (Blitzer and Boyer, 1978; Sztul et al., 1987) (Figure 4). This mechanism for sodium-dependent bile acid uptake is thus similar to the bile acid transport systems in intestine and kidney (Wilson, 1981; Wilson et al., 1981) and is analogous to that demonstrated for amino acids and sugars.

Although there is much evidence supporting the basolateral localization of Na^+,K^+-ATPase, Leffert et al. (1985) have provided immunologic evidence, using monoclonal antibodies against Na^+,K^+-ATPase, for its localization in the canalicular membrane as well. A recent report by Sutherland et al. (1988) suggests that the more rigid apical (canalicular) membrane prevents expression of Na^+,K^+-ATPase *in vitro* since increasing the fluidity of this membrane with agents such as 2-(2-methoxyethoxy)ethyl 8-(*cis*-2-*n*-octylcyclopropyl)octanoate (A_2C), benzyl alcohol, and Triton WR-1339 increased enzyme activity to 75% of that in basolateral membranes. The method used to prepare the canalicular membranes is apparently important to the detection of Na^+,K^+-ATPase activity. Sellinger et al. (1990) counter that use of free-flow electrophoresis to further purify canalicular liver plasma membranes originally separated by sucrose density centrifugation separated canalicular membranes into six fractions. Fractions enriched in canalicular markers were devoid of Na^+,K^+-ATPase activity, whereas this activity could be detected in fractions containing small amounts of canalicular markers but also containing glucagon-stimulated adenyl cyclase, a basolateral marker enzyme. Addition of A_2C to all subfractions did not reveal any further Na^+,K^+-ATPase activity. In contrast, free-flow electrophoresis of canalicular membranes isolated by the Mg^{2+}-precipitation technique used by Sutherland et al. (1988) could not separate Na^+,K^+-ATPase activity from canalicular membrane marker enzymes. Thus, Sellinger et al. (1990) argue that any Na^+,K^+-ATPase activity detected in canalicular membranes represents contamination by basolateral membranes.

Early studies in the dog showed competitive inhibition of taurocholate uptake by the co-administration of other bile acids (Glasinovic, 1975b). Co-administration of non-bile acid organic anions such as indocyanine green in

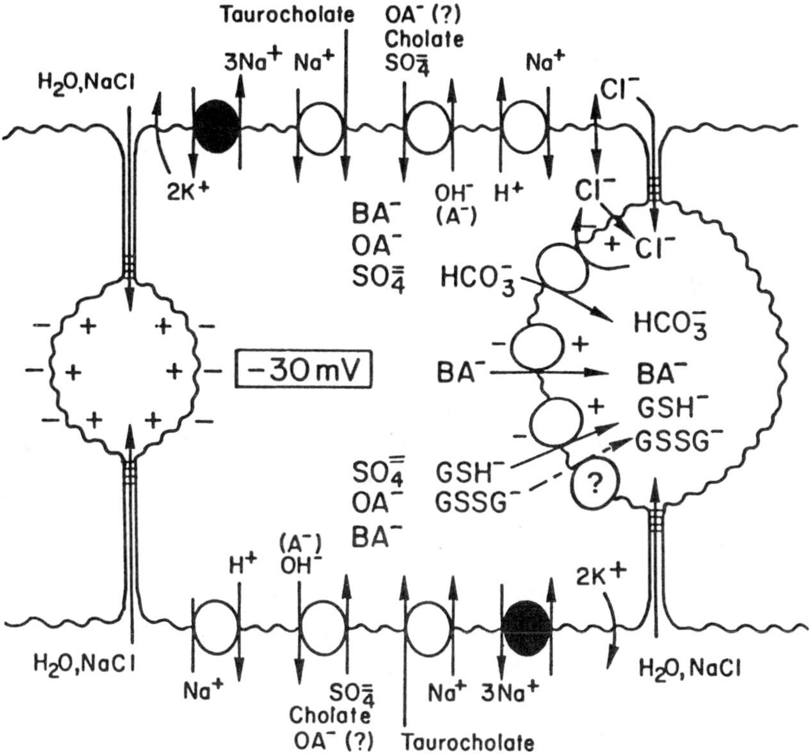

Figure 4. Schematic representation of some of the membrane transport systems in hepatocytes. See text for explanation. The solid circles represent the sodium pump, Na^+,K^+-ATPase. OA^-, organic anions; A^-, anions; BA^-, bile acids; GSH, reduced glutathione; GSSG, oxidized glutathione. (From Meier, 1988b, with permission.)

the rat, however, did not show inhibition (Paumgartner and Reichen, 1975). These data led to the suggestion that distinct carrier systems were responsible for the uptake of the bile acids and the non-bile organic anions. More recent data have shown that the substrate specificities of these carriers is much more complicated. Photoaffinity labeling studies utilizing a photoreactive diazirine derivative of taurocholate, (7,7-azo-3α,12α-dihydroxy-5β-cholan-24-oyl)-2-aminoethanesulfonic acid, identified two bile acid–binding proteins in the basolateral plasma membrane with molecular weights of 54,000 and 48,000 Da (von Dippe and Levy, 1983; Kramer et al., 1982; Wieland et al., 1984). The 48–49-kDa protein has been shown to represent the Na^+-dependent bile acid carrier (Ananthanarayanan et al., 1988), whereas the 54-kDa protein is postulated to represent a multispecific Na^+-independent organic anion carrier which is shared by taurocholate and organic anions such as bromosulfophthalein (BSP) and bilirubin (Reichen and Berk, 1979; Stremmel et al., 1983; Wolkoff and Chung, 1980; Stremmel and Berk, 1986), the nonconjugated bile acid, cholate, and other inorganic and organic anions such as sulfate, thiosulfate, oxalate, and succinate (Cheng and Levy, 1980; Ziegler et al., 1984; Hugentobler and Meier, 1986).

The means by which bile acids traverse the hepatocyte from the sinusoid to the canaliculus is not well understood. A number of intracellular bile acid–binding proteins have been postulated to serve in the intracellular transport of the bile acids. These include glutathione-*S*-transferases, Z protein, and the Y′ bile acid binders recently identified as 3α-hydroxysteroid dehydrogenases (Takikawa et al., 1987b and references therein). Bile acids have been shown to bind strongly to the 3α-hydroxysteroid dehydrogenases in the hepatocyte; inhibition of this binding caused a reflux of the bile acids out of the hepatocyte and delayed their excretion into bile (Takikawa 1987a, b). These data suggest that in addition to metabolizing the bile acids, these cytosolic proteins bind the bile acids and play an important role in mediating their transit across the hepatocyte. The bile acids are thought to diffuse through the cytosol across the hepatocyte both as the protein bound and the unbound monomer. Microtubules do not appear to be involved in the transport of the bile acids since agents that disrupt the microtubules have little effect on bile acid transport (Barnwell et al., 1984).

The secretion of bile acids across the canalicular membrane has also been shown to be a carrier-mediated process. The biliary excretion of the bile acids is saturable, exhibiting a Tm (transport maximum) of 10–12 μmol/min/kg in rats (Adler et al., 1977). The excretion of bile acids into the bile conforms to the definition of Tm except that the maximal transport rate does not remain constant, but declines upon increasing delivery of the bile acid at rates which exceed the Tm (Hardison et al., 1981). This is due to the cellular toxicity of the bile acids which results in a functional cholestasis. The term *secretory rate maximum,* or SRm, has therefore been used for bile acid biliary excretion instead of Tm. Hardison et al. (1981) have shown that the less toxic the bile acid, the more closely will the SRm approximate the Tm.

Like the bile acid uptake system on the sinusoidal membrane, the bile acid secretory carriers are thought to be distinct from those for the non-bile acid organic anions such as bilirubin and BSP. There is no evidence that bile acids inhibit the biliary excretion of non-bile acid organic anions, but rather infusion of bile acids *increases* the Tm for BSP and other anions (O'Maille et al., 1966; Gronwall and Cornelius, 1970). Further evidence for separate carrier systems comes from the mutant Corriedale sheep which has an inherited defect in the biliary excretion of organic anions such as BSP, whereas the biliary excretion of bile acids is normal (Alpert et al., 1969). More recently, a mutant rat strain (TM rats) was found which also has a selective defect in the biliary excretion of organic anions such as dibromosulfophthalein (DBSP) with no such defect for the biliary transport of taurocholate (Jansen et al., 1987).

Transport studies in isolated canalicular membrane vesicles have provided direct evidence that taurocholate efflux from the hepatocyte is carrier-mediated and is driven at least in part by the physiological intracellular negative electric potential of -30 to -40 mV (Inoue et al., 1984; Meier et al., 1984; Meier et al., 1987). This transport system thus differs from that in the sinusoidal membrane in that it is independent of sodium. Photoaffinity labeling studies using the taurocholate derivative, (7,7-azo-3α,12α-dihydroxy-5β-cholan-24-oyl)-2-aminoethanesulfonic acid, have identified a 100,000-Da glycoprotein in the canalicular, but not the sinusoidal, liver plasma membranes (Ruetz et al., 1987). Antibodies to the protein inhibited taurocholate transport in canalicular but not basolateral plasma membranes. Further studies have shown functional reconstitution of the 100-kDa glycoprotein inhibited electrogenic taurocholate transport in the reconstituted liposomes (Ruetz et al., 1988). These studies have provided conclusive evidence that this 100-kDa glycoprotein is the taurocholate carrier in the canalicular plasma membrane.

Fricker et al. (1989) have examined the effect of extrahepatic cholestasis (bile duct ligation) on the distribution and activity of this 100-kDa glycoprotein. After 50 h of bile duct ligation, there was an increased release of the 100-kDa protein (quantitated by polyclonal antibodies) into bile and an accumulation at the basolateral membrane and in intracellular vesicles. Functional studies demonstrated an increased electrogenic efflux of taurocholate in basolateral vesicles and a decreased efflux in canalicular vesicles in bile duct–ligated rats relative to sham controls. The authors postulate that this reversed polarity of the hepatocyte might serve as a protective mechanism to decrease intracellular accumulation of the toxic bile acids. Whether similar events occur during intrahepatic cholestasis is not known.

The transport of bile acids across the hepatocyte results in a concentration gradient of 100–1000 from bile to blood. The concentration of the bile acids ranges from 3.5 μ*M* in peripheral blood to 22 μ*M* in portal blood in humans (Ahlberg et al., 1977); in rats, bile acids in portal blood are in the range of 50–170 μ*M* (Olivecrona and Sjovall, 1959; Cronholm and Sjovall, 1967). Since bile acids are tightly bound to albumin and the major lipoproteins, the actual free concentration in blood is much lower. The overall concentration of bile acids in the hepatocyte is in the range of 50–100 μ*M*, however again, the bile salts are bound to intracellular proteins. The "activity" of the bile salts in the hepatocyte is thought to be less than 30 μ*M* due both to the intracellular binding and the rapid biliary excretion of the bile acids (Hofmann, 1988). The concentration of the bile acids in bile is much higher, in the range of 10–100 m*M*. Because of the micelle-forming properties of the bile acids, however, the actual free concentration is much lower. Nevertheless, it is clear that bile acids are transported against a concentration gradient; the step requiring active transport in this process appears to be that for uptake of the bile acids into the hepatocyte.

Bile Acid–Independent Flow

The concept of bile acid independent flow was developed to help explain the observation that linear extrapolation of the regression line relating bile flow (on the ordinate) and bile acid secretion (on the abscissa) resulted in a significant positive intercept on the ordinate, suggesting that a portion of bile flow was independent of bile acid secretion (Figure 3). Subsequent studies questioned the validity of linear extrapolation and suggested that the slope of the regression line increased substantially at low bile acid secretory rates (Balabaud et al., 1977; Baker et al., 1979). The distinction between bile acid–dependent and –independent flow is therefore not a precise one. Nevertheless, it is clear that a substantial portion of bile flow (about 50% in the rat and up to 40% in humans) is independent of bile acid secretion. The source of bile acid-independent flow is also generally accepted to be the result of transport of osmotically active solutes into the canalicular space with the subsequent passive movement of water. Several candidate organic and inorganic solutes have been proposed. Indirect studies in the isolated perfused rat liver suggested that inorganic anions such as chloride (Graf, 1983) or bicarbonate (Anwer and Hegner, 1983; Hardison and Wood, 1978; Van Dyke et al., 1982) might serve such a function. Recently, Meier et al. (1985) provided direct evidence for an electroneutral carrier-mediated chloride/bicarbonate exchange system as well as a chloride conductive leak pathway in canalicular membrane vesicles (Figure 4). These authors (Meier et al., 1985) propose that the canalicular excretion of chloride is driven by the intracellular negative potential resulting in a transcanalicular out to in chloride gradient which could then drive the electroneutral exchange for intracellular bicarbonate. Meier et al. (1987) also identified a bicarbonate/sulfate exchange system in canalicular membrane vesicles, which in conjunction with the chloride/bicarbonate exchange system is postulated to contribute to the canalicular excretion of cholephilic anions such as sulfate, or possibly other organic anions which might be substrates for this exchange system. Finally, glutathione has recently been postulated to be a significant driving force in the generation of bile acid–independent flow (Ballatori and Truong, 1989). Glutathione is found in bile at high concentrations (1–4 m*M*) in the rat and changes in its excretion in bile were significantly correlated with changes in bile flow (Ballatori and Truong, 1989). It seems likely that there are other solutes in bile which also contribute to the bile acid–independent component of bile flow.

Pathways of Fluid Movement

As indicated above, the bulk of fluid movement into bile is thought to occur primarily across the junctional complex from blood to bile via the paracellular pathway rather than across the canalicular membrane from the hepatocyte into bile via a transcellular pathway (Figure 1). The close correspondence between the osmolality of plasma and bile *in vivo* strongly suggests a direct equilibration of fluid and ions between plasma and bile rather than between the hepatocyte and bile. Such equilibration could occur at the level of the hepatocyte or downstream in the ductules and ducts. However, the low surface-to-volume ratio present in the ductules makes this latter possibility unlikely. Boyer (1980) used the red cell-free isolated perfused rat liver to show that the osmolality of bile faithfully followed that of the perfusate over a broad range (150–450 mosmol). This study ruled out equilibration at the level of the ducts since they are functionally inactive in the isolated liver perfused via the portal vein and do not permit the exchange of Na^+ or Cl^- (Graf and Peterlik, 1975). This occurs because the ducts receive their blood supply exclusively from the hepatic artery. The intercellular junction between hepatocytes is therefore the most likely site of fluid movement. This is consistent with observations that all epithelia that show iso-osmotic secretion are characterized by high ion conductivity across the junctional complexes (Schultz, 1977; Diamond, 1982). Freeze-fracture replica studies of the junctional complexes in the liver indicate that the number of interconnecting strands is intermediate between the classic "leaky" and "tight" epithelia (Chalcroft and Bullivant, 1970; Friend and Gilula, 1972; Goodenough and Revel, 1970). A number of studies indicate that the permeability of the junctional complex is not static but can vary in response to bile acids (Layden et al., 1978; Nemchausky et al., 1977; Reichen and Le, 1983), bile duct ligation (Metz and Bressler, 1979), hormones such as vasopressin (Lowe et al., 1988), and cholestatic agents such as phalloidin (Dubin et al., 1978) and α-naphthylisothiocyanate (Krell et al., 1982).

Studies designed to characterize the selective permeability of the junctional complexes using the charged and uncharged weight-matched solute pairs (carboxy-^{14}C)inulin and (methoxy-^{3}H)inulin and (^{14}C)ferrocyanide and

(^{3}H)sucrose showed that there is a barrier to negatively charged species (Bradley and Herz, 1978). Clearance values for sucrose and the much larger methoxyinulin were nearly identical, suggesting that these solutes move through similar channels or pores which are larger than those used by water and erythritol, which are readily diffusable smaller molecules. The validity of using markers such as sucrose and inulin to assess the permeability of the paracellular pathway has been questioned since sucrose (Graf, 1983) and inulin (Lorenzini et al., 1982) were identified in intracellular vesicles, indicating that they also reached the bile via a transcellular vesicular transport process. Lake et al. (1985) subsequently demonstrated that a significant portion of sucrose, inulin, and dextran transport from blood to bile occurs via a transcellular pathway. Jaeschke et al. (1987) have demonstrated that about 83% of the biliary clearance of sucrose occurs via the paracellular pathway in the untreated isolated perfused rat liver. Studies that use these inert fluid markers to measure the permeability of the paracellular pathway should therefore quantify the portion of clearance attributable to the paracellular pathway versus the transcellular pathway under all experimental conditions.

AGENTS WHICH INDUCE CHOLESTASIS

α-Naphthylisothiocyanate

α-Naphthylisothiocyanate (ANIT) (Figure 5) has been used as a model cholestatic agent for many years because it produces cholestasis in a reproducible and dose-dependent manner following acute administration and bile duct hyperplasia and biliary cirrhosis following chronic administration (see Plaa and Priestly, 1976 for an excellent review). In the rat, a single dose of ANIT results in retention of BSP after 2–4 h and hyperbilirubinemia at 12–24 h (Becker and Plaa, 1965; Indacochea-Redmond and Plaa, 1971; Drew and Priestly, 1976). Bile flow decreases within 16–24 h and remains depressed for about 5 days (Lock et al., 1982).

Doses of ANIT greater than 150 mg/kg result in complete cholestasis in the rat; the mouse was found to be more senitive to ANIT-induced cholestasis, the hamster less sensitive, and the dog and rabbit resistant (Indacochea-Redmond and Plaa, 1971). Metabolism of ANIT appears to be important for mediation of cholestasis. Cholestasis is increased by pretreatment of rats with phenobarbital, and this potentiation is diminished by concomitant treatment with inhibitors of protein synthesis such as actinomycin D and ethionine (Roberts and Plaa, 1965). Traiger et al. (1984, 1985) have shown that potentiation or inhibition of oxidative desulfuration of (^{35}S)-ANIT results in a corresponding potentiation or inhibition of the toxic effects of ANIT. Interruption of the enterohepatic circulation also markedly reduces the toxicity of ANIT, suggesting a role for intestinal metabolism as well (Roberts and Plaa,

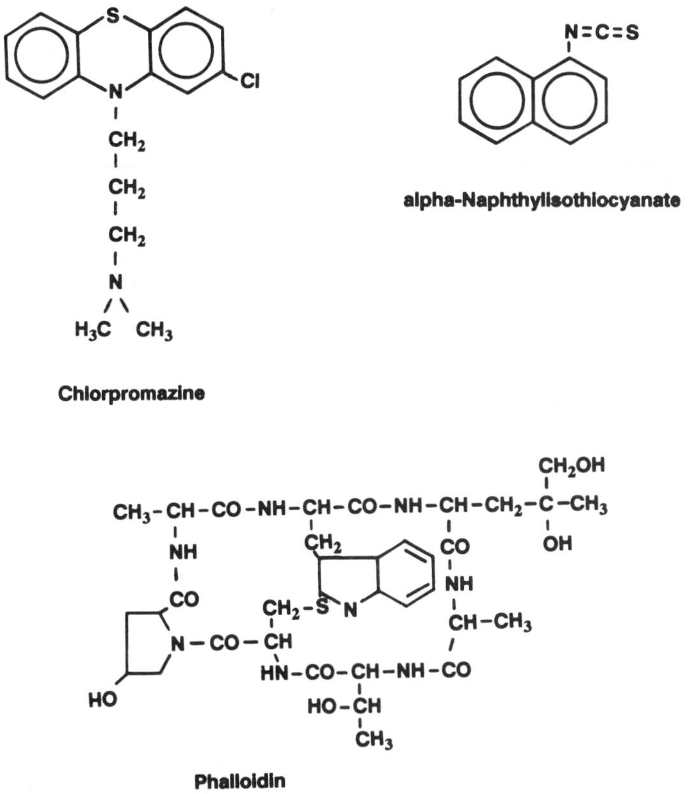

Figure 5. Structures of chlorpromazine, α-naphthylisothiocyanate and phalloidin.

1966). Although ANIT has been shown to be metabolized to intermediates which bind covalently to liver microsomes *in vitro* (El-Hawari and Plaa, 1977), there is a poor correlation between *in vivo* toxicity (cholestasis) and *in vitro* covalent binding. Sulfate, CO_2, and α-naphthylamine are the only identified metabolites of ANIT (Traiger et al., 1984; Mennicke et al., 1978). The nature of the metabolite(s) responsible for mediation of cholestasis remain unidentified. A recent preliminary report has shown that pretreatment of rats with L-buthionine-*S,R*-sulfoximine (BSO) protects against ANIT-induced cholestasis and hyperbilirubinemia, suggesting that a glutathione conjugate may be involved in ANIT-induced cholestasis (Dahm and Roth, 1989).

Although much of the focus of research of ANIT-induced cholestasis has focused on the hepatocyte, the initial lesions are seen in the cells lining the bile ducts. Connolly et al. (1988) have shown recently that within 4 h of dosing rats with 300 mg/kg ANIT, electron microscopy showed dilation of bile ducts, loss of microvilli from bile duct epithelial cells, and an apparent opening of the tight junctions between some bile duct epithelial cells. These changes were more pronounced after 6 h; by 24 h the majority of bile ducts were destroyed but by 48 h there was some evidence of regeneration. At the light microscopy level, no changes in parenchymal cells were seen at 4, 6, or 8 h; electron microscopy showed alteration in the tight junctions between hepatocytes at 6 h but none earlier. At 24 h there was evidence of parenchymal necrosis. These data are consistent with the observations of Krell et al. (1982) that ANIT treatment (250 mg/kg) of rats decreased the permeability barrier between the perfusate and the bile for (^{14}C)-sucrose and (^{32}P)-orthophosphate in the isolated perfused rat liver. These authors found that 7–11 h after treatment with ANIT, the equilibrium content of taurocholate and BSP in the perfusate was elevated approximately eightfold over that in control livers, with a comparable decrease in their biliary concentration, suggesting an exchange or regurgitation of solutes normally present in bile at high concentrations into the perfusate. Likewise, the bile:perfusate concentration ratios of (^{14}C)-sucrose and (^{32}P)-orthophosphate were increased dramatically 7–11 h after ANIT, indicative of an increased permeability to material normally restricted in its access to the bile. (The dramatic effects of ANIT on the permeability of the paracellular pathway are seen mainly in the Wistar rat; Fischer and Sprague-Dawley rats appear to be less sensitive; H. Jaeschke, personal communication). Jaeschke et al. (1987) examined the effect of ANIT on the transcellular and paracellular pathways of biliary sucrose clearance in the isolated perfused rat liver. In their model system, the liver was perfused with sucrose-containing medium (1 m*M*) until the biliary sucrose concentration reached an equilibrium. The liver was then perfused without sucrose-containing medium and the decline in the sucrose concentration in bile measured. Sucrose elimination from bile was found to follow biphasic first-order kinetics, showing a rapid initial decrease (half-life, 3.3 min), which was shown to represent the paracellular pathway, followed by a slower decrease (half-life, 29.4 min), which was shown to represent the transcellular pathway. These studies (Jaeschke et al., 1987) showed that ANIT decreased the half-life of the fast phase of the elimination of sucrose from the bile, indicating a reflux of sucrose into the perfusate via the paracellular pathway. The relative contribution of the transcellular pathway to the total clearance of sucrose was found to be 4% in ANIT-treated rats. ANIT has also been shown to increase the early appearance (5–7 min) of horseradish peroxidase (HRP) in bile following its infusion into the isolated perfused rat liver; this increase coincided with an increase in the biliary secretion of (^{3}H)-methoxyinulin which was coinfused with the HRP (Lowe et al., 1985). A second peak of HRP activity in bile was seen at 20–25 min and this coincided with the biliary secretion of ^{125}I-IgA, which was infused simultaneously with HRP. These data were interpreted to indicate that HRP enters bile via two pathways: the early peak represents a paracellular pathway and the later peak represents a transcytosis. These data are also consistent with an increased paracellular permeability induced by ANIT.

In summary, the cholestasis induced by ANIT is attributed to the increased permeability of the tight junctions between hepatocytes and bile duct epithelial cells, and the subsequent diffusion of osmotically active solutes from the bile into the plasma. The mechanism by which ANIT disrupts the tight junctions or the identity of the putative active metabolite which mediates this effect remain to be determined.

Agents that Act on Microfilaments and Microtubules

Phalloidin

Phalloidin (Figure 5) is a bicyclic heptapeptide derived from the poisonous mushroom *Amanita phalloides,* which causes vacuolization and necrosis of hepatocytes following acute administration of high doses to the rat. However, chronic treatment with low doses (500 μg/kg/day for 3 or more days) results in a striking accumulation of actin microfilaments in the pericanalicular region of the hepatocyte (Gabbiani et al., 1975) and a decrease in bile flow (Tuchweber and Gabbiani, 1976). Phalloidin has been shown to stabilize actin by binding to filamentous (F) actin and preventing its depolymerization (Lengsfeld et al., 1974). Phillips et al. (1975) postulated that the decrease in bile flow and the increased microfilaments were related; Dubin et al. (1978) subsequently showed that these changes occurred in parallel. The bile-to-plasma ratios of (^{3}H)-inulin and (^{14}C)-sucrose increased significantly in phalloidin-treated rats, suggestive of an increased permeability of the tight junctions (Elias et al., 1980). In support of this hypothesis, these authors also showed an increased penetration of ionic lanthanum into the zonula occludens and bile canaliculus in phalloidin-treated rats. Freeze-fracture replicas of the junctional complex indicated regions in which only a single element or strand separated the canaliculus from the lateral intercellular space. The bile acid concentration, however, was not decreased in phalloidin-treated rats (calculated from the data in Figure 1 and Table 2 in Elias et al., 1980) so that any increase in the permeability of the tight junction resulted in an increased penetration of sucrose and inulin from the plasma into the bile but did not result in a regurgitation of bile acids from the bile to the plasma. This is in contrast to the cholestasis seen following treatment with ANIT (*vide supra;* Krell et al., 1982). A decreased ability to transport bile acids into the canaliculus following phalloidin treatment has not been ruled out. Phalloidin has been shown to be taken up into hepatocytes by the bile acid transport system (Petzinger and Frimmer, 1980; Kramer et al., 1982; Wieland et al., 1984), thus explaining its concentration in and toxicity to the liver. An interaction with the bile acid carrier in the canalicular membrane has not been demonstrated.

Phillips et al. (1986, and references therein) have postulated that the microfilaments might subserve a contractile function to facilitate canalicular bile flow. This group has used isolated hepatocyte couplets in culture in which the canaliculus becomes sealed off from the culture media, and have shown that the bile canaliculi undergo spontaneous active contractions (Oshio and Phillips, 1981) which are stimulated by the bile acids. Phalloidin decreases the number of contractions and those that do occur appear slow and progressive or spastic (Watanabe et al., 1983). Cytochalasin B, which is also cholestatic when given in low doses (Phillips et al., 1975) and which acts to prevent the polymerization of actin (McLean-Fletcher and Pollard, 1980), also impairs contractions of the bile canaliculus in the isolated hepatocyte couplets (Phillips et al., 1983). Thus, these authors propose that canalicular contractions contribute an active force for the normal flow of bile in the canalicular system, and that motility disorders based on disruption of microfilament function can explain some forms of cholestasis (Phillips et al., 1986).

Cytochalasin B

Low-dose administration of cytochalasin also causes cholestasis, which is accompanied by dilatation of canaliculi and a loss of microvilli (Phillips et al., 1975). Cytochalasin has a number of effects, including detachment of microfilaments from the plasma membrane (Oda and Phillips, 1977) and prevention of the polymerization of actin (McLean-Fletcher and Pollard, 1980). Cytochalasin causes a reversible inhibition of canalicular contractions in the isolated hepatocyte couplets (Phillips et al., 1983), inhibits the transport of taurocholate (Reichen et al., 1981; Kacich et al., 1983), and the receptor-mediated transcytosis of IgA (Gebhart, 1984). The exact mechanism by which it inhibits bile flow is not known. As discussed above for phalloidin, disruption of microfilament function may explain the cholestasis.

Colchicine

Colchicine (and/or vincristine and vinblastine) inhibits the polymerization of tubulin into microtubules and disrupts various vesicle-mediated transport pathways. This includes the secretion of newly synthesized proteins such as fibrinogen from the hepatocyte into the blood (Redman et al., 1975; Feldman et al., 1975) and the biliary secretion of lipids (Gregory et al., 1978; Barnwell et al., 1984) and some proteins such as IgA (Mullock et al., 1980; Godfrey et al., 1982; Goldman et al., 1983) and horseradish peroxidase (Kacich et al., 1983; Bajwa and Fujimoto, 1983). Colchicine did not inhibit basal bile flow in the rat, but did inhibit bile acid-stimulated bile flow and bile acid secretion when administered alone and potentiated the inhibitory effect of phalloidin on these parameters (Dubin et al., 1980). Crawford et al. (1988) demonstrated that colchicine had no effect on the basal excretion of bile salts and phospholipids, but markedly inhibited the biliary excretion of bile salts following infusion of a bile acid load. Colchicine also inhibited slightly but significantly the excretion of a tracer dose of taurocholate in control animals and markedly inhibited the biliary secretion of a physiologic load of taurocholate in rats previously depleted of bile salts by 24 h biliary drainage. These authors concluded that the microtubule system plays a minimal role in the transport and excretion of bile salts under normal basal conditions, but that these processes become increasingly dependent on microtubular integrity as the transcellular flux and biliary excretion of bile salts increases.

Steroids

Estrogens

The ability of estrogens to induce cholestasis in humans and laboratory animals was recognized shortly after the observations that the oral contraceptives induced intrahepatic cholestasis in women who also developed intrahepatic cholestasis of pregnancy [see reviews by Kreek (1987), and Vore (1987), and references therein]. Intrahepatic cholestasis of pregnancy was first described in 1954 (Svanborg, 1954), and is characterized by pruritus, attributed to accumulation of bile acids, and jaundice, attributed to accumulation of bilirubin. Liver function tests show increased plasma levels of 5′-mononucleotidase, alkaline phosphatase, conjugated bilirubin, marked elevation of total bile acid, and BSP retention (Kreek et al., 1967; Kreek, 1987). Intrahepatic cholestasis of pregnancy is rare in most countries, with the highest prevalence occurring in Chile (12–22%) (Reyes et al., 1978), Bolivia (9%) (Reyes et al., 1979), and Sweden (2–3%) (Svanborg and Ohlsson, 1959). Studies in these countries suggest a genetic component (Dalen and Westerholm, 1974; Reyes et al., 1981). Estimates for the incidence of cholestasis in the United States and Great Britain are not available, but much lower incidences of 0.1–0.8% have been reported in Canada, France, Australia, and the Peoples Republic of China (Lunzer et al., 1986). Soon after the introduction of the oral contraceptive agents in 1960, intrahepatic cholestasis was also observed in a few women (Kreek, 1987). Approximately 50% of the women with intrahepatic cholestasis attributable to the oral contraceptives also had intrahepatic cholestasis of pregnancy (Furhoff and Hellstrom, 1973; Furhoff and Hellstrom, 1974); rechallenge experiments demonstrated that the estrogenic component (ethinylestradiol; Figure 6) was the active agent (Kreek et al., 1967). Mueller and Kappas (1964) had shown that administration of high doses of estradiol and estriol in humans for 7 days caused retention of BSP and decreased its transport maximum. Taken together, these data indicated that the estrogens decreased hepatic excretory function, and that in some "sensitive" individuals, this inhibition was sufficient to precipitate pruritus and jaundice.

Gallagher et al. (1966) were the first to characterize the structure-activity relationship of the estrogen-induced retention of BSP in an animal model, the rat, and showed that both synthetic and natural estrogens were effective. C18

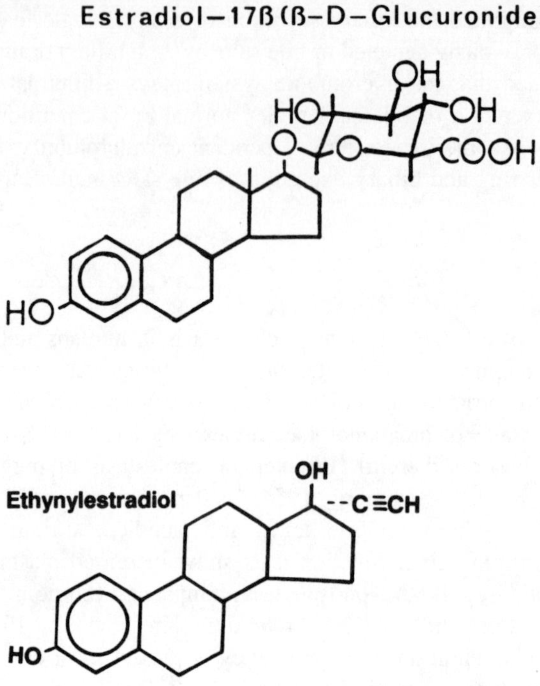

Figure 6. Structures of ethynylestradiol and estradiol-17β (β-D-glucuronide).

compounds with a phenolic A ring were the only natural steroids which regularly impaired BSP elimination. An oxygen substituent on the C or D rings was not essential, whereas modification of the phenolic A ring decreased or abolished activity. Diethylstilbestrol and mestranol were active, whereas C19, C21, and C24 steroids were inactive. Subsequent studies used ethinylestradiol or estrone at doses of 0.5–1 mg/day for 5–10 days and showed that estrogens decreased basal bile flow by about 50%; this inhibition was due primarily to a decrease in the so-called bile acid-independent fraction of bile flow (Kreek et al., 1969; Forker, 1969; Gumucio and Valdivieso, 1971). Basal bile acid secretion was not diminished, whereas the Tm, or SRm, was markedly inhibited by ethinylestradiol (Gumucio and Valdivieso, 1971).

A number of hypotheses have been put forward to explain the cholestasis induced by the estrogens. Forker (1969) noted that estrone (2.5 mg/day for 7 days) increased the bile:plasma ratio and the biliary clearance of sucrose and mannitol and decreased the biliary excretion and Tm for BSP. He concluded that estrone increased the permeability of the biliary tree resulting in a back-diffusion of BSP from bile to plasma, most likely via the paracellular pathway. Jaeschke et al. (1983) have argued, however, that the increased bile:plasma ratio of sucrose and inulin is due to a decrease in bile production, and that the high bile acid concentration in bile (greater than that in controls) is evidence against regurgitation of biliary contents into the plasma. These workers subsequently characterized the time course of changes induced by estradiol valerate at a dose of 1 mg/kg/week over a period of 12 weeks (Jaeschke et al., 1987). After 1 week, bile flow was significantly decreased and the taurocholate concentration in bile was significantly elevated. However, the permeability coefficients for sucrose and inulin were not elevated. After 3 weeks of treatment, bile flow remained depressed, but the concentration of taurocholate in bile was decreased and the permeability coefficients for sucrose and inulin were increased, consistent with a picture of increased permeability. Thus, use of a lower dose of estrogen permitted a dissociation of the decrease in flow which was detected at 1 week, from an increased permeability, which was not detected until after 3 weeks of treatment. These data offer strong evidence that an increased permeability is not the *cause* of the decrease in bile flow. Whether the increased permeability is the result of the decrease in bile flow or an independent effect of the estrogen is not known. Anderson et al. (1988) recently reported that treatment with ethinylestradiol (0.5 mg/d for 5 days) did not alter the immunohistochemical localization of ZO-1, a protein known to be a component of the tight junction. Control and estrogen-treated animals showed a similar continuous distribution of ZO-1 along the margins of the canaliculi, whereas bile duct ligation resulted in numerous breaks.

A decrease in the fluidity of the plasma membrane has also been proposed as a mechanism to explain the estrogen-induced cholestasis. Ethinylestradiol decreases profoundly the plasma cholesterol levels in rats due to a marked increase in the number of hepatic low-density-lipoprotein receptors and an increased hepatic clearance of plasma lipoproteins (Kovanen et al., 1979; Chao et al., 1979; Kovanen et al., 1981). This results in an increased cholesterol ester content of liver homogenate and plasma membranes and a decreased fluidity of these membranes (Davis et al., 1978a; Simon et al., 1980; Fewster et al., 1967; Davis et al., 1978b; Keeffe et al., 1979). Treatment of rats with estradiol or ethinylestradiol has been shown to decrease Na^+,K^+-ATPase activity in plasma membranes (Reichen and Paumgartner, 1977; Davis et al., 1978a; Heikel and Lathe, 1970) and the uptake of bile acids and other organic anions (steroid glucuronides) into isolated hepatocytes (Brock and Vore, 1984; Berr et al., 1984). Simon and co-workers have proposed that the decreased membrane fluidity is responsible for the decrease in bile flow and Na^+,K^+-ATPase activity, the decreased Tm for taurocholate and BSP secretion into bile and the decreased uptake of bile acids into the liver (Berr et al., 1984; Simon et al., 1980). This group has recently reported that ethinylestradiol selectively alters the lipid fluidity of the basolateral (sinusoidal) domain of the plasma membrane and that this decreased fluidity is not correlated with the increased cholesterol ester content of the membrane

(Rosario et al., 1988). These workers argue that the decreased bile acid Tm must therefore be the result of changes in the basolateral plasma membrane. Fricker et al. (1988) recently reported that ethinylestradiol treatment of rats abolished both Na^+-dependent taurocholate uptake in basolateral membrane vesicles and potential driven taurocholate efflux in canalicular membrane vesicles. Ethinylestradiol treatment also decreased phospholipids by 45 to 75% in *both* basolateral and canalicular membranes. The differences observed between these two studies is critical; i.e., Fricker et al. (1988) demonstrate an estrogen-induced lesion in the canalicular membrane, whereas Rosario et al. (1988) find changes only in the basolateral domain. The differences in the methods used to prepare the membrane fractions may account for the conflicting observations. Interestingly, Fricker et al. (1988) also noted that addition of *S*-adenosylmethionine to the vesicles *in vitro* restored taurocholate transport activity in both fractions of plasma membrane, presumably due to an increased conversion of phosphatidylethanolamine to phosphatidylcholine and the accompanying increase in membrane fluidity. These data are consistent with studies by Stramentinoli et al. (1981) and Boelsterli et al. (1983) demonstrating that administration of *S*-adenosylmethionine to rats increases plasma membrane fluidity and Na^+,K^+-ATPase activity and partly reverses the ethinylestradiol-induced decrease in bile flow.

Smith and Gordon (1988) recently compared the effects of two steroids, ethinylestradiol and spironolactone, on bile flow and liver plasma membrane fluidity. Although both agents significantly decreased membrane fluidity and Na^+,K^+-ATPase activity, ethinylestradiol decreased bile flow 42%, whereas spironolactone increased flow 22%. These authors concluded that changes in membrane fluidity cannot be the rate-determining step in cholestasis induced by ethinylestradiol.

Metabolism of the estrogens to cholestatic metabolites has been proposed as a third mechanism of estrogen-induced cholestasis. Studies in our laboratory have shown that following their intravenous administration to the rat, the glucuronide conjugates of the D ring of estradiol (Figure 6), estriol and ethinylestradiol cause an immediate, profound, dose-dependent and reversible cholestasis (Vore and Slikker, 1985; Meyers et al., 1980; Meyers et al., 1981; Vore et al., 1983). In contrast, administration of the glucuronide conjugates of the A ring under identical experimental conditions causes choleresis. The cholestatic glucuronides decrease bile acid secretion and the bile acid-independent component of bile flow (Meyers et al., 1980; Meyers et al., 1981). Studies in the rhesus monkey showed that estradiol-17β(β-D-glucuronide) at doses of 5.5–11 μmol/kg IV significantly inhibited indocyanine green clearance, whereas estradiol-3-(β-D-glucuronide) had no effect (Slikker et al., 1983). These doses of estradiol-17β(β-D-glucuronide) are similar in magnitude to the daily urinary excretion of D-ring glucuronides, primarily estriol-16α-glucuronide, of about 2 μmol/kg seen in the third trimester of normal pregnancy (Baker et al., 1979). These data suggest that the retention of BSP seen in normal pregnancy may be attributable to the high levels of D-ring glucuronides which occur in normal pregnancy.

The mechanism by which the cholestatic D-ring glucuronide conjugates of the estrogens induce cholestasis is not known. The parallel dose-response curves for the induction of cholestasis (Vore and Slikker, 1985) and the structure-activity relationship seen suggest their interaction at some "receptor" site. The structural similarities between the cholestatic glucuronides and the bile acids suggest that they may act at some common receptor site, such as the bile acid carrier. The cholestatic glucuronides do not inhibit Na^+-dependent taurocholate uptake in isolated hepatocytes (Brouwer et al., 1987) or in hepatic basolateral membrane vesicles (Zimmerli et al., 1989). Studies in the isolated perfused liver have shown that taurocholate protects against and reverses cholestasis induced by estradiol-17β(β-D-glucuronide) (Adinolfi et al., 1984; Durham and Vore, 1986). Adinolfi et al. (1984) have shown an increased bile:plasma ratio for (^{14}C)-sucrose and an increased sucrose clearance following estradiol-17β(β-D-glucuronide) administration in the isolated perfused liver, suggesting an increased permeability as a likely mechanism. However, the lack of a decreased bile acid concentration in bile argues against a significant change in permeability sufficient to cause a regurgitation of biliary contents into the perfusate. Kan et al. (1989) examined the paracellular permeability directly by measuring the biliary output of horseradish peroxidase administered as a single pulse in isolated perfused rat livers exposed to estradiol-17β(β-D-glucuronide). The cholestatic glucuronide increased the horseradish peroxidase detected in bile relative to vehicle controls about threefold in the absence of infused taurodehydrocholate and about sixfold when the bile acid was infused to restore bile flow to initial levels. These data provide evidence that the tight junctions are damaged by estradiol-17β(β-D-glucuronide). As discussed in subsequent correspondence by Jaeschke (1990) and Coleman and Kan (1990), it is still not clear if the increased paracellular permeability induced by estradiol-17β(β-D-glucuronide) is causally related to the cholestasis and if sufficient cholestatic estrogen glucuronides are formed during chronic estradiol treatment to be the cause of estrogen cholestasis.

Recent studies have shown that (^{3}H)-estradiol-17β(β-D-glucuronide) binds specifically to a high-affinity site in canalicular membranes; the cholestatic D-ring glucuronides and taurocholate also bind to this site, whereas the non-cholestatic A-ring glucuronides and BSP do not (Changchit et al., 1990). The nature of this site is not known, but could represent the canalicular bile acid carrier or some other site at which taurocholate acts to regulate bile flow.

In summary, the estrogens are thought to act by changing membrane fluidity and altering the mobility of bile acid carriers and perhaps other functions associated with the generation of bile flow, and/or by metabolism to potent cholestatic glucuronide conjugates. Since *in vivo* studies indicate that chronic estrogen treatment inhibits bile acid-independent flow, it would seem important to determine if estrogen treatment decreases the chloride-bicarbonate exchange system in canalicular membranes. Likewise, the effects of the cholestatic metabolites on bile acid transport and chloride-bicarbonate exchange in canalicular vesicles should be investigated. Finally, the possibility that hepatic estrogen receptors may mediate a change in the number or type of carriers has not been ruled out.

Anabolic Steroids

The cholestatic activity of the anabolic steroids has not been investigated to the same extent as that of the estrogens, even though the cholestatic jaundice induced by methyltestosterone (Figure 7) has been recognized since 1947 (Werner, 1947). Cholestasis has been observed primarily in patients treated with the anabolic steroids for Fanconi's and aplastic anemias, however, their misuse/abuse by athletes has also led to a number of hepatic complications [see Ishak and Zimmerman (1987) for a recent review]. Intrahepatic cholestasis usually develops after short-term use of anabolic steroids, whereas their long-term use may result in peliosis hepatis or the induction of benign or malignant tumors. The functional impairment most often demonstrated in humans has been the plasma retention of BSP, whereas appearance of jaundice, i.e., retention of bilirubin, appears in only a minority of patients (Plaa and Priestly, 1976; Ishak and Zimmerman, 1987). Structure-activity relationship studies have demonstrated that an alkyl group in the C-17 position is a key requirement for induction of jaundice and hepatic dysfunction in

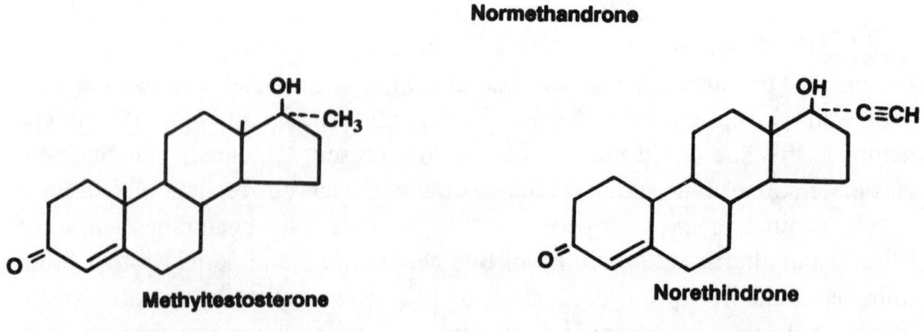

Figure 7. Structures of methyltestosterone, norethindrone, and normethandrone.

humans since testosterone and 19-nortestosterone have only rarely caused hyperbilirubinemia and did not induce BSP retention (deLorimier et al., 1965; Lennon, 1966a,b). The incidence of BSP retention was highest for normethandrone and norethindrone (Figure 7), moderate with methyltestosterone, fluoxymestrone, and methandriol, and low with ethisterone (deLorimier, 1965). A keto group at the C-3 position increases toxicity relative to a hydroxy substituent whereas saturation of the A ring decreases toxicity. Potency for retention of BSP appears unrelated to the anabolic, androgenic, or progestational activity of these steroids (deLorimier et al., 1965).

The mechanism by which the anabolic steroids decrease bile flow and induce retention of BSP is not known. They have been shown to decrease both bile acid-dependent and bile acid-independent flow (Despopoulus, 1971) and to inhibit Na^+,K^+-ATPase (Tuchweber et al., 1986). Phillips et al. (1978) have shown that norethandrolone may disrupt microfilaments resulting in motility disorders. Zimmerman has pointed out the structural similarities between the cholestatic estrogens, the anabolic steroids, and icterogenin, a cholestatic triterpine-pentacyclic alkaloid (Arias, 1963), and suggested that they may act via a similar mechanism (Zimmerman and Lewis, 1987). In this regard, it is interesting to note that the D-ring glucuronide conjugates of testosterone and dihydrotestosterone (Figure 8) are also cholestatic (Meyers et al., 1981; Vore et al., 1989). It would seem to be important to determine if the D-ring glucuronide conjugates of the 17-alkyl-substituted anabolic steroids are also cholestatic.

Bile Salts

Toxicity associated with bile has been of interest since the days of Hippocrates and Galen and the idea that yellow bile or black bile was the cause of human diseases (Palmer, 1972). More modern studies demonstrated that the bile salts are primarily responsible for the toxicity of bile. Holsti (1956) first showed that feeding of desiccated hog bile to rabbits induced hepatic cirrhosis and subsequently demonstrated that this effect could also be produced by lithocholate and glycolithocholate (Holsti, 1960). Lithocholic acid (Figure 8) is a naturally occurring bile acid, present in trace amounts in the bile of humans and other species, including the rat (Kakis and Yousef, 1978). It is a secondary, 3α-monohydroxy bile acid which is formed primarily in the colon by the bacterial 7α-dehydroxylation of the primary bile acid chenodeoxycholate. Recent studies of lithocholate-induced cholestasis have been stimulated by the use of chenodeoxycholate for the solubilization of cholesterol gallstones (Danziger et al., 1973).

The acute cholestatic effects of lithocholate were first noted by Ivy (1941) in the dog and, more recently, the cholestasis induced by taurolithocholate was noted in the rat (Javitt, 1966; Javitt and Emerman, 1968) and hamster

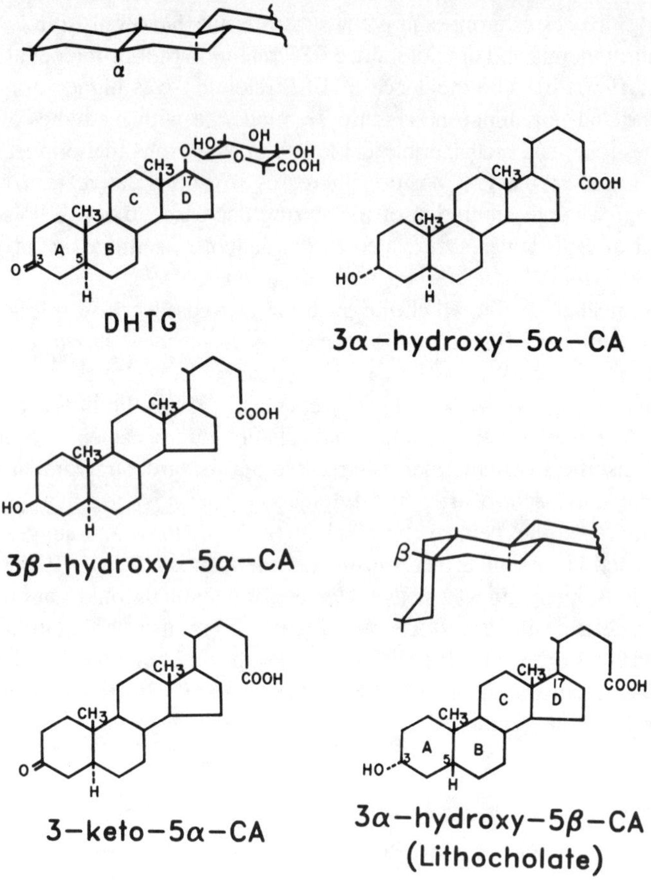

Figure 8. Structures of DHTG, some allo bile acids, and lithocholate. CA = cholanic acid. (From Vore et al, 1989, with permission.)

(Schaffner and Javitt, 1966). Javitt and Emerman (1968) originally postulated that the low solubility of lithocholate resulted in the formation of precipitates in the canaliculi with subsequent mechanical obstruction of the lumen. This hypothesis was supported by the observation that infusion of taurocholate or other micelle-forming bile acids at a 1:1 molar ratio protected against the cholestasis. Electron microscopy studies were unable to demonstrate a consistent relationship between the presence of precipitates and cholestasis, however, suggesting that precipitates were not a primary factor (Miyai et al., 1977). Numerous electron microscopic and freeze-fracture studies of lithocholate-induced cholestasis have shown specific changes in the bile canalicular membrane consisting of cytoplasmic evaginations and a bizarre lamellar transformation (Layden and Boyer, 1975; Miyai et al., 1975). Lithocholate increased the incorporation of cholesterol into the "bile canalicular" plasma membrane such that the cholesterol to phospholipid ratio increased from 0.31 to 1.94; taurolithocholate increased the ratio to 0.78 (Kakis and Yousef, 1978). The authors argue that these biochemical changes in the bile canalicular membrane may be responsible for the morphological changes and the physiological basis for the cholestasis. It should be pointed out that the "bile canalicular" membrane preparation used by these authors was enriched in both Na^+,K^+-ATPase and Mg^{2+}-ATPase, indicating the presence of membranes from both the sinusoidal and canalicular domain. Increased serum lithocholic acid levels have also been shown to increase the cholesterol content of the erythrocyte membrane (Cooper et al., 1972). Increased cholesterol content of membranes decreases membrane fluidity and permeability to water, small nonionic solutes, anions, and cations (Oelberg and Lester, 1986, and references therein). The decreased fluidity and permeability of the canalicular membrane has thus been proposed as a mechanism of lithocholate-induced cholestasis.

Lithocholate can be conjugated in the liver via amide bond formation with glycine or taurine and can be further metabolized by sulfation or glucuronidation at the 3-position. Sulfated glycolithocholic acid is the major metabolite of lithocholate in normal human bile (Cowen et al., 1975; Palmer and Bolt, 1971; Stiehl et al., 1980). Sulfation of lithocholate appears to decrease the cholestatic activity of lithocholate as taurolithocholate-sulfate had no effect on bile flow; sulfated-glycolithocholate was a potent cholestatic agent, however. Oelberg et al. (1984a) have identified lithocholate 3-*O*-β-glucuronide in the serum and urine of patients with cholestatic liver disease and have shown that it is also a potent cholestatic agent.

In their characterization of cholestasis induced by lithocholate glucuronide, Oelberg et al. (1984a) noted an opalescent material in bile which proved to be a lithocholic acid-glucuronide-calcium complex. These authors subsequently showed that in pure aqueous solution, calcium precipitated stoichiometrically with lithocholic acid glucuronide, and that the cholestatic bile salts lithocholate sulfate, lithocholate glucuronide, and lithocholate bound Ca^{2+} 10–60 times more avidly that did the noncholestatic bile acids taurocholate, glycocholate, and taurolithocholate sulfate (Oelberg et al., 1984b). Lester and colleagues have shown that the cholestatic bile acids act as Ca^{2+}-ionophores in phospholipid vesicles, and have postulated that these bile acids may induce cholestasis by increasing, at least transiently, the intracellular free calcium (Oelberg and Lester, 1986). Anwer et al. (1988) have recently shown that a number of bile acids increased the cytosolic Ca^{2+} activity of isolated rat hepatocytes within 10–30 s in a concentration-dependent manner (0.05–1.0 m*M*). The order of potency was lithocholate = taurolithocholate >

chenodeoxycholate = taurochenodeoxycholate = deoxycholate = tauro-deoxycholate; cholate and taurocholate were inactive in this regard. This order of potency is similar to that shown for the toxicity of the bile acids in isolated hepatocytes (Schölmerich et al., 1984). The initial increase in cytosolic Ca^{2+} was not due to cell damage, since bile-induced decreases in cell viability were not significant until 2 to 3 min. Combettes et al. (1988) have independently shown that lithocholate and taurolithocholate increased the cytosolic Ca^{2+} concentration in isolated rat hepatocytes with an ED_{50} of 25 μM and caused a net loss of Ca^{2+} from the cell; neither taurocholate nor cholate had any effect on intracellular free Ca^{2+}. These authors postulate that the endoplasmic reticulum serves as the source of the free cytosolic Ca^{2+}, whereas Anwer et al. (1988) argue that the source is extracellular Ca^{2+}. Thus, disturbances in the metabolism of intracellular free Ca^{2+} could initiate the inhibition of bile secretion induced by these agents.

The allo bile acids (Figure 8) are structural analogues of the natural bile acids in which the hydrogen at C-5 is alpha-oriented rather than beta-oriented as in the natural bile acids. Vonk et al. (1981) demonstrated that 3-β-hydroxy-5α-cholanic acid was also cholestatic in the rat and was approximately four times more potent than the 5-β analogue. We have recently compared the cholestatic potency of a series of allo-bile acids, 3α-hydroxy-5α-cholanic acid, 3-keto-5α-cholanic acid and 3β-hydroxy-5α-cholanic acid, with that of their 5β-analogues and dihydrotestosterone glucuronide (DHTG) (Figure 8) in order to characterize their structure-activity relationships with respect to (1) substituents at the 17 position of the D ring, (2) substituents at the 3-position of the A ring, and (3) the relationship of the A and B rings of the steroid nucleus with regard to their trans (where C-5 is α) vs. *cis* (where C-5 is β) configuration (Vore et al., 1989).

All agents caused a rapid and profound inhibition of bile flow which was dose-dependent following their intravenous administration in the rat. The logarithm of the dose versus the cholestatic response curves for DHTG, the allo-bile acids, and lithocholate were all parallel. DHTG was the most potent congener and was 2 times more potent than 3-keto-5α-cholanic acid and 5 times more potent than lithocholate. DHTG and 3-keto-5α-cholanic acid differ in structure only by the presence of the carboxylic acid side chain of the cholanic acids versus glucuronic acid at the 17 position. These data indicate that the glucuronic acid moiety contributes significantly but is not essential to the cholestatic activity, and that the trans configuration of the A and B rings of the steroid nucleus confers the greatest cholestatic potency.

As discussed above, bile acids other than lithocholate and its conjugates are hepatotoxic and will induce cholestasis if infused at rates which exceed the Tm. In general, the monohydroxy bile acids are more toxic than the dihydroxy bile acids which are more toxic than the trihydroxy bile acids (Billington et al., 1980; Vyvoda et al., 1977; Schölmerich et al., 1984).

Chenodeoxycholic and deoxycholic acids also induce hemolysis (Berliner and Schoenheimer, 1938), whereas ursodeoxycholate, the 7-β epimer of chenodeoxycholate, has only weak hemolytic activity (Palmer, 1979). The rank order of cholestatic potency of three bile acids following their intravenous infusion in the rat was taurodeoxycholate > taurochenodeoxycholate > taurocholate (Drew and Priestly, 1979). Hardison et al. (1981) obtained similar results, with tauroursodeoxycholate even less toxic than taurocholate. Hardison et al. (1981) have proposed that the SRm of a bile acid does not reflect saturation of the transport process, as required by the classical definition of Tm. Rather, the toxic bile acids become toxic to the transport process at rates less than the Tm. The less-toxic bile acids will therefore have SRm which most closely approximate the true Tm.

Bilirubin-Manganese

Witzleben et al. (1968) first noted that intravenous administration of manganese sulfate caused hepatic lesions similar to those seen in cholestasis, i.e., dilatation of canaliculi with swelling and loss of microvilli and a decreased Tm for bilirubin. Witzleben (1969) subsequently noted that manganese induced a slight decrease in bile flow that was markedly potentiated by infusion of bilirubin (Witzleben, 1971; Boyce and Witzleben, 1973). The extent of cholestasis depends both on the dose of bilirubin and the interval after manganese loading. Manganese apparently forms a complex of some sort with bilirubin, which could conceivably cause physical obstruction in the biliary tree. The nature of such a complex has not been identified, however, and the ultrastructural pathology does not support obstruction as a mechanism of cholestasis [see Plaa and Priestly (1976) for review]. More recent studies by Plaa and colleagues (De Lamirande and Plaa, 1978; De Lamirande and Plaa, 1979; De Lamirande et al., 1981; De Lamirande et al., 1982; Plaa et al., 1982; Ayotte and Plaa, 1985; Ayotte and Plaa, 1986) suggest that the canalicular membranes are physically altered, resulting in a decreased permeability to water. The bile canalicular membranes show an increased incorporation of cholesterol and phospholipid (Plaa et al., 1982) as well as incorporation of bilirubin and manganese. Ayotte and Plaa (1988) have recently shown that in Sprague-Dawley rats, unconjugated bilirubin and manganese are cholestatic, whereas in the homozygous male Gunn rat, only the combination of manganese-bilirubin ditaurate (a commercially available synthetic conjugate of bilirubin) was cholestatic. In neither group of rats was biliary excretion of bile salts impaired. These data are consistent with inhibition of the bile acid-independent component of bile flow. Further studies are needed to identify the physical-chemical nature of any manganese-bilirubin complexes and to investigate systematically the permeability of the paracellular pathway.

Phenothiazines

Chlorpromazine-induced jaundice has been considered a significant clinical problem with a reported incidence of 1–2% (Zimmerman and Lewis, 1987). However, Plaa and Priestly (1976) argue that this number may be inappropriately high since the number of cases reported for the entire 1965–1971 period (213) is remarkably less than 1% of the 1.5 million prescriptions for chlorpromazine dispensed in 1970 alone. The mechanism of cholestasis has been investigated in a number of model systems. Chlorpromazine (Figure 5) is a cation with detergent properties which binds to membrane phospholipids (Carey et al., 1976; Hruban et al., 1978; Di Francesco and Bickel, 1977). These properties are most likely responsible for its ability to alter membrane fluidity and inhibit Na^+,K^+-ATPase activity (Keeffe et al., 1980; Samuels and Carey, 1978). Studies in the isolated perfused rat liver have shown that chlorpromazine itself causes cholestasis; however, some of this effect could be attributed to the accompanying decrease in hepatic perfusion (Tavoloni and Boyer, 1980). In contrast, these authors showed that the 7,8-dihydroxy metabolite of chlorpromazine caused an irreversible inhibition of bile flow without any effect on hepatic perfusion, whereas the sulfoxide metabolite of chlorpromazine had no effect on either measure. These results are consistent with those of Samuels and Carey (1978) showing that the sulfoxide metabolite had no effect on hepatic plasma membrane Na^+,K^+-ATPase, whereas the hydroxylated metabolites were as toxic as chlorpromazine in this regard. Chlorpromazine and the cholestatic 7,8-dihydroxychlorpromazine have been shown to decrease Na^+,K^+-ATPase cation pumping in cultured rat hepatocytes in a dose-dependent fashion, whereas the noncholestatic chlorpromazine sulfoxide was shown to be inactive (Van Dyke and Scharschmidt, 1987). The decreased sodium pumping was associated with a decrease in the sodium electrochemical gradient and a modest decrease in the sodium-dependent transport of taurocholate that could contribute to chlorpromazine cholestasis. Chlorpromazine and its metabolites have also been shown to alter the polymerization and gelation of actin (Elias and Boyer, 1979).

Finally, there are numerous other drugs which have been reported to induce cholestasis. These have been summarized in an excellent and comprehensive review by Zimmerman and Lewis (1987). The mechanisms of cholestasis for these agents have not been investigated in the same detail as those described above and are not well understood.

ACKNOWLEDGMENTS

Preparation of this chapter and the work cited therein were supported in part by USPHS Grant HD13250. I thank Dr. Hartmut Jaeschke for his critical review and valuable suggestions.

REFERENCES

Adinolfi LE, Utili R, Gaeta GB, Abernathy CO, and Zimmerman JH: Cholestasis induced by estradiol-17β-D-glucuronide: mechanisms and prevention by sodium taurocholate. *Hepatology* 1984; 4:30–37.

Adler RD, Wannagat FJ, and Ockner RK: Bile secretion in selective biliary obstruction. Adaptation of taurocholate transport maximum to increased secretory load in the rat. *Gastroenterology* 1977; 73:129–136.

Ahlberg J, Anglelin B, Bjorkhem I, and Einarsson K: Individual bile acids in portal venous and systemic blood serum of fasting man. *Gastroenterology* 1977; 73:1377–1382.

Alpert S, Mosher M, Shanske A, and Arias IM: Multiplicity of hepatic excretory mechanisms for organic anions. *J Gen Physiol* 1969; 53:238–247.

Ananthanarayanan M, von Dippe P, and Levy D: Identification of the hepatocyte Na^+-dependent bile acid transport protein using monoclonal antibodies. *J Biol Chem* 1988; 263:8338–8343.

Anderson JM, Glade J, Stevenson BR, Boyer JL, and Mooseker MS: Altered immunohistochemical localization of the tight junction protein, ZO-1, in ethinyl estradiol and bile duct ligation-induced cholestasis. *Hepatology* 1988; 8:1227, A34.

Anwer MS and Hegner D: Role of inorganic electrolytes in bile acid-independent canalicular bile formation. *Am J Physiol* 1983; 244:G116–G124.

Anwer MS, Kroker R, and Hegner D: Cholic acid uptake into isolated rat hepatocytes. *Hoppe-Seyler's Z Physiol Chem* 1976; 357:1477–1486.

Anwer MS, Engleking LR, Nolan K, Sullivan D, Zimniak P, and Lester R: Hepatoxic bile acids increase cytosolic Ca^{++} activity of isolated rat hepatocytes. *Hepatology* 1988; 8:887–891.

Arias IM: Effects of a plant acid (icterogenin) and certain anabolic steroids on the hepatic metabolism of bilirubin and sulfobromophthalein (BSP). *Ann NY Acad Sci* 1963; 104:1014–1025.

Ayotte P and Plaa GL: Hepatic subcellular distribution of manganese in manganese and manganese-bilirubin induced cholestasis. *Biochem Pharmacol* 1985; 34:3857–3865.

Ayotte P and Plaa GL: Modification of biliary tree permeability in rats treated with a manganese-bilirubin combination. *Toxicol Appl Pharmacol* 1986; 84:295–303.

Ayotte P and Plaa GL: Biliary excretion in Sprague-Dawley and Gunn rats during manganese-bilirubin-induced cholestasis. *Hepatology* 1988; 8:1069–1078.

Bajwa RS and Fujimoto JM: Effect of colchicine and *S,S,S*-tributyl phosphorotrithioate (DEF) on the biliary excretion of sucrose, mannitol and horseradish peroxidase in the rat. *Biochem Pharmacol* 1983; 32:85–90.

Baker TS, Jennison KM, and Kellie AE: The direct radioimmunoassay of oestrogen glucuronides in human female urine. *Biochem J* 1979; 177:729–738.

Baker AL, Wood RAB, Moosa AR, and Boyer JL: Sodium taurocholate modifies the "bile acid independent" fraction of canalicular bile flow in the Rhesus monkey. *J Clin Invest* 1979; 64:312–320.

Balabaud C, Kron KA, and Gumucio JJ: The assessment of the bile salt nondependent fraction of canalicular bile water in the rat. *J Lab Clin Med* 1977; 89:393–399.

Ballatori N and Truong AT: Relation between biliary glutathione excretion and bile acid-independent bile flow. *Am J Physiol* 1989; 256:G22–G30.

Barnwell SG, Lowe PJ, and Coleman R: The effects of colchicine on secretion into bile of bile salts, phospholipids, cholesterol and plasma membrane enzymes: bile salts are secreted unaccompanied by phospholipids and cholesterol. *Biochem J* 1984; 220:723–731.

Becker BA and Plaa GL: Quantitative and temporal delineation of various parameters of liver dysfunction due to alpha-naphthylisothiocyanate. *Toxicol Appl Pharmacol* 1965; 7:708–718.

Berliner F and Schoenheimer R: Hemolytic and antihemolytic properties of bile acids and sterols in relation to structure. *J Biol Chem* 1938; 124:525–541.

Berr F, Simon FR, and Reichen J: Ethynylestradiol impairs bile salt uptake and Na-K pump function of rat hepatocytes. *Am J Physiol* 1984; 247:G437–G443.

Billington D, Evans CE, Godfrey PP, et al: Effects of bile salts on the plasma membranes of isolated rat hepatocytes. *Biochem J* 1980; 188:321–327.

Blitzer BL and Boyer JL: Cytochemical localization of Na^+,K^+-ATPase in the rat hepatocyte. *J Clin Invest* 1978; 62:1104–1108.

Blitzer B and Boyer J: Cellular mechanisms of bile formation. *Gastroenterology* 1982; 82:346–357.

Boelsterli UA, Rakhit G, and Balazs T: Modulation by *S*-adenosylmethionine of hepatic Na^+,K^+-ATPase, membrane fluidity and bile flow in rats with ethinyl estradiol-induced cholestasis. *Hepatology* 1983; 3:12–17.

Boyce W and Witzleben CL: Bilirubin as a cholestatic agent. II. Effect of variable doses of bilirubin on the severity of manganese-bilirubin cholestasis. *Amer J Pathol* 1973; 72:427–432.

Boyer JL: New concepts of mechanisms of hepatocyte bile formation. *Physiol Rev* 1980; 60:303–326.

Bradley SD and Herz R: Permselectivity of biliary canalicular membrane in rat: clearance probe analysis. *Am J Physiol* 1978; 235:E570–E576.

Brauer RW: Hepatic blood supply and the secretion of bile, in Taylor W (ed): *The Biliary System.* Oxford, Blackwell, 1965, pp 41–67.

Brauer RW, Leong GF, and Holloway RJ: Mechanics of bile secretion: effect of perfusion pressure and temperature on bile flow and secretion pressure. *Am J Physiol* 1954; 177:103–112.

Brock WJ and Vore M: The effect of pregnancy and treatment with estradiol-17β on the transport of organic anions into isolated rat hepatocytes. *Drug Metab Dispos* 1984; 13:695–699.

Brouwer KLR, Durham S, and Vore M: Multiple carriers for uptake of (^{3}H)estradiol-17β(β-D-glucuronide) in isolated rat hepatocytes. *Mol Pharmacol* 1987; 32:519–523.

Carey MC, Hirom PC, and Small DM: A study of the physicochemical interactions between biliary lipids and chlorpromazine hydrochloride. *Biochem J* 1976; 153:519–531.

Chalcroft JP and Bullivant S: An interpretation of liver cell membrane and junction structure based on observation of freeze-fracture replicas of both sides of the fracture. *J Cell Biol* 1970; 47:49–50.

Changchit A, Durham S, and Vore M: Characterization ^{3}H-estradiol-17β(β-D-glucuronide) binding sites in basolateral and canalicular liver plasma membranes. *Biochem Pharmacol* 1990; 40:1219–1225.

Chao YS, Windler EE, Chen GC, and Havel RJ: Hepatic catabolism of rat and human lipoproteins in rats treated with 17α-ethinyl estradiol. *J Biol Chem* 1979; 254:11360–11366.

Cheng S and Levy D: Characterization of the anion transport system in hepatocyte plasma membranes. *J Biol Chem* 1980; 255:2637–2640.

Coleman R: Biochemistry of bile secretion. *Biochem J* 1987; 244:249–261.

Coleman R and Kan KS: Oestradiol 17β-glucuronide and tight junctional permeability increase. *Biochem J* 1990; 266:622.

Combettes L, Dumont M, Berthon B, Erlinger S, and Claret M: Release of calcium from the endoplasmic reticulum by bile acids in rat liver cells. *J Biol Chem* 1988; 263:2299–2303.

Connolly AK, Price SC, Connelly JC, and Hinton RH: Early changes in bile duct lining cells and hepatocytes in rats treated with alpha-naphthylisothiocyanate. *Toxicol Appl Pharmacol* 1988; 93:208–219.

Cooper RA, Garcia FA, and Trey C: The effect of lithocholic acid on red cell membranes in vivo. *J Clin Invest* 1972; 79:7–18.

Cowen AD, Korman MG, Hofmann AF, and Cass OW: Metabolism of lithocholate in healthy man. I. Biotransformation and biliary excretion of intravenously administered lithocholate, lithocholylglycine, and their sulfates. *Gastroenterology* 1975; 69:59–66.

Crawford JM, Berken CA, and Gollan JL: Role of the hepatocyte microtubular system in the excretion of bile salts and biliary lipid: implications for intracellular vesicular transport. *J Lipid Res* 1988; 29:144–156.

Cronholm T and Sjovall J: Bile acids in portal blood of rats fed different diets and cholestyramine. *Eur J Biochem* 1967; 2:375–383.

Dahm LJ and Roth RA: L-Buthionine-*S,R*-sulfoximine (BSO) protects against alpha-naphthylisothiocyanante (ANIT)-induced hepatotoxicity. *FASEB J* 1989; 3:A1000.

Dalen E and Westerholm B: Occurrence of hepatic impairment in women jaundiced by oral contraceptives and in their mothers and sisters. *Acta Med Scand* 1974; 195:459–463.

Danziger RG, Hofmann AF, Thistle JL, and Schoenfield LJ: Effect of oral chenodeoxycholic acid on bile kinetics and biliary lipid composition in women with cholelithiasis. *J Clin Invest* 1973; 52: 2809–2821.

Davis RA, Kern F Jr, and Showalter R: Alterations of hepatic Na^+-K^+-ATPase and bile flow by estrogen: effects on liver surface, membrane lipid structure and function. *Proc Natl Acad Sci USA* 1978a; 75:4130–4134.

Davis RA, Showalter R, and Kern F: Reversal by Triton WR-1339 of ethynyl-oestradiol-induced hepatic cholesterol esterification. *Biochem J* 1978b; 174:45–51.

DiFrancesco C and Bickel MH: Membrane lipids as intracellular binders of chlorpromazine and related drugs. *Chem Biol Interact* 1977; 16:335–346.

Delacroix DL, Hodgson HJF, McPherson A, Dive C, and Vaerman JP: Selective transport of polymeric immunoglobulin A in bile. Quantitative relationships of monomeric and polymeric immunoglobulin A, immunoglobulin M, and other proteins in serum, bile and saliva. *J Clin Invest* 1982; 70:230–241.

De Lamirande E and Plaa GL: Role of manganese, bilirubin and sulfobromophthalein in manganese-bilirubin cholestasis in rats. *Proc Soc Exp Biol Med* 1978; 158:283–287.

De Lamirande E and Plaa GL: Dose and time relationships in manganese-bilirubin induced cholestasis. *Toxicol Appl Pharmacol* 1979; 49:257–263.

De Lamirande E, Tuchweber B, and Plaa GL: Hepatocellular membrane alteration as a possible cause of manganese-bilirubin-induced cholestasis. *Biochem Pharmacol* 1981; 30:2305–2312.

De Lamirande E, Tuchweber B, and Plaa GL: Morphological aspects of manganese-bilirubin induced cholestasis. *Liver* 1982; 2:22–27.

deLorimier AA, Gordan GS, Lowe RC, and Carbone JV: Methyltestosterone, related steroids and liver function. *Arch Intern Med* 1965; 116:289–294.

Despopoulos A: Hepatic and renal excretory metabolism of bile salts: a background for understanding steroid-induced cholestasis. *J Pharmacol Exp Ther* 1971; 176:273–283.

Diamond JM and Tormey JM: Studies on the structural basis of water transport across epithelial membranes. *Fed Proc* 1966; 25:1458–1463.

Diamond JM: Transcellular cross-talk between epithelial cell membranes. *Nature (London)* 1982; 300:683–685.

Drew R and Priestly BG: Choleretic and cholestatic effects of infused bile salts in the rat. *Experientia* 1979; 35:809–811.

Drew R and Priestly BG: Drug metabolism during alpha-naphthylisothiocyanate-induced cholestasis. *Toxicol Appl Pharmacol* 1976; 35:491–499.

Dubin M, Maurice M, Feldmann G, and Erlinger S: Phalloidin-induced cholestasis in the rat: relation to changes in microfilaments. *Gastroenterology* 1978; 75:450–455.

Dubin M, Maurice M, Feldmann G, and Erlinger S: Influence of colchicine and phalloidin on bile secretion and hepatic ultrastructure in the rat: possible interaction between microtubules and microfilaments. *Gastroenterology* 1980; 79:646–654.

Duffy MC, Blitzer BL, and Boyer JL: Direct determination of the driving forces for taurocholate uptake into rat liver plasma membrane vesicles. *J Clin Invest* 1983; 72:1470–1481.

Durham S and Vore M: Taurocholate and steroid glucuronides: mutual protection against cholestasis in the isolated perfused rat liver. *J Pharmacol Exp Ther* 1986; 237:490–495.

El-Hawari A and Plaa GL: alpha-Naphthylisothiocyanate (ANIT) hepatotoxicity and irreversible binding to rat liver microsomes. *Biochem Pharmacol* 1977, 26:1857–1866.

Elias E and Boyer JL: Chlorpromazine and its metabolites alter gelation and polymerization of actin. *Science* 1979; 206:1404–1406.

Elias E, Hruban Z, Wade JB, and Boyer JL: Phalloidin-induced cholestasis: a microfilament-mediated change in junctional complex permeability. *Proc Natl Acad Sci USA* 1980; 77:2229–2233.

Erlinger S: Secretion of bile, in Schiff L and Schiff E (eds): *Diseases of the Liver*, ed 6. JB Lippencott Company, 1987, pp 77–101.

Feldmann G, Maurice M, Sapin C, and Benhamou JP: Inhibition by colchicine of fibrinogen translocation in hepatocytes. *J Cell Biol* 1975; 67:237–243.

Fewster M, Pirrie RE, and Turner DA: Effect of estradiol benzoate on lipid metabolism in the rat. *Endocrinology* 1967; 80:263–271.

Forker EL: Bile formation in guinea-pigs: analysis with inert solutes of graded molecular radius. *Am J Physiol* 1968; 215:56–62.

Forker EL: The effect of estrogen on bile formation in the rat. *J Clin Invest* 1969; 48:654–663.

Forker EL: Hepatocellular uptake of inulin, sucrose and mannitol in rats. *Am J Physiol* 1970; 219:1568–1573.

French SW and Davies PL: Ultrastructural localization of actin-like filaments in rat hepatocytes. *Gastroenterology* 1975; 68:765–774.

Fricker G, Landmann L, and Meier PJ: Ethinylestradiol(EE) induced structural and functional alterations of rat liver plasma membranes and their reversal by *S*-adenosylmethionine (SAMe) in vitro. *Hepatology* 1988; 8:1224, A24.

Fricker G, Landmann L, and Meier PJ: Extrahepatic obstructive cholestasis reverses the bile salt secretory polarity of rat hepatocytes. *J Clin Invest* 1989; 84:876–885.

Friend DS and Gilula NB: Variations in tight and gap junctions in mammalian tissues. *J Cell Biol* 1972; 53:758–776.

Furhoff AK and Hellstrom K: Jaundice in pregnancy: a follow-up study of the series of women originally reported by L. Thorling. I. The pregnancies. *Acta Med Scand* 1973; 193:259–266.

Furhoff AK and Helstrom K: Jaundice in pregnancy: a follow-up study of the series of women originally reported by L. Thorling. II. Present health of the women. *Acta Med Scand* 1974; 196:181–189.

Gabbiani G, Montesano R, Tuchweber B, Salas M, and Orci L: Phalloidin-induced hyperplasia of actin filaments in rat hepatocytes. *Lab Invest* 1975; 53:562–569.

Gallagher TF, Mueller MN, and Kappas A: Estrogen pharmacology. IV. Studies on the structural basis for estrogen-induced impairment of liver function. *Medicine* 1966; 45:471–479.

Gebhardt R: Participation of microtubules and microfilaments in the transcellular biliary secretion of immunoglobulin A in primary cultures of rat hepatocytes. *Experientia* 1984; 40:269–271.

Glasinovic JC, Dumont M, Duval M, and Erlinger S: Hepatocellular uptake of taurocholate in the dog. *J Clin Invest* 1975a; 55:419–426.

Glasinovic JC, Dumont M, Duval M, and Erlinger S: Hepatocellular uptake of bile acids in the dog: Evidence for a common carrier-mediated transport system: an indicator dilution study. *Gastroenterology* 1975b; 69:973–981.

Godfrey PP, Lembra L, and Coleman R: Effects of colchicine and vinblastine on output of proteins into bile. *Biochem J* 1982; 208:153–157.

Goldman IS, Jones AL, Hradek GT, and Huling S: Hepatocyte handling of immunoglobulin A in the rat: the role of microtubules. *Gastroenterology* 1983; 85:130–140.

Goodenough DA and Revel JP: A fine structural analysis of intercellular junctions in the mouse liver. *J Cell Biol* 1970; 45:272–286.

Graf J: Canalicular bile salt-independent bile formation: concepts and clues from electrolyte transport in rat liver. *Am J Physiol* 1983; 244:G233–G246.

Graf J and Peterlik M: Mechanism of transport of inorganic ions into bile, in Taylor W (ed): *The Hepatobiliary System—Fundamental and Pathological Mechanisms*. New York, Plenum, 1975, pp 43–58.

Gregory DH, Vlahcevic ZR, Prugh MF, and Swell L: Mechanism of secretion of biliary lipids: role of a microtubular system in hepatocellular transport of biliary lipids in the rat. *Gastroenterology* 1978; 74:93–100.

Gronwall R and Cornelius CE: Maximal biliary excretion of sulfobromophthalein sodium in sheep. *Am J Dig Dis* 1970; 15:37–48.

Gumucio J and Valdivieso V: Studies on the mechanisms of ethinyl estradiol impairment of bile flow and bile salt excretion in the rat. *Gastroenterology* 1971; 61:339–344.

Hardison WGM, Hatoff DE, Miyai K, and Weiner RG: Nature of bile acid maximum secretory rate in the rat. *Am J Physiol* 1981; 241:G337–G343.

Hardison WGM and Wood GA: Importance of bicarbonate in bile salt independent fraction of bile flow. *Am J Physiol* 1978; 235:E158–E164.

Hatoff DE and Hardison WGM: Bile acids modify alkaline phosphatase induction and bile secretion pressure after bile duct obstruction in the rat. *Gastroenterology* 1980; 80:666–672.

Heikel RAJ and Lathe GH: The effect of 17-ethinyl substituted steroids on adenosine triphosphatases of rat liver plasma membrane. *Biochem J* 1970; 118:187–189.

Hofmann AF: Bile acids, in: Arias IM, Jakoby WR, Popper H, Schachter D, and Shafritz DA (eds): *The Liver: Biology and Pathobiology,* ed 2. New York, Raven Press, 1988.

Holsti P: Experimental cirrhosis of the liver in rabbits induced by gastric instillation of desiccated whole bile. *Acta Path Microbiol Scand* 1956; 112 (suppl):1–67.

Holsti P: Cirrhosis of the liver induced in rabbits by gastric instillation of 3-monohydroxycholanic acid. *Nature* 1960; 186:250.

Hruban Z, Tavoloni N, Reed JS, and Boyer JL: Ultrastructural changes during cholestasis induced by chlorpromazine in the isolated perfused rat liver. *Virchows Arch Cell Pathol* 1978; 26:289–305.

Hugentobler G and Meier PJ: Multispecific anion exchange in basolateral (sinusoidal) rat liver plasma membrane vesicles. *Am J Physiol* 1986; 251:G656–664.

Indacochea-Redmond N and Plaa GL: Functional aspects of alpha-naphthylisothiocyanate in various species. *Toxicol Appl Pharmacol* 1971; 19:71–80.

Inoue M, Kinne R, Tran T, and Arias IM: Taurocholate transport by rat liver sinusoidal membrane vesicles: evidence of sodium cotransport. *Hepatology* 1982; 2:572–579.

Inoue M, Kinne R, Tran T, and Arias IM: Taurocholate transport by rat liver canalicular membrane vesicles. Evidence for the presence of an Na^+-independent transport system. *J Clin Invest* 1984; 73:659–663.

Isak KG and Zimmerman JH: Hepatotoxic effects of the anabolic/androgenic steroids. *Sem Liv Dis* 1987; 7:230–236.

Ivy AC: The applied physiology of bile secretion and bile salt therapy. *JAMA* 1941; 117:1151–1154.

Jaeschke H: The pathophysiological significance of increased tight-junctional permeability during oestrogen cholestasis. *Biochem J* 1990; 266:620–621.

Jaeschke H, Krell H, and Pfaff E: No increase of biliary permeability in ethinylestradiol-treated rats. *Gastroenterology* 1983; 85:808–814.

Jaeschke H, Krell H, and Pfaff E: Quantitative estimation of transcellular pathways of biliary sucrose in isolated perfused rat liver. *Biochem J* 1987; 241:635–640.

Jaeschke H, Trummer E, and Krell H: Increase in biliary permeability subsequent to intrahepatic cholestasis by estradiol valerate in rats. *Gastroenterology* 1987; 93:533–538.

Jansen PLM, Groothuis GMM, Peters WHM, and Meijer DFM: Selective hepatobiliary transport defect for organic anions and neutral steroids in mutant rats with hereditary-conjugated hyperbilirubinemia. *Hepatology* 1987; 7:71–76.

Javitt NB: Cholestasis in rats induced by taurolithocholate. *Nature* 1966; 210:1262–1263.

Javitt NB and Emerman S: Effect of sodium taurolithocholate on bile flow and bile acid excretion. *J Clin Invest* 1968; 47:1002–1214.

Kacich RL, Renston RH, and Jones AL: Effects of cytochalasin D and colchicine on the uptake, translocation and biliary secretion of horseradish peroxidase and (^{14}C)-taurocholate in the rat. *Gastroenterology* 1983; 85:385–394.

Kakis G and Yousef IM: Pathogenesis of lithocholate- and taurolithocholate-induced intrahepatic cholestasis in rats. *Gastroenterology* 1978; 75:595–607.

Kan SK, Monte MJ, Parslow RA, and Coleman R: Oestradiol 17β-glucuronide increases tight-junctional permeability in rat liver. *Biochem J* 1989; 261:297–300.

Kaplan MM: Laboratory tests, in: Schiff L and Schiff ER (eds): *Diseases of the Liver,* ed 6. JB Lippencott, 1987, pp 239–246.

Kaplan MM and Righetti A: Induction of rat liver alkaline phosphatase: the mechanism of the serum elevation in bile duct obstruction. *J Clin Invest* 1970; 49:508–516.

Keeffe EB, Scharschmidt BJ, Blankenship NM, and Ockner RK: Studies on relationships among bile flow, liver plasma membrane NaK-ATPase and membrane microviscoscity in the rat. *J Clin Invest* 1979; 64:1590–1598.

Keeffe EB, Blankenship N, and Scharschmidt BF: Alteration of rat liver plasma membrane fluidity and ATPase activity by chlorpromazine hydrochloride and its metabolites. *Gastroenterology* 1980; 74:222–231.

Klaassen CD and Watkins JB III: Mechanisms of bile formation, hepatic uptake, and biliary excretion. *Pharmacol Rev* 1984; 36:1–67.

Kovanen PT, Brown MS, and Goldstein JL: Increased binding of LDL to liver membranes from rats treated with 17-ethinylestradiol. *J Biol Chem* 1979; 254:11367–11373.

Kovanen RP, Bilheimer DW, Goldstein JL, Jaramillo JJ, and Brown MS: Regulatory role for hepatic low density lipoprotein receptors in vivo in the dog. *Proc Natl Acad Sci USA* 1981; 78:1194–8.

Kramer W, Bickel U, Buscher HP, Gerok W, and Kurz G: Bile-salt-binding polypeptides in plasma membranes of hepatocytes revealed by photoaffinity labelling. *Eur J Biochem* 1982; 129:13–24.

Kreek MJ: Female sex steroids and cholestasis. *Semin Liver Dis* 1987; 7:8–23.

Kreek MJ, Peterson RE, Sleisenger MH, and Jeffries GH: Effects of ethinylestradiol-induced cholestasis on bile flow and biliary excretion of estradiol and estradiol glucuronide by the rat. *Proc Soc Exp Biol Med* 1969; 131:646–650.

Kreek MJ, Weser E, Sleisenger MH, and Jeffries GH: Idiopathic cholestasis of pregnancy. The response to challenge with the synthetic estrogen, ethinyl estradiol. *N Engl J Med* 1967; 277:1391–1395.

Krell H, Höke H, and Pfaff E: Development of intrahepatic cholestasis by alpha-naphthylisothiocyanate in rats. *Gastroenterology* 1982; 82:507–514.

Lake JR, Licko V, Van Dyke RW, and Scharschmidt BF: Biliary secretion of fluid-phase markers by the isolated perfused rat liver. Role of transcellular vesicular transport. *J Clin Invest* 1985; 76:676–684.

Layden TJ and Boyer JL: Scanning electron microscopy of the rat liver. Studies of the effect of taurolithocholate and other models of cholestasis. *Gastroenterology* 1975; 69:724–738.

Layden TJ, Elias E, and Boyer JL: Bile formation in the rat. The role of the paracellular shunt pathway. *J Clin Invest* 1978; 62:1375–1385.

Leffert HL, Schenk DB, Hubert JJ, Skelly J, Schumacher M, Ariyasu R, Ellisman M, Koch KS, and Keller GA: Hepatic (Na^+,K^+)-ATPase: a current view of its structure, function and localization in rat liver as revealed by studies with monoclonal antibodies. *Hepatology* 1985; 5:501–507.

Lengsfeld AM, Low I, Wieland T, Dancker P, and Hasselbach W: Interaction of phalloidin with actin. *Proc Natl Acad Sci USA* 1974; 71:2803–2807.

Lennon HD: Relative effects of a 17α-alkylated anabolic steroids on sulfobromophthalein (BSP) retention in rabbits. *J Pharmacol Exp Ther* 1966a; 151:143–150.

Lennon HD: Effects of various 17α-alkylsubstitutions and structural modifications of steroids on sulfobromophthalein (BSP) retention in rabbits. *Steroids* 1966b; 7:157–170.

Lock S, Lavigne J, and Plaa GL: The effect of alpha-naphthylisothiocyanate on bile secretion prior to and during the onset of cholestasis in the rat. *Toxicol Lett* 1982; 10:427–435.

Lorenzini P, Ilter P, Meier P, and Boyer JL: Taurochenodeoxycholic acid stimulates hepatic uptake of 3H-methoxyinulin (3MI) into membrane-bound compartments. *Hepatology (Baltimore)* 2:737a (abstract).

Lowe PJ, Kan KS, Barnwell ST, Sharma RK, and Coleman R: Transcytosis and paracellular movements of horseradish peroxidase across liver parenchymal tissue from blood to bile. *Biochem J* 1985; 229:529–537.

Lowe PJ, Miyal K, Steinbach JH, and Hardison WGM: Hormonal regulation of hepatocyte tight junctional permeability. *Am J Physiol* 1988; 255:G454–G461.

Lunzer M, Barnes P, Byth K, and O'Halloran M: Serum bile acid concentrations during pregnancy and their relationship to obstetric cholestasis. *Gastroenterology* 1986; 91:825–829.

McLean-Fletcher S and Pollard RD: Mechanisms of action of cytochalasin B on actin. *Cell* 1980; 20:329–341.

Meier PJ: Transport polarity of hepatocytes. *Sem Liver Dis* 1988a; 8:293–307.

Meier PJ: Liver plasma membrane transport systems and bile formation, in Davidson JS and Shaffer EA (eds): *Gastrointestinal and Hepatic Secretions: Mechanism and Control.* Calgary, The University of Calgary Press, 1988b, pp 1–7.

Meier PJ, St Meier-Abt A, Barrett C, and Boyer JL: Mechanisms of taurocholate transport in canalicular and basolateral rat liver plasma membrane vesicles. Evidence for an electrogenic canalicular organic anion carrier. *J Biol Chem* 1984; 259:10614–10622.

Meier PJ, St Meier-Abt A, and Boyer JL: Properties of the canalicular bile acid transport system in rat liver. *Biochem J* 1987; 242:465–469.

Meier PJ, Valantinas J, Hugentobler G, and Rahm I: Bicarbonate sulfate exchange in canalicular rat liver plasma membrane vesicles. *Am J Physiol* 1987; 253:G461–G468.

Meier PJ, Knickelbein R, Moseley RH, Dobbins JW, and Boyer JL: Evidence for carrier-mediated chloride/bicarbonate exchange in canalicular rat liver plasma membrane vesicles. *J Clin Invest* 1985; 75:1256–1263.

Mennicke WH, Krumbiegel G, and Görler K: Metabolism of isothiocyanates in rats. *Naunyn-Schmiedeberg's Arch Pharmakol Exp Pathol* 1978; 302:R13.

Metz J and Bressler D: Reformation of gap and tight junctions in regenerating liver after cholestasis. *Cell Tissue Res* 1979; 199:257–270.

Meyers M, Slikker W, and Vore M: Steroid D-ring glucuronides: characterization of a new class of cholestatic agents in the rat. *J Pharmacol Exp Ther* 1981; 218:63–73.

Meyers M, Slikker W, Pascoe G, and Vore M: Characterization of cholestasis induced by estradiol-17β-D-glucuronide in the rat. *J Pharmacol Exp Ther* 1980; 214:87–93.

Miyai K, Mayr WW, and Richardson AL: Acute cholestasis induced by lithocholic acid in the rat. A freeze-fracture replica and thin section study. *Lab Invest* 1975; 32:527–535.

Miyai K, Richardson AL, Mayr WW, and Javitt NB: Subcellular pathology of rat liver in cholestasis and choleresis induced by bile salts. I. Effects of lithocholic, 3β-hydroxy-5-cholenoic, cholic and dehydrocholic acids. *Lab Invest* 1977; 36:249–258.

Mueller MN and Kappas A: Estrogen pharmacology. I. The influence of estradiol and estriol on hepatic disposal of sulfobromophthalein (BSP) in man. *J Clin Invest* 1964; 43:1905–1914.

Mullock BM, Jones RS, Peppard P, and Hinton RH: Effects of colchicine on the transfer of IgA across hepatocytes into bile in isolated perfused rat livers. *FEBS Lett* 1980; 120:278–282.

Nemchausky BA, Layden TJ, and Boyer JL: Effects of chronic choleretic infusions of bile acids on the membrane of the bile canaliculus—a biochemical and morphologic study. *Lab Invest* 1977; 36:259–267.

O'Maille ERL, Richards TG, and Short AH: Factors determining the maximal rate of organic anion secretion by the liver and further evidence on the hepatic site of action of the hormone secretin. *J Physiol* 1966; 186:424.

Oda M and Phillips MJ: Bile canalicular membrane pathology in cytochalasin B induced cholestasis. *Lab Invest* 1977; 37:350–356.

Oelberg DG, Chari MV, Little JM, Adcock EW, and Lester R: Lithocholate glucuronide is a cholestatic agent. *J Clin Invest* 1984a; 73:1507–1514.

Oelberg DG, Dubinsky WP, Adcock EW, and Lester R: Calcium binding by lithocholic acid derivatives. *Am J Physiol* 1984b; 247:G112–G115.

Oelberg DG and Lester R: Cellular mechanisms of cholestasis. *Ann Rev Med* 1986; 37:297–317.

Olivecrona R and Sjovall J: Bile acids in rat portal blood. *Acta Physiol Scand* 1959; 46:284--290.

Oschman JL and Berridge MJ: The structural basis of fluid secretion. *Fed Proc* 1971; 30:49–56.

Oshio C and Phillips MJ: Contractility of bile canaliculi: implications for liver function. *Science* 1981; 212:1041–1042.

Palmer RH: Bile acids, liver injury, and liver disease. *Arch Intern Med* 1972; 130:606–617.

Palmer RH: Bile acid heterogeneity and the gastrointestinal epithelium: from diarrhea to colon cancer. *J Lab Clin Med* 1979; 94:655–660.

Palmer RH and Bolt MG: Bile acid sulfates. I. Synthesis of lithocholic acid sulfates and their identification in human bile. *J Lipids Res* 1971; 12:671–679.

Paumgartner G and Reichen J: Different pathways for hepatic uptake of taurocholate and indocyanine green. *Experientia* 1975; 31:306–307.

Petzinger E and Frimmer M: Comparative studies on the uptake of ^{14}C-bile acids and ^{3}H-demethylphalloin in isolated rat liver cells. *Arch Toxicol* 1980; 44:127–135.

Phillips MJ, Latham PS, and Poucell S: Electron microscopy of human liver diseases, in Schiff L and Schiff ER (eds): *Diseases of the Liver,* ed 6. JB Lippencott, 1987, pp 47–76.

Phillips MJ and Oda M: The bile canalicular web. *Fed Proc* 1974; 33:626 (abstract).

Phillips MJ, Oda M, and Funatsu K: Evidence for microfilament involvement in norethandrolone-induced intrahepatic cholestasis. *Am J Pathol* 1978; 93:729–744.

Phillips MJ, Oda M, Mak E, Fisher MM, and Jeejeebhoy KN: Microfilament dysfunction as a possible cause of intrahepatic cholestasis. *Gastroenterology* 1975; 69:48–58.

Phillips MJ, Oshio C, Miyairi M, and Smith CR: Intrahepatic cholestasis as a canalicular motility disorder: evidence using cytochalasin. *Lab Invest* 1983; 48:205–211.

Phillips MJ, Poucell S, and Oda M: Biology of disease. Mechanisms of cholestasis. *Lab Invest* 1986; 54:593–608.

Plaa GL, De Lamirande E, Lewittes M, and Yousef IM: Liver cell plasma membrane lipids in manganese-bilirubin-induced intrahepatic cholestasis. *Biochem Pharmacol* 1982; 31:3698–3701.

Plaa GL and Priestly BG: Intrahepatic cholestasis induced by drugs and chemicals. *Pharmacological Reviews* 1976; 28:207–273.

Popper H and Szanto PB: Intrahepatic cholestasis ''cholangiolitis''. *Gastroenterology* 1956; 31:683–700.

Redman CM, Banerjee D, Howell K, and Palade GE: Colchicine inhibition of plasma protein release from rat hepatocytes. *J Cell Biol* 1975; 66:42–59.

Reichen J and Berk PD: Isolation of an organic anion binding protein from rat liver plasma membrane fractions by affinity chromatography. *Biochem Biophys Res Commun* 1979; 91:484–489.

Reichen J, Berman MD, and Berk PD: The role of microfilaments and microtubules in taurocholate uptake by isolated rat liver cells. *Biochim Biophys Acta* 1981; 643:126–133.

Reichen J and Le M: Taurocholate, but not taurodehydrocholate, increases biliary permeability to sucrose. *Am J Physiol* 1983; 245:G651–G655.

Reichen J and Paumgartner G: Kinetics of taurocholate uptake by the perfused rat liver. *Gastroenterology* 1975; 68:132–136.

Reichen J and Paumgartner G: Uptake of bile acids by the perfused rat liver. *Am J Physiol* 1976; 231:734–742.

Reichen J and Paumgartner G: Relationship between bile flow and NaK adenosine triphosphatase in liver plasma membranes enriched in the bile canaliculi. *J Clin Invest* 1977; 60:429–434.

Reyes H, Gonzales MC, Ribalta J, Medina E, Aburto H, Matus C, Schramm G, and Katz R: Prevalence of intrahepatic cholestasis of pregnancy in Chile. *Ann Intern Med* 1978; 88:487–93.

Reyes H, Ribalta J, Gonzales MC, Segovia N, and Oberhauser E: Sulfobromophthalein clearance tests before and after ethinylestradiol administration, in women and men with familial history of intrahepatic cholestasis of pregnancy. *Gastroenterology* 1981; 81:226–232.

Reyes H, Taboada G,and Ribalta J: Prevalence of intrahepatic cholestasis of pregnancy in La Paz, Bolivia. *J Chronic Dis* 1979; 32:499–504.

Richards TG: The secretion of bile against pressure. *Gastroenterology* 1961; 40:705–707.

Roberts RJ and Plaa GL: Potentiation and inhibition of alpha-naphthylisothiocyanate-induced hyperbilirubinemia and cholestasis. *J Pharmacol Exp Ther* 1965; 150:499–506.

Roberts RJ and Plaa GL: The effect of bile duct ligation, bile duct cannulation and hypothermia on alpha-naphthylisothiocyanate induced hyperbilirubinemia and cholestasis in rats. *Gastroenterology* 1966; 50:768–774.

Rosario J, Sutherland E, Zaccaro L, and Simon FR: Ethinylestradiol administration selectively alters liver sinusoidal membrane lipid fluidity and protein composition. *Biochemistry* 1988; 27:3939–3946.

Ruetz S, Fricker G, Hugentobler G, Winterhalter K, Kurz G, and Meier PJ: Isolation and characterization of the putative canalicular bile salt transport system of rat liver. *J Biol Chem* 1987; 262:11324–11330.

Ruetz S, Hugentobler G, and Meier PJ: Functional reconstitution of the canalicular bile salt transport system of rat liver. *Proc Natl Acad Sci USA* 1988; 85:6147–6151.

Samuels AM and Carey MC: Effects of chlorpromazine hydrochloride and its metabolites on Mg^{++}- and Na^{+},K^{+}-ATPase activities of canalicular enriched rat liver plasma membranes. *Gastroenterology* 1978; 74:1183–1190.

Schaffner F and Javitt NB: Morphologic changes in hamster liver during intrahepatic cholestasis induced by taurolithocholate. *Lab Invest* 1966; 15:1783–1792.

Scharschmidt BF and Stephens JE: Transport of sodium, chloride and taurocholate by cultured rat hepatocytes. *Proc Natl Acad Sci USA* 1981; 78:986.

Schiff M: Gallenbildung, abhangig non der Augsaugung der Gallenstoffe. *Pfleugers Arch Ges Physiol* 1890; 3:598–613.

Schölmerich J, Becher M-S, Schmidt K, Schubert R, Kremer B, Feldhaus S, and Gerok W: Influence of hydroxylation and conjugation of bile salts on their membrane-damaging properties—studies on isolated hepatocytes and lipid membrane vesicles. *Hepatology* 1984; 4:661–666.

Schultz SG: The role of paracellular pathways in isotonic fluid transport. *Yale J Biol Med* 1977; 50:99–113.

Schwarz LR, Burr R, Schwenk M, Pfaff E, and Greim H: Uptake of taurocholic acid into isolated rat liver cells. *Eur J Biochem* 1975; 55:617–623.

Sellinger M, Barrett C, Malle P, Gordon ER, and Boyer JL: Cryptic Na^{+},K^{+}-ATPase activity in rat liver canalicular plasma membranes: evidence for its basolateral origin. *Hepatology* 1990; 11:223–229.

Simon RF, Gonzalez M, Sutherland E, Accatino L, and Davis RA: Reversal of ethinyl estradiol-induced bile secretory failure with Triton WR-1339. *J Clin Invest* 1980; 65:851–859.

Slikker W, Vore M, Bailey JR, Meyers M, and Montgomery C: Hepatotoxic effects of estradiol-17β-D-glucuronide in the rat and monkey. *J Pharmacol Exp Ther* 1983; 225:138–43.

Small DM: The physical chemistry of cholanic acids, in Nair PP and Kritchevsky D (eds): *The Bile Acids, Chemistry, Physiology and Metabolism.* New York, Plenum, 1971, pp 249–356.

Smith DJ and Gordon ER: Role of liver plasma membrane fluidity in the pathogenesis of estrogen-induced cholestasis. *J Lab Clin Med* 1988; 112:679–685.

Sperber I: Biliary secretion of organic anions and its influence on bile flow, in Taylor W (ed): *The Biliary System.* Oxford, Blackwell, 1965, pp 457–467.

Sperber I: Secretion of organic anions in the formation of urine and bile. *Pharmacol Rev* 1959; 11:109–134.

Stiehl A, Becker M, Czygan P, Frohling W, Kommerell B, Rothauwe HW, and Senn M: Bile acids and their sulphated and glucuronidated derivatives in bile, plasma, and urine of children with intrahepatic cholestasis: effects of phenobarbital treatment. *Eur J Clin Invest* 1980; 10:307–316.

Stramentinoli G, DiPadova C, Gualano M, Rabagnati P, and Galli-Kienle M: Ethynylestradiol-induced impairment of bile secretion in the rat: protective effects of *S*-adenosyl-*l*-methionine and its implication in estrogen metabolism. *Gastroenterology* 1981; 80:154–158.

Stremmel W and Berk PD: Hepatocellular uptake of sulfobromophthalein and bilirubin is selectively inhibited by an antibody to the liver plasma membrane sulfobromophthalein/bilirubin binding protein. *J Clin Invest* 1986; 78:822–826.

Stremmel W, Gerber MA, Glezerov V, Thung SN, Kochwa S, and Berk PD: Physiochemical and immunohistological studies of a sulfobromophthalein- and bilirubin-binding protein from rat liver plasma membrane. *J Clin Invest* 1983; 71:1796–1805.

Sutherland E, Dixon BS, Leffert HL, Skally J, Zaccaro L, and Simon FR: Biochemical localization of hepatic surface-membrane Na^+,K^+-ATPase activity depends on membrane lipid fluidity. *Proc Natl Acad Sci USA* 1988; 85:8673–8677.

Svanborg A: A study of recurrent jaundice in pregnancy. *Acta Gynecol* 1954; 33:434–444.

Svanborg A and Ohlsson S: Recurrent jaundice of pregnancy. *Am J Med* 1959; 27:40–49.

Sztul ES, Biemesderfer D, Caplan MJ, Kashgarian M, and Boyer JL: Localization of Na^+-,K^+-ATPase alpha-subunit to the sinusoidal and lateral but not canalicular membranes of rat hepatocytes. *J Cell Biol* 1987; 104:1239–1248.

Takikawa J, Stolz A, and Kaplowitz N: Cyclical oxidation-reduction of the C_3 position on bile acids catalyzed by rat hepatic 3α-hydroxysteroid dehydrogenase. I. Studies with the purified enzyme, isolated rat hepatocytes, and inhibition by indomethacin. *J Clin Invest* 1987a; 80:852–860.

Takikawa H, Ookhtens M, Stolz A, and Kaplowitz N: Cyclical oxidation-reduction of the C_3 position on bile acids catalyzed by 3α-hydroxysteroid dehydrogenase. II. Studies in the prograde and retrograde single-pass, perfused rat liver and inhibition by indomethacin. *J. Clin Invest* 1987b; 80:861–866.

Tavoloni N and Boyer JL: Relationship between hepatic metabolism of chlorpromazine and cholestatic effects in the isolated perfused rat liver. *J Pharmacol Exp Ther* 1980; 214:269–274.

Thureborn E: Human hepatic bile. *Acta Chir Scand Suppl* 1962; 303:1–63.

Traiger GJ, Vyas KP, and Hanzlik RP: The effect of thiocarbonyl compounds on alpha-naphthylisothiocyanate induced hepatotoxicity and the urinary excretion of (^{35}S)-alpha-naphthylisothiocyanate in the rat. *Toxicol Appl Pharmacol* 1984; 72:504–512.

Traiger GJ, Vyas KP, and Hanzlik RP: Effects of inhibitors of alpha-naphthylisothiocyanate induced hepatotoxicity on the in vivo metabolism of alpha-naphthylisothiocyanate. *Chem Biol Interact* 1985; 52:335–345.

Tuchweber B and Gabbiani G: Phalloidin-induced hyperplasia of actin microfilaments in rat hepatocytes, in Preisig R, Bircher J, and Paumgartner G (eds): *The Liver: Quantitative Aspects of Structure and Function*. Basel, S Karger, 1976, pp 84–90.

Tuchweber B, Weber A, Roy CC, and Yousef IM: Mechanisms of experimentally induced intrahepatic cholestasis. *Prog Liv Dis* 1986; 8:161–178.

Van Dyke RW and Scharschmidt BF: Effects of chlorpromazine on Na^+,K^+-ATPase pumping and solute transport in rat hepatocytes. *Am J Physiol* 1987; 253:G613–G621.

Van Dyke RW, Stephens JE, and Scharschmidt BF: Effect of ion substitution on bile acid-dependent and bile acid-independent bile formation by the isolated perfused rat liver. *J Clin Invest* 1982; 70:505–517.

von Dippe P and Levy D: Characterization of the bile acid transport system in normal and transformed hepatocytes. *J Biol Chem* 1983; 258:8896–8901.

Vore M: Estrogen cholestasis. Membranes, metabolites or receptors? *Gastroenterology* 1987; 93:643–649.

Vore M, Hadd H, and Slikker W: Ethynylestradiol-17β-D-ring glucuronide conjugates are potent cholestatic agents in the rat. *Life Sci* 1983; 32:2989–2993.

Vore M, Montgomery C, Durham S, Schlarman D, and Elliott WH: Structure-activity relationship of the cholestatic activity of dihydrotestosterone glucuronide, allo bile acids and lithocholate. *Life Sci* 1989; 44:2033–2040.

Vore M and Slikker W: Steroid D-ring glucuronides: a new class of cholestatic agents. *Trends Pharmacol Sci* 1985; 6:256–9.

Vonk RJ, Tuchweber B, Massé D, Perea A, Audet M, Roy CC, and Yousef IM: Intrahepatic cholestasis induced by allo monohydroxy bile acid in rats. *Gastroenterology* 1981; 81:242–249.

Vyvoda OS, Coleman R, and Holdsworth G: Effects of different bile salts upon the composition and morphology of a liver plasma membrane preparation. *Biochim Biophys Acta* 1977; 465:68–76.

Watanabe S, Miyairi M, Oshio C, Smith CR, and Phillips MJ: Phalloidin alters bile canalicular contractility in primary monolayer cultures of rat liver. *Gastroenterology* 1983; 85:245–253.

Werner SC: Clinical syndromes associated with gonadal failure in man. *Am J Med* 1947; 23:52–66.

Wheeler HO and Ramos O: Determinants of the flow and composition of bile in the unanesthetized dog during constant infusions of sodium taurocholate. *J Clin Invest* 1960; 39:161–170.

Wheeler HO, Ross ED, and Bradley SE: Canalicular bile production in dogs. *Am J Physiol* 1968; 214:866–874.

Wieland T, Nassal M, Kramer W, Fricker G, Bickel U, and Kurz G: Identity of hepatic membrane transport systems for bile salts, phalloidin, and antamanide by photoaffinity labeling. *Proc Natl Acad Sci USA* 1984; 81:5232–5236.

Wilson FA: Intestinal transport of bile acids. *Am J Physiol* 1981; 241:G83–G92.

Wilson FA, Burckhardt G, Murer H, Rumrich G, and Ullrich KJ: Sodium-coupled taurocholate transport in the proximal convolution of the rat kidney in vivo and in vitro. *J Clin Invest* 1981; 67:1141–1150.

Witzleben CL: Manganese-induced cholestasis: concurrent observations on bile flow rate and hepatic ultrastructure. *Am J Pathol* 1969; 57:617–625.

Witzleben CL: Bilirubin as a cholestatic agent: physiologic and morphologic observations. *Am J Pathol* 1971; 62:181–194.

Witzleben CL, Pitlick P, Bergmeyer J, and Benoit R: Acute manganese overload. A new experimental model of intrahepatic cholestasis. *Am J Pathol* 1968; 53:409–423.

Wolkoff AW and Chung CT: Identification, purification and partial characterization of an organic anion binding protein from rat liver plasma membrane. *J Clin Invest* 1980; 65:1152–1161.

Ziegler K, Frimmer M, and Fasold H: Further characterization of membrane proteins involved in the transport of organic anions in hepatocytes. *Biochim Biophys Acta* 1984; 769:117–129.

Zimmerli B, Valantinas J, and Meier PJ: Multispecificity of Na^+-dependent taurocholate uptake in basolateral (sinusoidal) rat liver plasma membrane vesicles. *J Pharmacol Exp Ther* 1989; 250:301–308.

Zimmerman HJ and Lewis JH: Drug-induced cholestasis. *Med Toxicol* 1987; 2:112–160.

Chapter
13

Hepatotoxic and Hepatocarcinogenic Effects of Chlorinated Ethylenes*

Jeffrey L. Larson[†] *and Richard J. Bull*
College of Pharmacy
Washington State University
Pullman, Washington

INTRODUCTION

The chlorinated ethylenes enjoy a wide variety of uses in industry. Vinyl chloride (VC) was used sporadically in industry from the 1930s to the 1950s, but is now widely used since in the manufacture of its polymer, polyvinyl chloride (PVC), the most used plastic industrial material in the world. VC is also used as a copolymer in the manufacture of food wrappings and plastics (Maltoni and Lefemine, 1974). Vinylidene chloride (VDC) is used extensively as a monomeric intermediate in the production of plastics as well (McKenna et al., 1977). The trans-isomer of dichloroethylene (*t*-DCE) is a universal solvent more widely used in industrial operations than the cis-isomer (*c*-DCE). Trichloroethylene (TCE) was once used extensively as an anesthetic and food extractant. TCE is now widely favored as an industrial solvent because of its nonflammability (Stewart et al., 1974). Perchloroethylene (PCE) is also used

*This Project is funded by Federal Funds (Air Force Grant No. AFSOR-86-0284 and by NIEHS Grant No. RO1 ESO4648-01A1).
[†]Current Address: Chemical Industry Institute of Toxicology, Research Triangle Park, NC 27709.

widely in industry as a general degreasant and as a solvent in the dry-cleaning industry. As a consequence of a lower vapor pressure, PCE is often favored for industrial applications over TCE.

The major human exposure to these chemicals are industrial. However, the widespread use of large quantities of these chemicals results in some exposure to the entire population. In the case of VC and VDC there is concern that unreacted monomers could be released from the plastics used in food wrappings and result in a relatively small carcinogenic risk in a very large population that would be difficult to detect (Haley, 1975). This potential risk is doubted by others who argue that the volatility of the compounds is so great that little or no contamination of food is expected (CEH, 1983). TCE and PCE are ubiquitous in air, with levels in the parts per trillion to parts per billion range. They are also found as common contaminants in ground and surface waters (EPA, 1984, 1987a). VCD, *t*-DCE, and *c*-DCE are relatively rare as contaminants of ground and surface waters (EPA, 1987b, 1987c, 1987d).

With the widespread exposure of humans to trace amounts of these chemicals in the environment, there has been considerable interest in their toxicological and, especially, their carcinogenic properties. Despite the fact that these compounds vary in structure by only the number and position of chlorine substitutions, it is becoming quite evident that these small differences in chemical structure substantially affect their toxicology, especially in the liver. The intent of this chapter is to place these differences in perspective and to point out the need to assess the risks that might be associated with each of the compounds independently. The key consideration is that the metabolites of these chemicals play differing roles in their hepatotoxic and hepatocarcinogenic effects.

For our purposes we will use these abbreviations: (VC) = 1-chloroethylene or vinyl chloride; (VDC) = 1,1-dichloroethylene or vinylidene chloride; (*t*-DCE) = *trans*-1,2-dichloroethylene; (*c*-DCE) = *cis*-1,2-dichloroethylene; (TCE) = 1,1,2-trichloroethylene; and (PCE) = 1,1,2,2-tetrachloroethylene or perchloroethylene.

HEPATOTOXICITY

Description of the Injury

The responses observed in the liver following acute poisoning with the individual ethylenes is given in Table 13–1. With the exception of VDC, the chlorinated ethylenes are similar in that most are relatively nontoxic to hepatocytes. The pattern of hepatocellular injury is a gradually evolving process such that death from acute exposures is more likely due to central nervous system (CNS) depression and respiratory failure than to total hepatic failure.

TABLE 13–1.
Description of Liver Damage Caused by Acute Chlorinated Ethylene Exposure

Damage to Liver	VC	VDC	1,2-DCE	TCE	PCE
Fatty infiltration	+	−	+	+	+
Necrosis	−	+	−	−	−
Endoplasmic reticulum	+	−	+	+	+
Chromatin clumping	−	+	−	−	−
Mitchondria damage	−	+	−	−	−
Plasma membrane	−	+	−	−	−

VDC is unique in that the most sensitive end point of short-term toxicity is liver damage (Chieco et al., 1982.) Table 13–1 shows that the pattern of liver damage is much different than that observed with the other compounds.

Acute exposures to the chlorinated ethylenes, other than VDC, produce characteristic changes in the endoplasmic reticulum and fat content of the hepatocytes. There is increased cytoplasmic disorder with random dispersion of organelles and degranulation and vacuolization of rough endoplasmic reticulum. The smooth endoplasmic reticulum is dilated and proliferation is observed. By 8 h after a 2-h exposure of rats to 1% TCE, the smooth endoplasmic reticulum contains electron-opaque regions, suggestive of membrane collapse (Reynolds and Moslen, 1977b).

The ability of the compounds to cause damage to the endoplasmic reticulum of hepatocytes can also be assessed by looking at detrimental effects on microsomal enzyme systems. TCE and VC both degrade the heme of microsomal P450, while PCE and VDC do not (Reynolds and Moslen, 1977b; Costa and Ivanetich, 1980). This degree of endoplasmic reticulum damage correlates well with the relative rates and extents of metabolism of the compounds (Costa and Ivanetich, 1980). The ability of the chemicals to cause endoplasmic reticulum damage is in the order TCE $>$ VC $>$ PCE $>$ VDC (Reynolds and Moslen, 1977a).

The degree and position of chlorine substitution also is important in affecting sensitivity of the animals to injury, although the pattern of injury remains the same. Hepatic structural derangements are observed in the rat following a 2-h exposure to 1% TCE, in contrast to a 6-h exposure to 5% VC vapor (Reynolds and Moslen, 1977b; Reynolds, 1977). Jenkins et al. (1972) compared the hepatotoxicities of *t*-DCE, *c*-DCE, and VDC following oral administration of the compunds to rats. VDC was much more potent in increasing the activities of serum alkaline phosphatase than either of the 1,2-isomers. The cis-isomer was slightly more hepatotoxic than the trans-isomer. This is expected in light of the fact that the rate of *c*-DCE metabolism is faster than that of *t*-DCE (Costa, 1983).

The ability of the compounds to produce fatty degeneration of the liver seems to be distinct from the ability to cause structural damage to the endoplasmic reticulum. Kylin et al. (1963) found that PCE was 10 times more effective in causing fatty degeneration in the livers of mice than was TCE. Acute inhalation exposures of 200 ppm PCE resulted in moderate fatty degeneration while concentrations of TCE as high as 3200 ppm produced only slight fatty degeneration. This difference in sensitivity was also observed following 6-week exposures (Kylin et al., 1965). Oral administration of the compounds shows the same differences in sensitivity. Buben and O'Flaherty (1985) administered either PCE or TCE by gavage to mice daily for 6 weeks. Liver triglycerides were markedly increased in mice administered as little as 100 mg/kg/day, while TCE-induced fatty changes were minimal even at doses of 3200 mg/kg/day. This study also reported that PCE was also a more effective hepatotoxin than TCE. Elevation of the serum enzyme glutamate-pyruvate transaminase was observed in mice given the 500 mg/kg PCE per day, in contrast to a dose of 2400 mg/kg/day of TCE to elicit a similar hepatotoxic response.

VDC, the most unsymmetrical of the chlorinated ethylenes, does not fit the pattern expected for chlorinated ethylene-induced liver damage. VDC is much more potent and faster acting than the other members of the group. The characteristics of the hepatotoxic response also differ, suggesting a different mechanism of action. Exposures of 200 ppm (0.02%) VDC vapor for 4 h resulted in massive hepatocyte death in fasted rats (Reynolds et al., 1980). The liver damage can be observed as prominent midzonal "stripes" of necrosis (Reynolds et al., 1980). Necrosis is not observed following acute exposures to the other chlorinated ethylenes.

The instability of the proposed epoxide intermediate of VDC (Greim et al., 1975) would suggest that this compound would produce the most severe damage to the endoplasmic reticulum. This is not the case. The endoplasmic reticulum is in fact undamaged acutely (Reynolds et al., 1975). Instead, the plasma and mitochondrial membranes of hepatocytes are the primary site of acute liver injury (Chieco et al., 1982). Mitochondria often appear swollen and the outer mitochondrial membranes are ruptured. Nuclear changes in the cell are also conspicuous, being characterized by a loss of perinucleolar chromatin and chromatin clumping.

The hepatotoxic response to VDC has been postulated to involve changes in cytosolic calcium. An increase in intracellular concentration of calcium occurs early in the hepatotoxic response to VDC (Long et al., 1989). Endonucleases are among the hepatic enzymes that can be activated by calcium (Burgoyne et al., 1970), and activation of endonucleases has been implicated in programmed cell death or "apoptosis" (Kerr et al., 1972). Long et al. (1989) investigated whether liver endonucleases were stimulated in rats administered VDC. Following both *in vivo* and *in vitro* exposure to VDC, derangement of intracellular calcium was observed, as reflected by endoplasmic reticulum calcium-pump inhibition. This event preceded cell death. The study was unable to find, however, any evidence of nucleolytic damage to DNA when calcium levels were first disrupted. It is concluded that nucleolytic destruction of DNA is not an early event in the hepatotoxicity produced by VDC, but secondary to destruction of the plasma membrane late in the course of heptocellular injury.

In contrast with the other chlorinated ethylenes, fatty infiltration is not observed in the livers of animals treated with a single hepatotoxic dose of VDC (Reynolds et al., 1980). This is surprising in light of the close correlation between fatty degeneration in the liver and hepatotoxicity observed for TCE and PCE (Buben and O'Flaherty, 1985).

The hepatotoxicities observed with longer exposures to the olefins tend to segregate into two distinct groups (Table 13–2). One class contains VC and probably VDC and the other class contains TCE and PCE. The changes observed with VC or VDC are comparable in some aspects to those observed with TCE or PCE. In all cases, fatty infiltration is observed, much to the same extent as in acute exposures. The noticeable difference is that continued VDC administration does lead to fatty degeneration and endoplasmic reticulum changes that are not evident with acute exposures (Prendergast et al., 1967; NTP, 1982).

Continued exposures to VC result in subcapsular fibrosis, areas of focal necrosis, and variable enlargement of the portal tracts. The fibrotic areas are also linked to areas of bile duct hyperplasia and cell hypertrophy. These pathological changes are observed in experimental animals following inhalation exposures to VC (Maltoni and Lefemine, 1974; Lee et al., 1978). The correlation of these pathological changes with VC-induced angiosarcomas in the liver led these investigators to suggest a role for fibroplasia in the he-

TABLE 13–2.
Description of Liver Damage Caused by Chronic Chlorinated Ethylene Exposures

Damage to Liver	VC	VDC	TCE	PCE
Fatty infiltration	+	+	+	+
Endoplasmic reticulum	+	+	+	+
Hepatocyte hypertrophy	−	−	+	+
Hepatocellular hyperplasia	−	−	−	−
Endothelial cell hyperplasia	+	+	−	−
Fibrosis	+	+	−	−
Focal necrosis	+	+	−	−
Hepatoangiosarcoma	+	+/−	−	−
Hepatocellular carcinoma	−	−	+	+

patocarcinogenic response. Liver biopsies from persons with a history of long-term industrial exposures to VC show excess fibrous connective tissue, portal tract enlargement, and bile duct hyperplasia (Thomas et al., 1975). Pathological changes similar to those induced by VC are seen in experimental animals following either oral or inhalation exposures to VDC (Prendergast et al., 1967; NTP, 1982; Quast et al., 1983). It is worth noting that the sole study showing an increased incidence of liver tumors in mice resulting from VDC administration showed these tumors to be of the same type as those associated with VC, hepatoangiosarcomas.

Chronic exposures of experimental animals to either TCE or PCE are not associated with fibrotic change or focal necrosis of endothelial cells in the livers. Rather, changes are observed in hepatocytes and are primarily limited to hyperplasia and hypertrophy (Elcombe et al., 1985; Odum et al., 1988). Clearly, the target cells of TCE and PCE differ from those of VC, and most likely VDC, when chronic exposures to the compounds is involved.

Role of Metabolism in the Toxic Response

Attempts have been made to account for the differing hepatotoxic potential of these compounds by examining the extent and location of chlorine substitution on the molecule (Bonse et al., 1975). Chlorine substitution causes electron delocalization and charge polarization of the compound. The most extreme case of electron delocalization is seen in VDC in which two chlorine atoms are attached to a single carbon. In this case, the charge polarization is the greatest of any in this particular class of olefins. The ethylene compounds TCE and VC are also unsymmetrical in the chlorine distribution on the molecule, resulting in charge polarization of the compounds. PCE and the 1,2-DCE's have a symmetrical distribution of chlorines; consequently the electron delocalizations are balanced and these olefins are more stable.

An additional effect of multiple chlorine substitutions is the enhancement of lipid solubility of the molecule. Increasing the number of chlorines on the molecule increases the lipid solubility, thus decreasing the ability of the compound to be exhaled and increasing the residence time of the compound in the body. This is an important consideration when studying the toxic potential of the compound. The increased residence time for TCE in comparison to VC accounts for the greater toxic potential of TCE (Reynolds and Moslen, 1977a).

Figure 1 illustrates the oxidative and conjugative reactions of the chlorinated ethylenes. The major metabolic pathway involves oxidative attack by cytochrome P450. This initial oxidative step is postulated to result in formation of either an epoxide or a catalytic oxo-iron complex as an intermediate. There is considerable disagreement as to the form of the actual intermediate. Bonse and Henschler (1976) argue that the initial oxidation step for the olefinic

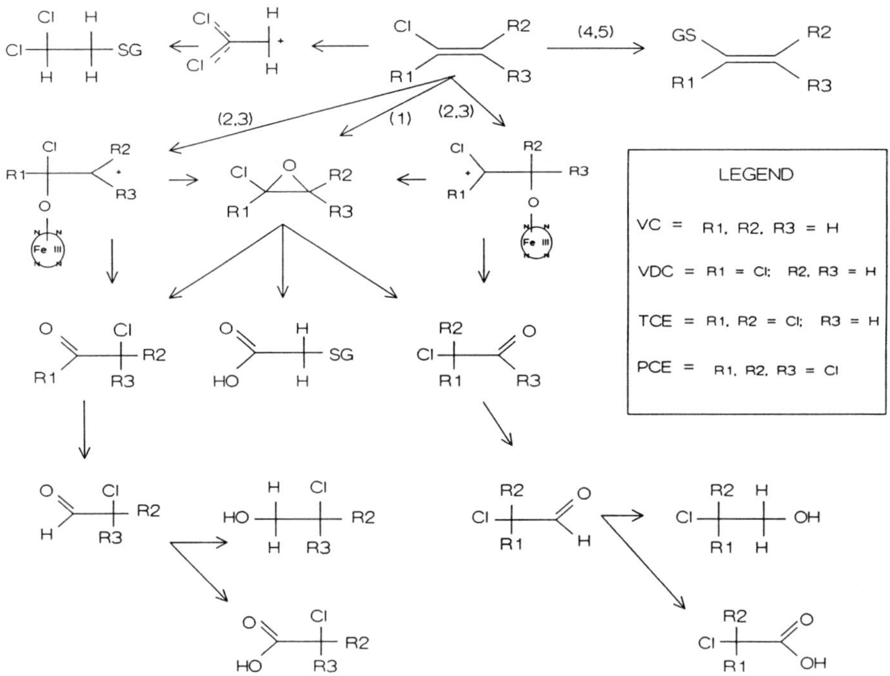

Figure 1. General scheme for the oxidative and conjugative metabolism of the chlorinated ethylenes. (1) Bonse and Henschler (1976); (2) Liebler and Guengerich (1983); (3) Miller and Guengerich (1983); (4,5) Dekant et al. (1986a, 1986b).

compounds involves formation of an epoxide. However, it has been demonstrated by Liebler and Guengerich (1983) that the epoxide does not rearrange to 2,2-dichloroacetaldehyde, a major metabolite of VDC, even in the presence of purified cytochrome P450. Miller and Guengerich (1983) also showed that the rearrangement of the TCE epoxide did not result in the formation of trichloroacetaldehyde, the precursor to the major urinary metabolites of TCE. Miller and Guengerich further demonstrated that the epoxide was not involved in covalent binding. Additions of the epoxide hydrolase inhibitor, 3,3,3-trichloropropylene oxide were ineffective at blocking microsomal binding of TCE. These investigators argue that the initial oxidation step is the formation of an oxo-iron complex which can rearrange one of several different ways, one of which is to the epoxide (Liebler and Guengerich, 1983; Miller and Guengerich, 1983).

It is known that the initial oxidation is followed by a preferential chlorine migration and subsequent formation of an aldehyde or an acid chloride, which can be further metabolized to an acid or an alcohol (Bonse and Henschler, 1976). The formation of an unstable reactive acid chloride is one reactive intermediate which binds cellular macromolecules. Dekant et al. (1987) identified the majority of alkylated macromolecules arising from PCE metabolism as *N*-trichloroacetylated phospholipids. Phospholipid adducts have also been identified following administration of TCE or VCD (Reichert et al., 1979; Dekant et al., 1984). Conjugation of reactive intermediates with glutathione is also known to occur for the chlorinated ethylenes and plays the major role in VDC toxicity, as will be discussed later (Allemand et al., 1978; Liebler et al., 1985).

A minor pathway of metabolism of the chlorinated ethylenes is via direct conjugation of glutathione with the chlorinated ethylene (Dekant et al., 1986a; Dekant et al., 1986b). This glutathione-substituted ethylene is further metabolized to the cysteine-ethylene conjugate by the enzymes of mercapturic acid biosynthesis in the renal tubules, yielding the cysteine-ethylene conjugate (Dekant et al., 1987). This metabolite is postulated to be responsible for the necrosis and induction of tumors in the kidney by TCE and PCE (Dekant et al., 1986a; Dekant et al., 1986b). The cysteine thioether may be acetylated to the mercapturic acid and excreted in the urine in a nontoxic pathway, but it is also a substrate for B-lyase which cleaves the thioether to pyruvate, ammonia, and a reactive, mutagenic fragment (Dekant et al., 1986c). The cysteine conjugate has been demonstrated to be toxic and mutagenic only upon further metabolism of the conjugates by renal B-lyase. Elfarra et al. (1986) and Dekant et al. (1986c) both demonstrated that the toxicity of the cysteine-ethylene conjugates could be prevented by either inhibitors of gamma glutamyl transpeptidase, the renal organic ion transport system or cysteine-conjugate B-lyase. The investigators conclude that the potential for damage is limited to the kidney due to the localization of the enzymes necessary to convert the conjugate to the toxic form.

It is readily apparent from studies done on the chlorinated ethylenes that metabolism is a prerequisite for the hepatotoxic response. Pretreatments which increase cytochrome P450, and presumably the metabolic activation of the parent compound to reactive intermediates, have been demonstrated to enhance the hepatotoxicity of VC, TCE, and PCE (Reynolds et al., 1975; Moslen et al., 1977).

Pretreatment of rats with phenobarbital or Aroclor 1254 potentiates the relative degree of TCE and PCE-induced liver injuries (Moslen et al., 1977). The magnitude of P450 induction correlated extremely well with the extent of liver injury in animals administered TCE, although liver injury was observed only in the Aroclor 1254-pretreated rats dosed with PCE.The induction of liver injury in the TCE-dosed animals was found to be closely associated with enhanced urinary excretion of total trichlorinated metabolites. Buben and O'Flaherty (1985) also report that the hepatotoxicity of both TCE and PCE is directly proportional to the extent of metabolism of the parent compound. Pretreatments of rats with compounds that are inducers of cytochrome P450 also enhanced the hepatotoxicity of VC, as observed by elevations of serum enzymes (Reynolds et al., 1975).

At the present time, a plausible hypothesis to explain the peculiar characteristics for the early hepatotoxic responses to VDC relative to the other chlorinated hydrocarbons is lacking. The pretreatment of rats with inducers of P450 does not enhance the toxicity of VDC (Reynolds et al., 1975). In fact, these animals were actually protected from the lethal hepatotoxic effects of VDC. Figure 1 contains a possible pathway for the nonenzymatic reaction of VDC with glutathione, as well as other cellular nucleophiles. This metabolic pathway might be expected from the character of chlorine substitution on the molecule. It is conceivable that the extent of unsymmetrical chlorine substitution on the molecule could lead to a sufficiently strong polarization of the compound such that it can act directly as an electrophile. This would explain the exquisite toxicity and be in agreement with the hypothesis that the chemical stability of the compound plays the major role in acute hepatotoxic response (Bonse and Henschler, 1976).

On the other hand, the study of Andersen and Jenkins (1977) is consistent with a role for microsomal oxidation in the activation of VDC to the hepatotoxic species. They were able to show that VDC hepatotoxicity, measured as increases in plasma transaminase levels, reached a plateau corresponding to the saturation of VDC metabolism. This argues for an oxidative activation of VDC to the reactive intermediate with subsequent conjugation with glutathione.

At this time, neither metabolic pathway seems to entirely explain the acute hepatotoxic response to VDC. If the molecule is depolarized to the extent that it acts as an electrophile (see Figure 1), the question remains as to the target organ specificity of the response observed, regardless of the route of exposure. Toxicity should not, in this case, be limited to the liver. On the other hand, cytochrome P450 activation to a reactive intermediate should result in the most unstable epoxide in this class of compounds (Greim et al., 1975). In this case, it is hard to reconcile the activation of VDC in the microsomes with the acute hepatocellular damage observed in the mitochondria and plasma membrane, especially since activation of the more stable ethylenes does lead to damage at the site of activation, the endoplasmic reticulum. Clearly, the mechanism of hepatotoxicity of VDC remains to be clarified.

It is known that the hepatotoxicity associated with VDC is dependent on the level of glutathione in the liver. Inhalation and oral exposures to VDC that are uniformly fatal to fasted rats are nonlethal and less hepatotoxic in fed rats (Jaeger et al., 1975; Andersen and Jenkins, 1977). The fasting of animals has been shown to markedly reduce liver concentration of glutathione. This decreased availability of hepatocellular GSH is reflected in a decrease in the amount of urinary metabolites of VDC associated with the GSH conjugation in fasted versus fed animals (Jaeger et al., 1974). Decreasing the hepatocellular level of GSH with diethyl maleate pretreatment also results in an enhancement of hepatotoxic response, while pretreatment of fasted rats with cysteine is able to protect rats from any hepatotoxic effects (Jaeger et al., 1974; Jaeger et al., 1975) Exposure to toxic quantities of VDC could easily overwhelm the supply of liver GSH with significant alkylation of tissue macromolecules ensuing. McKenna et al. (1977) present data consistent with a protective role for glutathione. In this study, appreciable covalent binding of VDC-derived radioactivity did not occur until hepatic GSH was depleted by 30%. The covalent binding preceded hepatocellular necrosis. It is reasonable to conclude then that the glutathione conjugation pathway is the major hepatocellular detoxification mechanism for VDC (Jaeger et al., 1974).

Earlier research attempted to associate the hepatotoxic and hepatocarcinogenic effects of the chlorinated ethylenes with reactive intermediates such as the epoxides and acid chlorides discussed above. While these intermediates may be important for VC, there is little question that most, if not all, of the hepatoxic effects of TCE can be accounted for by the formation of dichloroacetate (DCA) or trichloroacetate (TCA). Administration of DCA produces a marked enlargement of hepatocytes in B6C3F1 mice and to a lesser extent in rats (Bull et al., 1990). In mice these effects develop with 5 to 14 days of treatment, and by 14 days multiple areas of focal necrosis marked by massive accumulations of lymphocytes within the lesion are observed. A lesser degree of hypertrophy is produced with TCA, and there is no consistent evidence of focal necrotic lesions. In addition, prolonged treatment with TCA is marked by marked accumulations of lipid (Bull et al., 1990).

Species-Dependent Toxicity

The chlorinated ethylenes are metabolized qualitatively the same in mammals. The extent of metabolism of these compounds, however, is very different between species and appears to be a function of body surface area rather than body weight (Weiss et al., 1977). Consequently, one expects hepatic injury to vary between animal species. Indeed, the amount of chemical metabolized, or the "real exposure" of the animal to toxic intermediates, correlates well with acute hepatotoxicity. The extent of metabolism also correlates well with the hepatocarcinogenic response observed between experimental animal species.

VDC is more acutely hepatotoxic to mice than to rats. As with other chlorinated ethylenes, the rate of metabolism of VDC is more rapid in mice than it is in rats, and thus the production and concentrations of microsomal oxidation-derived reactive metabolites is greater (Jones and Hathaway, 1978). An enhanced production of VDC-derived reactive intermediates in mice is evidenced by a significantly greater amount of covalently bound radiolabeled metabolite in the livers of mice compared to rats (McKenna et al., 1977). It is then plausible that the covalent binding of reactive intermediates is responsible for the damage observed in VDC poisoning. However, as mentioned earlier, induction of cytochrome P450 protects rodents from the hepatoxicity of VDC. It is possible that this induction of cytochrome P450 induces detoxification pathways for VDC, i.e., GSH transferases, which could mask an increased production of reactive intermediates via microsomal oxidation.

The microsomal oxidation of TCE and PCE correlates well with the severity of the hepatotoxic response between species. TCE and PCE are metabolized to a greater extent in mice than in rats. Prout et al. (1985) found a similar percentage of TCE (greater than 90%) metabolized when mice and rats were administered single oral doses of 10 mg/kg TCE. Increasing the dose to 2000 mg/kg, however, markedly changed the degree of metabolism. While mice still eliminated only 14% of the dose as unchanged TCE, rats were found to eliminate 78%.

Peak blood concentrations of TCA four-fold higher in mice compared to rats, following an oral dose of 600 mg/kg TCE (Larson et al., 1990). At this dose the total exposure to TCA, measured as the area under the blood concentration versus time curve, was greater than two-fold higher in mice. After an oral dose of 200 mg/kg TCE, the real exposures between the species was approximately equal. Using the SIMUSOLV data analysis package (Dow Chemical) to fit the blood concentration data, the metabolic pathway for converting TCE was saturated at 130 mg/kg in rats, whereas this did not occur until 1100 mg/kg in the mouse. Thus, at high doses used in the NCI (1976) and NTP (1983) bioassays of TCE, the mouse is exposed to significantly greater amounts of TCA. DCA has also been identified in the urine of TCE-treated rodents (Green and Prout, 1985; Larson and Bull, 1989). As discussed later in this chapter, TCA, DCA, or subsequent metabolites of these chlorinated acetates appear to account for the hepatocarcinogenic effects of TCE in mice.

Mice are also more sensitive to the hepatoxic and hepatocarcinogenic effects of DCA and TCA administered directly in drinking water than are rats (Bull et al., 1990; Sanchez and Bull, 1990). Therefore, the net conversion of TCE and PCE to these metabolites is not the only factor accounting for the greater sensitivity of mice to the effects of TCE and PCE.

Schumann et al. (1980) administered PCE as either a 500-mg/kg oral dose or a 10–600-ppm 6-h inhalation exposure. In this study, mice metabolized 1.6 and 8.5 times more PCE than rats, following the 500 mg/kg and 10 ppm inhalation exposures, respectively. Since TCA is the major metabolite of PCE (Costa and Ivanetich, 1980), mice would be exposed to greater concentrations of TCA following equivalent PCE exposures than rats.

Metabolism of both TCE and PCE does result in binding of radiolabel to microsomal proteins (Van Duuren and Banarjee, 1976; Schumann et al., 1980). This covalent interaction of TCE or PCE-derived reactive may play a role in the hepatotoxic effects of TCE or PCE, although their role in hepatocarcinogenesis is doubtful, as will be discussed later. Alteration of the endoplasmic reticulum through covalent interactions of the membrane could result in the pathological changes observed. In addition, endoplasmic reticulum damage could affect the processing of triglycerides in the liver, resulting in the fatty degeneration. In keeping with the species sensitivity of the mouse compared to the rat, both Schumann et al. (1980) and Stott et al. (1982) demonstrated that covalent binding was much higher in the mouse than in the rat for either TCE or PCE.

HEPATOCARCINOGENESIS

Vinyl chloride is the only chlorinated ethylene for which a clear link with liver tumor induction in humans has been established. A relationship between angiosarcoma of the liver and long-term industrial exposure to vinyl chloride was first recognized by Creech and Johnson (1974) and Block (1974). Forman et al. (1985) found 115 deaths due to angiosarcoma of the liver among several thousand VC-exposed workers on the basis of reports from 10 countries during 1955–1984.

Description of Hepatocarcinogenic Response

Animal experiments have also clearly indicated VC as a carcinogen. Maltoni and Lefemine (1974) exposed Sprague-Dawley rats to inhalation exposures of 50 to 10,000 ppm VC for 12 months followed by observation periods to 135 weeks. Angiosarcomas of the liver were observed in up to 22% of the exposed rats. This type of liver tumor does not occur spontaneously in the rat. VC is also carcinogenic in Swiss Webster mice and Golden hamsters, inducing tumors at multiple sites thoughout the body, including angiosarcomas of the liver. The liver tumors were characterized by moderate-to-severe proliferation of endothelial cells lining the sinusoids, focal hemorrhage, invasion of the hepatic parenchyma with neoplastic cells, and mild-to-severe necrosis. These are the only tumors observed in the liver. Thus, as a carcinogen VC primarily targets endothelial cells in the liver.

Hepatic angiosarcomas were also observed in three mice treated with VDC in a study by Lee et al. (1978). Thus, VDC may also be targeting endothelial cells. It should be noted, however, that numerous other studies have been unable to show an increase in the incidence of these tumors in any animal species by VDC treatment (Maltoni et al., 1977; Quast et al., 1983). VDC has failed to induce hepatocellular tumors in any bioassay conducted to date.

In contrast to VC and VDC, TCE and PCE do not induce angiosarcomas of the liver, but induce hepatocellular adenomas and carcinoma instead. These tumors are observed only in B6C3F1 mice (NTP, 1986; 1988; 1990). At the least, the target cell for tumorigenesis in the liver is completely different with the more heavily chlorinated ethylenes, TCE and PCE.

To understand the potential for induction of human cancer by the chlorinated ethylenes, it is important to understand the mechanisms involved with the different chlorinated ethylenes. The remainder of the chapter focuses on VC and VDC and angiosarcoma development as one class of tumorigenic agents and response and the induction of hepatocellular carcinomas by TCE and PCE as the other.

Potential Mechanisms of Carcinogenesis

Vinyl Chloride and Vinylidene Chloride

The chlorinated ethylenes are examples of indirect-acting carcinogens. Just as the parent compounds need to be metabolized to elicit hepatotoxic response, so do the compounds require metabolism to the ultimate carcinogenic species.

A number of mechanisms have been proposed to explain the carcinogenicity of the chlorinated ethylenes. The ability of the reactive intermediates to interact with cellular DNA is one hypothesis. DNA adduct formation could result in a heritable somatic mutation and, hence, carcinogenesis. Current dogma holds that the initiation step of chemical carcinogenesis is a mutagenic event and is caused by alteration of DNA by the ultimate carcinogen intermediate (Miller and Miller, 1981). Vinyl chloride is mutagenic in bacteria upon metabolic activation (Rannug et al., 1976). By using different strains of bacteria, it was possible to show that the point mutations were due to base-pair substitutions, indicating that vinyl chloride possesses the ability to alkylate DNA and cause DNA mispairing of bases.

It has been suggested that the epoxide intermediate is the active carcinogen derived from VC (Van Duuren, 1975). VC has been shown to introduce a 2-oxoethyl group into nucleophilic sites in DNA (Osterman-Golker et al., 1977). The 2-oxoethyl group was found at the N-7 position of guanine in DNA in mice treated with VC. According to their calculations, the alkylating species was the epoxide and not chloroacetaldehyde.

Chloroethylene epoxide is one example of an alpha-chloroether, a class of potent carcinogens. The alpha-chloroethers are direct-acting DNA alkylating agents, and include the IARC Group 1 human carcinogenic compounds chloromethylmethyl ether and bis (chloromethyl) ether (IARC, 1987). The structural similarities of the other chlorinated ethylenes would seem to suggest that metabolism of these compounds to their epoxides would generate DNA-alkylating alpha-chloroethers as well. However, unlike VC, *in vivo* interactions of the other chlorinated ethylenes with DNA are minimal to nonexistent (Reitz et al., 1980; Parchman and Magee, 1982; Bronzetti et al., 1983).

VDC has been shown to be mutagenic in the Ames assay following addition of liver S-9 mix (Bartsch et al., 1975; Oesch et al., 1983). It has been suggested that the chemical instability of the nonsymmetrical VDC-epoxide would make this the most potent carcinogen of the chlorinated ethylene series (Greim et al., 1975). While it is clearly mutagenic, the carcinogenicity of VDC is open to question. Perhaps the instability of the epoxide makes it unlikely that it can diffuse from the site of metabolic activation to cellular DNA.

Reynolds and Moslen (1977a) suggest that epoxidation of VDC may be a relatively unimportant metabolic route. Instead, delocalization of electrons may activate the molecule towards a nucleophilic attack, with the nucleophile displacing a chlorine (see Figure 1). Consistent with this is the report that the bacterial mutagenicity of VDC (+S9) was not affected by the addition of microsomal epoxide hydrolase (Oesch et al., 1983). Additionally, the mutagenicity of VDC was reduced 50% by the addition of glutathione. These two observations suggest that the epoxide was not responsible for the mutagenicity, but that another metabolite subject to GSH-conjugation is the active form of VDC.

From the above discussion, it seems likely that VC-induced DNA alkylation is responsible for tumor initiation. The angioplastic response undoubtedly plays a role in making endothelial cells, as opposed to hepatocytes, the major target for tumor development. In contrast to VC, VDC does not appear to interact with DNA to any great extent *in vivo* (Reitz et al., 1980). Although VDC produces massive necrosis acutely at high doses, at the doses used in the carcinogenesis bioassays there was no evidence of an increased division of hepatocytes (Reitz et al., 1980). Thus, the small increase in angiosarcomas may have been primarily due to the angioplastic response to VDC.

Trichloroethylene and Perchloroethylene

The hepatocarcinogenic effects of TCE and PCE appear to be species and strain specific. TCE was shown to induce a dose-dependent increase in the incidence of hepatocellular carcinoma for both male and female B6C3F1 mice, but no significant increases in liver tumor yield in the Osborne-Mendel rats (NCI, 1976). Subsequent studies have confirmed the hepatocarcinogenicity of TCE in mice (NTP, 1990), and further demonstrated that TCE is incapable of inducing liver tumors in several strains of rat (NTP, 1988). TCE also fails to induce liver tumors in NMRI mice and Syrian hamsters (Henschler et al., 1980). Similarly, PCE was found to increase the incidence of hepatocellular carcinomas in B6C3F1 mice but not Osborne-Mendel rats (NCI, 1977). NTP (1986) and Mennear et al. (1986) again confirmed this species difference in Fisher 344 rats and B6C3F1 mice.

The potential for increased incidence of hepatocellular carcinoma induced

TABLE 13–3.
Relative Potency of Dichloroacetate, Trichloroacetate and Trichloroethylene as Hepatocarcinogens in Male B6C3F1 Mice

Chemical	Estimated Lifetime Dose Producing Extra Risk of 0.1 (mg/kg/day)		Administered Dose of TCE found in Urine (%)
	MLE	95% LCL	
TCE	156	122	—
DCA	18.3	9.4	<0.1
TCA	16.6	5.4	5

by PCE or TCE appears closely related to the concentration or amounts of stable metabolites produced in the different species. The recent identification of DCA and TCA as hepatocarcinogens in B6C3F1 mice suggests that the earlier "reactive" metabolites, i.e., acid chlorides or epoxides, are not necessary to explain the hepatocarcinogenic effects of TCE or PCE (Herren-Freund et al., 1987; Bull et al., 1990).

Bull et al. (1990) showed substantial differences in the dose-dependence of the hepatocarcinogenic response of the B6C3F1 mouse to TCA and DCA. The incidence of tumor formation in TCA-treated mice was linearly related to dose. On the other hand, the tumorigenic response to DCA displayed a very large increase when the dose was increased from 1 to 2 g/liter. This increased incidence of tumor formation at the high dose of DCA was associated with areas of focal necrosis and very high cell division in the immediate area surrounding these lesions (Sanchez and Bull, 1990). Such necrotic damage was not observed in the TCA-treated animals and was minimal at the low dose of DCA. Therefore, the tumorigenic response to DCA was largely attributed to repeated necrosis coupled with regenerative hyperplasia in those areas.

Table 13–3 compares the doses of TCE, DCA, and TCA that would increase the probability of cancer in mice by 0.1 in a lifetime. These figures were adjusted for the less than lifetime duration (52 weeks) of the study of DCA and TCA hepatocarcinogenicity by Bull et al. (1990). Based on the net excretion of TCA after administering TCE to animals and coupled with the fact that 65% of an administered oral dose of TCA is excreted in the urine (Larson, unpublished results), essentially all of the hepatocarcinogenic response to TCE can be accounted for by TCA. At the high doses used in the bioassays of TCE, DCA may also contribute to the hepatocarcinogenic response. The apparent conversion of TCE to DCA is much lower than to TCA. However, recovery of DCA in the urine after administration of an oral dose of DCA is less than 5% of the total dose. The more extensive metabolism of this compound suggests the possibility that significantly more might be formed than would be estimated for concentrations of DCA measured in blood or urine. Therefore, we cannot at the present time estimate the contribution of DCA to the tumorigenicity of TCE. Because of the sharp dependence of the tumorigenic response to DCA dose, it seems unlikely that it would contribute significantly to the induction of liver tumors at low TCE doses encountered in the general environment. DCA also does not appear to be a significant metabolite of PCE. Consequently TCA is probably responsible for the hepatocarcinogenicity of PCE (Odum et al., 1988).

It is also important to understand the tumorigenic response to DCA and the implications to the potential carcinogenicity of VDC. Dichloroacetaldehyde, derived from metabolism of VDC, is also a metabolic precursor to DCA (Jones and Hathaway (1978). A potential for inducing hepatocellular carcinomas with VDC exposures should exist. However, the data of Bull et al. (1990) demonstrate that the tumorigenic response to DCA is not a simple linear function of dose, but is heavily dependent on the dose of DCA being sufficient enough to cause cell necrosis. The doses received by mice consuming 1 and 2 g/liter DCA in the study resulted in doses of approximately 140 and 300 mg/kg/day, respectively, as estimated from water consumption. Chieco et al. (1982) report that the toxicity of 200 mg/kg/day VDC was massive in fasted animals. It is apparent that amounts of DCA required to produce a significant incidence of hepatocellular carcinomas (300 mg/kg/day) far exceed the maximum nonlethal doses of VDC that would be required to generate the DCA metabolically.

The question remains as to the mechanism of tumor induction by TCE, PCE, and their common metabolite TCA. Trichloroethylene, due to the unsymmetric nature of chlorine substitution, has been postulated to cause hepatocellular carcinoma via alkylation of DNA by an epoxide intermediate. However, several studies reveal that TCE is nonmutagenic, or only slightly so (Greim et al., 1975). Attempts to show an interaction of TCE with DNA *in vivo* have been unsuccessful (Parchman and Magee, 1982; Stott et al., 1982). Mirsalis et al. (1989) report that TCE was unable to induce unscheduled DNA synthesis in either Fischer 344 rat or B6C3F1 mouse hepatocytes following *in vivo* treatment. This indicates that the compound does not produce a repairable DNA adduct in vivo. PCE is nonmutagenic in *in vitro* mutagenicity assays (Greim et al., 1975). No reaction of perchloroethylene with hepatic DNA has been observed in either rats or mice (Schumann et al., 1980). TCA is apparently also incapable of producing point mutations (Andersen et al., 1972; Waskell, 1978). It would therefore appear that alternate mechanisms must explain the appearance of hepatocellular carcinomas in mice.

Studies by Schumann et al. (1980) and Elcombe et al. (1985) indicate that PCE and TCE increase cell division in the livers of mice at doses that produce hepatocellular tumors. Administration of TCE for 10 days at doses up to 1500 mg/kg/day by oral gavage resulted in a fivefold increase in ^{3}H-thymidine incorporation accompanied by an increased mitotic index in B6C3F1 mice. No such effect was seen in the rat (Elcombe et al., 1985). Administration of 100 mg/kg PCE for 11 days caused a twofold increase in DNA synthesis in the B6C3F1 mouse, as assessed by incorporation of tritiated thymidine into hepatocellular DNA, again with no effect in the rat (Schumann et al., 1980). It is important to note that no evidence of necrosis was observed in mice or rats, although Schumann et al. (1980) did report the occurrence of swollen centrilobular hepatocytes in the mouse following 11 days of treatment. Thus, the increase in mouse liver DNA replication following TCE or PCE dosing was apparently not due to regenerative hyperplasia, but did involve cell replication (Elcombe et al., 1985).

To some extent, TCA mimics the effects of TCE and PCE on mouse liver in terms of hepatocellular uptake of ^{3}H-thymidine. Sanchez and Bull (1990) report that administration of 1 and 2 g/L TCA in the drinking water of B6C3F1 mice for 14 days resulted in significantly increased ^{3}H-thymidine uptake in hepatocytes. However, this effect was not associated with increased cell division measured autoradiographically. It is not clear whether this increased incorporation of ^{3}H-thymidine reflects alterations in thymidine pool sizes or is the result of the induction of a DNA repair process.

Several authors have stated that the reason for the species specificity in tumorigenic response to TCE, PCE, and TCA lies in the ability of these compounds to cause peroxisome proliferation (Elcombe et al., 1985; Goldsworthy and Popp, 1987). A wide variety of hypolipidemic and plasticizer compounds are able to induce peroxisome proliferation and, subsequently, hepatocellular carcinoma. This association led Reddy et al. (1980) to propose a new class of carcinogens called peroxisome proliferators. For peroxisome proliferation to play a causal role in the carcinogenic response of TCE and PCE, rats and mice must respond differently, in terms of peroxisome proliferation and induction of peroxisomal B-oxidation activity.

Elcombe et al. (1985) administered TCE by gavage for 10 days to B6C3F1 mice and found that cyanide-insensitive palmitoyl-CoA oxidation (PCO) used to assay peroxisomal B-oxidation was increased 800%. The same treatment was without effect in the rat. Increases in peroxisome volume density in mice (1000% increase) were also reported. Again, the rat did not respond. Goldsworthy and Popp (1987) administered either TCE or PCE at oral doses of 1000 mg/kg for 10 days to both Fischer 344 rats and B6C3F1 mice. While increases in PCO above controls were observed in both species for TCE, and to a lesser degree PCE, the degree of elevations was three- to fourfold higher in the mice compared to rats for both compounds. Thus, while the ability to induce peroxisomal synthesis parallels the species sensitivity to these compounds. However, the induction of a greater level of peroxisome synthesis induced by TCE than its presumed active metabolite, TCA, in this study suggests some other factor is associated with TCE administration and was responsible for the carcinogenic response. Since the tumorigenic response to TCE appears to be accounted for by TCA, this would cast some doubt on a direct role for peroxisome proliferation in the hepatocarcinogenic effects of TCE.

Odum et al. (1988) reported no increase in either the number of peroxisomes or palmitoyl-CoA oxidation following inhalation exposures to PCE of 400 ppm for up to 28 days to Fischer 344 rats. Significant increases of peroxisomal activities were, however, observed in mice at this dose level.

Bull et al. (1990) found substantially greater amounts of autofluorescent lipofuscin in nontumorous portions of the liver in animals treated chronically with TCA than in control livers. This finding strongly suggests the involvement of radical mechanisms in the livers of these animals. The observation is consistent with increased peroxisomal activity, but other radical mechanisms, specifically a potential free radical arising from TCA metabolism, could well be responsible.

Recently, TCE and TCA were shown to inhibit gap junction communication in cultured hepatocytes from B6C3F1 mice (Klaunig et al., 1989). The compounds were without effect on intercellular communication in cultured cells from F344 rats. The inhibitory action of TCE was demonstrated to be the result of a metabolic product since treatment with SKF-525, an inhibitor of cytochrome P450, abolished the effect. The inhibition of intercellular communication caused by TCA was unaffected by SKF-525A pretreatment. Inhibition of gap junction–mediated intercellular communication has been associated with tumor promoters (Trosko and Chang, 1984). While the precise role of this effect in tumorigenesis is yet to be established, the study by Klaunig et al. (1989) indicates a clear difference between mouse and rat hepatocytes in terms of response to both TCE and TCA. This lead the authors to conclude that while the conversion to TCA may be critical to the hepatocarcinogenic response, there are intrinsic factors that have yet to be identified to explain the susceptibility of B6C3F1 mice to TCE and PCE-induced hepatocarcinogenesis.

SUMMARY

Although there is a superficial structural resemblance between the chlorinated ethylenes, it is clear that there are substantial differences in their toxicology and apparent mechanisms of action. At one extreme is VC, a chemical that apparently acts through a reactive intermediate, probably the epoxide, to produce damage to and tumors of endothelial cells in the liver. Its activity appears to depend largely on the induction of point mutations. At the other extreme are TCE and PCE which clearly do not act through epoxides or other reactive intermediates, but through subsequent chlorinated acetic acid metabolites, DCA and TCA. Although the mechanism by which these metabolites induce hepatic tumors in mice remains to be established, the target cell is clearly the hepatocyte.

REFERENCES

Allemand H, Pessayre D, Descatoire V, Degott C, Feldman G, and Benhamou JP: Metabolic activation of trichloroethylene into a chemically reactive metabolite toxic to the liver. *J Pharmacol Exp Therap* 1978; 204:714–723.

Andersen ME and Jenkins LJ Jr: Oral toxicity of 1,1-dichloroethylene in the rat: effect of sex, age, and fasting. *Environ Health Perspect* 1977; 21:157–163.

Andersen KJ, Leighty EG, and Takahashi MT: Evaluation of herbicides for possible mutagenic properties. *J Agr Food Chem* 1972; 20:649–656.

Bartsch H, Malaveille C, Montesano R, and Tomatis L: Tissue-mediated mutagenicity of vinylidene chloride and 2-chlorobutadiene in *Salmonella typhimurium*. *Nature* 1975; 255:641–643.

Berman JJ, Tong C, and Williams GM: Enhancement of mutagenesis during cell replication of cultured epithelial cells. *Cancer Lett* 1978; 4:277–283.

Block JB: Angiosarcoma of the liver following vinyl chloride exposure. *J Am Med Assoc* 1974; 229:53–54.

Bonse G and Henschler D: Chemical reactivity, biotransformation, and toxicity of polychlorinated aliphatic compounds. *CRC Crit Rev Toxicol* 1976; 10:395–409.

Bronzetti G, Bauer C, Corsi C, Del Carratore R, Galli A, Nieri R, and Paolini M: Genetic and biochemical studies on perchloroethylene *in vitro* and *in vivo*. *Mutat Res* 1983; 116:323–331.

Buben JA and O'Flaherty EJ: Delineation of the role of metabolism in the hepatotoxicity of trichloroethylene and perchloroethylene: a dose-effect study. *Toxicol Appl Pharmacol* 1985; 78:105–122.

Bull RJ, Sanchez IM, Nelson MA, Larson JL, and Lansing AJ: Liver tumor induction in B6C3F1 mice by dichloroacetate and trichloroacetate. *Toxicology* 1990; 63:341–359.

Burgoyne LA, Waqar NA, and Atkinson MR: Calcium-dependent priming of DNA synthesis in isolated rat liver nuclei. *Biochem Biophys Res Commun* 1970; 39:254–259.

CEH (Chemical Economics Handbook): Stanford Research Institute, Menlo Park, California, 1983.

Chieco P, Moslen MT, and Reynolds ES: Histochemical evidence that plasma and mitochondrial membranes are primary foci of hepatocellular injury caused by 1,1-dichloroethylene. *Lab Invest* 1982; 46:413–421.

Costa AK: The chlorinated ethylenes: their hepatic metabolism and carcinogenicity. *Diss Abstr Int [B]* 1983; 44(6):1791–B.

Costa AK and Ivanetich KM: Tetrachloroethylene metabolism by the hepatic microsomal cytochrome P-450 system. *Biochem Pharmacol* 1980; 29:2863–2869.

Creech JL and Johnson MN: Angiosarcoma of the liver in the manufacture of polyvinyl chloride. *J Occup Med* 1974; 16:150–151.

Dekant W, Metzler M, and Henschler D: Novel metabolites of trichloroethylene through dechlorination reactions in rats, mice and humans. *Biochem Pharmacol* 1984; 33:2021–2027.

Dekant W, Metzler M, and Henschler D: Identification of *S*-1,2,2-trichlorovinyl-*N*-acetylcysteine as a urinary metabolite of tetrachloroethylene. Bioactivation through glutathione conjugation as a possible explanation for its nephrocarcinogenicity. *J Biochem Toxicol* 1986a; 1:57–71.

Dekant W, Metzler M, and Henschler D: Identification of *S*-1,2-dichlorovinyl-*N*-acetyl-cysteine as a urinary metabolite of trichloroethylene. A possible explanation for its nephrocarcinogenicity in male rats. *Biochem Pharmacol* 1986b; 35:2455–2458.

Dekant W, Martens G, Vamvakas S, Metzler M, and Henschler D: Bioactivation of tetrachloroethylene. Role of glutathione *S*-transferase-catalyzed conjugation versus cytochrome P-450-dependent phospholipid alkylation. *Drug Metab Dispos* 1987; 15:702–709.

Dekant W, Vamvakas S, Berthold K, Schmidt S, Wild D, and Henschler D: Bacterial *b*-lyase mediated cleavage and mutagenicity of cysteine conjugates derived from the nephrocarcinogenic alkenes trichlorothylene, tetrachloroethylene and hexachlorobutadiene. *Chem Biol Interac* 1986c; 60:31–45.

Elcombe CR, Rose MS, and Pratt IS: Biochemical, histological and ultrastructural changes in rat and mouse liver following the administration of trichlorothylene: possible relevance to species difference in hepatocarcinogenicity. *Toxicol Appl Pharmacol* 1985; 79:365–376.

Elfarra AA, Lash LH, and Anders MW: Metabolic activation and detoxication of nephrotoxic cysteine and homocysteine S-conjugates. *Proc Natl Acad Sci* 1986; 83:2667–2671.

Environmental Protection Agency (EPA): Health Assessment Document for Trichloroethylene. EPA 600.8–82–006B. US Environmental Protection Agency, Washington, DC, 1984.

Environmental Protection Agency (EPA): PCE; occurrence in food, water and air. Office of Drinking Water, Washington, DC, 1987a.

Environmental Protection Agency (EPA): *Trans*-1,2-dichloroethylene; occurrence in food, water and air. Office of Drinking Water, Washington, DC, 1987b.

Environmental Protection Agency (EPA): *Cis*-1,2-dichloroethylene; occurrence in food, water and air. Office of Drinking Water, Washington, DC, 1987c.

Environmental Protection Agency (EPA): Dichloroethylene; occurrence in food, water and air. Office of Drinking Water, Washington, DC, 1987d.

Forman D, Bennett B, Stafford J, and Doll R: Exposure to vinyl chloride and angiosarcoma of the liver: a report of the register of cases. *Br J Ind Med* 1985; 42:750–753.

Freundt KJ, Liebaldt GP, and Lieberwirth E: Toxicity studies on trans-1,2-dichloroethylene. *Toxicology* 1977; 7:141–153.

Goldsworthy TL and Popp JA: Chlorinated hydrocarbon-induced peroxisomal enzyme activity in relation to species and organ carcinogenicity. *Toxicol Appl Pharmacol* 1987; 88:225–133.

Green T and Prout MS: Species differences in response to trichloroethylene. II. Biotransformation in rats and mice. *Toxicol Appl Pharmacol* 1985; 79:401–411.

Greim H, Bonse G, Radwan Z, Reichert D, and Henschler D: Mutagenicity *in vitro* and potential carcinogenicity of chlorinated ethylenes as a function of metabolic oxirane formation. *Biochem Pharmacol* 1975; 24:2013–2017.

Haley TJ: Vinylidene chloride: A review of the literature. *Clinical Toxicol* 1975; 8:633–643.

Hashimoto T: Individual peroxisomal B-oxidation enzymes. *Ann NY Acad Sci* 1982; 386:5–12.

Henschler D, Romen W, Elsasser H, Reichart D, Eder E, and Radwan Z: Carcinogenicity study of trichloroethylene by long-term inhalation in three animal species. *Arch Toxicol* 1980; 43:237–248.

Herren-Freund SL, Pereira MA, Khoury MD, and Olson G: The carcinogenicity of trichloroethylene and its metabolites, trichloroacetic acid and dichloroacetic acid, in mouse liver. *Toxicol Appl Pharmacol* 1987; 90:183–189.

International Agency for Research on Cancer (IARC): *Monographs on the Evaluation of Chemicals to Humans. Overall Evaluation on Carcinogenicity,* IARC Monographs, vol 1–42, suppl 7. IARC, Lyon, France, 1987.

Jaeger RJ, Trabulus MJ, and Murphy SD: Biochemical effects of 1,1-dichloroethylene in rats: dissociation of its hepatotoxicity from a lipoperoxidative mechanism. *Toxicol Appl Pharmacol* 1973; 24:457–467.

Jaeger RJ, Conolly RB, and Murphy SD: Effect of an 18-hr fast and glutathione depletion on 1,1-dichloroethylene-induced hepatotoxicity and lethality in rats. *Exp Mol Pathol* 1974; 20:187–194.

Jaeger RJ, Conolly RB, Reynolds ES, and Murphy SD: Biochemical toxicology of unsaturated halogenated monomers. *Environ Health Perspect* 1975; 11:121–128.

Jones DK and Hathaway DE: Differences in metabolism of vinylidene chloride between rats and mice. *Br J Cancer* 1978; 37:411–417.

Kerr JFR, Wyllie AH, and Currie AR: Apoptosis: a basic biological phenomenon with wide-ranging implications in tissue kinetics. *Br J Cancer* 1972; 26:239–251.

Klaunig JE, Ruch R, and Lin EC: Effects of trichlorothylene and its metabolites on rodent hepatocyte intercellular communication. *Toxicol Appl Pharmacol* 1989; 99:454–465.

Kylin BH, Reichard H, Sumegi I, and Yllner S: Hepatotoxicity of inhaled trichloroethylene, tetrachloroethylene and chloroform. Single exposure. *Acta Pharmacol Toxicol* 1963; 20:16–26.

Kylin B, Sumegi I, and Yllner S: Hepatotoxicity of inhaled trichloroethylene and tetrachloroethylene. Long-term exposure. *Acta Pharmacol Toxicol* 1965; 22:379–385.

Larson JL and Bull RJ: Effect of ethanol on the metabolism of trichloroethylene. *J Toxicol Environ Health* 1989; 28:395–406.

Larson JL, Hayton WL, and Bull RJ: Pharmacokinetic modeling of metabolism of trichloroethylene. *Toxicologist* 1990; 10:215.

Lee CC, Bhandari JC, Winston JM, House WB, Dixon RL, and Woods JS: Carcinogenicity of vinyl chloride and vinylidene chloride. *J Toxicol Environ Health* 1978; 4:15–30.

Liebler DC and Guengerich FP: Olefin oxidation by cytochrome P-450: evidence for group migration in catalytic intermediates formed with vinylidene chloride and *trans*-1-phenyl-1-butene. *Biochemistry* 1983; 22:5482–5489.

Liebler DC, Meredith MJ, and Guengerich FP: Formation of glutathione conjugates by reactive metabolites of vinylidene chloride in microsomes and isolated hepatocytes. *Cancer Res* 1985; 45:186–193.

Long RM, Moore L, and Schoenberg DR: Halocarbon hepatotoxicity is not initiated by Ca^{2+}-stimulated endonuclease activation. *Toxicol Appl Pharmacol* 1989; 97:350–359.

Lowenstein WR: Junctional intercellular communication and the control of growth. *Biochem Biophys Acta* 1979; 560:1–65.

Maltoni C: Recent findings on the carcinogenicity of chlorinated olefins. *Environ Health Perspect* 1977; 21:1–5.

Maltoni C, Cotti G, Morisi L, and Chico P: Carcinogenicity bioassays of vinylidene chloride: research plan and early results. *Med Lav* 1977; 68:241–262.

Maltoni C and Lefemine G: Carcinogenicity bioassays of vinyl chloride. I. Research plan and early results. *Environ Res* 1974; 7:387–405.

Marsman DS, Cattley RC, Conway JG, and Popp JA: Relationship of hepatic peroxisome proliferation and replicative DNA synthesis to the hepatocarcinogenicity of the peroxisome proliferators di(2-ethylhexyl)phthlate and [4-chloro-6-(2,3-xylidino)-2-pyrimidinylthio]acetic acid (Wy-14,643) in rats. *Cancer Res* 1988; 48:6739–6744.

McKenna MJ, Watanabe PG, and Gehring PJ: Pharmacokinetics of vinylidene chloride in the rat. *Environ Health Perspect* 1977; 21:99–105.

Mennear JH, Clarke W, Ragan H, and Miller R: Toxicology and carcinogenicity studies of inhaled tetrachloroethylene (PERC) in male and female F344 rats and B6C3F1 mice. *Toxicologist* 1986; 6:135.

Miller RE and Guengerich FP: Metabolism of trichloroethylene in isolated hepatocytes and reconstituted enzyme systems containing cytochrome P-450. *Cancer Res* 1983; 43:1145–1152.

Miller EC and Miller JA: Searches for ultimate chemical carcinogens and their reactions with cellular macromolecules. *Cancer* 1981; 47:2327–2345.

Mirsalis JC, Tyson CK, Loh EN, Steinmetz KL, Bakke JP, Hamilton CM, Spak DK, and Spalding JW: Induction of hepatic cell proliferation and unscheduled DNA synthesis in mouse hepatocytes following *in vivo* treatment. *Carcinogenesis* 1985; 6:1521–1524.

Mirsalis JC, Tyson CK, Steinmetz KL, Loh EK, Hamilton CM, Bakke JP, and Spalding JW: Measurement of unscheduled DNA synthesis and S-phase synthesis in rodent hepatocytes following *in vivo* treatment: testing of 24 compounds. *Environ Mol Mut* 1989; 14:155–164.

Moody CS and Hassan HM: Mutagenicity of oxygen free radicals. *Proc Natl Acad Sci* 1982; 79:2855–2859.

Moslen MT, Reynolds ES, and Szabo S: Enhancement of metabolism and hepatotoxicity of trichloroethylene and perchloroethylene. *Biochem Pharmacol* 1977; 26:369–375.

National Cancer Institute (NCI): *Carcinogenesis Bioassay of Trichloroethylene,* CAS No. 79–01–6, DHEW Publ no (NIH) 76–802, 1976.

National Cancer Institute (NCI): *Carcinogenesis Bioassay of Tetrachloroethylene for Possible Carcinogenicity,* DHEW Publ no (NIH) 77–813, 1977.

National Toxicology Program (NTP): *Carcinogenesis Bioassay of Vinylidene Chloride in F344 Rats and B6C3F1 Mice,* Cas no 75–35–4, DHEW publ no (NIH) 82–1784, 1982.

National Toxicology Program (NTP): Toxicology and carcinogenesis studies of trichloroethylene in four strains of rats (ACI, August, Marshall, Osborne-Mendel) (Gavage studies) NTP TR 273 DHHS, Public Health Service, 1988.

National Toxicology Program (NTP): Toxicology and carcinogenesis of tetrachloroethylene (perchloroethylene) in F344/N rats and B6C3F, mice (inhalation studies). NTP TR 311 DHHS, Public Health Service 1986.

National Toxicology Program (NTP): Carcinogenesis studies of trichloroethylene (without epichlorohydrine) in F344/N rats and B6C3F, mice (gavage studies) NTP TR 243, DHHS, Public Health Service 1990.

Newberne PM, Punyarit P, de Camargo J, and Suphakarn V: The role of necrosis in hepatocellular proliferation and liver tumors. *Arch Toxicol* 1987; suppl 10:54–67.

Odum J, Green T, Foster JR, and Hext PM: The role of trichloroacetic acid and peroxisome proliferation in the carcinogenicity of perchloroethylene in the mouse and rat. *Toxicol Appl Pharmacol* 1988; 92:103–112.

Oesch F, Protic-Sablijic M, Friedberg T, Klimisch HJ, and Glatt HR: Vinylidene chloride: changes in drug-metabolizing enzymes, mutagenicity and relation to its targets for carcinogenesis. *Carcinogenesis* 1983; 4:1031–1038.

Osterman-Golkar S, Hallmark D, Seperback D, Calleman CJ, Gothe R, Ehrenberg L, and Wachtmeister CA: Alkylation of DNA and proteins in mice exposed to vinyl chloride. *Biochem Biophys Res Comm* 1977; 76:259–266.

Parchman LG and Magee PN: Metabolism of [^{14}C] trichloroethylene to $^{14}CO_2$ and interaction of a metabolite with liver DNA in rats and mice. *J Toxicol Environ Health* 1982; 9:797–813.

Prendergast JA, Jones RA, Jenkins LJ, and Siegel J: Effects on experimental animals of long-term inhalation of trichloroethylene, carbon tetrachloride, 1,1,1-trichloroethane, dichlorodifluoromethane, and 1,1-dichloroethylene. *Toxicol Appl Pharmacol* 1967; 10:270–289.

Prout MS, Provan WM, and Green T: Species difference in response to trichloroethylene. I. Pharmacokinetics in rats and mice. *Toxicol Appl Pharmacol* 1985; 79:389–400.

Quast JF, Humiston CG, Wade CE, Ballard J, Beyer JE, Schwetz RW, and Norris JM: A chronic toxicity and oncogenicity study in rats and subchronic toxicity study in dog on ingested vinylidene chloride. *Fund Appl Toxicol* 1983; 3:55–62.

Rannug U, Gothe R, and Wachtmeister CA: The mutagenicity of chloroethylene oxide, chloroacetaldehyde, 2-chloroethanol and chloroacetic acid, conceivable metabolites ofvinyl chloride. *Chem Biol Interact* 1976; 12:251–263.

Reddy JK, Azarnoff DL, and Hignite CE: Hypolipidemic hepatic peroxisome proliferators form a novel class of chemical carcinogens. *Nature* 1980; 283:397–398.

Reichert D, Werner HW, Metzler M, and Henschler D: Molecular mechanism of 1,1-dichloroethylene toxicity. Excreted metabolites reveal different pathways of reactive intermediates. *Arch Toxicol* 1979; 42:159–169.

Reitz RH, Watanabe PG, McKenna MJ, Quast JF, and Gehring PJ: Effects of vinylidene chloride on DNA synthesis and DNA repair in the rat and mouse: a comparative study with dimethylnitrosamine. *Toxicol Appl Pharmacol* 1980; 52:357–370.

Reynolds ES: Liver endoplasmic reticulum: target site of halocarbon metabolites. *Advan Exptl Med Biol* 1977; 80:117–137.

Reynolds ES, Moslen MT, Szabo S, Jaeger RJ, and Murphy SD: Hepatotoxicity of vinyl chloride and 1,1-dichloroethylene. Role of mixed function oxidase system. *Am J Pathol* 1975; 81:219–236.

Reynolds ES and Moslen MT: Damage to hepatic cellular membranes by chlorinated olefins with emphasis on synergism and antagonism. *Environ Health Perspec* 1977a; 21:137–147.

Reynolds ES and Moslen MT: Hepatocellular structural and compositional derangements in trichloroethylene anesthetized animals. *Fed Proc* 1977b; 36:4153.

Reynolds ES, Moslen MT, Boor PJ, and Jaeger RJ: 1,1-Dichloroethylene hepatotoxicity. Time course of GSH changes and biochemical aberrations. *Am J Pathol* 1980; 101:331–344.

Ruch RJ and Klaunig JE: Effects of tumor promoters, genotoxic carcinogens and hepatocytotoxins on mouse hepatocyte intercellular communication. *Cell Biol Toxicol* 1986a; 2:469–483.

Sanchez IM and Bull RJ: Early induction of reparative hyperplasia in the liver of B6C3F1 mice treated with dichloroacetate and trichloroacetate. *Toxicology* 1990; 64:33–46.

Schumann AM, Quast JF, and Watanabe PG: The pharmacokinetics and macromolecular interactions of perchloroethylene in mice and rats as related to oncogenicity. *Toxicol Appl Pharmacol* 1980; 55:207–219.

Shen P: Tumors in control mice: literature tabulation. *Toxicol Appl Pharmacol* 1974; 30:337–359.

Stewart RD, Hake CL, and Peterson JE: Use of breathe analysis to monitor trichloroethylene exposures. *Arch Environ Health* 1974; 29:6–13.

Stott WT, Quast JF, and Watanabe PG: The pharmacokinetics and macromolecular interactions of trichloroethylene in mice and rats. *Toxicol Appl Pharmacol* 1982; 62:137–151.

Trosko JE and Chang CC: Role of intercellular communication on tumor promotion, in Slaga TJ (ed): *Mechanisms of Tumor Promotion,* vol 4. Boca Raton, CRC Press, 1984, pp 119–145.

Van Duuren BL: On the possible mechanism of carcinogenic action of vinyl chloride. *Ann NY Acad Sci* 1975; 246:256–267.

Van Duuren BL and Banarjee S: Covalent interaction of the carcinogen trichloroethylene in rat hepatic microsomes. *Cancer Res* 1976; 36:2419–2422.

Waskell L: A study of the mutagenicity of anesthetics and their metabolites. *Mut Res* 1978; 57:141–153.

Weiss M, Szeigoleit W, and Forster W: Dependence of pharmacokinetic parameters on the body weight. *Int J Clin Pharmacol* 1977; 15:572–575.

Chapter

14

Assessment of Hepatocarcinogenesis by Early Indicators

Norio Hirota
Department of Pathology
Jichi Medical School
Minamikawachi-machi
Kawachi-gun, Japan

Gary M. Williams
American Health Foundation
Valhalla, New York

INTRODUCTION

A variety of chemicals of diverse structure has been shown to cause liver cancer in experimental animals and humans. This review is concerned with assessment of chemically induced liver cell carcinogenesis by measurement of changes preceding the development of neoplasms.

In 1935, Sasaki and Yoshida reported the induction of liver tumors in rats with *o*-aminoazotoluene. In their sequential study of the liver changes, they described detailed histopathology of hyperplasia and nodular adenoma as well as carcinoma. In the following decades, diverse carcinogenic substances were synthesized and some naturally occurring carcinogens were discovered. Studies with a number of agents have corroborated the sequential events in the process of hepatocarcinogenesis. It is now well established that in the induction of hepatocellular carcinoma in rodents, several distinct cellular lesions precede the development of malignant tumors. Among those that have been postulated to be precursors of hepatocellular carcinoma are "oval cell" proliferation (Farber, 1956; Inaoka, 1967), hyperplastic (altered) foci (Firminger, 1955; Reuber, 1965; Newberne and Wogan, 1968; Bannasch, 1968; Friedrich-Freksa et al., 1969), and hyperplastic (neoplastic) nodules (Firminger, 1955; Reuber, 1965; Newberne and Wogan, 1968; Farber, 1973).

Oval cells can appear in rat liver within days of administration of a hepatocarcinogen. They are small, oval-shaped cells, as their name indicates, with scant cytoplasms and pale nuclei. They are usually concentrated in the periportal regions and have been thought to be of bile ductular origin (Grisham and Porta, 1964; Rubin, 1964). The findings of some hepatocellular properties in these cells have been interpreted as evidence that they may be related to hepatocytes, possibly as stem cells (Sell et al., 1981), although this must be taken cautiously because of the possibility of abberant gene expression in the oval cells. Nevertheless, several studies have provided evidence for the existence of oval cells as a subpopulation in normal liver (Grisham and Hartroft, 1961; Marceau et al., 1982; Furukawa et al., 1985).

The altered focus is the earliest-appearing distinct lesion of hepatocellular-type cells induced by liver carcinogens. Foci, as described in several species, are lesions smaller than a lobule and are composed of cells with abnormal staining properties (Reuber, 1965; Daoust, 1972; Williams, and Yamamoto, 1972; Bannasch, 1976) and a variety of phenotypic abnormalities (see below). Usually, the cells are different in size from normal hepatocytes and display nuclear abnormalities. The cells of foci are usually arranged in abnormal plates which still merge with those of the surrounding parenchyma. Ultrastructural studies have revealed distinct hepatocellular characteristics of the cells in foci (Timme, 1978; Hirota and Williams, 1982).

Nodular lesions that have been designated as hyperplastic nodules, neoplastic nodules, or adenomas appear later in carcinogenesis than foci. These nodular lesions are composed of cells similar to those in foci, but the nodules are larger and compress the surrounding parenchyma from which they are sharply demarcated. Cells in nodules possess some of the same phenotypic abnormalities as foci (Kitagawa, 1971; Williams et al., 1976). Nodules are considered by some investigators to be benign neoplasms and hence are referred to here as adenomas, following the original terminology of Sasaki and Yoshida (1935).

SIGNIFICANCE OF VARIOUS LIVER LESIONS CAUSED BY CHEMICALS

There are two general types of carcinogen, both of which affect the liver: DNA-reactive or genotoxic carcinogens and epigenetic or nongenotoxic carcinogens and epigenetic or nongenotoxic carcinogens (Weisburger and Williams, 1981). Most genotoxic carcinogens are carcinogenic only after biotransformation to highly reactive derivatives or ultimate forms which react with DNA. The liver is the most capable organ in such metabolism of chemicals, although other organs or tissues can generate the ultimate forms in some instances. In the liver, several genotoxic carcinogens have rapidly induced oval cell proliferation (refs., see below) and altered foci (Scherer et al., 1972; Williams and Watanabe, 1978). Some noncarcinogenic pathologic processes induce oval cell proliferation (Lopez and Massanti, 1955; Grisham and Hartroft, 1961) and, therefore this is not a specific effect of carcinogens. In contrast, of diverse chemical carcinogens, those that do not cause hepatocellular carcinoma do not induce altered foci, even after repeated exposures. Likewise, noncarcinogenic noxious agents do not induce foci. Both toxic agents and carcinogens can induce similar nonspecific hepatocellular changes such as focal degeneration and necrosis in the acute phase of their toxic effects. Altered foci, however, do not occur following the acute changes induced by noncarcinogens or carcinogens for other organs than the liver.

The nodular lesions, hyperplastic nodules, or adenomas, occur later in the carcinogenic process than foci, although a type of nodular lesion can be induced rapidly by the selective pressure of the Solt-Farber regimen (Solt et al., 1977) involving resistance to cytotoxicity (see below).

The pathogenesis of liver cancer induction by epigenetic agents, such as phenobarbital, estrogenic hormones, DDT, and peroxisome proliferators, is less well understood. Some of the same changes have been observed as with DNA-reactive carcinogens, but also there are differences (Rao et al., 1982, 1986).

Study of the various liver lesions caused by chemicals has several applications. Firstly, the carcinogen-induced altered populations are central in the pathogenesis of liver cancer. Of considerable practical importance, induction of these lesions can be used as a limited bioassay for carciongens (Weisburger and Williams, 1981; Pereira, 1982). Moreover, rapid induction of foci is an indication of genotoxicity of a chemical.

Also, foci can be used for early identification of liver tumor-promoting agents. Peraino and associates first demonstrated an increase in liver tumor incidence in rats given phenobarbital following the hepatocarcinogen 2-acetylaminofluorene (Peraino et al., 1971). Subsequently, it was shown that phenobarbital enhanced the persistence or progression of carcinogen-induced altered foci (Kitagawa and Sugano 1978; Pitot et al., 1978; Watanabe and Williams, 1978; Shinozuka et al., 1982; Peraino et al., 1984). This observation provided the basis for an approach to the rapid detection of promoting agents (Williams, 1982). Importantly, in the demonstration of tumor promotion with the quantitative or size-wise enhancement of the lesions, the test agent should be proven to be nongenotoxic and to produce the enhancement only after and not before the genotoxic carcinogen to exclude "summation" or "co-carcinogenic" effects (Williams, 1984). Otherwise, it is not feasible to discriminate promotion from an "initiation" effect by genotoxic carcinogens. Many liver tumor promoters have been detected in the two-stage experiments, including microsomal enzyme inducers, hypolipidemic agents, steroid hormones, diet, and miscellaneous exogenous factors (reviewed by Moore and Kitagawa, 1986). Current studies suggest that epigenetic carcinogens or promoters may affect the state or function of the cell membrane in such a way as to interfere with the transmission of regulatory factors, thereby liberating neoplastic cells for proliferation into tumors (Williams, 1981).

PRENEOPLASTIC AND PRECANCEROUS HEPATOCELLULAR LESIONS

The term *preneoplastic* or *precancerous* has been given various meanings. In the present context, it will be used to designate cellular populations that precede the development of neoplasms and display cellular and structural alterations consistent with their serving as precursors of neoplasms. Within such populations some cells may already be neoplastic, but at present there is no definite means to identify these.

Oval Cells

Oval cells often proliferate during early stages of hepatocarcinogenesis with different chemicals, particularly aminoazo dyes and aromatic amines in rats (Farber, 1956; Inaoka, 1967; Dempo et al., 1975; Guillouzo et al., 1981) and in mice (Jalanko and Ruoslahti, 1979).

The origin and fate of oval cells is controversial. Some investigators believe that they originate from bile duct cells (Inaoka, 1967; Iwasaki et al., 1972; Richards et al., 1982), while others claim that they constitute a distinct cell population fulfilling some of the criteria for liver stem cells (Sell and Leffert, 1982; Hayner et al., 1984) which could be the target of various carcinogens (Sell, 1983; Braun et al., 1987; Sell et al., 1987). Recently, in spite of the presence of heterogeneity in gene expression and biochemical function, more sophisticated fractionation studies (Hayner et al., 1984; Yaswen et al., 1984; Germain et al., 1985) have provided evidence that oval cells most probably originate from proliferation of bile ductular cells, supporting the conclusion of earlier morphological studies.

The fate of oval cells is not firmly established. They contain oncodevelopmental gene products such as α-fetoprotein and hybrids of fetal and adult forms of aldolase and pyruvate kinase as well as markers for hepatocyte-

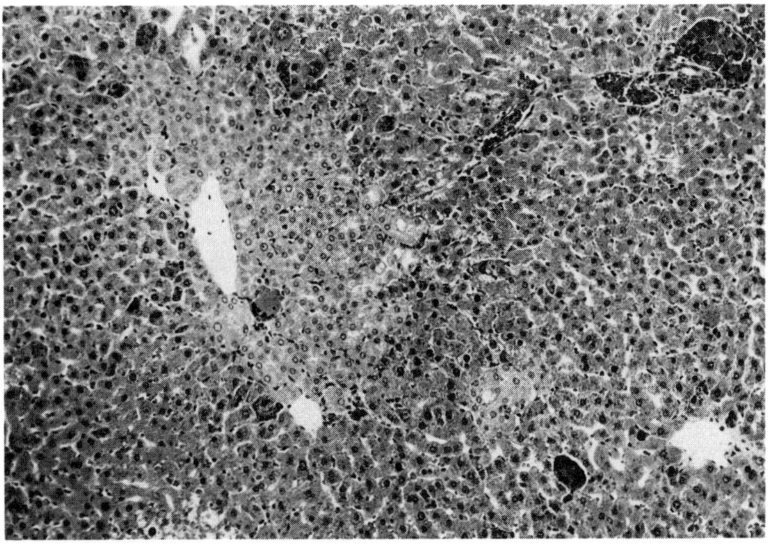

Figure 1. Eosinophilic cell altered focus. Formalin fixation, hematoxylin and eosin, magnification: × 85.

specific functions such as glucose-6-phosphatase and albumin mRNA, indicating the transitional nature of some of the oval cells into hepatocytes (Ogawa et al., 1974; Hayner et al., 1984; Yaswen et al., 1984; Germain et al., 1985).

Altered Foci

Most studies on the pathogenesis of foci of cellular alteration have been done in rat liver, although foci have been described in the mouse (Lipsky et al., 1979, 1980; Williams, and Numoto, 1984) and hamster (Stenback et al., 1986). Altered foci are generally distinguishable from surrounding normal liver tissue in sections stained with hematoxylin and eosin. Based on the staining properties of the cells in foci, they can be classified as eosinophilic (Figure 1), basophilic (Figure 2), and clear (Figure 3). Mixed types also occur. A variety of phenotypic changes has been identified in altered foci (Table 14-1). Foci also appear spontaneously in aged rats (Ogawa et al., 1981; Ward, 1981), but seem to differ in some respects from induced foci. For one, no elevation of γ-glutamyltranspeptidase (see below) was found in livers during spontaneous hepatocarcinogenesis in mice (Jalanko and Ruoslahti, 1979), and the enzyme was not found histochemically in most spontaneous basophilic foci in aged female rats (Mitaka and Tsukada, 1987). Also, naturally occurring basophilic foci of rats were not promotable with phenobarbital (Ward and Ohshima, 1985).

The numbers of foci are related to the dose of carcinogen (Emmelot and Scherer, 1980). They increase in number and size with continued carcinogen

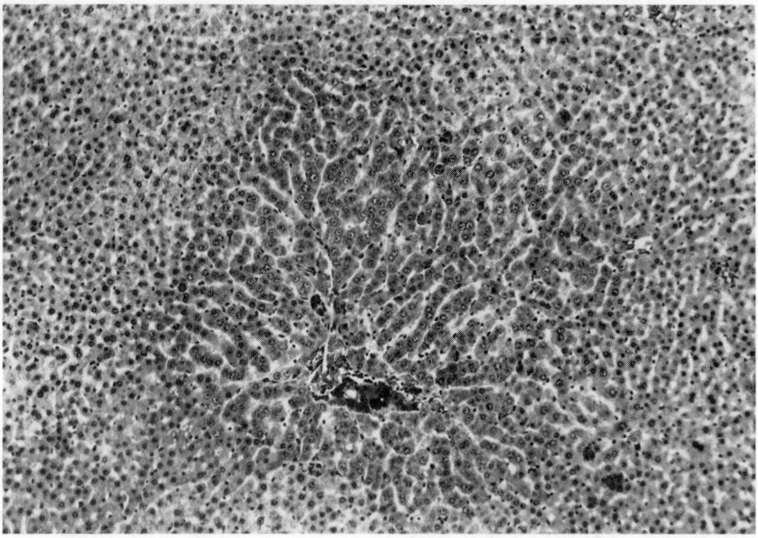

Figure 2. Basophilic cell altered focus. Formalin fixation, hematoxylin and eosin, magnification: × 85.

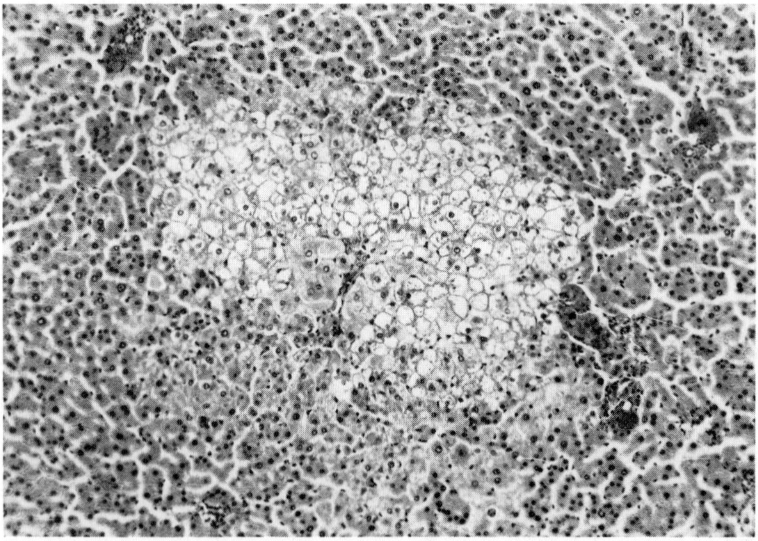

Figure 3. Clear cell altered focus. Formalin fixation, hematoxylin and eosin, magnification: × 85.

TABLE I
Phenotypic Abnormalities of Hepatocellular Altered Foci

Enzymes
Deficiency of glucose-6-phosphatase
Deficiency of adenosinetriphosphatase
Deficiency of L-pyruvate kinase
Appearance of placental form of glutathione transferase
Appearance of aldehyde dehydrogenase
Increase of nonspecific esterase
Increase of alkaline phosphatase
Increase of uridine diphosphate glucuronyltransferase
Excessive storage of glycogen
Resistance to iron accumulation
Production of α-fetoprotein
Decreased fibronectin
Expression of oncogenes
Resistance to cytotoxic effects of agents requiring metabolism

exposure (Williams, and Watanabe, 1978; Rabes and Szymkowiak, 1979) or with time after cessation of certain carcinogen exposures (Barbason and Betz, 1981; Moore et al., 1981).

The site of origin of foci in the hepatic lobules has not been studied in detail. In the Solt-Farber regimen of diethylnitrosamine/2-acetylaminofluorene-partial hepatectomy, foci were stated to have no paticular acinar distribution (Solt et al., 1977).

After some conditions of carcinogen exposure, certain phenotypic abnormalities in foci are quite persistent and stable (Pitot et al., 1978), whereas under other conditions, after cessation of carcinogen exposure, the expression of some phenotypic abnormalities in foci can diminish, a phenomenon referred to as phenotypic reversion (Williams and Watanabe, 1978). Whether the cells in reverting or regressing foci actually are lost, as suggested by some studies (Bursch et al., 1984), or whether they persist in an inapparent latent state (Williams, 1980), is still not resolved.

Rat liver foci apparently develop by the clonal expansion of an altered cell (Rabes et al., 1982). Their growth rate has been related to the complexity of phenotypic abnormalities in their cells (Pitot et al., 1978; Pugh and Goldfarb, 1978; Peraino et al., 1984). As a result of sustained growth, some foci progress into nodular lesions (Williams, 1976). Foci, as well as nodular lesions, have been considered by some investigators to be hyperplastic. But this position is not supported by observations that both are lesions with fixed modifications of DNA (Mori et al., 1982). Moreover, even under conditions in which some phenotypic reversion occurs after withdrawal of carcinogen exposure, they do not completely disappear (Williams, and Watanabe, 1978). Thus, they are not simply hyperplastic, but preneoplastic or even neoplastic in nature.

One proposal concerning the biological behavior of foci is that they have the potential to develop into either adenomas or carcinomas (Williams, 1980). In support of this suggestion, microspectrophotometric study of the DNA content of cells in foci, nodules, and carcinomas revealed that eosinophilic foci and nodules both had a wide range of DNA contents, whereas hyperbasophilic foci had an aneuploid DNA content like carcinomas (Mori et al., 1982). It is possible, therefore, that hyperbasophilic foci may be direct precursor lesions for carcinomas, as suggested from kinetic studies (Hirota and Williams, 1979a). The development of carcinomas from foci may require further genetic alterations in the cells of foci. Evidence has been provided that cells in rat liver foci have a lesser DNA-repair capacity than normal cells (Mori et al., 1985). This could render them more susceptible to genetic alteration during continued carcinogen exposure, even though they are less susceptible to toxic effects of activation-dependent carcinogens (Williams et al., 1976; Neal et al., 1981) as a consequence of reduced carcinogen activation and increased detoxification systems (Okita et al., 1976; Judah et al., 1977).

The demonstration that the progression of foci to neoplasm formation is enhanced by promoters (Pitot et al., 1978; Kitagawa and Sugano, 1978; Watanabe and Williams, 1978) provided important evidence that they are a source of neoplasms. Nevertheless, observations that foci can remain dormant or even undergo phenotypic reversion indicates that they are not inexorably destined for neoplastic development, but may be truly preneoplastic. The factors controlling growth or differentiation may be normal regulatory signals which are transmitted to foci through gap junctions (Williams, 1981).

Adenomas

Nodular lesions usually develop at a later stage of carcinogenesis, after induction of foci (Reuber, 1965; Williams and Yamamoto, 1972). These lesions (Figures 4, 5, 12) are characterized by a definite discontinuity with the surrounding hepatic plates, indicating a functional isolation of nodule cells from normal parenchyma (Williams, 1980; Hirota and Yokoyama, 1985). The plates of cells composing the nodules are more than one cell thick and irregular. Nodules can be categorized, like foci, according to their abnormal staining properties, but more often they display a mixture of cell types (Kitagawa, 1971; Hirota and Yokoyama, 1985). They appear variable rather than uniform in their properties as compared to foci.

For many years, nodules have been considered to be hyperplastic and to be precursors of carcinomas (Firminger, 1955; Reuber, 1965; Newberne et al., 1968; Farber, 1973). Support for the view that rat nodules are hyperplastic

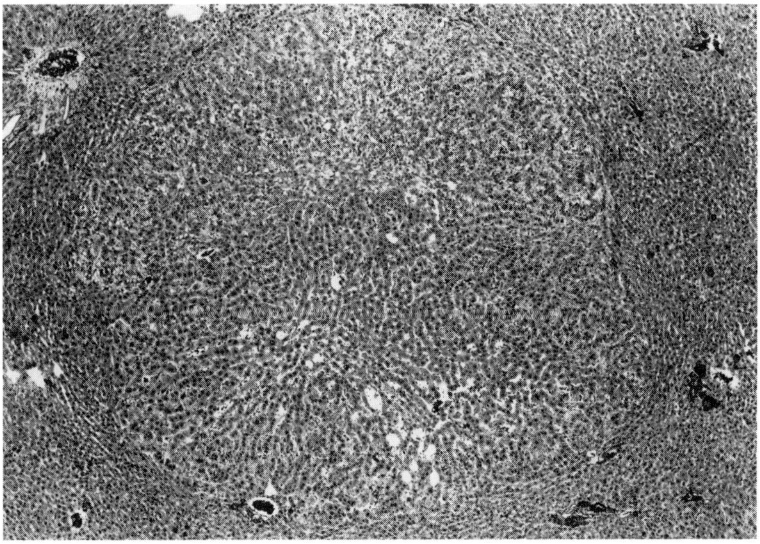

Figure 4. Adenoma (Neoplastic nodule). Formalin fixation, hematoxylin and eosin, magnification: × 43.

comes mainly from studies in which nodules have been reported to disappear following cessation of carcinogen exposure (Teebor and Becker, 1971). Recent experiments concerned with this possibility involve cytotoxic selection of lesions induced very early in carcinogenesis (Ogawa et al., 1980; Tatematsu et al., 1980). Without such a selective pressure, the lesions involved would probably be regarded as altered foci rather than nodules, and, therefore, the behavior of this type of nodular lesion can be interpreted as representing a special situation in which foci have become nodular because of rapid growth. As would then be expected, nodules of this type behave similarly to foci in their high frequency of phenotypic reversion or "remodeling" (Solt et al., 1977) once the selective pressure is removed.

The nature of nodules that are produced under conventional modalities of repeated or continuous exposure or that arise after cessation of carcinogen exposure is apparently different. Such nodules in rats are mostly persistent and progressively growing (Hirota and Williams, 1979a). Some nodules, however, tend to undergo regression with cystic change (Hirota and Williams, 1979a; Hirota and Yokoyama, 1985). The nodules with the regressive change are more distinctly recognizable in the siderotic liver (Figure 5) because of resistance to iron accumulation (see below) in the remnant of the nodule. The growth capability of nodules is apparently limited, since they have not displayed progressive growth in transplantation studies (Williams et al., 1977, 1980), even when recipients were treated with phenobarbital (Mori et al., 1983). Nevertheless, Wanson et al. (1981) reported that when nodule fragments were brought into contact with precultured embryonic chick heart fragments, they invaded the heart tissues. The above-described phenomena imply that the nodules are generally somewhat unstable in terms of independence or autonomy but that some nodule cells have a potential to progressive growth or malignant transformation. This is supported by the finding of decreased fibronectin in most nodules (see below).

Expert groups of pathologists have now recommended that carcinogen-induced hepatic nodules in rats be designated as neoplastic rather than hyperplastic (Squire and Levitt, 1975; Stewart et al., 1980). This is based upon experimental and biological evidence of properties of neoplasia and a general pathological concept that hyperplasia is defined as an increase in number of cells in a tissue or organ, excluding tumor formation. Thus, although the nature of nodules is controversial, we regard nodules as neoplastic rather than preneoplastic lesions and refer to them as adenomas.

It appears that there are several possible pathways for the development of liver cancer. One of these clearly involves the altered focus as a significant "preneoplastic" population. The remainder of this chapter will describe studies on the characteristics of foci and biochemical changes as early indicators for the assessment of liver cell carcinogenesis.

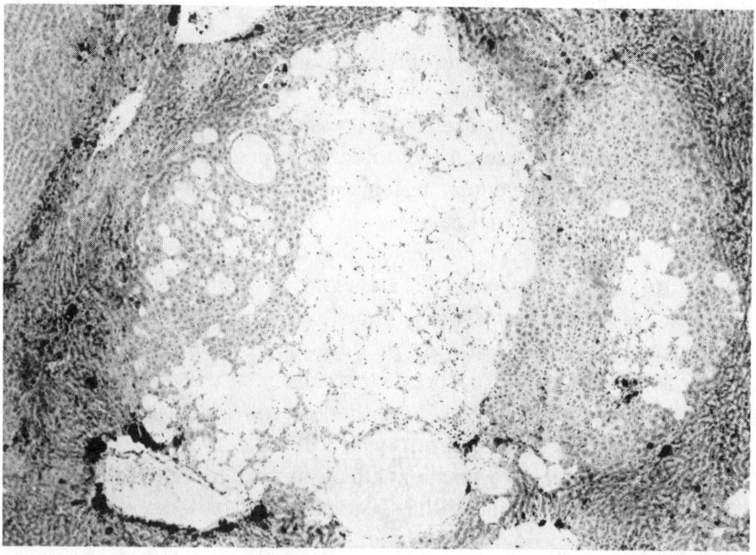

Figure 5. Adenoma with cystic change in the siderotic liver. Formalin fixation, iron reaction, magnification: × 48.

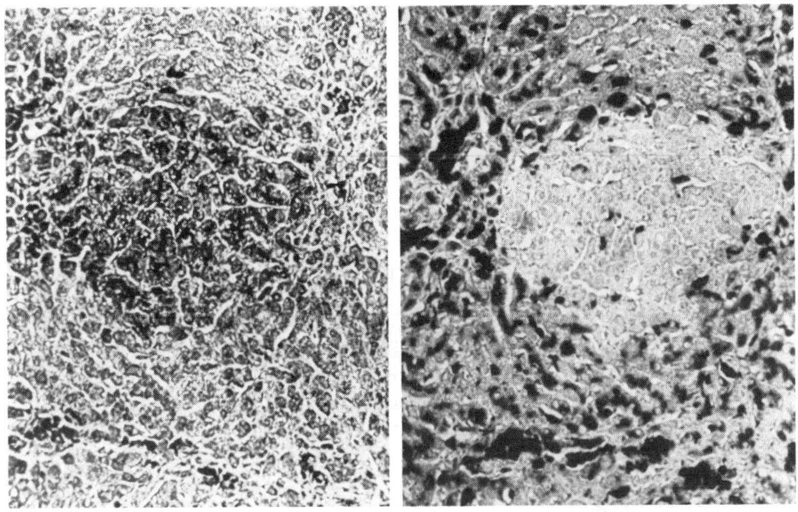

Figure 6. (a) GGT-positive focus. Acetone fixation, GGT reaction, magnification: × 126. (b) Iron-resistant focus. Serial section to Figure 6a, iron reaction, magnification: × 126.

Early Indicators for the Assessment of Liver Cell Carcinogenesis

Enzymatic Alterations

Foci are recognizable by abnormalities in histochemical activities of several enzyme. Carcinogen-induced foci frequently display enzymatic alterations such as deficiency of adenosinetriphosphatase (ATP) (Schauer and Kunze, 1968; Kitagawa, 1971; Rabes et al., 1972; Scherer et al., 1972) or glucose-6-phosphatase (G-6-P) (Gossner and Friedrich-Freksa, 1964; Schauer and Kunz, 1968; Bannasch, 1968; Friedrich-Freksa et al., 1969) and marked positivity for γ-glutamyl transpeptidase (GGT) (Kalengayi et al., 1975; Harada et al., 1976). These three enzymes have been demonstrated as histochemical markers for hepatic neoplasms in human cases as well (Gerber and Thung, 1980). GGT-positive foci (Figure 6a) correspond well to the foci identified by resistance to iron accumulation (Figure 6b) (see below) as well as deficiency of ATP (Figure 7a) and G-6-P (Figure 7b) (Hirota and Williams, 1979b).

Nonspecific esterase (NSE) has been used as a marker for detection of early foci following a single injection of diethylnitrosamine (Hirota et al., 1982). Therein, a very strong positive reaction for NSE (Figure 8) was occasionally shown in some foci, but it seems to be a less sensitive marker because of a rather mildly increased reaction in general. Under certain conditions, ATP and G-6-P exist in all hepatocytes, whereas NSE is primarily in hepatocytes in the central zone of the lobule. The former two are called

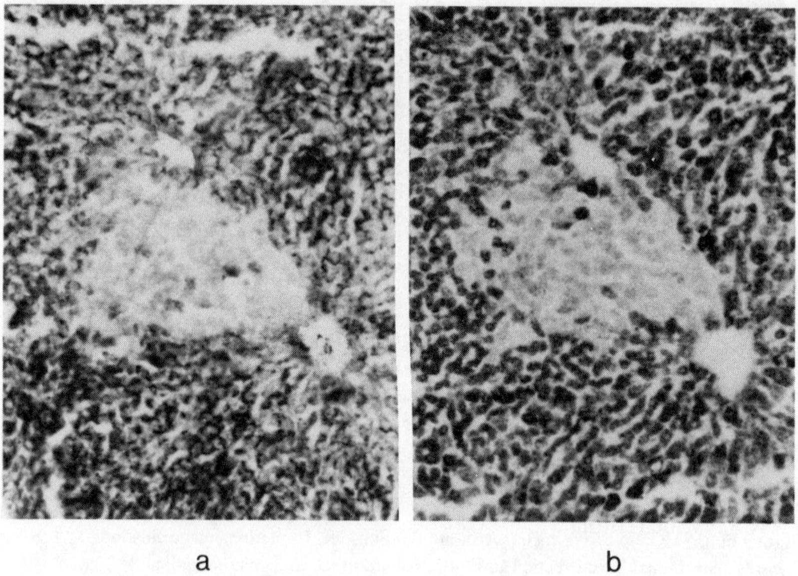

Figure 7. (a) ATP-deficient focus. Frozen section, ATP reaction, magnification: × 120. (b) G-6-P-deficient focus. Serial section to Figure 7a, G-6-P reaction, magnification: × 120.

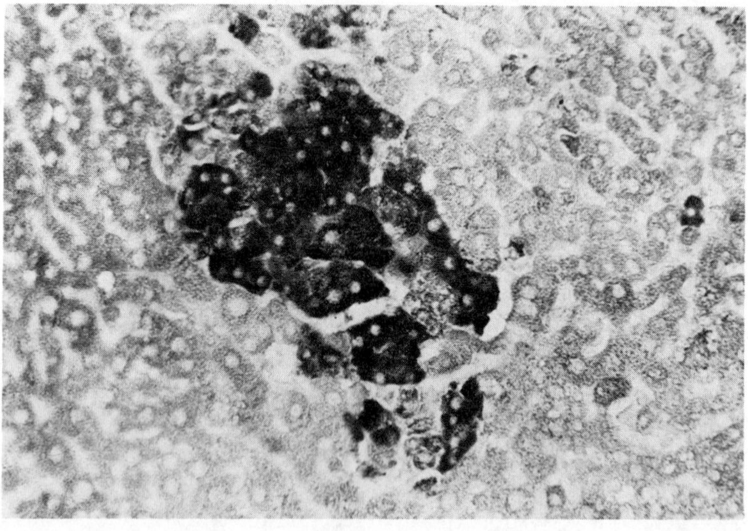

Figure 8. NSE-positive focus. Acetone fixation, NSE reaction, magnification: × 188.

negative markers for altered foci, and the latter is a positive marker as is GGT which is usually absent in normal hepatocytes in adult rats.

GGT, which is a membrane-bound enzyme in various mammalian tissues (Rosalki, 1975; Novogrodsky et al., 1976), plays an important role in hydrolysis of glutathione and its *S* derivatives (Curthoys and Hughey, 1979). GGT activity is high in fetal liver in rats and mice, but diminishes to low levels in the adult liver and is elevated again in liver tumors (Fiala and Fiala, 1970; Fiala et al., 1972). Histochemically, a diffuse positive reaction for GGT is found to be present in immature rat hepatocytes by 13 days of gestation (unpublished observations), but after birth hepatocellular activity of this enzyme gradually disappears and intrahepatic bile ducts become permanently positive for it. The expression of GGT in carcinogen-induced rat hepatocytes can be regarded as a carcinoembryonic feature or acquisition of fetal properties by the transformed cells (Harada et al., 1976; Uriel, 1976).

GGT-negative foci and nodules occur on occasion in the process of the development of carcinoma and they are found to be invariably basophilic or hyperbasophilic (Hirota and Yokoyama, 1985). Poorly differentiated liver cell carcinomas in rats and humans also show a negative reaction for GGT (Tanaka, 1974; Ida, 1977; Hirota and Yokoyama, 1985). Rao et al. (1982, 1986) reported that the absence of GGT activity was a common feature of hepatocarcinogenesis by peroxisome proliferating, hypolipidemic agents (nongenotoxic carcinogens) such as clofibrate. Hepatocytes in x-irradiation-induced altered foci were also negative for GGT (Kitagawa et al., 1985). These studies suggest that the mechanisms of GGT negativity are postulated to be related to the production of free radicals. Further, there is a speculation that the nondepletion of glutathione levels in preneoplastic hepatocytes induced by peroxisome proliferators is responsible for the lack of GGT activity (Hanigan and Pitot, 1985). Another possibility exists that high releasing rate of GGT from lipid layers of the plasma membrane in basophilic foci (see below) may result in its histochemical absence.

The appearance of abnormal enzymes such as aldehyde dehydrogenase (Jones et al., 1984) and a placental form of glutathione *S*-transferase (GST-P) (Sato et al., 1984; Kitahara et al., 1984; Sato, 1988) has been described. Immunohistochemical methods using anti-GST-P antibody demonstrated that GST-P was a more reliable marker than GGT (Tatematsu et al., 1985), although GST-P-negative foci were induced by peroxisome proliferators as were GGT-negative foci (Rao et al., 1986). All the phenotypic properties of enzymatic alterations are generally uniform in their expression within a focus, but often variable between foci (Kitagawa, 1971; Pitot et al., 1978; Pugh and Goldfarb, 1978; Hirota and Williams, 1979b).

Decrease of pyruvate kinase type L has immunohistochemically been demonstrated in ATP-deficient foci produced by different carcinogens (Fischer et al., 1987), indicating that alterations in carbohydrate metabolism may be significant in the process of hepatocarcinogenesis (see below).

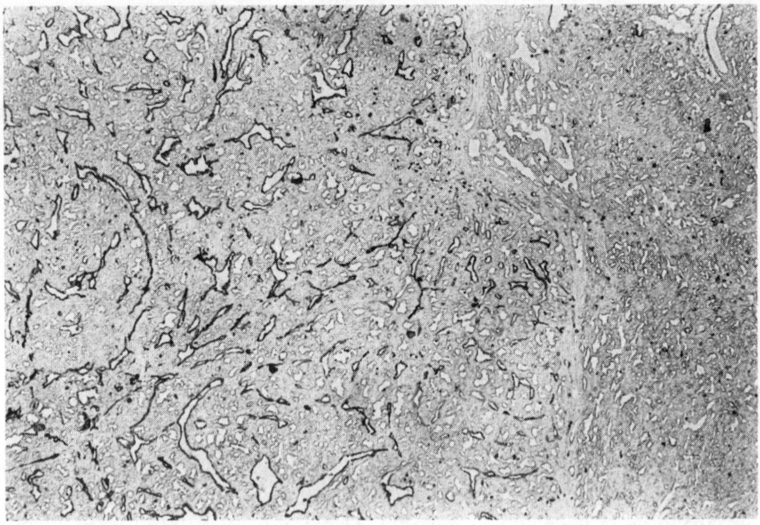

Figure 9. ALP-positive hepatocellular carinoma. Acetone fixation, ALP reaction, magnification: × 25.

Alkaline phosphatase (ALP), which is slightly detected in the bile canaliculi in the normal condition, appears on the plasma membrane of altered foci (Karasaki, 1975) and liver cell carcinomas with typical trabecular pattern (Figure 9), and it is of interest that the intestinal or intestinal-like isozyme of ALP in the carcinomas immunologically is increased in activity (Hirota et al., 1989a), while ALP isozymes of the normal liver consisted mainly of tissue-unspecific (liver/kidney/bone) type.

Serum GGT is elevated in rats exposed to either a hepatocarcinogen (Harada et al., 1976; Remandet et al., 1984) or a liver tumor promoter (Remandet et al., 1984) and, therefore, can be used as an index of hepatocarcinogenesis. Serum ALP "PI anchor" has been demonstrated both biochemically and histochemically in membrane-bound ALP (Hirota et al., 1988) like other cell-surface protein molecules which are covalently linked to the biomembrane via glycosylated PIs detected in a wide variety of eukaryotic cells (Cross, 1987; Low, 1988; Ferguson and Williams, 1988). GGT, on the other hand, was found to be devoid of PI anchor (Hirota et al., 1988); appearance of GGT in the bloodstream is possibly a result of protease digestion of the *N*-terminal residue of GGT and/or a detergent effect of bile. In any case, free lipids such as 1,2-diacylglycerol, which is an activator for protein kinase C (Nishizuka, 1986), resulting from liberation of ALP or GGT may be involved in intracellular signal transduction mechanisms during liver cell carcinogenesis. Thus, it seems worth noting that the elevation of serum levels of ALP or GGT caused by their accelerated release from the plasma membrane of preneoplastic or neoplastic cells may reflect the abnormal growth of those cells with production of 1,2-diacylglycerol on the biomembrane.

Glycogen Storage Abnormalities

Bannasch (1968) first reported that nitrosomorpholine-induced foci displayed extensive storage of glycogen and related the occurrence of hepatoma in human cases of glycogen storage disease to changes in carbonhydrate metabolism during liver carcinogenesis. Subsequently, glycogen-storing foci were demonstrated with other types of carcinogens (Williams et al., 1976). This abnormality is seen more distinctly as persistent retention of cellular glycogen after fasting for 24–36 h (Figure 10), appearing to be related at least in part to a deficiency of G-6-P in foci cells (Bannasch and Angerer, 1974; Hirota and Williams, 1979b; Hirota and Yokoyama, 1985). In a comparative study of glycogen-storage abnormality and other phenotypic changes during hepatocarcinogenesis by 2-acetylaminofluorene in rats, foci and nodules rich in glycogen storage account for a majority of all lesions over the course of experiment, while lesions devoid of glycogen storage gradually increased in number during maintenance. The glycogen-deficient lesions (Figure 11) displayed hyperbasophilia and often GGT negativity suggesting that they are important precursors for malignant transformation (Hirota and Yokoyama, 1985). The glycogen-deficient cells with the possible highest potential to malignancy are ultrastructurally characterized by numerous

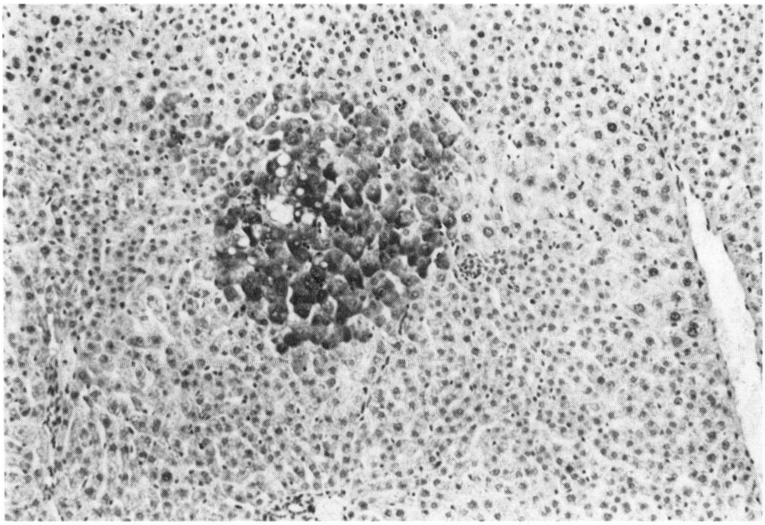

Figure 10. Altered focus with glycogen retention after 36-h fasting. Ethanol fixation, PAS, magnification: × 120.

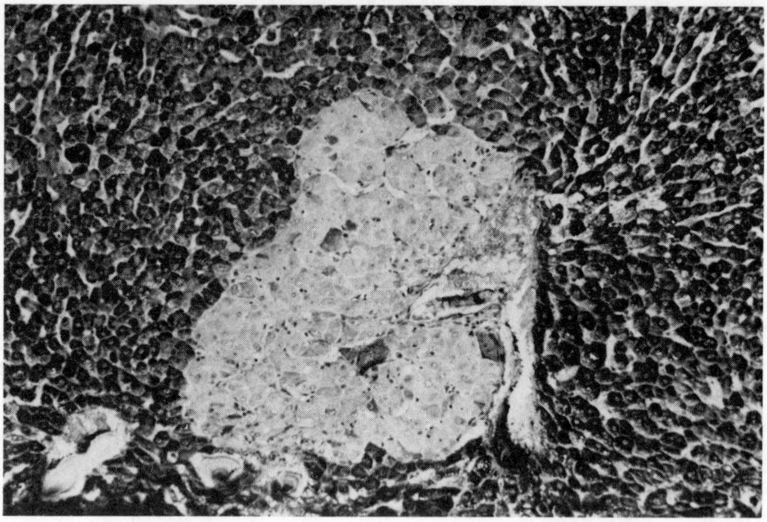

Figure 11. Glycogen-deficient altered focus. Acetone fixation, PAS, magnification: × 140.

polyribosomes lying free in the cytoplasmic matrix (Hirota and Williams, 1982), which is a general feature common to fast-growing populations of cells (Ghadially, 1975). This subcellular alteration accounting for cytoplasmic basophilia is compatible with acceleration of pentose-phosphate cycle with increased biochemical activity of glucose-6-phosphate dehydrogenase and glyceraldehyde-3-phosphate dehydrogenase in basophilic foci and nodules as well as carcinomas (unpublished data). The demonstration of increased glucose-6-phosphate dehydrogenase (Bannasch et al., 1984; Klimek et al., 1984) as a rather late alteration during liver cell carcinogenesis is indicative that early changes such as decrease in the activity of L-pyruvate kinase (Fischer et al., 1987) and increase of uridine diphosphate glucuronyltransferase (Fischer et al., 1983) may be largely shifted, with time, to pentose-phosphate shunt which fuels synthesis of RNA and DNA.

Iron Storage Abnormality

A deficiency in the accumulation of stainable cellular iron during either dietary or parenteral iron overload was shown by Williams and co-workers (Williams and Yamamoto, 1972; Williams, 1976; Williams et al., 1979) to characterize cells in foci (Figure 6b) and nodules (Figures 5, 12) or carcinomas in rat liver. This property provides a highly sensitive marker for the detection of altered foci during carcinogen administration, permitting quantitative studies of hepatocarcinogenesis (Williams and Watanabe, 1978; Hirota and Williams, 1979a; Hirota and Yokoyama, 1985). Following discontinuation of

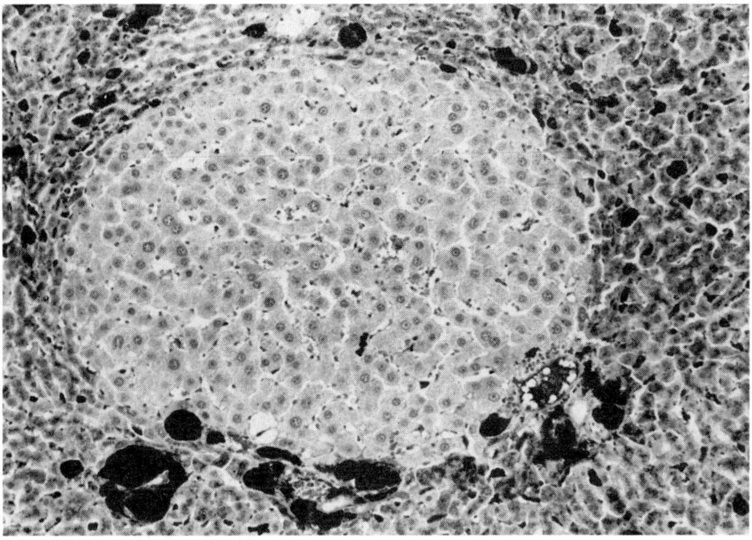

Figure 12. Adenoma with resistance to iron accumulation in the siderotic liver. Formalin fixation, iron reaction, magnification: × 132.

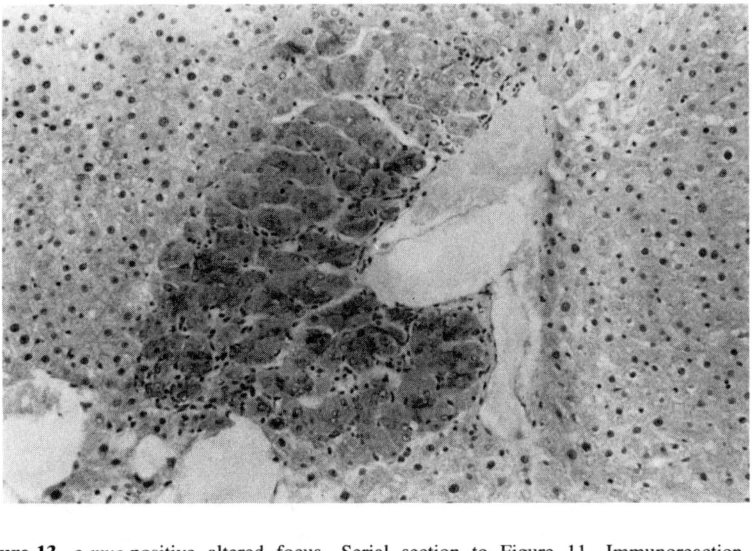

Figure 13. c-*myc*-positive altered focus. Serial section to Figure 11, Immunoreaction for c-*myc*, magnification: × 140.

carcinogen exposure, a phenotypic reversion of this property in foci can occur whereby the cells reacquire the ability to store cellular iron (Williams, and Watanabe, 1978; Furuya et al., 1984).

The basis for a deficiency in iron storage is not well defined. It has been reported that uptake of ^{59}Fe into hepatoma cells is actually increased (Yamashita, 1979), and that ferritin is also increased in carcinogen-induced altered hepatocytes (Beaumont et al., 1981) as well as in hepatoma cells *in vitro* (Lee et al., 1975) and *in vivo* (Lee and Richter, 1977). It is known that, unlike the export protein albumin which is produced on membrane-bound ribosomes, ferritin synthesis occurs on free polyribosomes of hepatocytes and is sensitive to the presence of iron in conformity to its role as an iron storage protein (Zahringer et al., 1977).The ferritin synthesis in normal hepatocytes stimulated by iron is considered to be at the level of translation rather than transcription, since ferritin mRNA is recruited from an inactive pool to free polyribosomes at which translation takes place (Zahringer et al., 1976). The translational response of ferritin synthesis may serve to protect hepatocytes against not only the potentially toxic effects of a rapid influx of iron but also against DNA damage by iron were there a direct interaction of the metal and a ferritin gene-promoter region in DNA (Shull and Theil, 1983). To reconcile the observations of increased iron uptake or ferritin production as described above and the absence of stainable iron in preneoplastic and neoplastic hepatocellular lesions, some explanations have been proposed. One of these involves crystal growth theory (Harrison et al., 1974; Hoy et al., 1974) that uptake and release of ferritin iron are demonstrated in not only carcinomas but also discrete areas of basophilic hepatocytes containing no hepatitis B surface antigen (Hirota et al., 1982).

Production of α-Fetoprotein

In 1963, Abelev made the discovery of the appearance of a fetal protein, α-fetoprotein (AFP), in the serum of adult mice with liver cancer (Abelev, 1971). Subsequently, production of this protein by both human and rat hepatocellular carcinomas was documented. Moreover, several laboratories (Watabe, 1971; Kitagawa et al., 1972; Kroes et al., 1972; Sell, 1978) independently found that an increase in serum AFP could be detected within a few weeks of carcinogen administration, long before the development of liver cancers. These studies demonstrated a biphasic pattern of the elevation of AFP; after early induction, if carcinogen exposure was discontinued, AFP diminished and reappeared only with the development of liver neoplasms. Thus, elevation of serum AFP was established as an indicator both of the early effects of liver carcinogens and of the presence of liver neoplasms.

The basis for the appearance of serum AFP early in hepatocarcinogenesis has been controversial. In several of the studies that first documented this phenomenon, AFP was associated with the proliferation of oval cells (Watabe, 1971; Kroes et al., 1973; Dempo et al., 1975). Subsequently, immunofluorescent and immunoperoxidase staining demonstrated the localization of AFP to these cells, but not those in foci (Tchipysheva et al., 1977; Kuhlmann, 1978; Sell, 1978; Guillouzo et al., 1981). Cytokeratin and AFP expression at an early stage of hepatocarcinogenesis by 3′-methyl-4-dimethylaminoazobenzene indicates the ductular origin of oval cells and a possible relation to immature hepatocytes of rats (Germain et al., 1985). But, it is not resolved whether AFP-positive cells possess a significant potential for neoplastic or malignant transformation. Another possibility for early AFP production suggested by Kroes et al. (1972) was the direct or indirect derepression by hepatocarcinogens of the gene or genes responsible for AFP production.

Thus, several mechanisms may exist for elevated AFP levels early in hepatocarcinogenesis. Although cellular localization of AFP is apparently not a useful marker for foci, elevated serum levels may be used as an indicator of hepatocarcinogenesis and is one of the few properties that can also be monitored in humans.

Disposition of Fibronectin

Fibronectin (FN), which is a high-molecular-weight glycoprotein in the extracellular matrix, is generally derived from fibroblasts, but it is, in some instances, produced by several types of epithelial cells (Ruoslahti et al., 1981). Normal hepatocytes secrete plasma FN and hepatocytes in vitro can secrete cellular FN (Martinez-Hernandez, 1984; Hynes, 1985; Schwarzbauer et al., 1985). However, whether the FN present in the space of Disse represents a plasma or cellular form or both is unknown, and it has not been well established that this is derived from hepatocytes or from endothelial cells (Martinez-Hernandez, 1988).

Immunohistochemical studies have examined the distribution of FN in the liver during hepatocarcinogenesis in rats (Sell and Ruoslahti, 1982; Maruyama et al., 1988), describing a significant decrease of FN content in preneoplastic and neoplastic lesions. Maruyama et al. (1988) reported that 40% of the altered foci and 90% of adenomas showed decreased FN content. They concluded that the decrease of FN in foci may be related to their development into adenomas. The cause of a decreased FN in neoplastic lesions is presently unclear (Martinez-Hernandez, 1988). Paucity of FN of cells in some foci and nodules with higher mitotic rates appears consistent with decrease of cell-surface associated FN before cell division demonstrated by Hynes (1973). However, some hepatocellular carcinomas secrete considerable amounts of FN (Sell and Ruoslahti, 1982; McCracken et al., 1984; Robinson et al., 1984).

The extracellular matrix is thought to be important in epithelial-mesenchymal interactions, influencing the processes of development or growth of neoplasms, as already established in tissue interactions during liver morphogenesis (LeDouarin, 1970). To substantiate this, further studies on extracellular substances, including FN, during liver carcinogenesis will be needed.

Glutamine Synthetase

This enzyme, which synthesizes glutamine, is confined to a ring of about three cells deep around the central veins. Gebhardt et al. (1989) showed that some hepatocarcinogens intially deplete activity after which foci arise which express the enzyme.

Expression of Oncogenes

Huebner and Todaro (1969) proposed the oncogene hypothesis that protooncogenes are present in all normal cells and that activation of these genes by retroviruses or carciongens leads to cancer. Since then, many studies on the expression of cellular (c-) oncogenes in a variety of normal tissues and neoplasms have been made. Of those, the c-*ras* and c-*myc* genes (Muller et al., 1982; Gonda et al., 1982) are of special relevance for liver pathology. In developing livers, c-*ki-ras,* c-*ha-ras,* and c-*myc* transcript levels are high at 17 days of gestation but decline to the low values characteristic of adult rat livers between 20 days of gestation and 3 days after birth (Yaswen et al., 1985). Increase in c-*ras* transcripts (up to eightfold of the normal level) is associated with normal proliferaterative processes such as regeneration in rat liver following partial hepatectomy (PH) (Goyette et al., 1983). The increase of such transcripts begins at about 18 h after PH but the level returns to normal by 72 to 96 h. Makino et al. (1984a) reported that marked increase in the expression of the c-*myc* oncogene found in rat liver at a very early stage of regeneration (up to ten- to 15-fold, within 1–3 h after PH) is transient. That is, the expression begins to decrease rapidly after 4 h and returns to less than double the normal level after 8 h. In reference to a possible regulatory mechanism of the c-*myc* gene, they suggested, taken together with the result of larger increase in c-*myc* transcription induced by inhibitory effect of cycloheximide on protein synthesis, that the c-*myc* is repressed by a short-lived protein that becomes abundant shortly after the onset of the proliferative process.

The changes that occur in the DNA of cells affected by genotoxic carcinogens, including point mutations, translocation, gene amplification, have been implicated in the activation of c-oncogenes to transforming activity (Balmain, 1985; Klein and Klein, 1985). Alteration in DNA methylation is also implicated in the activation of some genes (Razin and Riggs, 1980; Doerfler, 1983; Hoffman, 1984), and, in fact, their transcriptional activities are inversely related with their methylation status in normal or neoplastic livers of the mouse, rat, and human (Lapeyre et al., 1981; Vorce and Goodman, 1983, 1987; Kaneko et al., 1985; Nambu et al., 1987). Moreover, carcinogen-induced hypomethylation correlates with the increased transcriptional complexity of the mRNA of treated cells (Boehm and Drahovsky, 1983), indicating the initiation of transcription at sites previously inactive. Other oncogenes may be activated indirectly during the progression of liver carcinogenesis by either nontargeted or targeted genetic events in somatic cells. For example, some c-oncogenes such as c-*fos* (Greenberg and Ziff, 9184; Mitchell et al., 1985) and c-*myc* (Kelly et al., 1983; Bravo et al., 1985; Greenberg et al., 1985) are potential targets of protein kinase C which plays an important role in cell-cell communication (Nishizuka, 1986).

A feature of the activation of oncogenes reported thus far in liver adenomas or carcinomas is the high prevalence of activation of ras c-oncogenes in mice (Reynolds et al., 1986, 1987; Wiseman et al., 1986) and to a lesser extent in rats (McMahon et al., 1986). The reproducible activation of ras oncogenes in carcinogen-induced liver tumors has permitted studies of correlation between the activating mutations in the ras genes and the known mutagenic effects of certain chemical carcinogens (reviewed by Barbacid, 1987; Balmain and Brown, 1988). *H-ras*-1 oncogenes present in spontaneous liver tumors of aged B6C3F1 mice are activated by random mutations in the 61st codon (Fox and Watanabe, 1985; Fox et al., 1987; Reynolds et al., 1986, 1987), whereas the activating mutations conform to a specific pattern in the case of carcinogen-induced liver cell carcinomas (Wiseman et al., 1986; Reynolds et al., 1987), indicating ras oncogenes can be directly activated by the mutagenic properties of the initiating carcinogens.

Demonstration of five- to tenfold amplification of the c-*myc* gene and the overexpression of its mRNA in Morris hepatomas (Hayashi et al., 1984) implies that abnormal expression of the c-*myc* gene may be responsible for the progressive growth of liver carcinoma cells. A subsequent study documented that the c-*myc* gene was associated with rat liver carcinogenesis by 3′-methyl-4-dimethylaminoazobenzene, while c-*Ha-ras* gene was related to proliferation of hepatocytes, suggesting that activation of the c-*myc* gene might be caused by some irreversible change in its structure (Makino et al., 1984b). During hepatocarcinogenesis using a choline-deficient diet containing ethionine, c-*Ki-ras* and c-*myc* expressions increase in whole livers during the 35 weeks of the diet and in oval cells also, whereas c-*Ha-ras* transcripts increase transiently (Yaswen et al., 1985). Taking an established positive correlation between hypomethylation and gene activation, as described above, it is possible that the activation of those oncogenes may be associated with demethylation in DNA, because the reduction of methyl donors in the liver by feeding of choline-deficient diet may render liver cells more vulnerable to hypomethylation of DNA (Shinozuka, 1986).

Studies using immunohistochemical techniques have the advantage of allowing assessment of oncogene expression of individual cells in preneoplastic hepatocellular lesions, particularly in microscopic lesions which are difficult to isolate. This limitation is also overcome by the use of polymerose chain reaction amplification of DNA.

Recently, specific mutations in the p53 gene have been reported for human liver cancers (Hsu et al., 1991; Bressac et al., 1991). Interestingly, wild type p53 acts as a tumor suppressor gene, but some mutants function as dominant oncogenes.

Resistance to Cytotoxicity

Haddow (1938) first documented the resistance of neoplastic cells to the cytotoxic effects of carcinogens, and later investigators (Vasiliev and Guelstein, 1963) expanded on this feature. Farber and co-workers demonstrated that altered foci and hyperplastic nodules (Williams et al., 1976; Solt et al., 1977; Farber et al., 1979) in rat liver were similarly resistant to cytotoxicity. The resistance is not specific to the inducing agent; that is, lesions induced by one carcinogen are resistant to the cytotoxic effects of a chemically dissimilar agent (Laishes et al., 1978; Neal and Butler, 1978; Buckley et al., 1979). The resistance appears to result largely from alteration of drug-metabolizing enzymes leading to reduced xenobiotic activation and enhanced detoxification systems. In fact, microsomal cytochrome P450-dependent enzymes (Cameron et al., 1976) are decreased in 2-acetylaminofluorene-induced liver nodules of rats, while glutathione transferase (Neal et al., 1981; Kithara et al., 1983), uridine diphosphate-gluconyl-transferase (Bock et al., 1982; Fischer et al., 1983), and epoxide hydrolase (Novikoff et al., 1979; Kuhlmann et al., 1981) are increased in such nodules. Neal et al. (1981) reported the occurrence of a decreased covalent binding of aflatoxin B_1 to DNA following low-level feeding with the toxin which was due to increased detoxification of the agent by conjugation with reduced glutathione.

In the LEC rats, which have been established as a mutant strain displaying hereditary hepatitis and subsequent liver cancer (Sasaki et al., 1985; Yoshida et al., 1987), a large decrease in the content of cytochrome P450 and marked increase in the activities of epoxide hydrolase and GGT occur before the onset of fulminant hepatitis (Sugiyama et al., 1988). GST-P-positive foci appeared following hepatitis and their number and size increased with age (Oyamada et al., 1988).

The expression of a gene *mdr* associated with resistance to antineoplastic agents was elevated in hyperplastic nodules and hepatocellular carcinoma (Fairchild et al., 1987). Changes in the expression of several phase I and phase II drug-detoxifying enzymes might be closely related to the overexpression of *mdr*.

CONCLUSIONS

Hepato carcinogens typically induce large numbers of altered foci, although very few progress to neoplasm formation, indicating that there are biologically different types of foci. For instance, there seems to be differences between cryptogenic ''spontaneous'' foci and those induced by carcinogens. Studies of oncogene activation and expression inactivation of tumor suppressor genes may shed further light on the nature of different types of foci.

The induction of foci well in advance of carcionomas makes them a useful indicator of liver carcinogenesis, at least for genotoxic carcinogens. Liver neoplasm promoters enhance the expression of certain phenotypic abnormalities in foci and facilitate their growth. These effects can be used to assay for liver neoplasm promoters and, in deed, may become essential features for the definition of promoters.

The liver is the most versatile organ for biotransforming chemicals to react with DNA and hence much work using the liver as a model organ is being done. However, some basic biological and biochemical concepts pertaining to assessment of risk potential of either initiating or promoting agents for humans remain to be established.

REFERENCES

Abelev GI: *Adv Cancer Res* 1971; 14:295–358.
Balmain A: *Br J Cancer* 1985; 51:1–7.
Balmain A and Brown K: (1988). *Adv Cancer Res* 1988; 51:147–182.
Bannasch P: *Recent Results Cancer Res* 1968; 19:1–100.
Bannasch P: *Cancer Res* 1976; 36:2555–2562.
Bannasch P: *J Cancer Res Clin Oncol* 1984; 108:11–22.
Bannasch P and Angerer H: *Geschwulstforsch* 1974; 43:105–114.
Bannasch P, Hacker H-J, Klimek F, and Mayer D: *Adv Enz Regul* 1984; 22:97–121.
Barbacid M: *Annu Rev Biochem* 1987; 56:779–827.
Barbason H and Betz EH: *Br J Cancer* 1981; 44:561–566.
Beaumont C, Le Rumeur E, Guillouzo C, Latinier MF, Bourel M, and Guillouzo A: *Biochem Biophys Res Commun* 1981; 99:879–885.
Bock KW, Lilienblum W, Pfeil H, and Erikson LC: *Cancer Res* 1982; 42:3747–3752.
Boehm TLJ and Drahovsky D: *Recent Results Cancer Res* 1983; 84:212–225.
Braun L, Goyette M, Yaswen P, Thompson NL, and Fausto N: *Cancer Res* 1987; 47:4116–4124.
Bravo R, Burckhardt J, Curran T, and Muller R: *EMBO J* 1985; 4:1193–1197.
Bressac B, Kew M, Wands J, and Ozturk M: *Nature* 1991; 350:429–430.
Buckley JD, O'Connor PJ, and Craig AW: *Nature* 1979; 281:403–404.
Bursch W, Lauer B, Timmermann-Trosiener I, Barthel G, Schuppler J, and Schulte-Hermann R: *Carcinogenesis* 1984; 5:453–458.
Cameron R, Sweeny GD, Jones K, Lee G, and Farber E: *Cancer Res* 1976; 36:3888–3893.
Cater KC, Gandolfi AT, and Sipes IC: *Toxicol Pathol* 1985; 13: 3–9.

Cooper GM: *Science* 1982; 217:801–806.
Cross GAM: *Cell* 1987; 48:179–181.
Curthoys NP and Hughey RP: *Enzyme* 1979; 24:383–403.
Daoust R: *Cancer Res* 1972; 32:2502–2509.
Dempo K, Hisaka N, Yoshida Y, Kaneko A, and Onoe T: *Cancer Res* 1975; 35:1282–1287.
Doerfler W: *Annu Rev Biochem* 1983; 52:93–124.
Duesberg PH: *Nature* 1983; 304:219–226.
Emmelot P and Scherer E: *Biochim Biophys Acta* 1980; 605:247–304.
Fairchild CR, Ivy SP, Rushmore T, Lee G, Koo P, Goldsmith ME, Myers CE, Farber E, and Cowan KH: *Proc Natl Acad Sci USA* 1987; 84:7701–7705.
Farber E: *Cancer Res* 1956; 7:142–155.
Farber E: *Methods Cancer Res* 1973; 7:345–375.
Farber E, Parker S, and Gruenstein M: *Cancer Res* 1979; 36:3879.
Ferguson MAJ and Williams AF: *Annu Rev Biochem* 1988; 57:285–320.
Fiala S and Fiala AE: *Experientia* 1970; 26:889–890.
Fiala S, Fiala AE, and Dixon BJ: *J Natl Cancer Inst* 1972; 48:1393–1401.
Firminger HI: *J Natl Cancer Inst* 1955; 15(Suppl):1427–1442.
Fischer G, Altmannsberger M, Schauer A, and Katz N: *J Canc Res Clin Oncol* 1983; 106:53–57.
Fischer G, Ruschenburg I, Eigenbrodt E, and Katz NJ: *Cancer Res Clin Oncol* 1987; 113:430–436.
Fox TR and Watanabe DG: *Science* 1985; 228:596–597.
Friedrich-Freksa H, Papadopulu G, and Gössner W: *Z Krebsforsch* 1969; 72:240–253.
Furukawa K, Numoto S, Furuya K, Williams GM: *Cancer Res* 1985; 45:5011–5019.
Furuya K, Maeura Y, Williams GM: *Toxicol Pathol* 1984: 12:136–142.
Gebhardt R, Tanaka T, and Williams GM: *Carcinogenesis* 1989; 10:1917–192.
Germain L, Goyette R, and Marceau N: *Cancer Res* 1985; 45:673–681.
Ghadially FN: *Ultrastructural Pathology of the Cell* 1975; 1:530.
Gonda MA, Young HA, Elser JE, Rashed S, Talmadge CB, Nagashima K, Li CC, and Gilden RV: *J Virol* 1982; 44:520–529.
Gössner W and Friedrich-Freksa H: *Z Naturforsch* 1964; 19:862–863.
Goyette M, Petropoulos CJ, Shank PR, and Fausto N: *Science* 1983; 219:510–512.
Greenberg ME, Greene LA, and Ziff EB: *J Biol Chem* 1985; 260:14101–14110.
Greenberg ME and Ziff EB: *Nature* 1984; 311:433–438.
Grisham JW and Hartroft WS: *Lab Invest* 1961; 10:317–332.
Grisham JW and Porta EA: *Exp Mol Pathol* 1964; 3:242–261.
Guillouzo A, Weber A, Le Provost E, Rissel M, and Schapira F: *J Cell Sci* 1981; 49:249–260.
Haddow A: *Acta Uni Int Cancrum* 1938; 3:342–352.
Hanigan MH and Pitot HC: *Carcinogenesis* 1985; 6:165–172.
Harada M, Okabe K, Shibata K, Masuda H, Miyata K, and Enomoto M: *Acta Histochem* 1976; 9:168–179.
Harrison PM, Hoy TG, Macara IG, and Hoare RJ: *Biochem J* 1974; 143:445–451.
Hayashi K, Makino K, and Sugimura T: *Gann* 1984; 75:475–478.
Hayner NT, Braun L, Yaswen P, Brooks M, and Fausto N: *Cancer Res* 1984; 44:332–338.

Hirota N, Moryama S, and Yokoyama T: *J Natl Cancer Inst* 1982; 69(6):1299–1304.
Hirota N and Williams GM: *J Natl Cancer Inst* 1979a; 63:1257–1265.
Hirota N and Williams GM: *Am J Pathol* 1979b; 95:317–324.
Hirota N and Williams GM: *Cancer Res* 1982; 42:2298–2309.
Hirota N, Hamazaki M, and Williams GM: *Hepatogastroenterology* 1982; 29:49–51.
Hirota N and Yokoyama T: *Acta Pathol Jpn* 1985; 35:1163–1179.
Hirota N, Moriyama S, and Yokoyama T: *J Natl Cancer Inst* 1982b; 69:1299–1304.
Hirota N, Sakai T, and Komoda T: *Acta Histochem Cytochem* 1988; 21:606.
Hirota N, Sakai T, and Yokoyama T, and Komoda T: 1989a (in preparation).
Hirota N, Sakai T, and Williams GM: 1989b (in preparation).
Hoffman RM: *Biochim Biophys Acta* 1984; 738:48–87.
Hoy TG, Harrison PM, and Shabbir M: *Biochem J* 1974; 139:603–607.
Huebner RJ and Todaro GJ: *Proc Natl Acad Sci USA* 1969; 64:1087–1094.
Hsu IC, Metcalf RA, Sun T, Welsh JA, Wang NJ, and Harris CC: *Nature* 1991; 350:427–428.
Hynes RO: *Biochim Biophys Acta* 1973; 458:73–107.
Hynes RO: *Annu Rev Cell Biol* 1985; pp 67–90.
Ida M: *Acta Pathol Jpn* 1977; 27:647–655.
Inaoka Y: *Gann* 1967; 58:355–366.
Iwasaki T, Dempo K, Kaneko A, and Onoe T: *Gann* 1972; 63:21–30.
Jalanko M and Rouslahti E: *Cancer Res* 1979; 39:3495–3501.
Jones DE, Jr, Evces S, and Lindhal R: *Carcinogenesis* 1984; 5:1679–1687.
Judah DJ, Legg RF, and Neal GE: *Nature (London)* 1977; 265:343–345.
Kalengayi MMR, Rochi G, and Desmet VJ: *J Natl Cancer Inst* 1975; 55:579–588.
Kaneko Y, Shibuya M, Nakayama T, Hayshida N, Toda G, Endo Y, Oka H, and Oda T: *Jpn J Cancer Res (Gann)* 1985; 76:1136–1140.
Karasaki S: *Cancer Res* 1975; 35:482–491.
Kelly K, Cochran RN, Stiles CD, and Leder P: *Cell* 1983; 35:603–610.
Kitagawa T: *Gann* 1971; 62: 207–216.
Kitagawa T and Sugano H: *Gann* 1978; 69:679–687.
Kitahara A, Satoh K, Nishimura K, Ishikawa T, Ruike K, Sata K, Tsuda H, and Ito N: *Cancer Res* 1984; 44:2698–2703.
Kitahara A, Satoh K, and Sato J: *Biochem Biophys Res Commun* 1983; 112:20–28.
Kitagawa T, Nomura K, and Sasaki S: *Cancer Res* 1985; 45:6078–6082.
Kitagawa T, Yokochi T, and Sugano H: *Int J Cancer* 1972; 10:368–381.
Klein G and Klein E: *Nature* 1985; 315:190–195.
Klimek F, Mayer D, and Bannasch P: *Carcinogenesis* 1984; 5:265–268.
Kroes R, Williams GM, and Weisburger JH: *Cancer Res* 1972; 32:1526–1532.
Kroes R, Williams, GM, and Weisburger JH: *Cancer Res* 1973; 33:613–617.
Kuhlmann WD: *Int J Cancer* 1978; 21:368–380.
Kuhlmann WD, Krischan R, Kunz W, Guenther TM, and Oesch F: *Biochem Biophys Res Commun* 1981; 98:417–423.
Laishes BA, Roberts E, and Farber E: *Int J Cancer* 1978; 21:186–193.
Lapeyre JN, Walker MS, and Becker FF: *Carcinogenesis* 1981; 2:873–878.
LeDouarin N: in Wolff E (ed): *Tissue Interactions During Organogenesis*. New York, Gordon and Breach, 1970, pp 37–70.
Lee JCK, Lee SSC, Schlesinger KJ, and Richter GW: *Am J Pathol* 1975; 80:235–248.
Lee SSC and Richter GW: *J Biol Chem.* 1977; 252:2054–2059.

Lipsky MM, Hinton DE, Goldblatt PJ, Klaunig JE, and Trump BF: *Path Res Pract* 1979; 164:178–185.
Lipsky MM,Hinton DE, Klaunig JE, Goldblatt PJ, and Trump BF: *Carcinogenesis* 1980; 1:151–156.
Lopez M and Mazzanti L: *J Pathol Bacteriol* 1955; 69:243–250.
Low MG and Saltiel AR: *Science* 1988; 239:268–275.
Makino R, Hayashi K, and Sugimura T: *Nature* 1984a; 310:697–698.
Makino R, Hayashi K, Sato S, Sugimura T: *Biochem Biophys Res Commun* 1984b; 119:1096–1102.
Malvaldi G: *Eur J Cancer* 1981; 17:481–483.
Marceau N, Deschenes J, and Valet JP: *Oncodev Biol Med* 1982; 3:49–63.
Martinez-Hernandez A: *Lab Invest* 1984; 51:57–74.
Martinez-Hernandez A: *Lab Invest* 1988; 58:609–612.
Maruyama H, Tanaka T, Gebhardt R, Berghem L, and Williams GM: *Lab Invest* 1988; 58:630–635.
McCracken AA, Emmett M, Crowle AJ, Brown JL: *Hepatology* 1984; 4:715–721.
McMahon G, Davis E, and Wogan GN: *Proc Natl Acad Sci USA* 1987; 84:4974–4978.
McMahon G, Hanson L, Lee JJ, and Wogan GN: *Proc Natl Acad Sci USA* 1986; 83:9418–9422.
Mitaka T and Tsukada H: *Jpn J Cancer Res (Gann)* 1987; 78:785–790.
Mitchell RL, Zokas L, Schreiber RD, and Verma IM: *Cell* 1985; 40:209–217.
Moore MA and *Kitagawa*: *Int Rev Cytol* 1986; 101:125–173.
Moore MR, Drinkwater NR, Miller EC, Miller JA, and Pitot HC: *Cancer Res* 1981; 41:1585–1593.
Mori MA, Mayer D, and Bannasch P: *Carcinogenesis* 1982; 3:1429–1436.
Mori H, Furuya K, and Williams, GM: *J Natl Cancer Inst* 1983; 71:849–854.
Mori H, Sugie S, Ohbayashi F, Shima H, Yoshimi N, Takahashi M, and Williams, GM: *Carcinogenesis* 1985; 6:1087–1090.
Mori H, Tanaka T, Sugie S, Takahashi M, and Williams GM: *J Natl Cancer Inst* 1982; 69:1277–1282.
Muller R, Slamon DJ, Tremblay JM, Cline MJ, and Verma IM: *Nature* 1982; 299:640–644.
Nambu S, Inoue K, and Sasaki H: *Jpn J Cancer Res (Gann)* 1987; 78:695–704.
Neal GE and Butler WH: *Br J Cancer* 1978; 37:55–60.
Neal GE, Metcalfe SA, Legg RF, Judah DJ, and Green JA: *Carcinogenesis* 1981; 2:457–461.
Newberne PM and Wogan GN: *Cancer Res* 1968; 28:770–781.
Nishizuka Y: *Science* 1986; 233:305–312.
Novikoff AB, Novikoff PM, Stockert RJ, Becker FF, Yam A, Poruchynsky MS, Levin W, and Thomas PE: *Proc Natl Acad Sci USA* 1979; 76:5207–5211.
Novogrodsky A, Tate S, and Meister A: *Proc Natl Acad Sci USA* 1976; 73:2414–2418.
Ogawa K, Onoe T, and Tabeuchi J: *J Natl Cancer Inst* 1981; 67:407–412.
Ogawa K, Minase T, and Onoe T: *Cancer Res* 1974; 34:3379–3386.
Ogawa K, Solt DB, and Farber E: *Cancer Res* 1980; 40:725–732.
Okita K, Noda K, Fukumoto Y, and Takemoto T: *Gann* 1976; 67:899–902.
Oyamada M, Dempo K, Fujimoto Y, Takahashi H, Satoh MI, Mori M, Masuda R, Yoshida MC, and Satoh K, and Sato K: *Jpn J Cancer Res* 1988; 79:5–8.

Peraino C, Fry RJ, and Staffeldt E: *Cancer Res* 1971; 31:1506–1512.
Peraino C, Staffeldt EF, Carnes BA, Ludeman VA, Blomquist JA, and Vesselinovitsch SD: *Cancer Res* 1984; 44:3340–3347.
Pereira MA: *J Am Coll Toxicol* 1982; 1:101–118.
Pitot HC, Barsness L, Goldsworthy T, and Kitagawa T: *Nature (London)* 1978; 271:456–458.
Pugh TD and Goldfarb S: *Cancer Res* 1978; 38:4450–4457.
Rabes HM, Bucher T, Hartmann A, Linke I, and Dunnworld M: *Cancer Res* 1982; 42:3220–3227.
Rabes HM, Scholze P, and Jantsch B: *Cancer Res* 1972; 32:2577–2586.
Rabes HM and Szymkowiak R: *Cancer Res* 1979; 39:1298–1304.
Rao MS, Lalwani ND, Scarpelli DG, and Reddy JK: *Carcinogenesis* 1982; 3:1231–1233.
Rao MS, Subbarao M, and Reddy HK: *J Natl Cancer Inst* 1986; 77:951–956.
Razin A and Riggs AD: *Science* 1980; 210:604–610.
Remandet B, Gouy D, Berthe J, Mazue G, and Williams GM: *Fundam Appl Toxicol* 1984; 4:152–163.
Reuber MD: *J Natl Cancer Inst* 1965; 34:697–724.
Reynolds SH, Stowers SJ, Maronpot RR, Anderson MW, and Aaronson SA: *Proc Natl Acad Sci USA* 1986; 83:33–37.
Reynolds SH, Stowers SJ, Patterson RM, Maronpot RR, Aaronson SA, and Anderson MW: *Science* 1987; 237:1309–1316.
Richards WL, Tsukada Y, and Potter R: *Carcinogenesis* 1982; 3:1355–1359.
Robinson J, Viti M, and Hook M: *J Cell Biol* 1984; 98:946–953.
Rosalki SB: *Adv Clin Chem* 1975; 17:53–107.
Ruoslahti E, Engvall E, Hayman EG: *Coll Res* 1981; 1:95–128.
Rubin E: *Exp Mol Pathol* 1964; 3:279–286.
Sasaki T and Yoshida T: *Virchows Arch Pathol Anat* 1935; 295:175–200.
Sasaki M, Yoshida MC, Kagami K, Takeichi N, Kobayashi H, Dempo K, and Mori M: *Rat News Lett* 1985; 14:4–6.
Sato K: *Jpn J Cancer Res* 1988; 79:556–572.
Sato K, Kitahara A, Satoh K, Ishikawa T, Tatematsu M, and Ito N: *Gann* 1984; 75:199–202.
Schauer A and Kunze E: *Z Krebsforsch* 1968; 70:252–266.
Scherer E, Hoffmann M, Emmelot P, and Friedrich-Freska M: *J Natl Cancer Inst* 1972; 49:93–106.
Schwarzbauer JE, Paul JI, and Hynes RO: *Proc Natl Acad Sci USA* 1985; 82:1424–1428.
Sell S: *Cancer Res* 1978; 38:3107–3113.
Sell S: *Cancer Res* 1983; 43:1761–1767.
Sell S and Leffert HL: *Hepatology* 1982; 2:77–86.
Sell S, Leffert HL, Shinozuka H, Lombardi B, and Gochman N: *Gann* 1981; 72:479–487.
Sell S and Rouslahti E: *J Natl Cancer Inst* 1982; 69:1105–1114.
Shull GE and Theil EC: *J Biol Chem* 1983; 258:7921–7923.
Singer B and Kusmierek JT: *Annu Rev Biochem* 1982; 51:655–693.
Shinozuka H, Lombardi B, and Abanobi SE: *Carcinogenesis* 1982; 3:1017–1020.
Solt DB, Medline A, and Farber E: *Am J Pathol* 1977; 88:595–618.
Squire RA and Levitt MH: *Cancer Res* 1975; 35:3214–3233.

Stenback F, Mori H, Furuya K, and Williams, GM: *J Natl Cancer Inst* 1986; 76:327–333.
Stewart HL, Williams GM, Keysser CH, Lombard LS, and Montali RJ: *J Natl Cancer Inst* 1980; 64:177–207.
Stowers SJ, Wiseman RW, Ward JM, Miller EC, Miller JA, Anderson MW, and Eva A: *Carcinogenesis* 1988; 9:271–276.
Sugiyama T, Takeichi N, Kobayashi H, Yoshida MC, Sasaki M, and Taniguchi N: *Carcinogenesis* 1988; 9:1569–1572.
Tanaka M: *Acta Pathol Jpn* 1974; 24:651–665.
Tatematsu M, Mera Y, Ito N, Satoh K, and Sato K: *Carcinogenesis* 1985; 6:1621–1626.
Tatematsu M, Takano T, Hasegawa R, Imaida K, Nakanowatari J, and Ito N: *Gann* 1980; 71:843–855.
Tchipysheva TA, Guelstein VI, and Bannikov GA: *Int J Cancer* 1977; 20:388–393.
Teebor GW and Becker FF: *Cancer Res* 1971; 31:1–3.
Timme AH: *J Natl Cancer Inst* 1978; 61(2):407–409.
Uriel J, Bouillon D, and Aussel C: *Proc Nat Acad Sci* 1976; 73:1452–1456.
Vasiliev JM and Guelstein VI: *J Natl Cancer Inst* 1963; 31:1123–1143.
Vorce RL and Goodman JI: *Biochem Biophys Res Commun* 1985; 126:879–883.
Vorce RL and Goodman JI: *Toxicol Appl Pharmacol* 1987; 90:86–95.
Wanson JC, de Ridder L, and Mosselmans R: *Cancer Res* 1981; 41:5162–5175.
Ward JM: *Virchows Arch A Pathol Anat* 1981; 390:339–345.
Ward JM and Osshima M: *Carcinogenesis* 1985; 6:1255–1259.
Watabe H: *Cancer Res* 1971; 31:1192–1194.
Watanabe K and Williams GM: *J Natl Cancer Inst* 1978; 61:1311–1314.
Weisburger JH and Williams GM: *Science* 1981; 214:401–407.
Williams GM: *Cancer Res* 1976; 36:2540–2543.
Williams GM: *Biochim Biophys Acta* 1980; 605:167–189.
Williams GM: *Food Cosmet Toxicol* 1981; 19:577–583.
Williams GM: *Toxicol Pathol* 1982; 10:3–10.
Williams GM: *Fund Appl Toxicol* 1984; 4:325–344.
Williams GM, Hirota N, and Rice J: *Am J Pathol* 1979; 94:65–74.
Williams GM, Klaiber M, and Farber E: *Am J Pathol* 1977; 89:379–388.
Williams GM, Klaiber M, Parker SE, and Farber E: *J Natl Cancer Inst* 1976; 57:157–165.
Williams GM and Numoto S: *Carcinogenesis* 1984; 5:1689–1696.
Williams GM, Ohmori T, and Watanabe K: *Am J Pathol* 1980; 99:1–12.
Williams GM and Watanabe K: *J Natl Cancer Inst* 1978; 61:113–121.
Williams GM and Yamamoto RS: *J Natl Cancer Inst* 1972; 49:685–692.
Wiseman RW, Stowers SJ, Miller EC, Anderson MW, and Miller JA: *Proc Natl Acad Sci USA* 1986; 5825–5829.
Yamashita H: *Gann* 1979; 70:593–600.
Yaswen P, Goyette M, Shank PR, and Fausto N: *Mol Cell Biol* 1985; 5:780–786.
Yaswen P, Hayner NT, and Fausto N: *Cancer Res* 1984; 44:324–331.
Yoshida MC, Masuda R, Sasaki M, Takeichi N, Kobayashi H, Dempo K, and Mori M: *J Hered* 1987; 78:361–365.
Zahringer J, Baliga BS, Drake RL, and Munro HN: *Biochim Biophys Acta* 1977; 474:234–244.
Zahringer J, Baliga BS, and Munro HN: *Proc Natl Acad Sci USA* 1976; 73:857–861.

Chapter

15

Toxicological Implications of Peroxisome Proliferation

M. Sambasiva Rao and Janardan K. Reddy
Department of Pathology
Northwestern University
Medical School
Chicago, Illinois

Peroxisomes are single membrane-bound cytoplasmic organelles present in a wide variety of cells in both animals and plants including unicellular eukaryotes (Beevers, 1979, Bock et al., 1980; De Duve, 1983; Tolbert, 1981). These organelles were initially termed "microbodies" by the original discoverer Rhodin (1954), and later on De Duve (1965) coined the term "peroxisome" which reflects the presence in this organelle of enzymes that participate in peroxide-based reactions, i.e., generation of hydrogen peroxide by oxidases and degradation of H_2O_2 by catalase, the marker enzyme. Peroxisomes in fatty seedlings are called glyoxisomes (Breidenback and Beevers, 1967). Peroxisomes have been shown to contain more than 40 different enzymes, but the enzyme composition of peroxisomes varies depending upon the cell type (Tolbert, 1981). Peroxisomal enzymes participate in many metabolic reactions including respiration, gluconeogenesis, thermogenesis, purine catabolism, fatty acid oxidation, and plasmalogen synthesis (De Duve and Baudhuin, 1967; Hajra and Bishop, 1982; Lazarow, 1988).

Over the last 10 years interest and research in the field of peroxisomes has considerably increased. This is mainly due to recognition of (a) their role in the metabolism of long-chain fatty acids (Lazarow and De Duve, 1976); (b) identification of several chemicals that are of therapeutic and industrial

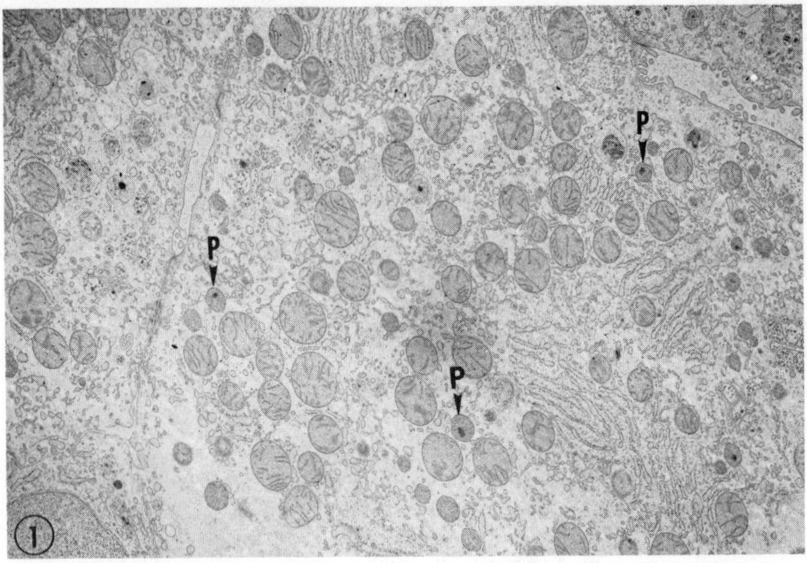

Figure 1. Electron micrograph of normal rat hepatocyte showing randomly distributed peroxisomes (P). Magnification: × 9,000.

value that cause marked proliferation of hepatic peroxisomes (Cohen and Grasso, 1981; Reddy et al., 1982a; Reddy and Lalwani, 1983); (c) discovery of carcinogenic effect of peroxisome proliferators in rodents (Reddy et al., 1976, 1980; Reddy and Lalwani, 1983); and (d) recognition of several human diseases associated with deficiency of peroxisomes (Goldfischer et al., 1973; Goldfischer and Reddy, 1984; Kaiser and Kramer, 1988). Several comprehensive reviews have been published recently on different aspects of peroxisomes, peroxisome proliferation, and peroxisome-associated diseases (De Duve and Baudhuin, 1967; Tolbert, 1981; Cohen and Grasso, 1981; Reddy and Lalwani, 1983; Goldfischer and Reddy, 1984; Lazarow and Fujiki, 1985; Kaiser and Kramer, 1988; Wilson et al., 1988). In this chapter we outline the morphology of peroxisomes and hepatic changes associated with the administration of peroxisome proliferators.

DISTRIBUTION AND MORPHOLOGY

Peroxisomes are present in all mammalian cells except in mature red blood cells (Hruban et al., 1972). Their size and number vary considerably between different species and between different tissues in the same species. Peroxisomes are comparatively abundant in parenchymal cells of liver and proximal tubular epithelial cells of kidney. In the mammalian liver there are about 500 to 1000 peroxisomes in each hepatocyte and occupy approximately 1.5 to 2% of the cytoplasmic volume (Figure 1) (Moody and Reddy, 1976;

Sternlieb, 1979; Gray and de la Iglesia, 1984). In the liver and kidney, peroxisomes measure approximately 0.3 to 1 μm in diameter (Rhodin, 1954; Hruban and Rechcigl, 1969). In other tissues, they are smaller in size, measuring about 0.15 to 0.25 μm (Hruban et al., 1972; Novikoff and Novikoff, 1972; Novikoff et al., 1973). The majority of peroxisomes are round to oval in shape, although other shapes such as rodlike and vermiform type have been identified (Shnitka, 1966; Hruban et al., 1972). Size and shape alterations are very common in peroxisomes that are induced by various chemicals (Hruban et al., 1974). Peroxisomes contain a finely granular or flocculent material of moderate electron density. Peroxisomes in liver cells of many species display urate oxidase containing nucleoid or crystalloid core consisting of parallel bundles of hollow tubules (Shnitka, 1966; Tsukada et al., 1966). Some mammals including humans, anthrapoid apes, and New World monkeys lack nucleoids, indicating the absence of urate oxidase. The presence of crystalloid core within peroxisomes usually correlates with the presence of urate oxidase (Figure 2) (Tsukada et al., 1966; Usada et al., 1988a). However, absence of a core does not rule out the existence of this enzyme (Vandenbosch, 1986). In addition to crystalloid cores, peroxisomes in some species also contain marginal plates. The significance and biochemical nature of these

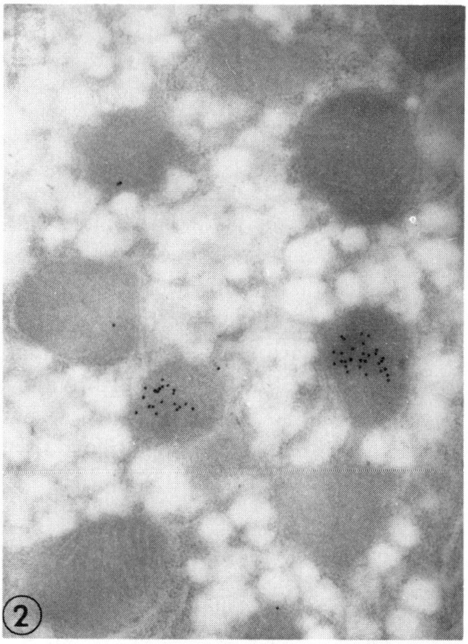

Figure 2. lowicryl embedded normal rat liver tissue stained for urate oxidase by Protein A-gold method. Urate oxidase is localized to nucleoid of peroxisomes (P). Magnification: × 32,000.

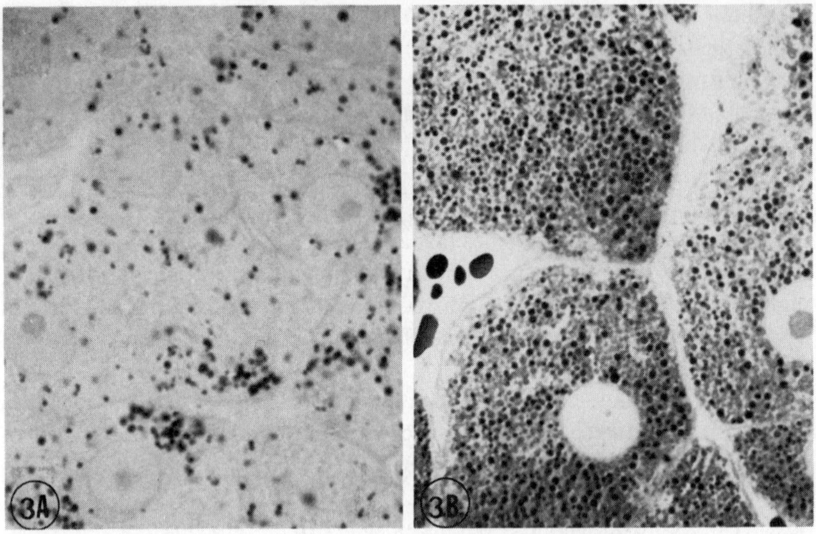

Figure 3. Photomicrograph of toluidine blue–stained 1-μm-thick section of epon-embedded rat liver tissue. (A) Control rat liver. (B) Ciprofibrate fed rat liver. Marked increase in the number of peroxisomes are present in the ciprofibrate treated rat. Magnification: × 1100.

marginal plates is not clear (Hruban and Rechcigl, 1969). Marginal plates are absent in liver peroxisomes of humans and rat.

Morphological identification of peroxisomes, which contain urate oxidase cores, is very easy under electron microscope. Even in the absence of cores, recognition of peroxisomes has become relatively simple because of cytochemical and recently developed immunocytochemical techniques. The cytochemical procedure developed by Novikoff and Goldfischer (1969) involves the demonstration of catalase, the marker enzyme for peroxisomes, using 3,3′-diaminobenzidine in an alkaline medium. By this procedure peroxisomes can be visualized both by light and electron microscope (Figure 3A). Development of specific antibodies against several peroxisomal proteins further facilitated the identification as well as specific distribution of peroxisomal proteins using immunofluorescence, immunoperoxidase, and protein A-gold methods (Figures 4A and 5A) (Reddy et al., 1981; Yokota and Fahimi, 1981; Bendayan and Reddy, 1982; Usuda et al., 1988b).

Peroxisomal Enzymes

At least 40 enzymes have been identified to be associated with peroxisomes (Tolbert, 1981). These include several classes such as catalase, oxidases, dehydrogenases, acyl and aminotransferases, reductases, hydrolases, and enzymes that participate in fatty-acid β-oxidation and glycerolipid syn-

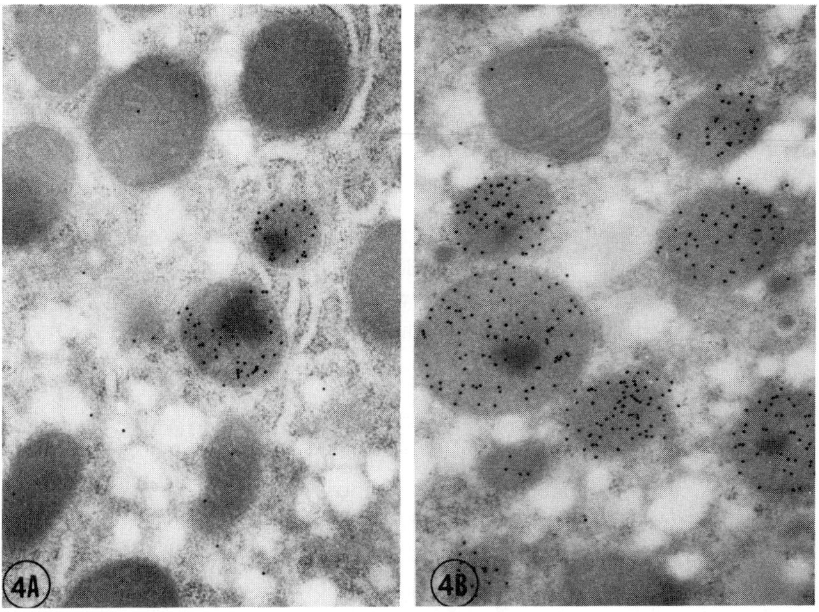

Figure 4. Immunocytochemical localization of fatty acyl-CoA oxidase by Protein A-gold method. (A) Normal rat liver. (B) Ciprofibrate treated rat liver. Peroxisomes of ciprofibrate treated animal shows increased number of gold particles. Magnification: × 32,000.

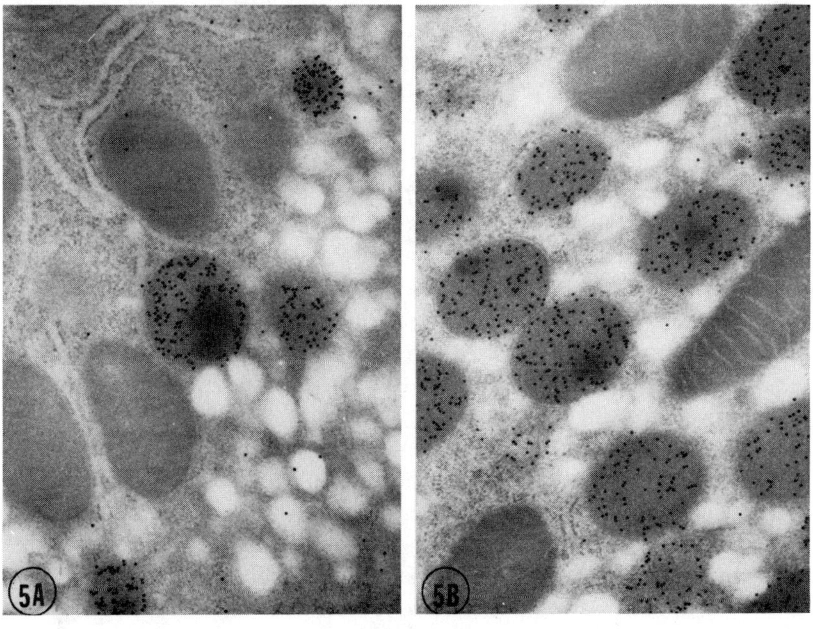

Figure 5. Immunocytochemical localization of catalase by Protein A-gold method. (A) Normal rat liver. (B) Ciprofibrate rat liver. The number of gold particles appear the same in both the control and treated animal. Magnification: × 32,000.

thesis. Of these different enzymes catalase and enzymes of fatty acid β-oxidation pathway have been well characterized. Catalase, an enzyme that protects cells from the deleterious effects of H_2O_2 generated by oxidases, is always present in peroxisomes of all cell types (De Duve, 1983), whereas the distribution of other enzymes may vary. In liver cells catalase constitutes about 15% of the total peroxisomal proteins (De Duve, 1983). Fatty acid β-oxidation enzyme system comprises four enzymes (acyl-CoA oxidase, enoyl-CoA hydratase, 3-hydroxy-acyl-CoA dehydrogenase and thiolase). These enzymes participate in the oxidation of long-chain fatty acids and dicarboxylic fatty acids. These enzymes have been purified from rat liver and are well characterized (Hashimoto, 1982). Fatty acyl-CoA oxidase the first enzyme in the peroxisomal β-oxidation pathway, catalyzes dehydrogenation of fatty acyl-CoA leading to generation of H_2O_2 by transfering two electrons to molecular oxygen. Since the energy conserving mitochondrial β oxidation of fatty acids is very active in all the cells, the role of peroxisomal β oxidation under normal physiological condition is not very clear. Recent studies have shown that peroxisomal β-oxidation enzymes, in addition to oxidizing long-chain fatty acids, also participate in the oxidation of cholesterol and formation of bile acids (Hagey and Krisans, 1982; Hayashi et al., 1984). These enzymes are present in the peroxisomes of several cell types (Small et al., 1981; Usuda et al., 1988b; Nemali et al., 1988).

Peroxisomal enzymes are synthesized on free polysomes in the cytosol in their final form (Goldman and Blobel, 1978; Fujiki et al., 1984; Lazarow, 1988). These newly synthesized proteins are selectively transported to and segregated within the preexisting peroxisome. The mechanism by which these cytosolic proteins are transferred posttranslationally to peroxisomes remains far from clear. Recent observations indicate that a highly conserved c-terminal tripeptide (e.g., *Ser-Lys-Leu;* or *Ser-Ayg-Leu*) serves as a peroxisomal transport signal (Gould et al., 1989). Whether this signal is recognized by a putative receptor on peroxisomal membrane remains to be elucidated. An understanding of the control mechanisms involved in selective transport of these proteins to peroxisomes is of great clinical significance since many of the peroxisomopathies are secondary to deficiency of peroxisomes (Wilson et al., 1988).

Peroxisomes are formed from preexisting peroxisomes. It has been shown that after the proteins are imported into the preexisting peroxisome, it enlarges in size and then divides (Lazarow and Fujiki, 1985). During embryonic development peroxisomes appear in rat liver on day 14 or 15 of gestation and increase in number just before and immediately after birth (Holmes and Masters, 1978; Stefanini, 1985). It is believed that peroxisomes are not formed *de novo* and are propagated from germ cells to the embryo (Lazarow and Fujiki, 1985). The half-life of peroxisomes is very short and is estimated to be about 1.5 to 3.5 days (Poole et al., 1969; Price et al., 1972; Miyazawa et al., 1980). The peroxisomes are decimated randomly by autophagy (De Duve and Baudhuin, 1967).

HEPATIC PEROXISOME PROLIFERATION AND PEROXISOME PROLIFERATORS

In liver cells peroxisomes are very few and are distributed randomly throughout the cytoplasm. Their distribution appears uniform in hepatocytes throughout different zones of the liver lobule. The number of peroxisomes in liver cells remains constant throughout the life-span of rats under normal physiological conditions (Schmucker, 1976). However, peroxisomes can be readily induced to proliferate in liver cells of rats and mice as a result of exposure to certain chemicals, and by altering the nutritional status of the animal (Table 15-1). Chemicals that are capable of inducing peroxisome proliferation are designated as "peroxisome proliferators" (Reddy et al., 1975). Recently, the administration of the anabolic steroid dehydroepiandrosterone to rats has been shown to induce peroxisome proliferation in liver cells.

In 1965, Hess and associates described the induction of peroxisome proliferation in hepatocytes of rats treated with clofibrate, a drug therapeutically used to lower serum triglycerides. Since then several structurally diverse groups of hypolipidemic drugs and some commercially used industrial agents such as phthalate ester plasticizers and solvent trichloroethylene have been shown to induce peroxisome proliferation in hepatocytes of several species of animals including primates (Svoboda and Azarnoff, 1966; Reddy and Krishnakantha, 1975; Moody and Reddy, 1978; Cohen and Grasso, 1981; Reddy et al., 1982b; Gray and de la Iglesia, 1984; Reddy et al., 1984; Elcombe et al., 1985; Lake and Gray, 1985; Lake et al., 1986). Although, peroxisome proliferative response to these chemicals is elicited in many species, the maximum response is observed in rats and mice, the two species that are extensively used in toxicological and carcinogenesis studies. Hypolipidemic drugs can be basically subdivided into two groups: group one consists of the prototype drug clofibrate and its structural analogs, and group two compounds are those that are structurally unrelated to clofibrate (Table 15-1). All peroxisome proliferators appear to induce similar types of qualitative changes in hepatocytes. However, quantitative differences are observed in the degree of proliferation of peroxisomes and the induction of peroxisome-associated enzymes (Lalwani et al., 1983b; Reddy et al., 1986a). Hypolipidemia caused by these chemicals has been linked to their ability to induce peroxisome proliferation (Reddy and Krishnakantha, 1975; Lazarow, 1977).

PEROXISOME PROLIFERATOR-INDUCED HEPATIC CHANGES

Administration of peroxisome proliferators results in predictable pleiotropic responses in the liver. These responses can be characterized as early

TABLE 15–1.
Compounds and Conditions That Cause Proliferation of Hepatic Peroxisomes*

Hypolipidemic Compounds				
Analogs of Clofibrate	Compounds Structurally Unrelated to Clofibrate	Industrial Agents	Nutritional Factors	Miscellaneous
Nafenopin	Tibric acid	di(2-Ethylhexyl)phthalate	High-fat diet	Triiodothyronine[11]
Methylclofenapate	Wy-14,643	di(2-Ethylhexyl)adipate	Vitamin E deficiency	Acetylsalicylic acid
SaH-42,348	BR-931	di(2-Ethylhexyl)Sebacate	Starvation[7]	Phenoxy acid heribicdes[12]
S-8527	Tiadenol	2-Ethylhexanol	Perfluorinated fatty acids and related compounds[8,9]	
AT-308	RMI-14,514	2-Ethylaldehyde	Riboflavin deficiency[10]	
Gemfibrozil	DG-5685	2-Ethylhexanoic acid		
Bezafibrate	DH-6463	2-(4-Dibenzofluopranyloxy)-2-methylpropionic acid		
Fenofibrate	LK-903	Trichloroethylene[4]		
LS-2265	DL-040[2]	Trichloroacetic acid[4]		
Ciprofibrate	LY171883[3]	Dimethrin[5]		
Simfibrate	Chlorcyclizine	di(2-Ethylhexyl)phophate[6]		
Clobuzarit[1]				

*Adapted from Reddy and Lalwani (1983) and updated with additional references. 1. Orton et al. (1984); 2. Lalwani et al. (1985); 3. Eacho et al. (1986); 4. Elcombe et al. (1985); 5. Hruban et al. (1974); 6. Lundgren and DePierre (1987); 7. Ishii et al. (1980); 8. Ikeda et al. (1985); 9. Ideka et al. (1988); 10. Reith (1973); 11. Fringes and Reith (1982); 12. Vainio et al. (1983).

TABLE 15–2.
Hepatomegalic Effect of Selected Hypolipidemic Compounds at Different Dose Levels in Rats and Mice*

Compound	Dose Level (% in diet)	Liver Weight/ 100 Gram b.wt.	Grams per Body Weight Mice
Clofibrate	0.125	—	6.4
	0.25	5.9	8.5
Nafenopin	0.125	8.1	24.6
	0.25	8.3	—
Ciprofibrate	0.001	4.8	—
	0.01	8.0	10.2
	0.02	8.2	12.3
Wy-14,643	0.125	7.6	15.4
	0.25	9.1	15.9
Tidenol	0.05	4.8	—
	0.1	6.9	—
	0.2	9.2	—
Di-(2-Ethylhexyl) phthalate	0.25	5.8	—
	0.5	6.0	—
	1.0	7.0	—
	2.0	8.0	11.8
	4.0	8.7	12.4
Control		3.9	5.7

*Adapted from: Reddy et al. (1976); Reddy and Krishnakantha (1975); Svoboda et al. (1967); Reddy et al. (1974); Lalwani et al. (1983b); Reddy et al. (1986a); Rao et al. (1988a).

"adaptive changes" and late "carcinogenic effects." The adaptive changes include hepatomegaly, proliferation of cytoplasmic organelles (particularly peroxisomes), and induction of certain peroxisomal, endoplasmic reticulum and cytosolic enzymes.

Adaptive Changes: Hepatomegaly

All peroxisome proliferators produce liver enlargement (Table 15-2). The magnitude of hepatomegalic effect is variable with different compounds and is dependent on the potency: the dose administered: and species, age, and sex of the animal used (Svoboda et al., 1967; Reddy et al., 1974; Reddy and Lalwani, 1983; Lalwani et al., 1983b; Elcombe et al., 1985; Lake and Gray, 1985; Kawashima et al., 1985; Dwivedi et al., 1989). Hepatomegalic effect is observed within a few days of administration of the compound and reaches a steady-state level within 2 weeks. The hepatomegaly persists for as long as the compound is administered (Reddy and Lalwani, 1983). After the treatment is discontinued, the liver begins to regress in size and reaches control weight within 2 weeks (Svoboda et al., 1967; Moody and Reddy, 1976; Reddy et

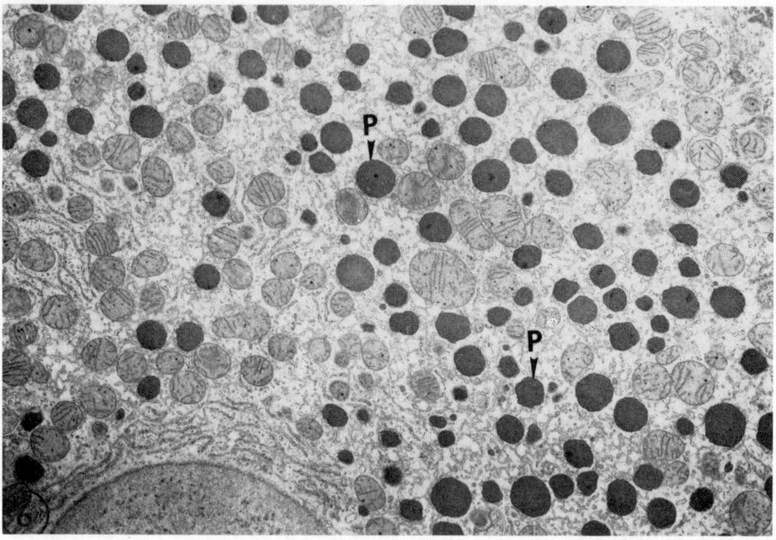

Figure 6. Hepatocyte from a rat fed ciprofibrate showing a large number of peroxisomes (P). Tissue was incubated in alkaline DAB medium before embedding. Reaction product is localized in the peroxisomes. Magnification: × 12,000.

al., 1978). The liver enlargement caused by peroxisome proliferators is secondary to hypertrophy and hyperplasia of hepatocytes.

In rats and mice, peroxisome proliferators cause a marked increase in the hepatocyte cytoplasmic volume (Moody and Reddy, 1976; Kolde et al., 1976; Anthony et al., 1978; Berman et al., 1983; Gray and de la Iglesia, 1984). This increase is principally due to increases in the number and size of peroxisomes (Figures 3B and 6). Peroxisome proliferative phenomenon is not unique to rats and mice (Cohen and Grasso, 1981; Reddy and Lalwani, 1983; Elcombe et al., 1985; Dwividi et al., 1989). Xenobiotic-induced peroxisome proliferation has been described in the liver parenchymal cells of hamsters, cat, chicken, dog, monkey, and human hepatocytes (Reddy et al., 1982b; Hanefeld et al., 1983; Reddy et al., 1984; Gray and de la Iglesia, 1984; Lalwani et al., 1985; Lake et al., 1986). Although, peroxisome proliferative response has been observed in several species, the magnitude of response appears to be species dependent.

In rats and mice, peroxisome proliferative response is dependent on the potency of the compound. For example, ciprofibrate, an analog of clofibrate, is 1000 times more potent than di(2-ethylhexyl)phthalate in inducing peroxisome proliferation in rats (Reddy et al., 1986a). Increase in the number of peroxisomes is evident within 24 h of administration of a xenobiotic (Hess et al., 1965) and by 2–3 weeks they occupy as much as 20 to 30% of the cytoplasmic volume as compared to less than 2% in normal hepatocytes

(Reddy and Lalwani, 1983; Gray and de la Iglesia, 1984). This increase in volume is both due to an increase in size and the number of peroxisomes. Peroxisome proliferation persists as long as the xenobiotic is administered, and the number of these organelles decreases precipitously after discontinuation of treatment reaching normal values within 7–10 days (Reddy et al., 1974; Moody and Reddy, 1976). In addition, peroxisome proliferators also cause mild to moderate increase in the amount of smooth endoplasmic reticulum (SER). The surface area of SER increases by about 50% over the control values during treatment (Moody and Reddy, 1976; Kolde et al., 1976; Anthony et al., 1978). Cessation of treatment results in a gradual decrease of SER volume (Svoboda et al., 1967; Moody and Reddy, 1976). Volume of rough endoplasmic reticulum decreases considerably, whereas mitochondria remain unaffected in the liver cells of rats treated with peroxisome proliferators.

The second factor that plays an important role in the enlargement of liver, at least during the initial period of peroxisome proliferator administration, is hepatocyte hyperplasia. In rats and mice, peroxisome proliferators cause enhanced DNA synthesis and mitotic activity within 24 h following administration and persist during the first 5–7 days of treatment. Once the hepatomegaly reaches a steady state, the increased cell proliferative response subsides even with continued treatment (Moody et al., 1977; Reddy et al., 1979; Elcombe et al., 1985; Smith-Oliver and Butterworth, 1987). Hyperplastic response in liver persists, albeit, at very low levels as long as the treatment with a peroxisome proliferator is continued. Moody et al. (1977), using a single injection of labeled thymidine in acatalasemic mice, have shown that labeling index after treatment with 0.125% nafenopin was 2.6%, 0.68%, and 0.45% at 5 days, 6 weeks, and 32 weeks, respectively. Recent studies using continuous labeling by osmotic pumps also showed maximum labeling at 1 week (34.8%) and declined to 2.9% by 5 weeks (Yeldandi et al., 1989). The induced hyperplasia is secondary to the mitogenic effect of peroxisome proliferator, rather than to compensatory hyperplasia, since there is no perceptible evidence of cell death at the dose levels used. Replicative DNA synthesis is also induced by peroxisome proliferators in hepatocytes maintained in culture (Bieri et al., 1984; Bieri et al., 1986). The morphological evidence of hyperplastic pattern was corroborated by increase and decrease in the levels of ornithine decarboxylase, an enzyme that plays an important role in the regulation of growth (Izumi et al., 1981; Fukami et al., 1986).

PEROXISOME PROLIFERATOR-INDUCED BIOCHEMICAL ALTERATIONS

Changes in the hepatic biochemical parameters in rats, mice, and hamsters treated with peroxisome proliferators have been well documented (Reddy et al., 1982a; Reddy and Lalwani, 1983; Orton et al., 1984; Lake et al., 1984; Elcombe et al., 1985; Dwivedi et al., 1989). The characteristic findings are disproportionate increases in the activities of several peroxisomal enzymes (Figures 4B and 5B). Levels of catalase increased by about twofold, whereas fatty acid β-oxidation enzymes increased by 20- to 30-fold (Lazarow, 1977; Miyazawa et al., 1980; Hashimoto, 1982; Reddy et al., 1986b). Urate oxidase which is localized exclusively in the peroxisomal crystalloids, is induced less than twofold (Usuda et al., 1988b; Reddy et al., 1988). The dissimilar changes in the levels of various peroxisomal enzymes is attributable to differential regulation of genes encoding peroxisomal enzyme. The mRNA levels of fatty acyl-CoA and bifunctional enoyl-CoA hydratase/3-hydroxy-acyl-CoA dehydrogenase, the first two enzymes of β-oxidation system, are increased by more than 20-fold, while catalase and urate oxidase mRNA levels increased less than twofold (Reddy et al., 1986b; Nemali et al., 1988; Nemali et al., 1989). Similar type of biochemical and molecular changes are also observed in cultured hepatocytes exposed to peroxisome proliferators (Lake et al., 1984; Feller et al., 1987; Bieri et al., 1988). Other peroxisomal enzymes such as carnitine acetyltransferase and carnitine octonoyltransferase also show increased activities (Moody and Reddy, 1974; Hajra et al., 1979; Farrell et al., 1984). Although several peroxisome proliferators induce qualitatively similar types of enzyme response, quantitatively there is marked variation. These quantitative differences generally reflect on the potency of the compound. In addition to peroxisomal enzymes, peroxisome proliferators also induce alterations in the levels of some ER, mitochondrial, and cytosolic enzymes. The activities in liver of laurate hydroxylase, NADPH cytochrome C reductase, epoxide hydrolase, and carnitine palmitoyltransferase are markedly increased (Gibson et al., 1982; Orton et al., 1982; Hammock and Ota, 1983; Wachter et al., 1984; Moody et al., 1985; Lake et al., 1986; Feller et al., 1987; Reddy et al., 1989). On the other hand, activities of two cytosolic enzymes, glutathione-*S*-transferase and glutathione peroxidase, are markedly decreased (Ciriolo et al., 1982; Awasthi et al., 1984; Furakawa et al., 1985; Perera et al., 1986).

The mechanism by which these structurally diverse peroxisome proliferators lead to peroxisome proliferation and the selective induction of some peroxisome-associated enzymes mainly in hepatocytes remains unclear. Three possible modes of action have been postulated. Those are (1) activation of specific genes by these xenobiotics either directly or through a receptor (Reddy and Rao, 1987), (2) mobilization of fatty acids from the fat depots to the liver by the peroxisome proliferators thus causing a substrate overload (Berge and Aarsland, 1985), and (3) peroxisome proliferators or their metabolites act as exogenous substrates for the peroxisomal β-oxidation enzymes (Albro et al., 1981). Currently available experimental data suggest the existence of peroxisome proliferator-binding protein (Lalwani et al., 1983a; Lalwani et al., 1987). However, further confirmation is necessary to establish that the cytosolic peroxisome proliferator-binding protein is a specific receptor responsible for peroxisome proliferator-induced pleiotropic responses.

CARCINOGENIC EFFECT OF PEROXISOME PROLIFERATORS

Various peroxisome proliferators that are used either as therapeutic compounds or industrial chemicals are shown to be complete hepatocarcinogens in rats and mice (Svoboda and Azarnoff, 1979; Reddy et al., 1980; Reddy et al., 1982b; Reddy and Lalwani, 1983; Rao et al., 1984a; Kluwe et al., 1985; Rao et al., 1987b; Rao et al., 1988a). These compounds are unequivocally established both as initiators and promotors (Table 15-3) (Rao and Reddy, 1987b). Chronic administration of peroxisome proliferators in the diet results in the development of hepatocellular carcinomas between 50 and 120 weeks. The latency period and the incidence of tumors appear to correlate well with the potency of the compounds to induce peroxisome proliferation. With potent peroxisome proliferators such as ciprofibrate and WY-14,643,

TABLE 15–3.
Peroxisome Proliferators That Are Proven To Be Hepatocarcinogens*

Species	Compound
Rats	Clofibrate
	Tibric acid
	Methylclofenapate
	Nafenopin
	Wy-14,643
	BR-931
	Ciprofibrate[1]
	Gemfibrozil[2]
	di(2-Ethylhexyl)phthalate[3]
Mice	BR-931
	Nafenopin
	Wy-14,643
	Ciprofibrate[4]
	di(2-Ethylhexyl)-adipate[3]
	di(2-Ethylhexyl)phthalate[3]
	Trichloroethylene[5]

*Adapted from Reddy and Lalwani (1983) and updated with additional references. 1. Rao et al. (1984a); 2. Fitzgerald et al. (1981); 3. Kluwe et al. (1985); 4. Rao et al. (1988a); 5. NTP (1983).

hepatocellular carcinomas (HCC) develop in nearly 100% of rats within 50 to 60 weeks, when these agents are administered in diet at dose levels ranging from 0.025 to 0.1% (Reddy et al., 1979; Rao et al., 1984b). In contrast, with a relatively less potent peroxisome proliferator, such as di(2-ethylhexyl)phthalate, tumors develop after 95 weeks of treatment at very high dietary concentrations (Kluwe et al., 1982; Rao et al., 1987b).

The pattern of development of hepatic lesions in rats treated with peroxisome proliferators is similar to that observed after treatment with genotoxic carcinogens. Carcinogenic treatment results in the appearance of altered areas and neoplastic nodules from which HCC arise (Farber and Sarma, 1987). In rats treated with 0.1% WY-14,643, altered hepatic foci appear around 22 weeks, neoplastic nodules between 26 and 50 weeks, followed by the emergence of HCC immediately thereafter (Rao et al., 1984b). All the potent peroxisome proliferators induce multiple hepatic tumors, which are distributed randomly in all lobes of liver. Histologically, a large percentage of hepatocellular carcinomas are well differentiated with trabecular pattern. Pulmonary metastases were observed in 20 to 40% of rats and mice with peroxisome proliferator-induced HCC (Rao and Reddy, 1987a).

The role of peroxisome proliferators as tumor promoters has been well established in rats and mice. Continuous dietary administration of peroxisome proliferators after diethylnitrosamine initiation led to a decrease in latency period and increased incidence of HCC (Reddy and Rao, 1978; Mochizuki et al., 1982; Ward et al., 1983). The promoting effect of Wy-14,643 on the number and size of diethylnitrosamine-induced altered hepatic foci was also demonstrated (Glauert et al., 1986). However, in some short term *in vivo* studies in which quantitation of γ-glutamyltranspeptidase positive foci were used as end point, no promoting effects were observed with some peroxisome proliferators (Stäubli et al., 1984; Numoto et al., 1985; Popp et al., 1985). The apparent lack of promotional effect is most likely due to the fact that GGT is not expressed in liver lesions and is not a useful marker to evaluate altered foci.

PHENOTYPIC PROPERTIES OF HEPATIC LESIONS INDUCED BY PEROXISOME PROLIFERATORS

After the initial observation of gammaglutamyltranspeptidase (GGT) negative property of hepatic lesions induced by Wy-16,643 by Rao et al. (1982) the preneoplastic and neoplastic lesions induced by different peroxisome proliferators have been well characterized (Rao et al., 1986a, b; Glauert et al., 1986; Rao et al., 1987). GGT is consistently negative in 70 to 90% of altered areas and more than 90% of neoplastic nodules and HCC. GGT negative property of HCC is persistent in metastatic lesions and in HCC established as transplantable tumors (Milano et al., 1987; Rao et al., 1988b). The absence of GGT expression appears to be due to failure of derepression of GGT gene as evidenced by lack of mRNA expression (Rao et al., 1988b). Glutathione-*S*-transferase placental form (GST-P) and α-fetoprotein are also not expressed in these lesions (Rao et al., 1988b). HCC induced by ciprofibrate are low in the activities of both drug-metabolizing and -detoxifying enzymes and epoxide hydrolase (Rao et al., 1987a). In addition, peroxisome proliferator-induced hepatic lesions are also deficient in glucose-6-phosphatase and adenosine triphosphatase and exhibit resistance to iron accumulation after overload (Rao et al., 1984b; Rao et al., 1986c). Although the morphological features of different hepatic lesions induced by peroxisome proliferators and genotoxic carcinogens are strikingly similar, there are considerable differences in the phenotypic properties between peroxisome proliferator-induced lesions and those induced by genotoxic carcinogens. Genotoxic hepatocarcinogen-induced lesions are strongly positive for GGT and GST-P. They also exhibit high levels of certain drug-detoxifying enzymes and epoxide hydrolase (Ogawa et al., 1980; Tatematsu et al., 1982; Tatematsu et al., 1985; Roomi et al., 1985; Wirth et al., 1987). The specific reason(s) for such a diversity in the phenotypic properties of liver lesions by these two groups of compounds is not clear. Probably, this may reflect on the difference in the mechanism of action of peroxisome proliferators and genotoxic carcinogens.

MECHANISM OF PEROXISOME PROLIFERATOR-INDUCED HEPATOCARCINOGENICITY

Peroxisome proliferators are considered as a novel class of chemical carcinogens because of their distinct properties (Reddy et al., 1980). As a group, these chemicals are nonmutagenic in prokaryotic and eukaryotic system (Warren et al., 1980; Fitzgerald et al., 1981; Glauert et al., 1984) and do not bind to and damage DNA (Butterworth et al., 1984; Gupta et al., 1985). Because of the nongenotoxic nature of these compounds, the mechanism by which they induce liver tumors remains intriguing. It has been postulated that the carcinogenicity of peroxisome proliferators is dependent on their biological actions, i.e., peroxisome proliferation (Rao and Reddy, 1987b; Reddy and Rao, 1987; Rao and Reddy, 1989) and hepatic cell hyperplasia (Grasso, 1985; Butterworth et al., 1987). Cell proliferation caused by peroxisome proliferators may not play a crucial role in hepatocarcinogenesis because hyperplasia is only transient (Cattley et al., 1987; Smith-Oliver and Butterworth, 1987; Rao and Reddy, 1989). Further, it has been shown that there is no increased incidence of tumors in the kidney, an organ in which there is continuous cell proliferation but minimal peroxisome proliferation after administration of peroxisome proliferators (Ward et al., 1988).

Peroxisome proliferation is associated with disproportionate increase in the levels of H_2O_2-generating and -degrading enzymes resulting in oxidative stress (Tomaszewski et al., 1986; Reddy et al., 1986b; Elliott et al., 1986; Reddy and Rao, 1987). Sustained oxidative stress leads to lipid peroxidation, and possibly DNA damage (Hassan, 1983; Fahl et al., 1984; Cerutti, 1985; Goel et al., 1986; Olson, 1988; Reddy and Rao, 1989). Recently Kasai et al. (1989) have shown increased levels of 8-hydroxydeoxyguanosine in the liver DNA of rats given ciprofibrate in diet for a protracted period of time, thus providing a direct evidence for oxidative stress-induced DNA damage.

CONCLUSIONS

Peroxisome proliferators are a diverse group of chemicals that are of either therapeutic or industrial value. Administration of these chemicals to rodents leads to pleiotropic response in livers. The early pleiotropic response is characterized by adoptive changes that include hepatic cell hyperplasia and proliferation of peroxisomes. The delayed component of pleiotropic response is the development of liver tumors. The mechanism by which peroxisome proliferators induces peroxisome proliferation and hepatocarcinogenesis is not clear. The current evidence suggests the presence of peroxisome proliferator-binding cytosolic protein which may mediate peroxisome proliferation. The carcinogenic effect of peroxisome proliferators appears to depend on the extent of peroxisome proliferation and the resulting oxidative stress. Peroxisome proliferators are nongenotoxic and cannot be identified by the commonly used *in vitro* and *in vivo* carcinogen screening tests. Evaluation of chemicals for peroxisome proliferation in hepatocytes either in intact animals and/or in cell culture systems should be very useful to screen potentially carcinogenic non-genotoxic carcinogens.

REFERENCES

Albro PW, Corbett JT, Schroeder J, Jordan S, and Mathews HB: The fate of di(2-ethylhexyl)phthalate in rats. *Toxicologist* 1981; 1:55.

Anthony LE, Schmucker DL, Mooney JS, and Jones AL: A quantitative analysis of fine structure and drug metabolism in livers of clofibrate-treated young adults and retired breeder rats. *J Lipid Res* 1978; 19:154–165.

Awasthi YC, Singh SV, Goel SK, and Reddy JK: Irreversible inhibition of hepatic glutathione *S*-transferase by ciprofibrate, a peroxisome proliferator. *Biochem Biophys Res Commun* 1984; 123:1012–1018.

Beevers H: Microbodies in higher plants. *Annu Rev Plant Physiol* 1979; 30:159–173.

Bendayan M, and Reddy JK: Immunocytochemical localization of catalase and heat-labile enoyl-CoA hydratase in the livers of normal and peroxisome proliferator-treated rats. *Lab Invest* 1982; 47:364–369.

Berge RK and Aarsland A: Correlation between the cellular level of long-chain acyl-CoA, peroxisomal β-oxidation, and palmitoyl-CoA hydrolase activity in rat liver. Are the two enzyme systems regulated by a substrate-induced mechanism? *Biochem Biohys Acta* 1985; 837:141–151.

Berman JS, Hiley CR, and Wilson AC: Comparison of the effects of the hypolipidaemic agents ICI53072 and clofibrate with those of phenobarbitone on liver size, blood flow and DNA content in the rat. *Br J Pharmacol* 1983; 78:533–541.

Bieri F, Bentley P, Waechter F, and Stäubli W: Use of primary cultures of adult rat hepatocytes to investigate mechanisms of action of nafenopin, a hepatocarcinogenic peroxisome proliferator. *Carcinogenesis* 1984; 5:1033–1039.

Bieri F, Bentley P, Waechter F, and Stäubli W: Mitogenicity of peroxisome proliferators in monolayers of adult rat hepatocytes. *Food Cosm Toxicol* 1986; 24:709.

Bieri F, Nemali MR, Muakkassah-Kelly S, Waechter F, Stäubli W, Reddy JK, and Bentley P: Increased peroxisomal enzyme mRNA levels in adult rat hepatocytes cultured in a chemically defined medium and treated with nafenopin. *Toxic in vitro* 1988; 2:235–240.

Bock P, Kramer R, and Pauelka M: Peroxisomes and related particles in animal tissues, in Altert M, Beerman W, Franke WW, Rudkin G, and Sitte P (eds): *Cell Biology Monograph,* vol 7. New York, Springer Verlag, 1980.

Breidenback RW and Beevers R: Association of the glyoxilate cycle enzymes in a novel subcellular particle from castor bean endosperm. *Biochem Biophys Res Commun* 1967; 27:462–469.

Butterworth BE, Bermudez E, Smith-Oliver T, Earle L, Cattley R, Martin J, Popp, JA, Strom S, Jirtle R, and Michalopoulos G: Lack of genotoxic activity of di(2-ethylexyl)phthalate (DEHP) in rat and human hepatocytes. *Carcinogenesis* 1984; 5:1329–1335.

Butterworth BE, Loury DJ, Smith-Oliver T, and Cattley RC: The potential role of chemically induced hyperplasia in the carcinogenic activity of the hypolipidemic carcinogens. *Toxicol Indust Health* 1987; 3:129–147.

Cattley RC, Conway JG, and Popp JA: Association of persistent peroxisome proliferation and oxidative injury with hepatocarcinogenicity in female F-344 rats fed di(2-ethylhexyl)phthalate for 2 years. *Cancer Lett* 1987; 38:15–22.

Cerutti PM: Prooxidant state and tumor promotion. *Science* 1985; 228:375–381.

Ciriolo MR, Mavelli I, Rotilio G, Borzatta V, Cristofari N, and Stanzani L: Decrease of superoxide dismutase and glutathione peroxidase in liver of rats treated with hypolipidemic drugs. *FEBS Lett* 1982; 144:264–268.

Cohen AJ and Grasso P: Review of the hepatic response to hypolipidemic drugs in rodents and assessment of its toxicological significance to man. *Food Cosmet Toxicol* 1981; 19:585–605.

De Duve C: Function of microbodies (peroxisomes). *J Cell Biol* 1965; 27:25A.

De Duve C: Microbodies in the living cell. *Sci Am* 1983; 248:74–84.

De Duve C and Baudhuin. Peroxisomes (microbodies and related particles). *Physiol Rev* 1967; 46:323–357.

Dwivedi RS, Alvares K, Nemali MR, Subbarao V, Reddy MK, Usman MI, Rademaker AW, Reddy JK, and Rao MS: Comparison of the peroxisome proliferator-induced pleiotropic response in the liver of nine strains of mice. *Toxicol Pathol* 1989; 17:16–26.

Eacho PI, Foxworthy PS, Johnson WD, Hoover DM, and White SL: Hepatic peroxisomal changes induced by a tetrazole-substituted alkoxyacetophenone in rats and comparison with other species. *Toxicol Appl Pharmacol* 1986; 83:430–437.

Elcombe CR, Rose MS, and Pratt JS: Biochemical, histological and ultrastructural changes in rat and mouse liver following the administration of trichloroethylene: possible relevance to species differences in hepatocarcinogenicity. *Toxicol Appl Pharmacol* 1985; 79:365–376.

Elliott BM, Dodd NJF, and Elcombe CR: Increased hydroxyl radical production in liver peroxisomal fractions from rats treated with peroxisome proliferators. *Carcinogenesis* 1986; 7:795–799.

Fahl WE, Lalwani ND, Watanabe T, Goel SK, and Reddy JK: DNA damage related to increased hydrogen peroxide generation by hypolididemic drug-induced liver peroxisomes. *Proc Natl Acad Sci USA* 1984; 81:7827–7830.

Farber E and Sarma DSR: Hepatocarcinogenesis: A dynamic cellular perspective. *Lab Invest* 1987; 56:4–22.

Farrell SO, Fiol CJ, Reddy JK, and Biebel LL: Properties of purified carnitine acyltransferases of mouse liver peroxisomes. *J Biol Chem.* 1984; 259:13089–13095.

Feller DR, Singh, Y, Shirhatti VR, Kocarek TA, Liu C, and Krishna C: Characterization of ciprofibrate and clofibric acid as peroxisomal proliferators in primary cultures of rat hepatocytes. *Hepatology* 1987; 7:508–516.

Fitzgerald JE, Sanyer JL, Schardein JL, Lake RS, McGuire EJ, and de la Iglesia FA: Carcinogen bioassay and mutagenicity studies with the hypolipidemic agent gemfibrozil. *J Natl Cancer Inst* 1981; 67:1105–1116.

Fringes B and Reith A: Time course of peroxisome biogenesis during adaptation to mild hyperthyroidism in rat liver. *Lab Invest* 1982; 47:19–26.

Fujiki Y, Rachubinski RA, and Lazarow PB: Synthesis of a major integral membrane polypeptide of rat liver peroxisomes on free polysomes. *Proc Natl Acad Sci USA* 1984; 81:7127–7131.

Fukami MH, Elkhom TS, Ekanger R, Flatmark T, and Nilsson A: Relation between induction of β-ornithine decarboxylase and specific gene expression in rat liver in response to the tumor promoter agent clofibrate. *Carcinogenesis* 1986; 7:1441–1446.

Furukawa K, Numoto S, Furuya K, Furukawa NT, and Williams, GM: Effects of hepatocarcinogen nafenopin, a peroxisome proliferator, on the activities of rat liver glutathione-requiring enzymes and catalase in comparison to the action of phenobarbitol. *Cancer Res* 1985; 45:5011–5019.

Gibson GG, Orton TC, and Tamburini PP: Cytochrome P-450 induction by clofibrate. *Biochem J* 1982; 203:161–168.

Glauert JP, Beer D, Rao MS, Schwarz M, Xu Y, Goldsworthy TL, Coloma J, and Pitot HC: Induction of altered hepatic foci in rats by the administration of hypolipidemic peroxisome proliferators alone or following a single dose of diethylnitrosamine. *Cancer Res* 1986; 46:4601–4606.

Glauert HP, Reddy JK, Kennan YS, Sattler GS, Subbarao V, and Pitot HC: Effect of hypolipidemic peroxisome proliferators on unscheduled DNA synthesis in cultured hepatocytes and on mutagenesis in Salmonella. *Cancer Lett* 1984; 26:147–156.

Goel SK, Lalwani ND, and Reddy JK: Peroxisome proliferation and lipid peroxidation in rat liver. *Cancer Res* 1986; 46:1324–1330.

Goldfischer S and Reddy JK: Peroxisomes (microbodies) in cell pathology. *Internat Rev Exp Pathol* 1984; 26:4–84.

Goldfischer S, Moore CL, Johnson AB, Spiro AJ, Valsamis MP, Wisniewski HM, Ritch RH, Norton WT, Rapin I, and Gartner LM: Peroxisomal and mitochondrial defects in the cerebro-hepato-renal syndrome. *Science* 1973; 182:62–64.

Goldman BA and Blobel G: Biogenesis of peroxisomes: intracellular site of synthesis of catalase and uricase. *Proc Natl Acad Sci USA* 1978; 75:5066–5070.

Gould SJ, Keller GA, Hosken N, Wilkinson J, and Subarmani SA: A conserved tripeptide sorts proteins to peroxisomes. *J Cell Biol* 1989; 108:1657–1664.

Grasso P: Peroxisome proliferation and hepatotoxicity in rodents. *Biochem Soc Trans* 1985; 13:861–862.

Gupta RC, Goel SK, Earley K, Singh B, and Reddy JK: ^{32}P-postlabeling analysis of peroxisomal proliferator—DNA adduct formation in rat liver *in vivo* and hepatocytes *in vitro*. *Carcinogenesis* 1985; 6:933–936.

Gray RH and de la Iglesia FA: Quantitative microscopy comparison of peroxisome proliferation by the lipid-regulating agent gemfibrozil in several species. *Hepatology* 1984; 4:520–530.

Hagey L and Krisans SJ: Degradation of cholesterol to propionic acid by rat liver peroxisomes. *Biochem. Biophys Res Comm* 1982; 107:834–841.

Hajra AK and Bishop JE: Glycerolipid biosynthesis in peroxisomes via the acyl dihydroxyacetone pathway. *Ann NY Acad Sci* 1982; 386:170–181.

Hajra AK, Burke CL, and Jones CL: Subcellular localization of acyl coenzyme A: dihydroxyacetone phosphate acyltransferase in rat liver peroxisomes (micro-bodies). *J Biol Chem* 1979; 254:10896–10900.

Hammock BD and Ota K: Differential induction of cytosolic epoxide hydralase, microsomal epoxide hydrolase and glutathione S-transferase activities. *Toxicol Appl Pharmacol* 1983; 32:355–367.

Hanefeld M, Kemmer C, and Kadner E: Relationship between morphologic changes and lipid-lowering actin of *P*-chlorophenoxyisobutyric acid (CPIB) on hepatic mitochondria and peroxisomes in man. *Atherosclerosis* 1983; 46:239–246.

Hashimoto T: Individual peroxisomal β-oxidation enzymes. *Ann NY Acad Sci* 1982; 386:5–12.

Hassan HM: Oxygen toxicity and mutagenesis in prokaryotes, in Cohen A and Greenwald RA (eds): *Oxygen Radicals and Their Scavenger Systems,* vol 1. New York, Elsevier Biomedical, 1983; pp 198–206.

Hayashi N, Fukami K, and Yamasaki F: Association of the liver peroxisomal fatty acyl-CoA beta-oxidation system with the synthesis of bile acids. I. *Biochemistry* 1984; 96:1713–1719.

Hess R, Staubli W, and Reiss W: Nature of the hepatomegalic effect produced by ethyl-chlorophenoxyisobutyrate in the rat. *Nature* 1965; 208:856–858.

Holmes RS and Masters CJ: Genetic control and ontogeny of microbody enzymes: a review. *Biochem Gen* 1978; 16:171–190.

Hruban Z, Gotoh M, Slesers A, and Chous S: Structure of hepatic microbodies in rats treated with acetylsalicylic acid, clofibrate, and dimethrin. *Lab Invest* 1974; 30:64–75.

Hruban Z and Rechcigl M, Jr: Microbodies and related particles. Morphology, biochemistry and physiology. *Internat Rev Cytol Suppl* 1969; 1:20–59.

Hruban Z, Vigil EL, Slesers A, and Hopkins E: Microbodies. Constituent organelles of animal cells. *Lab Invest* 1972; 27:184–191.

Ikeda T, Aiba K, Fukuda K, and Tanaka M: The induction of peroxisome proliferation in rat liver by perfluorinated fatty acids, metabolically inert derivatives of fatty acids. *J Biochem* 1985; 98:475–482.

Ikeda T, Ida-Enomoto M, Mori I, Fukuda K, Iwabuchi H, Komai T, and Suga T: Induction of peroxisome proliferation in rat liver by dietary treatment with 2,2,4,4,6,8,8-heptamethylnonane. *Xenobiotica* 1988; 18:1271–1280.

Ishii H, Horie S, and Suga T: Physiological role of peroxisomal β-oxidation in liver of fasted rats. *J Biochem* 1980; 87:1855–1858.

Izumi K, Reddy JK, and Oyasu R: Induction of hepatic ornithine decarboxylase by hypolipidemic drugs with hepatic peroxisome proliferative activity. *Carcinogenesis* 1981; 2:623–627.

Kaiser E and Kramer R: Clinical biochemistry of peroxisomal disorders. *Clinica Chem Acta* 1988; 173:57–80.

Kasai H, Okada Y, Nishimura S, Rao MS, Reddy JK: Formation of 8-hydroxydeoxyguanosine in liver DNA of rats following long term exposure to a peroxisome proliferator. *Cancer Res* 1989; 49:2603–2605.

Kawashima Y, Katoh H, Watanuki H, Takagishi M, and Kozuka H: Effect of long-term administration of clofibric acid on peroxisomal beta-oxidation, fatty acid binding protein and cytosolic long-chain acyl-CoA hydrolases in rat liver. *Biochem. Pharmacol* 1985; 34:325–329.

Kluwe WM, Huff JE, Mathews HB, Irwin R, and Haseman JK: Comparative chronic toxicities and carcinogenic potentials of 2-ethylhexyl-containing compounds in rats and mice. *Carcinogenesis* 1985; 6:1577–1583.

Kluwe WM, McConnell EE, Huff JE, Haseman JK, Douglas JF, and Hartwell WV: Carcinogenecity testing of phthalate esters and related compounds by the National Toxicology Program and the National Cancer Institute. *Environ Health Perspect* 1982; 45:129–133.

Kolde G, Roessner A, and Thermann H: Effects of clofibrate (alpha-*P*-chloro-phenoxyisobutyryl-ethyl-ester) on male rat liver. *Virch Arch B Cell Path* 1976; 22:73–87.

Lake BG and Gray TJB: Species differences in hepatic peroxisome proliferation. *Biochem Soc Trans* 1985; 13:859–861.

Lake BG, Gray TJB, and Gangolli SD: Hepatic effects of phthalate esters and related compounds—*in vivo* and *in vitro* correlations. *Environ Health Perspect* 1986; 67:283–290.

Lake BG, Gray TJB, Pelsrijcken WR, Beamand JA, and Gangolli SD: The effect of hypolipidaemic agents on peroxisomal β-oxidation and mixed function oxidase activities in primary cultures of rat hepatocytes. Relationship between induction of palmitoyl-CoA oxidation and lauric acid hydroxylation. *Xenobiotica* 1984; 14:269–276.

Lalwani ND, Alvares K, Reddy MK, Reddy MN, Parikh I, and Reddy JK: Peroxisome proliferator-binding protein: identification and partial characterization of nafenopin-, clofibric acid-, and ciprofibrate-binding proteins from rat liver. *Proc Natl Acad Sci USA* 1987; 84:5242–5246.

Lalwani ND, Fahl WE, and Reddy JK: Detection of a nafenopin binding protein in rat liver cytosol associated with the induction of peroxisome proliferation by hypolipidemic compounds. *Biochem. Biophys Res Commun* 1983a; 16:383–393.

Lalwani ND, Reddy NK, Ghosh S, Barnard SD, Molello JA, and Reddy JK: Induction of fatty acid β-oxidation and peroxisomal proliferation in the liver of Rhesus monkeys by DL-040, a new hypolipidemic agent. *Biochem. Pharmacol* 1985; 19:3473–3482.

Lalwani ND, Reddy NK, Qureshi SA, Sirtori CR, Abiko Y, and Reddy JK: Evaluation of selected hypolipidemic agents for the induction of peroxisomal enzymes and peroxisome proliferation in the rat liver. *Human Toxicol* 1983b; 2:27–48.

Lazarow PB: Three hypolipidemic drugs increase hepatic palmitoyl-coenzyme A oxidation in the rat. *Science* 1977; 197:580–581.

Lazarow PB: Peroxisomes, in Arias IM, Jakoby WB, Popper H, Schachter D, and Shafritz DA (eds): *The Liver: Biology and Pathobiology,* Ed 2. Raven Press, 1988, 241–254.

Lazarow PB and De Duve C: A fatty acyl-CoA oxidizing system in rat liver peroxisomes: enhancement by clofibrate, a hypolipidemic drug. *Proc Natl Acad Sci USA* 1976; 73:2043–2046.

Lazarow PB and Fujiki Y: Biogenesis of peroxisomes. *Annu Rev Cell Biol* 1985; 1:489–530.

Lundgren B and DePierre JW: Induction of xenobiotic-metabolizing enzymes and peroxisome proliferation in rat liver caused by dietary exposure to di(2-ethylhexyl)phosphate. *Xenobiotica* 1987; 17:585–593.

Milano M, Reddy JK, Prasad JD, and Rao MS: Absence of γ-glutamyl-transpeptidase activity in lung metastasis in rats with hepatocellular carcinomas induced by ciprofibrate, a peroxisome proliferator. *Cancer Lett* 1987; 38:65–71.

Miyazawa S, Furuta S, Osumi T, and Hashimoto T: Turnover of enzymes of peroxisomal β-oxidation in rat liver. *Biochem Biophys Acta* 1980; 630:367–374.

Mochizuki Y, Furukawa K, and Sawada N: Effects of various concentrations of ethyl α-*P*-chlorophenoxy isobutyrate (clofibrate) on diethylnitrosamine-induced hepatic tumor genesis in the rat. *Carcinogenesis* 1982; 3:1027–1029.

Moody DE and Reddy JK: Increase in hepatic carnitine acetyltransferase activity associated with peroxisomal (microbody) proliferation induced by the hypolipidemic drugs clofibrate, nafenopin and methyl clofenapate. *Res Commun Chem Pathol Pharmacol* 1974; 9:501–509.

Moody DE and Reddy JK: Morphometric analysis of the ultrastructural changes in rat liver induced by the peroxisome proliferator SaH 42-348. *J Cell Biol* 1976; 71:768–780.

Moody DE and Reddy JK: Hepatic peroxisome proliferation in rats fed plasticizers and related compounds. *Toxicol Appl Pharmacol* 1978; 45:497–504.

Moody DE, Loury DN, and Hammock BD: Epoxide metabolism in the liver of mice treated with clofibrate (ethyl-*x*-(*p*-chlorophenoxyisobutyrate)), a peroxisome proliferator. *Toxicol Appl Pharmacol* 1985; 78:351–362.

Moody DE, Rao MS, and Reddy JK: Mitogenic effect in mouse liver induced by a hypolipidemic drug, nafenopin. *Virch Arch B Cell Pathol* 1977; 23:291–296.

National Toxicology Program: Carcinogenesis studies of trichloroethylene (without epichloroxydrin) in F344/N rats and $B6C3F_1$ mice (gavage studies) DHEW 1983; 83:1799.

Nemali MR, Reddy MK, Usuda N, Reddy PG, Comeau LD, Rao MS, and Reddy JK: Differential regulation of peroxisome proliferation in identifying certain non-mutagenic carcinogens. *Toxicol Appl Pharmacol* 1989; 97:72–87.

Nemali MR, Usuda N, Reddy MK, Oyasu K, Hashimoto T, Osumi T, Rao MS, and Reddy JK: Comparison of constitutive and inducible levels of expression of peroxisomal β-oxidation and catalase genes in liver and extrahepatic tissues of rat. *Cancer Res* 1988; 48:5316–5324.

Novikoff AB and Goldfischer S: Visualization of peroxisomes (microbodies) and mitochondria with diaminobenzidine. *J Histochem Cytochem* 1969; 17:675.

Novikoff PM and Novikoff AB: Peroxisomes in absorptive mammalian small intestine. *J Cell Biol* 1972; 53:532–560.

Novikoff AB, Novikoff PM, Davis C, and Qunitana N: Studies on microperoxisomes. V. Are microperoxisomes ubiquitous in mammalian cells? *J Histochem Cytochem* 1973; 21:737–755.

Numoto S, Mori H, Furuya K, Levine WG, and Williams, GM: Absence of promotional or sequential carcinogenic effect in rat liver by carcinogenic hypolipidemic drug nafenopin given after *N*-2-fluorenylacetamide. *Toxicol Appl Pharmacol* 1985; 77:76–85.

Ogawa K, Solt DB, and Farber E: Phenotypic diversity as an early property of putative preneoplastic hepatocyte populations in the liver carcinogenesis. *Cancer Res* 1980; 40:723–733.

Olson MJ: DNA strand breaks induced by hydrogen peroxide in isolated rat hepatocytes. *J Toxicol Environ Health* 1988; 23:407–423.

Orton TC and Parker GL: The effect of hypolipidemic agents on the hepatic microsomal drug metabolizing enzyme system of the rat. *Drug Metab Dispos* 1982; 10:110–115.

Orton TC, Adam HK, Bentley M, Holloway B, and Tucker MJ: Clobuzarit: species differences in the morphological and biochemical response of the liver following chronic administration. *Toxicol Appl Pharmacol* 1984; 73:138–151.

Perera MI, Katyal SL, and Shinozuka H: Suppression of choline-deficient diet induced hepatocyte membrane lipid peroxidation in rats by peroxisome proliferators 4-chloro-6-(2,3-xylidino)-2-pyrimidinylthio (*n*-β-hydroxyethyl)acetamide and di(2-ethylhexyl)phthalate. *Cancer Res* 1986; 46:3304–3308.

Poole B, Leighton F, and DeDuve C: The synthesis and turnover of rat liver peroxisomes. II. Turnover of peroxisome proteins. *J Cell Biol* 1969; 41:536–546.

Popp JA, Garvey LK, Hamm TE, Jr, and Swenberg JA: Lack of hepatic promotional activity by the peroxisomal proliferating hepatocarcinogen di(2-ethylhexyl) phthalate. *Carcinogenesis* 1985; 6:144–146.

Price VE, Sterling WR, Tarantola VA, Hartley RW, Jr, and Rechcigl W, Jr: The kinetics of catalase synthesis and destruction *in vivo*. *J Biol Chem* 1972; 237:3468–3475.

Rao MS and Reddy JK: Phenotypic properties of preneoplastic and neoplastic hepatic lesions induced by peroxisome proliferators in rats, in Fahimi H and Sies H (eds): *Peroxismes in Biology and Medicine*. Heidelberg, Spring Verlag, 1987a, 263–272.

Rao MS and Reddy JK: Peroxisome proliferation and hepatocarcinogenesis. *Carcinogenesis* 1987b; 8:631–636.

Rao MS and Reddy JK: The relevance of peroxisome proliferation and cell proliferation in peroxisome proliferator-induced hepatocarcinogenesis. *Drug Metabol Rev* 1989; 21:103–110.

Rao MS, Dwivedi RS, Subbarao V, and Reddy JK: Induction of peroxisome proliferation and hepatic tumors in C57BL/6N mice by ciprofibrate, a hypolipidaemic compound. *Br J Cancer* 1988a; 58:46–51.

Rao MS, Kokkinakis DM, Subbarao V, and Reddy JK: Peroxisome proliferator-induced hepatocarcinogenesis: levels of activating and detoxifying enzymes in hepatocellular carcinomas induced by ciprofibrate. *Carcinogenesis* 1987b; 8:19–23.

Rao MS, Lalwani ND, and Reddy JK: Sequential histologic study of rat liver during peroxisome proliferator [4-chloro-6-(2,3-xylidino)-2-pyrimidinylthio] acetic acid (Wy-14,643)-induced carcinogenesis. *J Natl Cancer Inst* 1984b; 73:983–990.

Rao MS, Lalwani ND, Scarpelli DG, and Reddy JK: The absence of γ-glutamyl transpeptidase activity in putative preneoplastic lesions and heptaocellular carcinomas induced in rats by the hypolipidemic peroxisome proliferator Wy-14,643. *Carcinogenesis* 1982; 3:1231–1233.

Rao MS, Lalwani ND, Watanabe TK, and Reddy JK: Inhibitory effect of antioxidants ethoxyquin and 2(3)-tert-butyl-4-hydroxyanisole on hepatic tumorigenesis in rats fed ciprofibrate, a peroxisome proliferator. *Cancer Res* 1984a; 44:1072–1076.

Rao MS, Nemali MR, Usuda N, Scarpelli DG, Makino T, Pitot HC, and Reddy JK: Lack of expression of glutathione-*S*-transferase *P*, γ-glutamyl transpeptidase and α-fetoprotein messenger RNAs in liver tumors induced by peroxisome proliferators. *Cancer Res* 1988b; 48:4919–4925.

Rao MS, Subbarao V, and Reddy JK: Peroxisome proliferator-induced hepato-carcinogenesis: histochemical analysis of ciprofibrate-induced preneoplastic and neoplastic lesions for γ-glutamyltranspeptidase activity. *J Natl Cancer Instit* 1986a; 77:951–956.

Rao MS, Subbarao V, and Reddy JK: Increased resistance of peroxisome proliferator-induced hepatic lesions to iron overload in rats. *Cancer Lett* 1986c; 32:33–39.

Rao MS, Tatematsu M, Subbarao V, Ito N, and Reddy JK: Analysis of peroxisome proliferator-induced preneoplastic and neoplastic lesions of rat liver for placental form of glutathione *S*-transferase and α-glutamyltranspeptidase. *Cancer Res* 1986b; 46:5287–5290.

Rao MS, Usuda N, Subbarao V, and Reddy JK: Absence of γ-glutamyl transpeptidase activity in neoplastic lesions induced in the liver of male F-344 rats by di-(2-ethylhexyl)phthalate, a peroxisome proliferator. *Carcinogenesis* 1987; 8:1347–1350.

Reddy JK, Azarnoff DL, and Hignite CE: Hypolipidaemic hepatic peroxisome proliferators form a novel class of chemical carcinogens. *Nature* 1980; 283:397–398.

Reddy JK, Azarnoff DL, and Sirtori CR: Hepatic peroxisome proliferator induction by BR-931, a hypolipidemic analog of Wy-14,643. *Arch Internat Pharma* 1978; 234:4–14.

Reddy JK, Azarnoff DL, Svoboda DJ, and Prasad JD: Nafenopin-induced hepatic microbody (peroxisome) proliferation and catalase synthesis in rats and mice. Absence of sex differences in response. *J Cell Biol* 1974; 61:344–358.

Reddy JK, Goel SK, Nemali MR, Carrino JJ, Laffler TG, Reddy MK, Sperbeck SJ, Osumi T, Hashimoto T, Lalwani ND, and Rao MS: Transcriptional regulation of peroxisomal fatty acyl-CoA oxidase and enoyl-CoA hydrotase/3-hydroxyacyl-CoA dehydrogenase in rat liver by peroxisome proliferators. *Proc Natl Acad Sci USA* 1986b; 83;1747–1751.

Reddy JK and Krishnakantha TP: Hepatic peroxisome proliferation: induction by two novel compounds structurally unrelated to clofibrate. *Science* 1975; 190:787–789.
Reddy JK, Krishnakantha TP, Azarnoff DL, and Moody DE: 1-Methyl(-4-piperidyl-bis(*p*-chlorophenoxy)acetate. A new hypolipidemic peroxisome proliferator. *Res Commun Chem. Pathol Pharmacol* 1975; 10:589–592.
Reddy JK and Lalwani ND: Carcinogenesis by the hepatic peroxisome proliferators: evaluation of the risk of hypolipidemic drugs and industrial plasticizers to humans. *CRC Crit Rev Toxicol* 1983; 12:1–58.
Reddy JK, Lalwani ND, Qureshi SA, Reddy MK, and Moehle CM: Induction of hepatic peroxisome proliferation in nonrodent species, including primates. *Am J Pathol* 1984; 114:171–183.
Reddy MK, Lalwani ND, Qureshi SA, and Reddy JK: Induction of hamster hepatic peroxisomal β-oxidation and peroxisome proliferation associated 80,000 mol wt polypeptide by hypolipidemic drugs. *Human Toxicol* 1982b; 1:135–147.
Reddy JK, Lalwani ND, Reddy MK, and Qureshi SA: Excessive accumulation of autofluorescent lipofuscin in the liver during hepatocarcinogenesis by methyl clofenapate and other hypolipidemic peroxisome proliferators. *Cancer Res* 1982a; 42:219–266.
Reddy PG, Nemali MR, Reddy MK, Reddy MN, Yuan PM, Yuen S, Laffler TG, Shiroza T, Kuramitsu HK, Usuda N, Chisholm RL, Rao MS, and Reddy JK: Isolation and sequence determination of a cDNA clone for rat peroxisomal urate oxidase: liver specific expression in the rat. *Proc Natl Acad Sci USA* 1988; 85:9081–9085.
Reddy MK, Qureshi SA, Hollenberg PF, and Reddy JK: Immunocytochemical identity of peroxisomal enoyl-CoA hydratase with the peroxisome proliferation-associated 80,000 mol. wt. polypeptide in rat liver. *J Cell Biol* 1981; 89:406–417.
Reddy JK and Rao MS: Enhancement by Wy-14,643, a hepatic peroxisome proliferator of diethylnitrosamine-initiated hepatic tumorigenesis in the rat. *Br. J Cancer* 1978; 38:537–543.
Reddy JK and Rao MS: Peroxisome proliferators and cancer: mechanisms and implications. *Trends Pharmacol Sci* 1987; 7:438–443.
Reddy JK and Rao MS: Oxidative DNA damage caused by persistent peroxisome proliferation: its role in hepatocarcinogenesis. *Mut Res* 1989: 214:63–68.
Reddy JK, Rao MS, Azarnoff DL, and Sell S: Mitogenic and carcinogenic effects of a hypolipidemic peroxisome proliferator [4-chloro-6-(2,3-xylidino)-2-pyrimidinyl-thio] acetic acid (Wy-14,643) in rat and mouse liver. *Cancer Res* 1979; 39:152–161.
Reddy JK, Rao MS, and Moody DE: Hepatocellular carcinoma in acatalasemic mice treated with nafenopin, a hypolipidemic peroxisome proliferator. *Cancer Res* 1976; 36:1211–1217.
Reddy JK, Reddy MK, Usman MI, Lalwani ND, and Rao MS: Comparison of hepatic peroxisome proliferative effect and its implication for hepatocarcinogenicity of phthalate esters, di(2-ethylhexyl)phthalate, and di(2-ethylhexyl)-adipate with a hypolipidemic drug. *Environ Health Perspect* 1986a; 65:317–327.
Reddy JK, Usuda N, and Rao MS: Hepatic peroxisome proliferation: an overview. *Arch Toxicol* 1989; 12:207–216.
Reddy JK, Warren JR, Reddy MK, and Lalwani ND: Hepatic and renal effects of peroxisome proliferators: Biological implications. *Ann NY Acad Sci* 1982b; 386:81–110.

Reith A: The influence of triiodothyronine and riboflavin deficiency on the rat liver with special reference to mitochondria. *Lab Invest* 1973; 29:216–228.

Rhodin J: Correlation of ultrastructural organization and function in normal and experimentally changed proximal tubule cells of the mouse kidney. Doctoral thesis, Karolinska Institute, Stockholm, Akitbolaget, Godvil, 1954.

Roomi MW, Ho RK, Sarma DSR, and Farber E: A common biochemical pattern in preneoplastic hepatocyte nodules generated in four different models in the rat. *Cancer Res* 1985; 45:564–571.

Schmucker DL: Age-related changes in hepatic fine structure: a quantitative analysis. *J Gerantol* 1976; 31:135–143.

Shnitka TK: Comparative ultrastructure of hepatic microbodies in some mammals and birds in relation to species differences. *J Ultrastruct Res* 1966; 16:598–625.

Small GM, Hocking TJ, Sturdee AP, Burdett K, and Connock MJ: Enhancement by dietary clofibrate of peroxisomal palmityl-CoA oxidase in kidney and small intestine of albino mice and liver of genetically lean and obese mice. *Life Sci* 1981; 28:1875–1882.

Smith-Oliver T and Butterworth BE: Correlation of the carcinogenic potential of di(2-ethylhexyl)phthatale (DEHP) with induced hyperplasia rather than with genotoxic activity. *Mut Res* 1987; 188:21–27.

Stäubli W, Bentley P, Bieri F, Fröhlich E, and Wacchter F: Inhibitory effect of nafenopin upon the development of diethylnitrosamine-induced enzyme-altered foci within the rat liver. *Carcinogenesis* 1984; 5:41–46.

Stefanini S, Farrace MG, and Ceru Argento MP: Differentiation of liver peroxisomes in the fetal and newborn rat. Cytochemistry of catalase and *D*-amino acid oxidase. *J Embryol Exp Morphol* 1985; 88:151–163.

Sternlieb I: Electron microscopy of mitochondria and peroxisomes in human hepatocytes, in Popper H and Schaffner F (eds): *Progress in Liver Diseases,* vol 6. New York, Grune and Stratton, 1979, pp 81–104.

Svoboda DJ and Azarnoff DL: Response of hepatic microbodies to a hypolipidemic agent, ethylchlorophenoxybutyrate (CPIB). *J Cell Biol* 1966; 30:442–450.

Svoboda DJ and Azarnoff DL: Tumors in male rats fed ethylchlorophenoxyisobutyrate, a hypolipidemic drug. *Cancer Res* 1979; 39:3419–3428.

Svoboda D, Grady H, and Azarnoff D: Microbodies in experimentally altered cells. *J Cell Biol* 1967; 35:127–152.

Tatematsu M, Kaku T, Medline A, Eriksson L, Roomi W, Sharma RN, Muraya RK, and Farber E: Markers of liver neoplasia—real or fictional?, in Milman H and Sell S (eds): *Application of Biological Markers to Carcinogen Testing*. New York, Plenum, 1982, pp 25–42.

Tatematsu M, Mera Y, Ito N, Satoh K, and Sato K: Relativc merits of immunochemical demonstrations of placental A, B and C forms of glutathione *S*-transferase and histochemical demonstration of γ-glutamyl transferase as markers of altered foci during liver carcinogenesis in rats. *Carcinogenesis* 1985; 6:1621–1626.

Tolbert NE: Metabolic pathways in peroxisomes and glyoxisomes. *Annu Rev Biochem* 1981; 50:133–157.

Tomaszewski KE, Agarwal DK, and Melnick RL: In vitro steady-state level of hydrogen peroxide after exposure of male F-344 rats and female B6 $C3F_1$ mice to hepatic peroxisome proliferators. *Carcinogenesis* 1986; 7:1871–1876.

Tsukada H, Mochizuki Y, and Fujiwara S: The nucleotides of rat liver cell microbodies. *J Cell Biol* 1966; 28:449–460.

Usuda N, Reddy MK, Hashimoto T, Rao MS, and Reddy JK: Tissue specificity and species differences in the distribution of urate oxidase in peroxisomes. *Lab Invest* 1988a; 58:100–111.

Usuda N, Usman MI, Reddy MK, Hashimoto T, Reddy JK, and Rao MS: Immunocytochemical localization of urate oxidase, fatty acyl-CoA oxidase, and catalase in bovine kidney peroxisomes. *J Histochem Cytochem* 1988b; 36:253–258.

Vainio H, Linnainmma K, Kahonen M, Nickels J, Hietanen E, Mariemi J, and Peltoneu P: Hypolipidemic and peroxisome proliferation induced by phenoxyacetic acid herbicides in rats. *Biochem. Pharmacol* 1983; 32:2775–2779.

Vandenbosch KA: Light and electron microscopic visualization of uricase by immunogold labeling of sections of resin-embedded soybean nodules. *J Microsc* 1986; 143:187–196.

Wachter F, Bieri F, Stäubli W, and Bentley P: Induction of cytosolic and microsomal epoxide hydrolases by the hypolipidemic compound nafenopin in the mouse liver. *Biochem. Pharmacol* 1984; 33:31–34.

Ward JM, Hagiwara A, Anderson LM, Lindsey K, and Diwan BA: The chronic hepatic or renal toxicity of di(2-ethylhexyl)phthalate, acetaminophen, sodium barbital and phenobarbital in male B6C3F$_1$ mice: autoradiographic immunohistochemical and biochemical evidence associated for levels of DNA synthesis not associated with carcinogenesis or tumor promotion. *Toxicol Appl Pharmacol* 1988; 96:494–506.

Ward JM, Rice JM, Creasia D, Lynch P, and Riggs C: Dissimilar pattern of promotion by di(2-ethylhexyl)phthalate and phenobarbital of hepatocellular neoplasia initiated by diethylnitrosamine in B6C3F$_1$ mice. *Carcinogenesis* 1983; 4:1021–1029.

Warren JR, Simmon VF, and Reddy JK: Properties of hypolipidemic peroxisome proliferators in the lymphocyte [^{3}H]thymidine and Salmonella mutagenesis assays. *Cancer Res* 1980; 40:36–41.

Wilson GN, Holmes RD, and Hajra AK: Peroxisomal disorders: clinical commentary and future prospects. *Am J Med Genet* 1988; 30:771–792.

Wirth PI, Rao MS, and Evarts RP: Coordinate polypeptide expression during hepatocarcinogenesis in male F-344 rats: comparison of the Solt-Farber and Reddy models. *Cancer Res* 1987; 42:2839–2851.

Yeldandi AV, Milano M, Subbarao V, Reddy JK, and Rao MS: Evaluation of liver cell proliferation during ciprofibrate-induced hepatocarcinogenesis. *Cancer Lett* 1989; 47:21–27.

Yokota S and Fahimi D: Immunocytochemical localization of catalase in rat liver. *J Histochem Cytochem* 1981; 29:805–812.

Chapter

16

A Potential Role for Immunological Mechanisms in Halothane Hepatotoxicity

Andrea K. Hubbard
Toxicology Program
University of Connecticut
Storrs, Connecticut

Timothy P. Roth and A. Jay Gandolfi
Department of Anesthesiology
University of Arizona
Tucson, Arizona

INTRODUCTION

Hepatic injury is a recognized phenomenon following volatile anesthetic exposure, particularly after halothane (Figure 1) administrations and, to a lesser degree, other volatile anesthetics (Cousins et al., 1979, 1984). The hepatotoxicity associated with these anesthetics is sometimes fatal and always unexpected (Brown and Gandolfi, 1987). The sporadic nature of this injury has initiated extensive research to determine its mechanism, and major efforts are under way to produce a representative animal model.

Volatile halogenated anesthetics were originally thought to be chemically inert and refractory to biotransformation enzymes which would result in the majority of the anesthetic exhaled remaining unchanged (Cohen, 1978). However, subsequent studies demonstrated that volatile anesthetics undergo biotransformation (Van Dyke, Chenoweth, Van Poznak, 1964). This information immediately suggested that the hepatic injury was the result of the biotrans-

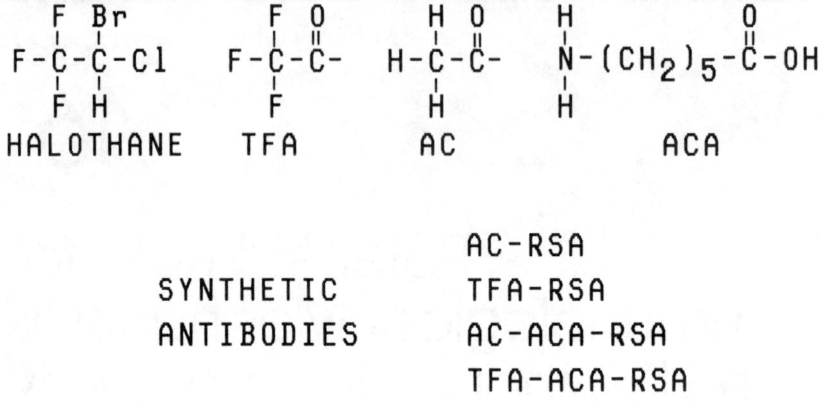

Figure 1. Chemical structure of the parent compound, halothane, and potential reactive intermediates/metabolites.

formation of these agents through the formation of reactive intermediates or toxic metabolites.

ANESTHETIC HEPATOTOXICITY STUDIES

All volatile halogenated inhalation anesthetics have been, at one time or another, associated with hepatic injury. Shortly after the introduction of halothane into clinical practice, case reports appeared of unexplained jaundice following halothane anesthesia (Virtue and Payne, 1958). The incidence of unexplained hepatitis following halothane anesthesia is estimated at one in 7,000 to 30,000 halothane administrations (Cousins et al., 1979, 1984).

Halothane-induced liver injury was initially thought reserved to middle-aged, slightly obese women who received multiple anesthesias (Brown and Gandolfi, 1987). However, cases of liver injury after single exposures in men, women, and children have now been reported (Walton, 1986; Hals et al., 1986; Zaric et al., 1986; Kenna et al., 1987a). Although use of halothane has curtailed enflurane, one of the two volatile anesthetics replacing halothane has also been linked to cases of unexplained postoperative jaundice (Lewis et al., 1983; Eger et al., 1986). In addition, isoflurane, the other anesthetic replacing halothane has been associated with 16% of the 45 reports of hepatic dysfunction following anesthesia with this agent (Stoelting et al., 1987). Often it is difficult to state unequivocally the presence or absence of anesthetic induced liver injury because of the anecdotal nature of the case reports and the spectrum of morphological and clinical responses (Benjamin et al., 1985).

A major difficulty in establishing the hepatotoxic potential of volatile anesthetics has been the lack of animal models which might determine factors responsible for toxicity. In all the preclinical studies conducted for halothane, there was no clear evidence suggesting it as a potential hepatotoxin (Gandolfi et al., 1988). The first successful animal models were based on the premise that halothane required bioactivation to be hepatotoxic (Brown and Sipes, 1977). Even though the major route of the metabolism of halothane is oxidative-producing trifluroacetic acid (Sipes et al., 1980), halothane can also be biotransformed reductively to reactive intermediates that covalently bound to liver tissue (Van Dyke and Gandolfi, 1976; Widger et al., 1976). For halothane to be metabolized reductively to any great extent *in vitro,* the hepatic biotransformation enzymes have to be induced and metabolism conducted in a hypoxic environment.

Based on these facts, two rat models for halothane hepatotoxicity were developed. One involved the pretreatment of rats with polychlorinated biphenyls followed by halothane exposure in 99% oxygen (Sipes and Brown, 1976). The other, referred to as the halothane-hypoxia model, required that rats be pretreated with phenobarbital and then exposed to halothane in 14% O_2 (Jee et al., 1980). Although these models have been confirmed by other investigators (Reynolds and Moslen, 1977; Cousins et al., 1979), both models have been questioned as to their relevance to the clinical situation and to the mechanism of toxicity as either a result of metabolism of the anesthetic or merely the conditions required to make the anesthetics toxic (Lind et al., 1985; Shingu et al., 1982).

It is a difficult task to separate potentially harmful physiological-pharmacological actions of an agent from its direct toxic effects. However, it was demonstrated that any liver injury produced by the pharmacological effects of the anesthetic (i.e., reduced hepatic blood flow) and the associated physiological state (i.e., hepatic ischemia/hypoxia) produced a distinctly different lesion and time course of injury (Lind et al., 1987, 1988).

This is in sharp contrast to the halothane-hypoxia model which requires phenobarbital pretreatment to produce hepatotoxicity, is limited to male rats (their biotransformation enzymes can be induced to higher levels than females), and has a time course of development and histopathology mimicking other classic hepatotoxins (Lind et al., 1985). A chemotoxic mechanism of injury in the halothane-hypoxia model has been supported by other laboratories as well (Gourlay et al., 1981; Hatano et al., 1985; De Groot and Noll, 1983).

The rat models for anesthetic hepatotoxicity have been under scrutiny as to their accuracy in representing the human situation (Gandolfi and Brown, 1983). In all the rat models of halothane-induced liver injury, multiple, sometimes stringent, pretreatments are required to produce a lesion which appears within hours, quickly resolves, does not progress, and is limited to males. Obviously, this is in marked contrast to the time course of events in the clinical situation (Benjamin et al., 1985; Goodman et al., 1984).

The reservations associated with the use of the rat models were, in part, solved with a model in the guinea pig (Luman et al., 1985; Lind et al., 1987). In this model, halothane produces in both sexes of selected strains a liver lesion which develops over a 2-day period and persists for over 7 days. The injury presents itself as either confluent centrilobular necrosis or scattered foci throughout the lobules. Neither phenobarbital pretreatment nor exposures in hypoxic atmospheres are required for the production of the hepatotoxicity. The injury is more prevalent in females and in heavier, older animals. In addition, biotransformation inhibitors or administration of deuterated halothane reduces the severity of the lesion. Although this animal model is a better representation of the clinical situation, it still does not account for the latent, progressive, often terminal liver injury seen in patients.

At this time, there are no animal models for enflurane and isoflurane hepatotoxicity beyond those that utilize hypoxic mechanisms.

IMMUNOLOGICALLY MEDIATED LIVER INJURY

Although anesthetic-induced hepatotoxicity may be initiated by direct or metabolite-mediate process following halothane exposure, this liver injury may also be the result of an immune-mediated process. However, the association between halothane-induced liver injury and an immune response directed toward a metabolite-altered hepatocyte protein has been under scrutiny for several years with ambiguous results. The following review will present evidence thus far published on the influence by the immune system in initiating or exacerbating this disease entity as substantiated by epidemiologic data, by humoral and cell-mediated mechanisms of injury, by the presence of halothane-induced antigens, by the relevance of immune complex levels in patients and by the association of HLA genotype in halothane patients to disease outcome. Finally, results will be described which suggest an immunologic cross-sensitization between halothane and enflurane, and the possible impact on current anesthesia.

Principles of Immunology

The immunologic system is a complex network composed of many different components which respond to substances recognized by the host as foreign. In defense against these foreign substances or antigens, the body commandeers two protective mechanisms known as humoral and cell-mediated immunity. Humoral immunity is dependent upon the synthesis and release of soluble antibodies by B lymphocytes into the blood and other body fluids. Cell-mediated immunity involves the production of antigen-sensitized T lymphocytes which are effectors of cellular defense.

Implicit in the effectiveness of immune defense are the concepts of memory and specificity. Through memory, immune cells have the capacity to respond more rapidly to an antigenic stimulus after an initial exposure. Both B and T lymphocytes have cell surface receptors for a specific antigen which will not recognize or bind to any other antigen. This specificity allows the appropriate immune response to a variety of environmental insults and allows the host to differentiate between self and non-self.

The characteristics which allow a substance to be recognized as antigenic include size, chemical composition, degree of foreigness, and route and dose of exposure. A hapten is a foreign substance which by itself is too small to be recognized as an antigen. However, if this hapten is bound covalently to a protein (self or non-self) of the appropriate size, the host will respond through an immune response directed, in part, toward the hapten. This concept is important in examining the role of the immune system in halothane-induced liver injury since small metabolites of halothane can serve as haptens in modifying hepatocyte proteins. These hepatocyte proteins would serve, then, as carriers for the haptens and would, in all probability, be rendered antigenic by the binding of metabolites. Early investigations demonstrated that during halothane metabolism in the liver, a biotransformation intermediate of this anesthetic could bind covalently to liver macromolecules, thereby serving as a hapten to alter the antigenicity of hepatocyte membrane proteins (Uehleke et al., 1973). The reactive metabolic intermediates of halothane responsible for modifying self-components in the hepatocyte membrane could include among others a trifluoroacetyl halide from the oxidative pathway, free radicals or carbene intermediates from reductive pathways (Gandolfi et al., 1980). Binding by halothane metabolites to liver macromolecules is evidenced in early studies which noted that ^{14}C halothane bound to the microsomal fraction of a halothane-perfused liver (Van Dyke and Wood, 1975). Further work noted that halothane was metabolized in the liver to reactive intermediates which bound irreversibly to microsomal lipids and proteins (Widger et al., 1976).

Epidemiologic Evidence

Epidemiologic data in humans supports a relationship between a halothane-induced immune response and the development of halothane hepatitis. Multiple administrations of halothane increase the incidence and intensity of hepatitis; the interval between anesthetic and jaundice is shorter after two or more exposures than after a single exposure (Inman and Mushin, 1978); and patients are reported to evidence rash, arthralgia, and peripheral and/or hepatic eosinophilia (Dienstag, 1980). In addition, early studies noted an increased frequency of antimitochondrial and anti-smooth muscle antibodies (Rodriguez et al., 1969), antithyroid antibodies, (Walton et al., 1976), and anti-nucleic acid antibodies (Moult and Sherlock, 1975) in these halothane hepatis (HH) patients. Although these reports did not constitute proof for immunologically mediated damage following halothane anesthesia, they led investigators to determine if the immune system could be involved and if the mechanism of liver injury could be attributed to humoral or cell-mediated immune events.

Immune Mechanism: Humoral Immunity

A role for antibody-mediated injury or humoral immunity in halothane hepatitis has been explored in both animal models as well as in patients with halothane-induced liver injury. These studies were based on the concept that the antigenic moiety was not the parent drug halothane, but instead a reactive biotransformation intermediate of halothane which alters the antigenicity of liver cell constitutents. The antigen, then, would be composed of the reactive intermediate (haptenic moiety) and an altered hepatocyte protein (carrier moiety). Early work determined that the oxidative metabolic intermediate of halothane, trifluoroacetyl halide (TFA), could be immunogenic whether conjugated to a non-self or heterologous carrier protein (Rosenberg and Wahlstrom, 1973) or to a self or homologous carrier protein (Mathieu et al., 1975a).

Initial work in our laboratory created a synthetic antigen through the conjugation of TFA to a carrier protein by using the method of Goldberger and Anfinsen (1962). This protein [rabbit serum albumin] (RSA) was of the same species (rabbit) used for eventual immunization and antibody generation. In this way, antibodies would be generated *in vivo* only toward the haptenic moiety since the carrier protein would not be recognized as foreign. The immunizing antigen or immunogen could then be used as the detecting antigen in an enzyme-linked immunosorbent assay (ELISA) for determining the specificity and titer of this anti-TFA antibody. In addition, modifications can be made of this detecting antigen by inserting a spacer arm (e.g., amino caproic acid) between the haptenic and carrier moieties (Hubbard et al., 1990). This insertion decreases potential steric hindrance of antibody reactivity with haptenic moieties. The level of conjugation (e.g., 20%, 40%, 80%, etc.) also enhances antibody reactivity. Both types of modifications ultimately influence the sensitivity of the ELISA in detecting hapten specific antibody. As the next step following the development of synthetic antigens, the native halothane-induced antigen(s) can be isolated from liver homogenates from halothane-exposed animals. These antigens can then be used as immunogens for generating antibodies specific for the natural antigens or as detecting antigens for monitoring levels of halothane-induced antibody from animals and humans alike.

Using trifluoroacetylated rabbit serum albumin (TFA-RSA) as the detecting antigen, the TFA moiety was found to be part of the immunogen responsible for generating a halothane-induced antibody response in rabbits (Callis et al., 1987) (Figure 2) or guinea pigs (Siadat-Pajouh et al., 1987) receiving multiple halothane exposures. Indirect evidence then suggests that a reactive metabolic intermediate of halothane could alter a hepatocyte protein to produce an immunogen composed in part of the intermediate. The subsequent anti-TFA antibody could potentially damage hepatocytes incorporating this TFA-altered liver protein. Although proof for this theory is still under investigation, it is somewhat contradicted by the fact that the presence of high levels of specific anti-TFA antibody in animals prior to halothane exposure did not exacerbate or alter the transient liver damage induced by administration of this anesthetic (Mathieu et al., 1975b; Callis et al., 1987).

Patients with liver complications following halothane anesthesia have also demonstrated a humoral immune response directed toward a hepatocyte protein altered by a reactive metabolic intermediate of halothane such as TFA. Rather than using TFA conjugated to a specific protein, early investigations used a more natural detecting antigen by examining reactivity of patient sera with hepatocytes from rabbits which had been exposed to halothane. These patients with fulminant halothane hepatitis (HH) evidence a circulating antibody which appeared to react with the surface of rabbit hepatocytes altered by halothane exposure (Vergani et al., 1980; Neuberger et al., 1983). Evaluated by two different methods for determining antibody-cell interaction, this antibody could be detected only in sera from halothane hepatitis patients and

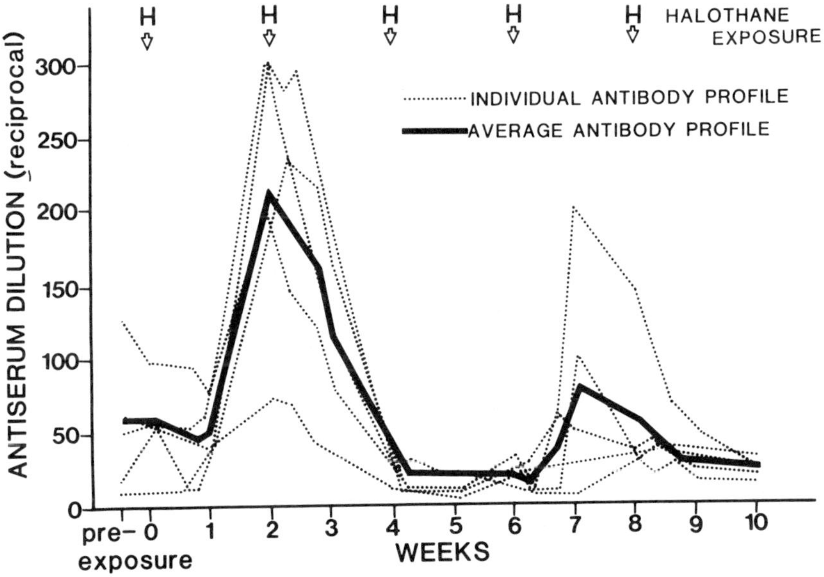

Figure 2. Halothane-induced antibody response specific for trifluoroacetylated-rabbit serum albumin. Rabbits were exposed five times at 2-week intervals to 1% halothane in 80% oxygen. At least two exposures were required for maximum anti-TFA antibody response. This response declined with successive exposures. Data are expressed as the reciprocal of the antibody dilution that yields an optical density of 0.5 (410/490 nm).

not from individuals with other liver complications. In addition, sera from two surgeons with hepatic damage following occupational exposure were also positive for this antibody (Neuberger et al., 1981a).

More recent studies using the sensitive and highly quantitative enzyme linked immunosorbent assay (ELISA) have noted that sera from HH patients will react with subcellular fractions of hepatocytes from halothane-exposed rabbits (Kenna et al., 1984) as well as with the halothane metabolite, TFA, conjugated to a carrier protein (Satoh et al., 1987). Indeed, recent work in our laboratory noted that five of six patients with HH have a high titer of circulating antibody that recognizes TFA, whereas ten patients exposed to halothane and not developing HH have minimal or no antibody (Hubbard et al., 1988). Presence of a halothane-induced antibody was determined in an ELISA using TFA-RSA and RSA (carrier control) as detecting antigens.

Speculation still remains as to the role this antibody may play in contributing to halothane-induced liver damage in these patients. However, detection of circulating antibody directed against altered liver proteins or reactive intermediates of halothane could provide two major advantages: (1) early diagnosis of this disease entity enabling potential therapeutic intervention and (2) defining the mechanism of immune-mediated hepatic damage by assessing the specificity of human antibodies.

Immune Mechanism: Cell-Mediated Immunity

Cell-mediated immunity by sensitized lymphocytes has also been studied as a potential mechanism in immunologically mediated halothane hepatitis. Sensitization of T lymphocytes to a specific antigen can be measured by cutaneous hypersensitivity, by induction of lymphocyte proliferation, or by release of lymphokines from T lymphocytes. Studies in animals revealed that the reactive metabolic intermediate, TFA, could serve as the haptenic moiety of an immunogen in inducing a cell-mediated immune response. This metabolite of halothane when coupled to a self or non-self protein could induce a specific delayed-type hypersensitivity response in animals as measured by cutaneous hypersensitivity (Mathieu et al., 1974) but did not induce potentially sensitized lymphocytes to proliferate (Mathieu et al., 1973). In addition, multiple halothane exposures produced a positive cutaneous hypersensitivity response to TFA-carrier protein in 4 of 29 animals exposed. However, there was no relationship between the degree of hepatic necrosis and positive skin reactions (Reves and McCracken, 1976b). Splenic leukocytes from rabbits receiving multiple exposures to halothane at 2-week intervals were found to respond by proliferation to TFA-RSA, but not RSA (Hubbard et al., 1988). Interestingly, splenic leukocytes from nonexposed animals also responded by proliferation suggesting that the hapten carrier complex, TFA-RSA, produced a nonspecific mitogenic effect. Peripheral blood leukocytes and leukocytes from the gut-associated lymph nodes from both exposed and nonexposed did not respond. Although, the potential role for sensitized lymphocytes in mediating halothane-induced damage is under investigation, there is evidence which does not support this as a mechanism in halothane-initiated damage. The induction of a delayed hypersensitivity response to the oxidative metabolite of halothane, TFA, prior to halothane exposure did not exacerbate halothane-induced liver damage in either guinea pigs (Mathieu et al., 1975); Reves and McCracken, 1976a) or rats (Ford et al., 1984).

A role for a cell-mediated mechanism in contributing to liver damage has also been investigated in halothane hepatitis patients. Initial studies *in vitro* explored a cell-mediated immune response by measuring the ability of lymphocytes from HH patients to respond to the potential antigen, halothane. If T lymphocytes from these patients were sensitized to halothane, they would respond by proliferation and/or release of lymphokines. Exposure of peripheral blood leukocytes from HH patients to halothane was reported to induce blast transformation or proliferation (Paronetto and Popper, 1970) as well as to evoke the release of the lymphokine, macrophage inhibitory factor (Price et al., 1977). However, later studies could not confirm the induction of such a lymphocyte response by the parent compound halothane (Walton et al., 1973; Moult et al., 1975). Based on the concept that the antigenic moiety was a metabolic intermediate and not the parent compound halothane, Vergani and colleagues (1978) used as antigen, liver homogenates containing metabolite-altered proteins from rabbits exposed to halothane. These liver homogenates were then incubated with leukocytes from HH patients and in 4 of 12 patients, lymphocytes were sensitized as measured by the release of a lymphokine.

Although a direct role for either antibodies or sensitized lymphocytes in mediating liver damage following halothane exposure has not been defined, several parallels exist between these two types of immune responses generated. (1) The oxidative metabolic intermediate of halothane, trifluoroacetyl halide, is part of the immunogen in immune responses generated directly through halothane exposures. (2) This metabolic intermediate can also serve as the haptenic portion of an immunogen in animals specifically immunized with TFA-carrier protein. (3) There is no evidence that the existence of specific antibody or sensitized lymphocytes prior to halothane exposure exacerbates any liver damage induced by anesthetic exposure.

Halothane-Induced Antigens

Studies over the past 10 years provide evidence that halothane exposure generates an immune response specific for a metabolic intermediate of this drug. The identity of the immunogenic moiety(ies) generated in the hepatocytes of halothane-exposed animals and humans responsible for this immune response is currently under investigation. The earliest data on the characteristics of this antigen note that sera from patients with fulminant hepatic failure following halothane anesthesia would react preferentially with hepatocytes from rabbits exposed to halothane under high oxygen tension (99%) rather than low (14%) (Neuberger et al., 1981b). These antibodies, then, displayed a specificity for an antigenic moiety generated under oxidative conditions of halothane metabolism. Recent studies by Satoh et al. (1985a,b,c) have identified an oxidative metabolic intermediate of halothane in the centrilobular area of liver sections and on the surface of hepatocytes from livers of rats exposed to halothane 2 h prior. These investigators have extended their observations and noted that a major target for the trifluoroacetyl halide intermediate is a microsomal esterase. Similar TFA adducts were also detected in liver homogenates from patients that had received halothane (Satoh et al., 1985b,c). Kenna et al. (1985) determined that sera from 11 of 14 patients with halothane hepatitis reacted with any of three specific polypeptides from liver homogenates of rabbits exposed once to halothane 18 h earlier. In rabbits receiving multiple halothane exposures, four novel liver proteins or antigens were recognized by populations of halothane-induced antibodies in a Western immunoblot assay (Roth et al., 1988a). Rabbits were exposed one to five times at two-week intervals to halothane and groups of rabbits were killed on a daily basis after each exposure. Liver homogenates were assessed for halothane-induced antigen in a Western immunoblot with anti-TFA antibody. Following one exposure, the greatest expression of antigen was during the first 2 days, whereas after five exposures the antigen expression was elevated for 4 days post-halothane. The greatest expression of these novel proteins occurred after the second and third exposure (Figure 3). Halothane-induced antigen expression could also be monitored in situ with an immunoperoxidase assay by examining liver sections from multiply exposed rabbits (Roth et al., 1988b). Antigen expression was localized to the centrilobular areas of the liver to a cell depth of 7 to 10 cells. This expression was apparent the first 2 days following one halothane exposure and during the first 4 days following five exposures. In guinea pigs exposed once to halothane, again antigen expression was localized to the centrilobular areas. This halothane-induced antigen could be detected as early as 6 h postexposure and was still apparent 90 h later.

These data suggest the following as a possible mechanism for the generation of a halothane-induced antigen(s) and subsequent immune response. Intracellular halothane is rapidly metabolized by the cytochrome P450 enzyme complex generating reactive metabolic species. One such reactive species, trifluoroacetyl halide, may react with intracellular and perhaps plasma membrane proteins altering their antigenicity. These altered proteins may be expressed on the cell surface or alternately be released from dying hepatocytes. Given the quantity and degree of foreigness of these proteins, an immune response (humoral and/or cell mediated) may ensue. Several populations of

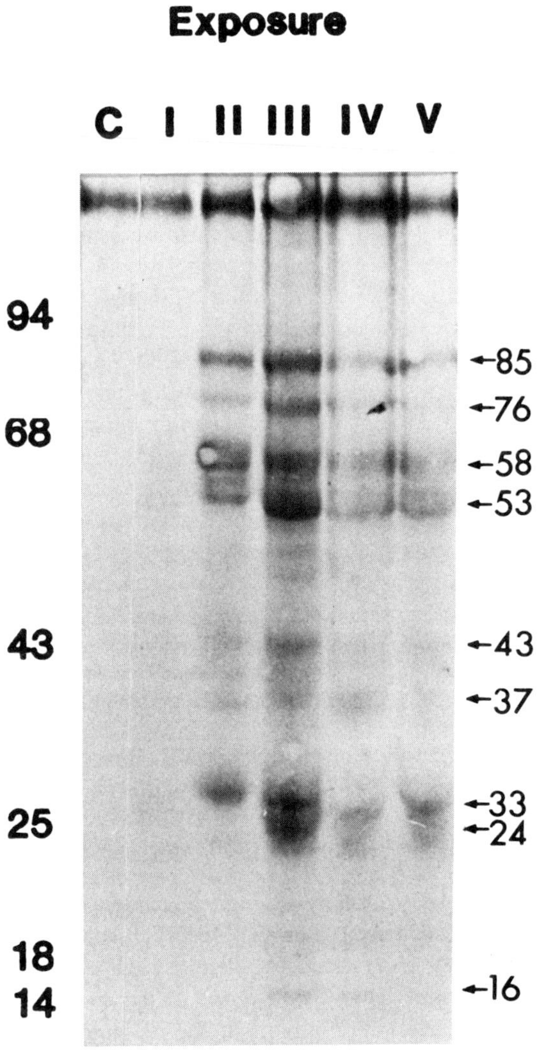

Figure 3. Halothane-induced antigen as detected by anti-TFA antibody in a Western analysis. Rabbits were exposed one, two, three, four, or five times to 1% halothane in 80% oxygen at 2-week intervals. Livers were homogenized, proteins electrophoretically separated and analyzed in a Western immunoassay. Protein bands, as detected by reactivity with anti-TFA antibody, were in highest concentration after exposures 2 and 3.

antibodies may be formed with specificities directed toward the hapten (TFA) or the carrier (altered hepatocyte proteins) or both. Antibodies may play a role in clearance of free antigen by the reticuloendothelial system with the formation of immune complexes or may lyse any hepatocytes which have altered membrane proteins.

Role of Immune Complexes

Reports of elevated immune complexes, antigen-antibody combinations, in HH patients support the theory that halothane exposures induce an antigen which subsequently will stimulate the production of specific antibodies. In addition, analysis of these immune complexes may reveal the identity of the potential immunogen and its role as a target in hepatic damage. A patient who developed hepatitis following multiple exposures to halothane developed an acute serum sickness syndrome and evidenced an elevated level of immune complexes in her serum and synovial fluid. Although unidentified, a metabolite of halothane was associated with these complexes (Williams, et al., 1977). Circulating immune complexes were also found elevated in all six patients with fulminant hepatic failure following halothane anesthesia. The authors speculate that an antibody produced in response to specific antigens generated by halothane, complexes with the corresponding antigen released from damaged liver cells. Since patients with paracetamol-induced hepatitis did not evidence elevated levels of immune complexes, the authors discount the theory that the immune complexes are simply a reflection of an immune response to released liver antigens as a result of hepatic necrosis (Canalese et al., 1981). Circulating immune complexes have been detected and isolated from sera from animal and HH patients (Roth et al., 1986). Although the specificity of these complexes is unknown, the chronology and level of these immune complexes correlate with the profile of anti-TFA antibody. The complexes appear just after peak levels of circulating antibody are detected (Osman et al., 1986).

Relationship of HLA Antigens with Halothane Hepatitis

Associations of specific genotypes with the tendency to develop halothane hepatitis may (1) provide a method for screening patients potentially at risk and (2) suggest a role for the immune system in mediating liver damage since immune response genes (HLA-DR) are part of the major histocompatibility complex and may reflect contributions of the immune response to disease. Although Gardner (1978) suggested HLA typing as a screening test for individuals potentially at risk for halothane hepatitis, later studies did not substantiate this concept. HLA A and B locus antigens were determined in 17 patients who had recovered from unexplained hepatitis following halothane anesthesia, and no statistical association could be found between HLA types and the development of halothane hepatitis (Eade et al., 1981). In contrast, the occurrence of severe hepatic necrosis after halothane anesthesia in three pairs of closely related women with a common ethnic origin suggested the possibility of familial susceptibility to halothane-induced fulminant hepatotoxicity (Hoft et al., 1981). Moreover, when all loci of HLA antigens, HLA A, B, C and DR were examined in patients recovering from halothane hepatitis, an association was noted between specific genotype and the development of post-halothane jaundice (Otsuka et al., 1985). The observation of genetic differences in halothane hepatotoxicity in different strains of rats and guinea pigs may also suggest a genetically linked association between genotype and halothane-induced liver disease (Gourlay et al., 1981; Lind et al., 1987). The contribution of the immune response to liver damage in this study might also be under immunogenetic influence.

Potential Cross-Reactivity with Other Anesthetics

The role for a specific anesthetic (halothane) -induced immune response in mediating liver damage initiated by other anesthetics (e.g., enflurane) is under investigation. Both enflurane and isoflurane will produce potential acetylating agents during their oxidative biotransformation (Burke et al., 1981; Hitt et al., 1974). Very recent evidence suggests a potential cross-sensitization between enflurane and halothane-modified hepatic proteins (Christ et al., 1988a,b). An antibody with specificity for the halothane intermediate, TFA, recognized and bound to microsomal proteins from rats treated with three consecutive intraperitoneal doses of either halothane or enflurane. These data suggest that individuals sensitized to halothane may produce antigenically similar moieties upon exposure to enflurane. This is one of the first potential, unifying themes for an anesthetic-induced liver injury which is latent, enhanced by multiple exposures, and progressive.

SUMMARY

The hypothesis that host immunologic mechanisms play a role in mediating anesthetic-induced liver damage can neither be accepted nor rejected. However, epidemiologic data from halothane hepatitis patients provide strong indirect support as does data generated over the past 5–8 years in animal models of halothane-induced immune reactivity (Table 16-1). Future research goals, employing highly sensitive immunologic techniques, will focus on discerning a more direct cause and effect relationship between cellular and humoral immune defenses and latent, progressive anesthetic-induced liver injury.

TABLE 16–1.
Animal Models of Halothane Induced Immune Response

	Halothane-Induced Antigen	Halothane-Induced Antibody	Increase in Liver Enzymes	Histopathologic Damage
Rat*	+	ND†	ND	ND
Rabbit‡	+	ND	ND	ND
Rabbit§ +	+	+	—	—
Guinea Pig§	+	+	+	+

*Pohl et al., NIH.
†Not determined.
‡Neuberger and Kenna, Kings College, England.

REFERENCES

Benjamin SB, Goodman ZD, Ishak KG, Zimmerman HJ, and Irey NS: The morphologic spectrum of halothane-induced hepatic injury: analysis of 77 cases. *Hepatology* 1985; 5:1163–1171.

Brown BR and Gandolfi AJ: Adverse effects of volatile anaesthetics. *Br J Anaesth* 1987; 59:14–23.

Brown BR Jr and Sipes IG: Biotransformation and hepatotoxicity of halothane. *Biochem Pharmacol* 1977; 26:2091–2094.

Burke T, Branchflower R, Lees D, and Pohl L: Mechanism of defluorination of enflurane. *Drug Metab Disp* 1981; 9:19–24.

Callis AH, Brooks SD, Waters SJ, Lucas DO, Gandolfi AJ, Sipes IG, Satoh H, and Pohl LR: Evidence for a role of the immune system in the pathogenesis of halothane hepatitis, in Roth SH and Miller KW (eds): *Molecular Cellular Mechanisms Anesthetics,* 1986.

Callis AK, Brooks SD, Roth TP, Gandolfi AJ, and Brown BR: Characterization of a halothane-induced humoral response in rabbits. *Clin Exp Immunol* 1987; 67:343–351.

Canalese J, Wyke RJ, Vergani D, and Eddleston ALWF: Circulating immune complexes in patients with fulminant hepatic failure. *Gut* 1981; 22:845–848.

Christ DD, Kenna JG, Satoh H, Pohl LR: Enflurane metabolism produces covalently-bound liver adducts recognized by antibodies from patients with halothane hepatitis. *Anesthesiology,* 1988a 69:833–838.

Christ DD, Satoh H, Kenna JG, and Pohl LR: Potential metabolic basis for enflurane hepatitis and the apparent cross-sensitization between enflurane and halothane. *Drug Metab Disp* 1988b; 16:1–6.

Cohen EN: Toxicity of inhalation anesthetic agents. *Br J Anaes* 1978; 59:665–675.

Cousins MJ, Sharp JH, Gourlay GK, Adams JF, Hayes DW, and Whitehead R: Hepatotoxicity and halothane metabolism in an animal model with application for human toxicity. *Anaesth Intens Care* 1979; 7:9–24.

Cousins MJ, Plummer JL, and Hall P: Toxicity of volatile anaesthetic agents. *Clinics Anesth* 1984; 2:551–575.

DeGroot H, Noll T: Halothane hepatotoxicity: Relation between metabolic activation, hypoxia, covalent binding, lipid peroxidation, and liver cell damage. *Hepatology* 1983; 3:601–606.

Dienstag JL: Halothane hepatitis. Allergy or idiosyncrasy? *N Engl J Med* 1980; 303:102–104.

Eade OE, Grice D, Krawitt EL, Trowell J, Albertini R, Festenstein H, and Wright R: HLAA and B locus antigens in patients with unexplained hepatitis following halothane anaesthesia. *Tissue Antigens* 1981; 17:428–432.

Eger EI, Smuckler EA, Ferrell LD, Goldsmith CH, and Johnson BH: Is enflurane hepatotoxic? *Anesth Analg* 1986; 65:21–30.

Ford DJ, Coyle DE, and Harrington JF: Effects of hypersensitivity to a halothane metabolite on halothane-induced liver damage. *Anesthesiology* 1984; 60:141–143.

Gandolfi AJ and Brown BK: Hypoxia and halothane hepatotoxicity. *Anesthesiology* 1983; 62:859–861.

Gandolfi AJ, Hubbard AK, and Nunn J: Viscerotoxicity of volatile anesthetics, in Nunn JF, Utting JE, and Brown BR (eds): *General Anesthetics,* ed 5. Sonehan, Butterworth, 1988.

Gandolfi AJ, White RW, Sipes IG, and Pohl LR: Bioactivation and binding of halothane *in vitro:* studies with ^{3}H and ^{14}C halothane. *J Pharmacol Exp Ther* 1980; 214:721.

Gardner LI: Halothane hepatitis. *Lancet* 1978; 2:791.

Goldberger R and Anfinsen C: The reversible masking of amino groups in ribonuclease and its possible usefulness in the synthesis of the protein. *Biochemistry* 1962; 1:401–405.

Goodman ZD, Benjamin SB, Ishar KG, Zimmerman HJ, and Irey NS: Pathology of halothane-associated liver injury. *J Lab Invest* 1984; 50:23A.

Gourlay GK, Adams JF, Cousins MJ, and Sharp JH: Time-course of formation of volatile reductive metabolites of halothane in humans and an animal model. *Br J Anaesth* 1980; 52:331.

Gourlay GK, Adams JF, Cousins MJ, and Hall P: Genetic differences in reductive metabolism and hepatotoxicity of halothane in three rat strains. *Anesthesiology* 1981; 55:96–103.

Hals J, Dodgson MS, Skulberg A, and Kenna JG: Halothane-associated liver damage and renal failure in a young child. *Acta Anaesthesiol Scand* 1986; 30:651–655.

Hatano H, Nomura F, Ohnishi K, Iijima T, Haysaka A, Iida S, Koen H, and Okuda K: Respective roles of hypoxia and halothane metabolism halothane-induced liver injury in rats. *Hepatology* 1985; 5:241–244.

Hitt BA, Mazze RI, Cousins MJ, Edmunds HN, Barr GA, and Trudell JR: Metabolism of isoflurane in Fischer 344 rats and man. *Anesthesiology* 1974; 40:62–67.

Hoft RH, Bunker JP, Goodman HI, and Gregory PB: Halothane hepatitis in three pairs of closely related women. *N Eng J Med* 1981; 304:1023–1024.

Hubbard AK, Levy JP, Roth TP, Gandolfi AJ: Use of structural alterations in the synthesis of halothane metabolite antigens to mimic halothane-induced immunogen. *Drug Chem Toxicol* 1990; 13:93–112.

Hubbard AK, Roth TP, and Gandolfi AJ: Elicitation of a cell mediated immune response to a reactive intermediate of halothane. *Toxicologist* 1988; 8:12.

Hubbard AK, Roth TP, Gandolfi AJ, and Brown BR: Halothane hepatitis patients generate an antibody response toward a metabolite of halothane. *Anesthesiology* 1988; 68:791–796.

Inman WHW and Mushin WW: Jaundice after repeated exposure to halothane: a further analysis of reports to the Committee on Safety of Medicines. *Br Med J* 1978; 2:1455–1456.

Jee R, Gandolfi AJ, Sipes IG, Braun BR: Factors influencing halothane hepatotoxicity in the rat hypoxic model. *Toxicol Appl Pharmacol* 1980; 52:267–277.

Kenna JG, Neuberger J, Mieli-Vergani G, Mowat AP, and Williams R: Halothane hepatitis in children. *Br Med J* 1987a; 294:1209–1211.

Kenna JG, Neuberger J, and Williams R: An enzyme-linked immunosorbent assay for detection of antibodies against halothane-altered hepatocyte antigens. *J Immunol Methods* 1984; 75:3–14.

Kenna JG, Neuberger J, and Williams R: Characterization of halothane-induced antigens by immunoblotting. *Biochem Soc Trans* 1985; 13:910–911.

Kenna JG, Neuberger J, and Williams R: Identification by immunoblotting of three halothane-induced liver microsomal polypeptide antigens recognized by antibodies in sera from patients with halothane-associated hepatitis. *J Pharmacol Exp Ther* 1987b; 242:733–740.

Lewis JH, Zimmerman HJ, Kshak KG, and Mullick FG: Enflurane hepatotoxicity. A clinico-pathologic study of 24 cases. *Ann Int Med* 1983; 98:984.

Lind RC, Gandolfi AJ, Brown BR, and Hall P LM: Halothane toxicity in the guinea pig. *Anesth Analg* 1987; 66:222–228.

Lind RC, Gandolfi AJ, Sipes IG, Brown BR: Comparison of the requirements for hepatic injury with halothane and enflurane in rats. *Anesth Analg* 1985; 64:955–963.

Lunam CA, Cousins MJ, and Hall PM: Guinea-pig model of halothane-associated hepatotoxicity in the absence of enzyme induction and hypoxia. *J Pharmacol Exp Ther* 1985; 232:802.

Mathieu A, DiPadua D, Kahan B, and Mills J: Lymphocyte transformation test, reaginic and complement fixing antibodies in guinea pigs sensitized to a metabolite of halothane and fluroxene. *Anesth Analg* 1973; 52:676–679.

Mathieu A, DiPadua D, Kahan BD, and Mills J: Humoral immunity to a metabolite of halothane, fluroxene, and enflurane. *Anesthesiology* 1975b; 42:612–616.

Mathieu A, DiPadua D, Mills J, and Kahan B: Experimental immunity to a metabolite of halothane and fluoroxene. Cutaneous delayed-type hypersensitivity. *Anesthesiology* 1974; 40:383–390.

Mathieu A, DiPadua D, Kahan BD, Galdabini JJ, and Mills J: Correlation between specific immunity to a metabolite of halothane and hepatic lesions after multiple exposures. *Anesth Analg* 1975a; 54:332–339.

Moult PJA and Sherlock S: Halothane-related hepatitis. *Q.J. Med* 1975; 44:99–103.

Moult PJA, Adjukiewicz AB, Gaylarde PM, Sarkany I, and Sherlock S: Lymphocyte transformation in halothane-related hepatitis. *Br Med J* 1975; 69–70.

Neuberger J, Gimson AES, Davis M, and Williams, R: Specific serological markers in the diagnosis of fulminant hepatic failure associated with halothane anesthesia. *Br J Anaesth* 1983; 55:15–19.

Neuberger J, Mieli-Vergani G, Tredger JM, Davis M, and Williams R: Oxidative metabolism of halothane in the production of altered hepatocyte membrane antigens in acute halothane induced necrosis. *Gut* 1981b; 22:669–672.

Neuberger J, Vergani D, Mieli-Vergani G, Davis M, and Williams R: Hepatic damage after exposure to halothane in medical personnel. *Br J Anaesth* 1981a; 53:1173.

Osman K, Roth TP, and Gandolfi AJ: *Circulating Immune-Complexes in Halothane-Induced Hepatic Injury in Rabbits*. Mountain West Society of Toxicology Regional Meeting, Tucson, 1986.

Otsuka S, Yamamoto M, Kasuya S, Ohtomo H, Yamamoto Y, Yoshida TO, and Akaza T: HLA antigens in patients with unexplained hepatitis following halothane anesthesia. *Acta Anaesth Scand* 1985; 29:497–501.

Paronetto F and Popper H: Lymphocyte stimulation induced by halothane in patients with hepatitis following exposure to halothane. *N Engl J Med* 1970; 282:277–280.

Pohl LR, Satoh H, Christ DD, and Kenna JG: The immunological and metabolic basis of drug hypersensitivities. *Annu Rev Pharmacol* 1988; 28:367–87.

Price C, Gibbs A, and Williams W: Halothane macrophage migration inhibition factor test in halothane-associated hepatitis. *J Clin Pathol* 1977; 30:312–316.

Reves J and McCracken L: Failure to induce hepatic pathology in animals sensitized to a halothane metabolite and subsequently challenged with halothane. *Anesth Analg* 1976a; 55:235–242.

Reves J and McCracken L: Hepatic pathology and skin test reactions to trifluoroacetylated autologous protein after repeated halothane anaesthesia in the guinea pig. *Br J Anaesth* 1976b; 48:419–424.

Reynolds E, Moslen MT: Halothane hepatotoxicity: enhanced by polychlorinated biphenyl pretreatment. *Anesthesiology* 1977; 47:19–27.

Rodriguez M, Paronetto F, Schaffner F, and Popper H: Antimitochondrial antibodies in jaundice following drug administration. *J Amer Med Assoc* 1969; 208:148–150.

Rosenberg PH and Wahlstrom T: Trifluoroacetic acid and some possible intermediate metabolites of halothane as haptens. *Anesthesiology* 1973; 38:224.

Roth TP, Callis (Hubbard) AK with Gandolfi AJ and Brown BR: Chronology of halothane induced antigen expression in halothane exposed rabbits. *Clin Exp Immunol* 1988a, 72:330–336.

Roth TP, Hubbard AK, and Gandolfi AJ: Localization of halothane-induced antigen in situ by specific anti halothane metabolite antibodies. *Toxicologist* 1988b; 8:11.

Roth TP, Hubbard AK, Gandolfi AJ, and Brown BR: Circulating immune complexes in halothane hepatitis patients correlate with anti-halothane antibody levels. *Pharmacologist* 1986; 28:194.

Satoh H, Davies HW, Takemura T, Maeda K, Gillette JR, and Pohl LR: An immunohistochemical approach to investigating the mechanism of halothane-induced hepatotoxicity, in Bridges et al. (eds): *Progress in Drug Metabolism,* vol 10. Taylor & Francis Ltd, pp 187–206, 1987.

Satoh H, Fukuda Y, Anderson DK, Ferrans VJ, Gillette JR, and Pohl LR: Immunogolical studies on the mechanisms of halothane-induced hepatotoxicity: immunohistochemical evidence of trifluoroacetylated hepatocytes. *J Pharma Exp Ther* 1985a; 233:857–862.

Satoh H, Gillette JR, Davies HW, Schulick RD, and Pohl LR: Immunochemical evidence of trifluoroacetylated cytochromes P-450 in the liver of halothane treated rats. *Mol Pharmacol* 1985c; 28:468–474.

Satoh H, Gillette JR, Takemura T, Ferrans VJ, Jelenich SE, Kenna JG, Neuberger J, and Pohl LR: Investigation of the immunological basis of halothane-induced hepatotoxicity, in Kocsis JJ, Jollow DJ, Witmer CM, Nelson JO, and Snyder R (eds): *Biological Reactive Intermediates III. Mechanisms of Action in Animal Models and Human Disease.* 1985, pp 657–673.

Satoh H, Gillette JR, Takemura T, Ferrans VJ, Jelenich SE, Kenna JG, Neuberger J, and Pohl LR: Investigation of the immunological basis of halothane-induced hepatotoxicity, in Kocsis JJ, Jollow DJ, Witmer CM, Nelson JO, and Snyder R (eds): *Biological Reactive Intermediates III. Mechanisms of Action in Animal Models and Human Disease.* 1985b, pp 657–673.

Shingu K, Eger El II, and Johnson BH: Hypoxia may be more important than reductive metabolism in halothane-induced hepatic injury. *Anesth Analg* 1982; 61:824–827.

Siadat Pajouh M, Hubbard AK, Roth TD, and Gandolfi AJ: Generation of halothane induced immune response in a guinea pig model of halothane hepatitis. *Anesth Analg* 1987; 66:1209–1214.

Sipes IG and Brown BR: An animal model of hepatotoxicity associated with halothane anesthesia. *Anesthesiology* 1976; 45:622–628.

Sipes IG, Gandolfi AJ, Pohl LR, Krishna G, and Brown BR Jr: Comparison of the biotransformation and hepatotoxicity of halothane and deuterated halothane. *J Pharmacol Exp Ther* 1980; 214:716–710.

Stoelting RK, Blitt CD, Cohen PJ, and Merin RG: Hepatic dysfunction after isoflurane anesthesia. *Anesth Analg* 1987; 66:147–153.

Uehleke H, Hellmer KH, Tarerelli-Poplawski S: Metabolic activators of halothane and its covalent binding in liver endoplasmic proteins *in vivo. Naunyn Schmiedebergs Arch Pharmacol* 1973; 279:39–51.

Van Dyke RA, Chenoweth MB, Van Poznak: Metabolism of volatile anesthetics. I. Conversion *in vivo* of several anesthetics to $^{14}CO_2$ and chloride. *Biochem Pharmacol* 1964; 13:1239–1250.

Van Dyke RA and Gandolfi AJ: Anaerobic release of fluoride from halothane: relationship of the binding of halothane metabolites to hepatic cellular constitutents. *Drug Metab Disp* 1976; 4:40–44.

Van Dyke RA and Wood CL: In vitro studies on irreversible binding of halothane metabolite to microsomes. *Drug Metab Dispos* 1975; 3:51–57.

Vergani D, Mieli-Vergani G, Alberti A, Neuberger J, Eddleston A, Davis M, and Williams R: Antibodies to the surface of halothane-altered rabbit hepatocytes in patients with severe halothane-associated hepatitis. *N Eng J Med* 1980; 303:66–71.

Vergani D, Tsantoulas D, Eddleston ALWF, Davis M, and Williams R: Sensitization to halothane-altered liver components in severe hepatic necrosis after halothane anesthesia. *Lancet* 1978; 2:801–803.

Virtue RW and Payne KW: Postoperative death after flurothane. *Anesthesiology* 1958; 19:562.

Walton B: Halothane hepatitis in children. *Anaesthesiology* 1986; 41:575–578.

Walton B, Dumonde D, Williams C, et al: Lymphocyte transformation. Absence of increased responses in alleged halothane jaundice. *JAMA* 1973; 225:494–498.

Walton B, Simpson BR, Strunin L, Doniach D, Perrin J, and Appleyard AJ: Unexplained hepatitis following halothane. *Br. Med J* 1976; 1:1171–1176.

Widger LA, Gandolfi AJ, and Van Dyke RA: Hypoxia and halothane metabolism *in vivo. Anesthesiology* 1976; 44:197–201.

Williams BD, White N, Amlot PL, Slaney J, and Toseland PA: Circulating immune complexes after repeated halothane anaesthesia. *Br Med J* 1977; 2:159–162.

Zaric D, Larsen SF, and Jacobsen E et al: Halothane hepatitis in a prospective study of postoperative complications. *Acta Anaesth Scand* 1986; 30:529–532.

INDEX

A

B

C

D

E

F

G

H

I

K

L

M

N

O

P

Q

R

S

T

U

V

W

X

Z

GPSR Compliance

The European Union's (EU) General Product Safety Regulation (GPSR) is a set of rules that requires consumer products to be safe and our obligations to ensure this.

If you have any concerns about our products, you can contact us on ProductSafety@springernature.com

In case Publisher is established outside the EU, the EU authorized representative is:

Springer Nature Customer Service Center GmbH
Europaplatz 3
69115 Heidelberg, Germany

Batch number: 10823827

Printed by Printforce, the Netherlands